ELECTRICAL ENGINEERING TEXTS

INDUSTRIAL ELECTRICITY

PART I

ELECTRICAL ENGINEERING TEXTS

INDUSTRIAL ELECTRICITY

PART I

BY

CHESTER L. DAWES, S. B., A. M.

*Associate Professor of Electrical Engineering, The Graduate School
of Engineering, Harvard University; Fellow, American
Institute of Electrical Engineers; Etc.*

SECOND EDITION

Printed in the United States of America

ISBN: 978-0-9851721-5-2

Digitally Reproduced in 2012 by
CONVERPAGE Digital Reproductions
23 Acorn Street
Scituate, MA 02066
www.converpage.com

PREFACE TO THE SECOND EDITION

Since 1924, when the first edition of this volume was published, the author has received from teachers and from students suggestions as to changes and additions that would make the book better adapted to their needs. Also, there have been many new developments in electrical engineering, particularly in measuring instruments, in the manufacture of electrical machinery, and in industrial applications. Hence, this revision has been undertaken with the object of improving the book as a text and of bringing industrial applications into accord with modern-day practice. For example, in storage-battery engineering, two new separator materials have been developed, "mipor" and "fiberglas," the latter being a fabric woven from glass fibers. Also, in the smaller sizes, the glass covers of glass-jar storage batteries are now sealed to the top of the jar, whereas in the earlier types they simply rested on top of the plates.

Instruments such as the Wheatstone bridge and potentiometer have been improved in design so that their accuracy and ease of manipulation are increased. In electrical machinery, the yokes, frames, and bases, instead of being a single casting, are now fabricated by welding from rolled sections. The description of automobile starting and lighting systems has also been changed to conform to latest practice.

Several readers have suggested that the order of the chapters be rearranged so that Permanent Magnets and Electromagnetism, formerly Chapters I and II, shall immediately precede Chapter VIII, The Magnetic Circuit. The three chapters on Magnetism accordingly have been arranged in this manner. The first chapter in the book is now devoted to the fundamental theory of current flow and electrical resistance and is followed by four chapters on electric circuits, batteries, and measurements. The characteristics of new magnetic materials such as "Alnico" have been added.

The text is in large measure devoted to the fundamentals of electrical engineering and the analyses of its industrial application based on these fundamentals. The fundamentals, of course, do not change, but in some instances they, together with the analyses, have been extended.

There is an entirely new set of problems and their number has been increased from 277 in the first edition to 322 in this edition.

The writer is indebted to H. W. Beedle of the Electric Storage Battery Company for his assistance in the preparation of the chapter on storage batteries; to A. L. Russell of the Franklin Union Technical Institute, Boston, for his assistance in the revision, especially in the preparation of the problems; and particularly to Professor H. E. Clifford, the Consulting Editor, for his interest and the important suggestions he has made in the preparation of this revision.

C. L. D.

HARVARD UNIVERSITY, CAMBRIDGE, MASS ,
   June, 1939.

PREFACE TO THE FIRST EDITION

The rapid increase in the industrial applications of electricity has resulted in the establishment of elementary electrical engineering courses in many technical high schools and in other schools not of collegiate grade. Because of the resulting demand for textbooks suitable for such courses, the author was requested to prepare such books. Since a considerable portion of the author's two volumes "A Course in Electrical Engineering" (Vol. I, "Direct Currents" and Vol. II, "Alternating Currents") is devoted to fundamental considerations, the Consulting Editor of these Electrical Engineering Texts and the author both agreed that some of this material could be utilized advantageously in the preparation of the present volumes. As the chapters in the author's Volume I, "Direct Currents" are arranged in the order which would naturally be followed by the majority of teachers giving courses in magnetism and direct currents, the arrangement and titles of chapters in the present volume conform very closely to those given in the previous work.

Throughout the text, the attempt has been made to explain in simple language the principles underlying electrical engineering and electrical-engineering apparatus and to give a bird's-eye view of electrical engineering and its problems to the student who is beginning to study the subject, either by himself or in courses of the grade of those given in the technical high schools.

With the possible exception of the incandescent lamp, nearly every important industrial application of electricity involves an electric circuit interlinked with a magnetic circuit. A knowledge of the underlying principles of these two circuits is necessary for an understanding of these industrial applications, and the first five chapters in this volume are arranged to give the student a good grounding in the elementary principles of magnetism and the electric circuit. A large number of illustrative problems and their solutions are given in order to show concrete applications of these principles.

Owing to the increasing industrial importance of batteries, due in large measure to their wide use for automobiles and radio apparatus, their underlying principles, their care and applications are given in some detail in Chapter VI. Industrial installations of electrical apparatus require a knowledge of conductor resistance, insulation resistance, volts, amperes, power, energy, etc. Chapter VII discusses various commercial methods of making electrical measurements and also considers the simple principles underlying electrical-measuring instruments. The author is convinced that ability to make electrical measurements accurately and intelligently is important to every person engaged in electrical work.

Sufficient theory and descriptive matter are given in the chapters on the magnetic circuit and on electrostatics to enable a student, pursuing an elementary course, to understand the phenomena which he may be called upon to analyze in industrial apparatus. The last four chapters are devoted to the construction and operation of electrical machinery. The ordinary operating characteristics are given and are briefly analyzed with reference to the underlying principles of the electric and magnetic circuits. Considerable attention is also given to the relation of these characteristics to the industrial applications of direct-current machinery. Since the electrical equipment of automobiles represents a very common industrial application of electricity, a typical ignition system is described in Chapter VIII, and a typical lighting and starting system is described in Chapter XIII. Direct-current power distribution is given in Part II, with alternating-current power transmission and distribution, since the two are so closely related.

The author is greatly indebted to Prof. H. E. Clifford, the Consulting Editor of these Electrical Engineering Texts, for his careful review of the manuscript and his many criticisms and suggestions.

C. L. D.

HARVARD UNIVERSITY, CAMBRIDGE, MASS.
August, 1924.

CONTENTS

CONTENTS

CHAPTER VI

MAGNETISM AND PERMANENT MAGNETS 130
   102. Magnets and Magnetism 130
   103. Magnetic Materials. 130
   104. Natural Magnets. 130
   105. Permanent (Artificial) Magnets 131
   106. Magnetic Field. 131
   107. Effect of Breaking a Bar Magnet. 132
   108. Weber's Theory 132
   109. Attraction and Repulsion of Magnetic Poles. 134
   110. Coulomb's Law of Magnetic Repulsion and Attraction . . . 134
   111. Lines of Force. 135
   112. Field Intensity. 135
   113. Flux Density 137
   114. Compass Needle. 137
   115. Magnetic Figures. 138
   116. Magnetic Induction. 139
   117. Law of the Magnetic Field 141
   118. Other Forms of Magnets 142
   119. Laminated Magnets 142
   120. Magnet Screens 143
   121. Magnetizing. 143
   122. Earth's Magnetism. 144

CHAPTER VII

ELECTROMAGNETISM. 145
   123. Magnetic Field Surrounding a Conductor 145
   124. Relation of Magnetic Field to Current 147
   125. Magnetic Field of Two Parallel Conductors 147
   126. Magnetic Field of a Single Turn 148
   127. Solenoid. 148
   128. Plunger Solenoid and Industrial Applications 150
   129. Telegraph Relay. 151
   130. Electric Bells, Buzzers 152
   131. Lifting Magnet. 153
   132. Magnetic Separator. 154
   133. Magnetic Circuit of Dynamos 155

CHAPTER VIII

THE MAGNETIC CIRCUIT. 157
   134. Magnetic Circuit. 157
   135. Magnetic Units 158
   136. Reluctance of the Magnetic Circuit. 159
   137. Permeability of Iron and Steel. 161
   138. Law of the Magnetic Circuit. 162
   139. Magnetization Curves for Iron and Steel 164

CHAPTER XIII

APPENDIX A

APPENDIX B

APPENDIX C

APPENDIX D

APPENDIX E

APPENDIX F

APPENDIX G

APPENDIX H

INDUSTRIAL ELECTRICITY

PART I

CHAPTER I

RESISTANCE

Introduction.—Electricity plays a most important part in both the social and economic sides of modern life. For example, without electrical energy, living in communities of even moderate size would be impossible, for such communities are dependent on electrical energy for transportation, communication, illumination, elevators, automobile starting, ignition, and water supply. Electrical energy owes its usefulness to several factors. It may be generated at the most favorable locations, such as near navigable water, so that fuel may be obtained economically and cooling water is available, and at sites where hydraulic energy is available. It can be transmitted economically over great distances. It can be converted efficiently into other forms of energy, such as mechanical, heat, light, chemical, and sound energy. Electrical energy itself involves no fumes or products of combustion. It is readily controlled and applied to multitudes of uses, such, for example, as the large motors for steel mills, the very small motors of vacuum cleaners, and the arcing tips of spark plugs. Without electrical energy for motor drives, cranes, welding, lighting, and heat-treatment, mass production, which has made such a large number of products available at low prices, could hardly exist.

Electricity in its applications to magnetism, generators, motors, power transmission, etc., follows certain definite laws, and it is only through a knowledge of these laws that the many applications of electricity to the service of man can be understood.

1. Nature of Electricity.—Modern science has shown that the atoms of all matter consist of a positively charged nucleus, or *proton*, around which revolve extremely small negatively charged particles, called *electrons*.

The mass of the atom is concentrated practically in the proton, and the protons for the different elements differ. The electrons are identical for all matter. (See Part II, Electron Tubes.) These electrons, or negative charges, are electricity. In good conductors of electricity, such as metals, the electrons are loosely held by the nucleus and are able to pass readily from atom to atom. With insulators, the electrons are held firmly by the nucleus, and a relatively high potential difference produces only a very small movement of electrons from atom to atom. When electrons are drawn to a body, the body becomes negatively charged. When electrons are withdrawn from a body, the body becomes positively charged.

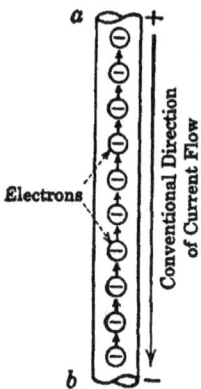

FIG. 1.—Electrons and direction of current flow.

Even with a moderate difference of potential applied to a conductor, the electrons move freely from atom to atom, constituting an electric current. Since the electrons are negative charges, the direction of the movement of the electrons is opposite to the conventional direction of the current. This is illustrated in Fig. 1 in which the potential of the end *a* of the conductor *ab* is positive with respect to that of the end *b*. The direction of movement of the electrons is from the negative end *b* to the positive end *a*.

The conventional direction of current is from the positive end *a* to the negative end *b*, the reverse of the direction of movement of the electrons.

When the electrons are not moving from atom to atom, the electricity is said to be *static*. When the electrons are moving from atom to atom, the electricity is said to be *dynamic*. Hence, static and dynamic electricity are the same, physically. (Also see Chap. IX.)

2. Electrical Resistance.—The current flowing in an electric circuit depends not only on the electromotive force impressed on the circuit, but on the circuit properties as well. For example, if a copper wire is connected across the terminals of a battery, a current will flow through the wire. This is shown in Fig. 2 (a), where the ammeter connected in the circuit indicates a current. If this copper wire is opened and a short length of resistor wire, of smaller cross-section than the copper wire, is introduced between two points, as *ab* (Fig. 2 (b)), two effects are noticed. The deflection of the ammeter pointer decreases, indicating that less current flows in the circuit. A perceptible amount of heat is developed in the resistor wire *ab*. If the resistor wire is removed and a small incandescent lamp substituted, the deflection of the ammeter pointer decreases further and the heat developed in the lamp filament is sufficient to bring it to incandescence.

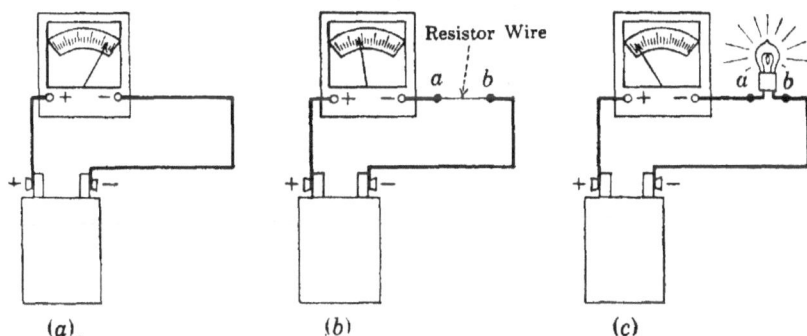

FIG. 2.—Effect of introducing resistance in an electric circuit.

It is clear, therefore, that the resistor wire and the incandescent lamp tend to reduce the flow of current and at the same time to cause heat to be developed within themselves. This property of tending to prevent the flow of current and at the same time causing heat dissipation is called *resistance*.

Resistance is explained on the electron theory by the fact that the electrons, in passing from atom to atom, must collide with the atomic nuclei as well as with other electrons. These collisions retard the motion of the electrons, thus reducing the current, and at the same time heat is generated.

Resistance in the electric circuit may be likened in its effect to friction in mechanics. For example, if a streetcar is running at a uniform speed on a straight, level track, friction tends to reduce the speed of the car. Likewise, resistance tends to reduce the value of the electric current. The energy which is used in maintaining the speed of the car is converted by friction into heat. Likewise, the energy expended in maintaining the current through

resistance is converted into heat.   Friction tends to impede the
flow of water in a pipe or a flume, some of the energy of the
water being expended in overcoming this friction.   The loss of
energy is represented by a loss of head.   This energy loss is
largely absorbed by the water, and careful measurements would
show a slight increase in its temperature.   It will be shown in the
next chapter that when an electric current flows through a resistance there is a loss of voltage or electric pressure, and also a loss
of energy, both these losses being directly proportional to the
value of the resistance.

All[1] substances have resistance, but the resistance of some substances is so great that it practically prevents the flow of current.
This leads to the classification of substances into conductors and
insulators.   No substance is a perfect conductor, and no substance
is a perfect insulator.   Silver, one of the best conductors, has
appreciable resistance, and glass and porcelain, among the best
insulators known, allow small currents to flow and therefore are
not perfect insulators.   The best conductors are the metals,
silver coming first and copper second.   Carbon and ordinary
water also may be classed as conductors.   Acid, alkaline, and
salt solutions are fair conductors.   Distilled or pure water, however, is a poor conductor.   Oils, glass, porcelain, silk, paper, cotton, ebonite, fiber, paraffin, and rubber may be considered as
nonconductors or good insulators.   Wood, either dry or impregnated with oil, is a good insulator, but wood containing moisture
is a partial conductor.

   3. Unit of Resistance.—The ohm is the practical unit of
resistance.   The unit as adopted in the United States is specifically defined by an act of Congress as follows:

   The *International Ohm* is the resistance offered to an unvarying electric current by a column of mercury at the temperature of melting ice,

---

[1] Professor Kamerlingh-Onnes of Leyden, in 1914, was able to produce a
circuit in which an electric current showed no diminution in strength 5 hr
after the emf had been removed.   The current was induced magnetically
in a short-circuited coil of lead wire at −270°C in the presence of liquid
helium and the inducing source then removed.   Liquid helium has the
lowest temperature known, being in the neighborhood of absolute zero
(−273°C).   This experiment indicates that the resistance of the lead was
practically zero at this extremely low temperature.

14.4521 grams in mass, of a constant cross-sectional area, and of a length of 106.300 centimeters.

One volt impressed across a resistance of *one ohm* causes a current of *one ampere* to flow. Also, *one ampere* flowing through *one ohm* for *one second* dissipates as heat *one joule*.

The resistance of insulating substances is ordinarily of the magnitude of millions of ohms, so that it is inconvenient to express this resistance in terms of a unit as small as the ohm. The *megohm*, equal to 1,000,000 (10^6) ohms, is the unit ordinarily used under these conditions. (The prefix "mega" means million.)

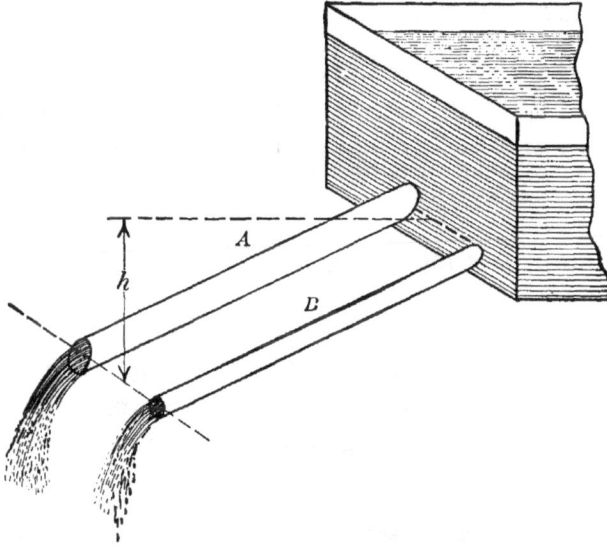

FIG. 3.—Water discharge through two pipes of equal lengths but having unequal cross-sections.

On the other hand, the resistance of bus-bars and of short pieces of metals may be so low that the ohm is too large a unit for convenient use. Under these conditions the *microhm* is used as the unit and is equal to 1/1,000,000 ohm (10^{-6} ohm). (The prefix "micro" means one-millionth.)

4. Resistance and the Geometry of Conductors.—The resistance of a body of given material depends on its size and shape and on the direction in which the electric current flows through it.

The flow of electricity through conductors is in many ways analogous to the flow of water through pipes.

Consider Fig. 3, in which is shown a reservoir to which two pipes *A* and *B* are connected. The two pipes are of the same length, but the cross-section of *A* is greater than that of *B*. There is the same difference of head (or pressure), *h*, between the ends of the pipes. It is obvious that pipe *A* discharges more water in a given time than pipe *B*, owing to the greater cross-sectional area of A.

The corresponding electrical analogue is given in Fig. 4. Two conductors *A* and *B* of equal lengths are connected to heavy cop-

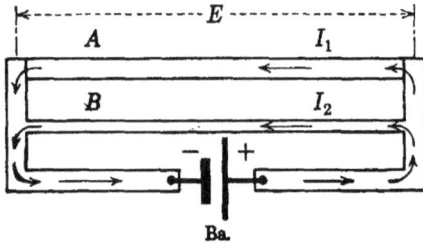

FIG. 4.—Flow of electricity through two conductors of equal lengths but having unequal cross-sections.

per bars. Conductor *A* has a larger cross-section than conductor *B*. The potential difference between the copper bars is maintained at *E* volts by the battery *Ba*. Therefore, the same electrical pressure or head of *E* volts exists between the ends of the two conductors. The quantity of electricity per second (I_1 amp) passing through conductor *A* is greater than the quantity of electricity per second (I_2 amp) passing through conductor *B*, because of the greater cross-sectional area of conductor *A*. Therefore, it may be said that the larger the cross-section either of water pipes or of electrical conductors, the greater the current flow with a given pressure. With a homogeneous

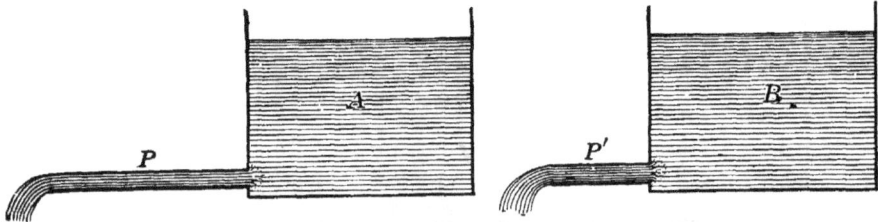

FIG. 5.—Water discharge through different-sized pipes.

electrical conductor of given length, the electrical resistance is *inversely* as its cross-section. This is not strictly true, however, of the frictional resistance of pipes to the flow of water.

Consider Fig. 5, in which two equal reservoirs *A* and *B* are to be emptied through pipes *P* and *P'* respectively. The pipe *P* is twice as long as the pipe *P'* but of one-half the cross-section. Therefore, both pipes have the same volume. It is evident that reservoir *B* will be emptied much quicker than *A*, because pipe *P'* has twice the cross-section of pipe P and therefore offers less resistance per unit of length. Further, the length of P' is only half that of P, and this again makes the total friction of P' half that of P, even if the cross-sections were equal.

In Fig. 6 are shown two conductors A and B of the same material. Conductor A has twice the length of conductor B, but only one-half the cross-section. Therefore, each conductor contains the same amount of material. It is evident, however, that the resistance per unit length of conductor A is twice that per unit length of B. Then, if conductors A and B were of the same length, conductor A would have twice the resistance of conductor B.

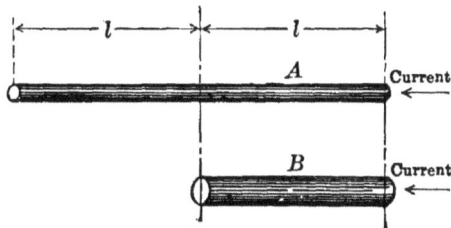

Fig. 6.—Two conductors of equal volume.

However, conductor A has twice the length of B and therefore must have 2×2, or 4 times the resistance of B.

5. Resistivity or Specific Resistance.—The deductions of Par. 4 may be summarized in the following rule for resistance:

The resistance of a homogeneous body of uniform cross-section varies directly as its length and inversely as its cross-section.

That is,

$$R = \rho \frac{L}{A} \tag{1}$$

where R is the resistance, L is the length in the direction of the current flow, A is the uniform area at right angles to the current flow, and ρ is a constant of the material known as its *resistivity* or *specific resistance*.

If L is 1 cm and A is 1 cm square, the body must have the form of a centimeter-cube, and

$$R = \rho \frac{1}{1 \times 1}$$

or

$$R = \rho.$$

In this case, ρ is called the *resistivity* or *specific resistance* of the substance per centimeter-cube. ρ may be expressed in terms of an inch-cube, a circular-mil-foot (p. 12), or in other units. The resistivity of copper is 1.724 microhms, or 1/580,000 ohm, per centimeter-cube at 20°C. It is evident that the resistance of a cube gives a perfectly definite resistivity, since the resistance between any two opposite faces is the same. The resistivities of various substances are given in Par. 15, p. 16. Knowing the specific resistance in terms of the centimeter-cube, the resistance of a wire, bar, etc., may be readily computed from Eq. (1).

Example.—Determine the resistance at 20°C of 2 km (1.24 miles) of solid aluminum wire having a cross-section of 4.0 sq mm (0.00620 sq in.). The resistivity of aluminum is 2.828 microhms per cm-cube at 20°C.

$$A = \frac{4.0}{100} = 0.040 \text{ sq cm.}$$
$$L = 2,000 \times 100 = 200,000 \text{ cm.}$$
$$\rho = 2.828 \times 10^{-6} = 0.000002828 \text{ ohm-cm-cube.}$$

Using Eq. (1),

$$R = 2.828 \times 10^{-6} \frac{200,000}{0.040} = 2.828 \times 10^{-6} \times 5,000,000 = 14.14 \text{ ohms. } Ans.$$

Example.—Determine the resistance at 20°C of 3,000 ft (915 m) of annealed 0000 copper wire, diameter of 0.460 in. (1.17 cm), the specific resistance of copper being 1.724 microhms (0.000001724 ohm) per cm-cube (20°C). (See Par. 15.)

$$3,000 \text{ ft} = 3,000 \times 12 \times 2.54 = 91,500 \text{ cm.}$$
$$\text{Cross-section} = \frac{\pi}{4}(0.460 \times 2.54)^2 = 1.07 \text{ sq cm.}$$
$$R = \rho\frac{L}{A} = (0.000001724) \times \left(\frac{91,500}{1.07}\right) = 0.1472 \text{ ohm. } Ans.$$

6. Volume Resistivity.—Since the volume of a body

$$V = LA$$

where L is its length and A its uniform cross-section, Eq. (1) may be written

$$R = \rho\frac{L}{A} = \rho\frac{L^2}{V} \qquad (2)$$
$$= \rho\frac{V}{A^2}. \qquad (3)$$

That is,

The resistance of a conductor varies directly as the square of its length when the volume is fixed.

The resistance of a conductor varies inversely as the square of its cross-section when the volume is fixed. (See Fig. 6, p. 7.)

Example.—A kilometer (3,280 ft) of wire having a diameter of 11.7 mm (0.460 in.) and a resistance of 0.031 ohm is drawn down so that its diameter is 5.0 mm (0.197 in.). What does its resistance become?

The original cross-section of the wire,

$$A_1 = \frac{\pi}{4}(11.7)^2 = 107.5 \text{ sq mm.}$$

The final cross-section,

$$A_2 = \frac{\pi}{4}(5.0)^2 = 19.64 \text{ sq mm.}$$

From Eq. (3),

$$R_1 = \rho \frac{V}{(107.5)^2} = 0.031 \text{ ohm;}$$

$$R_2 = \rho \frac{V}{(19.64)^2}.$$

Since the volume of the wire does not change during the drawing process and the resistivity constant ρ remains the same,

$$\frac{R_2}{R_1} = \frac{R_2}{0.031} = \frac{\rho\dfrac{V}{(19.64)^2}}{\rho\dfrac{V}{(107.5)^2}}.$$

$$R_2 = 0.031\frac{(107.5)^2}{(19.64)^2} = 0.031\frac{11,560}{386} = 0.93 \text{ ohm.} \quad Ans.$$

Since the area A of a circle is proportional to the square of its diameter D, $\left(A = \frac{\pi}{4}D^2\right)$, the ratio of the resistances of two cylindrical conductors of the same material is directly proportional to their lengths and inversely proportional to the *squares* of their diameters.

NOTE.—See Appendix F, p. 319, for Greek symbols.

If the resistances, lengths, and diameters of two cylindrical conductors of the same material are R_1, R_2, L_1, L_2, and D_1, D_2,

$$\frac{R_1}{R_2} = \frac{\dfrac{L_1}{D_1{}^2}}{\dfrac{L_2}{D_2{}^2}} = \frac{L_1D_2{}^2}{L_2D_1{}^2}. \tag{4}$$

Example.—A 1,000-ft (305-m) length of No. 14 (A.W.G.) copper wire (diameter of 64.0 mils—1.625 mm) has a resistance of 2.58 ohms at 25°C. Find the resistance at this temperature of 1,200 ft (366 m) of No. 16 wire (diameter 51.0 mils—1.296 mm).

Using Eq. (4) inverted,

$$\frac{R_2}{2.58} = \frac{1,200 \times \overline{64.0}^2}{1,000 \times \overline{51.0}^2}; \qquad R_2 = 2.58 \frac{1,200 \times 4,100}{1,000 \times 2,600} = 4.88 \text{ ohms.} \quad Ans.$$

7. Conductance.—Conductance is the reciprocal of resistance and may be defined as that property of a circuit or of a material which causes it to permit the flow of an electric current. The unit of conductance is the reciprocal ohm, or *mho*. Conductance is usually expressed by g.

$$g = \frac{1}{R};$$

also,

$$g = \gamma\frac{A}{L} \tag{5}$$

where γ is the *conductivity* or *specific conductance* of a substance, A the uniform cross-section, and L the length.

The conductivity of copper is 580,000 mhos per cm-cube at 20°C, 580,000 being equal to $1/(1.7241 \times 10^{-6})$.

Example.—Determine the conductance of an aluminum bus-bar 0.5 in. (1.27 cm) thick, 4 in. (10.16 cm) wide, and 20 ft. (6.10 m) long.

The conductivity of aluminum is 61 per cent that of copper, and copper has a conductivity of 580,000 mhos per cm-cube.

The conductivity of aluminum is

$$\gamma = 0.61 \times 580,000 = 354,000 \text{ mhos (centimeter-cube).}$$

The cross-section of the bus-bar,

$$A = 0.5 \times 4 \times 2.54 \times 2.54 = 12.9 \text{ sq cm.}$$

The length $L = 20 \times 12 \times 2.54 = 610$ cm.
The conductance, from Eq. (5),

$$g = 354,000 \times \frac{12.9}{610} = 7,490 \text{ mhos.} \quad Ans.$$

The resistance $r = \dfrac{1}{7,490} = 0.0001335$ ohm.

8. Circular Mil.—In the English and American wire systems, the *circular mil* is the standard unit of wire cross-section.

The term *milli* means one-thousandth; for example, a milli-volt is 1/1,000 volt. A *mil* is one-thousandth of an inch (0.0254 mm). A *square mil* is the area of a square, each side of which is one mil (0.001 in.), as shown in Fig. 7 (*a*). The area of a square mil is 0.001 × 0.001 = 0.000001 sq in. (0.000645 sq mm).

A *circular mil* is the area of a circle whose diameter is one mil (0.001 in.) (Fig. 7 (*b*)) and is usually written cir mil or C.M. As will be seen from an inspection of Fig. 7 (*c*), the circular mil is a smaller area than the square mil. The area in square inches

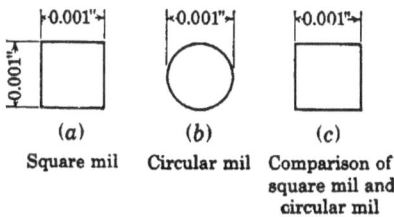

Fig. 7.—Square and circular mil.

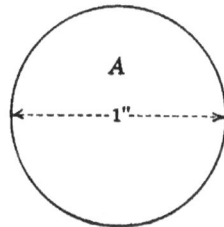

Fig. 8.—Cross-section of 1,000,000 cir. mils.

of a circular mil $(\pi/4)(0.001)^2 = 0.0000007854$ sq in. (0.000506 sq mm).

The circular mil is the unit in which the cross-section of wires and cables is measured, just as the square foot is the unit in which larger areas, such as floors or land, are measured. The advantage of the circular mil as a unit is that circular areas measured in terms of this unit bear a very simple relation to the diameters.

A, in Fig. 8, represents the cross-section of a wire having a diameter of 1 in. Its area

$$A = \frac{\pi}{4}(1)^2 \text{ sq in.}$$

The area of a circular mil, $a = (\pi/4)(0.001)^2$ sq in.

The ratio of the areas A/a gives the number of circular mils in A, or

$$\frac{A}{a} = \frac{\frac{\pi}{4}(1)^2}{\frac{\pi}{4}(0.001)^2} = \frac{1}{0.000001} = 1,000,000 \text{ cir mils.}$$

The general relation may be written:

$$\text{Cir mils} = \frac{D_1{}^2}{(0.001)^2} = 1,000,000(D_1)^2 = D^2 \qquad (6)$$

where D_1 is the diameter of the wire in *inches*.

D is the diameter of the wire in *mils*.

The matter may be summed up in two rules:

To obtain the number of circular mils in a solid cylindrical wire of given diameter, express the diameter in mils and then square it.

To obtain the diameter of a solid cylindrical wire having a given number of circular mils, take the square root of the circular mils and the result will be the diameter of the wire in mils.

Example.—The diameter of 00 wire (A.W.G.) is 0.3650 in. (9.26 mm). What is its circular milage?

$$0.3650 \text{ in.} = 365.0 \text{ mils.}$$
$$(365.0)^2 = 133,100 \text{ cir mils.} \quad \textit{Ans.}$$

Example.—The cross-section of a solid cylindrical wire is 52,640 cir mils. What is its diameter?

$$\sqrt{52,640} = 229.4 \text{ mils} = 0.2294 \text{ in. (5.83 mm).} \quad \textit{Ans.}$$

9. Circular-mil-foot.—Another convenient unit of resistivity, especially in the English system, is the resistance of a circular-mil-foot. This unit is the resistance of a wire having a cross-section of 1 cir mil and a length of 1 ft, as shown in Fig. 9. The resistance of

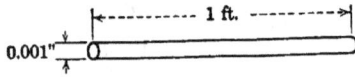

FIG. 9.—Circular-mil-foot.

a circular-mil-foot of copper at 20°C is 10.37 ohms. (In practical work, this resistance may frequently be taken as 10 ohms.) Knowing this resistivity, the resistance of any length and size of wire may be determined by Eq. (1), p. 7.

Example.—Determine the resistance at 20°C of a 750,000-cir-mil copper cable, 2,500 ft (762 m) long.

If the cross-section of the cable were 1 cir mil, the resistance would be 2,500 × 10.37 = 25,900 ohms. However, the cross-section is actually 750,000 cir mils; therefore,

$$R = \frac{25,900}{750,000} = 0.0346 \text{ ohm.} \quad \textit{Ans.}$$

Or Eq. (1), p. **7**, may be used directly, giving

$$R = 10.37 \frac{2,500}{750,000} = 0.0346 \text{ ohm.} \quad Ans.$$

When applying Eq. (1) under these conditions, the resistivity ρ must be expressed in *ohms per circular-mil-foot*, L in *feet*, and A in *circular mils*. (See Par. 15, p. 15.)

10. Temperature Coefficient of Resistance.—The resistance of the nonalloyed metals increases appreciably with the temperature. Since copper, as in the windings of electric machinery, often operates at temperatures much higher than that of the surrounding air, it is important to know the relation between temperature and resistance. The relation may be expressed as follows:

$$R_t = R_0(1 + at) \tag{7}$$

where R_t is the resistance at the temperature t, R_0 the resistance at 0°C, and a the *temperature coefficient of resistance* at 0°C. For copper, a is 0.00427, for aluminum, a is 0.0039, and for most of the unalloyed metals a is nearly 0.004. The temperature coefficient of resistance shows that with copper, for example, the resistance increases 0.427 of 1 per cent for each degree centigrade increase of temperature above 0°. Thus, if the resistance of a coil of copper wire is 100 ohms at 0°C, for every degree increase of temperature the resistance will increase

$$100 \times 0.00427 \text{ ohm, or } 0.427 \text{ ohm.}$$

At 40°C the increase of resistance will be $40 \times 0.427 = 17.08$ ohms, and the resistance at 40° will be $100 + 17.08 = 117.08$ ohms.

If the resistance at some definite temperature other than 0°C is known, ordinarily the resistance at 0°C must first be found before the resistance at other temperatures can be determined.

For this purpose, Eq. (7) may be transposed into the form

$$R_0 = \frac{R_t}{1 + at}. \tag{8}$$

Example.—The resistance of an electromagnet winding of copper wire at 20°C is 30 ohms. What is its resistance at 80°C?

The resistance at 0°C

$$R_0 = \frac{30}{1 + 0.00427 \times 20} = \frac{30}{1.085} = 27.65 \text{ ohms.}$$
$$R_{80} = 27.65(1 + 0.00427 \times 80) = 37.11 \text{ ohms.} \quad Ans.$$

The process of working back to 0° is a little inconvenient, although it is fundamental and easy to remember. Table 11 gives the temperature coefficients of resistance of copper at various initial temperatures. With this table available, the above problem involves but one computation.

Example.—From Table 11 the temperature coefficient of copper at 20° initial temperature is 0.00393. The rise in temperature = 80° − 20° = 60°.

Then, the resistance at 80°C

$$R_{80} = 30(1 + 0.00393 \times 60) = 37.07 \text{ ohms.} \quad Ans.$$

11.—Temperature Coefficients of Copper at Different Initial Temperatures

$$\left(\text{From formula } \frac{1}{(234.5 + t)} \right)$$

INITIAL TEMPERATURE	INCREASE IN RESISTANCE PER 1°C
0	0.00427
10	0.00409
20	0.00393
30	0.00378
40	0.00364

12. Resistance and Inferred Absolute Zero.—If the resistance of copper at temperatures between 0° and 100°C be plotted as a

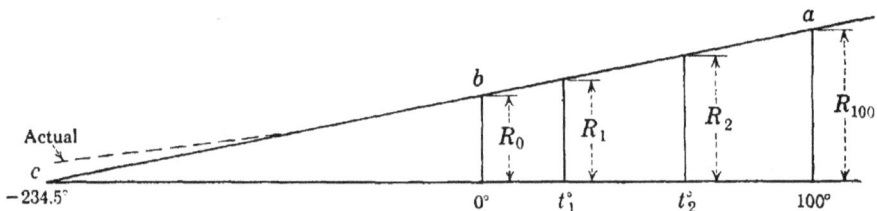

FIG. 10.—Variation of resistance of copper with temperature.

function of temperature, the result is practically a straight line *ba* (Fig. 10). If this line be extended, it will intersect the zero resistance line at −234.5°C (an easy number to remember), as shown in Fig. 10. This assumes that, between ordinary limits of temperature, copper behaves as if it had zero resistance at −234.5°C. [Actually, the curve bends at these extremely low temperatures, as shown by the dotted line (Fig. 10).] This gives a convenient method for determining temperature–resistance relations.

By the law of similar triangles (Fig. 10),

$$\frac{R_0}{234.5^0} = \frac{R_1}{234.5^0 + t_1}, \tag{9}$$

$$\frac{R_1}{234.5^0 + t_1} = \frac{R_2}{234.5^0 + t_2}, \tag{10}$$

where R_1 is the resistance at $t_1°$C and R_2 is the resistance at $t_2°$C.

Applying Eq. (10) to the previous example,

$$\frac{30}{234.5^0 + 20°} = \frac{R_{80}}{234.5_0 + 80°}.$$

$$R_{80} = 30\frac{234.5° + 80°}{234.5° + 20°} = 30\frac{314.5°}{254.5°} = 37.1 \text{ ohms.} \quad Ans.$$

13. Alloys.—Certain alloys, notably manganin and nickel-iron alloys, have practically zero temperature coefficients and are very useful as resistances for measuring instruments, where a change of resistance introduces error.

Table 14 gives the temperature coefficients at 20°C for various materials.

14.—Temperature Coefficients of Resistance

PER DEGREE CENTIGRADE AT 20°C

Aluminum	0.00388
Carbon (incandescent lamp)	−0.003
Graphite	−0.0006 to −0.0012
German silver	0.00031 to 0.00020
Iron and steel	0.0032 to 0.0056
Manganin	0.000011 to 0.000039
Nichrome	0.00015 to 0.00020
Platinum	0.00367

15. Resistivities.—The standard for the resistivity of copper of 0.15328 ohm per m-gram at 20°C, as recommended by the Bureau of Standards,[1] was adopted as the International Annealed Copper Standard by the International Electrotechnical Commission at Berlin, September, 1913. The corresponding resistivities in other units at 20°C are as follows:

[1] See "Copper Wire Tables," Circ. 31, U.S. Bureau of Standards, 1914

```
  0.15328   ohm (meter-gram)
875.20      ohms (mile-pound)
  1.7241    microhms (centimeter-cube)
  0.67879   microhm (inch-cube)
 10.371     ohms (circular-mil-foot)
  0.017241  ohm (meter-square millimeter)
```

Below are the resistivities at 20°C of some of the more common electrical conductors.

TABLE OF RESISTIVITIES AT 20°C

Metals	Cm-cube (microhms)	Cir-mil-ft (ohms)
Aluminum	2.828	17.01
German silver	33.3 to 48.2	200 to 290
Iron:		
Electrolytic	9.94	59.8
Cast	75 to 95	450 to 570
Manganin	44.9	270
Mercury	95.78	565
Nichrome	100 to 110	600 to 660
Platinum	10.0	60.2
Silver	1.5 to 1.7	9.0 to 10.2
Steel, soft	15.9	95.4

16. Resistivity Specifications.—The resistivities of commercial copper are usually slightly greater than the resistivity of the International Standard. In the standards of the American Institute of Electrical Engineers[1] it is recommended that the resistivity shall not exceed 891.58 ohms per lb-mile. Below are given the corresponding resistivities in other units.

```
 0.15614 ohm (meter-gram)
 1.75614 microhms (centimeter-cube)
 0.69150 microhm (inch-cube)
10.565   ohms (circular-mil-foot)
```

Example.—The resistance of a 20-ft length of No. 1 A.W.G. annealed solid copper wire is measured at 20°C and is found to be 0.00252 ohm. The diameter of the wire is 289.0 mils. Determine whether or not its resistivity is within the A.I.E.E. specifications.

[1] A.I.E.E. Standards No. 61, "Specifications for Soft or Annealed Copper Wire," adopted Dec. 16, 1927.

The circular mils = $(289.0)^2$ = 83,500.

The resistance per foot = 0.00252/20 = 0.000126 ohm.

The resistance per circular-mil-foot = 0.000126 × 83,500 = 10.52 ohms.

This value is slightly less than the value of 10.565 ohms specified by the A.I.E.E. and hence is within the specifications.

17. Resistances in Series and in Parallel.—If a number of resistances r_1, r_2, r_3, . . . , are connected in series (Fig. 11), that

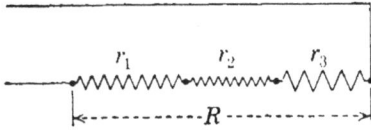

Fig. 11.—Resistances in series.

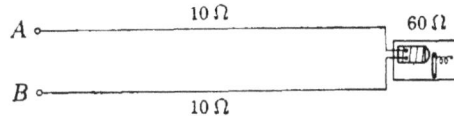

Fig. 12.—Sixty-ohm telegraph relay in series with a 20-ohm line.

is, end to end, the total resistance of the combination is

$$R = r_1 + r_2 + r_3 + \cdots \tag{11}$$

That is,

In a series circuit, the total resistance is the sum of the individual resistances.

Example.—A 60-ohm telegraph relay is connected to the end of a telegraph line (Fig. 12) each conductor of which has a resistance of 10 ohms. What is the resistance of the entire circuit?

The two line conductors and the relay form a series circuit, the resistance of which,

$$R = 10 + 60 + 10 = 80 \text{ ohms.} \quad Ans.$$

This is the resistance which would be measured between the ends A, B.

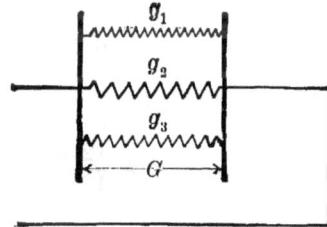

Fig. 13.—Conductances in parallel.

If a number of conductances, g_1, g_2, g_3, . . . , are connected in parallel (Fig. 13), the total conductance of this portion of the circuit must be equal to the sum of the individual conductances; that is,

$$G = g_1 + g_2 + g_3 + \cdots \tag{12}$$

Since

$$G = \frac{1}{R}, \qquad g_1 = \frac{1}{r_1}, \cdots$$

Eq. (12) may be written

$$\frac{1}{R} = \frac{1}{r_1} + \frac{1}{r_2} + \frac{1}{r_3} + \cdots \tag{13}$$

That is,

In a parallel circuit, the reciprocal of the total resistance is equal to the sum of the reciprocals of the individual resistances.

For a circuit of two resistances r_1 and r_2, in parallel, the joint resistance

$$R = \frac{r_1 r_2}{r_1 + r_2}. \tag{14}$$

Example.—A 12.0-ohm and a 7.00-ohm resistance are connected in parallel between points A and B (Fig. 14(a)). What is their joint resistance?

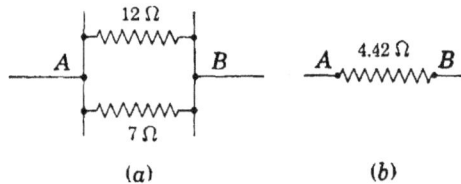

(a)   (b)

Fɪɢ. 14. —Two resistances in parallel and their equivalent resistance.

Using Eq. (13),

$$\frac{1}{R} = \frac{1}{12} + \frac{1}{7} = 0.0833 + 0.1429 = 0.2262 \text{ mho.}$$

$$R = \frac{1}{0.2262} = 4.42 \text{ ohms.} \quad Ans.$$

Using Eq. (14),

$$R = \frac{12 \times 7}{12 + 7} = \frac{84}{19} = 4.42 \text{ ohms (check).} \quad Ans.$$

This means that if the 12.0-ohm and 7.00-ohm resistances were replaced by a single resistance of 4.42 ohms (Fig. 14 (b)), the resistance between points A and B would be the same.

Example.—Determine the total resistance of a circuit having four branches, the individual resistances of which are 3.0, 4.0, 6.0, and 8.0 ohms.

$$\frac{1}{R} = \frac{1}{3.0} + \frac{1}{4.0} + \frac{1}{6.0} + \frac{1}{8.0} = 0.333 + 0.250 + 0.167 + 0.125$$

$$= 0.875 \text{ mho.}$$

$$R = \frac{1}{0.875} = 1.142 \text{ ohms.} \quad Ans.$$

18. American Wire Gage (A.W.G.).—The A.W.G. (formerly Brown & Sharpe gage) is based on a constant ratio of cross-section between wires of successive gage numbers. The following approximate relations make it a comparatively simple matter to determine the weight or resistance of any gage number without reference to the table: (1) No. 10 wire has a diameter of 0.1 in. and a resistance of 1 ohm per 1,000 ft. (2) The resistance of the wire doubles with every increase of 3 gage numbers. (3) Therefore, the resistance increases $\sqrt[3]{2} = 1.26$ ($1\frac{1}{4}$) times for each successive gage number and $(1.26)^2 = 1.6$ times for every two numbers. (4) The resistance is multiplied or divided by 10 for every difference of 10 gage numbers. (5a) The weight of 1,000 ft of No. 2 wire is 200 lb (90.7 kg). (5b) The weight of 1,000 ft of No. 10 wire is 31.4, or 10π lb (14.23 kg). (For constant length the weight varies inversely as the resistance.)

Example.—What are the resistance and weight of 1,000 ft (305 m) of 0000 wire?

The resistances will decrease as follows:

Gage number	10	7	4	1	000
Resistance	1	0.5	0.25	0.125	0.0625 (rules 1 and 2)

Resistance of 0000 = 0.0625/1.25 = 0.050 ohm (rule 3). *Ans.*

Weight of 1,000 ft No. 2 = 200 lb (rule 5a).
Weight of 1,000 ft   00 = 400 lb (rule 2).
Weight of 1,000 ft 0000 = 400 × 1.6 = 640 lb (290 kg) (rule 3). *Ans.*

The example might have been worked more quickly by rule 4.
Resistance of 1,000 ft of No.   10 = 1 ohm.
Resistance of 1,000 ft of No.   0 = 0.1 ohm (rule 4).
Resistance of 1,000 ft of No. 0000 = 0.050 ohm (rule 2). *Ans.*

Example.—Find the resistance, the circular milage, the diameter, and the weight of 3,500 ft (1,066 m) of No. 28 copper wire.

Resistance of 1,000 ft of No. 10 = 1 ohm (rule 1).
Resistance of 1,000 ft of No. 20 = 10 ohms (rule 4).
Resistance of 1,000 ft of No. 23 = 20 ohms (rule 2).
Resistance of 1,000 ft of No. 26 = 40 ohms (rule 2).
Resistance of 1,000 ft of No. 28 = 40 × 1.6 = 64 ohms (rule 3).

$$3.5 \times 64 = 224 \text{ ohms.} \quad Ans.$$

No. 10 contains 10,000 cir mils.
No. 20 contains 1,000 cir mils.

No. 23 contains 500 cir mils.

No. 26 contains 250 cir mils.

No. 28 contains $250/1.6 = 156$ cir mils.   *Ans.*

The diameter $D = \sqrt{156} = 12.5$ mils (0.318 mm).   *Ans.*

Weight of 1,000 ft of No. 10 $= 31.4$ lb.

Weight of 1,000 ft of No. 20 $= 3.14$ lb.

Weight of 1,000 ft of No. 23 $= 1.57$ lb.

Weight of 1,000 ft of No. 26 $= 0.785$ lb.

Weight of 1,000 ft of No. 28 $= 0.785/1.6 = 0.49$ lb.

$$0.49 \times 3.5 = 1.72 \text{ lb } (0.780 \text{ kg}).   Ans.$$

Appendices G and H, pp. 320 and 321, give the properties of copper wire. These tables were compiled by the Bureau of Standards and are accepted internationally as standards.

19. Stranded Wires and Cables.—Stranding increases the flexibility and is necessary for wires and cables whose cross-section is greater than 0000 A.W.G. It is often desirable for conductors of smaller cross-section. In order that cables may

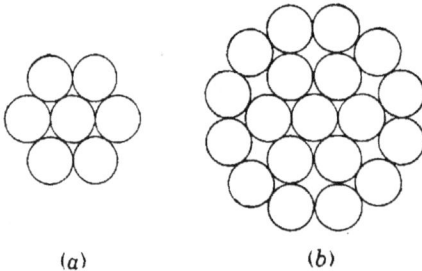

(a)   (b)

FIG. 15.—Make-up of a 7-strand and a 19-strand cable.

make up properly, the following number of cylindrical conductors, all of the same cross-section, must be used.

Six cylindrical conductors will just encircle a single one at the center (Fig. 15 (a)), making a 7-strand cable. Twelve more cylindrical conductors will just encircle the 7-strand cable (Fig. 15 (b)), making a total of 19 strands. The number of strands increases by six in each successive layer. Therefore, cables are ordinarily made up of 7, 19, 37, 61, 91, 127, etc., strands. Conductors made up in the foregoing manner are called *concentric-lay* wires and cables. Appendix H, p. 321, gives the properties of stranded wires and cables.

20. Conductors.—Although silver is a better conductor than copper, its use as a conductor is very limited because of its cost. In a few instances it is used where a delicate but highly conducting material is necessary, such as in the brushes and occasionally in the commutator of watt-hour meters. Silver contacts are frequently used in switches and circuit breakers since with silver the arcing is very much less than with copper and the circuit is therefore interrupted much more readily. Copper, because of its high conductivity and moderate cost, is used more extensively as a conductor than any other material. It has many good qualities, such as ductility and high tensile strength; it is not easily abraded; it is not corroded by the atmosphere; and it is readily soldered.

Aluminum has only 61 per cent of the conductivity of copper; but, for the same length and weight, it has about twice the conductance of copper. It is softer than copper, its tensile strength is much less, and it can be soldered only with difficulty. It is not affected by exposure to the atmosphere. The large diameter for a given conductance prohibits its use where an insulating covering is required. Aluminum is used extensively as a conductor for high-voltage transmission lines, where its lightness and large diameter are an advantage. It is used to some extent for low-voltage bus-bars, as it offers much greater radiating surface than copper of the same conductance.

Iron and steel have about nine times the resistance of copper for the same cross-section and length. The large cross-section for a given conductance prohibits their use where an insulating covering is necessary, and the increased weight prevents their use in most cases where the conductors must be placed on poles. They are commonly used as resistors in connection with rheostats and for third rails of electric railways. Iron and steel ordinarily must be protected from oxidation by galvanizing or by other protective covering. "Copperweld" consists of a steel wire surrounded by a layer of copper, fused or welded to the steel. The advantages are that it possesses the high tensile strength of steel, combined with the high conductivity of copper. Further, the copper protects the steel from corrosion. Its field is the transmission line conductor, where long spans make high tensile strength necessary. It is also used as an overhead ground wire on transmission lines, and for ground rods.

CHAPTER II

OHM'S LAW AND THE ELECTRIC CIRCUIT

Electric units are defined fundamentally in terms of the centimeter, the gram, and the second, giving the cgs or absolute systems of electrostatic and electromagnetic units. As the magnitudes of these cgs units differ considerably from the magnitudes ordinarily used in practice, the practical system of the ampere, volt, ohm, etc., has been adopted for commercial and scientific use.[1]

21. Practical Units. *Current.*—The unit of electric current is the *ampere* and represents the *rate of flow* of electricity. It corresponds in hydraulics to the rate of flow of water, expressed as cubic feet per second, gallons per minute, etc.

A current is said to have the strength of one absolute unit (one *absampere*) when its value is such that if one centimeter length of the circuit is bent into an arc of one centimeter radius, the current in it exerts a force of one dyne on a magnetic pole of unit strength placed at the center of the arc. The *ampere* is one-tenth of the absampere. In order that the value of the ampere may be readily determined by experiment, it is further defined by an act of Congress, 1894, as follows:

"The unit of current shall be what is known as the international ampere, which is one-tenth of the unit of current of the centimeter-gram-second system of electromagnetic units and is the practical equivalent of the unvarying current, which when passed through a solution of nitrate of silver in water in accordance with standard specifications, deposits silver at the rate of one thousand one hundred and eighteen millionths (0.001118) of a gram per second."

Quantity.—The unit of quantity is the *coulomb*. This is equal to the quantity of electricity conveyed by one ampere in one

---

[1] In June, 1935, the mks (meter-kilogram-second) system was adopted by the International Electrotechnical Commission. The volt, ampere. ohm, and other practical units remain the same.

second. The coulomb is analogous to the unit quantity of water in hydraulics, such as the cubic foot, the gallon, etc.

From this definition it is evident that an electric current may be expressed either in *coulombs per second* or in *amperes*.

Difference of Potential and Electromotive Force.—Difference of potential and emf (electromotive force) tend to cause a flow of electricity. The unit of potential difference and of emf is the *volt* and is defined as that potential difference or emf which, when impressed across the terminals of a resistance of one ohm, will cause a current of one ampere. The *international volt* is now more specifically defined as 1/1.01830 of the voltage of a normal Weston cell. (See p. 68.)

The mechanical analogue of potential is pressure. The difference in hydraulic pressure between the ends of a pipe causes or tends to cause the flow of water. The pressure of water behind the dam tends to cause water to flow through the penstock or through any leaks. The pressure in a boiler tends to cause steam to flow through the pipes, valves, etc. Likewise, electric pressure or difference of potential tends to cause a flow of electricity.

Resistance.—The *international ohm* is specifically defined as the resistance of a column of mercury at the temperature of melting ice (0°C), 14.4521 grams in mass, of a constant cross-sectional area and of a length of 106.300 cm. (See p. 4, Chap. I.)

22. Nature of the Flow of Electricity.—The flow of electricity through a circuit resembles in many ways the flow of water through a closed system of pipes. For example, in Fig. 16, water enters the mechanically driven centrifugal pump P at a pressure h_1 (represented by the length of a column of mercury) above the point of zero pressure shown by the line h_0. In virtue of the action of the pump blades, the pressure through the pump is increased from h_1 to h_2, representing a net increase of pressure H_1. The water then flows out along pipe F_1 to the hydraulic motor W. Because of the friction loss in the pipe F_1, the pressure at the motor terminals h_3 is slightly less than h_2. In other words, a pressure of $h_2 - h_3$ is required to overcome the frictional resistance of the pipe F_1. The distance between the horizontal line ac and the line ab shows the pressure drop along the pipe, this pressure drop being uniform.

In Fig. 17, the mechanically driven electrical generator G raises the potential of the current entering its negative terminal from v_1 to v_2 where v_1 and v_2 are measured from the earth, whose potential is ordinarily assumed as zero. (The various potentials are measured with voltmeters v'_1, v'_2, etc.) The generator, in raising the potential of this portion of the circuit from v_1 to v_2,

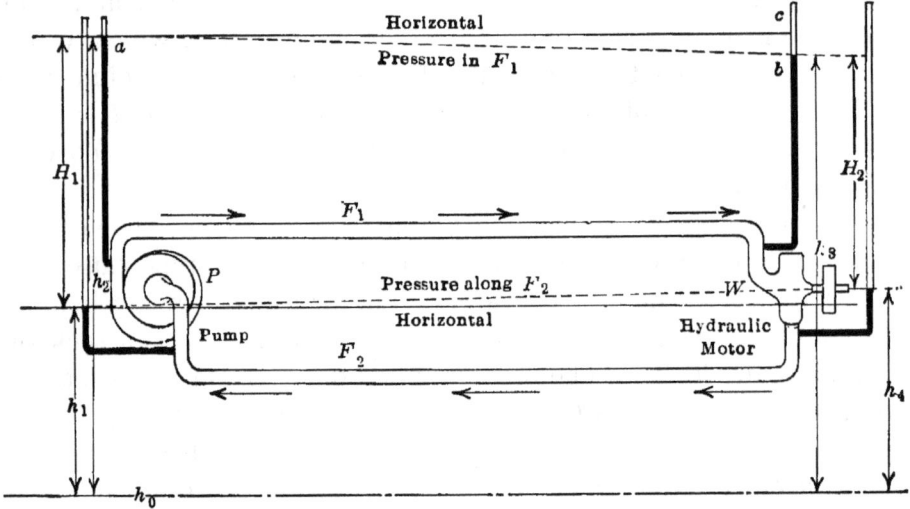

FIG. 16.—Flow of water through a hydraulic motor and pipe system.

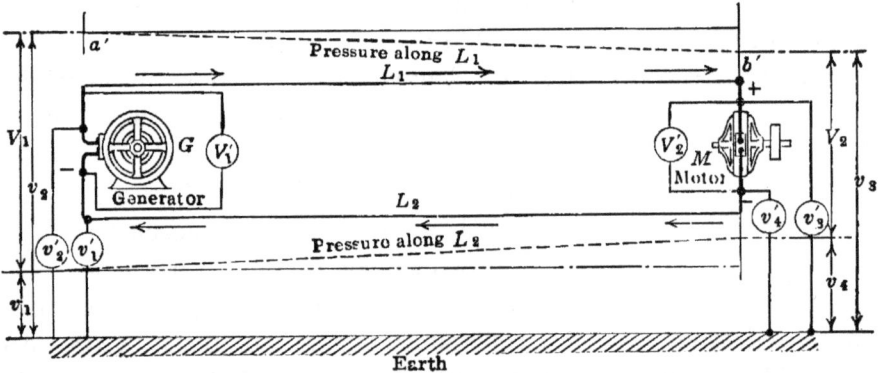

FIG. 17.—Flow of electric current through an electric motor and feeder system.

produces a net increase in electric pressure $v_2 - v_1 = V_1$. The current now flows out through the line L_1 to the + terminal of the motor M. Because of the line resistance, the potential drops from v_2 at the generator to v_3 at the motor in practically the same manner that the water pressure drops in pipe F_1 (Fig. 16). A voltage $v_2 - v_3$ is necessary to force the current through the line L_2. The line $a'b'$ shows the actual potential at each point along the wire, the distance of $a'b'$ from the ground line being proportional to the potential at each point. The potential drop is uniform.

Referring again to Fig. 16, the water enters the hydraulic motor W and, in overcoming the back pressure of the revolving blades, its pressure drops from h_3 to h_4, representing a net drop in pressure H_2. Pressure h_4 must necessarily be greater than h_1 in order that the water may flow back through the pipe F_2. The pressure $h_4 - h_1$ is necessary to overcome the friction loss in the pipe F_2. It is to be noted that H_2, the net pressure at the motor terminals, is less than the pressure H_1 at the pump, by the sum of the pressures necessary to overcome the friction in the *outgoing and return pipes* F_1 and F_2.

In a similar manner, the pressure of the electric current in flowing through the motor M drops from v_3 to v_4, representing a net drop in pressure V_2. A large percentage of this voltage V_2 is necessary to overcome the back emf of the motor. (See p. 269.) v_4 is necessarily greater than v_1, or the current could not flow along L_2 back to the negative terminal of the generator. It is to be noted that, as in the case of Fig. 16,

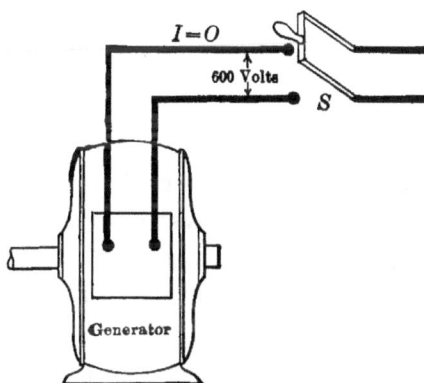

FIG. 18.—Illustrating the existence of potential difference without current.

the net potential difference V_2 at the motor M is less than the potential difference V_1 at the generator G by the drop in potential due to the resistance of both the *outgoing* and *return* wires.

Difference of potential is, therefore, the equivalent of pressure and *tends* to send current through a circuit; current is quantity of electricity per second. Potential difference may exist with no current flow, in the same manner that a boiler may have a very high steam pressure with no steam flow, owing to all the valves being closed. Likewise, a generator (Fig. 18) may have a very high potential difference at its terminals, yet no current flows because the switch S is open.

23. Difference of Potential.—In order that current may flow between two points, there must be a *difference of potential* between the two points, as shown in Fig. 17.   This is further illustrated in Fig. 19.   A large reservoir and a small tank are connected by a pipe *P*.   The water levels in the tank and in the reservoir are the same; that is, there is pressure in each, but there is no *difference in pressure* between them.   When the valve *V* is opened, no water

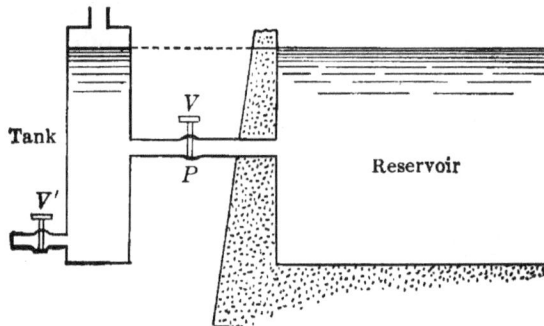

FIG. 19.—Tank and reservoir at the same pressure.

flows from the reservoir to the tank.   However, if the valve *V'* is opened, allowing the water level in the tank to fall, a difference in pressure results and water flows from the reservoir to the tank.

Figure 20 shows two batteries A_1 and A_2, each having an emf of 2 volts.   The positive terminal of A_1 has a potential of $+2$ volts above its negative terminal; likewise, the positive terminal

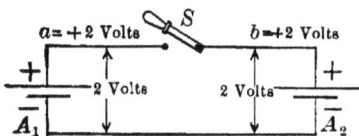

FIG. 20.—Two batteries having equal electromotive forces.

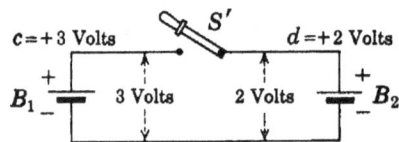

FIG. 21.—Two batteries having unequal electromotive forces.

of A_2 has a potential of $+2$ volts above its negative terminal. The negative terminals of both batteries are at the same potential because they are connected by a copper wire through which no current flows, and consequently there can be no potential difference between the ends of this copper wire.   Therefore, points *a* and *b* must each have the same potential of $+2$ volts. If now the switch *S* be closed, *no current* will flow from *a* to *b*, because there is *no difference of potential* between *a* and *b*.

In Fig. 21 the emf of battery B_1 is 3 volts, and therefore the potential of its positive terminal is 3 volts above that of its negative terminal. The emf of battery B_2 is 2 volts, and therefore the potential of its positive terminal is 2 volts above that of its negative terminal. The negative terminals are at the same potential. If this potential be assumed as zero, the point c is at a potential of +3 volts, whereas the potential of d is +2 volts. Therefore, the point c is at a potential of 3 − 2, or 1 volt higher than d. When switch S' is closed, a current will flow from c to d, in virtue of c being at a higher potential than d.

24. Measurement of Voltage and Current.—Voltage or potential difference is measured ordinarily with a voltmeter. The

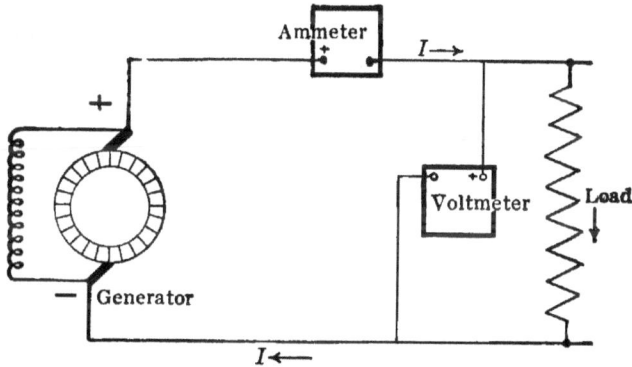

Fig. 22.—Proper method of connecting a voltmeter and an ammeter.

voltmeter, therefore, should be connected *across* or *between* the wires whose difference of potential is to be measured, as shown in Fig. 22.

Current is measured ordinarily with an ammeter. As current is the *quantity of electricity per second* passing in the wire, the ammeter must be so connected that only the current to be measured flows through it. This is accomplished by opening one of the wires of the circuit and inserting the ammeter, just as a water meter is inserted in a pipe when it is desired to measure the flow of water in the pipe. When the ammeter is so connected, the current flowing through it to the load is measured by the ammeter. (See Fig. 22.) *An ammeter should never be connected across the line.*

25. Ohm's Law.—Ohm's law states that, for a steady current, the current in a circuit is *directly* proportional to the total emf acting in the circuit and is *inversely* proportional to the total resistance of the circuit.

The law may be expressed by the following equation if the current I is in *amperes*, the emf E is in *volts*, and the resistance R is in *ohms*:

$$I = \frac{E}{R} \text{ amp.} \qquad (15)$$

That is, the current in amperes in a circuit is equal to the emf of the circuit in volts divided by the resistance of the circuit in ohms. Potential difference may be represented by either the letter V or the letter E, V usually signifying terminal voltage and E emf or induced voltage.

Example.—The resistance of the field winding of a shunt motor is 41 ohms. What current flows through the winding when it is connected across 220-volt mains?

$$I = \frac{E}{R} = \frac{220}{41} = 5.37 \text{ amp.} \quad \textit{Ans.}$$

By transformation, Eq. (15) becomes

$$E = IR \text{ volts.} \qquad (16)$$

That is, the voltage across any part of a circuit is equal to the product of the current in amperes and the resistance in ohms, provided the current is steady and there are no sources of emf within this part of the circuit.

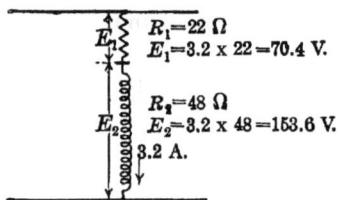

$R_1 = 22\ \Omega$
$E_1 = 3.2 \times 22 = 70.4$ V.
$R_2 = 48\ \Omega$
$E_2 = 3.2 \times 48 = 153.6$ V.
3.2 A.

Fig. 23.—Voltage drops across a generator field and its rheostat.

Example.—The resistance of the field winding of a shunt generator is 48.0 ohms and the resistance of its rheostat is 22.0 ohms. (See Fig. 23.) If the field current is 3.2 amp, what is the voltage across the field winding, the voltage across the rheostat, and the voltage at the generator terminals?

$E_1 = IR_1 = 3.2 \times 22.0 = \underline{70.4}$ volts across rheostat.
$E_2 = IR_2 = 3.2 \times 48.0 = 153.6$ volts across field winding.
 Total $= \overline{224.0}$ volts at generator terminals.

Also,

$E = I(R_1 + R_2) = 3.2(22.0 + 48.0) = 224.0$ volts (check). *Ans.*

Again, if Eq. (15) be solved for the resistance, the result is

$$R = \frac{E}{I} \text{ ohms.} \qquad (17)$$

That is, the resistance of a circuit or part of a circuit is equal to the voltage divided by the current, provided that the current is steady and there are no sources of emf within the part of the circuit under consideration. This formula is very useful in resistance measurements. (See p. 102.)

Example.—The voltage across the terminals of a generator field is 230 volts, and the field current is 7.0 amp. What is the resistance of the field circuit?

$$R = \frac{E}{I} = \frac{230}{7} = 32.9 \text{ ohms.} \quad Ans.$$

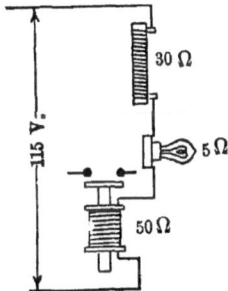

FIG. 24.—Relay circuit.

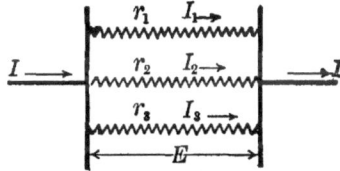

FIG. 25.—Parallel circuit.

26. Series Circuit.—As is stated in Par. 17, p. 17, if several resistances are connected in series, the total resistance is the sum of the individual resistances. That is,

$$R = r_1 + r_2 + r_3 + \cdots \qquad (18)$$

Therefore, the current in such a series circuit

$$I = \frac{E}{R} = \frac{E}{r_1 + r_2 + r_3 + \cdots}. \qquad (19)$$

Example.—A 50-ohm relay is connected in series with a 30-ohm resistance tube and with a small pilot lamp having a resistance of 5.0 ohms, as shown in Fig. 24. The operating voltage is 115 volts. What current flows in this relay circuit?

$$I = \frac{115}{50 + 30 + 5.0} = \frac{115}{85} = 1.35 \text{ amp.} \quad Ans,$$

27. Parallel Circuit.—In Par. 17, p. 17, the relation of total resistance to the component resistances in a parallel circuit was proved by transforming conductances into resistances. Equation (13), p. 18, giving the relation among parallel resistances, may be proved by Ohm's law as follows: Consider the circuit of Fig. 25, consisting of resistances r_1, r_2, and r_3 in parallel across the voltage E. Let I_1 be the current in resistance r_1, I_2 the current in r_2, and I_3 the current in r_3.

Then

$$I_1 = \frac{E}{r_1}, \qquad I_2 = \frac{E}{r_2}, \qquad I_3 = \frac{E}{r_3}. \qquad \text{[Eq. (15)]}$$

Adding,

$$I_1 + I_2 + I_3 = \frac{E}{r_1} + \frac{E}{r_2} + \frac{E}{r_3} = E\left(\frac{1}{r_1} + \frac{1}{r_2} + \frac{1}{r_3}\right).$$

Let the total current be $I = I_1 + I_2 + I_3$.
Let the equivalent resistance be R.

$$I = \frac{E}{R}.$$

Substituting I for $I_1 + I_2 + I_3$,

$$I = \frac{E}{R} = E\left(\frac{1}{r_1} + \frac{1}{r_2} + \frac{1}{r_3}\right),$$

and

$$\frac{1}{R} = \frac{1}{r_1} + \frac{1}{r_2} + \frac{1}{r_3}. \tag{20}$$

That is,
The reciprocal of the equivalent resistance of a parallel circuit is the sum of the reciprocals of the individual resistances.

If but two resistances are involved,

$$R = \frac{r_1 r_2}{r_1 + r_2}. \tag{21}$$

Example.—Four lamps, having hot resistances of 50, 40, 25, and 15 ohms, are connected in parallel across the mains of a 32-volt lighting system

(Fig. 26).   What is the equivalent resistance of the combination, and what is the total current?

Applying Eq. (20),

$$\frac{1}{R} = \frac{1}{50} + \frac{1}{40} + \frac{1}{25} + \frac{1}{15} = 0.0200 + 0.0250 + 0.0400 + 0.0667$$
$$= 0.1517 \text{ mho.}$$

$$R = \frac{1}{0.1517} = 6.59 \text{ ohms.}\quad Ans.$$

$$I = \frac{32}{6.59} = 4.86 \text{ amp.}\quad Ans.$$

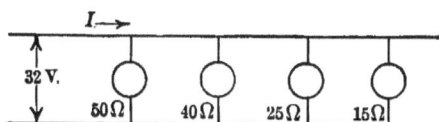

FIG. 26.—Circuit consisting of four lamps in parallel.

28. Division of Current in a Parallel Circuit.—In Fig. 27, two resistances R_1 and R_2 are connected in parallel across the voltage E.   The current I_1 flows in the resistance R_1, and the current I_2

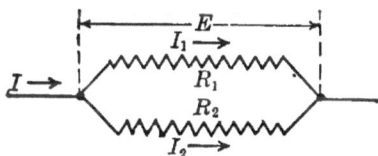

FIG. 27.—Division of current in a two-branch parallel circuit.

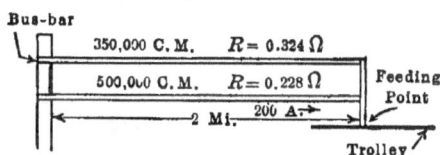

FIG. 28.—Trolley feeders in parallel.

flows in the resistance R_2.   It is desired to obtain the ratio of I_1 to I_2.

$$I_1 = \frac{E}{R_1}, \qquad I_2 = \frac{E}{R_2},$$

and

$$\frac{I_1}{I_2} = \frac{E/R_1}{E/R_2} = \frac{R_2}{R_1}. \tag{22}$$

That is,

In a parallel circuit of two branches, each of which contains resistance only, the currents are inversely as the resistances.   (This does *not* apply to the division of current between the field and armature of a shunt motor *when the motor is running.*)

Example.—A 2-mile length of 350,000 cir-mil trolley-feeder having a total resistance of 0.324 ohm and a 2-mile length of 500,000 cir-mil feeder having a total resistance of 0.228 ohm are connected in parallel between the station bus-bars and the trolley feeding point (Fig. 28). If a current of 200 amp flows in the 500,000 cir-mil feeder, what current flows in the 350,000 cir-mil feeder?

Let I_1 be the current in the 350,000 cir-mil feeder and I_2 be the current in the 500,000 cir-mil feeder. Applying Eq. (22),

$$\frac{I_1}{200} = \frac{0.228}{0.324}, \qquad I_1 = 200\frac{0.228}{0.324} = 141 \text{ amp. } Ans.$$

Since both feeders are connected across the same difference of potential, $I_1R_1 = I_2R_2$.
Hence,

$$141 \times 0.324 = 45.6 \text{ volts.} \qquad 200 \times 0.228 = 45.6 \text{ volts (check).}$$

That is, 45.6 volts are lost in overcoming the joint resistance of the two feeders or the resistance of either separately.

Example.—A current of 12.0 amp. divides between two paths in parallel, part passing through a branch having a resistance of 8.0 ohms, the other branch having a resistance of 12.0 ohms. How much current flows in each branch?

Let I_1 be the current in the 8.0-ohm branch and I_2 be the current in the 12.0-ohm branch.

$$\frac{I_1}{I_2} = \frac{12.0}{8.0} \text{ (Eq. (22)).} \tag{I}$$

$$I_1 + I_2 = 12. \tag{II}$$

$$I_1 = I_2\frac{12.0}{8.0} = I_2\frac{3}{2} \text{ (Eq. (I)).}$$

Substituting in (II),

$$I_2\frac{3}{2} + I_2 = 12.0.$$

$$\frac{5I_2}{2} = 12.0, \qquad I_2 = 4.8 \text{ amp. } Ans.$$

$$I_1 = 4.8\frac{3}{2} = 7.2 \text{ amp. } Ans.$$

29. Series-parallel Circuit.—A circuit may consist of groups of parallel resistances in series with other resistances or groups of resistances, as shown in Figs. 29 and 30. When such is the case, each group of parallel resistances is first combined into its equivalent single resistance by Eq. (20), p. 30 and the whole is then treated as a series circuit.

Example.—Two heating units having resistances of 8.0 and 12 ohms are connected in parallel and are supplied from constant-voltage, 115-volt bus-bars over a 100-ft run of No. 10, A.W.G. wire, each conductor of which has a resistance of 0.10 ohm (Fig. 29). What is the total current, and what is the current in each heating unit?

The 8.0-ohm and 12-ohm resistances are combined into an equivalent resistance R_1 by Eq. (20).

$$\frac{1}{R_1} = \frac{1}{8.0} + \frac{1}{12} = 0.1250 + 0.0833 = 0.2083 \text{ mho.}$$

$$R_1 = \frac{1}{0.2083} = 4.80 \text{ ohms.}$$

$$I = \frac{115}{0.10 + 0.10 + 4.80} = \frac{115}{5.00} = 23.0 \text{ amp.} \quad Ans.$$

The voltage across the units

$$V = 23.0 \times 4.80 = 110.4 \text{ volts.}$$

The current in 8.0 ohms, $I_1 = \frac{110.4}{8.0} = 13.8 \text{ amp.} \quad Ans.$

The current in 12 ohms, $I_2 = \frac{110.4}{12} = 9.2 \text{ amp.} \quad Ans.$

Example.—Determine the total current in the circuit shown in Fig. 30; the voltage across each part of the circuit; the current in each resistance.

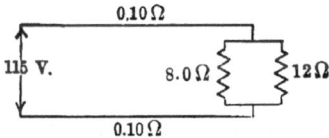

FIG. 29.—Heating units in parallel and in series with line resistance.

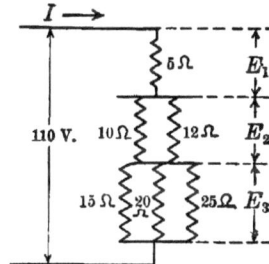

FIG. 30.—Series-parallel circuit.

Combine first the 10-ohm and 12-ohm resistances into a resistance R_1.

$$\frac{1}{R_1} = \frac{1}{10} + \frac{1}{12} = 0.10 + 0.0833 = 0.1833 \text{ mho.}$$

$$R_1 = \frac{1}{0.1833} = 5.45 \text{ ohms.}$$

Likewise, combining the group of three resistances into R_2,

$$\frac{1}{R_2} = \frac{1}{15} + \frac{1}{20} + \frac{1}{25} = 0.0667 + 0.050 + 0.040$$
$$= 0.1567 \text{ mho.}$$

$$R_2 = \frac{1}{0.1567} = 6.38 \text{ ohms.}$$

$$I = \frac{110}{5 + 5.45 + 6.38} = \frac{110}{16.83} = 6.53 \text{ amp.}$$

$E_1 = 6.53 \times 5.0 \quad = \quad 32.7$ volts.

$E_2 = 6.53 \times 5.45 \quad = \quad 35.6$ volts.

$E_3 = 6.53 \times 6.38 \quad = \quad 41.7$ volts.

$\qquad\qquad\qquad$ Total $= \overline{110.0}$ volts (check).

Current in 10 ohms $= \dfrac{35.6}{10} = 3.56$ amp.

Current in 12 ohms $= \dfrac{35.6}{12} = 2.97$ amp.

$\qquad\qquad\qquad$ Total $= \overline{6.53}$ amp. (check).

Current in 15 ohms $= \dfrac{41.7}{15} = 2.78$ amp.

Current in 20 ohms $= \dfrac{41.7}{20} = 2.09$ amp.

Current in 25 ohms $= \dfrac{41.7}{25} = 1.67$ amp.

$\qquad\qquad\qquad$ Total $= \overline{6.54}$ (check).

30. Electrical Power.—The unit of electrical power is the *watt* and is defined as follows: "The *International Watt* is the energy expended per second by an unvarying electric current of one international ampere under the pressure of one international volt." The power in watts is, therefore, equal to the product of the volts and the amperes. Thus the power

$$P = EI \text{ watts.} \qquad\qquad (23)$$

Since $E = IR$ in a circuit containing resistance only (Eq. (16), p. 28, Eq. (23) may be written

$$P = (IR)I = I^2R. \qquad\qquad (24)$$

Substituting for I its value ($I = E/R$) in Eq. (23),

$$P = E\left(\frac{E}{R}\right) = \frac{E^2}{R}. \qquad\qquad (25)$$

Equation (23) is useful when the volts and the amperes are known; Eq. (24) is useful when the current and the resistance are known; and Eq. (25) is useful when the voltage and the resistance are known.

Example.—The resistance of a 150-scale voltmeter is 12,000 ohms. What power is consumed by this voltmeter when it is connected across a 125-volt circuit?

Since the voltage and the resistance are known, Eq. (25) is most convenient.

$$P = \frac{(125)^2}{12,000} = 1.30 \text{ watts. } Ans.$$

This may be verified by Eq. (23).

$$I = \frac{125}{12,000} = 0.0104 \text{ amp.}$$
$$P = 125 \times 0.0104 = 1.30 \text{ watts (check).}$$

Example.—A motor takes 40 amp at 220 volts (Fig. 31). The field resistance is 110 ohms and the armature resistance R_a is 0.25 ohm. (*a*) How much power does the motor take? (*b*) How much power is consumed by the field? (*c*) How much power does the armature take? (*d*) What power is lost as heat in the armature?

(*a*) Applying Eq. (23), $P_1 = 220 \times 40 = 8,800$ watts. *Ans.*

(*b*) Applying Eq. (25), $P_2 = (220)^2/110 = 440$ watts. *Ans.*

(*c*) The field current $I_f = {}^{220}\!/_{110} = 2.0$ amp.

The armature current $I_a = 40.0 - 2.0 = 38.0$ amp.

The armature power $P_a = 220 \times 38.0 = 8,360$ watts. *Ans.*

(*d*) The armature heating loss is found by using Eq. (24).

$$P_c = (38.0)^2 \times 0.25 = 361 \text{ watts. } Ans.$$

The watt is often too small a unit for commercial use, and the *kilowatt* (1,000 watts) is used when large amounts of power are considered. It is often necessary to transform from mechanical horsepower to electrical power, and conversely; and a knowledge of the relation of the two is, therefore, useful.

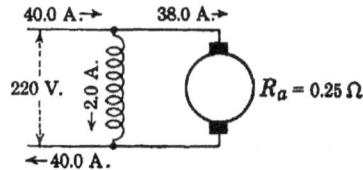
Fig. 31.—Power taken by a shunt motor.

$$746 \text{ watts} = 1 \text{ hp,} \tag{26}$$
$$0.746 \text{ kw} = 1 \text{ hp,} \tag{27}$$

and

$$1 \text{ hp} = \tfrac{3}{4} \text{ kw, very nearly,} \tag{28}$$
$$1 \text{ kw} = \tfrac{4}{3} \text{ hp very nearly.} \tag{29}$$

(Also see Appendix, p. 317.)

Example.—An electric motor takes 28 amp at 550 volts and has an efficiency of 89 per cent.   What horsepower does it deliver?

Input = 28 × 550 = 15,400 watts.
Output = 15,400 × 0.89 = 13,700 watts.
$$\frac{13,700}{746} = 18.35 \text{ hp at the pulley.} \quad Ans.$$

31. Electrical Energy.—Power is the *rate of doing work* or the *rate of expenditure of energy.*   For example, the definition of the international watt, p. 34, states that a watt is equal to the energy expended *per second* by an unvarying electric current of one international ampere under the pressure of one international volt.   That is, power is *rate* of expenditure of energy.   The energy expended in one second by one international ampere under the pressure of one international volt is one *joule.*   Hence, a joule is equal to a *watt-second.*   Electrical energy is given by the product of electrical power and time.   The unit of electrical energy is therefore the *watt-second,* or *joule.*

$$W = EIt \text{ watt-sec} \qquad (30)$$

if t is in seconds, E is in volts, and I is in amperes.

The watt-second is ordinarily too small a unit for commercial purposes, and so the larger unit, the *kilowatt-hour* (kw-hr) is commonly used.   1 kw-hr = 1,000 × 60 × 60 = 3,600,000 = (3.6 × 10⁶) joules or watt-sec.

The distinction between power and energy (or work) should be clearly kept in mind.   Power is *rate* of doing work, just as velocity is rate of motion.   On the other hand, energy is the total work done and is equal to the power multiplied by the time during which it acts, just as distance covered is the velocity or rate of motion multiplied by the time.   To speak of a train traveling at a rate of 40 miles per hr gives no information as to the total distance which the train travels.   Likewise, to speak of 50 kw does not state the amount of energy that is involved.   The statement "electricity is sold for so many cents per kilowatt" is incorrect.   The correct expression is "electrical *energy* is sold for so many cents per kilowatt-*hour.*"

Example.—If energy is sold for 5 c per kw-hr, how many kilowatts may be purchased for 20 c?

This question as it stands cannot be answered, since the *time* is not given. If, however, it is assumed that the power is to be used for 1 hr,

$$\frac{20\ c}{5\ c} = 4\ \text{kw-hr available.}$$

$$\frac{4\ \text{kw-hr}}{1\ \text{hr}} = 4\ \text{kw.} \quad Ans.$$

If used in 0.5 hr, $\dfrac{4\ \text{kw-hr}}{0.5\ \text{hr}} = 8\ \text{kw.} \quad Ans.$

If used in 0.001 hr, $\dfrac{4\ \text{kw-hr}}{0.001\ \text{hr}} = 4{,}000\ \text{kw.} \quad Ans.$

So the 20 c could purchase *any number* of *kilowatts*, depending on the time during which the power is supplied.

In a similar way, horsepower is *rate of doing work* and is equivalent to 33,000 ft-lb *per min* and *not* to 33,000 ft-lb. A motor developing ⅛ hp could do 33,000 ft-lb of work if allowed 8 min in which to do it. When speaking of *work* in connection with horsepower, the *horsepower-hour* is the unit ordinarily used.

Example.—How many watt seconds are supplied by a motor developing 2 hp for 5 hr?

$$2 \times 5 = 10\ \text{hp-hr}$$
$$10\ \text{hp-hr} \times 746 = 7{,}460\ \text{watt-hr}$$
$$7{,}460 \times 3{,}600 = 2.68 \times 10^7\ \text{watt-sec.} \quad Ans.$$

32. Heat and Energy.—It is well known that heat may be converted into mechanical and electrical energy and, conversely, that electrical and mechanical energy may be converted into heat. The complete cycle of energy transformation is well illustrated by a steam power plant. The energy is brought to the plant in the coal, as *chemical energy*. The ingredients of the coal combine with the oxygen of the air, thus converting the chemical energy into *heat energy*. A certain percentage of this heat is transferred to the boiler and produces steam. The expansion of the steam in the engine cylinders, or through the buckets and blades of the turbine, converts the heat energy of the steam into *mechanical energy*. This mechanical energy drives the generator, which converts a large percentage of this energy into *electrical energy*. A portion of this electrical energy is transformed into heat in the wires, bus-bars, etc. Finally, the remainder is used to supply lamps, propel electric cars, and operate motors, and some may be used for chemical processes. Ultimately, all the energy is converted into heat or chemical or other forms of energy.

The following table shows approximately what becomes of each 100 heat units, existing initially in the coal, in the most efficient modern power plants, using superheaters, condensers, and large generating units.

EFFICIENCY OF ENERGY CONVERSION

	Form of energy	Efficiency (percentage)	Heat units converted
Coal..............................	Chemical	......	100.0
Boiler............................	Heat	85	85.0
Turbine...........................	Mechanical	30	25.5
Generator.........................	Electrical	96	24.5
Distribution system (to point of utilization).......................	Electrical	80	19.6
Motors (large)....................	Mechanical	85 (av.)	16.7
Lamps.............................	Light	3.5	0.69

Figure 32* shows graphically the flow of energy in a modern electric power system, from the boiler to the point of utilization, the efficiencies of energy conversion being those given above.

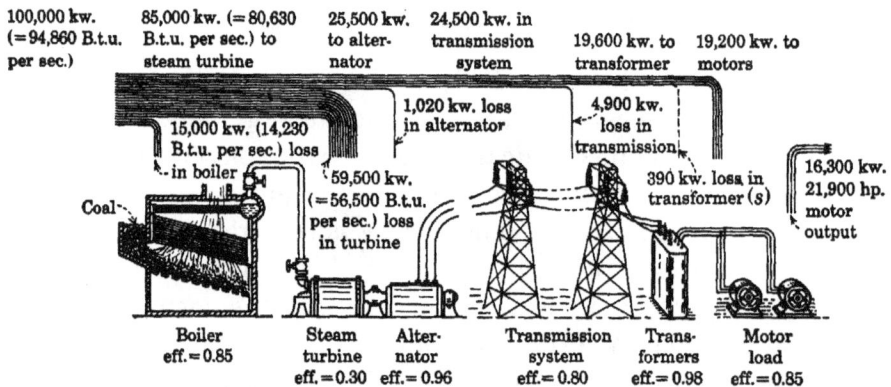

FIG. 32.—Energy flow in typical power system.

It is apparent that even in the most modern power plants the over-all efficiency is low.

33. Thermal Units.—The unit of heat in the English system is the Btu (British thermal unit) and is equal to the amount

* This diagram is based on one by R. A. Philip; see *Trans.*, A.I.E.E., Vol. 34, p. 781, 1915.

of heat required to raise the temperature of one pound of water by 1°F. It is equal to 778 ft-lb (called the "mechanical equivalent" of heat). That is, if all the heat energy of a Btu could be converted into useful work it could raise 1 lb a distance of 778 ft.

In the cgs system the heat unit is the gram-calorie[1] and is the amount of heat required to raise the temperature of one gram of water by 1°C. A gram-calorie is equal to 4.184 watt-sec, or joules.

By Joule's law the heat developed in a circuit is

$$W = \frac{1}{4.184} I^2Rt = 0.2390 I^2Rt \text{ gram-cal} \qquad (31)$$

where t is in seconds, I in amperes, and R in ohms.

Example.—It is desired to raise the temperature of 1 liter of water from 20°C to the boiling temperature 100°C, with an immersion heater which takes 3.50 amp at 110 volts. Neglecting losses: (*a*) How much time is required? (*b*) At 6 c per kw-hr, what does it cost to heat the water?

Since 1 liter = 1,000 cc, 1,000 cc × 80° = 80,000 gram-cal required.

$$80,000 \times 4.18 = 334,000 \text{ joules or watt-sec.}$$

(*a*) Rate of energy supply, 110 × 3.50 = 385 watts.

$$\frac{334,000}{385} = 868 \text{ sec, or } 14.5 \text{ min.} \quad Ans.$$

(*b*) $$\frac{334,000}{3,600,000} = 0.093 \text{ kw-hr.}$$

$$0.093 \times 6 \text{ c} = 0.56 \text{ c.} \quad Ans.$$

Example.—Ten horsepower is delivered by a pump circulating 400 gal of water per min through a certain cooling system. How many degrees Fahrenheit per minute would the temperature of the water be raised by the action of the pump?

$$10 \text{ hp} = 10 \times 33,000 = 330,000 \text{ ft-lb per min.}$$
$$\frac{330,000}{778} = 424 \text{ Btu per min.}$$
$$400 \text{ gal weigh } 400 \times 8.34 = 3,336 \text{ lb.}$$
$$\frac{424}{3,336} = 0.13°F. \quad Ans.$$

34. Potential Drop in a Feeder Supplying One Concentrated Load.—Figure 33 shows a feeder (consisting of a positive and a

[1] See Appendix A, p. 317.

negative wire) supplying a motor load. The feeder is connected
to bus-bars having a constant potential difference of 230 volts.
The feeder is 1,000 ft long and consists of two 250,000 cir-mil
conductors. The maximum load on the feeder is 250 amp. It is
required to determine the voltage at the motor terminals and the
efficiency of transmission.

As was stated in Par. 22, p. 23, the voltage at the motor must
be less than that at the bus-bars because of the voltage lost
in supplying the resistance drop in the feeder.

From Appendix H, p. 321, the resistance of 1,000 ft of 250,-
000 cir-mil cable is 0.0431 ohm. As was shown in Par. 22, the net

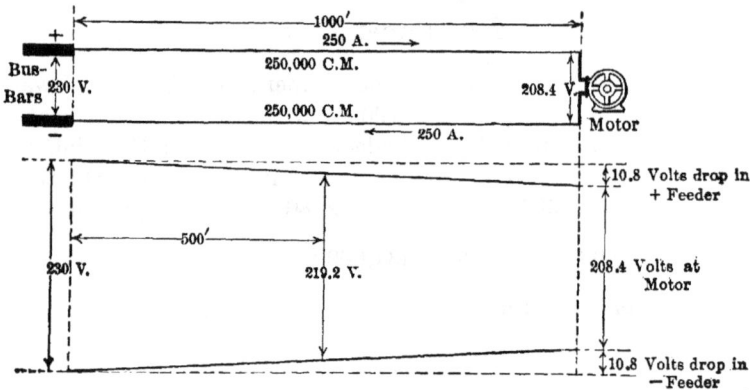

Fig. 33.—Voltage drop in a feeder due to a single load.

voltage at the receiving end of the line is less than the voltage at
the sending end by the voltage loss in both the *outgoing* and the
return wires. Therefore, the drop in 2,000 ft of cable must be
taken into consideration, the total resistance being 0.0862 ohm.

The current is 250 amp.

By Eq. (16), p. 28, the voltage drop in the line

$$E' = 250 \times 0.0862 = 21.55 \text{ volts.}$$

Therefore, the voltage at the motor terminals is

$$230 - 21.6 = 208.4 \text{ volts.} \quad Ans.$$

In Fig. 33 the voltage drop along the line is shown graphically.
The voltage at the sending end of the line is 230 volts, and there
is a uniform drop in each wire, this drop increasing uniformly
to 10.8 volts, making a total voltage loss of 21.6 volts. The

potential difference between the two wires 500 ft from the sending end will be $230 - 10.8 = 219.2$ volts as shown.

The power delivered to the motor $= 208.4 \times 250$ watts.

The power delivered to the line $= 230 \times 250$ watts.

The efficiency of transmission $= \dfrac{\text{output}}{\text{input}} = \dfrac{208.4 \times 250}{230 \times 250} = \dfrac{208.4}{230}$, or 90.6 per cent.

With one concentrated load, the efficiency of transmission is given by the voltage at the load divided by the voltage at the sending end of the line.

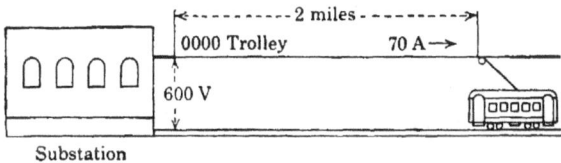

Fig. 34.—Substation supplying a trolley car.

The power loss in each wire

$$P_c = I^2R = (250)^2 \times 0.0431 = 2{,}690 \text{ watts.}$$
$$\text{Total power loss} = 2 \times 2{,}690 = 5{,}380 \text{ watts.}$$

Percentage power loss $= \dfrac{5{,}380}{230 \times 250}100 = \dfrac{5{,}380}{57{,}500}100 = 9.4$ per cent.

The percentage power loss also equals

$$\frac{I^2R}{EI}100 = \frac{IR}{E}100 = \frac{21.55}{230}100 = 9.4 \text{ per cent.}$$

That is, with a single concentrated load, the percentage power loss is equal to the ratio of the IR drop in the line to the sending-end voltage.

Example.—A trolley car, 2 miles from the substation (Fig. 34), takes 70 amp over a 0000 hard-drawn, copper, trolley wire having a resistance of 0.270 ohm per mile. The rail and ground return have a resistance of 0.06 ohm per mile. If the substation voltage is 600 volts: (a) What is the voltage at the car? (b) What is the efficiency of transmission?

Resistance of trolley $= 2 \times 0.270 = 0.540$ ohm.

Resistance of rail return $= 2 \times 0.06 = 0.120$ ohm.

Total resistance $= 0.660$ ohm.

Total IR drop $= 70 \times 0.660 = 46.2$ volts.

(a) Voltage at car $= 600 - 46.2 = 553.8$ volts. *Ans.*

(b) Efficiency of transmission $= 553.8/600 = 0.923$, or **92.3** per cent. *Ans.*

CHAPTER III

BATTERY ELECTROMOTIVE FORCE— KIRCHHOFF'S LAWS

35. Battery Electromotive Force and Resistance.—If a voltmeter be connected across the terminals of a battery (Fig. 35), the switch S being open, the instrument will indicate a voltage E. If the switch S be closed, allowing a current I to flow, the instrument will indicate a voltage V which is less than E.

The voltage E, measured when the battery delivers no current, is the *internal voltage* or the emf of the battery; the voltage V,

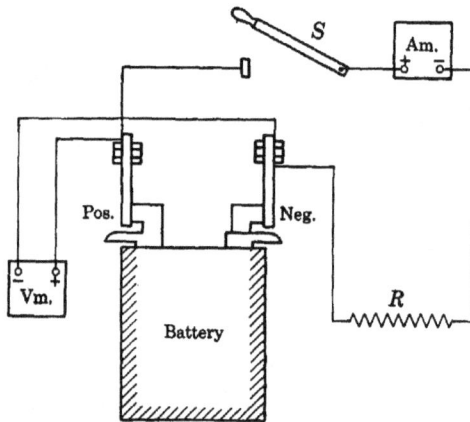

Fig. 35.—Connections for measuring battery resistance.

measured when a current I flows, is the *terminal voltage* of the battery for that particular current value.

The difference between the open-circuit voltage E and the voltage V, measured when current is being taken from the battery, is the *voltage drop* in the battery due to the passage of current through the battery resistance. Every cell has resistance, lying for the most part in the electrolyte but partly in the battery plates and terminals. When the external circuit is closed so that current flows, voltage is required to send this current through the

42

battery resistance, just as voltage is required to send current through an external resistance.

If the voltage E, measured at the battery terminals when the circuit is open, drops to V when the circuit is closed, the voltage $e = (E - V)$ is the voltage drop through the cell due to the passage of the current I. Let the cell resistance be r. Then, by Ohm's law,

$$E - V = e = Ir \text{ (Eq. (16), p. 28)}$$

or

$$r = \frac{e}{I} = \frac{E - V}{I} \text{ (Eq. (17), p. 29).} \tag{32}$$

That is, the internal resistance of the battery is equal to the open-circuit voltage minus the closed-circuit terminal voltage divided by the closed-circuit current.

Also, by transformation,

$$E = V + Ir. \tag{33}$$
$$V = E - Ir. \tag{34}$$

When a battery delivers current, the terminal voltage is equal to the emf minus the resistance drop within the battery.

Example.—The open-circuit voltage of a storage cell is 2.20 volts. The terminal voltage measured when a current of 12 amp flows is found to be 1.98 volts. What is the internal resistance of the cell?

The voltage drop through the cell

$$E - V = 2.20 - 1.98 = 0.22 \text{ volt.}$$

Then

$$r = \frac{0.22}{12} = 0.0183 \text{ ohm.} \quad Ans.$$

In making a measurement of this character, it must be remembered that under open-circuit conditions the ordinary voltmeter takes some current. If the cell capacity is small (as with a Weston cell), the voltmeter current alone may reduce the terminal voltage to a value one-half, or even less, of the open-circuit voltage. Under these conditions the ordinary voltmeter cannot be used to measure the emf of the cell.

Moreover, it is impossible to measure directly the internal voltage of the battery when the battery delivers current, for the voltage drop occurs *within* the cell itself. Figure 36 represents these conditions so far as their effect on the internal circuit is concerned. A battery cell B is enclosed in a sealed box. Its resistance r is considered as removed from the cell itself and con-

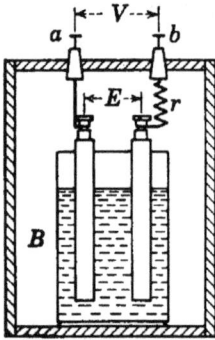

FIG. 36.—Internal resistance of a cell.

nected external to the cell, but within the sealed box. The cell may be considered as having no resistance, its resistance having been replaced by r. The connections are brought through bushings in the box to terminals a and b. When no current is being delivered by the cell, if a voltmeter be connected across the terminals a and b, the instrument will measure the internal emf E. If, however, a current I flows, the terminal voltage will drop from E to V, owing to the voltage drop in the resistance r. Under these conditions, it is impossible to measure E directly when current is flowing, since the voltmeter can be connected only outside the resistance through which the voltage drop occurs.

Example.—A voltmeter, connected across the terminals of a dry cell, reads 1.40 volts when the cell is open-circuited. The voltmeter reads 1.02 volts when the cell delivers a current of 3.0 amp. What is the internal resistance of the cell?

The voltage drop within the cell $e = 1.40 - 1.02 = 0.38$ volt. By Ohm's law,

$$r = \frac{0.38}{3.0} = 0.127 \text{ ohm.} \quad Ans.$$

Or, by applying Eq. (32) directly,

$$r = \frac{1.40 - 1.02}{3.0} = \frac{0.38}{3.0} = 0.127 \text{ ohm.}$$

The voltage E and the resistance r are seldom constant but are more or less dependent on the current. They are also affected by temperature, change in specific gravity of the electrolyte, polarization, etc.

36. Battery Resistance and Current.—As is shown in Par. 35, the resistance within the battery tends to reduce the flow of current. If, in Fig. 35, the switch S be closed, the cell emf E will be acting on a circuit consisting of the internal resistance

r of the cell and the resistance R of the external circuit. The resistances r and R being in series, the total resistance in the circuit is their sum. The current is

$$I = \frac{E}{r + R} \text{ amp.} \tag{35}$$

The total power developed within the battery

$$P = EI \text{ watts.} \tag{36}$$

The power lost in the battery

$$p = I^2 r \text{ watts.}$$

The power delivered to the external circuit by the battery,

$$P' = VI \text{ watts} \tag{37}$$

where V is the terminal voltage.

If the cell is short-circuited, R is zero and $I = E/r$. Under these conditions all the electrical energy developed by the cell is converted into heat within the cell itself.

Example.—A battery cell having an emf of 2.2 volts and an internal resistance of 0.03 ohm is connected to an external resistance of 0.10 ohm. What current flows, and what is the efficiency of the battery as used? If the battery is short-circuited, what current flows?

$$I = \frac{2.2}{0.03 + 0.10} = \frac{2.2}{0.13} = 16.9 \text{ amp.} \quad Ans.$$

Power lost in the battery

$$p = (16.9)^2 \times 0.03 = 8.57 \text{ watts.}$$

The useful power

$$P' = (16.9)^2 \times 0.10 = 28.6 \text{ watts.}$$

P' is equal to the total power P developed by the battery minus the battery loss.

$$P = 2.2 \times 16.9 = 37.2 \text{ watts.}$$
$$P' = 37.2 - 8.6 = 28.6 \text{ watts (check).}$$
$$\text{Eff.} = \frac{28.6}{28.6 + 8.6}, \text{ or 76.9 per cent.} \quad Ans.$$

The short-circuit current

$$I' = \frac{2.2}{0.03} = 73.3 \text{ amp.} \quad Ans.$$

From the foregoing, the following rule may be deduced:

The current in a circuit is equal to the total emf acting in the circuit divided by the total resistance of the circuit.

37. Batteries Receiving Energy.—If a resistance load be connected across a battery, current will flow from the positive terminal of the battery through the load resistance and will return to the battery through the negative terminal. (See Fig. 49, p. 61.) As has been pointed out, the battery terminal voltage will be less than its open-circuit value, owing to the current flowing through the internal resistance of the battery. Under these conditions, the battery is a source of energy and is acting as a generator; that is, it *delivers* energy.

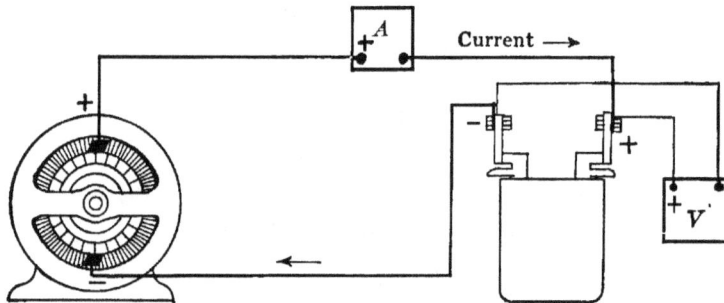

Fig. 37.—Generator charging a battery.

If current is forced to *enter* the battery at its *positive* terminal, the battery will no longer be supplying but will be receiving energy. This energy must be supplied from some other source, as from another battery or, as is more common, from a generator. The cell shown in Fig. 37 has an emf of 2 volts, and a voltmeter *V*, connected across its terminals, indicates 2 volts when no current flows. If another source of electrical energy, such as a d-c generator, supplies a potential difference of just 2 volts, and its + terminal is connected to the + terminal of the battery and its − terminal is connected to the − terminal of the battery, as shown in the figure, the voltmeter *V* will still read 2 volts and the ammeter *A* will read zero. That is, the battery neither delivers nor receives energy, and no effect is noted other than those observed when the battery is open-circuited. Under these conditions the battery is said to be "floating." If, however, the terminal voltage of the generator be raised slightly, the ammeter *A* will indicate a current flowing from the + terminal of the generator *into* the + terminal of the battery, a direction opposite to that which the current has when the battery *delivers* energy. The voltmeter will no longer read 2 volts but will indicate a potential difference somewhat greater than 2 volts.

The foregoing may be illustrated by a mechanical analogue (Fig. 38) which shows a car standing on the track. A force of 400 lb is necessary to overcome the standing friction of the car on the track. At one end of the car a force F is applied. Before the force F can move the car, its value must at least equal 400 lb. When F is exactly 400 lb, the car will not move, just as no current flows into the battery when the generator terminal voltage is just equal to the emf of the battery. When the force F exceeds 400 lb, however, the car will move, the force effective in

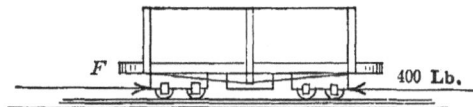

FIG. 38.—Force necessary to start a car.

starting the motion being the amount by which F exceeds 400 lb. Thus, if $F = 450$ lb, 400 lb is utilized in overcoming the 400 lb opposing force due to friction, and 50 lb is effective in moving the car.

With the battery, no current flows until a terminal voltage in excess of the 2 volts is produced by the generator. Thus, if the generator terminal voltage be raised to 2.4 volts, 2.0 volts of this is utilized to "buck" the 2.0 volts of the cell and 0.4 volt is effective in sending current into the cell. If the resistance of the leads be negligible and the cell resistance be 0.1 ohm, the current will be

$$I = \frac{2.4 - 2.0}{0.1} = \frac{0.4}{0.1} = 4.0 \text{ amp.}$$

Therefore, if E is the emf of a battery, r its resistance, and V the terminal voltage when current enters its positive terminal,

$$I = \frac{V - E}{r}. \tag{38}$$

$$V = E + Ir. \tag{39}$$

$$E = V - Ir. \tag{40}$$

That is, the emf of the cell is less than the terminal voltage by the amount of the resistance drop in the cell itself. These equations should be compared with Eqs. (32), (33), and (34), p. 43.

Under these conditions, the cell is *receiving* electric energy, as when a storage battery is being charged.

Example.—A 6-volt ignition battery, consisting of three lead storage cells in series, has a total emf of 6.2 volts and an internal resistance of 0.08 ohm. What is the terminal voltage of the battery when it is being charged at the 15-amp rate? How much energy is stored chemically per hour?

The resistance drop in the battery

$$e = 15 \times 0.08 = 1.2 \text{ volts.}$$

The terminal voltage

$$V = 6.2 + 1.2 = 7.4 \text{ volts.} \quad \textit{Ans.}$$

The energy stored in an hour is

$$15 \times 6.2 \times 1.0 = 93.0 \text{ watt-hr.} \quad \textit{Ans.}$$

38. Battery Cells in Series.—Strictly speaking, a battery consists of more than one unit or cell. However, the term "battery" may also mean a single cell, when this cell is not acting in conjunction with others. It is important to know the behavior of cells with respect to the external circuit, when the cells are connected in series and in parallel.

When cells are connected in series, their emfs are added to obtain the total emf of the battery, and their resistances are added to obtain the total resistance of the battery.

Thus, if several cells, having emfs E_1, E_2, E_3, E_4, . . . and resistances r_1, r_2, r_3, r_4, . . . are connected in series to form a battery, the total emf of the combination is

$$E = E_1 + E_2 + E_3 + E_4 + \cdots \tag{41}$$

and the total resistance is

$$r = r_1 + r_2 + r_3 + r_4 + \cdots \tag{42}$$

Equation (41) assumes that the cells are all connected + to − so that their emfs add. If any cell be connected so that its emf opposes the others, its emf in Eq. (41) must be preceded by a minus sign.

If an external resistance R is connected across the terminals of this battery, then by Eq. (35), p. 45, the current is

$$I = \frac{E}{r + R} = \frac{E_1 + E_2 + E_3 + E_4 + \cdots}{r_1 + r_2 + r_3 + r_4 + \cdots + R}. \tag{43}$$

Example.—Four dry cells, having emfs 1.30, 1.30, 1.35, 1.40 volts and resistances 0.3, 0.4, 0.2, 0.1 ohm, are connected in series to operate a relay having a resistance of 10 ohms. What current flows in the relay?

$$I = \frac{1.30 + 1.30 + 1.35 + 1.40}{0.3 + 0.4 + 0.2 + 0.1 + 10} = \frac{5.35}{11.0} = 0.486 \text{ amp.} \quad Ans.$$

A battery consisting of n equal cells in series has an emf n times that of one cell but has the current capacity of one cell only.

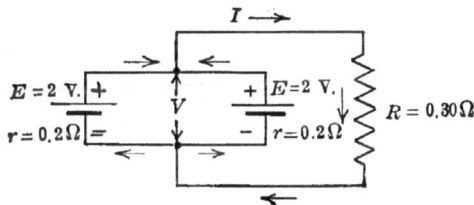

FIG. 39.—Parallel connection of two equal cells.

39. Equal Batteries in Parallel.—To operate satisfactorily in parallel, all the batteries should have the same emf. The behavior of batteries having unequal emfs can be treated as a special problem (see Par. 41).

Figure 39 shows two cells in parallel, each having an emf of 2.0 volts and a resistance of 0.2 ohm. It is clear that the emf of the entire battery is no greater than the emf of either cell. The current, however, has two paths through which to flow. Therefore, for a fixed external current, the voltage drop in each cell is one-half that occurring when the entire current passes through one cell. If the internal resistance of each cell is 0.2 ohm, the resistance of the battery must be 0.2/2 = 0.10 ohm.

Example.—Find the external current in Fig. 39, when a resistance of 0.30 ohm is connected across the battery terminals.

The battery resistance = 0.20/2 = 0.10 ohm.

$$I = \frac{2.0}{0.10 + 0.30} = 5.0 \text{ amp.} \quad Ans.$$

The current in each cell is 2.5 amp.

Considering a single cell only, the terminal voltage

$$V = 2.0 - 2.5 \times 0.2 = 1.5 \text{ volts.}$$

Considering the battery as a whole, the terminal voltage

$$V = 2.0 - 5.0 \times 0.1 = 1.5 \text{ volts (check).}$$

$$\text{The current } I = \frac{1.5}{0.30} = 5.0 \text{ amp (check).}$$

40. Kirchhoff's Laws.—By means of Kirchhoff's laws it is possible to solve circuit networks that would otherwise be difficult of solution. The two laws may be stated as follows:

1. *In any branching network of wires, the algebraic sum of the currents in all the wires that meet at a point is zero.*

2. *In any complete electrical circuit, the sum of all the emfs and all the resistance drops, taken with their proper signs, is zero.*

The first law is obvious. It states that the total current *leaving* a junction is equal to the total current *entering* the junction. If this were not so, electricity would either accumulate or disappear at the junction, an obvious impossibility.

The law is illustrated in Fig. 40. Four currents I_1, I_2, I_3, I_4 meet at the junction O. The first three currents flow *toward* the junction and so have plus signs, as they *add* to the quantity of electricity at the point O. The current I_4 flows *away* from the junction and so has a minus sign, as it subtracts from the quantity at the point O. Then

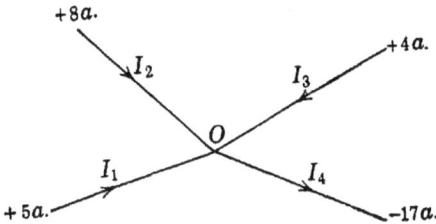

Fig. 40.—Illustrating Kirchhoff's first law.

$$I_1 + I_2 + I_3 - I_4 = 0 \text{ (first law).} \qquad (44)$$

Assume that $I_1 = 5$ amp, $I_2 = 8$ amp, and $I_4 = 17$ amp. Then

$$5 + 8 + I_3 - 17 = 0$$

and

$$I_3 = +4 \text{ amp,}$$

the plus sign indicating that the current I_3 flows toward the junction.

The second law is but another application of Ohm's law (Eq. (16), p. 28). The basis of the law is obvious; if one starts at a certain point in a circuit and follows continuously around the paths of the circuit until the starting point is reached, he must come to the same potential with which he started. Therefore, the sources of emf encountered in this passage must necessarily be equal to the voltage drops in the resistances, every voltage being given its proper sign.

This second law is illustrated by the following example.

Two batteries (Fig. 41 (a)), having emfs of 10 and 6 volts and internal resistances of 1 and 2 ohms, are connected in series oppos-

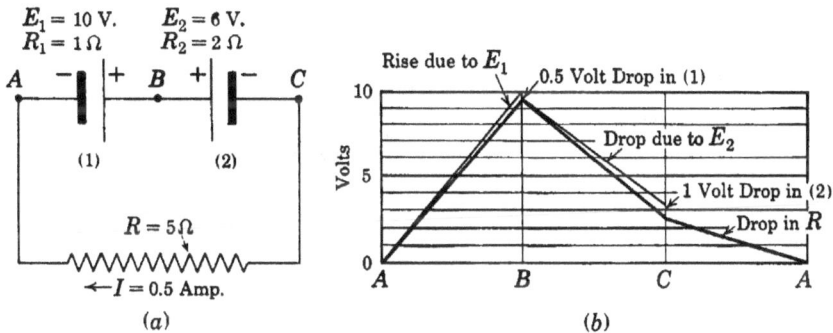

FIG. 41.—Voltage relations in a circuit

ing (their + terminals connected together) and in series with an external resistance of 5 ohms. Determine the current and the voltage at each part of the circuit.

Since the two batteries act in opposition, the net emf of the two batteries is $10 - 6 = 4$ volts.

The current

$$I = \frac{10 - 6}{1 + 2 + 5} = \frac{4}{8} = 0.5 \text{ amp.}$$

When a current flows through a resistance, there is always a *drop* of potential in the direction of the current. This is illustrated by Fig. 17, p. 24, where the line $a'b'$ shows that the voltage drops in the direction of the current. This is further illustrated by a stream of water, such as a river. When one travels in the direction of the flow of water, he finds that the level of the water is continually dropping with respect to some fixed level on the earth's surface.

It also follows that when a current flows through resistance there is always a *rise* of potential in the direction opposite to the current. Similarly, when one travels along a river against the current, he finds that the level of the water is continually rising with respect to some fixed level on the earth's surface.

Referring again to Fig. 41, consider the point *A* as being at reference or zero potential. In passing from *A* to *B* one goes from the negative to the positive terminal of battery (1). Therefore, there is a 10-volt *rise* in potential due to the emf of battery (1). Since the current flows in the direction from *A* to *B*, there occurs a simultaneous 0.5-volt $(10-9.5)$ drop of potential due to the current of 0.5 amp flowing through the 1-ohm resistance of battery (1). Therefore, the net potential at *B* is but 9.5 volts greater than that at *A*, as is shown in Fig. 41 (*b*). In going from *B* to *C* there is a drop of 6 volts, due to going from the + to the − terminal of battery (2), and there is also a further drop of 1 volt, due to the current of 0.5 amp flowing through the 2-ohm resistance of battery (2). This makes the net potential at *C* equal to

$$9.5 - 6 - 1 = +2.5 \text{ volts.}$$

In going from *C* to *A* there is a drop in potential of 2.5 volts, due to the current of 0.5 amp flowing through the 5-ohm resistance. When point *A* is reached, the potential has dropped to zero.

Therefore, the sum of all the emfs and *IR* drops in the circuit, each with its proper sign, is equal to zero. This is illustrated as follows:

Electromotive Forces	*IR* Drops
Cell (1) = + 10 volts	Cell (1) = $-0.5 \times 1 =$ − 0.5 volt
Cell (2) = − 6 volts	Cell (2) = $-0.5 \times 2 =$ − 1.0 volt
Total = + 4 volts	5-ohm res. = $-0.5 \times 5 =$ − 2.5 volts
	Total = − 4.0 volts

$$+4 + (-4) = 0$$

41. Simple Applications of Kirchhoff's Laws.—In the application of Kirchhoff's second law to specific problems, the question of algebraic signs may be troublesome and is a frequent source of error. If the following rules be kept in mind, no difficulties should occur:

A rise in potential should be preceded by a + sign.

A drop in potential should be preceded by a − sign.

For example, in going *through a battery* from the − terminal to the + terminal, the potential *rises* owing to the battery emf, so that this emf should be preceded by a + sign. On the other hand, when going *through the battery* from the + terminal to the − terminal, the potential *drops* due to the battery emf, so that a − sign should precede this emf. These relations are illustrated by Fig. 41.

When going through a resistance in the *direction* of the current, the voltage drops, so that this voltage should be preceded by a − sign. When going through a resistance in the direction *opposite* to the current, the voltage rises, so that this voltage should be preceded by a + sign. These two rules apply to the internal resistances of batteries as well as to external resistances.

The foregoing rules are further illustrated by the electric circuit shown in Fig. 42. Two batteries, having emfs E_1, E_2, and internal resistances r_1, r_2, are connected in parallel, with positive terminal to positive terminal and negative terminal to negative terminal. An external resistance R is connected across the battery terminals *ab*. Assume the current in the left-hand battery to be I_1, the current in the right-hand battery

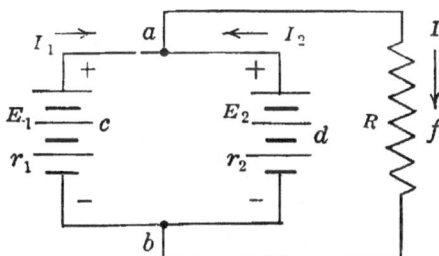

Fig. 42.—Application of Kirchhoff's laws to battery circuits.

I_2, and the external current I. The arrows show the *assumed* direction of current flow. Since there are three unknown quantities I_1, I_2, I, three independent equations are necessary.

First, considering the path *bcafb* and applying Kirchhoff's second law,

$$+E_1 - I_1 r_1 - IR = 0. \qquad (I)$$

Another equation may be obtained by applying Kirchhoff's second law to path *bdafb*.

$$+E_2 - I_2 r_2 - IR = 0. \qquad (II)$$

A third equation may be obtained by considering path *bcadb;* but when this equation is combined with (II) it gives (I), and when combined with (I) it gives (II). Hence, this third equation, when combined with either (I) or (II), will not give a solution. The third equation must involve Kirchhoff's *first* law. For example, Kirchhoff's first law may be applied to junction *a*.

$$+I_1 + I_2 - I = 0. \tag{III}$$

I_1 and I_2 are preceded by $+$ signs, since they add to the quantity of electricity at *a*, and I is preceded by a $-$ sign, since it subtracts from the quantity at *a*.

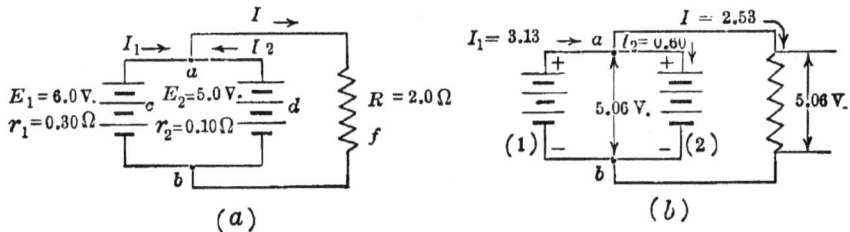

FIG. 43.—Determination of currents in batteries by Kirchhoff's laws.

Equations (I), (II), and (III), when solved simultaneously, give the three currents, I_1, I_2, I. This may be illustrated by giving numerical values to the emfs and resistances shown in Fig. 43 (a).

For path *bcafb*,

$$+6.0 - I_1(0.30) - I(2.0) = 0. \tag{I}$$

For path *bdafb*,

$$+5.0 - I_2(0.10) - I(2.0) = 0. \tag{II}$$

At junction *a*,

$$I_1 + I_2 - I = 0, \quad \text{or} \quad I_1 + I_2 = I. \tag{III}$$

Substituting the value of I from (III) in (I) and (II),

$+6.0 - 0.30I_1 - 2.0(I_1 + I_2) = 0,$ or
$$+6.0 - 2.3I_1 - 2.0I_2 = 0; \tag{IV}$$
$+5.0 - 0.10I_2 - 2.0(I_1 + I_2) = 0,$ or
$$+5.0 - 2.0I_1 - 2.1I_2 = 0. \tag{V}$$

Multiplying (IV) by 2.1, (V) by 2.0, and subtracting,

$$12.6 - 4.83\,I_1 - 4.2\,I_2 = 0$$
$$10.0 - 4.00\,I_1 - 4.2\,I_2 = 0$$
$$\overline{2.6 - 0.83\,I_1 = 0}$$
$$I_1 = +3.13 \text{ amp.} \quad Ans.$$

Substituting the value of I_1 in (V),

$$5.0 - 6.26 - 2.1\,I_2 = 0.$$
$$2.1\,I_2 = 5.0 - 6.26 = -1.26.$$
$$I_2 = -0.60 \text{ amp.} \quad Ans.$$

The negative sign means that the current I_2 flows in a direction *opposite* to that *assumed*.

From (III)

$$I = 3.13 - 0.60 = 2.53 \text{ amp.} \quad Ans.$$

As a check,

$$V_{ab} = 6.0 - 3.13 \times 0.3 = 6.0 - 0.94 = 5.06 \text{ volts.}$$
$$V_{ab} = 5.0 - (-0.60 \times 0.1) = 5.06 \text{ volts.}$$
$$V_{ab} = 2.53 \times 2.0 = 5.06 \text{ volts.}$$

Figure 43 (*b*) shows the same circuit as that of Fig. 43 (*a*) but gives the numerical values of voltage and current and the actual directions of the currents as well. It will be noted that battery (1) supplies 3.13 amp. Of this current, 2.53 amp flow to the external resistance R and 0.60 amp flows into battery (2). Battery (2) is, therefore, receiving energy. If it is a storage battery, it is being charged.

42. Assumed Direction of Current Flow.—In solving for the three unknown currents in the problem of Par. 41, a negative value of the current I_2 was required to satisfy the three simultaneous equations. This negative value of the current I_2 means that the current I_2 flows in a direction opposite to the *assumed* direction, as is shown by comparing the directions of the arrows in Figs. 43 (*a*) and (*b*). Consequently, in writing the equations for a circuit, it is immaterial whether or not the arrows are shown as pointing in the actual direction of flow of current. The signs preceding the answers show whether or not the assumed directions are correct. If the value of current found by solving the equations is preceded by a positive sign, the assumed direction is correct. If the value of current found by solving the equations is preceded by a negative sign, the actual direction of the current is opposite to the direction assumed. After having once assumed the direction of current, however, it is necessary that the quantities in the equations be preceded by signs consistent with these assumed directions.

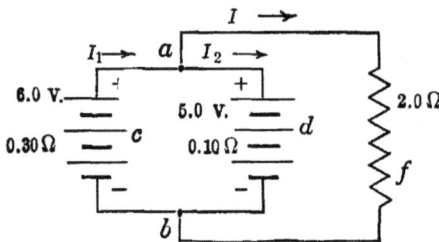

Fig.   44.—Application   of   Kirchhoff's laws to parallel batteries.

As an example, let it be required to solve the problem of Par. 41, with the flow of the current I_2 taken in the direction *adb*, its actual direction of flow as determined by solving the equations in Par. 41. The circuit is shown in Fig. 44.

For path *bcafb*,

$$+6.0 - 0.30\,I_1 - 2.0\,I = 0. \tag{I}$$

For path *bdafb*,

$$+5.0 + 0.10 I_2 - 2.0 I_1 = 0. \tag{II}$$

At junction (*a*),

$$I_1 - I_2 - I = 0, \quad \text{or} \quad I = I_1 - I_2. \tag{III}$$

Substituting this value of I in (I) and (II).

$6.0 - 0.30\,I_1 - 2.0(I_1 - I_2) = 0$ or
$$6.0 - 2.3\,I_1 + 2.0 I_2 = 0: \quad \text{(IV)}$$
$5.0 + 0.10\,I_2 - 2.0(I_1 - I_2) = 0$ or
$$5.0 - 2.0\,I_1 + 2.1 I_2 = 0. \quad \text{(V)}$$

Multiplying (IV) by 2.1 and (V) by 2.0 and subtracting.

$$
\begin{array}{l}
12.6 - 4.83I_1 + 4.2I_2 = 0 \\
10.0 - 4.00I_1 + 4.2I_2 = 0 \\
\hline
2.6 - 0.83\,I_1 = 0
\end{array}
$$

$$I_1 = +\frac{2.6}{0.83} = +3.13 \text{ amp.} \quad \textit{Ans.}$$

Substituting the value of I_1 in (V),

$$5.0 - 6.26 + 2.1I_2 = 0,$$
$$2.1I_2 = +1.26, \qquad I_2 = +0.60 \text{ amp.} \quad Ans.$$

The + sign indicates that the *assumed* direction is the *actual* direction of current flow.

The fact that when writing the simultaneous equations it is immaterial whether or not the assumed direction of the current be correct is again illustrated by the network shown in Fig. 45. The three unknown currents I_1, I_2, I are all *assumed* to be flowing *toward* the junction a, a condition which obviously cannot exist. The fact that the assumed direction of at least one of these

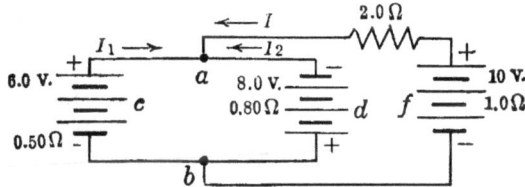

FIG. 45.—Application of Kirchhoff's laws to a general network.

currents is incorrect is shown by a negative sign appearing before one of the currents in the solutions of the equations.

Example.—Find the magnitude and direction for each of the three currents I_1, I_2, I shown in the network, Fig. 45.

Applying Kirchhoff's second law to path *bcafb*,

$$+6.0 - 0.50\,I_1 + 2.0\,I - 10 + 1.0\,I = 0. \qquad \text{(I)}$$

For path *bcadb*,

$$+6.0 - 0.50\,I_1 + 8.0 + 0.80\,I_2 = 0. \qquad \text{(II)}$$

Applying Kirchhoff's first law to the junction a,

$$I_1 + I_2 + I = 0, \qquad \text{(III)}$$
$$I = -I_1 - I_2.$$

substituting this value of I in (I) and solving,

$$+6.0 - 0.50\,I_1 - 3.0\,I_1 - 3.0\,I_2 - 10 = 0,$$
$$-4.0 - 3.50\,I_1 - 3.0\,I_2 = 0, \qquad \text{(IV)}$$
$$\text{(II)} \qquad +14.0 - 0.50\,I_1 + 0.80\,I_2 = 0,$$
$$-4.0 - 3.50\,I_1 - 3.0\,I_2 = 0 \qquad \text{(IV)}$$

Multiplying (II) by 7, and subtracting from (IV),

$$+98.0 - 3.50 I_1 + 5.60 I_2 = 0 \qquad\qquad (V)$$
$$\overline{-102.0 - 8.60 I_2 = 0}$$
$$I_2 = -11.86 \text{ amp.} \quad Ans.$$

Substituting this value of I_2 in (IV),

$$-4.0 - 3.50 I_1 + 35.58 = 0,$$
$$3.50 I_1 = 31.58,$$
$$I_1 = +9.02 \text{ amp.} \quad Ans.$$

From (III),

$$+9.02 - 11.86 + I = 0,$$
$$I = +2.84 \text{ amp.} \quad Ans.$$

Therefore, I_2 flows away from the junction a, being opposite to the direction assumed. The currents I_1 and I flow toward the junction, as assumed.

The foregoing results may be checked by computing the voltage across ab (V_{ab}), using each value of current.

$$V_{ab} = +6.0 - (9.02 \times 0.50) = 1.49 \text{ volts.}$$
$$V_{ab} = -8.0 - (-11.86 \times 0.80) = 1.49 \text{ volts.}$$
$$V_{ab} = +10 - (2.84 \times 3.0) = 1.48 \text{ volts (check).}$$

43. Further Applications of Kirchhoff's Laws.—In solving problems involving networks, the number of unknowns, and

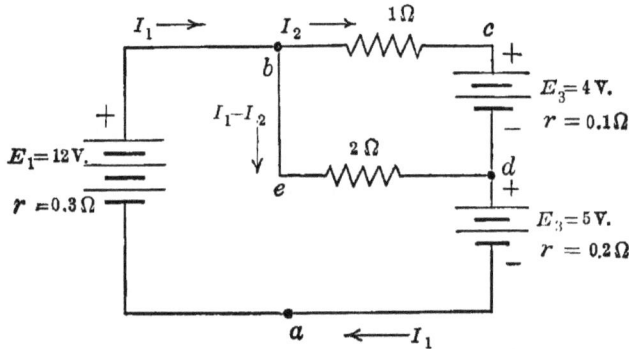

FIG. 46.—Further application of Kirchhoff's laws.

hence of independent equations, may sometimes be reduced by writing one of the currents at a junction as the sum or difference of the other currents at that junction. For example, in Fig. 43, p. 54, the current in the external circuit may be denoted by $(I_1 + I_2)$. This reduces the number of unknowns and of equations from three to two. Likewise, in Fig. 44, p. 56, $(I_1 - I_2)$ may be used to designate the current in the branch f, and in

Fig. 45, $(-I_1 - I_2)$ may be used to designate the current I. That is, the currents by Kirchhoff's first law are indicated on the diagram, rather than being expressed in a separate equation. This principle is illustrated by the following example:

Example.—Find the currents in all three branches of the network shown in Fig. 46, the assumed directions of currents being as shown.

The current in branch *bcd* is designated as $I_1 - I_2$, rather than by a distinct symbol such as I_3.

Applying Kirchhoff's second law to path *abcda*,

$$+12 - 0.3\,I_1 - 1\,I_2 - 4 - 0.1\,I_2 - 5 - 0.2\,I_1 = 0,$$

or

$$+3 - 0.5\,I_1 - 1.1\,I_2 = 0. \tag{I}$$

For path *abeda*,

$$+12 - 0.3\,I_1 - (I_1 - I_2)2 - 5 - 0.2\,I_1 = 0,$$

or

$$+7 - 2.5\,I_1 + 2\,I_2 = 0. \tag{II}$$

Multiplying (I) by 2.0 and (II) by 1.1,

$$+6 - I_1 - 2.2\,I_2 = 0, \tag{III}$$
$$+7.7 - 2.75\,I_1 + 2.2\,I_2 = 0. \tag{IV}$$

Adding (III) and (IV),

$$+13.7 - 3.75\,I_1 = 0,$$
$$I_1 = +3.65. \quad Ans.$$

Substituting the value of I_1 in (III),

$$+6 - 3.65 - 2.2\,I_2 = 0,$$
$$2.2\,I_2 = 2.35,$$
$$I_2 = +1.07. \quad Ans.$$

To check, voltage V_{ba} is found,

$$12 - (3.65 \times 0.3) = 12 - 1.095 = 10.905 \text{ volts.}$$

For path *adcb*,

$$+5 + (3.65 \times 0.2) + 4 + (1.07 \times 1.1) = 9 + 1.907 = 10.907 \text{ volts (check).}$$

CHAPTER IV

PRIMARY AND SECONDARY BATTERIES

44. Principle of Electric Batteries.—If two copper strips or plates be immersed in a dilute sulphuric acid solution (Fig. 47) and be connected to the terminals of a voltmeter, no appreciable deflection of the voltmeter will be observed. This shows that no appreciable difference of potential exists between the copper strips. If, however, one of the copper strips be replaced

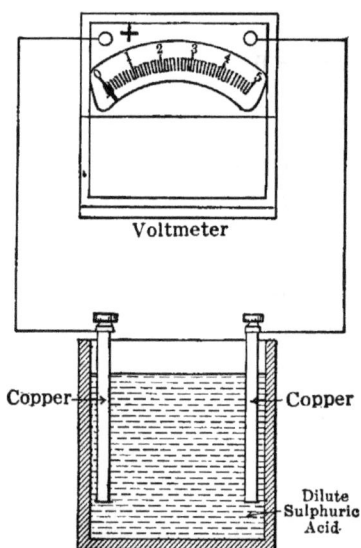

FIG. 47.—Primary cell with copper electrodes.

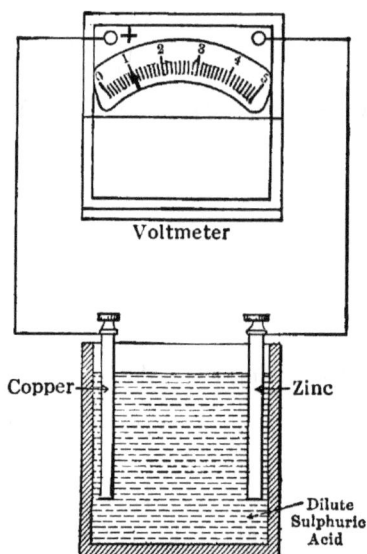

FIG. 48.—Primary cell with copper and zinc electrodes.

by a zinc strip (Fig. 48), the voltmeter needle will deflect and will indicate approximately 1 volt, showing that a potential difference now exists. It will be necessary to connect the copper to the + terminal of the voltmeter and the zinc to the − terminal in order that the voltmeter may read up scale. This shows that, so far as the *external* circuit is concerned, the copper is positive to the zinc.

The preceding experiment may be repeated with various metals. For example, carbon or lead may be substituted for the copper and a potential difference will be found to exist between each of these and the zinc, although it will not be of the same value as it is for the copper-zinc combination. Likewise, other metals may be substituted for the zinc, and potential differences will be found to exist.

Furthermore, it is not necessary that sulphuric acid be used for the solution. Other acids, such as hydrochloric, or chromic, may be substituted for the sulphuric; or salt solutions, such as common salt (sodium chloride), ammonium chloride (sal ammoniac), copper sulphate, or zinc sulphate, may be used.

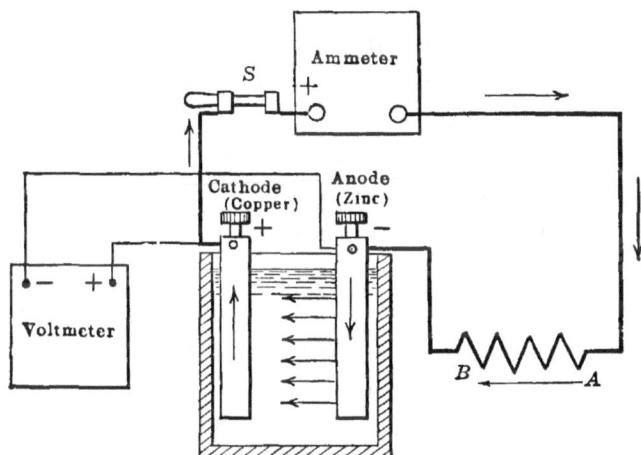

Fig. 49.—Current-flow in a single cell.

In order to obtain a difference of potential between the two metal plates, but two conditions are necessary:

1. The plates or electrodes must be of different materials.

2. They must be immersed in some electrolytic solution, such as an acid, alkali, or salt.

Again, if current be taken from the cell shown in Fig. 48, by connecting a resistance across its terminals (Fig. 49), current will flow from the copper through the resistance *AB* and into the cell through the zinc. Inside the cell, however, the current will flow *from* the *zinc* through the solution *to* the *copper*, as shown in Fig. 49. Since current flows *from zinc to copper within the cell*, zinc is said to be electrochemically positive to copper. Therefore, when considering such an electrochemical cell, the copper is positive to the zinc when the external circuit is considered, but the zinc is *electropositive to* the copper when the plates and the solution alone are considered.

45. Definitions.—An electrochemical cell which delivers electrical energy in virtue of the difference of potential between its electrodes is called a *galvanic cell*. An electrochemical cell in which electrolytic and chemical reactions are caused to take place by electrical energy received from an external source is called an *electrolytic cell* (see p. 87, Par. 75). The metal strips or plates of a cell are *electrodes*. The electrode at which current enters the solution (as the zinc, Fig. 49) is the *anode*, and the electrode at which current leaves the solution (as the copper, Fig. 49) is the *cathode*.

The solution used in a cell is the *electrolyte*.

When the cell delivers energy, the zinc plate diminishes in weight. Not only is this true for this particular cell; but with *primary* cells in general, the flow of current is accompanied either by a loss in weight of at least one of the plates or by a reduction of the plate materials to a compound of lesser chemical energy. Energy is stored in the cell *chemically*, and the electrical energy is delivered at the expense of the plate which goes into solution. That is, one plate is converted into another chemical compound, this change being accompanied by a decrease in the available chemical energy of the system. Therefore, *chemical energy* is converted into *electrical energy* when the cell delivers current.

Hence,

A galvanic cell or battery is a device for transforming chemical energy into electrical energy.

Such cells or batteries are divided into two classes: *primary cells* and *secondary cells*.

In a *primary cell*, it is necessary from time to time to renew the electrolyte and the electrode which goes into solution, by fresh solution and new plates.

In a *secondary cell*, the electrolyte and the electrodes which undergo change during the process of supplying current are restored electrochemically by sending a current through the cell in the reverse direction.

46. Primary Cells.—Although it is stated in Par. 44 that there are many combinations of metals and solutions capable of generating an emf and so forming a cell, only a limited number of such combinations are commercially useful. The general requirements of a good cell are as follows:

1. There must be practically no local action, that is, little or no wastage of the materials when the cell is not delivering current.

2. The emf must be of such a magnitude as to enable the cell to deliver a reasonable amount of energy with a moderate current.

3. Frequent replacement of materials must not be necessary, and such materials must not be expensive.

4. The internal resistance and the polarization effects must not be too great, otherwise the output of the battery will be seriously impaired.

As an illustration, the cell shown in Fig. 48 would not be useful, because both the copper and the zinc would waste away even were the battery delivering no current. Polarization (see Par. 48) would be too great, and hence the battery would be capable of delivering a comparatively small current only.

47. Internal Resistance.—As is pointed out in Chap. III, every cell or battery has internal resistance, which reduces the magnitude of the current and causes the terminal voltage to drop when current is taken from the cell. A small portion of this resistance lies in the electrodes, but the greater portion lies in the electrolyte itself and at the contact surface between the electrolyte and the electrodes. The resistance of a cell may be reduced, therefore, by increasing the areas of the electrodes in contact with the electrolyte and by reducing the distance between the positive and negative electrodes.

Increasing the size of the cell *does not increase its emf.* This emf depends only on the material of the two electrodes and the electrolyte. Figure 50 shows two gravity cells, made up of the same materials, but differing in size. The cells are in opposition, that is, their + terminals are joined and their − terminals are joined. A galvanometer *G* connected in one of the leads reads zero, indicating that no current flows from the larger to the smaller cell. Hence, their emfs must be equal.

48. Polarization.—When a fixed resistance is connected across a simple galvanic cell (Fig. 49), the current decreases in value with time. This is due both to a decrease in the emf of the cell and to an increase in its internal resistance. These changes are caused by *polarization*. Small bubbles of hydrogen form on the positive plate or cathode. These bubbles, acting in conjunction with the cathode, produce an emf in opposition to that of the cell

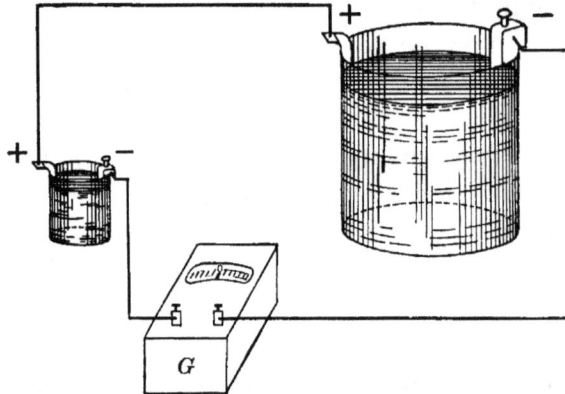

FIG. 50.—Equality of electromotive forces in cells of unequal size.

thus decreasing the net emf of the cell. They also increase the resistance of the cell, thus decreasing the current.

A common method of reducing polarization is to bring some oxidizing agent, such as chromic acid or manganese peroxide,

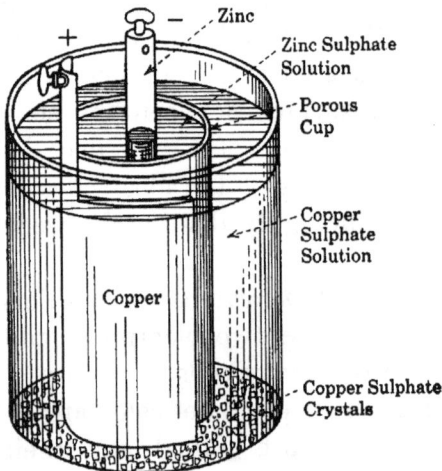

FIG. 51.—Daniell cell.

into intimate contact with the cathode. The hydrogen combines readily with the oxygen of these compounds to form water (H_2O). (Also see Pars. 52 and 53.)

49. Daniell Cell.—This cell (Fig. 51) is a two-fluid cell having copper and amalgamated zinc as electrodes. It consists of a glass jar, inside of which is a porous cup containing zinc sulphate solution or a solution of zinc sulphate and sulphuric acid. The zinc anode or negative electrode is immersed in this electrolyte. The porous cup is placed in a solution of copper sulphate with copper sulphate crystals in the bottom of the jar. The copper plate, which is cathode, surrounds the porous cup. The porous cup keeps the two solutions separated. As the copper is in a copper sulphate solution, there is no polarization. This cell is designed for use in a circuit which is closed continually; if the cell is left idle, the electrodes waste away. After the cell is taken out of service for some time, the electrodes should be removed and the porous cup should be thoroughly washed. The emf of this cell is about 1.1 volts.

50. Gravity Cell.—The gravity cell (Fig. 52) is similar to the Daniell cell, except that gravity, rather than a porous cup, keeps the electrolytes separated. The cathode, which is of copper, is made of strips riveted together and placed in the bottom of the cell together with copper sulphate crystals. A solution of copper sulphate is then poured to within a few inches of the top of the jar. The connection to the copper is usually an insulated copper wire fastened to the copper and carried through the solution to the top of the jar.

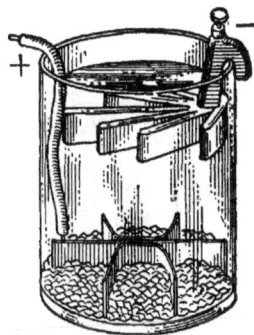

Fig. 52.—Gravity cell.

The anode is zinc, is usually rather massive, and is cast in the form of a crow's foot and hung on the top of the jar. This is surrounded by a zinc sulphate solution. The solutions are kept separated by gravity. The copper sulphate, having the greater specific gravity, tends to remain at the bottom of the jar. In the operation of the cell, the zinc goes into solution as zinc sulphate, and metallic copper comes out of the copper sulphate solution and is deposited on the copper electrode. The cathode, therefore, gains in weight, whereas the anode loses in weight.

The gravity cell is a *closed-circuit* battery, and the circuit should be kept closed, therefore, for the best results. The cell has been found very useful in connection with railway signals, fire-alarm systems, and telephone exchanges, all closed-circuit work, although the storage battery has replaced it in many instances. Its emf is approximately 1.09 volts.

51. Edison-Lalande Cell.—The Edison-Lalande cell (Fig. 53) has a cathode of copper oxide suspended between two zinc plates which form the anode. The electrolyte is caustic soda (NaOH), one part by weight of soda to three of water. To prevent the soda being acted on by the air, the electrolyte is covered with a layer of mineral oil. The emf is about 0.95 volt, and when the cell delivers nominal current the terminal voltage drops to 0.75 volt. There is little or no local action in this cell, and it can be used on both open-circuit and closed-circuit work.

52. Le Clanché Cell.—The Le Clanché cell is perhaps the most familiar type of primary battery. The cathode is molded carbon, and the anode is amalgamated zinc. The electrolyte is sal ammoniac (ammonium chloride). This type of cell is suited only

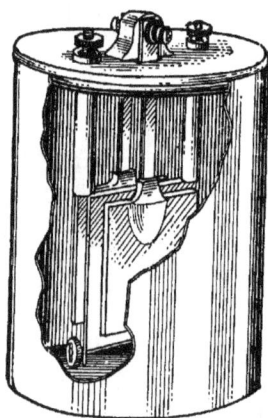

Fig. 53.—Edison-Lalande cell or Edison primary battery.

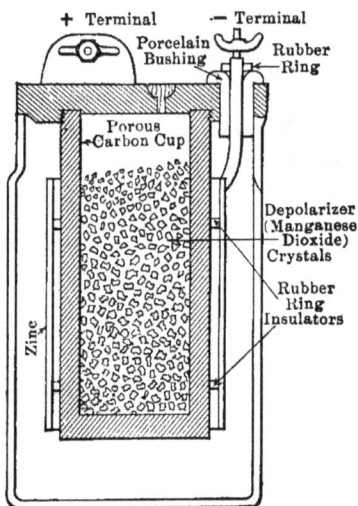

Fig. 54.—Porous-cup Le Clanché cell.

to open-circuit work because of the rapidity with which it polarizes. The emf is 1.4 volts, and the terminal voltage is approximately 1 volt when the cell is in normal operation. The most common method of reducing polarization in the cell is to bring manganese dioxide into intimate contact with the carbon. This gives up oxygen readily, which unites with the hydrogen bubbles to form water.

In one type of Le Clanché cell, a pencil zinc is suspended in the center of a hollow cylinder of carbon and manganese dioxide. An improved type, the porous-cup cell, is shown in Fig. 54. In this form, a hollow carbon cylinder is filled with manganese dioxide, and the zinc, in cylindrical form, surrounds the carbon cylinder.

The solution should consist of 3 oz (85 grams) of sal ammoniac to 1 pt (0.47 liter) of water. To prevent the solution "creeping," the top of the cell is dipped in paraffin and the top of the carbon is covered with a black wax.

Because of its simplicity and the fact that it contains no injurious acids or alkalies, this type of cell was once used for intermittent work, such as ringing doorbells, telephone work, and open-circuit telegraph work. The dry cell (Par. 53), because of its more convenient form, has to a large extent replaced this type of cell.

53. Dry Cell.—Dry cells are a modification of the Le Clanché cell and, as they are light, portable, and convenient, have practically displaced other types of primary cells. The term "dry cell" is a misnomer, for no cell that is actually dry will deliver an appreciable current.

A cross-section of a typical dry cell is shown in **Fig. 55.** The anode is sheet zinc, in the form of a cylinder, and acts as the container of the cell. The zinc is lined with some nonconducting material, such

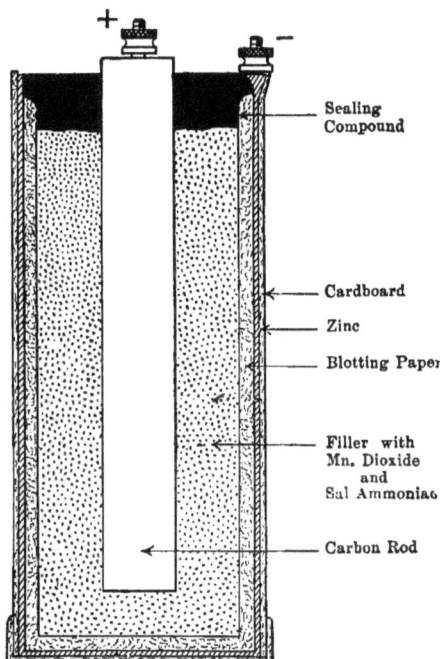

Fig. 55.—Dry cell—sectional view.

as blotting paper or plaster of paris. The anode consists of a carbon rod with the mixture of coke, carbon, etc., which surrounds this rod. The rod is located axially in the zinc container. The depolarizing agent, powdered manganese dioxide, is mixed with finely crushed coke and pressed solidly into the container between the carbon rod and the nonconducting material which lines the zinc. Sal ammoniac, with perhaps a little zinc sulphate, is added and the cell is then sealed with wax or some tar compound. The cell is always set in a close-fitting cardboard container.

The emf of a dry cell is about 1.5 volts when new but drops to about 1.4 volts with time, even though the cell remains idle. A cell is practically useless after a year to 18 months, even if not used meanwhile. The internal resistance of the cell is about 0.1 ohm when new and increases to several times this value with time. A method for testing the condition of a cell is to short-circuit it through an ammeter, when it should deliver an instantaneous current of 1.5/0.1 or 15 amp, if in good condition. When delivering appreciable current, the terminal voltage is very nearly 1 volt.

One of the chief causes of a cell's becoming useless is the using up of the zinc as a result of electrochemical actions within the cell. This allows the solution to leak out and to dry up, and the cell then becomes worthless.

Dry cells can supply moderate currents only intermittently, but they are capable of supplying small currents, of the magnitude of 0.1 amp, continuously. They are used extensively for electric bells, buzzers, telephones, telegraph instruments, gas-engine ignition, flash lamps, A, B, and C batteries for radio receiving sets, and for many other purposes.

54. Weston Standard Cell.—It is essential that reproducible standards of current, resistance, and voltage be available. The ampere may be determined by weighing the amount of silver deposited in a standard voltameter (see Par. 21, p. 22). This method is not satisfactory as a working method, since it requires skill and care in making the measurement, and the cell materials must have a high degree of purity. The standard or international ohm involves a column of mercury of fixed dimensions (see Par. 21, p. 23), and extremely accurate secondary standards, usually of manganin, are readily obtainable. The international volt is defined as 1/1.0183 of the emf of a Weston *normal* cell at 20°C. This standard is readily reproducible to within a few parts in 100,000. Since the working standards of resistance and voltage are readily obtainable and do not require unusual skill in their use, these two are almost universally used, and the current is obtained by Ohm's law. (See p. 27.)

A view of an unmounted Weston cell is shown in Fig. 56. The cathode is mercury, located at the bottom of one leg of an H-tube. Above this is mercurous sulphate paste. A porcelain tube extends to the top of the cell and acts as a vent for any gases that are formed. In the bottom of the other leg of the H-tube is the anode, of cadmium amalgam. Another porcelain tube acts as a vent. The electrolyte is cadmium sulphate. Platinum-wire leads from the cathode and the anode are sealed into the tubes at the bottom. The top of the cell is sealed with cork, paraffin, and wax. The entire cell is mounted in a metal case with binding posts at the top.

In the *unsaturated cell*, or working standard (Fig. 56), the solution is saturated at 4°C, and, as no crystals are left in the solution, its concentration is substantially constant at other temperatures. Such cells have practically zero temperature coefficient. A

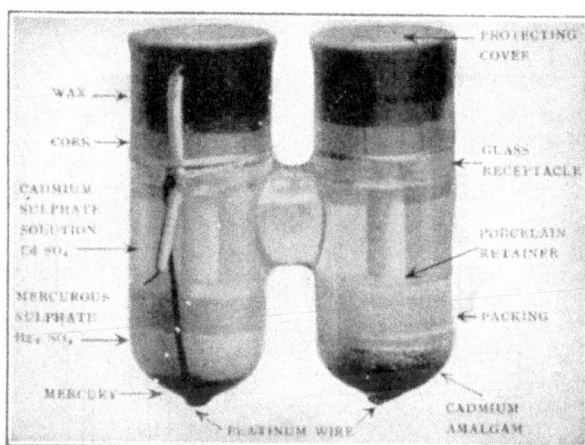

FIG. 56.—Unmounted Weston unsaturated cell.

certificate should accompany each cell, giving its emf, which usually is about 1.0186 volts.

As the resistance of a Weston cell is about 200 ohms, its emf cannot be measured by any method requiring appreciable current, such as the use of a voltmeter. By means of the Poggendorff method, described in Par. 93, p. 116, the cell is used without delivering current.

STORAGE BATTERIES

55. Storage Batteries.—A storage or secondary cell (sometimes called an accumulator) involves the same principles as a primary cell, but the two differ from each other in the manner in which they are renewed. The materials of a primary cell, which are used up in the process of delivering current, are replaced by new materials, whereas, in the storage cell, the cell materials are restored to their initial condition by sending a current through the cell in a reverse direction. There are but two types of storage cells i ι common use, the *lead-lead-acid* type and the *nickel-iron-alkali* type. In both of these cells *the active materials do not leave the el ctrodes.*

THE LEAD-LEAD-ACID BATTERY

56. Lead-lead-acid Cell.—The principle underlying the lead cell may be illustrated by a simple experiment. Two plain lead

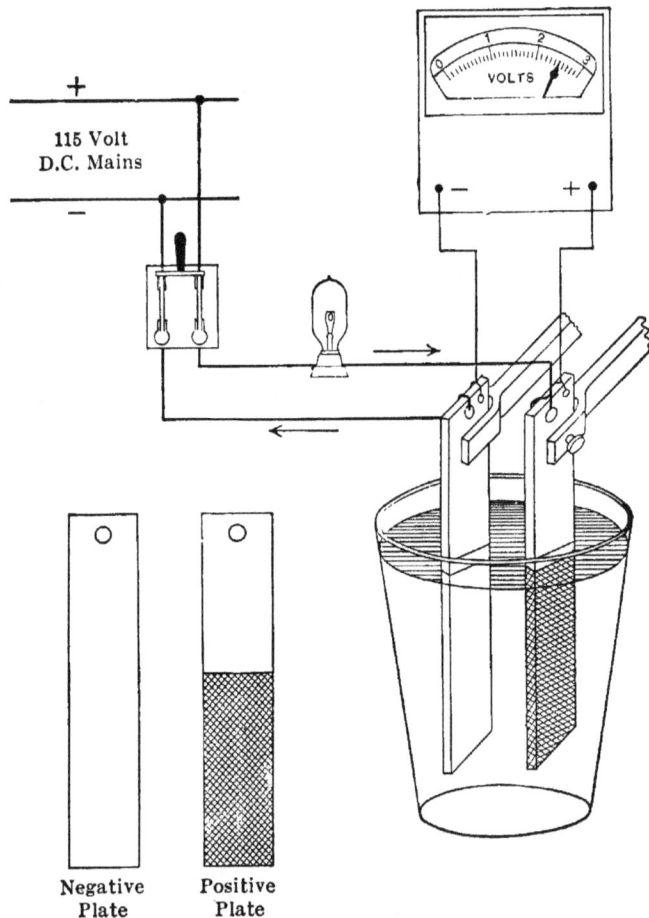

Fig. 57.—Forming the plates of an elementary lead storage cell.

strips (Fig. 57) are immersed in a glass of dilute sulphuric acid. These are connected in series with an incandescent lamp and supplied from a 115-volt d-c source. When current flows through the cell, bubbles of gas will be given off from each plate. but it will be found that a much greater number are given off by one plate than by the other. After a short time one plate will be observed to have changed to a dark chocolate color, and the other apparently will not have changed its appearance. A careful examination, however, will show that the metallic lead at the surface of the latter plate has started to change from solid metallic lead to spongy lead.

When the current is flowing as shown in Fig. 57, the voltmeter connected across the cell indicates about 2.5 volts. If the current be interrupted by opening the switch, the voltmeter reading will fall to about 2.1 volts, and the cell will now be found to be capable of delivering a small current, though the amount of energy that such a cell can deliver is very limited. As the cell discharges, the terminal voltage drops off slowly to about 1.75 volts, after which it drops more rapidly until it becomes zero and the cell is apparently exhausted. The color of the dark-brown plate will now have become lighter. After a short rest the cell will recover slightly and will again deliver current for a brief period.

The plate which is a dark chocolate color in this experiment is the positive plate or cathode, and the one which is partially converted to spongy lead is the negative plate or anode. The bubbles which are noted come mostly from the negative plate and are free hydrogen gas. When current is passed through such a cell, the metallic lead of the positive plate is converted into lead peroxide, whereas the negative plate, though not changed chemically, is converted from solid lead into the spongy or porous form. When the cell is discharged, the lead peroxide of the positive plate is changed to lead sulphate and the spongy lead of the negative plate is changed to lead sulphate, so that both plates tend to become electrochemically the same.

The principle of the cell is the same as that of the primary cell. When the two lead plates are the same electrochemically, that is, when both are lead sulphate, there is no emf. When the positive is converted to the peroxide and the negative to spongy lead by the action of an electric current, the two plates become electrochemically dissimilar and an emf results. This emf is about 2.1 volts, the excess of 0.4 volt observed in charging the cell being necessary to overcome the internal resistance and the polarization effects. This simple experiment illustrates the principle underlying the operation of lead storage cells.

The chemical reactions which take place in a lead storage cell are as follows:

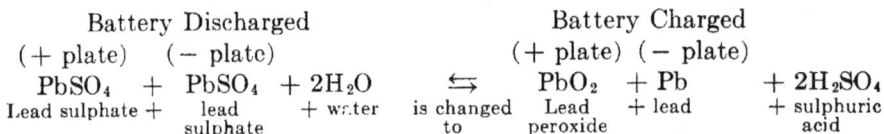

Battery Discharged				Battery Charged		
(+ plate)	(− plate)			(+ plate)	(− plate)	
$PbSO_4$ +	$PbSO_4$	+ $2H_2O$	$\leftrightarrows$	PbO_2	+ Pb	+ $2H_2SO_4$
Lead sulphate +	lead sulphate	+ water	is changed to	Lead peroxide	+ lead	+ sulphuric acid

When read from left to right, the equation shows the reactions which occur in the battery on charge. When read from right to left, it shows the reactions which occur on discharge.

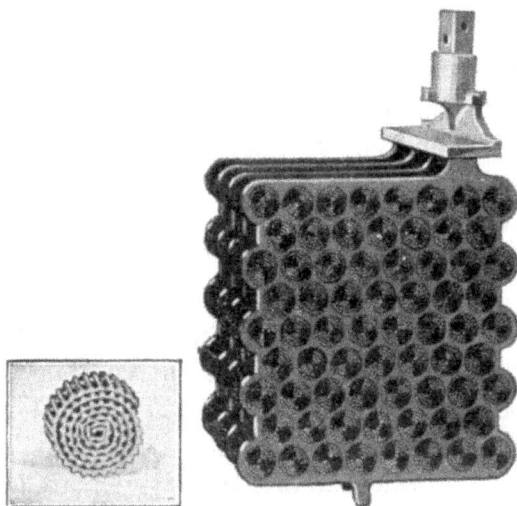

Fig. 58.—Planté (Manchester) positive group and button.

It will be noted that, when the battery is being charged, the change which takes place in the electrolyte is that water is converted into sulphuric acid. This accounts for the rise of specific gravity on charge. On discharge, the sulphuric acid is dissociated and reacts with the lead peroxide to form water. Therefore, the specific gravity of the electrolyte decreases when the cell is discharging. When charging, free hydrogen is given off at the negative plate and oxygen at the positive plate. Because of the explosive nature of hydrogen, *no flame should be allowed to come near a storage battery while the battery is charging.*

It would not be practicable to construct storage cells of plain lead sheets, such as were used in this experiment. The current capacity of the cell would be so small that the cell could not deliver currents of commercial value for any length of time, unless the cell were made prohibitively large in order to secure the necessary plate area.

If the charging of the elementary cell (Fig. 57) were carried further, the dark lead peroxide of the positive plate would be observed to fall off in flakes and drop to the bottom of the cell. Therefore, in a commercial cell, provision must be made to minimize this flaking of the active material.

57. Planté Plates.—There are two methods of increasing the active plate area, the Planté process and the Faure process. In the Planté process, the active material on the plates is formed from the metallic lead by passing a current through the cell first in one direction and then in the reverse direction. This procedure works the lead on the surface of the plates into active material.

The Manchester plate, shown in Fig. 58, is made by this process. A grid made of lead and antimony is perforated. The active material consists of corrugated lead ribbon, which is coiled in

FIG. 59.—Skeleton for pasted plate.

spirals and pressed into the perforations of the grid. The peroxide has a greater volume than the lead from which it is derived. Hence, when the cell is charged, these spirals expand and become more firmly embedded in the plate.

58. Faure or Pasted Plate.—This type of plate consists of a lead-antimony latticework or skeleton (Fig. 59) into which lead oxide is applied in the form of a paste. The battery is then charged. The paste on the positive grid is converted into peroxide and that on the negative grid into spongy lead.

The chief advantage of the pasted plate is its high energy capacity, especially for short periods, together with its lesser size, weight, and cost for a given discharge rate. It is, therefore, very useful where lightness and compactness are necessary, as with electric-vehicle batteries, ignition and starting batteries for internal-combustion motors, etc.   The pasted type of positive has a shorter life than the Planté type, due to a more rapid shedding of the active material.

In all storage batteries there is one more negative than positive plate.   This allows all the positives to be worked on *both* sides. Were any of the positives to be worked on one side only, the expansion of the active material, which occurs when it is converted to the peroxide on charge, would produce buckling.

59. Exide-Ironclad Cell.—Where it is desired to combine ruggedness and long life, as with vehicle batteries, the Exide-Ironclad Cell is frequently used.   The positive consists of a lead-antimony frame which supports a number of slotted, hard-rubber tubes.   In the center of each tube there is an irregular lead-antimony core.   Lead peroxide is pressed into the tubes forming the active material.   The perforations are so small that shedding of the active material is practically eliminated. An ordinary pasted plate is used for the negative.   This type of cell, although somewhat more expensive than the Planté and pasted types, is more rugged and has a life from two to three times that of the flat pasted-plate type.

Storage batteries are divided into two general classes, *stationary batteries* and *portable batteries.*

60. Stationary Batteries.—The plates of this type of battery may be either of the Planté type or of the pasted type, depending on the nature of the service.   For merely regulating duty, involving only moderate, though continual, charging and discharging, the Planté plate is preferable.   Where a battery is installed for emergency service, to carry very large overload for a short period during a temporary shutdown of the generating apparatus, the pasted plate, because of its high discharge rate, is preferable. As such batteries are usually located in thickly settled city districts where ground area is valuable, the high discharge rate of the pasted plate is of great advantage.

61. Tanks.—The containing tanks for stationary batteries are of two types: glass, and lead-lined wooden tanks.   Glass jars are used only for cells of small capacity, as they are expensive and have not the requisite mechanical strength in the larger sizes. Wooden tanks are lined with sheet lead.   The seams of the lead lining must be sealed by burning the lead with a nonoxidizing flame. Solder should never be used. The wood should be painted with an acid-resisting paint, such as asphaltum.

When glass jars are used, the plates are suspended by projecting lugs which rest on the edges of the jar. (See Fig. 64, p. 78.) In the lead-lined tanks, the plates are similarly suspended on two glass slabs, which rest on the bottom of the tank (Fig. 60). The plates of like polarity are burned to a heavy lead strip or busbar to which the current-carrying lug is either burned or bolted. There should always be a liberal space between the plates and the bottom of the tank to allow the red lead peroxide to accumulate without short-circuiting the plates. All types of stationary batteries should have a glass cover to reduce evaporation and to intercept the fine acid spray which is produced during the charging period.

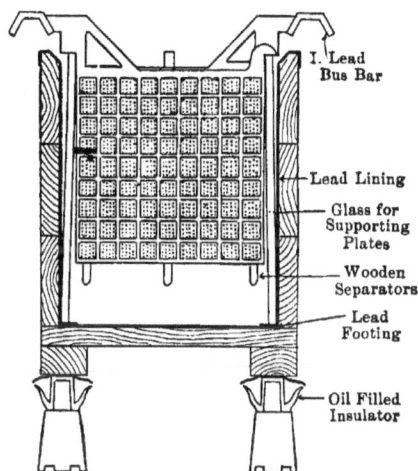

Fig. 60.—Lead-lined wooden-tank storage cell.

62. Separators.—To prevent the positive and negative plates from coming in contact with one another, separators are necessary. For a number of years thin sheets of wood have been used almost universally for separators. They act as complete barriers between plates and do not add appreciably to the resistance of the cell. The separators are usually grooved vertically to permit the circulation of the electrolyte (Fig. 60; also see Fig. 64, p. 78). The wood is specially treated to remove ingredients that would be detrimental to the electrolyte. These separators should never be allowed to become dry, as then they decompose readily.

A perforated, hard-rubber sheet between the wood separator and the positive plate is used in some assemblies to ensure maximum life. Such a sheet interposes resistance, depending on the degree of perforation, and thus reduces somewhat the effectiveness of the battery at high rates of discharge.

A rubber separator having microscopic pores—hence the trade name Mipor—has recently been developed by the Electric Storage Battery Co. This separator is grooved in the same manner as wood separators, has all the advantages of the wood separator, does not add any serious restriction to cell action, and is not affected by the electrolyte or by heat. Another recent development is the use of Fiberglas for separators. This consists of a fabric woven from thread spun from very fine glass fibers.

63. Electrolyte.—The electrolyte should be chemically pure sulphuric acid. When fully charged, the specific gravity should be 1.210 for Planté plates and not higher than 1.300 for pasted plates. This solution may be made from concentrated acid (oil of vitriol, sp. gr. 1.84) by *pouring the acid into water* in the following ratios:

PARTS WATER TO ONE PART ACID

Specific Gravity	Volume	Weight
1.200	4.3	2.4
1.210	4.0	2.2
1.240	3.4	1.9
1.280	2.75	1.5

Considerable heat is evolved when sulphuric acid and water are mixed. This results in the generation of a large amount of steam, if the water is added to the acid. This should be avoided, as it may scatter the acid and break the container and may cause personal injury.

The specific gravity of a solution may be determined directly by the use of a hydrometer (Fig. 61). This consists of a weighted bulb and graduated tube which floats vertically in the liquid whose specific gravity is to be measured. The specific gravity is read at the point where the surface of the liquid intercepts the tube. Such a tube may be left floating permanently in stationary batteries in a representative cell called a *pilot cell*.

The small amount of liquid and the inaccessibility of vehicle and starting batteries make the use of such a hydrometer impossible. To determine the specific gravity with such batteries, the syringe hydrometer is used (Fig. 62). The syringe contains a small hydrometer; and when sufficient liquid is drawn into the syringe tube, the small hydrometer floats and may be read directly.

Figure 63 shows the change in specific gravity during charge and discharge.

64. Specific Gravity.—When the battery is charged, hydrogen is given off at the negative plate and oxygen is given to the positive plate to convert it into the peroxide. The electrolyte gives up water, which means that the solution becomes more concentrated (see chemical equation, p. 72). The specific gravity will rise from the complete discharge value of 1.160 to 1.210 when fully charged, as shown in Fig.

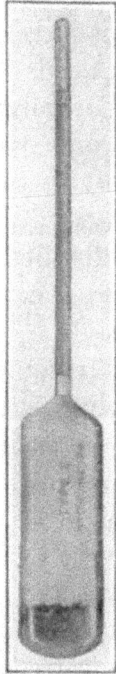

Fig. 61.—
Simple
hydrometer.

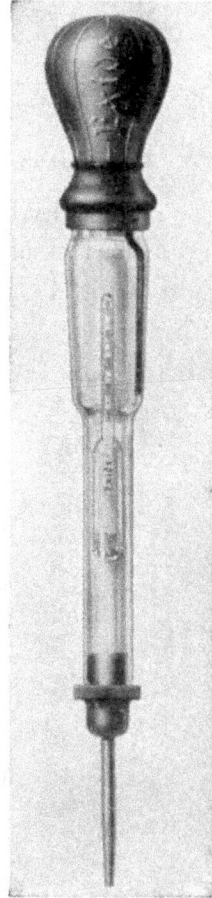

Fig. 62.—Syringe
hydrometer.

63. At *a*, the gassing point, the specific gravity drops slightly, owing to the presence of hydrogen bubbles in the electrolyte. On *discharge*, the specific gravity decreases even after the battery has ceased to deliver current, as the dilute acid in the pores of the active material diffuses into the solution. The specific gravity is such a good indication of the state of charge of the battery that the hydrometer reading is generally used to determine how nearly charged or discharged the battery may be.

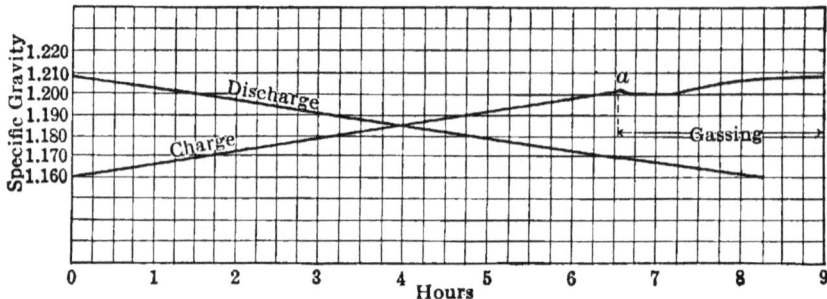

FIG. 63.—Variation of specific gravity in a stationary battery.

As the hydrogen and oxygen which escape from the battery during the charge and discharge periods are, ordinarily, only dissociated water, nothing but water needs to be added to replace the electrolyte. Acid must be added only when an actual loss of electrolyte takes place, such as occurs with a leaky tank. As a rule, distilled water is used to replace the evaporation of the electrolyte.

The freezing temperature of the electrolyte is considerably reduced with increasing specific gravity. For example, the freezing temperature is $-6°F$ when the specific gravity is 1.180, and $-90°F$ when the specific gravity is 1.280. Hence, there is little danger of a battery's freezing in the temperate zone, if it is well charged.

FIG. 64.—Open-type cell on glass sand tray.

65. Installing in Service.—The sealed type of battery is shipped complete with electrolyte in the cells and ready for service. The battery should be given a freshening charge before using. The plates, tanks, electrolyte, and containers of batteries of the open type are packed separately when shipped. When received, the separators should be placed immediately where they can be kept wet. The jars should be set in sand trays, as shown in Fig. 64. The plates should be handled carefully. Before the battery is ready for service, the active material on the plates, which is more or less converted into lead salts during exposure to the atmosphere, must be reduced electrically. Therefore, the battery should be given an initial charge at the rate and for the time recommended by the manufacturer.

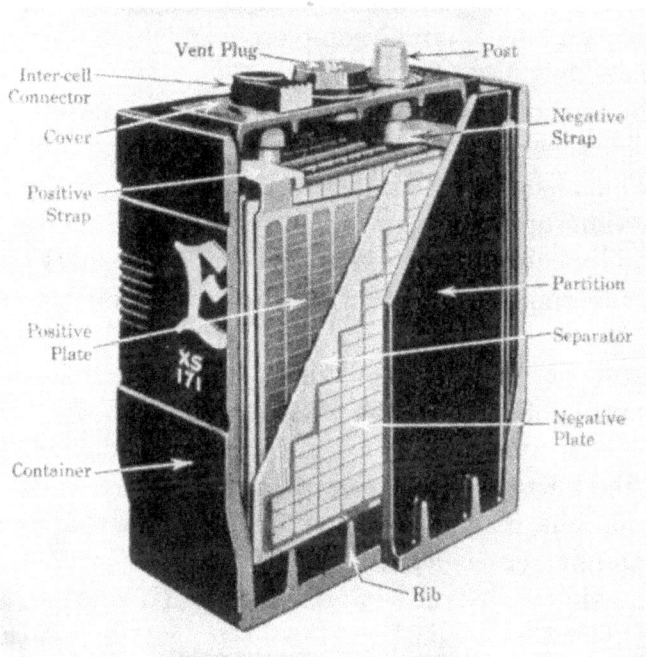

FIG. 65.—Cutaway of cell of portable battery.

If the battery stands for a long time without being used, the active material becomes more or less converted into inactive lead sulphate. Therefore, a battery if idle should be charged occasionally.

66. Portable Batteries.—In the design of batteries for propelling vehicles and for automobile starting, it is necessary to obtain a high discharge rate with minimum weight and size. Therefore, pasted plates are used for both positives and negatives. These are made extremely thin and are insulated from one another by very thin separators. They are then packed tightly in a hard-rubber jar (Fig. 65). The jar is sealed with an asphaltum compound to prevent the liquid splashing out. There is a hole in the top of the jar which is closed with a cap.   This permits the replenishing of the electrolyte.   A vent in the cap allows the gases to escape.   The discharge rates of this type of battery may be high, for example, as when doing starting duty.   Furthermore, the ampere-hour capacity of the battery for its weight and size must be high.   Hence, the volume of electrolyte is small and the specific gravity must vary between wide limits.   When the battery is fully charged, the specific gravity is as high as 1.280 and 1.300; and when it is completely discharged, the specific gravity is as low as 1.100.

Portable batteries are usually shipped assembled, charged, and complete with the electrolyte, so that they are ready for service when received.   However, a freshening charge is advisable.   As the space for the electrolyte in vehicle batteries is limited and as considerable gassing occurs, the level of the electrolyte falls rapidly, so that frequent additions of water are necessary.

67. Rating of Batteries.—Practically all batteries have a nominal rating based on the 8-hr rate of discharge.   Thus, if a Planté battery can deliver a current of 40 amp continuously for 8 hr, the battery will have a rating of $40 \times 8 = 320$ amp-hr. The normal charging rate of such a battery would be 40 amp. Even if the battery is capable of delivering 40 amp for 8 hr, it will not be able to deliver 64 amp for 5 hr (320 amp-hr), but only 88 per cent of this, or 56.4 amp for 5 hr; 56.4 amp is called the 5-hr rate.

Below is given the percentage capacity with different discharge rates:

	Hr				Min	
Discharge rate.....................	8	5	3	1	20	6
Percentage of capacity at 8-hr rate:						
Planté type..................	100	88	75	55.8	37	19.5
Pasted type..................	100	93	83	63	41	25.5

This falling off in capacity with higher rates of discharge is due to the inability of the free solution to penetrate the pores of the active material.   After such a battery has stood for a short time, it will be found to have recovered its capacity to some extent and is therefore capable of delivering more current, even after apparently having become exhausted. (See p. 71.) This is due to the final penetration of the pores of the active material by the free solution (see p. 77).

Batteries are able to discharge at very high rates for very short intervals. For example, a starting battery having an 8-hr rating of 10 amp is often called upon to supply 450 amp when doing starting duty.

68. Charging.—The following rule may be observed in charging a lead battery: The charging rate in amperes always may be made

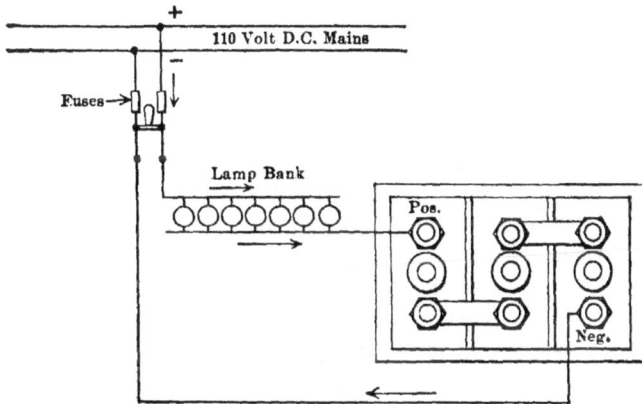

Fig. 66.—Charging a starting battery from 110-volt mains.

equal to the number of ampere-hours that have been discharged by the battery. For example, if 200 amp-hr are out of the battery, a charging rate of 200 amp may be used. As the battery charges, the ampere-hours out of the battery decrease, and the charging rate must correspondingly decrease. The rate should never be such that violent gassing occurs. Gassing represents a loss of energy, since a considerable portion of the charging energy is used in merely breaking up the water into hydrogen and oxygen. In addition, gassing causes the battery to become heated, the acid is carried out in the form of fine bubbles, and active material may be carried away by the erosive action of the bubbles. When a battery is fully charged, any rate will produce gassing, but the rate may be reduced to such a low value that it is practically harmless. This rate is called the *finishing rate*.

A battery may be charged in 5 hr by beginning at a rate several times the finishing rate and tapering off to the finishing rate as the charge progresses (constant-voltage method). On the other hand, a constant current of moderate value may be used over a much longer period, even as long as 16 hr (constant-current method).

A common example of the constant-current method is the charging of batteries from 110-volt mains. This is illustrated by Fig. 66, which shows the charging of a 6-volt battery. The emf of the battery is so small in comparison with the 110 volts of the mains that the current is determined almost entirely by the resistance of the lamp bank or any similar resistance. Hence, the current remains essentially constant irrespective of any small changes in battery emf. This method, although convenient, is inefficient from the energy point of view, as will be seen from the following.

Example.—When a 6-volt starting battery is charged from 110-volt d-c mains at the 15-amp rate, its terminal voltage is 7.4 volts. (*a*) How much power is being delivered to the battery? (*b*) How much power is consumed in the series resistance? (*c*) What percentage of the power delivered by the line is received by the battery?

(*a*) $7.4 \times 15 = 111$ watts.   *Ans.*

(*b*) $(110 - 7.4)15 = 1,539$ watts.   *Ans.*

$$(c) \quad \frac{7.4 \times 15}{110 \times 15}100 = \frac{7.4}{110}100 = 6.7 \text{ per cent.}   Ans.$$

The efficiency of this method is increased by increasing the number of batteries connected in series. Therefore, it is desirable to charge at one time as many batteries as possible.

Example.—What percentage of the power delivered by the line is utilized if two batteries are charged at the 15-amp rate in the foregoing problem?

$$\frac{2 \times 7.4 \times 15}{110 \times 15}100 = \frac{14.8}{110}100 = 13.4 \text{ per cent.}   Ans.$$

The percentage of power utilized is doubled by charging two batteries in series.

The constant-potential method of charging is frequently to be preferred, as the charging current automatically tapers off owing to the rise in cell emf as the cell approaches the completely charged condition, and little or no further attention is required. The rise in terminal voltage during charge is shown in Fig. 67. The terminal voltage is about 2.2 volts at the beginning of charge and rises to 2.6 volts, with little further rise in voltage. The diminishing rate of increase of voltage near the end of charge occurs in the gassing period and indicates that the cell is nearing the completion of charge.

With 2.3 volts per cell and no series resistance, the current at the beginning of charge is usually too large. Hence, it is advisable to use a small series resistance. If a series resistance is used, a voltage of 2.5 or 2.6 volts per cell is desirable, as otherwise adjustments must be made during the charging period.

A battery cannot be charged from a-c supply without the use of a rectifying device.[1] The battery must be connected so that

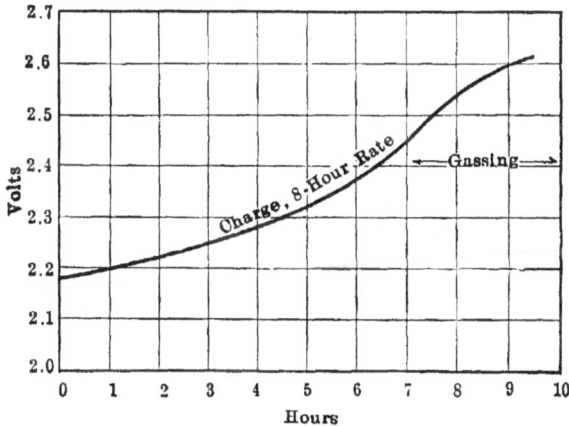

FIG. 67.—Variation in cell voltage during charge.

current enters its positive terminal from the positive line (Fig. 66). If doubt exists as to the polarity of either the line or the battery, use a voltmeter, if one is available. If a voltmeter is not available, dip the ends of two wires which are connected to either the battery or the circuit into slightly acidulated water. Bubbles form about the negative wire (also see p. 70).

EDISON NICKEL-IRON-ALKALINE CELL

69. Construction.—This type of storage cell was invented by Thomas A. Edison in 1901. Nickel and iron are used for the electrodes, and the electrolyte is alkaline, consisting of a 21 per cent potassium hydroxide solution. The cell is light and durable owing mainly to the employment of steel in the construction of the electrodes and container.

[1] See Tungar and Dawes, "Industrial Electricity," 2d ed., "Copper-Oxide Rectifier," Part II, Chap. X.

The positive plate consists of a nickeled-steel grid, holding nickeled-steel tubes which contain the positive active material. When inserted in the tubes, the active material is in the form of nickel hydrate but changes to an oxide of nickel after the formation treatment. To give the electrolyte free access to the active material, the tubes are perforated. To obtain improved electrical conductance, the active material is alternated with layers of pure metallic nickel flake at the time that it is tamped into the tubes. The tubes are either $3/16$ or $1/4$ in. (0.476 or 0.635 cm) inside diameter and about 4 in. (10.16 cm) long and are reinforced by eight encircling seamless steel rings spaced equidistantly over the tube length.

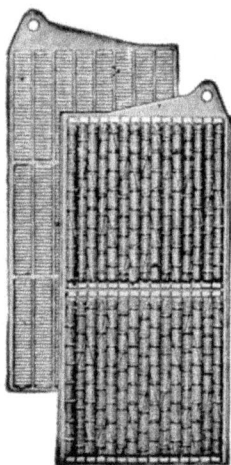

The negative plate is generally similar in construction to the positive plate except that a finely divided oxide of iron is used as active material and is contained in rectangular perforated nickeled-steel pockets instead of in tubes. Positive and negative plates are shown in Fig. 68.

70. Electrochemical Principle.—The electrolyte, which consists of a 21 per cent solution of potassium hydroxide to which has been added a small amount of lithium hydroxide, does not attack the steel tubes, container,

FIG. 68.—Positive and negative plates of Edison storage cell.

etc., but actually preserves them. The composition of the electrolyte does not change appreciably on charge or on discharge so that its specific gravity and conductivity remain essentially constant.

The chemical reactions in the cell are complex but may be summarized briefly as follows. On charge, the active material of the negative plate, iron oxide (FeO), is reduced to iron (Fe). The active material of the positive plate, nickel oxide (NiO), is oxidized to nickel dioxide (NiO_2). On discharge, the reverse process occurs, the negative plate being oxidized to iron oxide and the positive plate being reduced to nickel oxide. Throughout charge and discharge, the solution consisting of $2KOH + 2H_2O$ remains unchanged, both chemically and in concentration. Hence, unless evaporation takes place, the specific gravity does not change with either charge or discharge.

71. Assembly.—By means of a steel connecting rod which passes through holes at the top of the grids, the positive and negative plates are assembled into positive and negative groups. Steel spacing washers between adjacent plates on the connecting rod ensure proper plate spacing. Positive and negative plate groups are intermeshed to form complete elements, separation between alternate positive and negative plates being accomplished by vertical hard-rubber grids and pins.

The assembled elements are placed in a corrugated nickeled-steel container (see Fig. 69), after which the steel cover is welded in position. Projecting from the cell cover is the filling aperture,

FIG. 69.—Five Edison storage cells mounted in a tray.

on which is mounted a hinged filler cap which is held positively either open or closed by a clip spring. Suspended from the filler cap is a hard-rubber valve which seats by gravity when the cap is closed, thus excluding external air and reducing evaporation, yet permitting the escape of gas.

The individual cells are connected by connector lugs which are steel forgings bored to fit the taper of the projecting lugs. All lugs, links, and nuts are nickel-plated. The cells are mounted in wooden trays to form a battery (Fig. 69).

72. Discharge and Charge Characteristics.—The rated capacity of the Edison cell is based on a normal 5-hr discharge until the voltage becomes 1 volt per cell for A, B, C, and N types and on a normal $3\frac{1}{3}$-hr discharge to 1 volt per cell for G and L types. Figure 70 shows normal charge and discharge curves for the Edison A-type cell. It will be noted that the average discharge voltage is about 1.2 volts per cell. At discharge rates other than normal, voltage values will vary above and below this average.

73. Advantages and Applications.—The extensive use of steel in the construction of the Edison cell makes it rugged and capable of withstanding the vibration and shock incidental to commercial service. Moreover, the use of steel for tubes, pockets, and grids makes possible a plate construction which securely retains the active materials and which eliminates buckling and warping. The cell is light per watt-hour of capacity and is capable of withstanding electrical abuse such as overcharge, overdischarge,

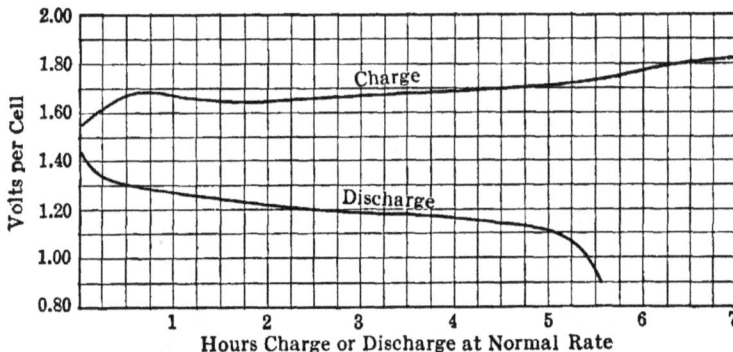

Fig. 70.—Voltage during charge and discharge of Edison cell.

accidental short circuit, charge in the reverse direction, being left in a discharged condition, all without injury.

The Edison cell finds its greatest usefulness in storage-battery-propelled trucks, tractors, mine and industrial locomotives, railway-passenger-car lighting and air conditioning, multiple-unit car control, railway signaling, and emergency lighting, and in isolated electric plants. The types now manufactured are not adapted to automobile starting service and are not sold for that purpose.

74. Efficiency of Storage Batteries.—The efficiency of a storage battery is the ratio of the watt-hour output to the watt-hour input.

Example.—A fully charged cell is discharged at a uniform rate of 38 amp for 6 hr at an average terminal voltage of 1.95 volts. The cell is now charged at a uniform rate of 40 amp for 6 hr, bringing it back to the fully charged condition. The average terminal voltage during charge is 2.3 volts. What is the efficiency of the cell?

Watt-hours   input = 40 × 2.3   × 6 = 552.
Watt-hours output = 38 × 1.95 × 6 = 445.
Eff. = $445/552$, or 80.7 per cent.

The *ampere-hour efficiency* of a storage battery is the ratio of the ampere-hours output to the ampere-hours input. In the preceding example the ampere-hour efficiency may be found as follows:

Ampere-hours input = 40 × 6 = 240.
Ampere-hours output = 38 × 6 = 228.
Ampere-hour eff. = $228/240$, or 95 per cent.

The much lower watt-hour efficiency is due to the difference between the voltage of charge and that of discharge, as shown in Figs. 63 and 70.

75. Electroplating.[1]—Electroplating consists for the most part in plating the baser metals with protective coverings of such

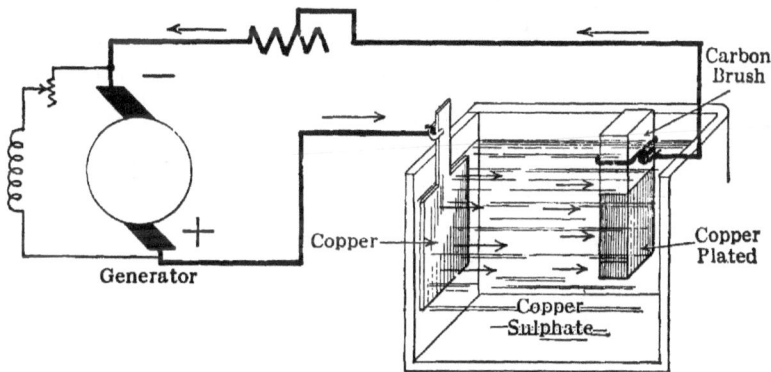

FIG. 71.—Copper-plating bath.

metals as copper, nickel, chromium, gold, and silver and is an important industry. As a simple illustration, assume that it is desired to copper plate a dynamo carbon brush. The portions of the brush to be plated are immersed in a solution of copper sulphate (Fig. 71). A copper strip is also immersed in the solution and is connected to the positive terminal of a dynamo or some other source of d-c supply. The article to be plated is connected to the negative terminal of the supply. Under these conditions the current will carry copper from the solution and deposit it on the carbon brush. The copper which leaves the solution is replaced by copper which is carried from the copper strip (the anode) into solution, and so there is no change in the

[1] See "Standard Handbook," 6th ed., Sec. 23, Pars. 136 to 180, for a more complete discussion.

solution itself.   The current density used depends on the metal being deposited.

It is not necessary that the anode be of the metal which it is desired to deposit.   Other metals may be used.   Under these conditions, however, the solution in time becomes contaminated by the going into solution of the anode.   If an inert substance such as carbon be used as anode, acid is formed in the solution.

The *only opposing* emf in the bath just described is the IR drop in the solution.   Hence, electroplating baths are naturally low-voltage devices, and, when practicable, several are connected in series.   A low-voltage, high-current generator is generally used for plating purposes.   In practice, there are many refinements to be observed.

Electrotyping is another common example of electroplating. An impression is made in wax with the type or object to be reproduced.   The surface of the wax is made conducting by applying a thin coating of graphite.   Copper is then plated on this surface.   It is later backed by type metal to give it the necessary mechanical strength.

CHAPTER V

ELECTRICAL INSTRUMENTS AND
ELECTRICAL MEASUREMENTS

76. Principle of Direct-current Instruments.—It is a fundamental law that a conductor carrying current and placed in a magnetic field tends to move at right angles to the direction of the field and to the direction of the current. An extension of this law is that a coil carrying a current and placed in a magnetic field tends to turn so as to make the direction of the field of the coil the same as that of the magnetic field in which the coil is placed. Most d-c measuring instruments operate on this principle. Moreover, this is the basic principle on which all motors operate, which is considered in detail later. (See Par. 212, p. 267.)

Consider Fig. 72 (a), which shows a rectangular coil pivoted so that it may rotate about an axis perpendicular to the plane of the paper. This coil is placed in a uniform magnetic field. Current is led into the coil through flexible connections. The direction of current is downward in the lower left-hand side and upward in the upper right-hand side of the coil, as is indicated by the small circles with cross and dot. The coil tends to assume such a position that the flux of the system is a maximum (Par. 117, p. 141). This position is shown in Fig. 72 (b), in which the coil has its plane perpendicular to the magnetic field. The ampere-turns of the coil produce a flux whose direction is from left to right and thus in conjunction with the field due to the magnet. For a given current in the coil, the flux of the system is obviously a maximum when the coil is in this position.

Also, the tendency of the coil to turn may be explained by the laws of attraction of magnetic poles. The coil itself has a north pole n and a south pole s. The s-pole of the coil is attracted by the n-pole of the magnet, and the n-pole of the coil is attracted by the s-pole of the magnet, as shown by the heavy arrows in (a).

Moreover, in Fig. 72 (*a*) the conductors in the lower side of the coil tend to move downward at right angles to the field, and those in the upper side tend to move upward at right angles to the field.

Hence, under the conditions of Fig. 72 (*a*), the coil tends to rotate in a counterclockwise direction until its plane is perpendicular to the magnetic field in which the coil finds itself, and its own field acts in conjunction with this magnetic field.

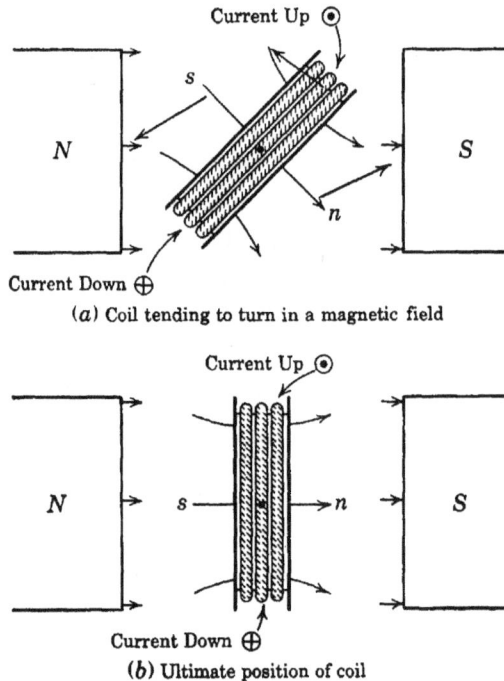

(*a*) Coil tending to turn in a magnetic field

(*b*) Ultimate position of coil

FIG. 72.—Turning moment of an instrument coil.

This behavior of a coil carrying a current and placed in a magnetic field must be thoroughly understood, for it is the underlying principle of most measuring instruments and is also the principle on which all electric motors operate.

77. D'Arsonval Galvanometer.—A galvanometer is a sensitive instrument used for detecting as well as measuring small electric currents. The D'Arsonval galvanometer, which operates on the principle of a coil turning in a magnetic field, is the most common type of galvanometer. Figure 73 shows the principle of its construction. A coil of very fine wire is suspended between the poles of a permanent magnet by means of a filament, usually a flat strip of phosphor bronze.

A soft-iron core is usually placed between the poles of the magnet (Fig. 73). This core, by reducing the length of the air-gap, increases the flux linking the coil. The flux also tends to enter the core radially, which makes the deflections of the galvanometer nearly directly proportional to the current flowing in the galvanometer coil.

Electrical connections to the coil are made through the phosphor-bronze suspension and a very flexible spiral filament fastened to the bottom of the coil (Fig. 73).

Any rotation of the coil produces torsion in the filament, which opposes the turning of the coil. The torsion is called the *restoring force*. When the moment of the restoring force and the turning moment due to the current are equal, the galvanometer assumes a steady deflection.

There are two common methods of reading the deflection of a galvanometer. A plane mirror is mounted on the coil system (Fig. 73), and a scale and telescope are mounted about 0.5 m from the galvanometer (Fig. 74). The reflection of the scale in the mirror can be seen in the telescope and the deflection is read by means of a crosshair in the telescope.

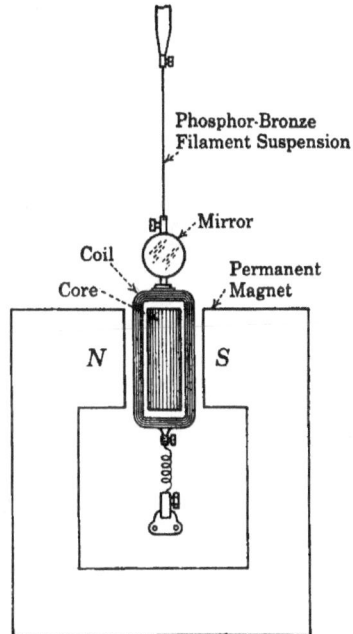

FIG. 73.—Principle of the D'Arsonval galvanometer.

A second method is to use a concave mirror on the galvanometer moving system. A lamp filament is placed some distance from the mirror, and its image is focused on a ground glass to which a scale, graduated in centimeters, is fastened.

Damping.—If a galvanometer coil, which is hung freely, starts to swing, it will continue swinging for some time unless it is retarded or damped. If the coil be wound on a metal bobbin, the motion of the bobbin through the magnetic field induces eddy currents in the bobbin, and these are in such a direction as to put an electric load on the moving system, as in an electric generator, which opposes the motion. The same result may be obtained by binding short-circuited copper coils on the main coil, or by shunting the galvanometer externally with a resistance (see Ayrton shunt), or even by short-circuiting the galvanometer.

78. Galvanometer Shunt.—In order to measure currents greater than the rated currents of the galvanometer and to

FIG. 74.—Telescope and scale method of reading a galvanometer.

protect the galvanometer from excessive currents, galvanometers are provided with shunts. There are two common types of shunt. One type (Fig. 75) consists of three or four separate resistances which are plugged in parallel with the galvanometer, one at a time. Ordinarily, these are so adjusted in value that, with a given current to be measured, the successive galvanometer currents are in the ratio of 10 to 1. Referring to Fig. 75, when the plug is entirely removed, no shunt is connected across the galvanometer and the entire line current I flows through the galvanometer. When the plug is in the uppermost ($\frac{1}{9}$) position, one-tenth of the line current I goes through the galvanometer and nine-tenths goes through the shunt. When the plug is in the

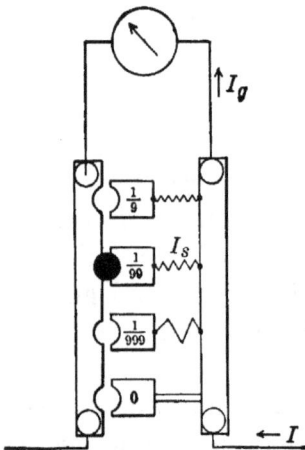

FIG. 75.—Galvanometer shunt.

position shown in the figure, one-hundredth of the line current I goes through the galvanometer; when in the second position from the bottom ($\frac{1}{999}$), one-thousandth of the line current goes through the galvanometer; when in the lowest position (0), the galvanometer is short-circuited and practically none of the line current goes through the galvanometer. The values of the shunt resistances necessary for obtaining the foregoing shunting ratios are determined as follows:

If, for example, it is desired that the galvanometer current be one-tenth of the line current, nine-tenths of the current must be shunted, and the shunt resistance must be one-ninth of the galvanometer resistance, as shown by the upper resistance (Fig. 75). If it is desired that the galvanometer current be one-hundredth of the line current, the shunt resistance must be one ninety-ninth of the galvanometer resistance. This may be proved as follows:

Let R_g and I_g be the galvanometer resistance and current, I_s the shunt current, and I the line current. By Eq. (22), p. 31,

$$\frac{I_g}{I_s} = \frac{I_g}{I - I_g} = \frac{R_s}{R_g}.$$

If $I = 10I_g$,

$$\frac{R_s}{R_g} = \frac{I_g}{10I_g - I_g} = \frac{1}{9}.$$

If $I = 100I_g$,

$$\frac{R_s}{R_g} = \frac{I_g}{100I_g - I_g} = \frac{1}{99}.$$

Example.—The resistance of a galvanometer is 600 ohms. What resistances should shunt it in order to reduce its deflections in the ratio of 10 to 1 and 100 to 1?

$$R_1 = {}^{600}\!/_9 = 66.7 \text{ ohms.} \quad Ans.$$
$$R_2 = {}^{600}\!/_{99} = 6.06 \text{ ohms.} \quad Ans.$$

79. Ayrton Shunt.—The circuit connections of the *Ayrton shunt* are shown in Fig. 76. A permanent resistance AB is connected across the galvanometer terminals. One line terminal is permanently connected to one end of this resistance, and the other line terminal C is movable and can be connected to various points along AB. With a fixed line current, the maximum galvanometer current is obtained when C is at B. If C be moved to a, where resistance Aa is one-thousandth of the total resistance AB, the galvanometer current will be one-thousandth of its maximum value. If C be moved to b, where Ab is one-hundredth of the resistance AB, the galvanometer current will be one-hundredth of its maximum value, etc.

The advantages of the Ayrton shunt are:

1. The shunt is applicable to any galvanometer, regardless of the galvanometer resistance.

2. A fixed resistance is shunted across the galvanometer, which gives a constant value of damping in open-circuit ballistic measurements (see p. 204).

The maximum sensitivity of the galvanometer is reduced by the addition of the shunt. The maximum sensitivity of the galvanometer *in combination with its shunt* is obtained when C is at B. The galvanometer is then shunted by the resistance AB, so that the entire line current does not go through the galvanometer. That is, the sensitivity of the galvanometer in combination with its shunt is less than its sensitivity when used alone. If the resistance of the shunt is only five times that of the galvanometer, the sensitivity will be reduced in the ratio of 6 to 5, which is not objectionable usually.

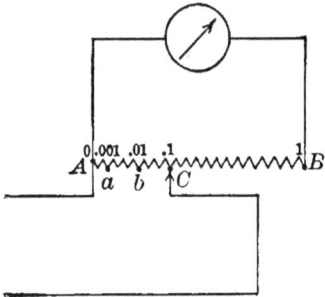

Fig. 76.—Ayrton shunt.

The *multiplying power* of a shunt is the ratio of the line current to the galvanometer current. The product of the galvanometer current and the multiplying power gives the line current.

80. Weston Type Instrument.—For the measurement of direct current and voltage, the Weston type instrument, developed by Edward Weston, is practically in universal use. The instrument is based on the principle of the D'Arsonval galvanometer, but it is so constructed that it is portable and is provided with a pointer and scale for indicating the deflections of the moving coil.

The essential parts of the instrument are shown in **Fig. 77.** As in the D'Arsonval galvanometer, a permanent magnet is necessary, being made in horseshoe form. Two soft-iron pole pieces are fitted to the magnet poles, and a soft-iron, cylindrical core is held between these pole pieces by a strip of brass. This gives a short, uniform air-gap and a radial field. The moving coil consists of very fine, silk-covered, copper wire wound on an aluminum bobbin. The aluminum bobbin, besides supporting the coil mechanically, makes the instrument highly damped. This damping is due to the currents induced in the aluminum, because of its cutting the magnetic field.

The moving system is supported at the top and bottom by hardened steel pivots turning in cup-shaped jewels, usually sapphire. This method of supporting the moving coil is almost frictionless and makes the instrument portable, whereas the D'Arsonval galvanometer is not readily portable. The current is led into and out of the coil by two flat spiral springs, one at the top and the other at the bottom of the coil. These springs also serve as the controlling device for the coil, any tendency of the

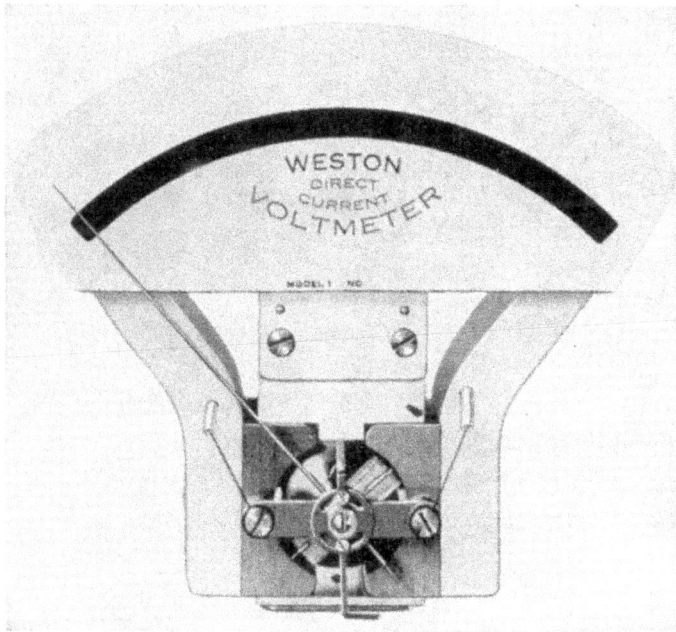

FIG. 77. –Movement of Weston instrument.

coil to turn being opposed by these two springs. The springs are coiled in opposite directions so that the effect of change of temperature, which causes a spiral spring to coil or uncoil, will not cause the needle to change its zero position. A very light and delicate aluminum pointer is attached to the moving element, to indicate the deflection of the coil. The pointer moves over a graduated scale, which may be marked in volts or in amperes, as the case may be. Because of the radial field, the deflection of the moving coil in this type of instrument is practically proportional to the current in the moving coil, so that the scale of the instrument has substantially uniform graduations, which is desirable. The internal connections of a Weston intrument are shown in **Fig. 78.**

Instruments of this construction having weak springs are often used for portable galvanometers (Fig. 79). Although

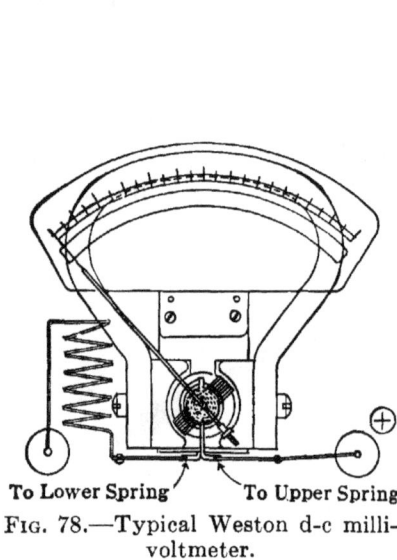

FIG. 78.—Typical Weston d-c milli-voltmeter.

FIG. 79.—Weston portable galvanometer and micro-ammeter.

lacking the extreme sensitivity of the suspended type, they can be made sufficiently sensitive for many kinds of work, and their ruggedness and portability make them very useful.

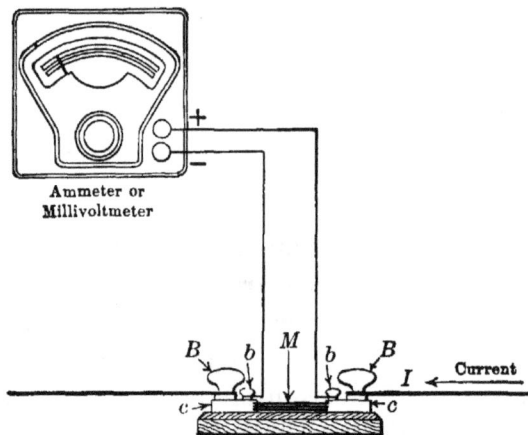

Ammeter or
Millivoltmeter

FIG. 80.—Ammeter with external shunt.

81. Ammeters.—An ammeter is an instrument which measures the current in an electrical circuit. The moving coil of Weston portable instruments deflects to full-scale value with currents of 0.01 to 0.05 amp in the coil. The instrument may be used directly, therefore, for measurement of these small currents. In order to measure currents greater than this, the larger portion of the current must be diverted from the moving coil by a *shunt*. The shunt is merely a low resistance, usually made of manganin strip *M* brazed to comparatively heavy copper blocks *c* (Fig. 80.) Two sets of binding nuts are fastened to the copper blocks. The heavy wing nuts *BB* carry the main current through the shunt. The small posts *bb* connect with the ammeter leads. The ammeter is, in reality, a voltmeter which measures the voltage drop across a resistance.

The voltage drop across the shunt is

$$V_{sh} = I_{sh}R_{sh},$$

where I_{sh} and R_{sh} are the shunt current and the shunt resistance. If R_{sh} is constant, the voltage drop across the shunt is proportional to the current in the shunt, so that the instrument readings are proportional to the current in the shunt. For this reason the ammeter itself (Figs. 78, 80) is often marked "millivoltmeter." For full-scale deflection the shunt is adjusted

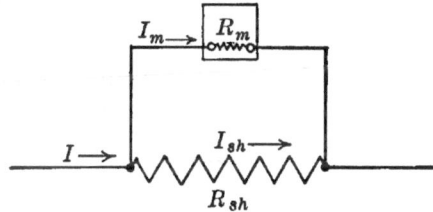

FIG. 81.—Division of current between ammeter and shunt.

usually to give a drop of about 50 millivolts. The current taken by the instrument itself is usually from about 0.02 to 0.05 amp, so that it is usually negligible compared with the main current. Therefore, in most cases the line current practically equals the shunt current.

An ammeter and its shunt may also be considered as a divided circuit. In Fig. 81, let R_{sh} and I_{sh} be the shunt resistance and the shunt current, and let R_m and I_m be the instrument resistance and the instrument current. By the law of divided circuits (Eq. (22), p. 31):

$$\frac{I_{sh}}{I_m} = \frac{R_m}{R_{sh}}.$$

That is, the currents in the shunt and in the instrument are inversely as their resistances.

Example.—Assume that the resistance of an instrument is 2.5 ohms, the resistance of the shunt 0.0005 ohm, and that the line current is 90 amp. What is the value of the instrument current?

As the current in the line differs from the shunt current by a very small amount, the two may be assumed equal. Then,

$$\frac{90}{I_m} = \frac{2.5}{0.0005}.$$
$$I_m = 0.018 \text{ amp. } Ans.$$

For accuracy, the current must always divide between the instrument and the shunt in the same ratio. Therefore, the resistance of the shunt and the resistance of the instrument must not change, or both must change in the same ratio. Hence, the leads with which the instrument is calibrated should always be used to connect the shunt to the instrument. The lugs and binding-post contacts should be kept clean from oxide and dirt.

An ammeter with an external shunt may be made to have a large number of scales or ranges. Assume, in the example just given, that the instrument gives full-scale deflection when the instrument current is 0.02 amp. The potential drop across the instrument terminals and hence across the shunt is

$$0.02 \times 2.5 = 0.050 \text{ volt,}$$

or 50 millivolts. Dividing this voltage by the shunt resistance, the shunt current is

$$I = \frac{0.05}{0.0005} = 100 \text{ amp,}$$

or the instrument deflects full scale with practically 100 amp in the line.

If a shunt having a resistance of 0.005 ohm be substituted, the 50 millivolts drop across the shunt may be obtained with 10 amp (10 × 0.005 = 0.050). Therefore, a 10-scale ammeter results. By the choice of suitable shunts, the same instrument may be made to give full-scale deflection with 1 amp, and with 5,000 amp.

In the smaller sizes of instruments up to 50 amp, and where only one scale is desired, the shunt is placed usually within the instrument. For ranges between 50 and 100 amp, the use of an internal or an external shunt is optional. Above 100 amp, it is usual to have the shunt external to the instrument, because of its size and its heating loss.

An ammeter can usually be distinguished from a voltmeter by the fact that its binding posts are heavy and are of bare metal, except in the case of an instrument having an external shunt. The posts of millivoltmeters and voltmeters are of much lighter construction and are covered with hard rubber or bakelite for insulation purposes.

Any instrument when connected in a circuit should disturb the circuit conditions as little as possible. An ammeter shunt, since it is connected in series with the line, should have as low a resist-

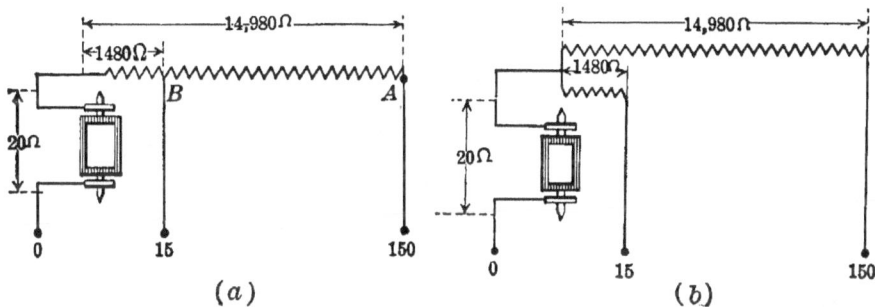

Fig. 82.—Methods of connecting resistance in a voltmeter.

ance as is practicable, so that very little additional resistance is introduced into the circuit. To protect ammeters from excessive currents, provision may be made for *short circuiting* the ammeters while readings are not being taken.

82. Voltmeters.—The Weston type instrument, when used as a voltmeter, is essentially the same as when used as an ammeter, in so far as the movement and magnet are concerned. (See Fig. 77.) The moving coil of the voltmeter is usually wound with more turns and of finer wire than that of the ammeter and so has a higher resistance. It therefore requires less current for a given deflection. The principal difference, however, lies in the manner of connecting the instrument to the circuit. As a voltmeter is connected directly across the line to measure the voltage, it is desirable that the voltmeter take as little current as is practicable. Because of its comparatively low resistance, the moving coil of the voltmeter cannot be connected directly across the line, as it would take ordinarily an excessive current and hence would be burnt out. Therefore, it is necessary to connect a high resistance in series with the moving coil, as shown in Fig. 82. By Ohm's law, the current through the instrument is proportional to the voltage, so that the instrument scale can be graduated in volts.

The series resistance required is easily determined. Assume that an instrument gives full-scale deflection with 0.01 amp in the moving coil, and that the coil resistance is 20 ohms. If it is desired that the instrument indicate 150 volts, full scale, the total resistance of the instrument circuit must be

$$R = \frac{V}{I} = \frac{150}{0.01} = 15,000 \text{ ohms.}$$

As the resistance of the instrument itself is 20 ohms, 14,980 ohms additional are necessary.

If it be desired that this same instrument also have a full-scale deflection with 15 volts, the resistance of 14,980 ohms may be tapped so that the resistance *OB* (Fig. 82 (*a*)) = 15/0.01 = 1,500 ohms, and this tap can be brought to a binding post. Another method of securing the same result is shown in Fig. 82 (*b*). A separate resistance of 1,500 − 20 = 1,480 ohms is connected from a binding post to the junction of the 14,980-ohm resistance and the moving coil. This last method is advantageous, as it permits independent adjustment of each resistance; injury or repair in one resistance does not affect the other.

83. Multipliers or Extension Coils.—The range of a voltmeter may be increased by the use of external resistance connected in series with the instrument.

Example.—The resistance of a 150-scale voltmeter is 17,000 ohms. What external resistance should be connected in series with it so that its range is (*a*) 300 volts? (*b*) 600 volts?

a. The deflection of the voltmeter depends on the current flowing through its moving coil. Hence, in order to maintain the same current through the instrument at 300 volts and at 150 volts, the resistance of the circuit must be doubled.

Therefore,

17,000 × 2 = 34,000 ohms are necessary.

As the resistance of the voltmeter itself is 17,000 ohms, the added resistance will be

34,000 − 17,000 = 17,000 ohms. *Ans.*

b. The total resistance must now be

$$600/150 \times 17{,}000 = 68{,}000 \text{ ohms.}$$

As 17,000 ohms is already within the instrument, $68{,}000 - 17{,}000 = 51{,}000$ ohms must be added external to the instrument. *Ans.*

External resistances used in this manner are called *multipliers*, or sometimes *extension coils*. Their multiplying power M is given by

$$M = \frac{R_z + R_m}{R_m} \qquad (45)$$

where R_z is the resistance of the multiplier and R_m the resistance of the instrument.

Example.—In the above example (*b*), the multiplying power of the multiplier is

$$M = \frac{51{,}000 + 17{,}000}{17{,}000} = 4.$$

84. Vacuum Thermocouple.—The deflection of the Weston type instrument depends on the force developed on an electric current in a magnetic field. Current may also be measured by its heating effect, and in the Hartmann and Braun ammeter the deflections depend on the thermal expansion of a resistor heated by the current.[1] The vacuum thermocouple operates on the principle that when two dissimilar metals, in contact at a junction, are heated an emf is developed. Either this emf or the resulting current may be measured with a galvanometer or a sensitive indicating instrument.

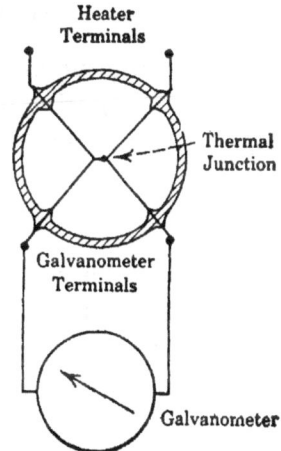

FIG. 83.—Vacuum thermocouple and galvanometer.

In Fig. 83, a very small thermal junction in contact with a small resistor or heater is enclosed in a small evacuated bulb. The terminals of the thermal junction are connected to a galvanometer or to a sensitive d-c indicating instrument. The current to be measured, which, if too large, may be shunted, flows in the heater circuit, raising the temperature of the thermal

[1] See DAWES, C. L., "A Course in Electrical Engineering," 3d ed., Vol. I, Par. 111, p. 156.

junction, thus causing an emf to be developed.   The deflections
of the indicating instrument are nearly proportional to the square
of the current in the heater.   Since an alternating current is
defined by its heating effect (see Chap. I, Part II), such an
instrument can be used as a transfer instrument to measure alternating current in terms of direct current.   Also, since inductive
and capacitive effects are practically nil, the thermocouple may
be used to measure audio- and radiofrequency currents.   The
heater circuit is delicate and is limited to very small currents, and
care is necessary in its use.   For larger currents, a shunt may be
used.   With a series resistance, the thermocouple may be used
to measure voltage.

ELECTRICAL MEASUREMENTS

MEASUREMENT OF RESISTANCE

85. Voltmeter-ammeter Method.—The resistance of any por
tion of an electric circuit which does not contain a source of emf
is, by Ohm's law,

$$R = \frac{V}{I}$$

where V is the voltage across that portion of the circuit and I is
the steady current flowing in that portion of the circuit.   The
voltage V may be measured with a voltmeter, the current I
with an ammeter, and the resistance R may be computed.

Let it be required to determine the resistance R in the circuit
shown in Fig. 84.   The source of power is the 110-volt supply.
The resistance R is comparatively small in value and, if connected directly across 110 volts, would take an excessive current.
Therefore, it is necessary to insert a resistance R' in series with
R to limit the current.   The voltmeter, however, must be connected directly across R, as it is desired to know the resistance
of this portion of the circuit only.

Example.—The voltmeter (Fig. 84) reads 19.0 volts when the ammeter
reads 24.0 amp.   What is the value of the resistance R?

$$R = \frac{19.0}{24.0} = 0.792 \text{ ohm.}   \textit{Ans.}$$

As a matter of interest, let it be required to determine the resistance of
R'.   The voltmeter terminals are transferred from across R to across R'

(probably a higher range voltmeter will be necessary). Under these conditions, the voltmeter reads 91.0 volts and the ammeter still reads 24.0 amp. Therefore,

$$R' = \frac{91.0}{24.0} = 3.79 \text{ ohms.} \quad Ans.$$

It is sometimes desired to measure resistances of such low value that, if a voltmeter were connected directly across their terminals, the contact resistance, which may be comparatively large, would introduce considerable error and might even exceed in magnitude the resistance which it is desired to measure. To eliminate this error due to contact resistance, the voltmeter terminals are connected well inside the terminals *BB* (Fig. 85)

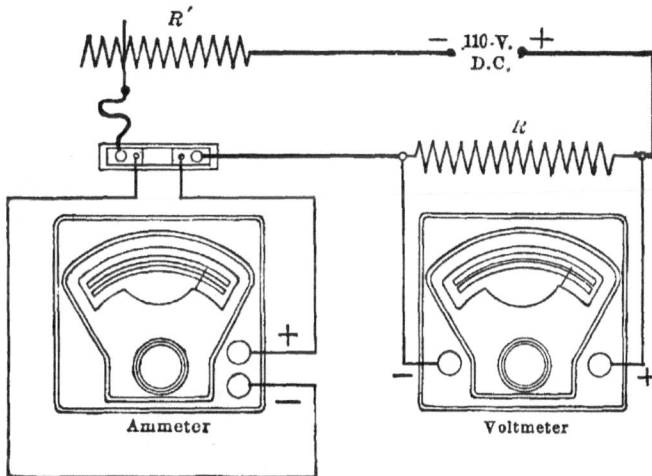

Fig. 84.—Voltmeter-ammeter method of measuring resistance.

through which the current is led to the specimen whose resistance is being measured. As the voltmeter takes only a very small current, small, sharp-pointed contacts *CC* may be used. Since the resistance of the voltmeter is comparatively high, it is only necessary that the contact resistances at *CC* be negligible compared with the resistance of the instrument. This condition is easily met. As these contacts are small and sharp, the points of contact on the specimen can be determined very accurately.

Example.—When the ammeter (Fig. 85) reads 50 amp, the millivoltmeter indicates 40 millivolts. The contacts *CC* are 23 in. apart. What is the resistance per inch length of the rod?

The resistance for 23 in. is

$$R = \frac{0.040}{50} = 0.00080 \text{ ohm.}$$

The resistance per inch

$$r = \frac{0.00080}{23} = 0.0000348 \text{ ohm.} \quad Ans.$$

86. Voltmeter Method.—It is possible to measure resistance by means of a voltmeter alone, provided the resistance

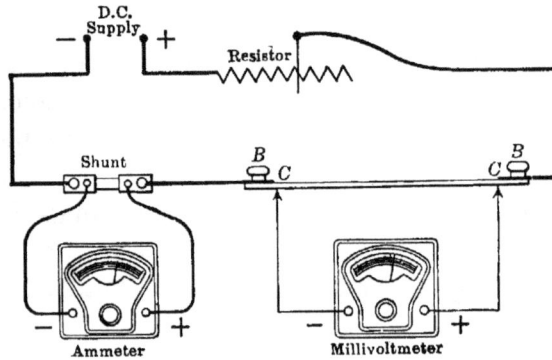

Fig. 85.—Measuring the resistance of a metal rod.

to be measured is comparable with that of the voltmeter. In Fig. 86 (a), let it be required to measure the resistance R. The voltmeter is first connected across the source of supply and a reading V_1 taken. It is then transferred so that the resistance

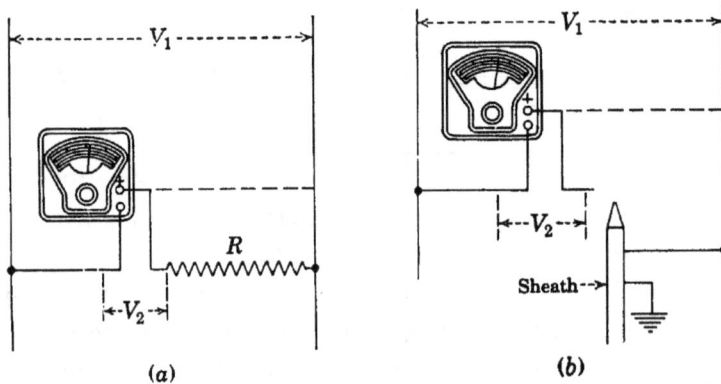

(a)   (b)

Fig. 86.—Measurement of resistance by the voltmeter method.

R is in series with it across the source of supply, and the voltmeter reading is again taken. Let this reading be V_2.

As V_1 is the total circuit voltage and V_2 is the voltage across the instrument, the voltage across the unknown resistance R is obviously $V_1 - V_2$. When the voltmeter is in series with R, the same current i must flow through each, so the voltages are as follows:

$$V_2 = iR_v, \tag{I}$$
$$V_1 - V_2 = iR, \tag{II}$$

where R_v is the resistance of the voltmeter.

Dividing (II) by (I) and solving for R,

$$R = R_v \frac{V_1 - V_2}{V_2}. \tag{46}$$

This method of measuring resistance is particularly useful in determining insulation resistance of dynamo windings, cables, etc. As such resistances are very high, they are usually expressed in megohms (1 megohm = 1,000,000 ohms). It will be seen from Eq. (46) that, the greater the value of R_v, the greater the resistance that can be measured by this method. For this reason, special 150-scale voltmeters, having resistances of 100,-000 ohms ($\frac{1}{10}$ megohm), are available. These give a sensitivity about six times as great as can be obtained with the ordinary 150-scale voltmeter.

Figure 86 (*b*) shows the application of this method to the measurement of the insulation resistance of a cable.

Example.—When a 100,000-ohm voltmeter is connected across a d-c line, it reads 125.0 volts. One terminal of the voltmeter is then connected to the core of a lead-covered cable, and the sheath of the cable is connected to the other side of the line (Fig. 86 (*b*)). The voltmeter now reads 8.0 volts. What is the insulation resistance of the cable?

$$x = 0.1\frac{125.0 - 8.0}{8.0} =$$

1.46 megohms. *Ans.*

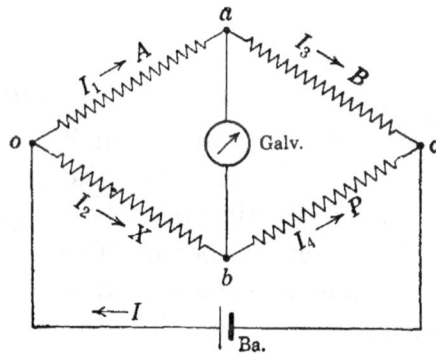

Fig. 87.—Wheatstone bridge circuit.

87. Wheatstone Bridge.—In

distinction to the foregoing methods of measuring resistance, the Wheatstone-bridge method is one in which the unknown resistance is balanced against other known resistances. The bridge in its simplest form is shown in Fig. 87. Three known resistances A, B, P and the unknown resistance X are connected in the form of a diamond. Connections are made from a battery Ba to the two opposite corners o and c of the diamond. A galvanometer is connected across the other two corners, a and b.

To make a measurement, the two arms A and B are each set at some fixed value of resistance, usually 1, 10, 100, 1,000, etc., ohms. The arm P is then adjusted until the galvanometer does not deflect. If the galvanometer does not deflect, no current flows through it and therefore the two points a and b must be at the *same potential*. Also, the currents $I_1 = I_3$ and $I_2 = I_4$, as no current flows through the galvanometer.

If the points a and b are at the same potential, the voltage drop $oa =$ the voltage drop ob, and

$$I_1 A = I_2 X. \tag{I}$$

Also, the voltage drop $ac =$ the voltage drop bc, and

$$I_3 B = I_4 P.$$

Since

$$I_1 = I_3 \quad \text{and} \quad I_2 = I_4,$$
$$I_1 B = I_2 P. \tag{II}$$

Dividing (I) by (II),

$$\frac{I_1 A}{I_1 B} = \frac{I_2 X}{I_2 P} \quad \text{or} \quad \frac{A}{B} = \frac{X}{P}.$$

$$X = \frac{A}{B} P, \tag{47}$$

which is the equation of the Wheatstone bridge. A and B are called the ratio arms and P the balance or rheostat arm. Obviously, the battery and the galvanometer may be interchanged without affecting the relation given in Eq. (47).

The many types of Wheatstone bridge found in practice do not differ in principle from that shown in Fig. 87. The differences lie in the positions of the arms A, B, P on the bridge, as well as in the manner in which the coils in these arms are cut in and cut out of circuit.

88. Dial-type Bridge.—In the commercial Wheatstone bridge, the resistances A, B, P are made up of individual resistance units or coils. For a number of years the resistance coils of such bridges were connected across transverse gaps cut in heavy brass or composition mounted on the hard-rubber or bakelite top of the bridge box. The coils were removed from circuit by short-circuiting the gaps with tapered, brass plugs having hard-rubber tops. An objection to such bridges is that, in time, dirt and oxide accumulate on the plugs and their contact resistance becomes sufficiently high to cause error. Also, in such bridges, the manipulation is slow owing to the necessity of inserting and removing plugs, which must be done with a twisting action. However, many such bridges are still in use. Also, this type of bridge is preferred for standards, because with clean plugs and careful use the contact resistance may be reduced to negligible

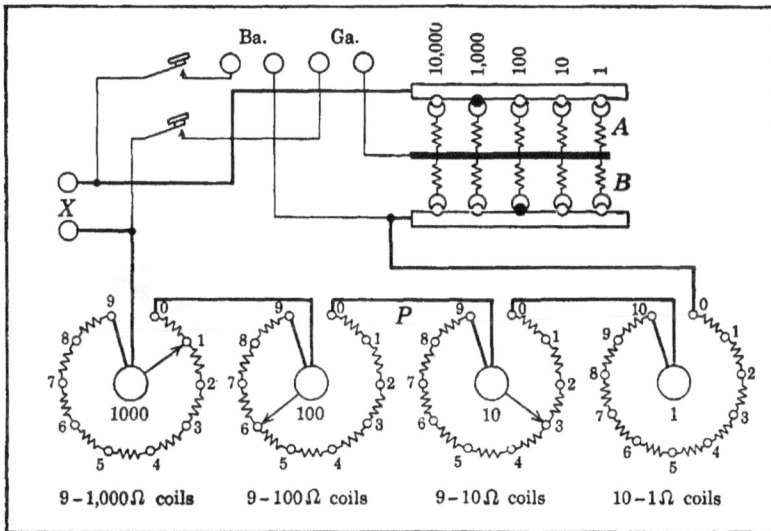

Fig. 88.—Diagram of dial bridge.

values because of the high pressure that can be obtained. In modern bridges, however, the dial type using decades (or nine resistances per dial) has superseded the plug type. The dials can be quickly manipulated to give a balance, the wiping action of the dial brushes over the contact surfaces tends to keep the contacts clean, and at most there are but two contact resistances per dial.

The diagram of a simple dial bridge is shown in Fig. 88. The rheostat arm P consists of four dials in series, one of nine 1,000-ohm coils, one of nine 100-ohm coils, one of nine 10-ohm coils, and one of ten 1-ohm coils. The ratio arms A and B are so arranged that by means of a single plug in each arm it is possible to connect in circuit a 10,000-ohm, a 1,000-ohm, a 100-ohm, a 10-ohm, or a 1-ohm resistance. The dials in the rheostat arm P can be manipulated to select very rapidly any resistance between

1 ohm and 10,000 ohms.   This permits a balance to be obtained quickly.

Ordinarily, the battery and galvanometer are external to the bridge, being connected to the terminals *Ba* and *Ga*.   Sometimes the galvanometer is incorporated within the bridge box itself. The unknown resistance is connected across the terminals *X*.

To increase the range, bridges are frequently provided with additional dials, such as one dial consisting of nine 10,000-ohm

Fig. 89.—Dial-type Wheatstone bridge.   (*Leeds and Northrup Company.*)

coils, and a second dial consisting of ten 0.1-ohm coils.   The connections of the bridge in Fig. 88 are identical with the simplified diagram of Fig. 87.

A Leeds and Northrup dial bridge is shown in Fig. 89.   The connections are essentially those of Fig. 88, except that provision is made for interchanging arms *A* and *B* by changing the positions of the two plugs shown.   In bridges, the dial contacts as well as the plugs should be cleaned occasionally.

89. Procedure in Balancing Bridge.—In using the bridge, much time may be saved if a systematic procedure is followed in obtaining a balance.   Assume that it is desired to measure an unknown resistance to four significant figures.   Connect the resistance across terminals *X* (Fig. 88) and connect the battery and galvanometer to terminals *Ba* and *Ga*.   The galvanometer should be shunted to prevent violent deflections when the bridge is considerably out of balance.   First, make the ratio arms *A* and B each 1,000 ohms, a 1 to 1 ratio. With the galvanometer well shunted and all the dials in P in the zero positions (res. = 0), depress first the battery key and then the galvanometer key. The galvanometer is observed to deflect to the left. Now, move the 9,000-ohm dial to its maximum value, and the galvanometer deflects to the right. From these observations, three facts are determined. First, the unknown resistance is less than 9,000 ohms; second, the galvanometer deflects to the left if the resistance in P is too small; and, third, it deflects to the right if the resistance in P is too large. The resistance in the 9,000-ohm dial is gradually reduced, the battery and galvanometer keys being depressed at each position. The galvanometer continues to deflect to the right until the dial is in the zero position, when the galvanometer deflects to the left. This shows that the unknown resistance is less than 1,000 ohms. The 900-ohm dial is now turned from zero to the 100-ohm position, and the galvanometer continues to deflect to the right, showing that the unknown resistance is less than 100 ohms. This dial is returned to the zero position, the 90-ohm dial is turned to the 10-ohm position, and the galvanometer continues to deflect to the right. This dial is also returned to the zero position, and the 10-ohm dial is turned from zero upward. The galvanometer deflects to the left at 2 ohms and to the right at 3 ohms. This shows that the unknown resistance lies between 2 and 3 ohms. With the 1-to-1 ratio chosen, the unknown resistance can be determined to only one significant figure, and four significant figures are desired. To obtain four significant figures, ratio arm A is set at 1 ohm and ratio arm B remains at 1,000 ohms. The 9,000-ohm dial is set at 2,000 ohms. By successive trials, at the same time reducing the shunting of the galvanometer, a final balance is obtained at 2,761 ohms in P. Then,

$$X = \frac{A}{B} P = \frac{1}{1,000} \, 2,761 = 2.761 \text{ ohms.}$$

In obtaining a balance, the battery key should always be depressed before the galvanometer key, so that the current in the bridge has time to reach a constant value, as otherwise the emf of self-induction may introduce an error.

90. Slide-wire Bridge.—The slide-wire bridge is a simplified Wheatstone bridge, in which the balance is obtained by means of a slider which moves over a resistance-wire usually of manganin. A typical slide-wire bridge is shown in Fig. 90. The resistance-wire AB, 100 cm long, is stretched tightly between two heavy copper blocks CD, 100 cm apart. A meter scale is placed along this wire. A contact key K' is movable along the scale; and when the key K' is depressed, a knife-edge makes contact with the wire. The rest of the bridge consists of a heavy copper bar E, a known resistance R, and the unknown resistance X. R is connected

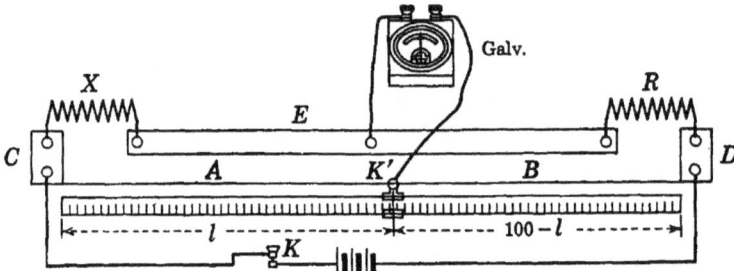

Fig. 90.—Slide-wire bridge.

between D and E, and X is connected between C and E, although the positions of R and X are interchangeable.

The galvanometer is connected between the key K' and E, and the battery terminals are connected to C and D. A balance is obtained by moving K' along the wire until the galvanometer shows no deflection.

Let l be the distance in centimeters from the left-hand end of the scale to K' when a balance is obtained. Then $100 - l$ is the distance from K' to the right-hand end of the scale. Let r be the resistance per unit length of the wire. Then the resistance of l is lr, and that of the remainder of the wire is $(100 - l)r$.

By the law of the Wheatstone bridge,

$$\frac{X}{lr} = \frac{R}{(100 - l)r}. \tag{I}$$

r cancels, and (I) becomes

$$X = R\frac{l}{(100 - l)}. \tag{48}$$

Eq. (48) may also be written

$$\frac{X}{R} = \frac{l}{(100 - l)}. \tag{49}$$

That is, when a balance is obtained, the slide-wire is divided into two parts which are to each other as X is to R.

The slide-wire is not so accurate as the coil bridge, because the slide-wire may not be uniform; the solder at the points of contact at C and D makes the length of the wire uncertain; the slide-wire cannot be read so accurately as the resistance units of a bridge can be adjusted.

Example.—Assume that R (Fig. 90) equals 10 ohms and that a balance is obtained at 74.6 cm from the left-hand end of the scale. Find the unknown resistance X.

From Eq. (48),

$$X = 10\frac{74.6}{100 - 74.6} = 10\frac{74.6}{25.4} = 29.37 \text{ ohms.} \quad Ans.$$

CABLE TESTING

91. Murray Loop.—The slide-wire bridge offers a convenient method for locating grounds in cables and wires. Figure 91

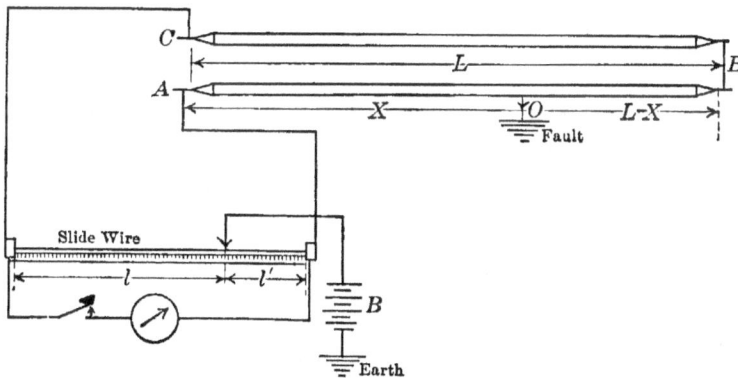

Fig. 91.—Murray-loop test.

shows a cable AB which has become grounded at the point O, owing to a defect in the insulation. CB is the return conductor and is similar to AB, except that it has no ground or "fault." The two conductors are connected at B, the far end of the two conductors, which may be at some power station, telephone exchange, etc.

The slide-wire is then connected to the home ends of the cable as shown. It will be noted that the battery and the galvanometer are not in the positions shown in Fig. 90 but have been interchanged. This is done in order that earth currents shall not enter the galvanometer circuit and thus produce false balances. Also, if the resistance of the ground is high, the emf of the battery B may be increased until sufficient current to operate the bridge is sent through the ground resistance. The resistance of the fault to ground does not produce any error in the measurement so long as the conductor is not broken. If the conductor is broken, with both ends lying on the ground, the resistance of the conductor is increased and a false location of the fault may result.

In Fig. 91, the distance X to the fault may be found by means of Eq. (49).

$$\frac{X}{l'} = \frac{L + (L - X)}{l} \tag{50}$$

where L is the length of the cable.

The slide-wire is divided into two sections which are to each other as the two lengths of cable on each side of the fault.

Solving Eq. (50) for X,

$$X = \frac{2Ll'}{l + l'}. \tag{51}$$

This assumes that the resistance per foot of both conductors is the same and is uniform. The jumper tying the conductor ends together at B should make good connection, as contact resistance at this point may introduce an appreciable error. A ratio and a rheostat arm of a bridge box may obviously be used instead of the slide-wire.

Example.—A cable 2,000 ft long consists of two conductors. One conductor is grounded at some point between stations A Murray-loop test, with a 100-cm, slide-wire bridge, is connected as in Fig. 91 to locate the fault. A balance is obtained at 85 cm. How far from the station is the ground?
From Eq. (51) in which

$$L = 2{,}000, \qquad l' = 15, \qquad l = 85,$$

$X = (4{,}000 \times 15)/100 = 600$ ft. from the end at which the measurement is made.

92. Insulation Testing.—In practice it is necessary to measure the insulation resistance of cables, both at the factory and after the cable is installed. A low value of insulation resistance may indicate that the insulation is of an inferior grade. A low insulation resistance after installation may indicate improper handling or faulty installation. The voltmeter method described in Par. 86, p. 104 is applicable in many cases; but where the insulation resistance is high, even a high-resistance voltmeter is not sufficiently sensitive.

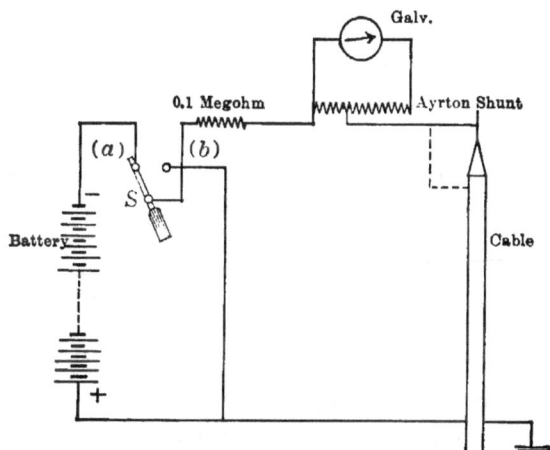

Fig. 92.—Measurement of insulation resistance of cable.

To make the measurement, a sensitive galvanometer is used. A source of potential, of moderately high voltage, 100 to 500 volts, is also necessary. Such potential may be secured from d-c mains, although dry cells, silver chloride cells, and test-tube batteries connected in series are more satisfactory. A simple diagram of connections is shown in Fig. 92.

The method is one of substitution. A known resistance, usually 0.1 megohm (100,000 ohms), is first connected in the circuit and the galvanometer deflection noted. The unknown resistance X is then substituted and the galvanometer reading again noted. As the currents in the two cases are inversely proportional to the circuit resistances, the unknown resistance can be determined, the galvanometer deflections being used rather than actual values of current. Let D_1 be the deflection with the 0.1 megohm and D_2 be the deflection with the unknown resistance.

$$\frac{X}{0.1} = \frac{D_1}{D_2}.$$

$$X = 0.1\frac{D_1}{D_2}. \tag{52}$$

Under ordinary circumstances it would not be possible to obtain accurate results under these particular conditions, because the unknown resistance may be in the hundreds of megohms and the known resistance is but 0.1 megohm. This would make the deflection D_2 so many times smaller than D_1 that it would not be readable.

This difficulty is overcome by the use of the Ayrton shunt, described in Par. 79, p. 93. When the 0.1 megohm only is in circuit, the galvanometer sensitivity is such that ordinarily it would deflect off scale unless the galvanometer were shunted.

Therefore, the shunt is adjusted to some low value, as 0.0001. Let this setting of the shunt be S_1 and the galvanometer deflection D_1. The multiplying power of the shunt is $M_1 = 1/S_1$. (See p. 94.) The cable is now substituted for the 0.1 megohm and the shunt adjusted until a reasonable deflection is obtained. Let this deflection be D_2 and the value of the shunt S_2. Its multiplying power is now $M_2 = 1/S_2$.

The ratio of the currents in the circuit in the two cases is

$$\frac{I_1}{I_2} = \frac{M_1 D_1}{M_2 D_2}.$$

Therefore, the unknown resistance, from Eq. (52), is

$$X = 0.1\frac{I_1}{I_2} = 0.1\frac{M_1 D_1}{M_2 D_2}. \tag{53}$$

The galvanometer acts merely as an ammeter to measure the current leaking through the insulation. If the battery voltage and the galvanometer reading in amperes were known, the insulation resistance could be computed directly. By the substitution of the 0.1 megohm, the insulation resistance may be determined directly from the galvanometer readings.

In practice, instead of substituting the cable for the 0.1 megohm, the cable is first short-circuited by the wire shown dotted (Fig. 92), and the value of $M_1 D_1$ determined. This wire is then removed, placing the cable in circuit. The 0.1 megohm is left permanently in circuit to protect the galvanometer in case of accidental short circuit of the cable. The 0.1 megohm is usually negligible compared to the insulation resistance of the cable, so that no correction is necessary.

A switch or key S is usually provided. When in position (a), the circuit is closed through the cable. When thrown over to (b), the cable, which is charged electrostatically, discharges through the galvanometer.

When the switch S is first closed at (a), there is a rush of current which charges the cable electrostatically (see p. 195). Owing to dielectric absorption, it takes time to charge the cable. Therefore, the charging current flows for some time, decreasing continuously. As it is often inconvenient to wait for the galvanometer to reach a steady deflection, it has been agreed arbitrarily to take the deflection at the end of 1 min as the value to be used in determining insulation resistance.

When the switch S is thrown to (b), the electrostatic charge in the cable rushes out through the galvanometer in the reverse direction. Owing to absorption, it requires considerable time for the cable to become totally discharged.

In making insulation-resistance measurements, precautions must be taken to insulate thoroughly the apparatus itself. Hard-rubber or bakelite posts should be used for supports; and, wherever possible, the leads should be carried through the air rather than allowed to rest on the ground. The variation of insulation resistance with temperature is very large, so that the temperature at which the measurements are made should be carefully determined and stated.

Example.—A cable is tested for insulation resistance, the connections given in Fig. 92 being used. When the cable is short-circuited, the galvanometer deflection, with the shunt set at 0.0001, is 24 cm, the resistance of the circuit being 0.1 megohm. When the short circuit is removed from the cable, a deflection of 17 cm is obtained after the cable has been electrified for 1 min and with the shunt set at 1.0. The cable is 1,800 ft long.

(a) What is its insulation resistance? (b) What is the insulation resistance of a mile length of the cable?

$$M_1 = \frac{1}{0.0001} = 10,000.$$

$$M_2 = \frac{1}{1.0} = 1.$$

$$D_1 = 24 \text{ cm.} \quad \text{and} \quad D_2 = 17 \text{ cm.}$$

From Eq. (53)

$$X = 0.1 \frac{10,000 \times 24}{1 \times 17} = 1,410 \text{ megohms. } \textit{Ans.}$$

(*b*) The insulation resistance of a mile length will be *less* than that of the 1,800-ft length, because the amount of leakage current is directly proportional to the length of the cable. Therefore, the resistance of this leakage path is inversely proportional to the length of cable. The cross-sectional area of the leakage path for the mile length is greater than it is for the 1,800-ft length. Therefore, the insulation resistance of a mile length

$$R = 1,410 \frac{1,800}{5,280} = 481 \text{ megohms. } \textit{Ans.}$$

93. Potentiometer.—The potentiometer is an instrument for making accurate measurements of voltage. Its standardization depends primarily on the Weston standard cell. (See p. 68.) The principle is as follows:

Assume in Fig. 93 (*a*) that a standard cell *S* has an emf of exactly 1 volt. Let a storage cell *Ba* supply current to a wire *AB* through a rheostat *R*. Let the wire *AB* be divided into 15 divisions each of 1 ohm resistance, making the total resistance of *AB* equal to 15 ohms. The standard cell is connected with its negative terminal to the negative terminal of the storage cell and its positive terminal is connected to the tenth 1-ohm coil *C* through a galvanometer and key. If 0.1 amp flows through the wire *AB*, the voltage drop through each division of *AB* will be 0.1 volt and the voltage drop across *AC* will be 1.0 volt. If the key be depressed, no current will flow through the galvanometer, as the standard-cell emf is equal and opposite to this 1-volt drop. If, however, the current in *AB* is not exactly 0.1 amp, current will flow through the standard-cell circuit, owing to the voltage drop *AC* being either greater or less than 1 volt. If the current is less than 0.1 amp, the galvanometer deflects in one direction; and if it is greater than 0.1 amp, the galvanometer deflects in the reverse direction. Obviously, it is possible to adjust the current in *AB* by means of the rheostat *R* to such a value that the galvanometer deflection is zero. Under these conditions, the current in *AB* is exactly 0.1 amp and the potential drop across each resistance in *AB* is 0.1 volt. Therefore, *AB* may be marked in volts as shown.

Let it be required to measure some unknown emf E whose value is known to be less than 1.5 volts. Its negative terminal is connected to the end A of the wire AB (Fig. 93 (*b*)). The positive terminal of the emf is connected through the galvanom-

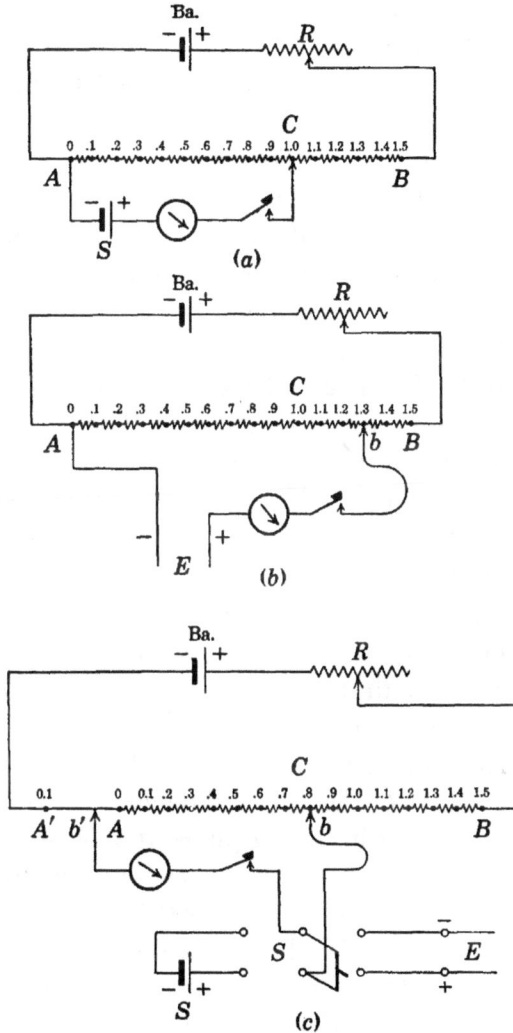

Fig. 93.—Simple potentiometer.

eter and key to a movable contact *b*. It is assumed that the current in AB has been adjusted to exactly 0.1 amp. Contact *b* is moved along AB until the galvanometer deflection is zero. This means that the emf E is just balanced against an equal drop in the wire AB. As AB is calibrated in volts, the value of E may be read directly on AB. This method of measuring voltage is the *Poggendorff method* and is the fundamental principle of the potentiometer.

The transfer from the standard cell to the emf to be measured may be accomplished by means of a double-pole, double-throw (D.-P. D.-T.) switch *S* shown in Fig. 93 (*c*). The standard-cell emf actually is slightly greater than 1 volt. In Fig. 93 (*c*) an extra slide-wire *A'A*, having a resistance of 1.0 ohm, is added to the potentiometer wire *AB*. This makes the fractional parts of 0.1 volt readily obtainable. To make the standard-cell adjustment, the contacts *b* and *b'* are set to values corresponding to the emf of the standard cell, and the potentiometer is then balanced with the switch to the left. The switch is then thrown to the right, and the potentiometer is balanced against the unknown emf *E* by moving the contacts *bb'*.

94. Leeds and Northrup Potentiometer.—In commercial potentiometers, the circuit is essentially that of Fig. 93 (*c*), but with a few accessories added, so that measurements may be made quickly and conveniently. The low-resistance potentiometer, manufactured by Leeds and Northrup, is used quite generally. The connections are shown in Fig. 94.

A 2-volt storage battery is connected to the terminals +BA. and −BA. The working circuit of the potentiometer is +BA., C, P, O, the shunted dial B, the dial D, the shunted dial H, and −BA. The resistance and rheostats, COARSE, FINE, MEDIUM, permit adjustment of the working current and correspond to the rheostat *R* (Fig. 93). The value of the resistance in COARSE is adjusted by inserting plugs at 1, 2, 4.

The potentiometer resistance corresponding to 0.1 volt is 5 ohms, so that the working current is $0.1/5 = \frac{1}{50}$ or 0.02 amp. The total resistance of dial D, corresponding to 1.5 volts, is 75 ohms. The dial H, which consists of a slide-wire wound spirally on a marble cylinder, corresponds to 0.11 volt; hence the equivalent resistance of the dial and shunt is 0.11/0.02 or 5.5 ohms.

The connection of the negative terminal of the standard cell is made through the galvanometer to the 0.5-volt contact on dial D. The resistance between this contact and the 1.5-volt contact corresponds to 1 volt. The emf of an unsaturated Weston standard cell (Par. 54, p. 68) is somewhat greater than 1 volt but may vary slightly for different cells. The ordinary value is about 1.0186 volts. In order to provide for the emf in excess of 1 volt, the dial B with its series and shunt resistances is in series with the working circuit. This corresponds to the slide-wire $A'A$ (Fig. 93 (c)). The positive terminal of the standard cell is connected to the contactor T, and the dial is graduated in standard-cell volts. Hence, when using the potentiometer, it is merely necessary to set T to the emf of the standard cell being used.

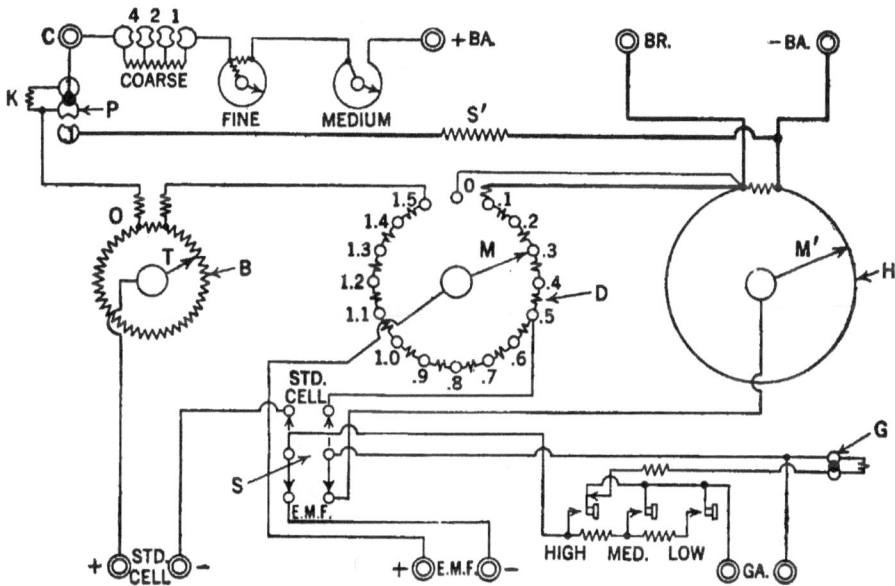

FIG. 94.—Connections of Leeds and Northrup low-resistance potentiometer.

M and M' are movable contacts which are adjusted to balance the unknown emf M moves over the 15 0.1-volt contacts, giving a range of 1.5 volts. M' moves over the spirally wound slide-wire H, giving a range of 0.11 volt which may be read to hundredths, thousandths, etc., of a volt. A D.-P., D.-T. switch S, corresponding to S in Fig. 93 (c), connects either the standard cell or the emf to the potentiometer circuit. There are three galvanometer keys marked HIGH, MED., LOW. The key HIGH should first be depressed since it inserts considerable resistance in series with the galvanometer. As balance is approached, the key MED. and then LOW should be depressed, there being no series resistance with this last key. If necessary, the plug G may be used to insert resistance in series with the galvanometer so that the total resistance may correspond more nearly to the critical damping resistance of the galvanometer. The binding post BR. is used only when it is desired to measure the potentiometer resistances when checking their values. A

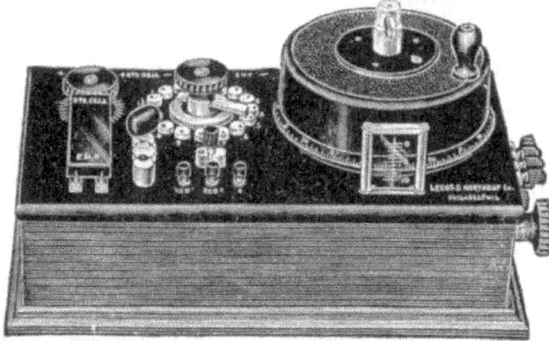

FIG. 95.—Leeds and Northrup potentiometer without accessories.

resistance S′ shunts nine-tenths of the current from the potentiometer system when the plug P is changed. A resistance K is automatically put in circuit, keeping the total potentiometer resistance, and therefore the load on the battery, constant. By this arrangement, the potentiometer readings are all one-tenth their normal values.

A view of the potentiometer is shown in Fig. 95.

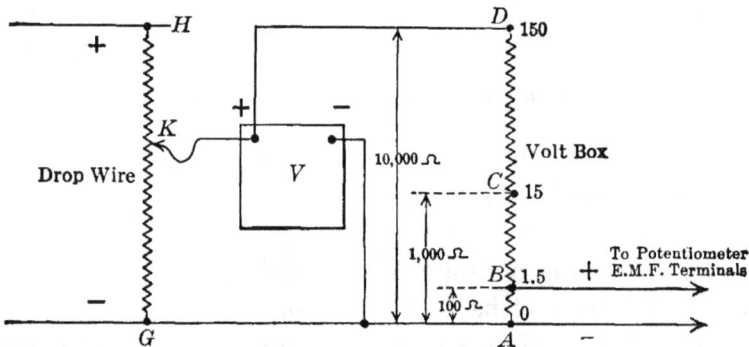

FIG. 96.—Volt-box and drop-wire connections.

95. Voltage Measurements with Potentiometer.—The maximum range of a potentiometer is ordinarily from zero to 1.6 volts. For the measurement of potentials in excess of this value, a *volt box* is necessary. A volt box is merely a high resistance from which suitable taps are brought out. This is illustrated by the resistance AD (Fig. 96). Assume that the resistance AD is

10,000 ohms and AB is 100 ohms. *If no current leaves the wire at B,* the voltage drop across AB will be $100/10,000 = \frac{1}{100}$ of that across AD. If leads be carried from AB to the potentiometer, the potentiometer will measure $\frac{1}{100}$ of the voltage across AD, since no current is taken from B, the potentiometer principle being a null method. Therefore, when a voltmeter V is being calibrated, it should be connected in parallel with AD. If the potentiometer reads 1.142 volts, the true line voltage across the voltmeter will be $1.142 \times 100 = 114.2$ volts. If the voltmeter reads 113.8 volts, the error is -0.4 volt and the correction is $+0.4$ volt.

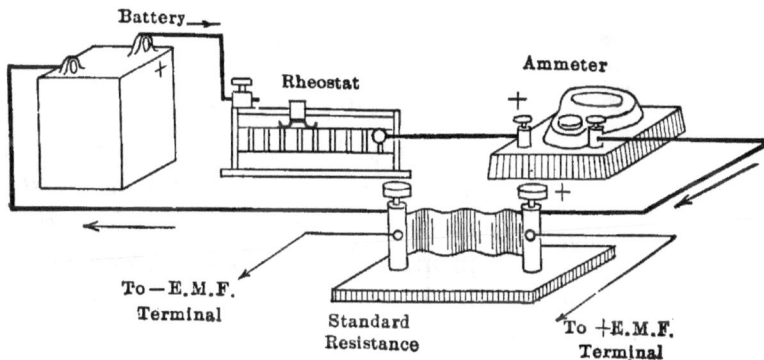

FIG. 97.—Connections for calibrating an ammeter.

In a similar manner, voltages from 1.5 to 15 volts are connected across AC, the multiplying factor in this case being 10.

The Drop-wire.—GH is a resistance connected directly across the line. One voltmeter terminal and one terminal of the volt box are connected to the end G of this resistance. The other terminal of the voltmeter and the remaining terminal of the volt box are connected to a movable contact K. By sliding K along GH, any desired voltage up to that across the drop-wire may be obtained. When used in this manner, GH is called a *drop-wire* or a *potentiometer wire*. It is not essential to the operation of the volt box and is merely a convenient means for adjusting the voltage.

96. Measurement of Current with Potentiometer. A potentiometer is designed primarily to measure *voltage*. By applying Ohm's law it may also be used to measure current. Let an unknown current I flow through a known resistance R. If E, the voltage drop across R be measured, the current I is determined, since, for this part of the circuit, both the voltage and the resistance are known. Therefore,

$$I = \frac{E}{R}.$$

In order to determine the errors of the ammeter, it is necessary to know the exact current through the instrument. The method of making the measurement is shown in Fig. 97. The ammeter is connected in series with a battery, the standard resistance, and a rheostat for controlling the current. Standard resistances are provided with four terminals as a rule, two heavy ones for current and two smaller binding posts for potential. (See Fig. 98 and also Fig. 85, p. 104.) The two potential binding posts are connected to the potentiometer, the proper polarity being observed. The voltage across the standard resistance is then measured by means of the potentiometer.

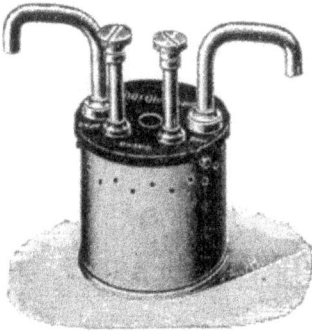

Fig. 98.—Standard resistance. 0.01 ohm.

Standard resistances are usually adjusted to decimal values, such as 10, 1, 0.1, 0.01, etc., ohms. They are ordinarily rated

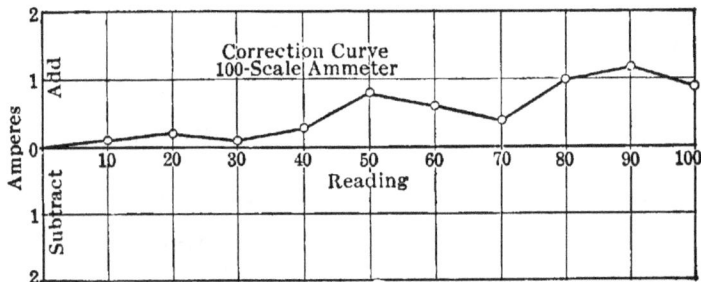

Fig. 99.—Ammeter correction curve.

to carry a current that will give 1.0 volt drop. Thus, the 1 ohm can carry 1 amp; the 0.001 ohm, 1,000 amp; etc. To keep the resistances cool, they are often immersed in oil.

Knowing that the potentiometer is limited to 1.6 volts, it is a simple matter to select the proper standard resistance. An instrument having a range of 100 amp would require 1.6/100 = 0.016 ohm, approximately; 0.01 ohm would be used.

Likewise, a 15-scale instrument would require $1.6/15 = 0.107$ ohm, approximately. A 0.1-ohm resistance would be used.

When instruments are calibrated, they should be checked at 10 or 15 points throughout the scale. The corrections are plotted as ordinates with the corresponding scale readings as abscissas. (The instrument readings are plotted as abscissas.) As an instrument scale is subject to scale errors, etc., it is customary to connect successive points of the plot by straight lines, as shown in Fig. 99. For example, in Fig. 99, the correct current when the instrument reads 50 amp is $50 + 0.8 = 50.8$ amp.

MEASUREMENT OF POWER

97. Voltmeter-ammeter Method.—Since power is equal to the product of volts and amperes (see p. 34), the power taken by a circuit or a device may ordinarily be measured by an ammeter in series and a voltmeter in parallel with the circuit. The power

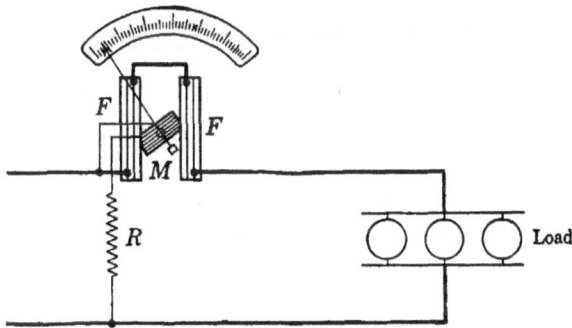

FIG. 100.—Indicating wattmeter.

is the product of the two readings. By using calibrated instruments, the power may be measured with a precision as good as one-half of 1 per cent. With commercial uncalibrated instruments, the precision is less, usually 1 to 2 per cent.

98. Wattmeter.—The wattmeter measures power directly. It consists of fixed coils FF, and a pivoted coil M free to rotate within the magnetic field produced by coils FF, as shown in Fig. 100. The coils FF are wound with comparatively few turns of heavy wire and are capable of carrying the entire current of the circuit. The moving coil M is wound with fine wire, and the current is led into it through two control springs in the same manner that current is led into the coil of a Weston instrument. The fixed coil is connected in series with the load in the same manner as an ammeter is connected. The moving coil is connected across the line in series with a high resistance R in the same manner as a voltmeter coil is ordinarily connected.

The field produced by the coils FF is proportional to the current, and the current in the coil M is proportional to the voltage. Therefore, the turning moment is proportional to the power of the circuit and also depends on the angular position of M with respect to FF, which is taken into consideration when the scale is marked.

Owing to the high degree of precision obtainable by the use of the voltmeter and ammeter, the wattmeter is seldom used for d-c measurements. As it is affected by stray fields, reversed readings should be taken; that is, both the current and the voltage should be reversed, and the average of the two readings used. The wattmeter is used extensively with alternating current. A more complete description, together with its uses, is found in Part II, Chap. III.

MEASUREMENT OF ENERGY

99. Watthour Meter.—The watthour meter is a device for measuring *energy* (see p. 36). As energy is the product of power and time, the watthour meter must take into consideration both of these factors. As power is usually sold on an energy basis, many dollars may depend on the accuracy of such a meter. Therefore, a proper understanding of its mechanism and method of adjustment is essential.

In principle, the watthour meter is a small motor whose instantaneous speed is proportional to the *power* passing through it, and the total revolutions in a given time are proportional to the total *energy* or watt-hours delivered during that time.

Referring to Fig. 101, the line is connected to two terminals on the left-hand side of the meter. The upper terminal is connected to two coils FF in series. These coils are wound with wire sufficiently large to carry the maximum current taken by the load and are so connected that their magnetic fields act in conjunction. The armature rotates in the magnetic field so produced.

The other line wire runs straight through the meter to the load. A shunt circuit is tapped to the upper line on the left-hand side. It runs first to the armature, through the silver brushes B, which rest on the small commutator *C*. From the brushes the line passes through coil *F'*, and through a resistance *R* to the lower line wire. This resistance *R* is omitted in certain types of meters.

As the load current flows through *FF* and there is no iron in circuit, the magnetic field produced by these coils is proportional to the *load current*. As the armature, in series with resistance, is connected directly across the line, the current in the meter armature is proportional to the *line voltage*. Neglecting the small voltage drop in *FF*, the torque acting on the armature

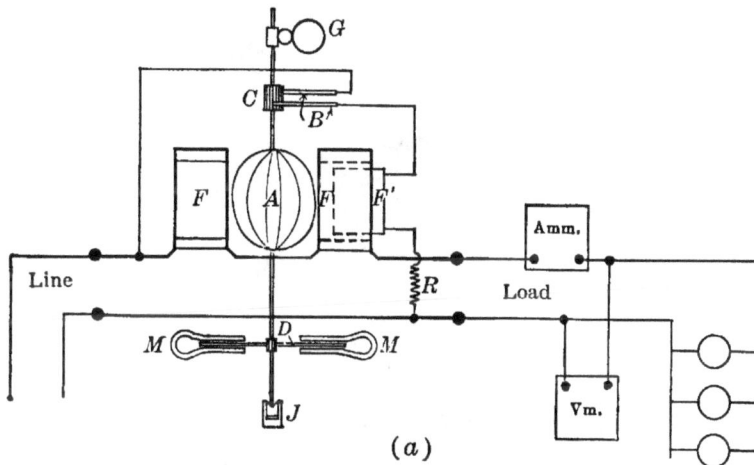

FIG. 101.—Connections of watthour meter.

must be proportional to the product of the load current and the load voltage or, in other words, proportional to the power passing through the meter to the load.

It can be proved that, if the meter is to register correctly, there must be a retarding torque acting on the moving element which is proportional to its speed of rotation. To meet this condition, an aluminum disc *D* is pressed on the motor shaft. This disc rotates between the poles of two permanent magnets *MM*. In cutting the field produced by these magnets, eddy currents are induced in the disc, retarding its motion. As the strength of these currents is proportional to the angular velocity of the disc and they are acting in conjunction with a magnetic field of constant strength, their retarding effect is proportional to the speed of rotation.

Friction cannot be entirely eliminated in the rotating element. Near the rated load of the meter, the effect of the friction torque is practically negligible, but at light loads the friction torque, which is nearly constant at all loads, is a much greater proportion of the load torque. As the ordinary meter may operate at light loads during a considerable portion of the time, it is desirable that the error due to friction be eliminated. This is accomplished by means of coil F' connected in series with the armature. F' is so connected that its field acts in the same direction as that due to

FIG. 102.—Thomson watthour meter.   (*General Electric Company.*)

coils FF. Therefore, it assists the armature A to rotate. Being connected in the shunt circuit, it acts continuously. The coil is movable, and its position can be so adjusted that the friction error is eliminated.

To reduce friction and wear, the rotating element of the meter is made as light as possible. The element rests on a jewel bearing J, which is a sapphire in the smaller types and a diamond in the larger types of meter. The jewel is supported on a spring. A hardened steel pivot rests in the jewel. In time, the pivot becomes dulled and the jewel roughened, which increases friction and causes the meter to register low unless F' is readjusted.

The moving element turns the clockwork of the meter dials through a worm and the gears G.

Figure 102 shows the interior of a Thomson watthour meter.

100. Adjustment of Watthour Meter.—Even if the initial adjustment be accurate, the registration of a watthour meter may, in time, become incorrect. This is due to many causes, such as pitting of the commutator, roughening of the jewel, wear on the pivot, or change in the strength of the retarding magnets. As the cost of energy to consumers is based on the registration of such meters, it is important that they be kept in adjustment, as a small error in the larger sizes may ultimately mean a difference of many dollars one way or the other.

To adjust the meter, it may be loaded as shown in Fig. 101. The power taken by the load is measured by a calibrated voltmeter and ammeter. The revolutions of the disc D are counted over a period of time which is measured with a stop watch. The relation between watt-hours and the revolutions of the *disc*, in most meters, is as follows:

$$\frac{W \times t}{3,600} = K \times N \tag{54}$$

where W is in watts, t is in seconds, K is the "meter constant," and N is the revolutions of the disc in time t. The meter constant is usually marked on the disc.

When the meter is tested, the voltmeter and ammeter are read intermittently while the revolutions of the disc are being counted. A run of about a minute gives good results.

Let the average watts determined from the corrected voltmeter and ammeter readings be W_1.

The average watts, as indicated by the meter during the same period, are, from Eq. (54),

$$W = \frac{K \times N \times 3,600}{t}. \tag{55}$$

The percentage accuracy of the meter is

$$\frac{100W}{W_1}.$$

Example.—In the test of a 10-amp watthour meter, having a constant of 0.4, the disc makes 40 revolutions in 53.6 sec. The average volts and amperes during this period are 116 volts and 9.4 amp. What is the percentage accuracy of the meter at this load?

Average standard watts $W_1 = 116 \times 9.4 = 1,090$.

Average meter watts, from Eq. (55),

$$W = \frac{0.4 \times 40 \times 3,600}{53.6} = 1,074.$$

$$\text{Accuracy} = \frac{1,074}{1,090} \, 100 = 98.5 \text{ per cent.} \quad Ans.$$

This means that the meter is 1.5 per cent slow and should be speeded up slightly. With calibrated indicating instruments and careful adjustment, a meter may easily be brought within 0.5 per cent of accurate registration.

There are two adjustments to be made. Near full load, if the meter is running slow, the magnets are moved nearer the

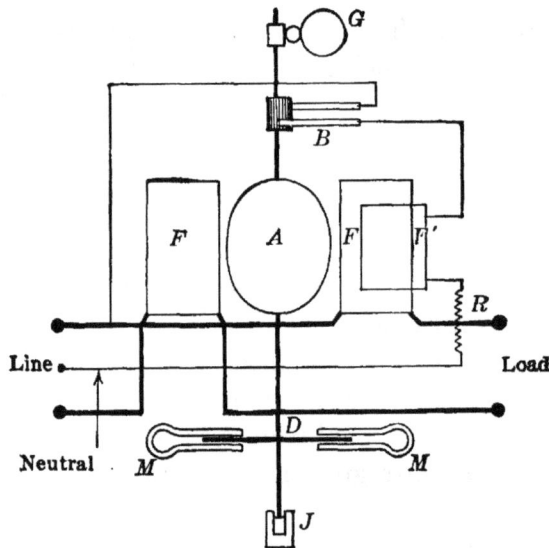

Fig. 103.—Diagram of a three-wire watthour meter.

center of the disc where the effect of the retarding currents is reduced; and if the meter is running fast, the magnets are moved farther from the center. If the meter has been correctly adjusted near full load and is found to be in error near light load, the error is obviously due to friction. The light-load adjustment (made at 5 to 10 per cent rated load) is effected by moving the friction-compensating coil F'. If the meter is slow, the coil F' is moved in nearer the armature; and if the meter is fast, it is pulled out farther from the armature. This adjustment of F' may affect the full-load adjustment slightly so that the meter should be rechecked at full load and then again at light load.

101. Three-wire Watthour Meter.—The three-wire meter is designed to register energy on a three-wire system. It does not differ materially from the meter shown in Fig. 101 except that the two coils *FF* are connected in opposite sides of the line, as shown in Fig. 103. The armature circuit may be connected to the neutral as shown, or it may be connected across the outer wires. If the armature circuit is connected across the outer wires, the neutral connection to the meter is omitted. If connected to neutral, the meter does not register accurately unless the voltages between the two outer lines and neutral are equal, although the error is usually small.

CHAPTER VI

MAGNETISM AND PERMANENT MAGNETS

102. Magnets and magnetism are involved in the operation of practically all electrical apparatus. Therefore, an understanding of their underlying principles is essential to a clear conception of the operation of such apparatus.

Magnets may be divided into two general classes: *permanent magnets*, which have the property of retaining their magnetism indefinitely and which require no exciting ampere-turns; *electromagnets*, whose magnetism depends on the magnetic action of electric currents (Chap. VII). Permanent magnets are made of hardened steel and its alloys. Electromagnets are made of soft iron and steel, which are highly responsive to changes in the magnetizing effects of electric currents.

103. Magnetic Materials.—Iron (or steel) is far superior to all other metals and substances as a magnetic material and, with the exception of a few of its alloys, is practically the only metal used for magnetic purposes. Cobalt and nickel (and some of their alloys) possess magnetic properties which are far inferior to those of iron. Liquid oxygen is also attracted to the poles of magnets. Certain alloys of iron may have highly desirable magnetic properties. For example, *Permalloy*, an alloy of 80 per cent nickel and 20 per cent iron, has extremely high permeability at low flux densities. Alloys of iron and cobalt make excellent permanent magnets. *Alnico*, a recently developed iron-aluminum-nickel-cobalt alloy, has unusually good permanent magnetic properties.

104. Natural Magnets.—Magnetic phenomena were first noted by the ancients. Certain stones, notably at Magnesia, Asia Minor, were found to have the property of attracting bits of iron; hence, the name *magnets* was given to those magic stones. The fact that such stones had the property of pointing north and south, if suspended freely, was not discovered until the tenth or twelfth century. The use of such a stone in navigation gave it the name *Lodestone*, or leading stone. Natural magnets are composed of an iron ore known in metallurgy as magnetite, having the chemical composition Fe_3O_4.

105. Permanent (Artificial) Magnets.—If a piece of hardened steel be magnetized, either by coming in contact with another magnet or by means of an electric current, it will be found to have acquired a substantial amount of magnetism, which it will retain indefinitely. Such a steel magnet is called a *permanent magnet*. Permanent magnets ordinarily derive their initial excitation from an electric current, as will be shown later. If a

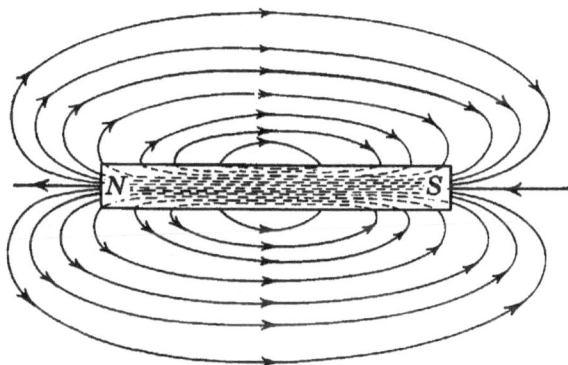

FIG. 104.—Magnetic field about a bar magnet.

piece of soft steel or soft iron be similarly treated, it retains but a very small portion of the magnetism initially imparted to it.

These properties make it desirable to use hardened steel, or its alloys, when a permanent magnet is desired and to use soft iron or steel when it is essential that the magnetism respond closely to changes of magnetizing force. It is found that even hardened steel *ages* or loses some of its magnetism with time. Where a high degree of permanency is desired, as in electrical instruments or in magnetos, the magnets are aged artificially.

106. Magnetic Field.—It is found that magnetism manifests itself as if it existed in lines called *lines of magnetism* or *lines of induction*. These magnetic lines taken as a whole are called the *magnetic flux*. The region or space in which these lines exist is called the *magnetic field*. Further, if the lines of induction of a field due to a magnet be determined experimentally, it is found that they emerge from one region of the magnet and enter at some other region, as shown in the bar magnet, Fig. 104. These regions are called the *poles* of the magnet.   The two poles are distinguished by the position which they seek if suspended freely.   The one which points north is called the *north-seeking pole,* or *north pole* for short, and the other the *south-seeking pole,* or *south pole.*   In practice it is assumed that the lines of induction leave the magnet at the north pole and reenter the magnet at the south pole.   Within the magnet the lines of induction continue from the south to the north pole so that each line of induction forms a closed loop, as shown in Fig. 104.   The plane midway between the poles is the *neutral zone* or *equator* of the magnet.

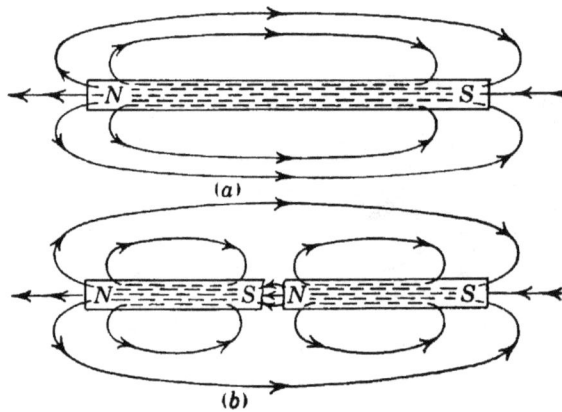

(a)

(b)

Fig. 105.—Effect of breaking a bar magnet.

The entire path through which the lines of induction pass is called the *magnetic circuit.*

107. Effect of Breaking a Bar Magnet.—Neither a north pole nor a south pole can exist alone.   For every north pole there exists an equal south pole.   If an ordinary bar magnet be broken at the middle, or at various points, each fragment will constitute a bar magnet having its north and its south pole lying in the same respective directions as those of the original magnet, as is illustrated in Fig. 105.   This phenomenon is easily explained by noting that the lines of induction still continue to pass from one fragment to the next adjacent one, and in so doing constitute north and south poles, as shown in (b), Fig. 105.   Experimentally, this phenomenon may be illustrated by magnetizing a highly tempered steel knitting needle and breaking it at various points.

108. Weber's Theory.—An explanation of the appearance of north and south poles on breaking a magnet, and of other phenomena associated with the magnetization of iron, is offered by Weber's theory, which has been expanded by Ewing. Each molecule of a magnet is assumed to be a very small magnet, as shown in Fig. 106 (a). Under ordinary conditions the small magnets are arranged in a haphazard way, as shown at (a), so that the various north and south poles neutralize one another, and no external effect is produced. Upon the application of a magnet-izing force, however, the small magnets tend to arrange them-

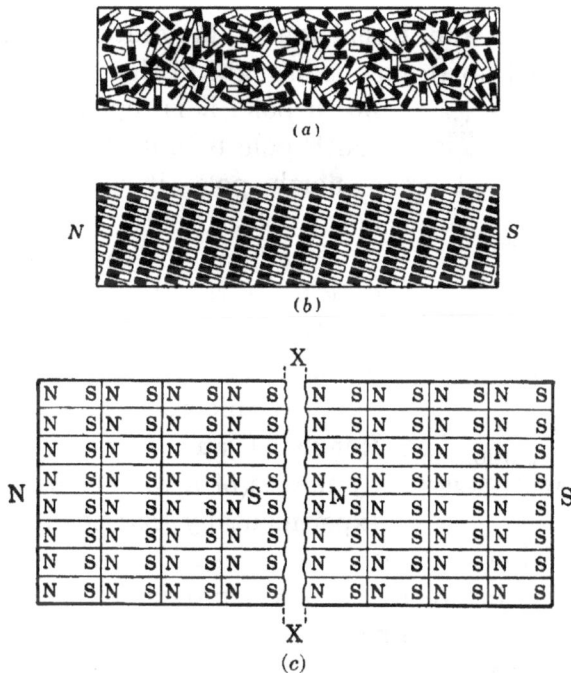

(a)

(b)

(c)

FIG. 106.—Weber-Ewing theory and cutting of magnet.

selves in such a manner that their axes are parallel and their north poles are all pointing in the same general direction as the magnetizing force. This is shown in Fig. 106 (b). Figure 106 (c) shows a section of bar magnet to which a magnetizing force has been applied. At one end of the magnet the molecular *N*-poles combine to form the north pole of the magnet. Likewise, at the other end of the magnet the molecular *S*-poles combine to form the south pole of the magnet. Within the magnet, how-ever, each *N*-pole is neutralized by the adjacent *S*-pole. If the magnet be broken at *XX*, the molecular *N*- and *S*-poles no longer neutralize one another along the sections thus produced, so that a new *N*-pole is formed on the right-hand section and a new *S*-pole on the left-hand section.

This theory is further substantiated by grinding a permanent magnet into very small particles. Each of the small particles possesses the properties of the bar magnet, each having its own north and south poles. Further, the theory offers a rational explanation of saturation, hysteresis, etc., occurring in iron subjected to a magnetizing force. These phenomena will be considered later.

109. Attraction and Repulsion of Magnetic Poles.—When a freely suspended north pole is brought in the vicinity of another north pole, it is repelled, whereas, if a south pole is brought in the vicinity of a north pole, it is attracted to the north pole. South poles are also found to repel one another. From this it may be stated that *like poles repel one another and unlike poles attract one another.*

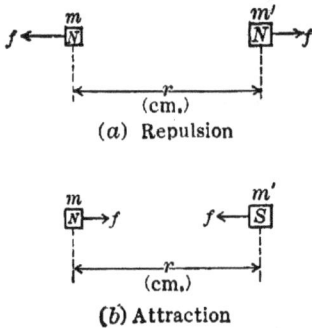

(a) Repulsion

(b) Attraction

Fig. 107.—Repulsion and attraction between magnetic poles.

110. Coulomb's Law of Magnetic Repulsion and Attraction.—Coulomb, by means of a torsion balance, proved experimentally that the force of attraction (or repulsion) between two poles is inversely as the square of the distance between the poles, provided the dimensions of the poles are small compared with the distance between them. *A unit magnetic pole is one of such strength that if placed at a distance of one centimeter in free space from a similar pole of equal strength will repel it with a force of one dyne.*

Pole strength is measured by the number of unit poles which, if placed side by side, would be equivalent to the pole in question.

The force *f*, existing between poles in air, is

$$f = \frac{mm'}{r^2} \text{ dynes} \tag{56}$$

where *m* and *m'* are the pole strengths (in terms of unit pole) of two magnetic poles, spaced a distance *r* cm apart, as shown in Fig. 107. This force may be attraction or repulsion, according as the poles are unlike or like.

Example.—Two north poles, one having a strength of 500 units and the other a strength of 150 units, are placed 4.0 in. (10.16 cm) apart in air. What is the force in grams acting between these poles, and in what direction does it act?

$$4.0 \text{ in. } = 4.0 \times 2.54 = 10.16 \text{ cm.}$$

$$f = \frac{500 \times 150}{(10.16)^2} = \frac{75,000}{103.2} = 727 \text{ dynes.}$$

$$\frac{727}{981} = 0.742 \text{ gram.} \quad \text{Poles repel each other.} \quad Ans.$$

111. Lines of Force.—Thus far the magnetic field has been considered with respect to the lines of magnetism or induction only. If a small unit north pole be placed in such a field, two effects will be observed:

1. This pole will be urged along the lines of induction.

2. The force urging this pole will be greatest where the lines of induction are most dense, and, moreover, the force at a point will be proportional to the number of lines per unit area at the point taken perpendicular to the lines in the field in which the pole finds itself.

From these statements it can be seen that *lines of force*, similar to lines of induction, can be drawn to represent the force at the various points in the magnetic field. In much of the literature of the subject, lines of induction and lines of force are used indiscriminately. The fallacy of so doing is immediately apparent upon considering a solid bar magnet. The lines of induction pass completely through the solid metal of the magnet, whereas the lines of force terminate at the poles. To be sure, a magnetic force does exist within the magnet; but this force can be determined only by making a cavity of special form in the magnet, and the force acting under these conditions is quite distinct from that indicated by the number of lines of induction passing through the bar. In air, however, the lines of force and lines of induction coincide. For example, the magnetic lines *external* to the bar magnet (Fig. 104) may also be considered as lines of force.

112. Field Intensity.—It has been stated that the force acting on a magnetic pole placed in a magnetic field is proportional to the number of lines of force per square centimeter at that point. *Unit field intensity is defined as the field strength which will act on a unit pole with a force of one dyne.* Field intensity is represented by the symbol H, the unit of field intensity being the *oersted*.

The field intensity in oersteds is therefore equal to the dynes per unit pole. It follows that one *line of force* perpendicular to and passing through a square centimeter represents unit field intensity. It is evident that if a pole of m units be placed in a field of intensity H, as shown in Fig. 108, the force acting on this pole will be

$$f = m \times H \text{ dynes.} \tag{57}$$

For this to hold, a pole placed in a field must be of such small strength that it will have no appreciable disturbing effect on the magnetic field.

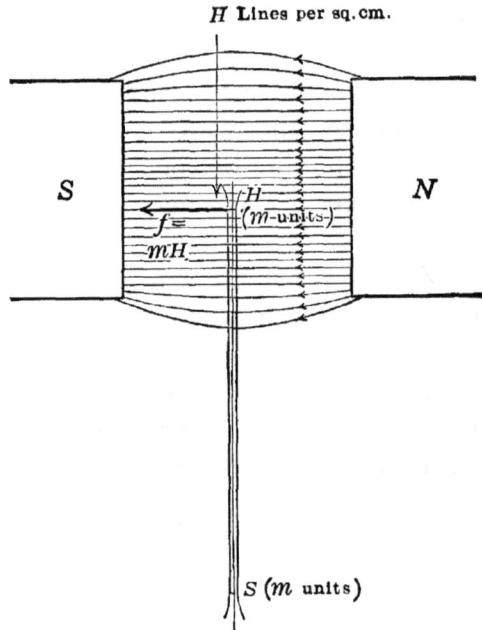

Fig. 108.—Force acting on a magnetic pole in a magnetic field.

Example.—A total flux of 150,000 lines exists in air between two parallel pole faces, each 10 cm square. The field is uniformly distributed. Determine: (a) the field intensity in oersteds; (b) the force in grams acting on a small pole of 200 units pole strength, if placed in this magnetic field.

(a) Flux density = $150,000/(10 \times 10)$ = 1,500 lines per sq cm, or 1,500 gausses. Being in air, this value of flux density also equals the field intensity H. Hence,

$$H = 1,500 \text{ oersteds.} \quad Ans.$$

(b)
$$f = m \times H = 200 \times 1,500 = 300,000 \text{ dynes.}$$

$$\frac{300,000}{981} = 306 \text{ grams.} \quad Ans.$$

113. Flux Density.—Flux density is the number of lines of induction per unit area taken perpendicular to the induction. In free space, flux density and field intensity are numerically the same; but within magnetic material the two are entirely different. The two should not be confused. The unit of flux density in the cgs system is one line per square centimeter and it is named the *gauss*. The expression "lines per square centimeter" is frequently used. Also, in practical work where the English system of units is used, flux density is expressed in lines per square inch.

114. Compass Needle.—The compass consists of a hardened-steel needle or small bar, permanently magnetized and accurately

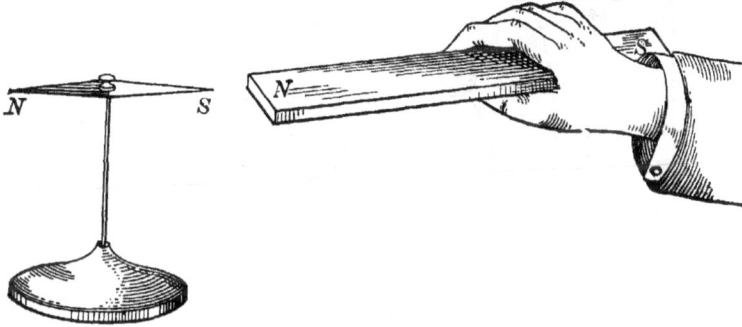

FIG. 109.—Compass needle and bar magnet.

balanced on a sharp pivot. The north-seeking end or north pole points north, and the south-seeking end points south. The north pole of the needle is usually colored blue or given some distinguishing mark. With the exception of a few used for lecture purposes, the needle is enclosed in an airtight case for mechanical protection. Mariner's compasses are mounted carefully upon gimbals, so that they always hang level. Unless the compass is compensated, large errors may be introduced into mariner's compasses by the magnetic effects of the steel hull and the masses of iron in the machinery of the ship. This compensation is accomplished by placing heavy iron balls near the compass, which neutralize the magnetic effect of the ship itself.

By means of the compass, the polarity of a magnet is readily determined. The *south* pole of the compass points to the *north* pole of the magnet, as shown in Fig. 109. Likewise, the *north* pole of the compass points to the *south* pole of the magnet.

This action of the compass needle follows immediately from the law that like poles repel and unlike poles attract each other. The compass is, therefore, very useful in practical work, for it enables one to determine the polarity of the various poles of motors and generators, thus showing whether or not the exciting coils are connected correctly.

Further, the compass needle always tends to set itself in the direction of the magnetic field in which it finds itself, the north end of the needle pointing in the direction of the lines of force or magnetic lines. This is illustrated in Fig. 110. By placing a small compass at various points in the neighborhood of a magnet and drawing an arrow at each point, the arrow pointing in the

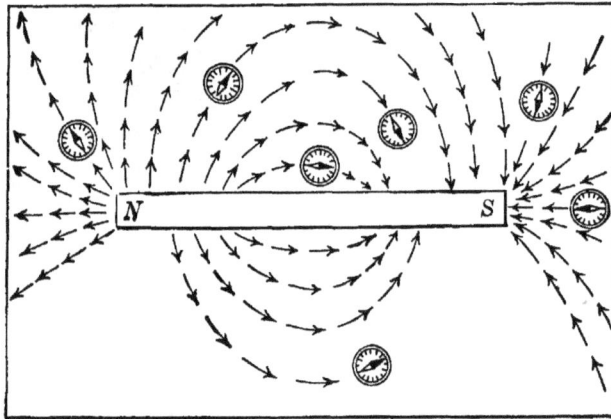

FIG. 110.—Exploring the field about a bar magnet with a compass.

same direction as the needle, the field around the magnet may be mapped, as shown in Fig. 110. In mapping a field in this manner, it must be remembered that the earth's field may exert considerable influence on the compass needle in addition to the effect of the field being studied.

115. Magnetic Figures.—If a card be placed over a magnet and iron filings be sprinkled on the card, a magnetic figure is obtained. The filings at each point set themselves in the direction of the line of force at that point, and the resulting figure shows in close detail the character of the magnetic field. Figure 111 shows the magnetic field due to two bar magnets placed side by side, with unlike poles adjacent. On the other hand, Fig. 112 shows the field due to these same bar magnets when like poles are adjacent. It will be noted in Fig. 111 that the lines of force seem like elastic bands stretched from one pole to the other, acting to pull the unlike poles together.   In **Fig. 112**, the lines of

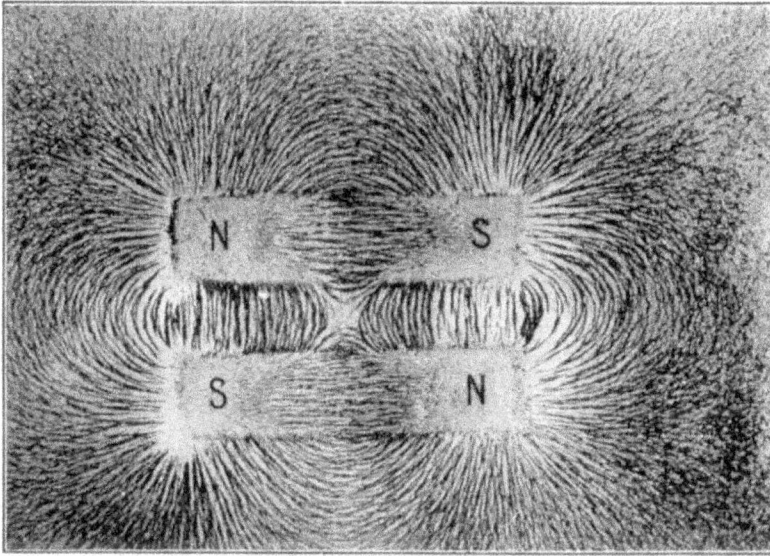

FIG. 111.—Magnetic figure, unlike poles adjacent.

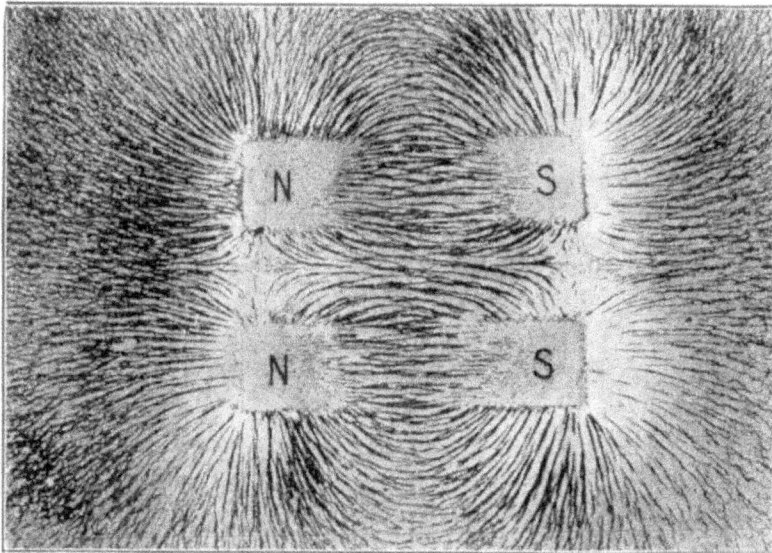

FIG. 112.—Magnetic figure, like poles adjacent.

force from the two like poles appear to repel one another, indicating a state of repulsion between the poles.

116. Magnetic Induction.—If a magnet is brought near a piece of soft, nonmagnetized iron, the piece of iron becomes

magnetized by *induction*.   If the north pole of the magnet is
brought near the iron, a south pole is induced in that part of the
iron nearest the inducing magnet; and if the south pole of the
magnet is brought near the iron, a north pole is similarly induced.

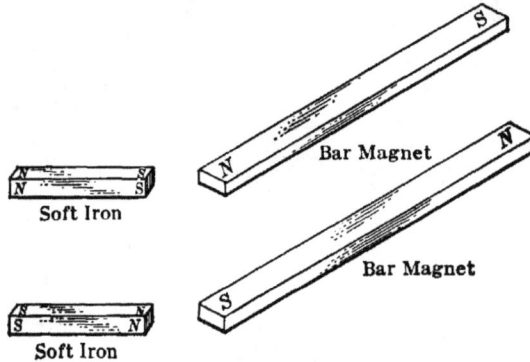

FIG. 113.—Poles induced by magnetic induction.

This is illustrated in Fig. 113.   The reason that poles are induced
in soft iron is indicated in Fig. 114, where a bar of soft iron is
brought in proximity to a north pole.   A considerable number of
the lines of induction leaving this north or inducing pole enter
the adjacent end of the soft-iron bar readily and leave the soft-

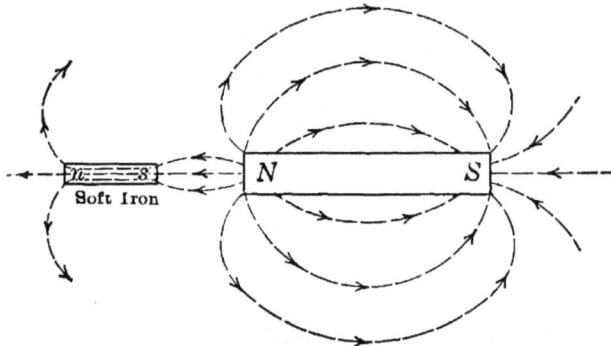

FIG. 114.—Relation of induced poles and magnetic lines.

iron bar at its further end, because the magnetic conductivity
of the iron is considerably greater than that of the air.   This
makes the end of the soft-iron bar nearest the inducing pole
a south pole and its further end a north pole.   From the foregoing, the ability of magnets to attract soft iron is readily understood.   An opposite pole to that of the magnet is induced in the
adjacent portions of the iron, and these two poles, being of unlike
polarity, attract each other.

It is sometimes noticed that if a comparatively weak north pole be brought in the vicinity of a strong north pole, attraction between the two results, rather than the repulsion which might be expected. This is no viola-tion of the laws governing the attraction and repulsion of magnetic poles but comes from the fact that the strong north pole induces a south pole, which overpowers the existing weak north pole and results in attraction. It is easy to

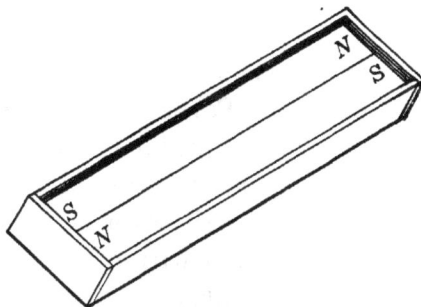

Fig. 115.—Proper method of "keeping" bar magnets.

reverse the polarity of a compass needle by holding one end too close to a strong magnetic pole of the same polarity.

For a similar reason, when two bar magnets are put away in a box, the adjacent ends should be of opposite polarity, as shown in Fig. 115. They will retain their magnetism better under

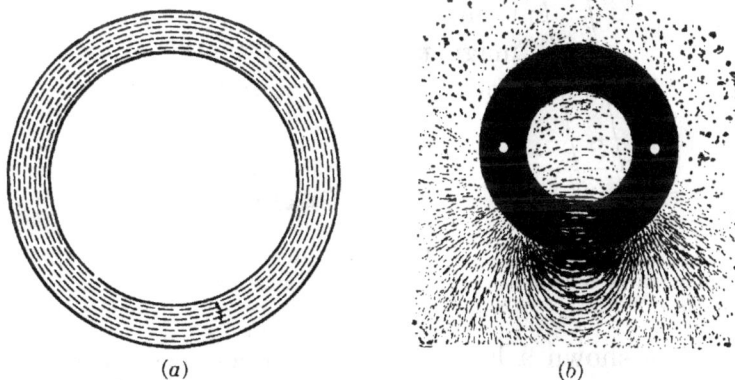

(a)   (b)

Fig. 116.—Field of circular instrument magnet.

these conditions. When a horseshoe magnet is not in use, a "keeper" of soft iron should be placed across the poles.

117. Law of the Magnetic Field.—*The magnetic field always tends to conform itself so that the maximum amount of flux is attained.* This offers further explanation of the attraction of iron to poles of magnets. The iron is drawn toward the magnet so that the magnetic lines may utilize the iron as a part of their circuit, since iron conducts these lines much better than air.

This is illustrated in the horseshoe magnet of Fig. 117. The armature is drawn toward the poles of the magnet, and the portion of the magnetic circuit in air is materially shortened, so that the number of magnetic lines is materially increased. The maximum flux exists when the armature is in contact with the poles.

118. Other Forms of Magnets.—The simple bar magnet frequently is not suitable for practical work. For the same amount of material, other forms are more powerful and more compact. Figure 116 (*a*) shows a closed ring magnet. All the magnetic flux is contained within the ring and little or no external effect exists so that this form of magnet is not very useful. If the ring be cut however, a north and a south pole are formed and the flux in the gap can be utilized. Such a ring magnet, which is employed in the Weston-type instrument, is shown in Fig. 116 (*b*). Ring magnets of this type are very efficient.

The horseshoe magnet, shown in Fig. 117, is very useful, for two reasons. Since the two poles are near each other, a comparatively strong field exists between them. Further, if the function of the magnet is to exert a pull on an armature, both poles are equally effective. In Fig.

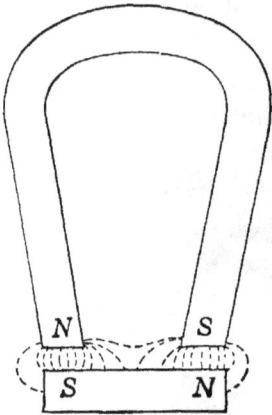

Fig. 117.—Horseshoe magnet attracting a soft-iron armature.

Fig. 118.—Compound or laminated bar magnet.

77, p. 95, is shown a horseshoe magnet as used in Weston d-c instruments.

119. Laminated Magnets.—It is found that thin steel magnets are stronger in proportion to their weight than thick ones. For a given amount of material, a magnet made up of several laminations, as shown in Figs. 118 and 119, is more powerful than one made of a solid piece of metal. This is due in large measure to the fact that during heat-treatment the material within the thicker magnets is not properly tempered throughout. Figure 119 shows the form of horseshoe magnet generally used for telephone and ignition magnetos.

120. Magnet Screens.—There is no known insulator for magnetic flux. No appreciable change in the flux or in the pull of a magnet is noticed if glass, paper, wood, copper, or other such material be placed in the magnetic field. However, it is often desirable to shield galvanometers and electrical measuring instruments from the earth's field and from stray fields due to generators, conductors carrying currents, etc. This is done by surrounding the instrument with an iron shell, as shown in Fig. 120. This shell by-passes practically the entire flux and thus prevents it from affecting the sensitive portions of the instrument. The smaller the openings in the shell, the more effective the screening becomes.

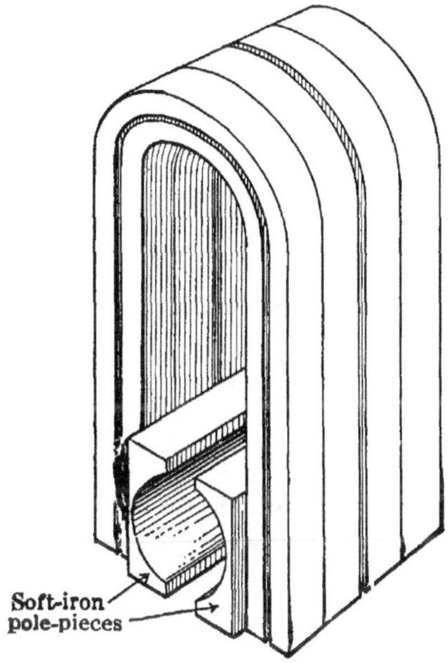

Soft-iron pole-pieces

FIG. 119.—Compound horseshoe magnet used in magnetos.

121. Magnetizing.—A magnet may be magnetized by merely rubbing it with another magnet. The resulting polarity at any point is opposite to that of the last pole which came in contact with this point. Stronger mag-

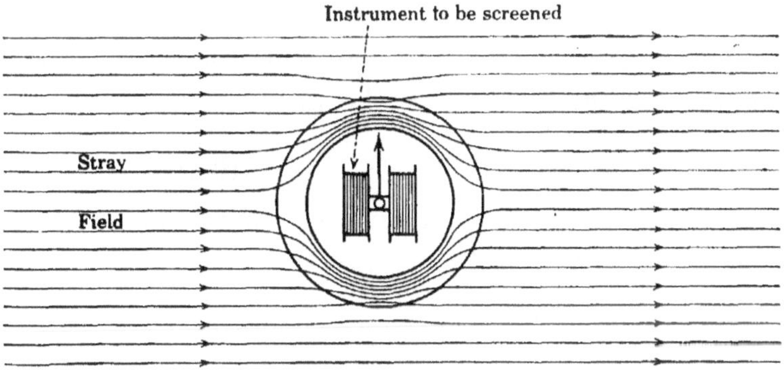

Instrument to be screened

Stray

Field

FIG. 120.—Magnetic screen.

nets may be obtained by placing them between the poles of a very powerful electromagnet. Figure 121 shows this method of magnetizing a horseshoe magnet. An armature or "keeper" should be placed across the poles of the horseshoe magnet before removing it from the electromagnet. The most common method of magnetizing permanent magnets, however, is to insert the magnets in a suitable exciting coil and to cause a large current to flow in the coil.

122. Earth's Magnetism.—The earth behaves as if it were a huge bar magnet, the poles of which are not far from the geographical poles. The north magnetic pole (corresponding to the south pole of a magnet) is situated in Boothia Felix, about 1,000 miles from the geographical north pole. The south magnetic pole has never been located, but experiment points to the existence of two south poles. Due to the noncoincidence of the geographical and magnetic poles and to the presence of magnetic materials in the earth, the compass points to the true north in only a few places on the earth's surface. The deviation from the true north is called the *declination*, and magnetic maps are provided showing the declination at various parts of the earth. At New York it is about 9° west.

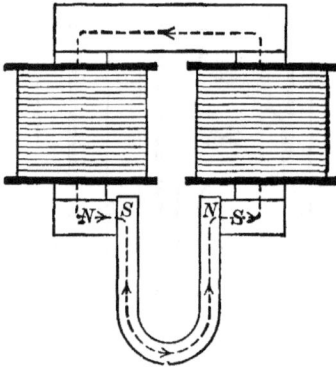

Fig. 121.—Magnetizing a horseshoe magnet with an electromagnet.

A freely suspended and balanced needle does not take up a position parallel to the earth's surface, when under the influence of the earth's magnetism alone, but assumes a position making some angle with the horizontal. This angle is called the *dip* of the needle. At New York it is about 70° north.

The field intensity (total, not horizontal) of the earth's field at New York is about 0.61 cgs units or oersteds, although this value varies slightly with time.

CHAPTER VII

ELECTROMAGNETISM

123. Magnetic Field Surrounding a Conductor.—It had long been suspected that some relation existed between electricity and magnetism, but it remained for Oersted in 1819 to show that this relation not only existed but that it was a definite relation.

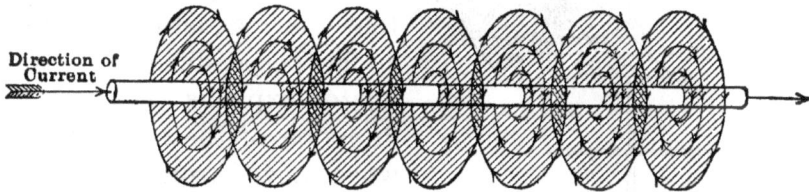

Fig. 122.—Magnetic field about a straight conductor.

If a compass be brought in the neighborhood of a single conductor carrying an electric current, the needle deflects, indicating the presence of a magnetic field. It is further observed that the needle always tends to set itself at right angles to the

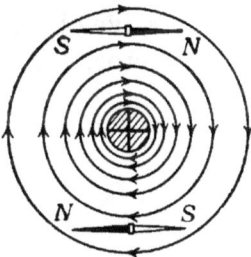

Fig. 123.—Lines of flux surrounding a straight cylindrical conductor—current inwards.

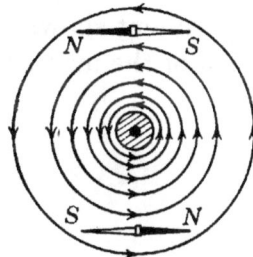

Fig. 124.—Lines of flux surrounding a straight cylindrical conductor—current outwards.

current. When it is held above the conductor, the needle points in a direction opposite to that which it assumes when held beneath the conductor. Further investigation shows that with a straight cylindrical conductor, the magnetic flux exists in circles about the conductor (if there is no other magnetic field in the vicinity), as shown in Figs. 122, 123, and 124. These circles have their centers at the axis of the conductor, and their planes are perpendicular to the conductor. If the current in the conductor be reversed, the direction in which the compass needle is deflected will also reverse, showing that the direction of this magnetic field depends on the direction of the current. The relation is shown in Fig. 122. The fact that the magnetic flux exists in circles perpendicular to the conductor explains the reversal in direction of the compass needle when moved from a point above the conductor to a point beneath it, since the direction of the field above the conductor must be opposite to that beneath the conductor. This is illustrated

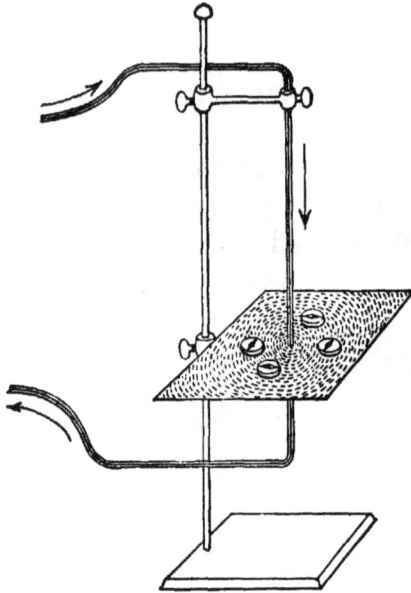

Fig. 125.—Investigation of the magnetic field surrounding a conductor.

in Figs. 123 and 124.[1]

The experiment shown in Fig. 125 is illustrative of this concentric relation of the magnetic flux to the conductor. A conductor

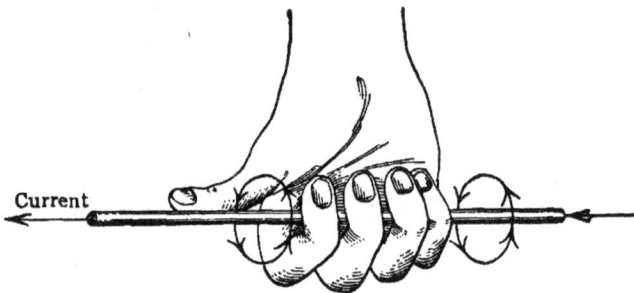

Current

Fig. 126.—Hand rule.

carrying a current is brought vertically down through a horizontal sheet of cardboard. Iron filings sprinkled on the cardboard form

[1] A circle having a cross inside ($\oplus$) indicates that the current is flowing into the paper, the cross representing the feathered end of an arrow. A circle having a dot at the center ($\odot$) indicates that the current is flowing out of the paper, the dot representing the tip of an arrow.

concentric circles. (A current of about 100 amp is necessary to obtain distinct figures.) If four or more compasses are arranged as shown in Fig. 125, they will indicate, by the direction in which the needles point, that the magnetic lines are circles having the axis of the wire as centers.

124. Relation of Magnetic Field to Current.—A definite relation exists between the direction of the current in a conductor and the direction of the magnetic field surrounding the conductor. There are two simple rules by which this relation may be remembered.

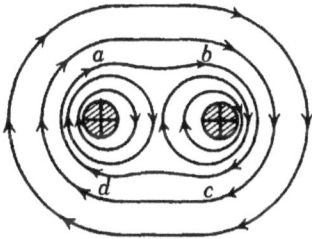

FIG. 127.—Magnetic field about two straight parallel conductors—current in same direction.

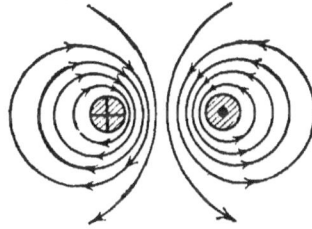

FIG. 128.—Magnetic field about two straight parallel conductors—current in opposite directions.

Hand Rule.—Grasp the conductor in the *right* hand with the thumb pointing in the direction of the current. The fingers will point in the direction of the lines of flux (Fig. 126).

Corkscrew Rule.—The direction of the current and that of the resulting magnetic field are related to each other as the forward travel of a corkscrew and the direction in which it is rotated.

This last rule is probably the most common and the most easily remembered. It must not be inferred from this rule, however, that the magnetic field exists in spirals about the conductor. It exists actually in planes perpendicular to the conductor.

125. Magnetic Field of Two Parallel Conductors.—When each of two parallel conductors carries an electric current, flowing in the same direction, there is a tendency for the two conductors to be drawn together. The reason for this is obvious. In Fig. 127, the lines of force encircle each conductor in the same direction (corkscrew rule), and the resultant field is an envelope of lines tending to pull the conductors together. Further reason for this attraction is given by the rule of Par. 117, stating that the magnetic field tends to conform itself so that the number of magnetic lines is a maximum. The drawing together of the conductors reduces the length of the magnetic path *abcd*. In any given plane, perpendicular to the conductors, the field *due to each conductor separately* is still circular in form, but the *resultant magnetic field* is no longer circular, as is shown in Fig. 127.

In Fig. 128 is shown the field which results when two parallel conductors carry current in opposite directions. In any given plane, perpendicular to the conductors, the magnetic lines are circles, but these circles are not concentric either with one another or with the conductor. The lines are crowded between the conductors and therefore tend to push the conductors farther apart. Again, when the conductors separate, the area is increased, so that the electric circuit in this case tends to conform itself so that the magnetic flux is a maximum.

From the foregoing, the following rules may be formulated:

Conductors carrying current in the same direction attract one another; conductors carrying current in opposite directions repel one another.

All electric circuits tend to take such a position as will make their currents parallel and flowing in the same direction.

This phenomenon is especially pronounced in modern, large-capacity power systems. Bus-bars have been wrenched from their clamps, transformer coils have been pulled out of place, and transformers wrecked by the forces produced by the extremely large currents arising under short-circuit conditions.

126. Magnetic Field of a Single Turn.—If a wire carrying a current be bent into a loop, a field similar to that shown in Fig. 129 results. The direction of the field is determined by the corkscrew rule. The single turn has a north pole and a south pole, which possess all the properties of similar poles of a short permanent magnet. A compass needle placed in this field assumes the direction shown, the north pole pointing in the direction of the magnetic field.

127. Solenoid.—An electric conductor wound in the form of a helix and carrying current is called a *solenoid*. A simple solenoid and the magnetic field produced within it are shown in Fig. 130. The solenoid may be considered as consisting of a large number of single turns, such as the turn in Fig. 129, placed together. The solenoid winding may consist of several layers, as shown in Fig. 132. The solenoid, like the bar magnet, has a north and a south pole as shown in Fig. 130.

The relation of the direction of the flux within the solenoid to the direction in which the current flows in the helix may be

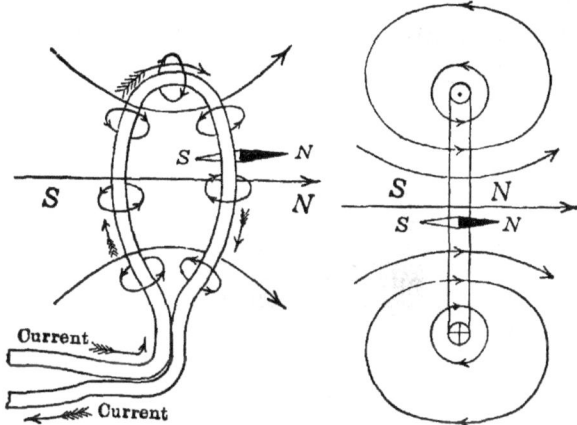

FIG. 129.—Magnetic field produced by a single-turn carrying current.

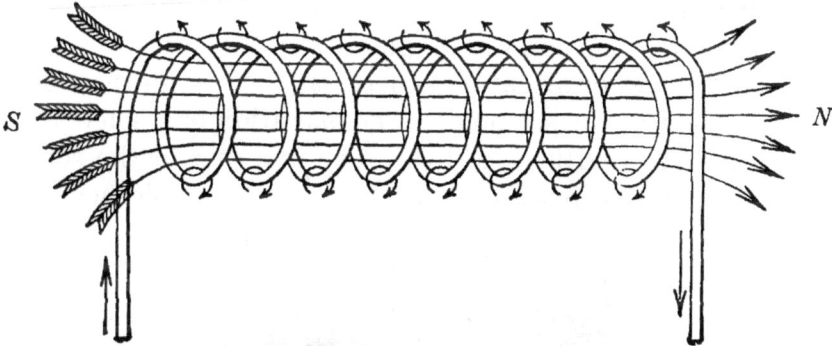

FIG. 130.—Magnetic field produced by a helix or solenoid.

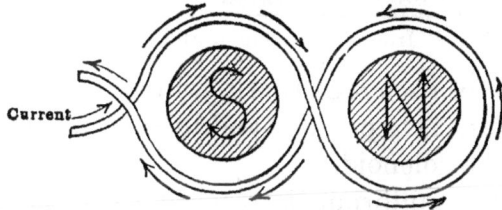

FIG. 131.—Relation of magnetic poles to direction of exciting current.

determined by the hand rule or by the corkscrew rule of Par. 124. Another simple method is shown in Fig. 131, where the arrows at the ends of the *N* and the *S* show the direction of current in the coil. For example, when looking down upon a north pole, the current direction in the coil will be counterclockwise, as shown by the N; when looking down upon a south pole, the direction of the exciting current will be clockwise, as shown by the S.

128. Plunger Solenoid and Industrial Applications.—In practice, the solenoid is used for tripping circuit breakers, for operating contactors in automatic motor starters, for operating voltage-

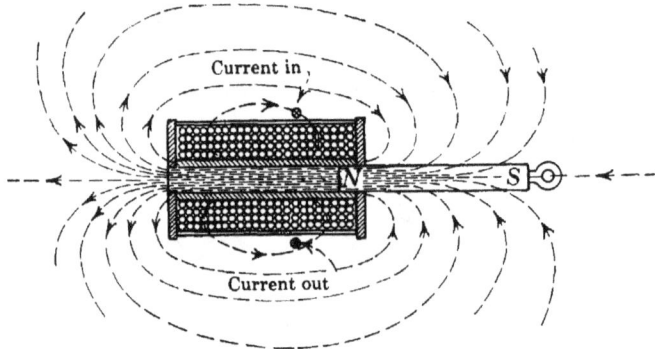

Fig. 132.—Simple solenoid and plunger.

regulating devices, for arc lamp feeds, for operating valves, and for many other purposes. In practically all cases a soft-iron (or steel) plunger or armature is necessary to obtain the tractive effort required of the solenoid. The operation of a solenoid and plunger is indicated in Fig. 132. The flux due to the solenoid produces magnetic poles on the plunger. The pole nearer the

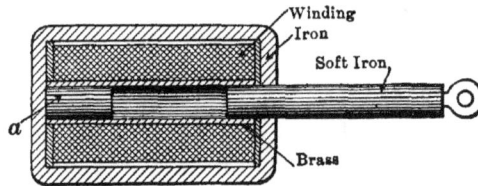

Fig. 133.—"Ironclad" solenoid and plunger with stop.

solenoid will be of such sign that it will be urged along the lines of force (see Par. 111, p. 135), and in such a direction as to be drawn within the solenoid.

A position of equilibrium is reached when the center of the plunger reaches the center of the solenoid (Fig. 132). Figure 133 shows an "ironclad" solenoid commonly used for tractive work. The ironclad feature increases the range of uniform pull and produces a very decided increase of pull as the plunger approaches the end of the stroke. When a stop a is used, the solenoid becomes a *plunger electromagnet.*   This changes the characteristics of the solenoid, in that the maximum pull now occurs when the end of the plunger is near the stop.

An important practical application of the solenoid occurs in the braking of elevators and cranes. When the power is removed from the lifting motor or when the power is interrupted due to a broken wire or other accident, the brake must be applied immediately. One method of accomplishing this is shown in Fig. 134. When the power, for any reason, is interrupted, the plunger *P* of the solenoid *A* drops, owing partly to gravity and partly to the action of the springs *S*. The springs *S* immediately force the levers *L* against the brake bands *B*, pressing these against the brake drum *D*, thus

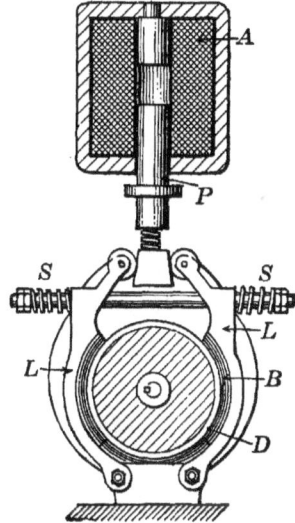

Fig. 134.—Plunger electromagnet operating a crane brake.

effecting the braking action. When the power is applied to the lifting motor, the plunger *P* is pulled up, thus releasing the brake. A plunger electromagnet is most suitable for this purpose because the stroke is short and the pull is positive.

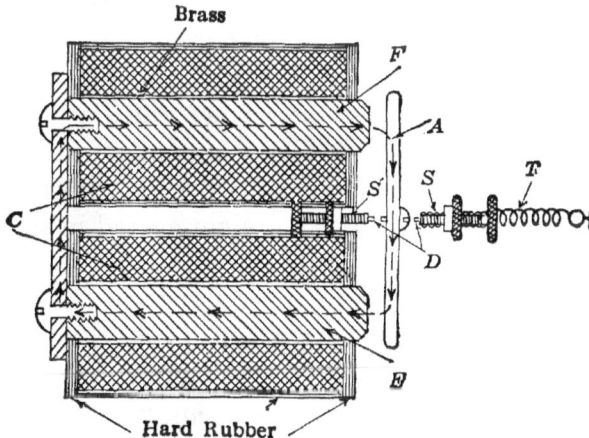

Fig. 135.—Telegraph relay.

129. Telegraph Relay.—A common use of the solenoid is illustrated by the relay or the sounder used in telegraphy. To increase the pull upon the armature, two solenoids are used.

each being placed on one of the legs of a horseshoe or U-shaped magnet. When the coils C (Fig. 135) are energized, the iron armature A is attracted because of the tendency of the magnetic lines to make their path of minimum length. As a rule, the armature A is not allowed to close the magnetic circuit completely; for under these conditions the magnetic lines still exist after the excitation is removed, thus preventing rapid release of the armature. The stop S' prevents the armature making contact with the cores FF and thus completely closing the magnetic circuit. The two sets of contacts D close any secondary circuit that the relay may be operating. The spring T draws the armature back against a stop S when the energizing ceases.

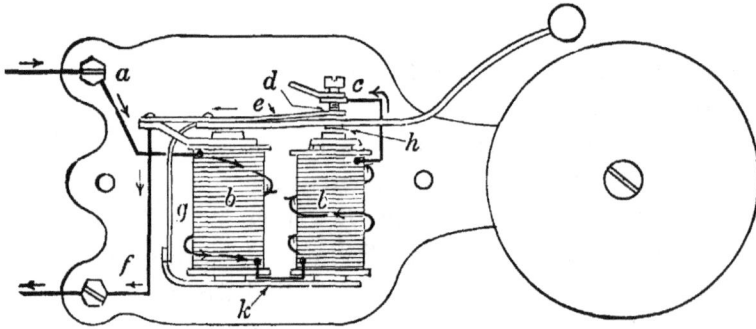

FIG. 136.—Electric bell.

130. Electric Bells, Buzzers.—Electric bells and buzzers also utilize the principle of the electromagnet in their operation. A diagram of a typical electric bell is given in Fig. 136. The path of the current is indicated by the arrows, the upper binding post a being assumed positive. Current leaving the binding post a passes through the two solenoids b, b, which are fastened to the iron yoke k. The solenoids b, b are connected in series and their magnetic fields act in opposite directions. The current next passes to the stud c, then through the contact point d (usually of tungsten or platinum) to the contact spring e, which is fastened to the iron clapper arm, and so to the binding post f. The spring g holds the clapper away from the gong and holds the spring e against contact d. The current causes the solenoids b, b to attract the clapper arm. The clapper strikes the gong, and the contact at d is broken, interrupting the current. The solenoids thus being de-energized, the spring g is able to restore the clapper arm to its initial position, again closing the contact at *d*. The cycle is then repeated. A more or less rapid striking of the gong results. The small copper stop *h*, which prevents the clapper arm from coming in contact with the core of the solenoid and sticking, should be noted.

131. Lifting Magnet.—Lifting magnets are used commercially to handle iron and steel in various forms. A considerable saving of time and labor is effected by their use, because chains and slings for holding the load are unnecessary. They are useful for handling steel billets in rolling mills, although the

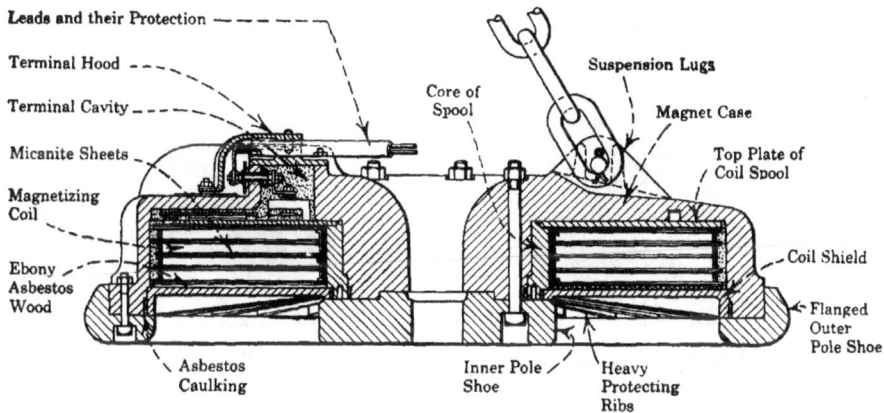

FIG. 137.—Cross-section of Cutler-Hammer lifting magnet.

billets cannot be picked up when red hot, as they lose their magnetic properties at high temperatures. Lifting magnets effect a great saving of labor when small pieces of iron, such as scrap iron, are handled, for they will pick up large quantities at every lift. Without such a magnet it would be necessary to handle each individual piece by hand. Figure 137 shows in cross-section a typical Cutler-Hammer lifting magnet.

The magnet is disc-shaped, and the magnetizing coil is concentric with the magnet, encircling the inner pole shoe. The rim of the magnet constitutes the outer pole shoe. The coil shield is hardened nonmagnetic steel plate and protects the coil structure from the impact of the load. (Also see Fig. 182 *A*, p. 209.)

Figure 138 shows a lifting magnet in operation.

It should be understood that the magnet itself does little or no work in the actual lifting but, like a hook and chain, merely serves as a holding device. The actual work is performed by the engine or motor which operates the steel ropes or chains attached to the magnet.

Fig. 138.—Cutler-Hammer lifting magnet handling car wheels.

132. Magnetic Separator.—Another important application of magnetic principles is found in the magnetic separator, shown in Fig. 139. It is specially designed to remove steel and iron

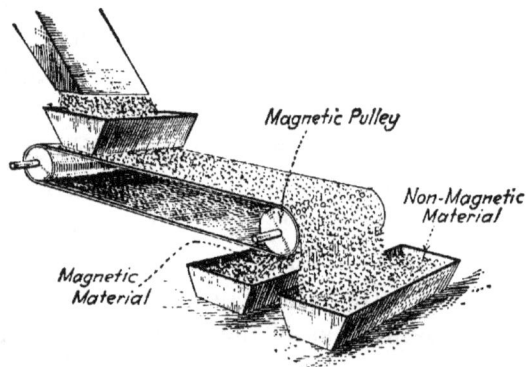

Fig. 139.—Magnetic separator.

from coal, rock, ore, etc., but it may be used for separating steel shot from molding sand, iron chips from machine-shop turnings, etc. The material is fed on an endless belt running at a speed of about 100 ft per min. The belt passes over a magnetized pulley. The nonmagnetic material immediately drops into a hopper, but the magnetic material is held by the pulley until the belt leaves the pulley, when this material drops into a second hopper. The pulley is magnetized by concentric exciting coils, to which current is carried by means of slip-rings.

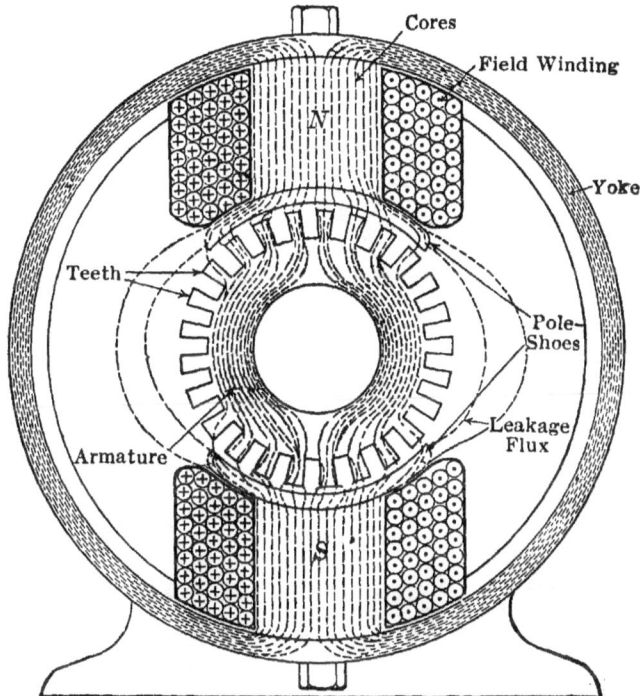

Fig. 140.—Magnetic circuit and field windings of a modern bipolar generator.

133. Magnetic Circuits of Dynamos.—One of the most important uses of electromagnets is in the magnetic circuits of generators and motors.

The magnetic circuit of a bipolar generator of modern design is shown in Fig. 140. Some of the flux, called *leakage flux*, passes between the field cores and shoes without entering the armature. In this type of dynamo the long air-path existing between the pole shoes reduces this leakage flux to a minimum. It is to be noted that the flux in the cores divides as it passes into the armature and into the yoke. Ordinarily the cross-section of the armature and of the yoke needs be practically only one-half the cross-section of the field cores. Direct-current dynamos of the bipolar type are made in small units usually.

Figure 141 shows the more complex magnetic circuits of a multipolar generator having eight poles, the poles being alternately north and south. The flux in the field cores divides both on reaching the yoke and on reaching the armature iron, and here again the cross-section of the yoke and of the armature iron needs to be only approximately one-half that of the cores. In both Figs. 140 and 141 the magnetic leakage is materially reduced

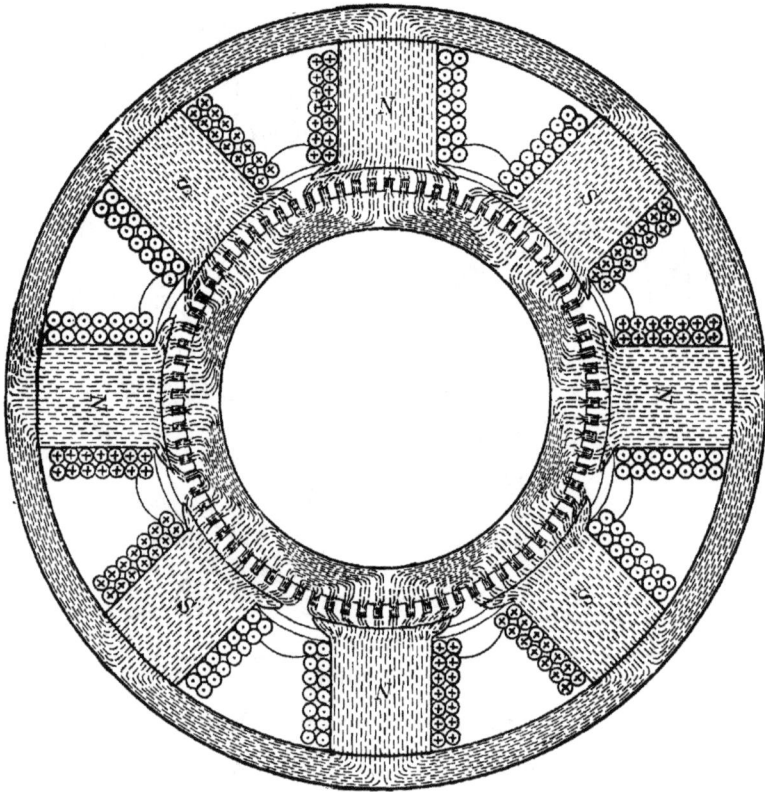

Fig. 141.—Magnetic circuits of a multipolar generator.

by placing the exciting ampere-turns on the pole cores as near the armature as possible.

It should be understood that magnetic leakage of itself does not directly lower the power efficiency of a dynamo, since to maintain a *constant* magnetic field *does not require an expenditure of energy.* (See p. 175.) However, both the yokes and the cores must have sufficiently large cross-sections to carry the leakage flux in addition to the useful armature flux. This increases both the amount of iron and the amount of field copper and hence increases the weight and the cost of the dynamo.

CHAPTER VIII

THE MAGNETIC CIRCUIT

The general nature of magnetic phenomena is considered in Chaps. VI and VII, though the quantitative relations existing among the magnetic flux, the ampere-turns, and the magnetic circuit itself are not given. As with the electric circuit, it is possible to determine the values of magnetic quantities in a circuit, provided sufficient data are given.

134. Magnetic Circuit.—The magnetic circuit resembles the electric circuit in many ways. For example, magnetic flux is equal to the magnetomotive force divided by the magnetic resistance (reluctance); drop in magnetic potential is equal to the magnetic flux multiplied by the reluctance. Also, Kirchhoff's two laws are applicable to the magnetic circuit as well as to the electric circuit.

On the other hand, magnetic calculations and measurements cannot be made with the same high degree of precision with which electric calculations can be made. Electric currents are usually confined to definite paths whose geometry is accurately known. Magnetic flux cannot be confined to definite paths, since there is no known insulator for magnetic flux. Air itself is a fairly good magnetic conductor.

Ordinarily, magnetic paths are short and have large cross-sections as compared with their lengths. Usually their geometry is not simple. Again, the reluctance of iron and steel varies over wide limits depending both on the previous magnetic history and on the flux density. The reluctance of a given piece of steel may increase fifty to one hundred times with increase of flux density.

The foregoing factors make it extremely difficult to calculate magnetic quantities with a high degree of precision. However, with an understanding of the magnetic circuit, it becomes possible to make computations that are sufficiently close for most practical applications.

135. Magnetic Units. *Ampere turns.*—The ampere-turns (*NI*) acting on a magnetic circuit are given by the product of the turns linked with the circuit and the amperes flowing in these turns. For example, 10 amp flowing in 150 turns give 1,500 amp-turns. The same result is produced by 15 amp flowing in 100 turns. If any ampere-turns act in opposition, they must be subtracted.

Magnetomotive Force.—Magnetomotive force (mmf; also, *F*) tends to produce a flux in the magnetic circuit and corresponds to emf in the electric circuit. It is directly proportional to the ampere-turns of the circuit, the constant of proportionality being $0.4\pi = 1.257$. That is, $F = 0.4\pi NI = 1.257NI$.

The mmf of a circuit is measured by the work done in carrying a unit north pole once around the entire circuit.

The unit of mmf is the *gilbert*. The gilberts acting on a circuit are obtained by multiplying the ampere-turns by 0.4π or 1.257.

Reluctance.—Reluctance ($\mathfrak{R}$) is resistance to the passage of magnetic flux and corresponds to resistance in the electric circuit. The unit of reluctance is the reluctance of a centimeter-cube of air. This unit has not been given a name as yet and is referred to as the "cgs unit of reluctance."

Permeance.—The permeance ($\mathcal{P}$) of a circuit is the reciprocal of the reluctance ($\mathcal{P} = 1/\mathfrak{R}$) and may be defined as that property of the circuit which permits the passage of magnetic flux or of lines of induction. Permeance corresponds to conductance in the electric circuit.

Permeability.—The permeability (μ) of a material is the ratio of the flux or of the number of lines of induction existing in that material to the flux or the number of lines of induction which would exist if that material were replaced by air, the mmf acting on the material remaining unchanged. The permeability of air is taken as unity; and with the exception of iron, steel, nickel, liquid oxygen, and certain iron oxides, practically all materials may be considered as having a permeability unity. The permeability of commercial iron and steel ranges from 50 and even lower to about 2,000. In special investigations, vacuum-treated iron has attained a permeability of 5,000 and even greater.

Flux.—The magnetic flux (ϕ) is equal to the total number of lines of induction existing in the circuit and corresponds to current in the electric circuit. The unit of flux is the *maxwell*, but "line of induction" or simply "line" is generally used.

Flux Density.—The flux density (*B*) is the number of max-wells or of lines of induction per unit area, the area being taken perpendicular to the direction of the flux. The unit of flux density in the cgs system is one line per square centimeter and is called the *gauss*. Flux density is usually expressed in "lines per square centimeter" or "lines per square inch." (Also see p. 137.)

$$B = \frac{\phi}{A} \text{ gausses}$$

where *A* is the area in square centimeters and ϕ the flux in maxwells through and normal to this area.

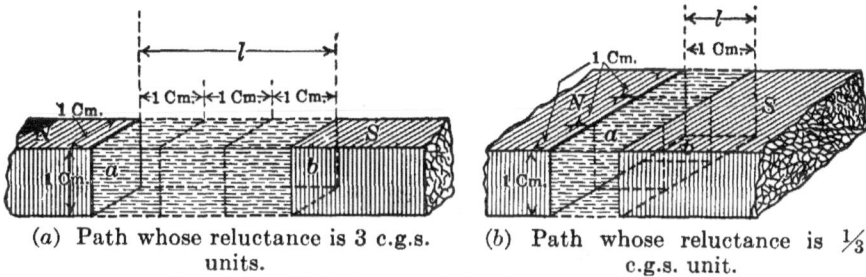

(a) Path whose reluctance is 3 c.g.s. units. (b) Path whose reluctance is $\frac{1}{3}$ c.g.s. unit.

Fig. 142.—Reluctance of simple magnetic paths.

136. Reluctance of the Magnetic Circuit.—The unit reluctance is defined as that of a centimeter-cube of air. If the portion of a magnetic circuit between pole faces *a* and *b* (Fig. 142 (a))[1] consists of a path in air having a length of 3 cm and a cross-section of 1 sq cm, this path is equal to 3 cm-cubes placed in series. As the total flux must pass through each cube in turn, it is evident that the total reluctance is 3 cgs units. The reluctance is proportional to the length of the flux path.

On the other hand, if the path has a length of 1 cm and a cross-section of 3 sq cm (Fig. 142 (b)), the reluctance of the path is one-third that of 1 cube alone, or $\frac{1}{3}$ cgs unit. The reluctance is inversely proportional to the cross-section of the path.

Moreover, if these paths were in iron, having a permeability μ, the flux would be μ times its value in air, provided the same

[1] The actual flux path between small pole faces taken by themselves would not be as shown in Fig. 142 (a), but the flux would "fringe," as shown in Figs. 108 and 116 (b), pp. 136 and 141.

mmf were maintained between the two pole faces.   This means a lower reluctance.   *The reluctance of any portion of a magnetic circuit is proportional to its length, inversely proportional to its uniform cross-section, and inversely proportional to the permeability of the material.*   If the centimeter is used as the unit of length, the constant of proportionality is unity, since the reluctance of a path in air *one* centimeter long and *one* square centimeter cross-section is *one* cgs unit.   Hence,

$$\mathcal{R}_1 = \frac{l_1}{A_1\mu_1} \text{ cgs unit.} \tag{58}$$

where l_1 is the length in centimeters of that part of the circuit under consideration, A_1 is the uniform cross-section in square centimeters of that portion of the circuit, and μ_1 is the permeability of that portion of the circuit.

Example.—A portion of a magnetic circuit consists of a cylindrical steel casting 3.0 in. (7.62 cm) diameter, 8.0 in. (20.3 cm) long, whose permeability is 1,300 at the flux density at which it is to operate.   Find the reluctance and the permeance of this portion of the circuit.

The cross-section

$$A = \frac{\pi(7.62)^2}{4} = 45.6 \text{ sq cm.}$$

The reluctance

$$\mathcal{R} = \frac{20.3}{45.6 \times 1,300} = 0.000342 \text{ cgs unit.}   \textit{Ans.}$$

The permeance

$$\mathcal{P} = \frac{1}{0.000342} = 2,920 \text{ units.}   \textit{Ans.}$$

If a magnetic circuit consists of several reluctances in series, the total reluctance is

$$\mathcal{R} = \mathcal{R}_1 + \mathcal{R}_2 + \mathcal{R}_3 + \mathcal{R}_4 + \cdots \tag{59}$$

$$= \frac{l_1}{A_1\mu_1} + \frac{l_2}{A_2\mu_2} + \frac{l_3}{A_3\mu_3} + \frac{l_4}{A_4\mu_4} + \cdots \text{ cgs units.}$$

Figure 143 shows four reluctances in series.

Permeances in parallel are added together to find the total permeance, just as conductances in parallel are added together to find the total conductance.

The total permeance

$$\mathcal{P} = \mathcal{P}_1 + \mathcal{P}_2 + \mathcal{P}_3 + \mathcal{P}_4 + \cdots \qquad (60)$$

Reluctances in parallel are combined in the same manner as resistances in parallel.

$$\frac{1}{\mathcal{R}} = \frac{1}{\mathcal{R}_1} + \frac{1}{\mathcal{R}_2} +$$
$$\frac{1}{\mathcal{R}_3} + \frac{1}{\mathcal{R}_4} + \cdots \qquad (61)$$

137. Permeability of Iron and Steel.—The permeability of iron and steel depends on the quality of the material, the flux density, and the previous magnetic history.

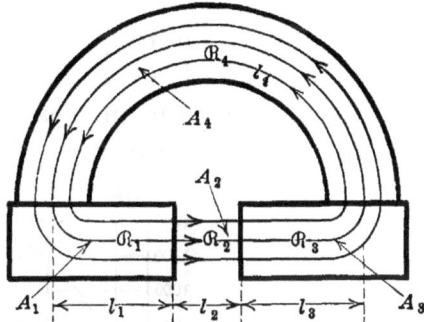

FIG. 143.—Reluctances in series.

If the permeability of iron and steel were constant, a graph plotted with flux density B as ordinates and mmf per unit length H as abscissas would be a straight line, since the reluctance $\mathcal{R}$ is constant so that the flux density is proportional to the mmf

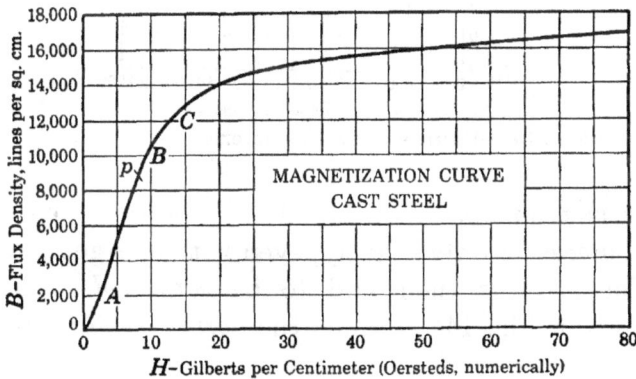

FIG. 144.—Magnetization curve for cast steel.

per unit length. This relation is shown by Eq. (62), p. 163, in which $\phi = B$ when a unit cross-section is considered.

As the permeability varies widely, the relation of the flux density B to the mmf per unit length H is not a straight line, but a curve (Fig. 144) whose equation is not simple. A curve which shows the relation of flux density B to mmf per unit length H is called a *magnetization curve*. Figure 144 gives the magnetization curve for one grade of cast steel. Magnetomotive force gradient H in gilberts per centimeter is plotted as abscissa and the corresponding flux density B in gausses is plotted as ordinate. The gilberts per centimeter are equal to $0.4\pi nI$, where nI is the ampere-turns per centimeter length of the material.

From the origin to point A, the curve is slightly concave upward. From A to B the curve is nearly a straight line. Beyond B the rate of increase of flux density decreases and the steel approaches saturation. The point C, where the bend in

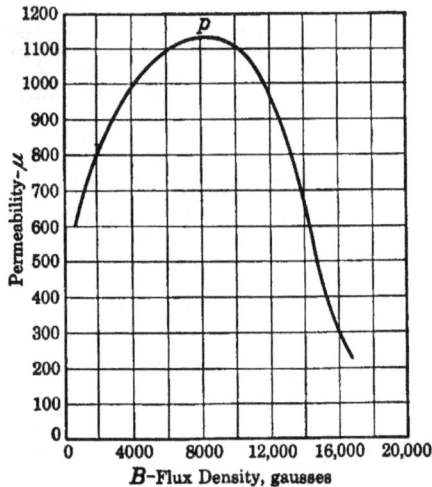

FIG. 145.—Permeability characteristic of cast steel.

the curve is *pronounced*, is the "knee" of the curve. Beyond C the flux increases only slightly, even with a large increase in the mmf. The steel is then said to be *saturated*. The type of curve shown in Fig. 144 is called the *normal* saturation or induction curve.

Figure 145 shows the permeability curve for the steel of Fig. 144. Each ordinate is obtained by dividing B by H for each point of the curve in Fig. 144. It will be noted that the permeability varies over a wide range. It begins at a comparatively low value, increases to a maximum at the point p, and decreases to about one-fifth its maximum value.

138. Law of the Magnetic Circuit.—The relation among flux, mmf, and reluctance, for the magnetic circuit, is identical with the relation among current, emf, and resistance for the electric circuit. (See Par. 25, p. 27.)

$$\phi = \frac{F}{\mathcal{R}} \text{ maxwells.} \tag{62}$$

The flux in maxwells is equal to the mmf in gilberts divided by the reluctance of the magnetic circuit in cgs units.

If the magnetic circuit consists of several parts in series having reluctances $\mathcal{R}_1$, $\mathcal{R}_2$, $\mathcal{R}_3$, $\cdots$ and mmfs F_1, F_2, F_3, $\ldots$ then, from Eqs. (59) and (62),

$$\phi = \frac{F_1 + F_2 + F_3 + \cdots}{\dfrac{l_1}{A_1\mu_1} + \dfrac{l_2}{A_2\mu_2} + \dfrac{l_3}{A_3\mu_3} + \cdots}$$

$$= \frac{0.4\pi(N_1I_1 + N_2I_2 + N_3I_3 + \cdots)}{\dfrac{l_1}{A_1\mu_1} + \dfrac{l_2}{A_2\mu_2} + \dfrac{l_3}{A_3\mu_3} + \cdots} \text{ maxwells.} \tag{63}$$

Example.—The ring magnet (Fig. 146) is wound with 250 turns of wire, in which a current of 1.5 amp flows. The permeability of the iron at the

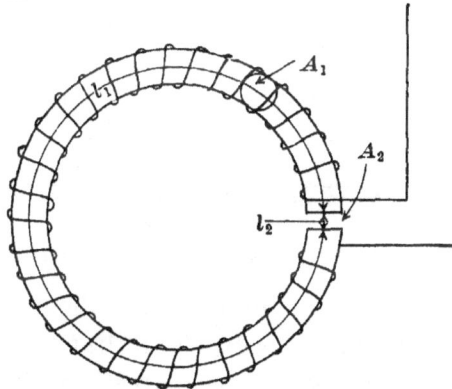

Fig. 146.—Ring-type electromagnet. $l_1 = 18.0$ in. (45.7 cm), $A_1 = A_2 = 0.20$ sq in. (1.29 sq cm), $l_2 = \frac{3}{16}$ in. (0.476 cm)

operating flux density is 800. Neglecting fringing, determine the flux in the ring and also the flux density.

$F = 0.4\pi \times 1.5 \times 250 = 471$ gilberts.
$l_1 = 18$ in. $= 18 \times 2.54 = 45.7$ cm.
$l_2 = \frac{3}{16}$ in. $= \frac{3}{16} \times 2.54 = 0.476$ cm.
$A_1 = A_2 = 0.2$ sq in. $= 0.2 \times 2.54 \times 2.54 = 1.29$ sq cm.

From Eq. (63),

$$\phi = \frac{471}{\dfrac{45.7}{1.29 \times 800} + \dfrac{0.476}{1.29 \times 1.0}} = \frac{471}{0.0443 + 0.369} = 1{,}140 \text{ maxwells. } Ans.$$

The flux density

$$B = \frac{1,140}{1.29} = 884 \text{ lines per sq cm (gausses)}$$

or

5,700 lines per square inch.   *Ans.*

139. Magnetization Curves for Iron and Steel.—Typical magnetization curves for commercial iron and steel are shown in Fig. 147. It will be noted that silicon steel and dynamo sheet steel are superior at the lower flux densities and that cast steel is slightly superior at the higher densities. The eddy-current and

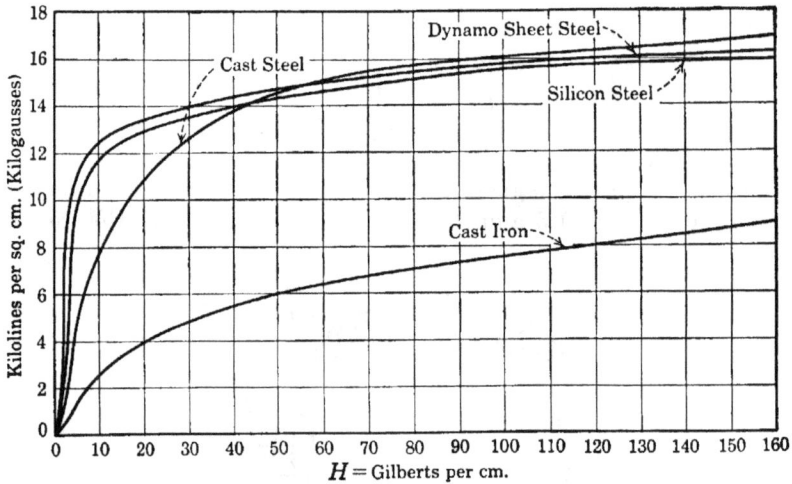

FIG. 147.—Typical magnetization curves.

hysteresis losses per unit volume in the silicon steel are so much less than in the dynamo sheet steel that silicon steel is used almost exclusively for transformer cores (see Par. 141 and Chap. VII, Part II), where the iron is subjected to rapid reversals in direction of the flux. On the other hand, dynamo sheet steel is used for dynamo cores (see p. 227), because it is cheaper than silicon steel and the iron losses are of less importance than in transformers. Cast iron has about half the permeability of cast steel and can be used only where weight is not important.

The value of B in gausses (Fig. 147) is found by multiplying the ordinate by 1,000.

The permeability for any given flux density is found by dividing the value of B at that point by the corresponding value of H, as described on p. 162.

140. Hysteresis.—If the mmf acting on an iron sample forming a closed magnetic circuit begins at zero and increases, the relation between mmf gradient and flux density will be similar to that shown by curve *Oa* (Fig. 148). This curve is called the *normal saturation* or magnetization curve and has already been discussed.

If the mmf gradient is now decreased, the flux density will *not* decrease along the line *aO* but will decrease less rapidly along *ab*. When point *b* is reached, the mmf gradient is zero, but

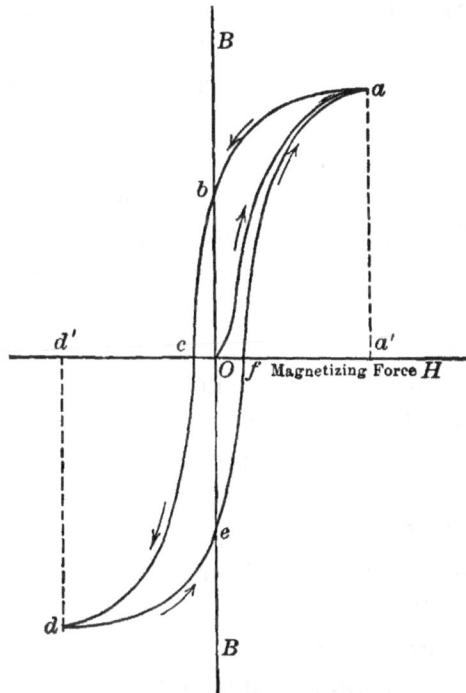

Fig. 148.—Hysteresis loop.

the magnetic induction has not reached zero, having the value *Ob*. The flux density *Ob* is called the *remanence*. To reduce the flux density to zero, the magnetizing force must be reversed in direction. That is, it requires a negative magnetizing force *Oc* to reduce the flux density to zero. The magnetizing force *Oc* is called the *coercive force*.

If now the magnetizing force is increased in the negative direction to *d'*, where *Od'* = *Oa'*, the flux density will be carried to a negative maximum *d'd*. The negative maximum flux density *d'd* is equal to *a'a*. If the magnetizing force is increased toward zero, the curve will pass through point *e* when the magnetizing force is again zero and the negative remanence $Oe = Ob$. A positive coercive force $Of = Oc$ is necessary to bring the flux density again to zero. When the magnetizing force again becomes Oa', the flux density will return to its original maximum positive value at a, closing the loop.

In Fig. 148, it will be noted that the magnetization *lags* behind the magnetizing force. Starting with any degree of magnetization, if the magnetizing force is increased and then decreased, Fig. 148 shows that the values of magnetization which are associated with increasing values of the magnetizing force are different from the values which are associated with corresponding decreasing values of the magnetizing force. Likewise, if the magnetizing force is decreased and then increased, the values of magnetization which are associated with decreasing values of the magnetizing force are different from the values which are associated with corresponding increasing values of magnetizing force. That is, with iron, the magnetization *lags* behind the magnetizing force. For example, at point b the magnetizing force is zero and the flux does not become zero until later at point c. This lag is called *hysteresis*, which means "to lag."

The cycle of magnetization shown in Fig. 148 is called a *hysteresis loop*.

141. Hysteresis Loss.—The lag of magnetization behind the magnetizing force is attributed to molecular friction (see Weber's theory, p. 132). Therefore, it must require an expenditure of energy to carry the iron through a cycle. The expenditure of energy per cycle is proportional to the area of the hysteresis loop. The energy involved is called *hysteresis loss*. In silicon steel, this loss is less than in ordinary steel, which is one reason for using silicon steel for transformer cores.

With most iron and steel, hysteresis loss varies practically as the 1.6 power of the flux density. Quantitatively, hysteresis loss

$$W_h = \eta B^{1.6} \text{ ergs per cc per cycle} \tag{64}$$

where η is a constant and B is the maximum flux density in gausses. Below are given a few typical values of η.

Hard cast steel	0.025	Sheet iron	0.004
Forged steel	0.020	Silicon sheet steel	0.0010
Cast iron	0.013	Silicon steel	0.0009

Example.—Determine the ergs loss per cycle in a core of sheet iron, the net volume of which is 40 cc, when the maximum flux density is 8,000 gausses.

$$W_h = 0.004 \times 8,000^{1.6} \text{ ergs.}$$
$$\log 8,000 = 3.9031.$$
$$1.6 \times 3.9031 = 6.2449.$$
$$\log 1,757,000 = 6.2449.$$
$$W_h = 0.004 \times 1,757,000 = 7,028 \text{ ergs per cubic centimeter per cycle.}$$
$$\text{Total loss } W = 7,028 \times 40 = 281,000 \text{ ergs per cycle. } Ans.$$
$$= 281,000 \times 10^{-7} \text{ joules per cycle.}$$

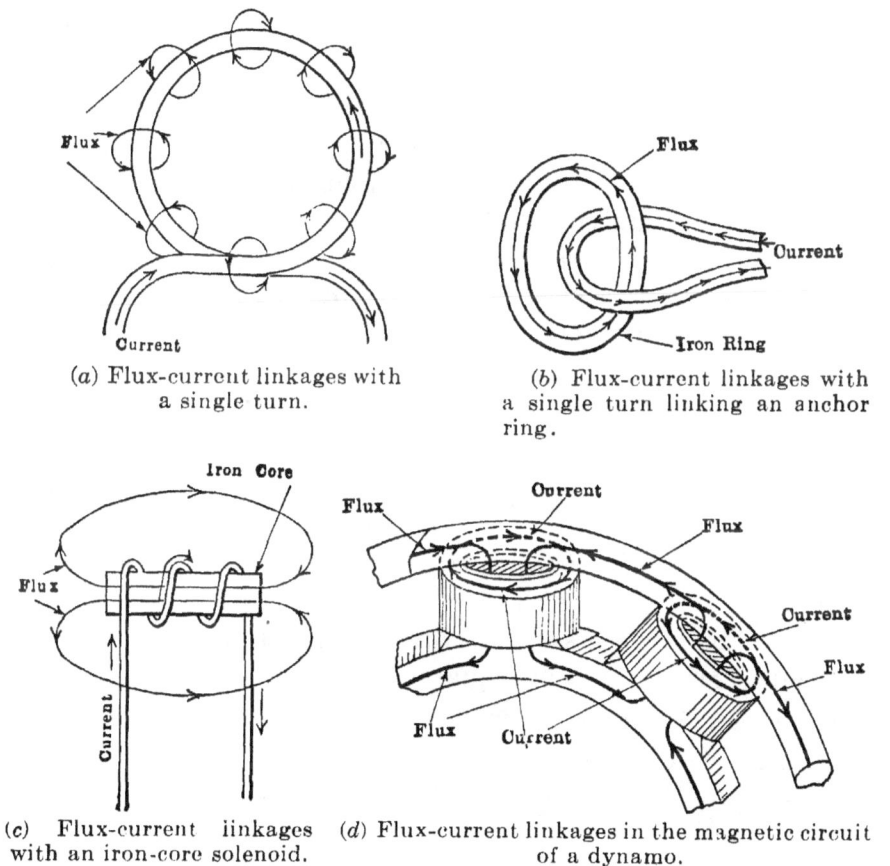

(a) Flux-current linkages with a single turn.

(b) Flux-current linkages with a single turn linking an anchor ring.

(c) Flux-current linkages with an iron-core solenoid.

(d) Flux-current linkages in the magnetic circuit of a dynamo.

Fig. 149.—Typical examples of flux-current linkages.

INDUCTANCE

142. Magnetic Linkages.—If a current flows in a conductor, a magnetic flux is set up about the conductor. *This magnetic flux completely encircles the conductor and the current in the conductor completely encircles the flux* (see Chap. VII, pp. 145 to 150).

Some familiar examples of this are given in Fig. 149, where the currents and associated fluxes are shown. As a current and the resulting flux always completely encircle each other, they are said to *link* each other. This is shown particularly well in Fig. 149 (*b*), where a conductor carrying a current is linked with an anchor ring.

The product of the turns of conductor and the number of lines of flux linking these turns is called the *linkages* of the circuit.

Example.—There are two exciting coils, each of 700 turns, on a certain electromagnet. When the current is 8.0 amp, the flux in the magnetic circuit is 3,000,000 maxwells. What are the magnetic linkages?

$$1{,}400 \times 3{,}000{,}000 = 4.2 \times 10^9 \text{ amp-maxwell linkages.} \quad \textit{Ans.}$$

The number of these *linkages per unit current* in a circuit is called the "self-inductance" of the circuit and is represented by the symbol *L*, implying linkages. The practical unit of self-inductance is the *henry*, which is defined as *the induction in a circuit when the electromotive force induced in this circuit is one international volt while the inducing current varies at the rate of one ampere per second.*

Since, from definition, self-inductance is equal to the linkages per ampere,

$$L = \frac{N\phi}{I \times 10^8} \tag{65}$$

where *L* is the self-inductance in henrys, ϕ is the flux in maxwells linking *N* turns, and *I* is the current in amperes.

Note.—It is necessary to divide by 10^8 because 10^8 cgs magnetic lines or maxwells equal one line in the practical system of volts, amperes, etc.

Example.—What is the self-inductance of the foregoing circuit?

$$L = \frac{4.2 \times 10^9}{8.0 \times 10^8} = 5.25 \text{ henrys.} \quad \textit{Ans.}$$

In Eq. (65) it is assumed that the reluctance of the magnetic circuit remains constant. In magnetic computations this assumption is usually made for it is difficult to express variation of reluctance by an equation.[1]

[1] See Dawes, C. L., "A Course in Electrical Engineering," 3d ed., Vol. I, Chap. XII, Direct Currents, Par. 198, p. 295, for an analysis of self-inductance with varying reluctance.

It should be noted in Eq. (65) that if the reluctance of the magnetic circuit and the current remain constant, the flux ϕ is proportional to the turns N, that is, $\phi = 0.4\pi NI/\mathfrak{R}$. Hence, with constant reluctance, the self-inductance varies as the *square of the turns.* (See Prob. 200, p. 358.)

143. Induced Electromotive Force.—If the terminals of an insulated coil (Fig. 150 (*a*)) be connected to a galvanometer and a magnetic field be established through this coil, either by thrusting a bar magnet into the coil or by some other means, the galvanometer will be observed to deflect momentarily and then to return to its zero. This shows that an emf has been temporarily induced

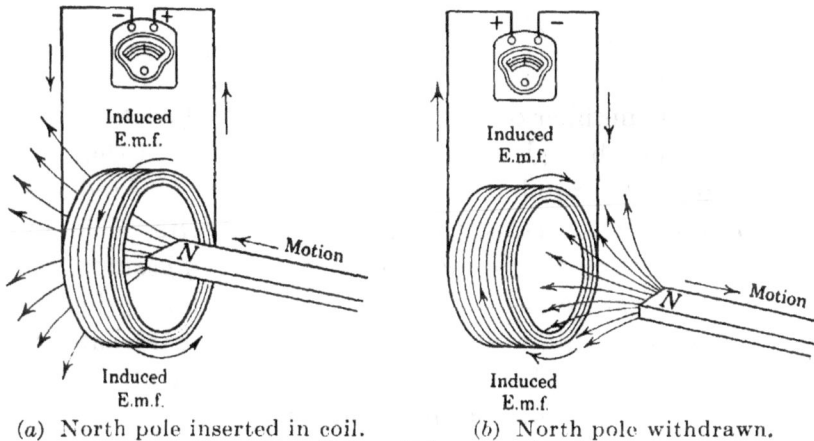

(*a*) North pole inserted in coil.   (*b*) North pole withdrawn.
Fig. 150.—Induced electromotive force.

in the coil. When the flux through the coil has ceased to change, this emf also ceases. If investigation be made, it will be found that the direction of this induced emf is that shown in the figure and that this direction is such that if the emf produces a current, this current will tend to push the bar magnet *out* of the coil or will oppose its entering the coil. A study of Fig. 150 (*a*) shows that the induced current flows in a counterclockwise direction. Hence, it produces an mmf acting outward to the right and therefore opposes the entrance of the N-pole of the bar magnet.

If the magnet be withdrawn from the coil (Fig. 150 (*b*)), the galvanometer will be observed to deflect momentarily as before, but the deflection is opposite to its direction in the first case. The direction of the induced emf is now such that if it causes current to flow, this current will tend to prevent the magnet from being withdrawn from the coil.   That is, in Fig. 150 (*b*), the current would flow in a clockwise direction and produce an mmf, hence a flux, whose direction is inward to the left.   Since a north pole tends to move in the direction of the lines of force, this mmf will oppose the withdrawal of the bar magnet.   The emf in each case is momentary and ceases when the *change of flux* linking the coil ceases.

If careful measurements be made, the value of this emf will be found to depend on: (1) the number of turns in the coil, (2) the rate at which the flux linked with the coil changes.

The average emf in volts is given by

$$e = -\frac{N\phi 10^{-8}}{t} \text{ volts,} \qquad (66)$$

where N is the number of turns in the coil, ϕ is the *total change* of flux in maxwells linked with the N turns, and t is the time in seconds required to produce this change of flux. 10^{-8} reduces the flux ϕ to practical units, so that e is given in volts.   The minus sign indicates that the induced emf acts in *opposition* to the effect which produces it.   That is, if current flows, due to the emf, its direction will be such as to oppose either the bar magnet entering the coil or its being withdrawn from the coil.

ϕ/t is the average rate of change of flux, so that the induced emf may be said to be proportional to the *number of turns* and the *rate of change of flux* linked with these turns.

Example.—When an armature coil of a d-c generator is directly under a north pole, 2,400,000 maxwells link the coil.   When the coil has moved to such a position that its center line is midway between poles, the flux linking it is zero.   The coil has eight turns, and the time required for the coil to move half the distance between poles is $\frac{1}{180}$ sec.   What is the average emf induced in the coil?   (See Fig. 182, p. 209.)

Applying Eq. (66),

$$e = -\frac{8 \times 2,400,000 \times 10^{-8}}{\frac{1}{180}}$$
$$= -8 \times (2.4 \times 10^{6}) \times (1.8 \times 10^{2}) \times 10^{-8} = -34.6 \text{ volts.}   \textit{Ans.}$$

The fact that the currents produced by induction *oppose* the motion producing them should be carefully noted, for this principle is manifest in practically all types of electric machinery.

This principle was first formulated by Lenz, in a form known as Lenz's law, which in effect says:

In all cases of electromagnetic induction, the induced emfs have a direction such that the currents which they produce oppose the effect which produces them.

This law is also based on the law of the conservation of energy. That is, the induced currents are produced at the expense of the mechanical energy required to push the magnet into the coil *against their opposition*, or the energy required to withdraw the magnet against the opposition of the induced currents, which try to prevent the withdrawal.

144. Electromotive Force of Self-induction.—If a coil be connected to a battery and a switch *S* be closed (Fig. 151), current

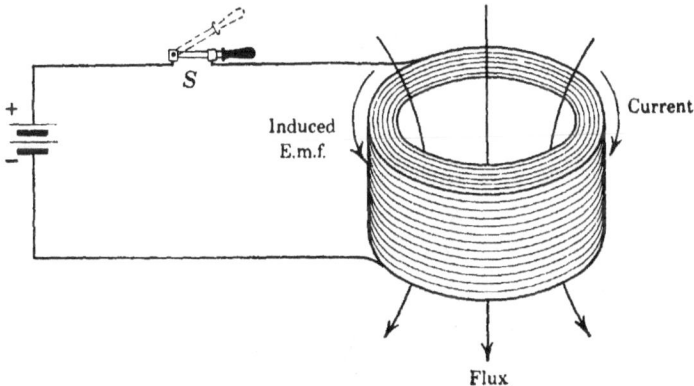

Fig. 151.—Relation of electromotive force of self-induction to current.

begins to flow in the coil. This current produces a flux which links the coil. As this flux increases, it must induce an emf in the coil, the magnitude of which depends on the number of turns in the coil and the rate at which the flux increases. By Lenz's law and also from a consideration of Fig. 150 (*a*), the emf thus induced must have such a direction as to oppose the increase in the flux linking the coil. That is, in Fig. 151, the current, by the corkscrew rule, produces a flux acting downward, which is practically equivalent to thrusting the *N*-pole of a bar magnet downward through the coil. The direction of the induced emf must be such that, if it produced a current, this current would oppose the increase of flux. Hence, the induced emf must act in a counterclockwise direction. This induced emf, therefore, must *oppose* any increase of current. Hence, the current cannot reach

its maximum value at once but is retarded in its rise by the opposing emf of self-induction.

In Fig. 152 is shown graphically the rise of current in a circuit
containing resistance only, the impressed voltage being 10 volts
and the resistance 20 ohms.   When the switch S is closed, the
current at once reaches its maximum or Ohm's law value of
0.5 amp.

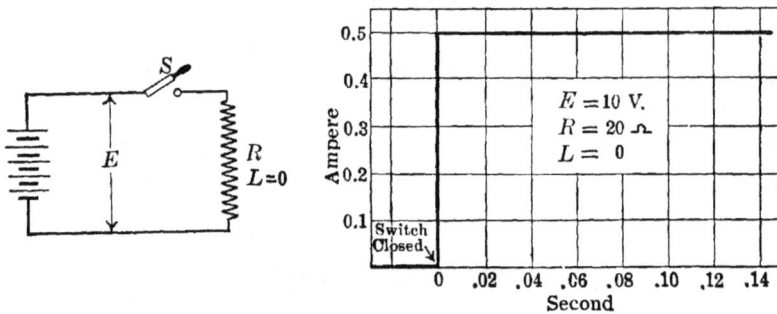

Fig. 152.—Rise of current in a noninductive circuit.

With the inductive circuit, however, the current gradually
approaches its Ohm's law value.   To be exact, it takes an infinite
time for the current to reach its Ohm's law value, although in a
comparatively short time it reaches substantially this value.
Figure 153 shows the rise of current in a circuit in which the

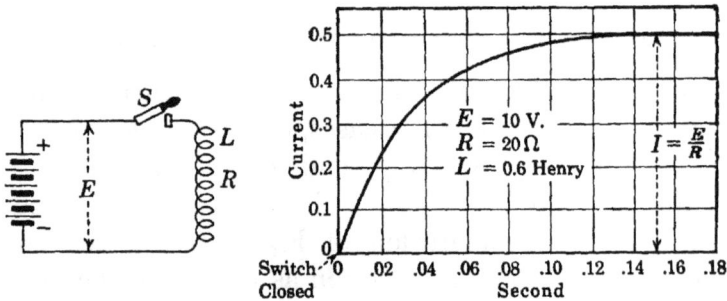

Fig. 153.—Rise of current in an inductive circuit.

impressed voltage is 10 volts, the resistance 20 ohms, and the
self-inductance 0.6 henry.

It will be noted that at the instant the switch is closed the
current begins to rise at a comparatively high rate.   Its rate
of increase, however, diminishes with time, until the rate of
increase becomes practically zero and the current equals its

Ohm's law value.[1]   This curve should be compared with that in Fig. 152, in which the circuit has the same impressed voltage and the same resistance but has no inductance.

This delayed rise of current in a circuit due to self-inductance should be carefully kept in mind, since it accounts for some of the time lag observed in relays, trip coils, etc.   The effect of inductance is also one of the controlling factors in the initial current rush on short circuit.

145. Decay of Current in Inductive Circuit.—If an inductive circuit, carrying current, be short-circuited, the current does not

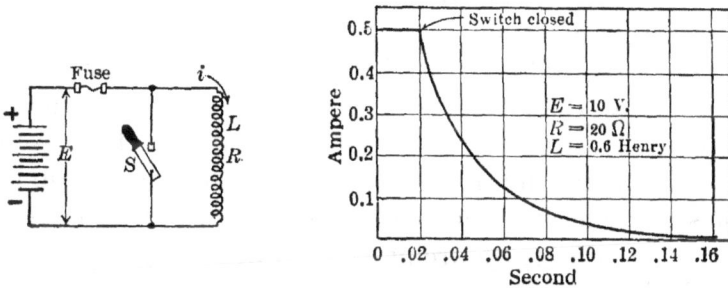

FIG. 154.—Decay of current in an inductive circuit.

cease immediately, as it does in a noninductive circuit under similar conditions, but continues to flow and does not become zero until an appreciable time after the instant of short circuit. This is due to the emf of self-induction.   The flux linking the coil is due to the current; and when the current decreases, this flux also decreases.   In decreasing, the flux induces an emf in the coil. In the same way that the current due to the induced emf tends to prevent the flux being withdrawn in Fig. 150 (b), so now the emf of self-induction tends to *prevent* the decrease of the current.

A curve showing the decrease of current with time is given in Fig. 154.   The circuit has the same constants as the circuit shown in Fig. 153.   It is usually advisable to fuse the battery so that it will not be injured, since short-circuiting the inductive

---

[1] The equation of this curve is

$$i = \frac{E}{R}(1 - \epsilon^{-\frac{Rt}{L}}) \text{ amp}$$

where i is the current at a time t sec after the closing of the switch. ϵ is the Napierian logarithmic base = 2.7183.

circuit also short-circuits the battery, as is shown in Fig. 154.*

146. Effects of Self-inductance.—It thus appears that the effect of inductance is always to oppose any change in circuit conditions. If the current tends to increase, inductance *opposes* the increase; if the current tends to decrease, inductance *opposes* the decrease. Inductance corresponds to inertia. The effect of inertia is to oppose any change in the motion of a body. If a body is at rest, inertia opposes its being set in motion; if a body is in motion, inertia opposes its coming to rest.

After having established the current in the circuit of Fig. 153, if the switch S be opened, a noticeable arc will appear at the switch contacts. This arc will be much greater in magnitude than that formed at the contacts of the switch in the circuit of Fig. 152, with resistance only in the circuit, although the current and circuit voltage are the same in each case. This arc is due to the emf of self-induction, which in some circuits may have such a value as to cause severe arcing at the switch contacts. In fact, this emf has been known to reach such values in alternator fields as to puncture their insulation when the field circuit is suddenly opened. To protect the field from puncture, a field discharge switch is ordinarily used (see Part II).

A voltmeter should never be allowed to remain connected across the field circuit of a dynamo while the switch is opened. The emf of self-induction, resulting from the large rate of change of flux linking the large number of field turns, may be many times the normal field voltage. Voltmeters have been ruined by neglecting this precaution.

147. Calculation of the Electromotive Force of Self-induction. From Eq. (65), p. 168, the average emf induced in a coil, due to a change in the flux linking the coil, is

$$e = -N\frac{\phi}{t}10^{-8} \text{ volts}$$

* The equation of this curve is

$$i = I_0 \epsilon^{-\frac{Rt}{L}} \text{ amp}$$

where I_0 is the current at the instant of short circuit. If the current has reached a steady value as in Fig. 154, $I_0 = E/R$.

where N is the number of turns, and ϕ/t is the average rate at which the flux changes.

Remembering that

$$L = \frac{N\phi}{I}10^{-8} \quad \text{or} \quad N\phi10^{-8} = LI \text{ (Eq. (65), p. 168)}$$

and also that the emf of self-induction opposes the change in current, its value may be written

$$e = -\frac{N\phi10^{-8}}{t} = -L\frac{I}{t} \text{ volts.} \tag{67}$$

The emf of self-induction is equal to the product of the self-inductance and the rate of change of current with respect to time. The minus sign indicates that this emf opposes the change of current.

From the definition, p. 168, and also from Eq. (67), it is obvious that a current changing at the rate of one ampere per second in a circuit of inductance one henry induces an emf of one volt.

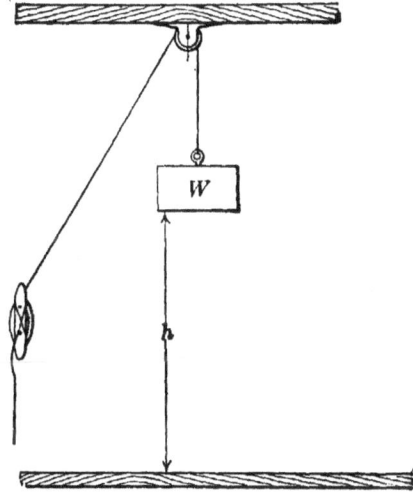

FIG. 155.—Energy of a suspended weight.

Example.—The inductance of the field circuit of a generator is 6 henrys. If the field current of 12 amp is interrupted in 0.05 sec, what is the average induced emf in the field winding?

$$e = -6\frac{12}{0.05} = -1{,}440 \text{ volts.} \quad Ans.$$

148. Energy of the Magnetic Field.—To establish a magnetic field requires the expenditure of energy. This is obvious from the fact that the current, in building up, flows against a counter emf. However, to *maintain* a *constant field* does *not* require an expenditure of energy, even in electromagnets. The energy lost in the exciting coils of electromagnets is accounted for as heat in the copper and is not concerned with the energy of the magnetic field itself.

That is, if the exciting coils be removed entirely from the iron cores of the magnet and rewound so that they are practically noninductive and hence produce no field, the energy loss for a given constant direct current will be the same as when the coil is associated with the magnetic field.

The energy of the magnetic field is stored or potential energy and is similar to the energy of a suspended weight (Fig. 155). Work is performed in raising the weight to its position, *but no expenditure of energy is required to maintain the weight in this position.* The energy of the weight due to its position is Wh ft-lb, where W is the weight in pounds and h the height in feet through which the weight has been raised. This energy is available and can be utilized in many ways.

In the same way the energy stored in the magnetic field is available and may make itself manifest in many ways, as, for example, the arc at the switch contacts. In an a-c circuit all this energy may be returned to the circuit.

The energy of the field in joules, or watt-seconds, is

$$W = \frac{1}{2}LI^2 \tag{68}$$

where L is the circuit inductance in henrys and I is the current in amperes.

Example.—Find the energy stored in the generator field in the problem of Par. 147. What is the average power during the time that the circuit is being interrupted?

$$W = \frac{1}{2} \times 6 \times 12^2 = 432 \text{ watt-sec.} \quad Ans.$$

$$P = \frac{432}{0.05} = 8{,}640 \text{ watts} = 8.64 \text{ kw.} \quad Ans.$$

Equation (68) shows that the energy of the magnetic field is proportional to the *square of the current* when the inductance is fixed. Therefore, if the current can be reduced by a suitable resistance to one-half its initial value before opening a highly inductive circuit, the energy of the arc at the switch contacts can be reduced to one-fourth of its value. This fact should be remembered when opening the field circuit of a dynamo.

The low-tension or make-and-break system of ignition, sometimes used with internal-combustion engines, is an example of the utilization of the emf of self-induction and the energy stored in the magnetic field. A diagram of the system is shown in Fig. 156. The emf of the battery *Ba* is usually 6 volts, and the battery may consist of six dry cells in series or a 3-cell storage battery. *U* is a coil wound with a number of turns of insulated wire, and the core consists either of laminated iron or of iron wires. *K* is a cam rotating in synchronism with the engine shaft, which actuates a follower *F* on the lever *L* pivoted on the shaft *P*. The shaft *P* extends through the cylinder wall, and the end *L'* of the lever and the sparking points *C* are within the cylinder. The upper point *C* is mounted rigidly on an insulated stem. The spring *M* holds the follower *F* against the cam; and as the cam

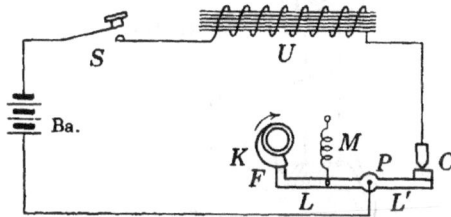

Fig. 156.—Make-and-break ignition system.

advances farther, *F* drops over the lip of the cam and the lever *L* jumps upward causing the sparking points *C* to open suddenly. When the switch *S* is closed and the cam action causes the sparking points *C* to close the circuit, a magnetic field is built up in the coil *U*. Accordingly, magnetic energy ($\frac{1}{2}LI^2$) is stored. When the sparking contact *C* is suddenly opened, this stored energy is released as heat at the contacts and ignites the charge within the cylinder. The phenomenon may also be considered from the point of view of induced emf. When the cam closes the circuit, a flux is built up in the core of the coil *U*. When the circuit is suddenly interrupted, an emf $e = -N(\phi/t)10^{-8}$ volts (Eq. (66), p. 170) is induced, where *N* is the number of turns in the coil and *t* is the time of interruption of the circuit. This emf is of considerable magnitude and causes the spark at the contacts. The advantage of this system of ignition is that there are no high-tension wires. It is used with 2-cycle motors, particularly in small open motorboats and also with some stationary engines. The high-tension or jump-spark system is much more widely used and is described on p. 180. This latter system also depends on stored magnetic energy for its operation.

149. Mutual Inductance.—In Fig. 157 are shown two coils *A* and *B*. Coil *A* is connected to a battery through a switch *S*. Coil *B* is not connected to any source of voltage but is connected to a galvanometer. Coil *B* is placed so that its axis is nearly coincident with that of *A* and the two coils are close together. When the switch *S* is closed, current flows in coil *A*, building up a field which links the coil. The position of *B* with respect to *A* results in a considerable part of the magnetic flux produced by *A* linking *B*. Therefore, if the current in *A* be interrupted by opening the switch *S*, or if it be altered in magnitude, a *change of flux simultaneously* occurs in *B*, inducing an emf in *B*. This emf is detected by the galvanometer connected across the terminals of *B*. Upon closing the switch *S* the galvanometer will deflect momentarily and then return to zero. Upon opening the switch *S* the deflection will reverse and again return to zero. This shows that the induced emf on opening the circuit of *A* is opposite in direction to the induced emf on closing the circuit. Because coil *B* is in such a relation to *A* that an emf is induced in *B*, due to a

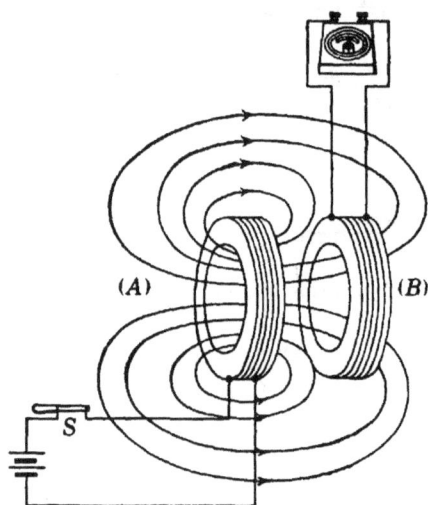

FIG. 157.—Mutual inductance between two coils.

change of current in *A*, these two coils are said to possess *mutual inductance*. The induced emf in coil *B* is an *emf of mutual induction*.

Even though coils *A* and *B* be brought close together, all the flux ϕ_1 produced by coil *A* does not link coil *B*. Only a certain fraction *K* of ϕ_1 links *B*, *K* being less than unity. *K* is called the *coefficient of coupling* of the circuits *A* and *B*.

If the current in coil *A*, when changing at the rate of one ampere per second, causes one volt to be induced in coil *B*, the coils have a mutual inductance of *one henry*. If the two coils have a mutual inductance of one henry, the current in coil *B*, when changing at the rate of one ampere per second, causes one volt to be induced in coil *A*.

From the foregoing, it is obvious that the emf induced in coil B, due to a rate of change of current I_1/t in coil A, is

$$c_2 = -M\frac{I_1}{t} \text{ volts.} \tag{69}$$

Likewise, the emf induced in coil A, due to a rate of change of current I_2/t in coil B, is

$$e_1 = -M\frac{I_2}{t} \text{ volts} \tag{70}$$

where M is the mutual inductance in henrys and t is the time in seconds.

Example.—The mutual inductance of coils A and B (Fig. 157) is 0.3 henry. A current of 2 amp in coil A is interrupted in 0.04 sec. What is the average emf induced in coil B?

$$e_2 = -0.3\frac{2}{0.04} = -15 \text{ volts.} \quad Ans.$$

Likewise, if a current of 2 amp in coil B is interrupted in 0.04 sec, -15 volts will be induced in coil A. The minus sign signifies that current, flowing as a result of this induced emf, opposes the flux producing this emf. (See Lenz's law, p. 171.)

It may be shown that the mutual inductance

$$M = K\sqrt{L_1 L_2} \tag{71}$$

where K is the coefficient of coupling, L_1 is the self-inductance of one circuit, and L_2 is the self-inductance of the second circuit.

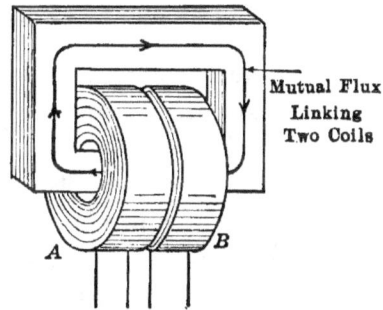

Fig. 158.—Effect of iron core on mutual inductance.

The mutual inductance of two circuits may be substantially increased by linking the circuits with an iron core. Thus, if two coils, similar to those shown in Fig. 157, be placed on an iron core (Fig. 158), the coefficient of coupling K may be made very nearly unity. That is, practically all the flux linking coil A also links coil B. (Fig. 158 shows a transformer.)

150. Induction Coil.—A common example of mutual inductance occurs in the induction coil (Fig. 159). A primary winding P, of comparatively coarse wire and few turns, is wound on a laminated iron core C. This winding is connected to a battery B.   The primary current is interrupted by the contact D, in series with the battery and the primary.   The circuit is closed by the armature A being pulled against the contact D by the spring.

When the core C is magnetized by the primary current, the armature A is drawn toward it and away from D, opening the circuit and causing the flux in the core to drop practically to zero. The spring then pulls the armature A against the contact D again, and the cycle is repeated.   By this process the flux in the core C

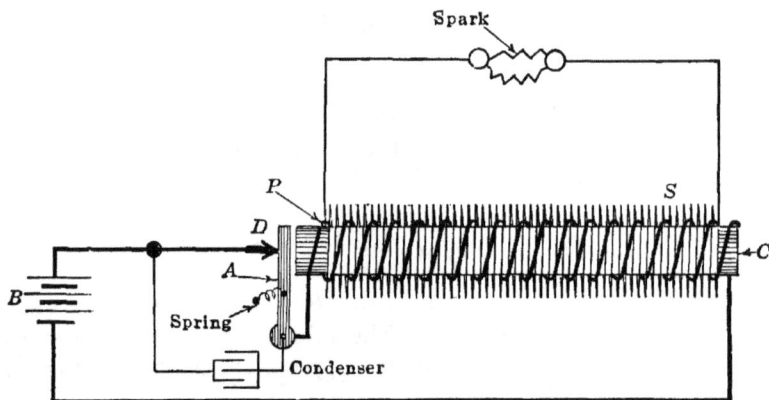

Fig. 159.—Induction coil.

is continually being established and then destroyed.   A condenser is shunted across the contact D in order to reduce sparking when the circuit is opened.

On the same core is placed a secondary winding S, consisting of many turns of fine wire.   This winding is thoroughly insulated from the primary winding; but as it is wound on the same core as P, the two coils have a high coefficient of coupling.   Because of the change of flux in the core, due to the interruptions of the primary current, a large alternating emf is induced in the secondary winding.   This induced emf may be considered as due to the mutual inductance existing between the primary and secondary coils.   The induction coil has many practical applications.   For example, in automobile radio sets an induction coil is used to convert the 6-volt storage-battery emf to 300 to 400 volts alternating.   This alternating emf is rectified to give the plate voltage and the grid-bias voltage.

151. Battery-ignition Systems.—A common example of the combined use of mutual inductance and the energy stored in the magnetic field is in battery-ignition systems which are used with internal-combustion engines.

The wiring diagram of a typical system is shown in Fig. 160. The ignition coil itself does not differ materially from the induction coil. It consists of an iron core made of either laminations or wires, a primary winding of comparatively few turns, and a secondary winding of a comparatively large number of turns. The primary and secondary windings are insulated from each other, although each circuit is grounded at one point. The pri-

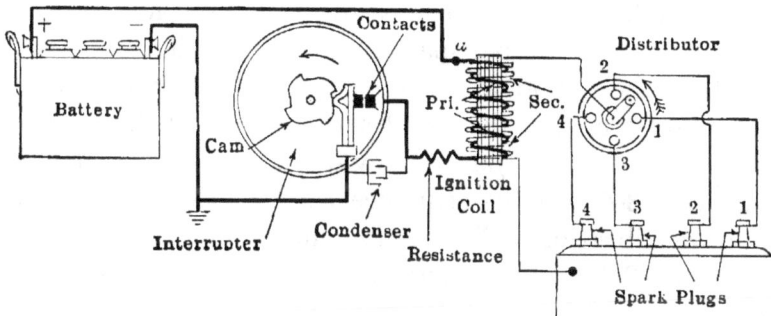

FIG. 160.—Typical battery-ignition system.

mary current of the ignition coil, however, is not interrupted by a vibrator, as it is with the induction coil, but by a set of tungsten contacts, actuated by a cam driven by the engine crankshaft through gears. The cam ordinarily has the same number of projections as the engine has cylinders and runs at half the engine speed, since in a four-cycle engine each cylinder fires only once in two revolutions of the crankshaft. One of the contacts is mounted on a spring, which tends to hold the contacts open. As the cam rotates, the projections close the contacts and hence close the primary circuit. Because of the inductance of the primary, the current builds up slowly (see Fig. 153, p. 172). Therefore, the induced secondary emf is not sufficient to cause a spark to jump at the spark plugs. The cam allows the contacts to remain closed a sufficient time to permit the primary current to attain practically its Ohm's law value. When the cam projection passes the follower, the spring causes the contacts to open quickly, thus suddenly interrupting the primary current. The magnetic flux in the core of the ignition coil suddenly collapses and induces a high voltage in the secondary. This high voltage causes a spark to jump across the gap in the spark plug.

The secondary ground terminal is usually connected to the primary terminal of the ignition coil at *a*. The induced emf of the primary and the condenser charge then contribute to the energy of the spark.

The distributor is usually made of molded insulation, such as bakelite, to give it good insulating and mechanical properties. A number of equally spaced metal contacts are set in the insulation, there being one contact for each cylinder. A rotating arm, which is well insulated, is connected to the secondary of the ignition coil. This arm passes over the contacts and conducts the current to the proper spark plug. The order of the contacts is such that the cylinders are fired in the proper sequence. In a four-cylinder engine the usual arrangement of the cranks on the crankshaft is such that the cylinders must fire in either the sequence 1-3-4-2 or the sequence 1-2-4-3. These sequences also minimize vibration. The engine shown in Fig. 159 fires in the sequence 1-2-4-3. The distributor is usually incorporated with the interrupter.

A condenser is shunted across the primary contacts in order to reduce sparking when they open. A resistance of high-temperature-coefficient material is connected in series with the primary of the ignition coil in order to limit the current, particularly if the ignition switch is inadvertently left closed when the engine is not running.

The operation of this ignition system may also be considered from the energy point of view. When the primary contacts are closed, a magnetic field is built up in the core of the ignition coil. When the primary circuit is opened suddenly, this energy is suddenly released. A considerable portion of the energy appears at the spark plug. Naturally, some of the energy is lost at the interrupter contacts and also energy is stored in the condenser, which is later dissipated when the contacts reclose.

It is evident that with this system but one spark is obtained each time a cylinder is fired. With the induction coil, a number of sparks are obtained whenever the primary and interrupter circuits are supplied with voltage. This is an advantage when starting an engine, particularly if it is cold. On the other hand, the induction coil is not well adapted to high-speed automobile and airplane engines, because of the comparative sluggishness in the action of the vibrator. (Also see p. 314.)

CHAPTER IX

ELECTROSTATICS: CAPACITANCE

Thus far, the electric current, electricity in motion, has been considered. Electricity in motion is called *dynamic* electricity. Electricity may also be stationary and under these conditions is called *static* electricity. There is no difference in the fundamental nature of static and dynamic electricity. Static elec-

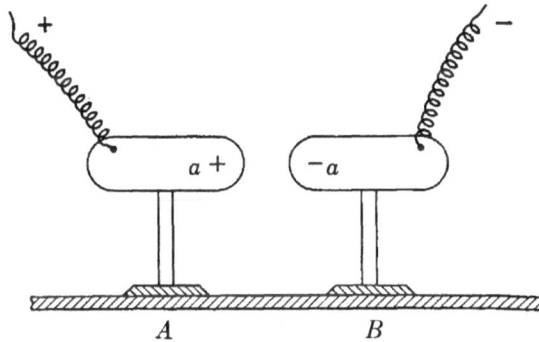

FIG. 161.—Electrostatic charges on insulated conducting bodies.

tricity usually appears different because usually the potential is extremely high and the quantity is small.

152. Electrostatic Induction.—In Fig. 161 are shown two similar elongated conducting bodies *A* and *B*, each of which is insulated from ground by a nonconducting support. The terminals of an influence machine (Par. 157) are connected, one to each body. The body *A*, connected to the positive terminal of the influence machine, is charged with positive electricity; and the body *B*, connected to the negative terminal of the influence machine, is charged with negative electricity. The charges will distribute themselves over the entire surface of the bodies, but the density of the charge will be greatest on the ends of the bodies which are adjacent, owing to the fact that the positive and negative charges attract each other.

If the two wires from the electrostatic machine be disconnected, the two charges will not be affected. In time, the charges will leak away through the insulating supports.

This experiment illustrates the fact that positive and negative charges attract each other. Also, electricity is stored in the two conducting bodies and accordingly they form an electrostatic condenser.

153. Free and Bound Charges.—In Fig. 162 (*a*) and (*b*) are shown two elongated conducting bodies or electrodes A and B, similar to those in Fig. 161; and these electrodes are in close proximity to each other. Initially, neither body has any charge. Then in (*a*) a positive charge is placed on A. A negative charge b will be found on the end of B nearest A. As B did not hold any charge initially and it is assumed to be perfectly insulated, no electricity can have left B and none can have reached it from external sources, so that the net charge on B must still be zero. Therefore, a positive charge b' equal to b must also appear on

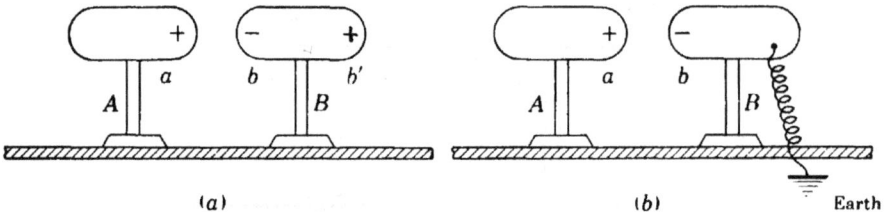

Fig. 162.—Electrostatic induction.

B at the end farthest from A. Since these two charges are of opposite sign, the net charge on B is still zero. The negative charge b is as near as possible to the positive inducing charge a as in Fig. 161, whereas the positive charge b' is as far away as possible from the positive charge a, owing to the fact that unlike charges attract each other and like charges repel each other.

Charges a and b are *bound charges*, whereas charge b' is a *free charge*. This may be proved by connecting B to ground (see Fig. 162 (*b*)). The charge b' will escape to ground, whereas the two charges a and b remain bound by their mutual attraction. Charge b' will seek a position as far away from a as possible.

If a were a negative charge, b would be a positive charge.

154. Electricity by Friction.—If a glass rod or a rod of hard rubber, amber, or other resinous material be rubbed with a dry woolen cloth, the surface of the rod will acquire a charge of electricity.

If the rod be held near a suspended pith, cork, or even an insulated metal ball, such as a (Fig. 163), the ball at first will deflect toward the rod and take some such position as b (Fig. 163). If the ball be allowed to touch the rod, it will remain in contact for a short time and then will be repelled to some such position as c.

This experiment indicates the presence of electricity on the rod. The reasons for the foregoing phenomena are: Rubbing the rod with the woolen cloth usually produces negative elec-

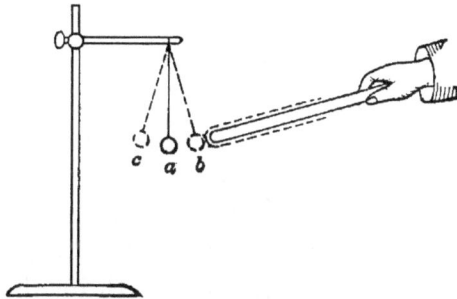

FIG. 163.—Suspended ball attracted and repelled by electrostatic charge.

tricity on the rod and positive electricity on the cloth. Being an insulator, the rod allows electricity to leak away only very slowly. When the negatively charged rod is held near the pith ball, a positive charge is induced on the side of the pith ball nearer the rod (Fig. 164). The total charge on the ball was zero

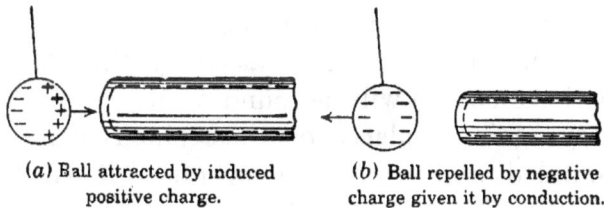

(a) Ball attracted by induced positive charge.

(b) Ball repelled by negative charge given it by conduction.

FIG. 164.—Attraction and repulsion of insulated ball.

before the rod was brought near it; and since the ball is well insulated, its *net* charge cannot have changed appreciably. Hence, there must be equal quantities of negative and positive electricity on the ball. The negative electricity is repelled by the negative charge on the rod and moves to those parts of the ball which are farthest removed from the inducing charge on the rod, as in Fig. 162 (a).

Unlike charges attract and like charges repel each other. The negative charge on the rod attracts the positive charge on the ball and repels the negative charge on the ball. The positive charge being nearer the inducing charge, attraction prevails over repulsion and the ball moves toward the rod.

If the ball be allowed to touch the rod, part of the negative charge on the rod flows to the ball. This not only neutralizes the positive charge on the ball, but the ball acquires further negative charge from the rod. The negative charge which the ball thus acquires is repelled by the negative charge on the rod, and the ball takes position *c* (Fig. 163).

155. Electrophorus.—Figure 165 shows a flat metal dish *a*, filled with some resinous material *b*, such as a mixture of resin and wax—or hard rubber may be used. If the resinous material be rubbed with either a woolen cloth or a catskin, it will acquire a negative charge by friction. If a metal disc *c* with insulating handle *d* be laid on the resinous material in the dish, it will touch it at only very few points. The negative charge on the resinous material will *induce* a positive charge on the under side of the disc, and a negative charge must obviously appear on the upper side of the disc, in the same manner as in Fig. 162 (*a*). If the disc be touched or grounded, the *free* negative charge flows to ground, leaving only the *bound* positive charge on the disc (Fig. 165). If the disc be raised by the glass handle *d*, a strong positive charge will be found on it. After discharging, the disc may again be placed on the resinous material, its top surface grounded, and again on being raised it will be found to have a strong positive charge. The operation may be repeated many times before the negative charge on the resinous material leaks away.

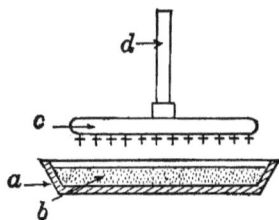

Fig. 165.—The electrophorus.

The positive charge on the disc represents energy. This energy is obviously not acquired at the expense of the negative charge on the surface of the resinous material, since the decrease in the charge on this material is due wholly to leakage. The energy is due to the work done in raising the disc against the force of attraction between the inducing negative charge on the resin and the induced positive charge on the disc.

A similar principle is involved in the operation of influence machines which generate static electricity (see p. 188).

156. Gold-leaf Electroscope.—An electroscope is a sensitive instrument for detecting small electric charges. It consists of a wide-mouthed glass jar (Fig. 166), closed with a stopper of insulating material in which is fitted a varnished glass tube. A stiff wire or rod passes through the tube. At the top of the rod there is either a metal ball or a disc, and at the bottom of the rod two pieces of gold leaf are suspended. When a charge is brought in the vicinity of the electroscope (Fig. 166), a charge of opposite sign is induced on the metal ball and a charge of the same sign as the inducing charge appears on the two pieces of gold leaf (Fig. 162 (a)). Since the charges on the two pieces of gold leaf are of like sign, they repel each other. As an example, a negatively charged glass rod is shown in the vicinity of the electroscope (Fig. 166). A positive charge is induced on the knob, and a negative charge appears on the two pieces of gold leaf.

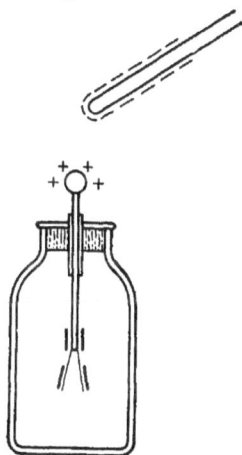

FIG. 166.—Gold-leaf electroscope.

The positive charge is a *bound charge*, as the negative inducing charge prevents its leaving the knob, even if a conducting path to ground be provided. The negative charge is not so held, is a *free charge*, and if a conducting path be provided, it will flow to ground. (See Par. 153.)

If, therefore, the knob be touched, or grounded with the negatively charged rod still in position, the negative charge on the two pieces of gold leaf flows to ground, as is shown by the pieces of gold leaf coming together. If the inducing charge be removed by withdrawing the rod, the bound positive charge becomes free and distributes itself over the knob, rod, and gold leaves, and the pieces of gold leaf again repel each other.

The polarity of a charge may be determined by means of the electroscope. Assume that the electroscope is charged negatively throughout by *touching* the knob with a rod of hard rubber which has been rubbed with flannel or silk. If the unknown charge be brought in the vicinity, it will induce on the knob a charge of opposite polarity, and on the gold leaves a charge of the same polarity as that of the unknown charge. Therefore, if the unknown charge is negative, the gold leaves will spread farther apart; if it is positive, they will come closer together.

157. Influence Machines.—There are various types of machines for generating so-called "static" electricity. Some depend on friction for the generation of electricity, but this type is not satisfactory. The machines which give the best results, and which are in common use, depend on electrostatic induction or influence for their operation. The principle is fundamentally that of the electrophorus. Small conducting strips called "carriers," usually of tin foil, are brought consecutively under the influence of a charged body at a time when the charge on the

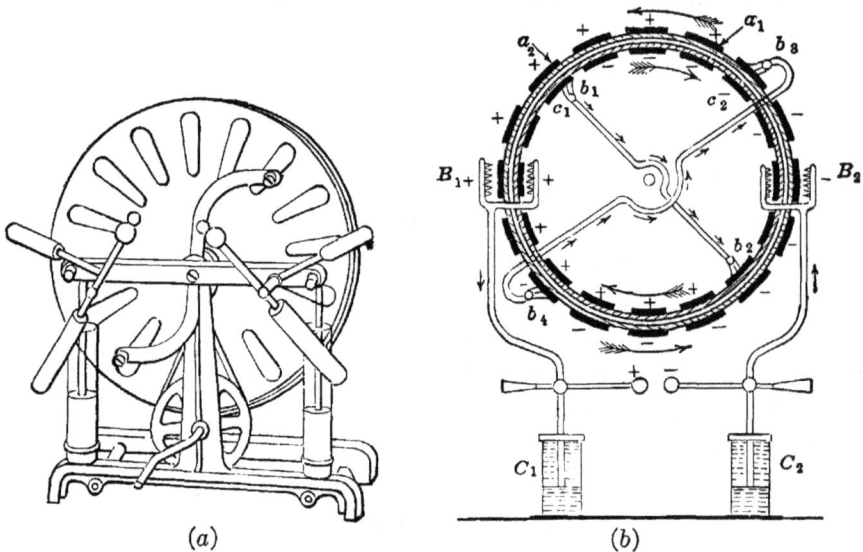

(a)   (b)

FIG. 167.—Wimshurst influence machine.

carriers is zero. Simultaneously, each carrier is touched by a conductor which allows the "free" charge (Par. 153) to escape. The "bound" charge is then carried to a collecting brush. With the electrophorus, a conducting plate is brought near the charged resinous material and grounded to permit the free charge to escape. The bound charge is then carried away with the disc.

A common type of influence machine is the Wimshurst machine (Fig. 167 (a)). Two circular glass or ebonite plates are mounted on a common shaft with small clearance between them. The two plates rotate at the same speed, but in opposite directions. A number of carriers, usually of tin foil, are mounted on the outer surface of each plate. The machine requires a pair of neutralizing brushes and a pair of collecting combs for each plate.

The details of operation of the Wimshurst machine are illustrated in Fig. 167 (*b*), in which, for clearness, the two flat circular plates are represented by two concentric glass or ebonite cylinders. The back plate is represented by the outer cylinder, shown as rotating in a counterclockwise direction, and the front plate is represented by the inner cylinder, shown as rotating in a clockwise direction. The tin-foil sectors of the actual machine are represented by black segments. The neutralizing brushes are shown as b_1, b_2 and b_3, b_4. The sharp-pointed collecting combs B_1 and B_2 are also shown.

The action of the machine is as follows: Assume that some segment a_1 on the back plate acquires in some manner a positive charge, which may be very small. This charge is carried from right to left. When it reaches position a_2 directly opposite the neutralizing brush b_1, a segment c_1 on the front plate comes directly under its influence, being touched at the same time by brush b_1. This results in a negative charge, which is a *bound* charge, being induced on this front segment c_1, and a *free* positive charge also being induced on c_1 (see Pars. 152 and 155). The positive charge, which is a free charge, flows away through the brush b_1. This leaves only the bound negative charge on the segment c_1.

The segment c_1 now carries this negative charge from left to right. When it reaches position c_2, a segment on the back plate comes directly under its influence, being touched at the same time by brush b_3. This results in a positive bound charge appearing on this back segment, the free negative charge escaping through neutralizing brush b_3. In this manner the positive charges on the segments of the back plate moving from right to left are increased. They, in turn, increase the charges on the segments of the front plate. As this action is cumulative, the machine will build up its charge.

The positive charge on a_2 will move directly under the sharp points of the positive collecting brush B_1, where, in virtue of brush discharge (see Fig. 171, p. 195), it is taken off. When this segment a_2 reaches the neutralizing brush b_4, it comes under the influence of a positive charge. Hence, it obtains a bound negative charge. Its free positive charge neutralizes the free negative charge coming from brush b_3.

Throughout the machine the foregoing reactions occur. These reactions are all of such character as to cause an increase in the generated energy.

The electrical energy generated by such a machine is of extremely high voltage. The current must necessarily be very small, since its value is determined by the rate at which the carriers can bring the charges to the collecting brushes. The capacity of the machine may be increased by increasing the number of pairs of plates.

A sphere gap, shown in Fig. 167 (*a*) and (*b*), is usually connected across the terminals of such a machine. A series of small, thin, and rapidly occurring sparks ordinarily jump the gap. By using two Leyden jars $C_1 C_2$ (see p. 199), whose outer coatings are connected together, large, vigorous sparks are obtained, occurring at less frequent intervals.

The electrical energy which this machine generates is obtained from the mechanical work done in separating positive and negative charges.

This is the same principle as in the electrophorus, where the free charge is allowed to escape and the electrical energy is obtained in virtue of the mechanical work done in separating the positive induced charge from the negative inducing charge.

158. Dielectric or Electrostatic Field.—The preceding experiments show that a condition of stress exists in the medium surrounding an electrostatic charge. For example, the fact that unlike charges attract each other and like charges repel each other indicates a condition of stress in the intervening medium. This condition of stress is analogous to that existing in the magnetic field, although the two fields are quite different physically.

If a charged body or a neutral body be placed in the region surrounding an electric charge, a force will be found to act on the body. As with the magnetic field, this condition of stress may be represented by lines of force, the number of lines per square centimeter, taken normal to their direction, representing the force in dynes exerted on a unit plus charge. These lines are called *dielectric* or *electrostatic* lines of force.

These lines have the following properties: They originate on a positive charge and terminate on a negative charge. They must always leave and enter a conducting surface normal to the surface, as otherwise there would be a component of force or stress tangential to the surface which would result in a flow of current. They can never intersect, since the *resultant* of any number of separate electrostatic forces can have but a single direction at

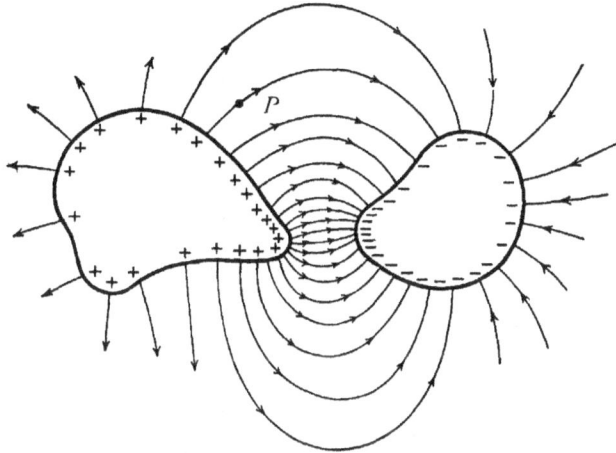

Fig. 168.—Electrostatic field due to charged conducting bodies.

a point in space. They cannot exist within a conducting body, since there can be no force without a resulting flow of current in a conductor. A positive charge P, free to move (Fig. 168), if placed at the positive end of an electrostatic line of force will

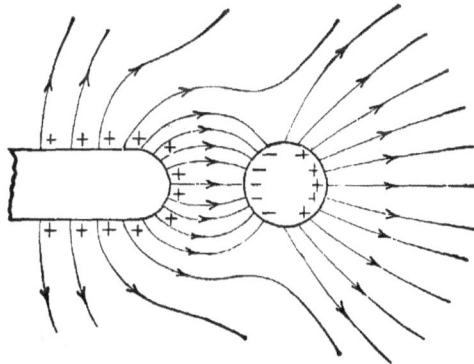

Fig. 169.—Electrostatic induction explained by dielectric lines.

move along the line until it reaches the negative end. Figure 168 shows a typical electrostatic field between two irregular, charged, conducting bodies.

The properties of electrostatic lines are almost identical with those of magnetic lines of force except that they act on dielectric rather than on magnetic media. The dielectric lines adjust themselves so that the reluctance of the field is a minimum and they tend to contract like elastic bands.

Electrostatic induction (or influence) may be explained by means of these lines. Figure 169 shows a positively charged rod and a pith ball. Some of the lines leaving the positively charged rod terminate at the negative induced charge on the pith ball (Fig. 164 (*a*), p. 185). Likewise, lines leave the free positive charge on the pith ball and go out into space. In contracting, the lines between the charged rod and the pith ball tend to pull the pith ball nearer the rod, whereas the lines leaving the pith ball tend to pull it away from the rod. Since the former lines terminate on positive and negative charges which are quite close together, their effect predominates (compare Fig. 169 with Fig. 114, p. 140).

159. Force Between Charged Bodies (Coulomb's Law).—By means of a very sensitive torsion balance, Coulomb, after a series of experiments conducted in the years 1785 to 1789, proved that *the force between two small charged bodies in air is proportional to the product of the charges and inversely proportional to the square of the distance between them.* If the charges q_1 and q_2 are given in terms of unit electrostatic charge (Par. 160), and the distance r between them is given in centimeters, in vacuum or in air the force

$$F = \frac{q_1 q_2}{r^2} \text{ dynes.} \tag{72}$$

If the charges are in a medium having a dielectric constant κ (see p. 198), the force

$$F = \frac{q_1 q_2}{\kappa r^2} \text{ dynes.} \tag{73}$$

For air and vacuum, κ is equal to unity. For most other dielectrics, κ is greater than unity. Hence, with fixed *distance* and fixed *charges*, the force between charged bodies is decreased if the medium is other than air or vacuum.

If the charges are not concentrated at points and if the distance between them is comparatively small, Eqs. (72) and (73) do not hold.

A charge on a sphere which is remote from all other charged bodies distributes itself uniformly on the *surface* of the sphere.

The electrostatic lines of force or flux must leave radially and uniformly from such a charged sphere, as if they came from its center. It follows that the effect of a charged sphere at all points in space external to the sphere is the same as if the entire charge on the sphere were concentrated at its center.

Example.—Two equal spheres A and B, each of which has a radius of 0.5 cm, are spaced 30 cm between centers in air. If sphere A is given a charge of 100 positive electrostatic units (statcoulombs) and sphere B a charge of 150 negative units (statcoulombs), find: (*a*) the force (acting on a unit positive charge) just outside the surface of each sphere, neglecting the effect of the other sphere; (*b*) the force of attraction between the two spheres.

a. The force acting just outside the surface of sphere A, from Eq. (72),

$$F_A = + \frac{100 \times 1}{(0.5)^2} = 400 \text{ dynes, radially } outward. \quad Ans.$$

At the surface of sphere B

$$F_B = - \frac{150 \times 1}{(0.5)^2} = 600 \text{ dynes, radially } inward. \quad Ans.$$

b. The force acting between the two spheres, from Eq. (72),

$$F = \frac{(+100) \times (-150)}{(30)^2} = 16.67 \text{ dynes, } attraction. \quad Ans.$$

160. Unit Electrostatic Charge.—*Unit electrostatic charge is that charge which, when placed at a distance of one centimeter in air from a similar and equal charge, repels it with a force of one dyne.* Thus in Eq. (72) when F is 1 dyne, r is 1 cm, and q_1 and q_2 are equal, q_1 and q_2 must each be equal to unity. The relation is illustrated in Fig. 170, which shows two unit positive charges $+q$, 1 cm apart and repelling each other with a force of 1 dyne.

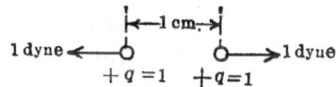

Fig. 170.—Unit electrostatic charge.

In the electrostatic system, unit charge is called the *statcoulomb* and also the *esu* (electrostatic unit).

161. Fundamental Electrostatic Laws.—The preceding discussions and experiments are illustrative of the following fundamental electrostatic laws.

For every positive charge there must exist an equal negative charge.

Charges of unlike sign attract each other, and charges of like sign repel each other.

A positive charge will induce a negative charge on a body near it.

A negative charge will induce a positive charge on a body near it.

An electrostatic line originates at a positive charge and terminates at a negative charge.

Electrostatic lines distribute themselves so that the electrostatic reluctance of the system is a minimum.

162. Dielectrics.—If electrostatic phenomena are being considered, the medium between two conductors is called a *dielectric*. This is in distinction to the properties of the same medium as an insulator, which relates to electrical *conduction*. The dielectric properties of a medium relate to the number of electrostatic lines which it permits to pass through it with a given potential difference and also to the number of electrostatic lines per unit area taken normal to their direction which it can permit without being ruptured. On the other hand, the insulating properties of a medium relate to the current which it conducts with a given potential difference.

For example, air is not a particularly good dielectric so far as preventing sparkover is concerned. It does not permit the passage of electrostatic lines as readily as most media, and it ruptures when the potential gradient (see Par. 163) reaches 30,000 volts per cm or 75,000 volts per in. On the other hand, it is one of the best insulators.

163. Dielectric Strength.—There is one difference, however, between electrostatic lines, on the one hand, and either magnetic lines of induction or electric-current streamlines on the other. No matter how much current flows in a conductor, the conductor is not injured mechanically, provided it be kept cool. Neither is a magnetic conductor injured, no matter how many magnetic lines exist in it. But there is a limit to the number of electrostatic lines which may exist in a medium. If the lines become too concentrated, the medium cannot withstand the stresses which result, and it is ruptured or "breaks down." This breakdown may be followed by a dynamic arc, which increases the injury to the medium by burning.

The ability of a substance to resist electrostatic breakdown is called its *dielectric strength*. This is expressed in volts per unit thickness when the substance is placed between flat electrodes having rounded corners. For example, the dielectric strength of air is approximately 3,000 volts per mm, or 75 volts per mil.

Rubber and varnished cambric have a much greater dielectric strength than air, that of rubber being about 16,000 volts per mm, or 400 volts per mil, and that of varnished cambric being about twice that for rubber.

The volts per unit thickness impressed across a dielectric is called the *voltage gradient*. For example, if 24,000 volts are impressed across 30 mils of dielectric between flat parallel electrodes, the gradient is 24,000/30, or 800 volts per mil.

An example of air being ruptured under dielectric stress is seen by connecting a gap having needles or other sharp-pointed electrodes (Fig. 171) across the terminals of an influence machine or of a high-voltage transformer. A bluish discharge occurs around the points, due to the air being

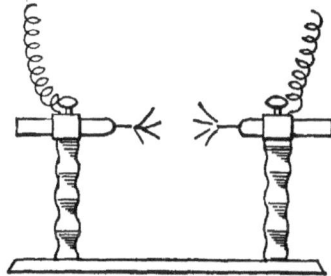

FIG. 171.—Brush or corona discharge from needle points.

ruptured by the concentration of electrostatic lines at the points. This bluish brush discharge is often called *corona*. It occasionally appears on high-voltage transmission lines (see Part II).

164. Capacitance.—A condenser consists of two conductors separated by a dielectric.

The electrodes of Figs. 161 and 162, pp. 183 and 184, form condensers.

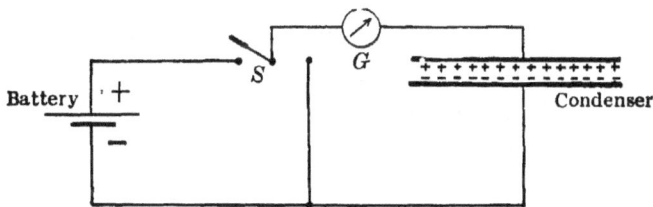

FIG. 172.—Charging and discharging a condenser.

Figure 172 shows two conducting plates connected to a battery, the plates being separated by a dielectric, such as air or glass. There is also a single-pole, double-throw (S.-P. D.-T.) switch S and a galvanometer G in the circuit. If the switch S be closed to the left, the upper electrode of the condenser is connected through the galvanometer to the positive terminal of the battery, and the lower electrode of the condenser is connected directly to the negative terminal of the battery. The galvanometer deflects momentarily and then comes back to zero. This indicates that, when the switch is closed, a quantity of electricity flows through the galvanometer and that the current ceases to flow almost immediately. This current flows for a time sufficient to charge the condenser. After the condenser has become fully charged, the current ceases, because the emf of the condenser is equal and opposite to that of the battery. As this condenser emf opposes the current entering the condenser, it may be considered as a back emf. Any current which may flow after the condenser has become fully charged is a leakage current flowing through the insulation. If the switch S be opened for a short

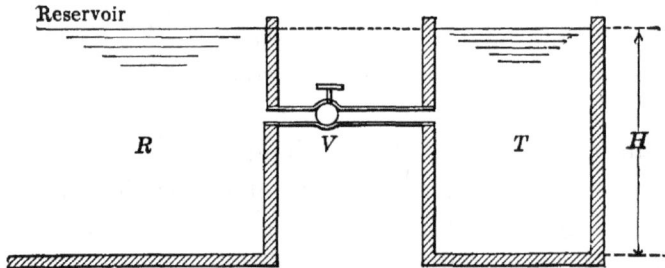

FIG. 173.—Reservoir and connected tank.

time and then closed again, no deflection of the galvanometer will be noted unless there is leakage through the insulation.

This phenomenon of charging a condenser from a battery is not unlike the filling of a tank T from a reservoir R (Fig. 173). When the valve V is first opened, water will rush through the connecting pipe and will continue to flow at a diminishing rate until the level H of the water in the tank T is equal to the level of the water in the reservoir. If the tank does not leak, no water flows through the pipe after the water levels have become equal. In the same way the condenser (Fig. 172) takes current until its potential is equal to that of the battery, when current ceases to flow.

To prove that electricity has actually been stored in the condenser (Fig. 172), the switch S may be closed to the right. This short-circuits the condenser through the galvanometer. The galvanometer now deflects momentarily in a direction opposite to that on charge, showing that the current now flows *out* of the positive plate. The condenser now becomes discharged, as is shown by there being no longer any deflection of the galvanometer

If the voltage of the battery (Fig. 172) be increased, the galvanometer deflection on charge and on discharge will increase also. This is due to the fact that the charge given to the condenser is *proportional to the voltage* across its terminals, just as the amount of water in the tank will be proportional to its height *H* (Fig. 173). The relation between the voltage and the charge in a condenser may be expressed by

$$Q = CE. \tag{74}$$

That is, the quantity of charge in a condenser is equal to the voltage multiplied by a constant *C*. This constant *C* is called the *capacitance* of the condenser. The practical unit of capacitance is the *farad*. If *C* is in farads and *E* in volts, *Q* is in coulombs or ampere-seconds.

The farad is too large a unit for practical purposes, since a condenser having a capacitance of 1 farad would have a prohibitively large volume. The capacitance of the earth, as an isolated sphere, is less than one-thousandth of a farad. The *microfarad* (μf), equal to one-millionth of a farad, is the unit of capacitance ordinarily used.

By transposition, Eq. (74) may be written

$$C = \frac{Q}{E}. \tag{75}$$

That is, the capacitance of a condenser in *farads* is the ratio of its charge in *coulombs* to the potential difference between its terminals in *volts*.

Also,

$$E = \frac{Q}{C}. \tag{76}$$

That is, the potential in volts across a condenser is the ratio of its charge in *coulombs* to its capacitance in *farads*.

Example.—A condenser having a capacitance of 200 μf is connected across 600-volt mains. If the current is maintained constant at 0.1 amp, how long must it flow before the condenser is fully charged?

The quantity in the condenser, when fully charged, is $Q = 0.000200 \times 600 = 0.12$ coulomb, or amp-sec.

$$0.12 = 0.1t.$$
$$t = 1.2 \text{ sec.} \quad Ans.$$

165. Dielectric Constant or Permittivity.—A parallel-plate condenser (Fig. 174 (*a*)) with air as a dielectric has a measured capacitance C_1. If a slab of glass or of hard rubber be inserted between the plates so as to fill the intervening space completely (Fig. 174 (*b*)) and the capacitance of the condenser again be measured, it will be found to be greater than its previous value. Let this new value be C_2. The increase in capacitance must be due to the presence of the glass or of the hard rubber.

The ratio $C_2/C_1 = \kappa$ is called the *dielectric constant* or *permittivity* (sometimes *specific inductive capacity*) of the material

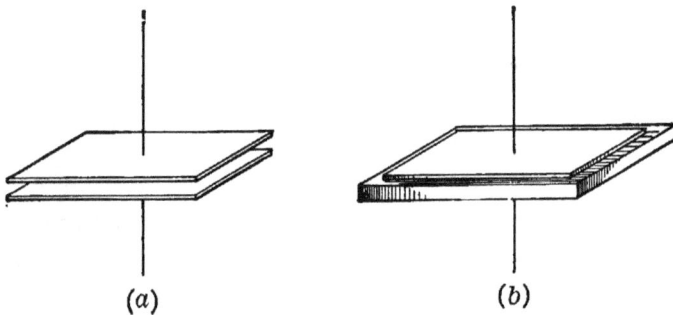

(a)   (b)

Fig. 174.—Plate condenser having air and then glass as a dielectric.

between the condenser plates. The dielectric constant of air is assumed to be unity, just as the magnetic permeability of air is assumed to be unity.

In this table are given the dielectric constants of some of the more common dielectrics:[1]

Bakelite............	4.5 to 7.5	Paraffin................	1.9 to 2.3
Glass..............	5.5 to 10	Rubber compounds.....	3.0 to 7.5
Ice...............	86.4	Hard rubber..........	1.9 to 3.5
Mica..............	2.5 to 7.1	Transformer oils........	2.3 to 2.6
Paper.............	1.7 to 2.6		

166. Types of Condensers.—There are several types of condensers. A simple and common type is the Leyden jar (Fig. 175 (*a*)). The base and lower portion of a glass jar are covered both inside and outside with a thin, metallic coating, such as tin foil. Contact with the inner coating is made through a loose chain, the lower end of which touches the inner coating and the upper end of which is fastened to a metallic rod. The rod passes

[1] For more complete data, see "Standard Handbook," 6th ed., Sec. 4.

through the insulating cover and is often terminated at the top by a round, metallic knob. The inner and outer coatings of the jar constitute the condenser plates, and the glass of the jar is the dielectric. Such condensers have capacitances of the order of 0.003 μf.

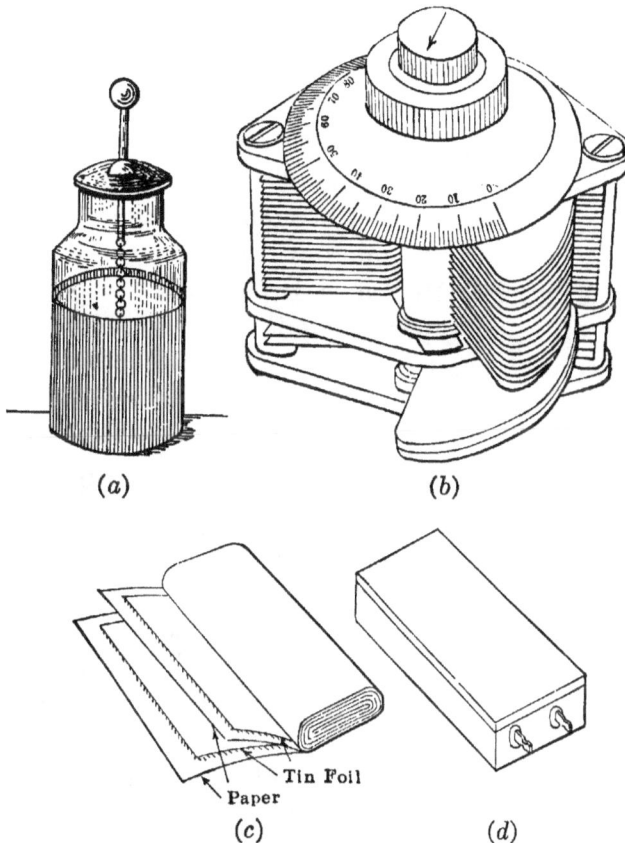

FIG. 175.—Types of condensers: (a) Leyden jar; (b) variable radio air condenser; (c) telephone condenser (internal view); (d) telephone condenser (assembled).

A common type of variable air condenser is shown in Fig. 175 (b). It consists of a set of equally spaced, semicircular, fixed plates, which are connected together, and a similar set of equally spaced, semicircular, movable plates, which are fastened rigidly together but mounted as a unit on an axis free to rotate. The mounting is such that the movable plates move into the spaces between the fixed plates. The fixed set of plates and the movable set of plates are insulated from each other. The capacitance is varied by turning the movable set of plates so that more or less of their plate area is between the fixed plates. Even with the movable plates entirely outside the spaces between the fixed plates, there is still a small capacitance between the edges of the fixed and movable plates. With the exception of the insulating supports, this type of condenser has air as the dielectric. At radio frequencies, losses in dielectrics other than air are excessive. The capacitance of such condensers is nearly proportional to the number of plates. The capacitance of a condenser of 11 plates is of the order of 0.00025 μf. A vernier on the top serves usually to indicate the condenser setting rather than its capacitance. The capacitance of such condensers is so small that it is usually expressed in micromicrofarads ($\mu\mu$f). One micromicrofarad is equal to 10^{-6} μf or 10^{-12} farad.

The two foregoing types of condensers have very low capacitance. Telephone condensers, on the other hand, are made to have very large capacitance per unit volume. They are made by rolling together very tightly two long strips of very thin foil, with two strips of thin paraffined paper, one strip of paper being between the strips of foil and the other being outside one of the strips of foil (Fig. 175 (c)). The width of the paper is slightly greater than that of the foil. The roll is then sealed in a tin box and the two terminals brought out through the top (Fig. 175 (d)). As much as 2 μf may be obtained with a box 1 by $2\frac{1}{4}$ by $4\frac{1}{2}$ in. (2.54 by 5.7 by 11.4 cm). Such condensers will withstand voltages as great as 600 volts, direct current, without puncturing.

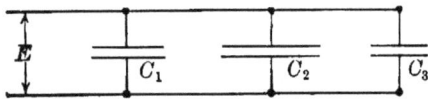

Fig. 176.—Capacitances in parallel.

For power work, such as for power-factor correction, and in lightning generators, condensers are made which in a single unit will withstand up to 60,000 volts alternating. Such condensers or capacitors are insulated with paper impregnated with high-grade oils and are sealed in a sheet-metal container.

167. Equivalent Capacitance of Condensers in Parallel.—Let it be required to determine the capacitance C of three condensers in parallel, the condensers having capacitances C_1, C_2, C_3. This arrangement of condensers is shown in Fig. 176. Let the common voltage across the condensers be E and the total resulting charge Q. Obviously,

$$Q = CE$$

and

$$Q_1 = C_1E, \qquad Q_2 = C_2E, \qquad Q_3 = C_3E. \tag{I}$$

The total charge

$$Q = Q_1 + Q_2 + Q_3 = CE. \tag{II}$$

Substituting from (I) in (II),

$$CE = C_1E + C_2E + C_3E.$$
$$CE = (C_1 + C_2 + C_3)E$$
$$\therefore C = C_1 + C_2 + C_3. \tag{77}$$

That is,

If condensers are connected in parallel, the resulting capacitance is the sum of the individual capacitances.

This is analogous to the grouping of conductances in parallel in the electric circuit.

Example.—Three condensers, having capacitances 5, 10, 12 μf, are connected in parallel across 600-volt mains. (*a*) What single condenser would replace the combination? (*b*) What is the charge on each condenser?

(*a*) $\qquad C = 5 + 10 + 12 = 27$ μf. *Ans.*

(*b*) $\qquad Q_1 = 5 \times 600 = 3,000$ microcoulombs.

$\qquad Q_2 = 10 \times 600 = 6,000$ microcoulombs.

$\qquad Q_3 = 12 \times 600 = 7,200$ microcoulombs. *Ans.*

Total charge = 16,200 microcoulombs = 27×600 microcoulombs (check).

168. Equivalent Capacitance of Condensers in Series.—In Fig. 177, three condensers, having capacitances C_1, C_2, C_3, are connected in series across the voltage E. It is desired to determine the capacitance of an equivalent single condenser. Let E_1, E_2, E_3 be the potential differences across the condensers C_1, C_2, C_3. When the voltage E is applied to the system, there will be $+Q$ units of charge on the positive plate of C_1 and, by the law of electrostatic induction, $-Q$ units must be induced on the negative plate of C_1.

Now consider the region a, which consists of the negative plate of C_1, the positive plate of C_2, and the connecting lead. This system is insulated from all external potentials, since it is assumed that the condensers have perfect insulation. Before the voltage was applied to the system of condensers, no charge existed in the region a. After the application of the voltage, the net charge in this region must still be zero, as no charge can flow through the insulation. Therefore, $+Q$ units must come into existence in order that the net charge in the region a may remain zero. $(+Q + (-Q) = 0)$. This charge of $+Q$ units will go to the plate of C_2, since it is repelled by the $+$ charge on C_1, just as the charge b' (Fig. 162 (a), p. 184) took a position on the end of the electrode as far as possible from the positive inducing charge a. The same reasoning holds for the region b, between C_2 and C_3, and accordingly there will be a negative charge $-Q$ on the lower electrode of C_3. Therefore, each of

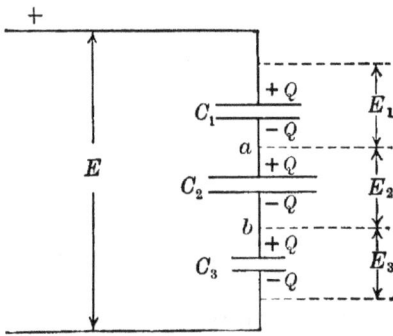

FIG. 177.—Capacitances in series.

the three condensers in series has the same charge Q. (This is analogous to resistances in series, each of which must carry the same current, if no leakage exists.)

Consider the voltages, E_1, E_2, E_3.

$$E_1 = \frac{Q}{C_1}, \qquad E_2 = \frac{Q}{C_2}, \qquad E_3 = \frac{Q}{C_3} \text{ (from Eq. (76), p. 197).} \quad \text{(I)}$$

The sum of the three condenser voltages must equal the line voltage,

$$E_1 + E_2 + E_3 = E. \qquad (II)$$

From (I) and (II),

$$E = \frac{Q}{C_1} + \frac{Q}{C_2} + \frac{Q}{C_3}.$$

Also $E = Q/C$, as by definition the equivalent condenser C must have a charge Q.

Substituting this value for E,

$$\frac{Q}{C} = \frac{Q}{C_1} + \frac{Q}{C_2} + \frac{Q}{C_3}.$$

$$\frac{1}{C} = \frac{1}{C_1} + \frac{1}{C_2} + \frac{1}{C_3}. \qquad (78)$$

That is,

The reciprocal of the equivalent capacitance of a number of condensers in series is equal to the sum of the reciprocals of the capacitances of the individual condensers.

Equation (78), giving the equivalent capacitance of condensers in series, is similar to Eq. (13), p. 18, giving the equivalent resistance of resistances in parallel.

In assuming for condensers connected in series that, with direct current, the potential across each condenser is inversely proportional to its capacitance, the factor of leakage is absolutely neglected. If the condensers are even slightly leaky, however, a current flows through the series and eventually the potential distributes itself according to Ohm's law.

$$E_1 = IR_1, \qquad E_2 = IR_2, \qquad E_3 = IR_3$$

where I is the leakage current and R_1, R_2, R_3 are the ohmic resistances of the three condensers.

Example of condensers connected in series.

Assume that the three condensers of Par. 167, p. 201, having capacitances of 5, 10, 12 μf, are connected in series across 600-volt mains. Determine: (a) the equivalent capacitance of the combination; (b) the charge on each condenser; (c) the potential across each condenser, assuming no leakage.

(a)
$$\frac{1}{C} = \frac{1}{5} + \frac{1}{10} + \frac{1}{12} = 0.383.$$

$$C = \frac{1}{0.383} = 2.61 \ \mu f. \quad Ans.$$

(b) $Q = 2.61 \times 600 = 1{,}566$ microcoulombs, on each condenser. *Ans.*

(c) From Eq. (76), p. 197,

$$E_1 = \frac{1{,}566 \times 10^{-6}}{5 \times 10^{-6}} = 313 \text{ volts},$$

$$E_2 = \frac{1{,}566 \times 10^{-6}}{10 \times 10^{-6}} = 157 \text{ volts},$$

$$E_3 = \frac{1{,}566 \times 10^{-6}}{12 \times 10^{-6}} = 130 \text{ volts}. \quad Ans.$$

169. Energy Stored in Condensers.—When a charge is stored in a condenser, a difference of potential exists between the positive and negative plates so that energy must be stored. The existence of this energy is shown by the spark resulting from short-circuiting the condenser plates. The energy in joules, or watt-seconds, is

$$W = \tfrac{1}{2}QE. \tag{79}$$

Since $Q = CE$ and $E = Q/C$, (79) may be written

$$W = \tfrac{1}{2}CE^2, \tag{80}$$

$$W = \frac{1}{2}\frac{Q^2}{C}. \tag{81}$$

The similarity in form of (80) to the equation for the energy stored in the magnetic field should be noted (see Eq. (68), p. 176). The energy stored in the *electrostatic* field is proportional to the square of the *voltage*, whereas the energy stored in the *electromagnetic* field is proportional to the square of the *current*.

Example.—Determine the stored energy in each of the condensers in series of Par. 168 and the total stored energy.

Using Eq. (81),

$$W_1 = \frac{1}{2}\frac{(1,566 \times 10^{-6})^2}{5 \times 10^{-6}} = 0.2453 \text{ joule.}\quad Ans.$$

$$W_2 = \frac{1}{2}\frac{(1,566 \times 10^{-6})^2}{10 \times 10^{-6}} = 0.1225 \text{ joule.}\quad Ans.$$

$$W_3 = \frac{1}{2}\frac{(1,566 \times 10^{-6})^2}{12 \times 10^{-6}} = 0.1020 \text{ joule.}\quad Ans.$$

The total energy $W = \tfrac{1}{2}\,(1,566 \times 10^{-6} \times 600) = 0.4698$ joule. *Ans.*

170. Measurement of Capacitance.—There are two common methods of measuring capacitance, the d-c or ballistic method and the a-c or bridge method.

Ballistic Method.—In the ballistic or d-c method a galvanometer, used ballistically, is employed. It can be shown that if the moving coil of the ordinary galvanometer has considerable inertia and is properly damped, its maximum throw, due to the impulse produced by the sudden passage of a current through the moving coil, is proportional to the total quantity of electricity passing through the galvanometer. This assumes that the entire charge passes through the coil before the coil begins to move. Let D be the maximum galvanometer or ballistic deflection in centimeters. Then

$$Q = KD \text{ coulombs} \tag{82}$$

where Q is the quantity and K is the galvanometer constant.

To make the measurement, the apparatus may be connected as shown in Fig. 178. A battery B supplies the current for the apparatus. The measurement may be made on either the charge or the discharge of the condenser, or check measurements may be made using both charge and discharge. If the condenser is at all leaky, the discharge method is preferable.

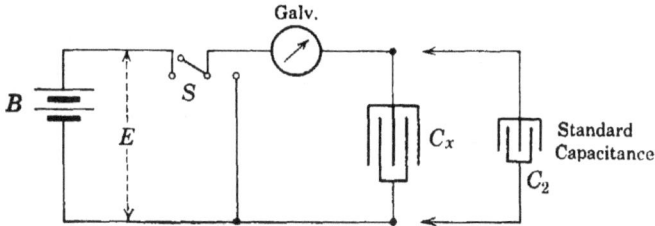

FIG. 178.—Ballistic method of measuring capacitance.

When the switch S is closed to the left, the condenser C_x is charged through the galvanometer and the maximum deflection of the galvanometer is read. The galvanometer should return immediately to zero. If it shows a steady deflection, it indicates a leaky condenser. In a corresponding manner the ballistic deflection of the galvanometer may be read on discharge by closing switch S to the right after charging. Let D_1 be the deflection of the galvanometer, on either charge or discharge, when the unknown capacitance C_x is connected.

A standard capacitance C_2 is now substituted for the unknown capacitance. Let D_2 be the corresponding galvanometer deflection. Then,

$$\frac{C_x}{C_2} = \frac{D_1}{D_2},$$

or

$$C_x = C_2 \frac{D_1}{D_2}. \tag{83}$$

$\frac{C_2}{D_2}$ is the galvanometer constant.

It is often desirable to use an Ayrton shunt in such measurements, as it gives the apparatus greater range. When such a shunt is used, proper correction must be made for its multiplying power (see p. 93).

Bridge Method.—In the bridge method, two capacitances form adjacent arms of a Wheatstone bridge and two resistances form the other two arms (Fig. 179). An alternating current having a frequency of about 1,000 cycles is desirable, since a telephone receiver at this frequency makes a very sensitive detector. The secondary of an induction coil or a vacuum-tube oscillator may be used as the source of power.

Let C_x be the unknown capacitance and C_2 a standard capacitance, which may or may not be adjustable. R_1 and R_2 are two known resistances, one of which should be adjustable unless C_2 is so.

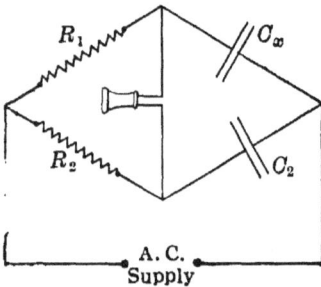

Fig. 179.—Bridge method of measuring capacitance.

Either C_2 or one of the resistances is adjusted until there is no sound in the telephone, showing that the bridge is in balance. Under these conditions,

$$\frac{C_x}{C_2} = \frac{R_2}{R_1}.$$

$$C_x = C_2 \frac{R_2}{R_1}. \qquad (84)$$

Compare this equation with Eq. (47), p. 106. Capacitances in the electrostatic circuit correspond to conductances in the electric circuit.

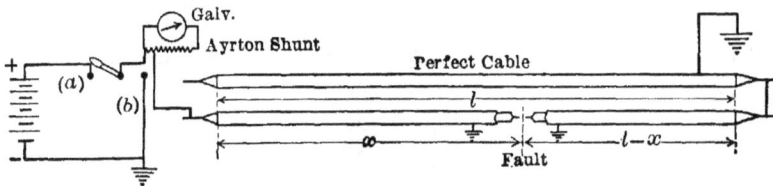

Fig. 180.—Locating an "open" in a cable.

In both of the foregoing measurements, it is assumed that there is little if any leakage in the condensers.

171. Fault Location with a Total Disconnection.—In Chap. V it is shown that a grounded fault in a cable can be located by suitable resistance measurements, as, for example, the Murray-loop test. If a cable be totally disconnected and its broken ends remain insulated, such loop tests are impossible. The distance to the fault may now be determined by capacitance measurements. The connections are shown in Fig. 180. The capacitance C_1 of the length x to the fault is first measured, usually by the ballistic method. If a *similar* perfect cable parallels the faulty cable, the two are looped at the far end and the capacitance C_2 of the combined length l of the perfect cable and the length $l - x$ of the faulty cable is measured. Let D_1 and D_2 be the galvanometer deflections. If the cables are uniform, the capacitances are proportional to the lengths.

Therefore,

$$\frac{C_1}{C_2} = \frac{x}{2l - x} = \frac{D_1}{D_2}.$$

$$x = l\frac{2D_1}{D_1 + D_2} \tag{85}$$

CHAPTER X

THE GENERATOR

A generator is a machine which converts mechanical energy into electrical energy. This is accomplished by means of an armature, carrying conductors on its surface, acting in conjunction with a magnetic field. Electromotive force is induced by the relative motion of the armature conductors and the magnetic field; and when current is delivered to an external circuit, electrical energy is given out by the armature.

In the d-c generator the field is usually stationary and the armature rotates. In most types of a-c generators the armature is stationary and the field rotates. In either case the rotating member is driven by mechanical power applied to its shaft.

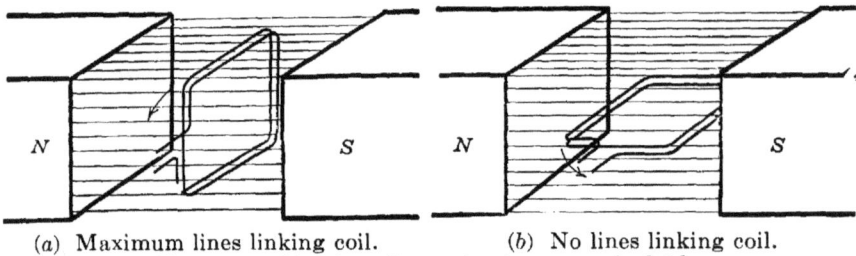

(a) Maximum lines linking coil.   (b) No lines linking coil.
Fig. 181.—Simple coil rotating in magnetic field.

172. Induced Electromotive Force in Generator.—It is shown in Chap. VIII that if the flux linking a coil is varied in any manner an emf is *induced* in the coil. The action of the generator is based on this principle. The flux linking the armature coils is varied by the relative motion of the armature and field.

In Fig. 181 a coil revolves in a uniform magnetic field produced by a north and a south pole. In (a) the coil is perpendicular to the magnetic field and in this position the maximum possible flux links the coil. Let this flux be ϕ.

If the coil be rotated counterclockwise a quarter of a revolution, it will lie in the position shown in (b). As the plane of the coil is parallel to the direction of the flux, no flux links the coil in this

208

position. Therefore, in a quarter revolution the flux which links the coil has been decreased by ϕ. Since the flux linking the coil has changed in the quarter revolution, an emf is induced in the coil.

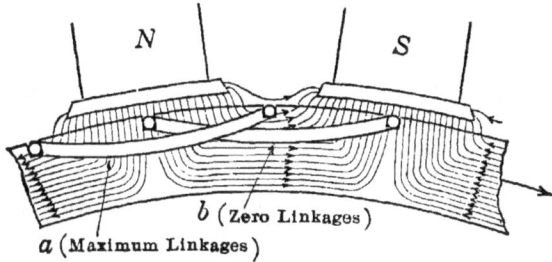

FIG. 182.—Change of coil linkages in generator.

The change in flux linkages of an armature coil, brought about by the change in relative position of armature coil and flux, is illustrated in Fig. 182 for a multipolar generator. In position a the flux from a north pole links the coil. In position b no flux *links* the coil, the flux merely entering and leaving the coil without linking it. When the coil moves to a position directly under the south pole, the flux to the south pole links it. The change in the flux linking the coil during the time required for the coil to move through one pole pitch is 2ϕ where ϕ is the

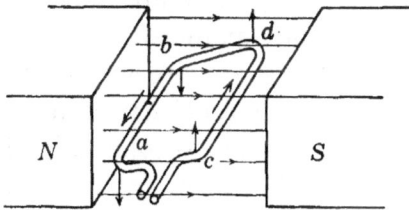

FIG. 183.—Electromotive force induced in individual coil sides.

flux entering the armature from the north pole. Therefore, an emf is induced in the coil.

173. Electromotive Force Induced in Single Conductor.— When an emf is induced by the relative motion of a coil and a magnetic field, it is usually more convenient, for purposes of analysis, to consider the emf as due to the *cutting* of the magnetic flux by the individual conductors forming the coil sides, rather than to the change in flux linkages of the coil itself. For example, in Fig. 183, the single-turn coil may be considered as composed of two active conductors ab and cd, connected by the inactive conductors ac and bd. (The conductor ac is closed through the external circuit.) No emf is induced in the inactive conductors ac and bd because of their geometrical relation to the flux, as will be shown later. The coil (Fig. 183) is shown as rotating counterclockwise. The direction of the induced emf in conductor *ab* will be outward and that in *cd* will be inward by Fleming's right-hand rule (see Par. 174, p. 211). Considering the emf of the coil as a whole, the two emfs in *ab* and *cd* are additive even though their space directions are opposite.

The method of considering the emf as due fundamentally to the cutting of the flux by the individual conductors gives the same numerical result that is obtained by considering the emf as due to change in flux linkages.

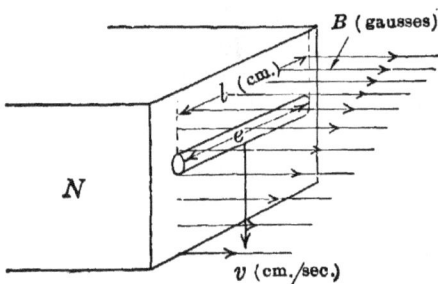

FIG. 184.—Electromotive force induced in single conductor.

The emf in volts generated by a single conductor which cuts a magnetic field is given by the fundamental relation

$$e = Blv10^{-8} \text{ volts} \qquad (86)$$

where B, l, and v are mutually perpendicular.

B is the flux density of the field in gausses and is assumed uniform over the entire length l of the conductor; l is the length of the conductor in *centimeters;* and v is the velocity of the conductor in centimeters per second (Fig. 184).

That the emf in any single conductor is not large is illustrated by the following example.

Example.—The average flux density over the pole face of the ordinary generator is approximately 45,000 lines per sq in.; hence,

$$B = \frac{45,000}{(2.54)^2} = 7,000 \text{ gausses.}$$

The pole faces cover only about 0.7 the peripheral surface of the armature. Hence, the average flux density over the entire peripheral surface is $7,000 \times 0.7 = 4,900$ gausses. The peripheral speed of the ordinary armature is of the order of 100 ft per sec, or approximately 3,000 cm per sec. As a rule, the axial length of the armature of such a generator will not exceed 2 ft. Hence, the length $l = 24 \times 2.54 = 61$ cm.

The induced emf per single conductor is

$$e = 4,900 \times 61 \times 3,000 \times 10^{-8} = 9 \text{ volts.}$$

It is necessary, therefore, to connect a number of such conductors in series in order to obtain the usual commercial voltages of 125, 250, and 600 volts.

174. Direction of Induced Electromotive Force—Fleming's Right-hand Rule.—A definite relation exists among the direction of the flux, the direction of motion of the conductor, and the direction of the emf induced in the conductor, just as a definite relation exists between the direction of current and the direction of the flux which it produces.

A convenient method for determining this relation is the *Fleming right-hand rule.* In this rule the fingers of the *right* hand are utilized as follows:

Set the forefinger, the thumb, and the middle finger of the right hand at right angles to one another (Fig. 185). If the *forefinger points along the lines of flux and the thumb in the direction*

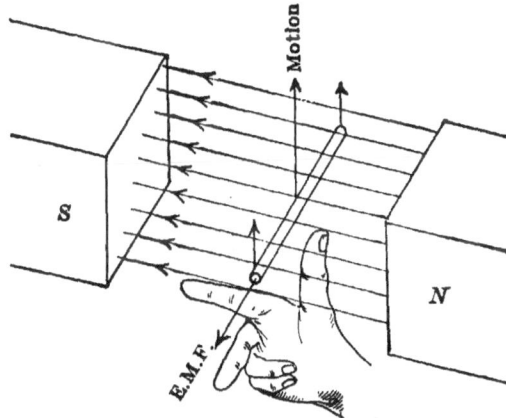

Forefinger along lines of flux; thumb in direction of motion; middle finger gives direction of induced emf.
FIG. 185.—Fleming's right-hand rule.

of motion of the conductor, the middle finger will point in the direction of the induced emf. (The fact that the *forefinger* points along the lines of *flux* assists in memorizing this rule.)

This rule is illustrated by Fig. 185.

175. Electromotive Force Generated by the Rotation of a Coil. A coil of a single turn is shown in Fig. 186 (*a*). The coil rotates in a counterclockwise direction at a uniform speed in a uniform magnetic field. As the coil assumes successive positions, the emf induced in it changes. When it is in position (1) the emf generated is zero, for in this position neither active conductor is *cutting* magnetic lines but is actually moving parallel to these lines. When the coil reaches position 2 (shown dotted), its

conductors are cutting across the lines obliquely and the emf
has a value indicated at (2) in Fig. 186 (*b*). When the coil
reaches position (3), the conductors are cutting the lines *perpendicularly* and are therefore cutting at the maximum possible

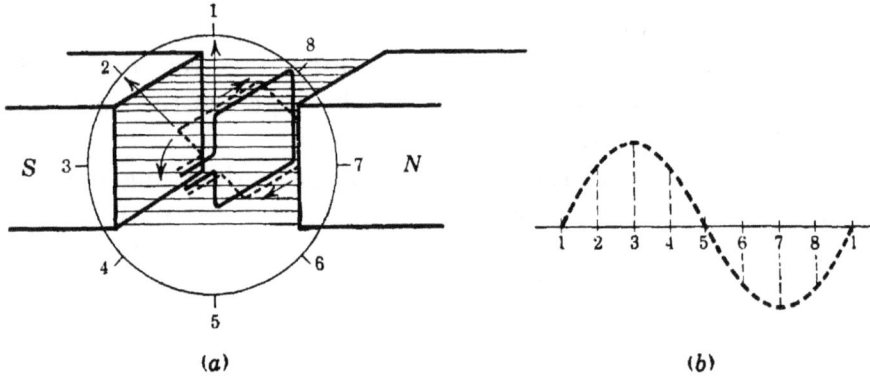

Fig. 186.—Emf induced in coil rotating at constant speed in uniform magnetic
field.

rate. Hence, the emf is a maximum when the coil is in this
position (Fig. 186 (*b*)). At position (4) the emf is less, owing to a
lesser rate of cutting. At position (5) no lines are being cut and,
as in (1), there is no emf. In position (6), the direction of the
emf in the conductors will have reversed, as each conductor is

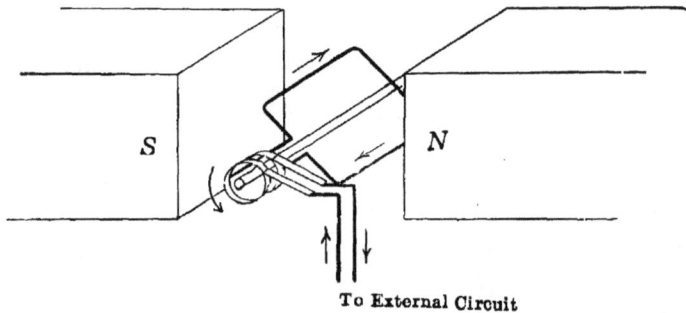

To External Circuit

Fig. 187.—Current conducted from rotating coil by means of slip-rings.

under a pole of opposite sign to that for positions (1) to (5).
The emf increases to a negative maximum at (7) and then
decreases negatively until the coil again reaches position (1).
After this the coil merely repeats the cycle. The values of
induced emf for each position of the coil as it makes one revolution are plotted in Fig. 186 (*b*).

This induced emf is alternating, and an emf varying in the manner shown is called a *sine wave*[1] of emf. This alternating emf may be impressed on an external circuit by means of two *slip-rings* (Fig. 187). Each ring is continuous and insulated from the other ring and from the shaft. A metal or a carbon brush rests on each ring and conducts the current from the coil to the external circuit.

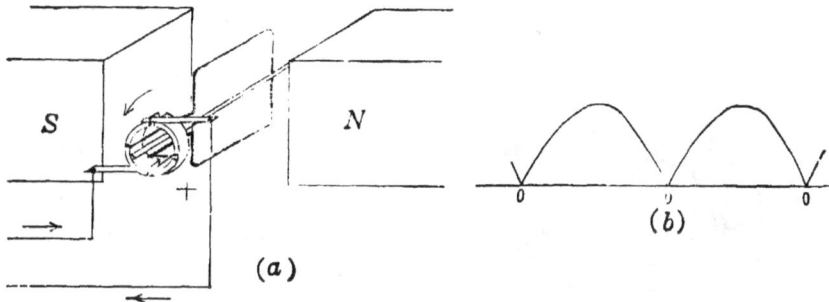

FIG. 188.—Rectification of current by split-ring or commutator.

If a unidirectional or *direct current* is desired, that is, one whose direction of flow in the external circuit is always the same, such rings cannot be used. A direct current must always flow into the external circuit in the *same direction*. As the coil current must necessarily be alternating, since the emf which produces it is alternating, as has just been shown, this current must be rectified before it is allowed to enter the external circuit. This rectification can be accomplished by using a split-ring, as shown in Fig. 188. Instead of using two rings, as in Fig. 187, one ring only is used. This is split by saw cuts at two points diametrically opposite each other. Each of the two ends of the coil is connected to one of the sections or segments so produced.

Consideration of Fig. 188 shows that, as the direction of the current in the coil reverses, its connections to the external circuit are simultaneously reversed. Therefore, the direction of flow of current to the external circuit is not changed. Values of current to the external circuit for different positions of the coil are plotted in Fig. 188 (*b*). The brushes pass over the cuts in the ring when the coil is perpendicular to the magnetic field or when it is in the so-called *neutral* plane and is generating no emf, as shown in Fig. 186. These neutral points are marked 0-0-0 in Fig. 188 (*b*).

[1] See Part II, Chap. I, for a more detailed discussion.

By comparing Fig. 186 (*b*) with Fig. 188 (*b*), it will be seen that so far as the *external* circuit is concerned the negative half of the wave has been reversed, and so made positive.

The split-ring shown in Fig. 188 is a simple *commutator*. The commutator causes the alternating current generated in the individual armature coils to flow to the external circuit as a unidirectional or direct current. In commercial dynamos there is a large number of segments or commutator bars.

A voltage with a zero value twice in each cycle, as shown in Fig. 188 (*b*), could not be used commercially for d-c service. Also, a single-coil generator would have a small output for its size and weight. By employing a large number of coils and commutator segments, the emf becomes practically steady and the output of the generator per unit weight becomes sufficiently large to make the machine commercially economical.

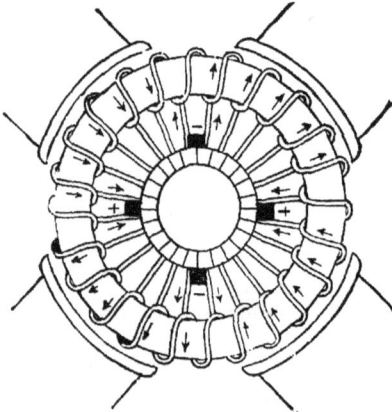

Fig. 189.—Gramme-ring winding.

176. Gramme-ring Winding.—This type of winding in diagrammatic form (Fig. 189) consists of insulated wire wound spirally around a hollow cylinder of iron with taps taken from the winding at regular intervals and connected to commutator segments. This winding is simple and has the advantage that a single winding is adapted to any number of poles, if the voltage limitations do not prevent. The portions of the conductors which lie inside the cylinder or ring cut practically no flux and act merely as connectors for the active portions of the conductors which lie on the outer surface of the cylinder or armature. Because of the small proportion of active conductor in such a winding, a relatively large amount of copper is required. In generators of small diameter, there is not sufficient room to carry the inactive conductors back through the armature core. In a Gramme-ring winding, formed coils cannot be used, and this makes the winding expensive.

It will be noted in the Gramme-ring winding (Fig. 189) that one-fourth of the turns on the armature are connected in series between each pair of positive and negative brushes. Hence, in such a winding, the emf between brushes is the sum of the emfs of all the coils that lie between brushes. When one coil passes a brush, another coil moves forward to take its place. Figure 190 shows the emf between brushes due to four coils, it being assumed that the voltage curve for each coil is a sine wave. The emf of each coil is plotted separately. These emfs do not all have their

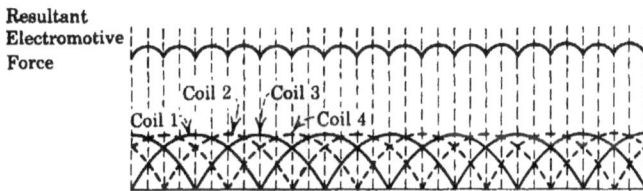

Fig. 190.—Resultant emf due to four series-connected coils between brushes.

zero value at the same time, nor do they reach their maximum value at the same time, owing to the different positions of the individual coils. The resultant emf at any point is the sum of these individual emfs at that point. This voltage should be compared with the emf obtained with the single-coil winding shown in Fig. 188 (*b*). It will be noted that a fairly smooth resultant emf is obtained with as few as four coils even, the "ripples" being noticeable but comparatively small in magnitude.

177. Drum Winding.—The objections to the ring winding are overcome by the use of the drum winding. All the conductors of the drum winding lie on the surface of the armature and are connected to one another by front and back connections or coil ends. (*ad* and *bc*, Fig. 191, are coil ends. Also see Fig. 183, p. 209.) With the exception of these end connections, all the armature copper is "active"; that is, it cuts flux and so is active in generating emf.

The sides of each coil should be approximately one pole pitch (the peripheral distance between centers of adjacent poles) apart. If one conductor is under a north pole, the other is then under a south pole; and as both move in the same direction, but under opposite poles, the space directions of the emfs of the two conductors will be opposite (Fig. 191). However, owing to the manner in which these conductors are connected at their ends, their emfs are additive.   (Also see Fig. 183, p. 209.)

In most Gramme-ring windings, and in the earlier drum-wound dynamos, the surface of the armature core was smooth.   The conductors were held in position partly by projecting pins and

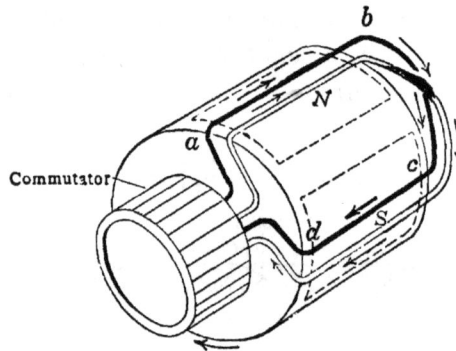

Fig. 191.—Two coils in place on a four-pole, drum-wound armature.

were prevented by binding wires from flying out under the action of centrifugal force.   The smooth-core armature has been superseded by the "ironclad" or slotted armature in which the conductors are embedded in slots, as indicated in Fig. 194, p. 218.

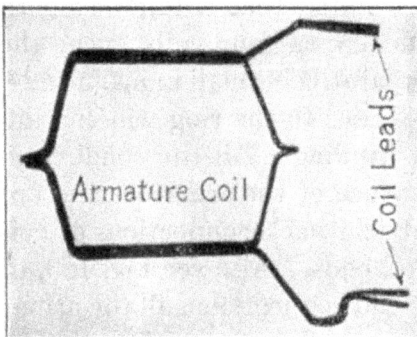

Fig. 192.—Formed armature coil.

The slots are lined with insulation, and the conductors are held firmly in the slots by wooden or nonconducting wedges in the larger machines and by binding wires in the smaller machines (see Fig. 203, p. 229).   The slotted construction is much better mechanically than the smooth-core construction and also permits a much shorter air-gap.

178. Simplex Lap Winding.—As a rule, d-c armatures are wound with form-made coils (Fig. 192).   These coils are usually wound by machines with the necessary number of turns and are taped with cotton or mica tape.   They are then bent into proper shape by another machine.   The two ends of the coil are left bare so that later they may be soldered to the commutator bars. The span of the coil, called the *coil pitch*, should be equal or nearly equal to the pole pitch, so that when one side of the coil is under the center of a north pole the other side is under a south pole. The span of the coil may be less than the pole pitch, in which case the winding is called a *fractional-pitch winding*. A fractional pitch as low as eight-tenths is sometimes used. Since the span of the coil is less than the pole pitch, the induced emfs in the two coil sides do not reach their maximum values simultaneously and the coil emf is reduced. With the ordinary values of fractional pitch, the induced emf is reduced by only a small amount. Some saving of copper results from the shorter end connections.

Usually, d-c windings are *two-layer* windings. That is, with the single-coil type of winding, each slot contains two coil sides (Fig. 194). With multiple-coil windings (Fig. 196, p. 220) the coil sides lie in two layers in the slots, one layer along the bottom

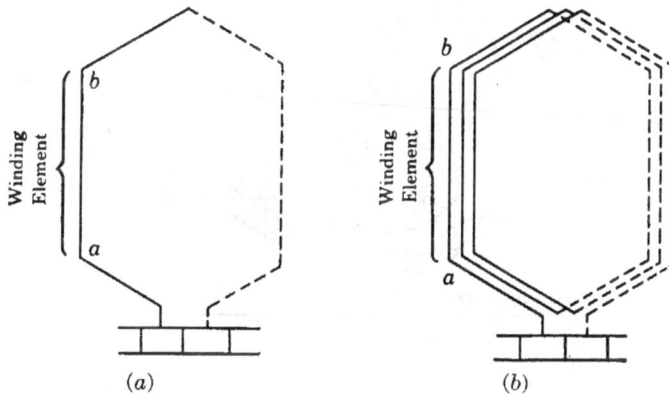

FIG. 193.—Single-turn coil and three-turn coil of an armature winding.

and the other layer along the top of the slots. The coils are so placed on the armature that one side of each coil occupies the top of one slot and its other side occupies the bottom of another slot spaced approximately one pole pitch away (Fig. 194). With this arrangement, the coils overlap one another, not unlike shingles on a roof, and the winding is made to fit readily on the armature. Also, with this type of winding, the end connections are easily made, as the coil ends can be bent around one another in a systematic manner, passing from the bottom to the top layer by means of a peculiar twist in the ends of the coils.

If an armature is simplex *lap-wound* and each coil has a single turn, the two ends of the coil are connected to adjacent commutator segments (Fig. 193 (a)). If each coil consists of several turns, the coil is taped up as a unit (Fig. 192) and the two ends of the coil are connected to adjacent commutator segments

(Fig. 193 (*b*)). The coil sides which are thus taped together constitute a *winding element* (Fig. 193 (*b*)).

In the simplest form of two-layer winding, two coil sides occupy a single slot (Fig. 194). It is customary in designing such a winding to number the coil sides, designating the coil side or the winding element at the top of a slot as 1, that directly under it in the same slot as 2, the one at the top of the adjacent slot to the right as 3, etc., as shown in Figs. 194 and 195. The number of elements that a coil connection spans on the *commutator end* of the armature is called the *front pitch*, and the number of elements that

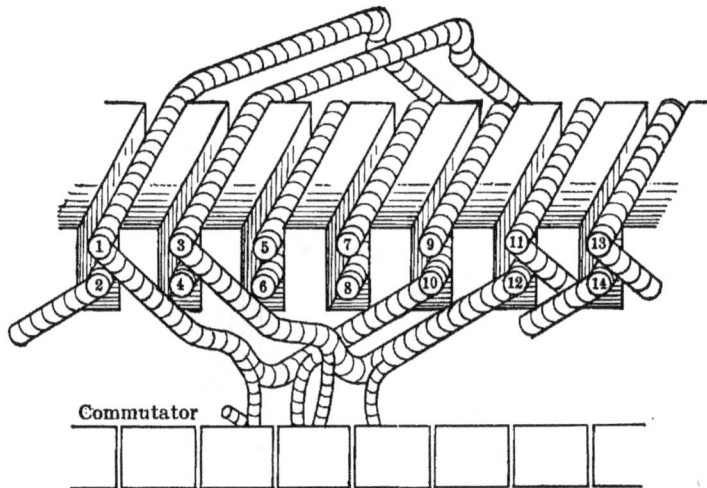

FIG. 194.—Simplex lap winding having back pitch 9 and front pitch 7.

a coil connection spans at the *back of the armature* is called the *back pitch*. The average of the front and back pitch is called the *average pitch*. For example, in Fig. 194, the element 3 connects with element 10 at the front of the armature through a commutator segment; hence, the front pitch is 7. The element 1 connects with the element 10 at the back of the armature; hence, the back pitch is 9. This is also shown in Fig. 195. *Since all connections are made from elements at the top of the slots to other elements at the bottom of other slots, both the front pitch and the back pitch must be odd.* Also, the pitch must be such that if one side of a coil is under a north pole the other side is under a south pole.

A study of Figs. 194 and 195 shows that the connection on the back of the armature, whose span equals the back pitch, joins the two sides of any one coil together. The coil connection on the front of the armature, whose span equals the front pitch and which is ordinarily made through a commutator segment, is the means by which the armature coils are connected in series.

The winding in Fig. 195, a part of which is shown in Fig. 194, is designed for a four-pole, 18-slot dynamo. There are, therefore, 36 elements. In order that the two sides of any coil may span a distance approximately equal to the pole pitch, the average pitch must be in the neighborhood of $\frac{36}{4}$, or 9. In this winding the

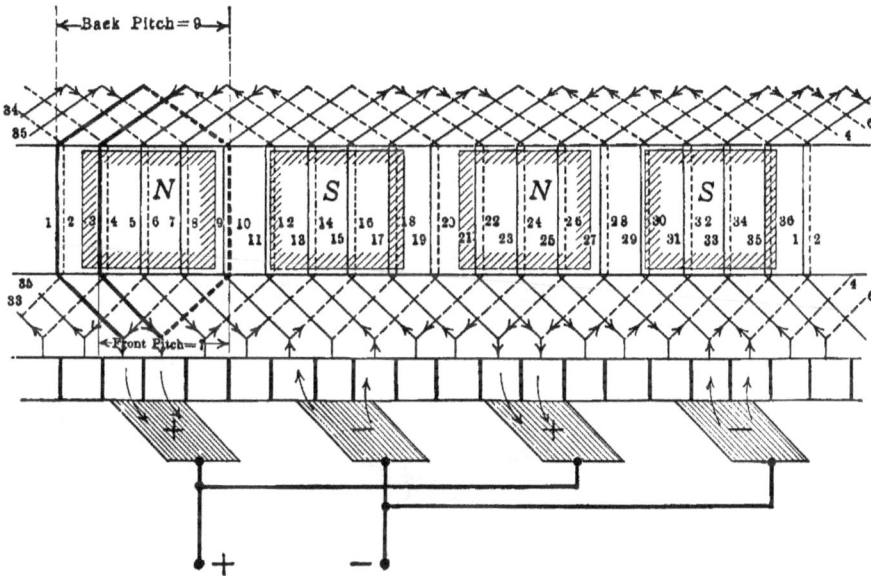

Fig. 195.—Development of a four-pole lap winding.

back pitch is chosen as 9 and the front pitch as 7, giving an average pitch of 8. Since the back pitch is 9, the coil must span a peripheral distance on the armature equal to the pole pitch. Hence, this back pitch of 9 gives a *full-pitch winding*, as the emfs in the two sides of the coil are additive at every instant of time. The fact that the average pitch is 8 and the front pitch is 7 does *not* mean that this is a fractional-pitch winding, since the front connection is merely the means by which adjacent coils are connected in series. Figure 195 shows the development of the entire winding.

If the back pitch is greater than the front pitch, as in Figs. 194 and 195, the winding advances clockwise when viewed from the commutator end and is called a *progressive* winding. If the back pitch is less than the front pitch, the winding advances counterclockwise when so viewed and is called a *retrogressive* winding.

It is obvious that the number of commutator segments is one-half the number of winding elements.

179. Multiple Coils.—In the larger sizes of dynamos, it is often necessary to place several coil sides or elements in one slot, usually 4, 6, or 8. For a given armature, if two elements per slot were used, the number of slots would be excessive and the slots would be too small. Figure 196 illustrates the placing of six elements in each slot. These elements are taped together and placed in the slot as a unit. The coil so formed is called a

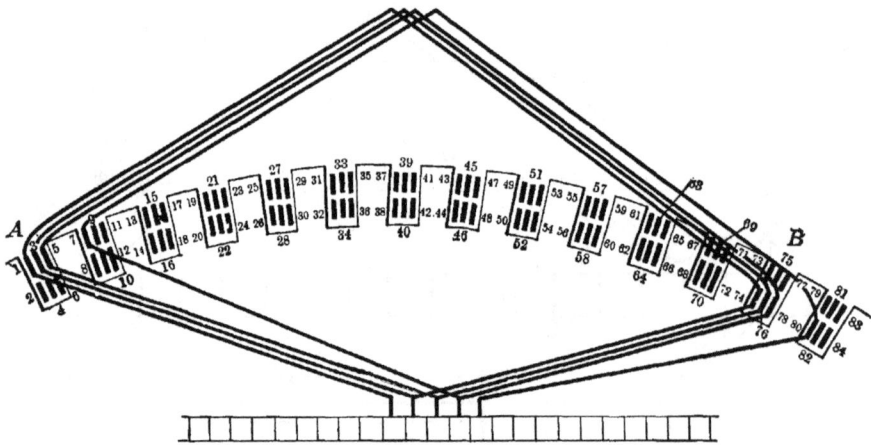

FIG. 196.—Connections of a triple coil.

multiple coil. The elements are numbered (Fig. 196) as if there were two elements per slot. The pitch, however, must be chosen so that the elements 1, 3, and 5 lie in the top of a single slot, and 74, 76, and 78 of the same coil lie at the bottom of another single slot. Thus, in Fig. 196, the back pitch is 73 so that elements 1, 3, 5, constituting one side of the triple coil, lie in the top of one slot and elements 74, 76, 78, constituting the other side of the triple coil, lie in the bottom of another slot.

180. Paths through an Armature.—If four batteries, each having an emf of 2 volts and a current rating of 10 amp, be connected in parallel (Fig. 197 (a)), there will be *four* paths for the current to follow in going through the batteries. The voltage of the combination will be 2 volts and the ampere rating will be 40 amp, making a total power rating of 80 watts. If now these same batteries be arranged in two groups of two in series (Fig. 197 (b)) there will be only two paths for the current to follow, but the voltage is now 4 volts. The current rating is now 20 amp, and the power rating is $4 \times 20 = 80$ watts, the same as before.

Similarly, the conductors in an armature may be connected so that certain groups of conductors are in series, and these groups may then be connected so that there are two or more paths in parallel. As with the batteries, the power rating of the dynamo is not changed by connecting the conductors in series and parallel groups, but the voltage and ampere ratings are changed. To determine the number of such parallel paths, start at one of the

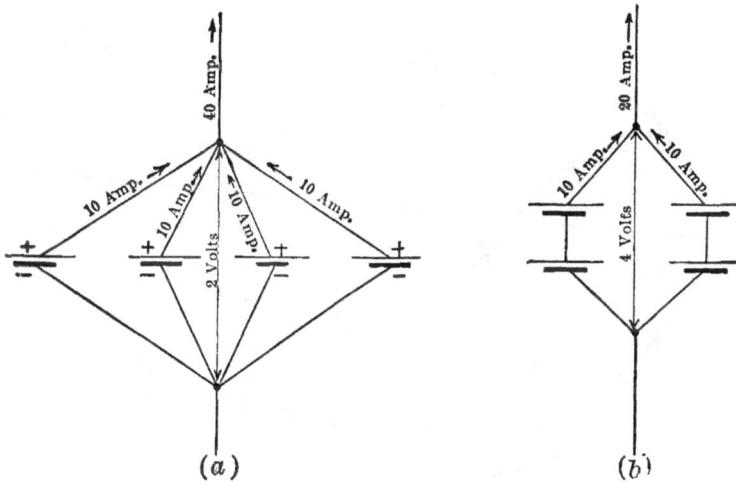

FIG. 197.—Parallel and series-parallel arrangement of batteries.

dynamo terminals, as, for example, the negative, and see how many different paths through the armature it is possible to follow in reaching the positive terminal.

To illustrate for a drum-wound armature having a lap winding, the 18-slot winding of Figs. 194 and 195, developed in circular form, is shown in Fig. 198. Remembering that the two negative brushes are connected together to form the negative terminal of the dynamo and the two positive brushes are connected together to form the positive terminal of the dynamo, there are four distinct paths between the negative and positive terminals of the dynamo. For the sake of clearness, two paths are shown with heavy lines, one from brush a to brush b, and the other from brush c to brush d. By tracing through the lighter lines, two more paths may be found, one between brushes c and b and the other between brushes a and d, making four paths in all.

In all simplex lap windings there are as many paths through the armature as there are poles.

From the foregoing, it is clear that with a given dynamo having a fixed number of armature conductors, the current-carrying capacity of the armature is directly proportional to the number of paths through the armature, and the voltage between the

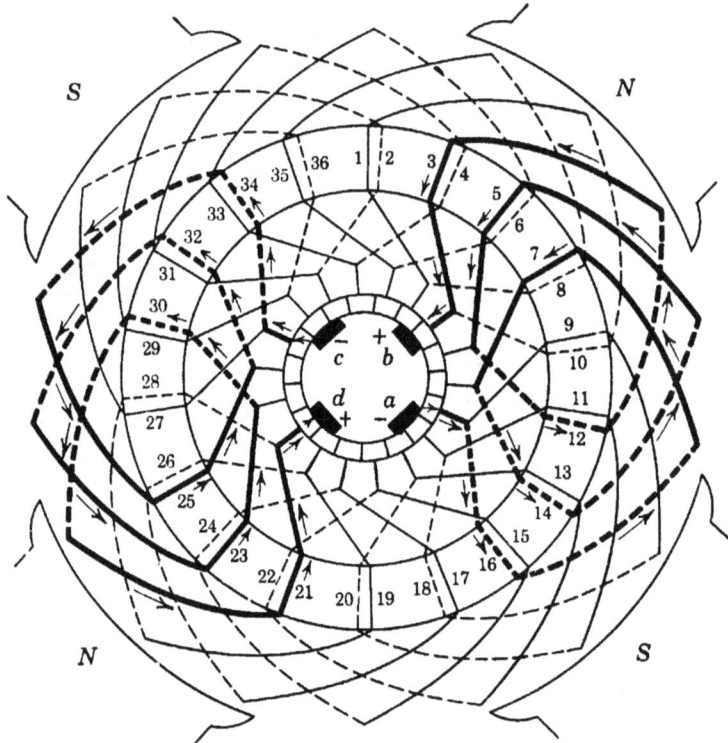

FIG. 198.—Heavy lines show two of the four parallel paths of a lap winding.

terminals of the machine is inversely proportional to the number of paths through the armature. The kilowatt or power rating remains unchanged.

181. Simplex Wave Winding.—It has been shown that, with the *lap winding*, a conductor under one pole is connected directly to a second conductor which occupies a nearly corresponding position under the next pole. This second conductor is then connected *back* again to a conductor under the *original* pole, but removed two or more conductors from the initial conductor. This is shown in Fig. 199 (*a*), where conductor *ab* under a north pole is connected to conductor *cd* having a corresponding position under the next south pole. Conductor *cd* is then connected back to *ef*, which is adjacent to *ab* under the original north pole. Obviously, it would make no difference, so far as the direction and magnitude of the induced emf in the winding are concerned, if the connection, instead of returning back to the same north pole, advanced *forward* to the next north pole, as shown in Fig. 199 (*b*). When the connection is made in this manner, the winding passes successively every north and south pole before it returns again to the original pole, as shown at *a'b'* in Fig. 199 (*b*). The winding, after passing once around the armature, reaches conductor *a'b'* lying under the same pole as the initial

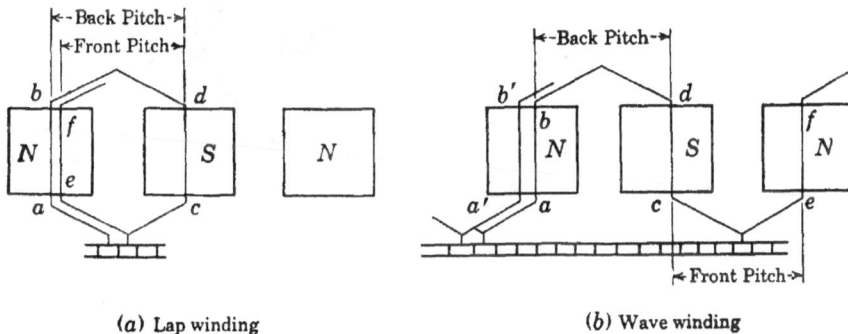

(*a*) Lap winding   (*b*) Wave winding

Fig. 199.—Lap and wave windings.

conductor *ab*. When a winding advances from pole to pole in this manner, it is called a *wave winding*. The number of units spanned by the end connections on the back of the armature is called the *back pitch* (Fig. 199 (*b*)). This corresponds to the back pitch in the lap winding (Fig. 199 (*a*)). The number of elements which the end connections span on the commutator end of the armature is the *front pitch* (Fig. 199 (*b*)). This should be compared with the front pitch in Fig. 199 (*a*). In the wave winding, as in the lap winding, *the front pitch and the back pitch must both be odd in order that one side of a coil may lie in the top of a slot and the other side in the bottom of a slot.* Unlike the lap winding, the front pitch may equal the back pitch in the wave winding. The average of the front pitch and the back pitch gives the *average pitch*.

The wave winding is much more restricted than the lap winding in the front and back pitch that can be used and in the number of slots that may be used with a given number of poles. This is illustrated in Fig. 200, which shows a six-pole dynamo. In (*a*) there are 34 commutator segments and slots; hence, there are 68 elements. The average pitch should be $^{68}\!/_6$, or 11, approximately. If both the front pitch and the back pitch are 11, and if element *ab* is numbered 1, the connections are made 1–12–23–34–45–56–67 or *a'b'*, which is two elements to the left of *ab* and in the top of the adjacent slot. This winding may be continued in this manner, until every slot is occupied by two elements and the winding closes on itself. Since this winding, after each passage

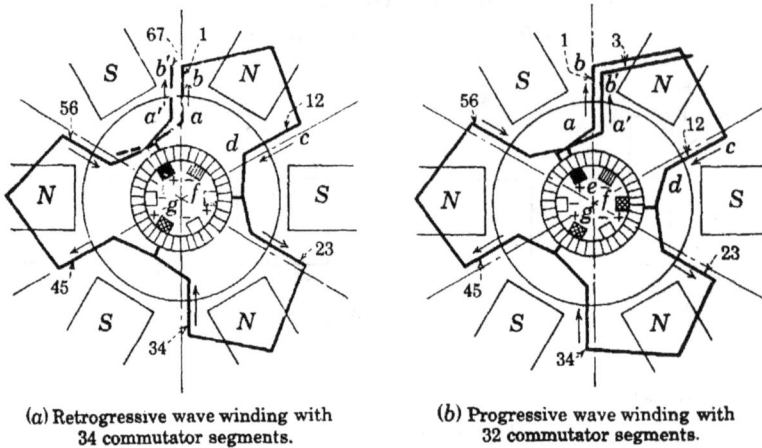

(*a*) Retrogressive wave winding with 34 commutator segments.

(*b*) Progressive wave winding with 32 commutator segments.

FIG. 200.—Wave windings.

around the armature, falls in a counterclockwise direction from its starting point, it is a *retrogressive* winding.

In Fig. 200 (*b*), there are 32 commutator segments and 32 slots; hence, there are 64 winding elements. Let conductor *ab* be numbered 1 and let both front and back pitch be 11. The winding connections are then 1–12–23–34–45–56–3 or *a'b'*, which is two elements to the right of *ab* and in the top of the adjacent slot. Again, this winding will close on itself after every slot on the armature is filled with two elements. Since this winding, after each passage around the armature, falls in a clockwise direction from its starting point, it is a *progressive* winding.

If this armature had 33 slots or 66 winding elements, the connections would be 1–12–23–34–45–56–67 (or 1). Hence, the winding would close on itself after one passage around the armature and would therefore not be a possible winding. That is, the winding would close on itself when only *one* slot under each pole contained only a *single* element, whereas the winding should close on itself only after *every* slot contains *two* elements. Again, if the winding closed on itself after a single passage around the armature, three coils, all of which are generating emf (Fig. 200), would form a closed circuit as shown above. That is, the coils would be short-circuited, and a large current would circulate within the armature coils.

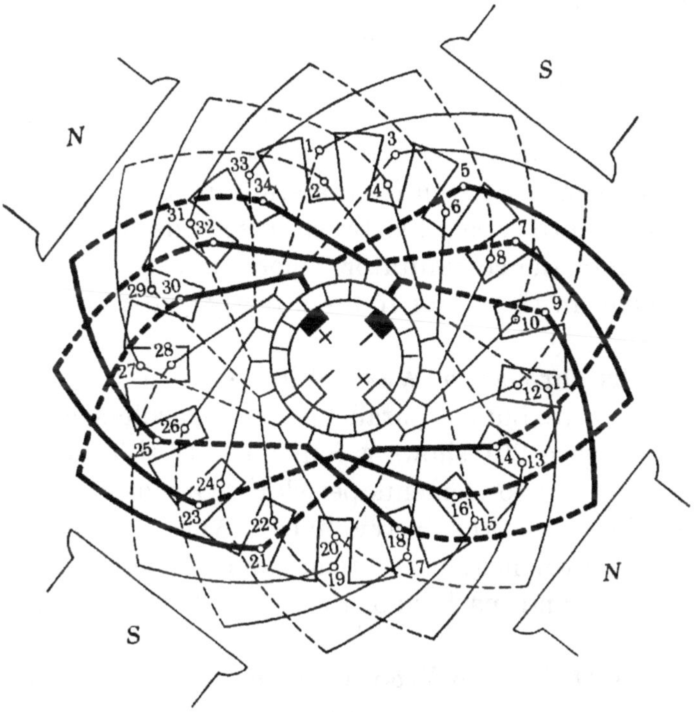

Fig. 201.—Seventeen-slot, four-pole, simplex wave winding: back pitch = 9, front pitch = 7; one of two parallel paths shown with heavy lines; the other with light lines including coils short-circuited by brushes.

If a wave winding were necessary under the same conditions of 33 slots, a *dummy* or *idle* coil would be used. This consists merely of one of the regular coils which is *not* connected to the commutator, but has its ends taped. It merely acts as a filler. The removal of the coil reduces the active winding elements to 64, which, as has just been shown, makes a possible winding.

A complete wave winding for a four-pole, 17-slot dynamo is shown in Fig. 201.

182. Number of Brushes with Wave Winding.—In Fig. 200 (*a*), three positive brush sets *e*, *f*, *g* are shown. Each rests at the instant shown on commutator segments connected to the conductors *ab*, *cd*, etc.   At this instant, these conductors lie practically in the neutral plane of the machine and there is practically no emf induced in them.   Therefore, brush sets *e*, *f*, and *g* are connected together through the armature by *idle* conductors.   Consequently, but one positive brush set is necessary to carry the current from the armature, since any one positive brush set is connected by idle conductors to all the points in the winding from which current should be collected (also see Fig. 201).

All three brush sets would be used ordinarily, since more current could be collected with a commutator of given length.

The foregoing applies equally well to the number of negative brush sets, as may be seen from a study of Fig. 201.

In railway motors, only two brush sets are used, and are located near the top of the commutator, where they are accessible from the handholes in the motor frame directly under the floor of the car.

183. Paths through a Wave Winding.—In a simplex wave winding, there are but *two* parallel paths, regardless of the number of poles.   Figure 201 shows a four-pole, 17-slot, simplex wave winding, having two elements per slot.   One of the two parallel paths is shown by the heavy lines.   Approximately half the winding is shown heavy, the remainder of the winding constituting the second path and the coils short-circuited by the brushes.

184. Uses of the Two Types of Winding.—With a given number of poles (except for two poles) and armature conductors, the wave winding gives a higher emf than the lap winding, since there are fewer parallel paths.   This type of winding is used, therefore, in small dynamos, especially in those designed for 600-volt circuits, where a lap winding would require a very large number of small conductors.   This, in turn, would involve a higher winding cost and less efficient utilization of the space in the slots.

The wave winding has the additional advantage that the emf in each path is produced by series-connected conductors, which lie under successive north and south poles.   Any magnetic unbalance, therefore, due to such causes as air-gap variation and difference in pole strength, does not produce cross-currents, because the corresponding conductors of each and every path are moving by the same poles at every instant and the effect of such unbalancing will be the same in each path.

When large currents are required, the lap winding is more satisfactory, since it gives a large number of parallel paths. This is particularly true of large, engine-driven, multipolar generators.

DYNAMO CONSTRUCTION

185. Armature.—The cross-section of a typical dynamo is shown in Fig. 202. The armature iron consists of circular sheet-steel stampings or laminations such as are shown at (*b*), which are assembled and clamped over a cast-iron spider shown in (*a*) and (*b*). The spider is keyed to the shaft. Laminated construction is necessary for an armature, as otherwise the eddy-current loss due to the armature's rotating in the magnetic field would be prohibitive. The laminations are usually separated at intervals by spacers to give ventilating ducts which permit the passage of air into the spider and radially outward, thus cooling the iron and the winding. In the generator shown in Fig. 202 there is only one such duct.

Figure 203 shows an armature in the process of winding. The slots are first lined with fish paper, which is a thin, tough fiber, and the coils, already formed and taped, are then forced into the slots. The coils are held in the slot by either wooden or fiber wedges or with banding wire. The coils are usually impregnated with an insulating varnish or with asphaltum.

186. Yoke or Frame.—The yoke or frame of a dynamo has two functions. It forms a part of the magnetic circuit, and it acts as a mechanical support for the machine as a whole. In small dynamos where weight is of little importance, the yoke is frequently a steel or iron casting. However, in the modern dynamo it is the practice to fabricate, so far as possible, the parts of the dynamo from steel plate. Thus, in the dynamo shown in Fig. 204, the yoke is made by rolling a steel plate about a cylindrical mandrel and then welding. Either the feet are of steel stampings riveted or welded to the frame or else they are fabricated from steel pieces welded together and then welded to the frame (Fig. 204). In large dynamos the yoke is frequently made in two sections which are bolted together. The top section is easily removed, facilitating shipment and permitting easy removal of the armature.

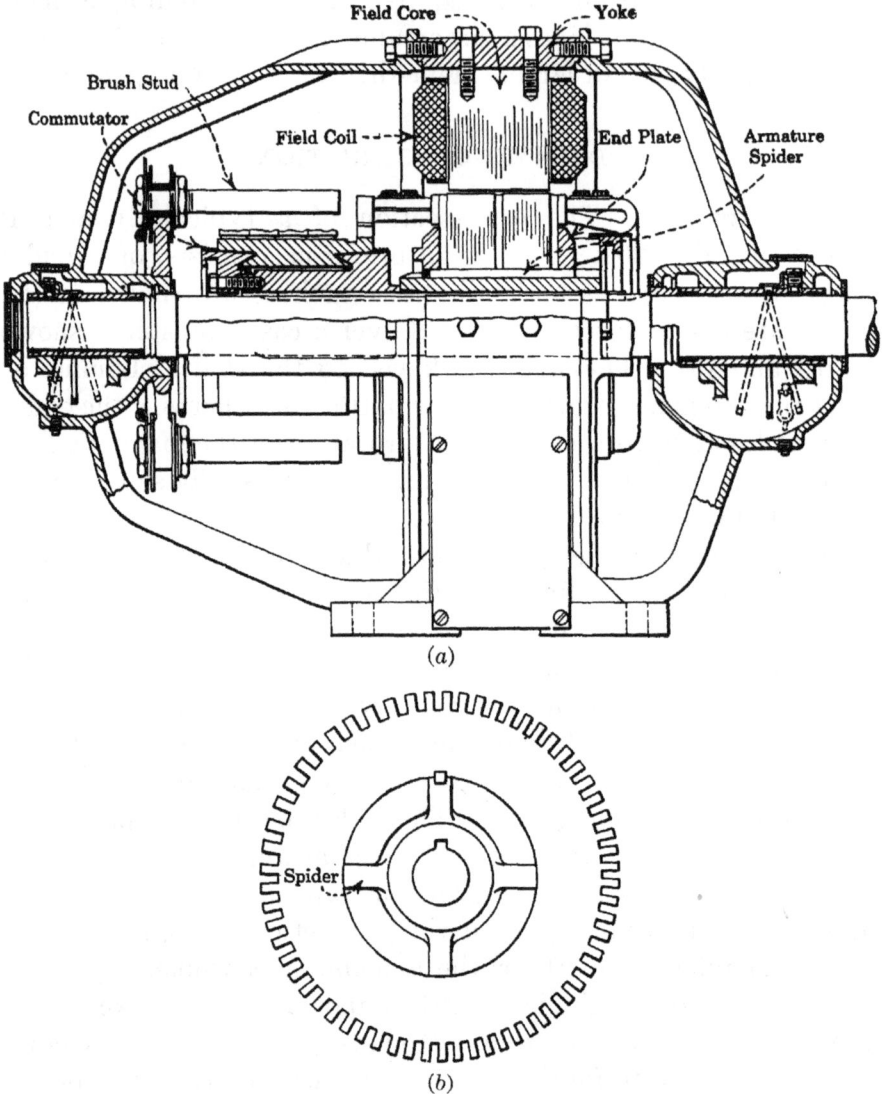

Fig. 202.—(a) Direct-current generator with shield bearings, longitudinal section; (b) armature stamping.   (*General Electric Company.*)

187. Field Cores.—Field cores are made of forged steel and cast steel. In the moderate and larger sizes of dynamos it is now the practice to make the field cores of steel laminations (Fig. 205). When made of forged or cast steel, the cores are usually circular in cross-section, as such a section permits the

minimum length of turn for a given core cross-section. The
advantages of the laminated construction are that eddy-current

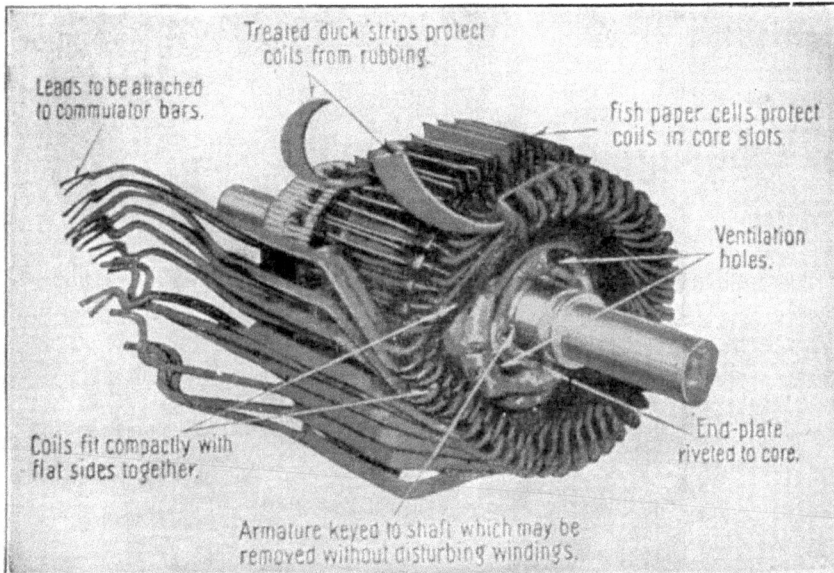

Leads to be attached to commutator bars.

Treated duck strips protect coils from rubbing.

fish paper cells protect coils in core slots.

Ventilation holes.

Coils fit compactly with flat sides together.

End-plate riveted to core.

Armature keyed to shaft which may be removed without disturbing windings.

FIG. 203.—Partly wound armature, showing method of assembling coils. (*West-inghouse Electric & Manufacturing Company.*)

FIG. 204.—Type SK d-c motor with fabricated yoke and feet. (*Westinghouse Electric & Manufacturing Company.*)

losses in the pole face are reduced, and in addition there
are mechanical advantages. In some designs the laminations

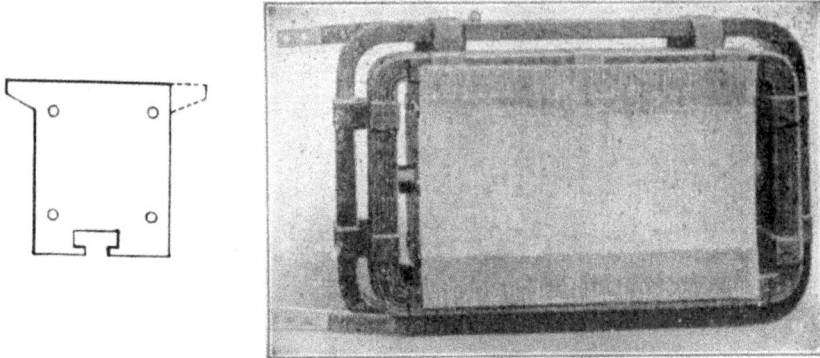

Fig. 205.—Field-core lamination and pole piece assembled—Westinghouse d-c motor.

Fig. 206.—Shunt-field coil and edgewise series winding.

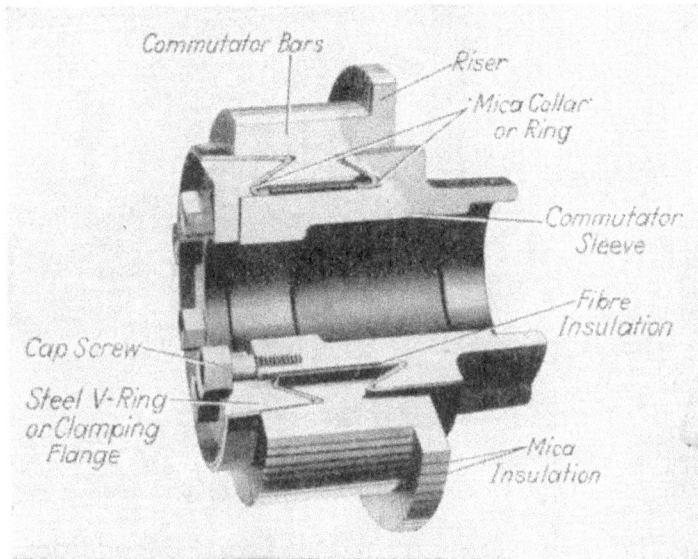

Fig. 207.—Commutator construction. (*Reliance Electric and Engineering Company.*)

have only one pole tip (Fig. 205) and they are stacked so that the pole tip comes alternately on one side and the other. This results in there being but half the iron in a pole-tip cross-section, thus producing a saturated pole tip, which assists commutation. After being stacked to the proper thickness, the laminations are

Fig. 208.—Field structure and brush rigging for low-speed, commutating-pole generator. (*General Electric Company.*)

riveted together. The poles are usually held to the yoke by bolts (Figs. 204 and 208).

188. Field Coils.—The field coils are usually wound with cotton-covered enamel wire. The coils are dried in a vacuum and are then impregnated with an insulating compound. The outer cotton insulation is often protected by tape or cord. In the larger dynamos, an air-space is often left between layers of the winding for ventilating purposes. The coils are also wound on metal spools (Fig. 206). An edgewise series winding, set some distance from the shunt winding, is shown also in Fig. 206.

189. Commutator.—The commutator is made of wedge-shaped segments of hard-drawn or drop-forged copper, insulated from one another by thin layers of mica. The segments are held together by steel V-rings or clamping flanges (Fig. 207), which pull the segments inward when the flanges are drawn together by

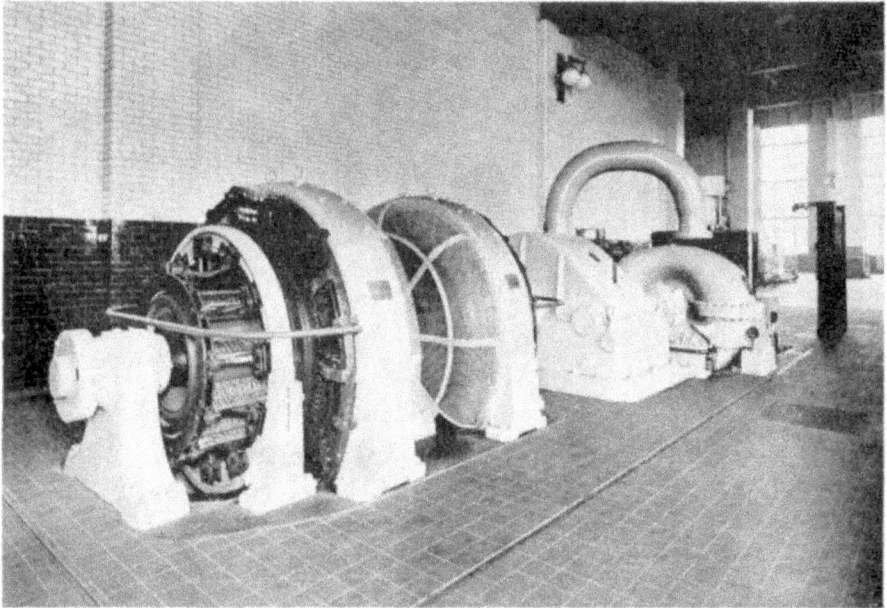

Fig. 209.—Turbine-driven 3,000-kw d-c generator unit. (*General Electric Company.*)

the through bolts. These flanges are prevented from short-circuiting the segments by two cone-shaped collars or rings of built-up mica (Fig. 207). This construction is illustrated by the commutator of the machine shown in Fig. 202, p. 228.

The leads from the armature coils may be soldered into small, longitudinal slits in the ends of the commutator segments; or the segments may have risers (Figs. 202 and 207), to which these leads are soldered.

190. Brushes.—The function of the brushes is to conduct the current from the commutator to the external circuit. They are made usually of carbon, although in very low-voltage machines they may be made of copper gauze or of patented metal compounds. The brush should be free to slide in its holder, in order that it may follow any irregularities in the commutator. The brush is made to bear down on the commutator by a spring. The pressure should be from 1 to 2 lb per sq in. To decrease the electrical resistance, the upper portion of the brush is copper-plated (see p. 87) and this plating is connected to the brush holder by a pigtail made of copper ribbon. Figure 208 shows the field structure and brush rigging of a low-speed, commutating-pole generator, designed for direct connection to a steam or internal-combustion engine. The copper rings which conduct the current from the brushes to the flexible cables should be noted.

Figure 209 shows a General Electric Company 3,000-kw, turbine-driven, d-c generator unit. The unit consists of two generators on the same shaft driven from the turbine through reduction gears.

CHAPTER XI

GENERATOR CHARACTERISTICS

The principles involved in the generation of emf in an armature
and the method by which direct current is obtained from an alternating emf are discussed in Chap. X. It remains to consider
the induced emf from the *quantitative* point of view and to analyze
the behavior of the various types of generator in operation.

191. Electromotive Force Induced in Armature.—In Chap. X
it is shown that if a conductor *cuts* a magnetic field an emf is
generated or induced in the conductor. It is also shown that
if conductors are placed on the surface of an armature and are
properly connected, their induced emfs add. Further, if unidirectional or direct current is desired from such an armature,
the conductors on the armature surface must be connected to a
commutator. From Eq. (86), p. 210, ($e = Blv10^{-8}$ volts), the
magnitude of the emf between brushes can be computed. Equation (86), however, involves the flux density, the length of
conductor, and its linear velocity. For a generator, these three
factors can be determined only after some computation, whereas
the total flux per pole and the rotational speed of the armature
are usually known or can be determined readily. Therefore, by
proper substitutions in Eq. (86), an equation may be derived
which is easily applied to a generator armature.

In a generator, the total flux ϕ per pole is proportional to the
average flux density B under the pole, and the total flux cut by a
conductor in each revolution is proportional to the number of
poles P. The length l of the active conductors is a constant,
and the linear velocity v of the conductor is proportional to the
speed of the armature in rpm. If all the conductors Z on the
armature surface were connected in series, the emf induced in
the armature would be proportional to Z. However, as shown in
Chap. X, there are two or more parallel paths P' through the
armature, so that the emf between brushes is proportional to
Z/P'.

By substituting these values in Eq. (86), the following equation for the total induced emf between brushes is obtained, where ϕ is the total flux in maxwells leaving a north pole and entering the armature, S is the speed of the armature in rpm, P is the number of poles, Z is the total number of conductors on the surface of the armature, and P' is the number of parallel paths through the armature.

$$E = \frac{\phi SPZ}{60P'10^8} \text{ volts.}^* \tag{87}$$

As S is given in rpm the quantity 60 in the denominator is necessary to give rps.

With a simplex lap winding, $P' = P$; with a simplex wave winding, $P' = 2$.

Example.—A 900-rpm, six-pole generator has a simplex lap winding. There are 300 conductors on the armature.

The poles are 10 in. square, and the average flux density under the poles is 50,000 lines per sq in.   What is the emf induced between brushes?

$\phi = 10 \times 10 \times 50,000 = 5,000,000$ lines.
$S = 900$ rpm
$P = 6.$
$P' = 6$ (see Par. 180, p. 222).
$E = \dfrac{5,000,000 \times 900 \times 6 \times 300}{60 \times 6 \times 10^8} = 225$ volts.   *Ans.*

192. Saturation Curve.—Equation (87) may be written as follows:

$$E = \left(\frac{PZ}{60P'10^8}\right)\phi S. \tag{88}$$

For a given generator, the quantity within the parenthesis is constant and may be denoted by K.

Hence,

$$E = K\phi S. \tag{89}$$

The induced emf in a generator, therefore, is *directly proportional to the flux and to the speed.*

* The complete derivation of this equation is given in C. L. DAWES, "A Course in Electrical Engineering," 3d ed., Vol. I, Chap. XII, Direct Currents.

If the speed be kept constant, the induced emf is directly proportional to the flux ϕ.

The flux is produced by the field ampere-turns, and as the turns on the field remain constant, the flux depends on the field current. It is not directly proportional to the field current because of the varying permeability of the magnetic circuit.

Figure 210 shows the relation existing between the field ampere-turns and the flux per pole. Ordinarily, the flux is

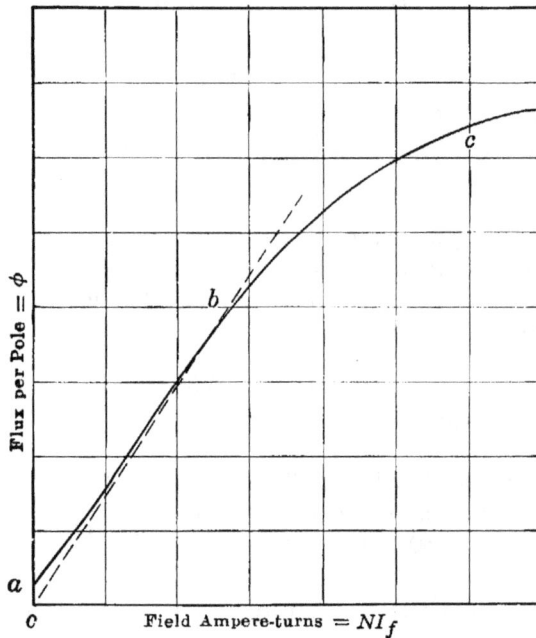

FIG. 210.—Saturation curve.

not zero when the field current is zero but has some value, as *oa*, because of the residual magnetism in the magnetic circuit. At first the line *ab* is practically straight, since most of the reluctance of the magnetic circuit is in the air-gap. At the point *b* the effect of saturation begins and the curve falls away from the straight line.

From *b* to *c* and beyond, the iron is close to saturation and the flux increases very slowly with further increase in field current. In fact, beyond point *c*, it is practically impossible to obtain an appreciable increase in flux without a very large increase in field current. (Compare this curve with the curves of Fig. 147, p. 164.)

From Eq. (89), the induced emf is proportional to the flux, if the speed is maintained constant. Therefore, if the induced emf be plotted as ordinates with field current as abscissas, a curve similar to that of Fig. 211 is obtained. In Fig. 211, a saturation

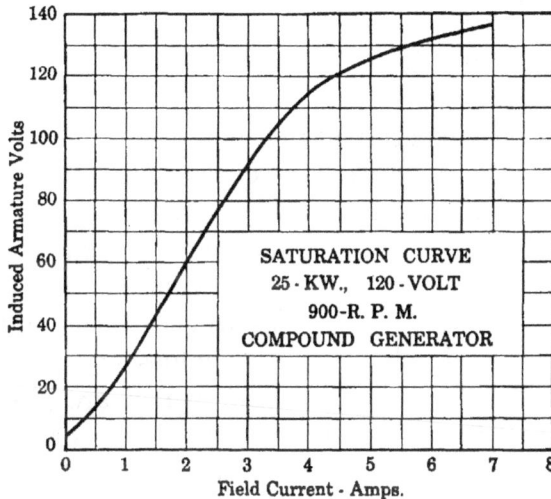

FIG. 211.—Saturation curve of 25-kw, 120-volt, 900-rpm generator.

curve is shown for a 25-kw, 120-volt, 900-rpm, compound generator. This curve shows that with a shunt-field current of 3.3 amp, an emf of 100 volts is induced in the armature when the speed is 900 rpm. With a field current of 4.4 amp, 120 volts is induced in the armature at 900 rpm.

The saturation curve has a considerable effect on the operation of all generators and motors, as will be shown later.

193. Hysteresis in Dynamos.—The saturation curve *Oab* (Fig. 212) is determined for *increasing* values of field current. If, when point *b* is reached, the field current be *decreased*, the curve *baO* will not be followed. For any given field current, the corresponding

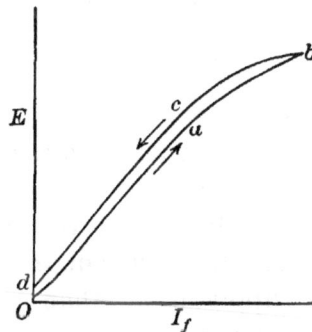

FIG. 212.—Hysteresis loop of magnetic circuit of dynamo.

induced emf will now be greater than it was for *increasing* field current, as is shown by the curve *bcd*. This is due to hysteresis in the iron of the magnetic circuit. (See Par. 140, p. 165.)

For a given value of field current, there is no single value of flux. The value of flux for a given field current depends on whether the field current was *increased* until it reached the given value or whether it was *decreased*. This characteristic of the magnetic circuit should be carefully kept in mind, for the operating characteristics of both generators and motors are affected to a considerable degree by hysteresis in the magnetic circuit.

194. Determination of Saturation Curve.—To determine the saturation curve experimentally, connect the field, in series with an ammeter, across a d-c source of power. A voltmeter should be connected across the armature terminals. The ammeter measures the field current, values of which are plotted as abscissas; the voltmeter reads the values of induced armature emf, which are plotted as ordinates. The connections are shown in Fig. 213. As the voltage drop due to the voltmeter current is negligible within the armature, the terminal volts and the induced emf under these conditions are equal. During the experiment, the speed should be determined each time the ammeter and voltmeter readings are taken. If the speed cannot be maintained constant, corrections can be made for any variation, since the induced emf is directly proportional to the speed (Eq. (89), p. 235).

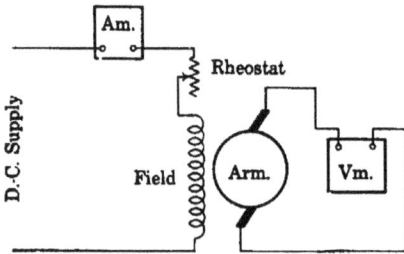

Fig. 213.—Connections for obtaining saturation curve.

In determining the saturation curve experimentally, the field current should be varied continuously in *one direction*, either with increasing values or with decreasing values, as shown in Fig. 212. Otherwise, hysteresis effects will be introduced.

The field current in this experiment should be obtained from a supply other than the generator itself, for two reasons. If the generator excited its own field, the emf and the field current would be interdependent so that it would be difficult to adjust the field current without the emf in turn changing this adjustment. Also, a voltage drop would exist in the armature due to the field current, and the voltmeter would not give the true induced emf, although the error from this cause would be small.

195. Types of Generator.—There are three general types of generator, the shunt, the compound, and the series. In the shunt type the field circuit is connected across the armature terminals, in shunt with the load (Fig. 214). A shunt-field rheostat is usually connected in series with the shunt field. The shunt field must have a comparatively high resistance in order that it may not take too great a proportion of the generator current.

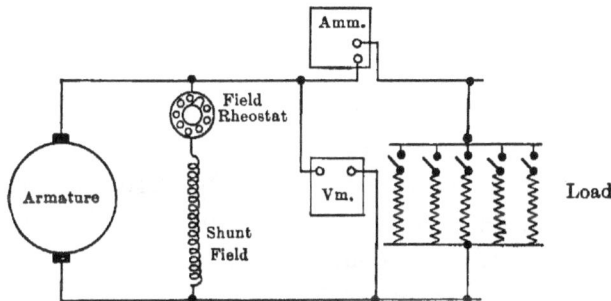

FIG. 214.—Shunt-generator connections.

The compound generator is similar to the shunt type, but has an additional field winding connected in series with the armature or with the load. The series winding is ordinarily connected so that it aids the shunt-field winding and hence increases the flux with increase of load (Fig. 227, p. 257). The series generator is excited by a winding of comparatively few turns connected in series with the armature or load. (See p. 258.)

SHUNT GENERATOR

196. Building Up of Shunt Generator.—The shunt generator excites its own field. When the generator is started, there is no voltage across its terminals; hence, there is no field current. Ordinarily, however, there is some residual magnetism or flux in the magnetic circuit of the machine due to the retentivity of the iron. This is shown by the ordinate *oa* (Fig. 210, p. 236). As the generator comes up to speed, a small induced emf appears at its terminals due to the cutting of this residual flux by the armature conductors. Therefore, a small current flows through the field. If the field resistance is sufficiently low and the connections of the field to the armature terminals are such that the small field current due to this small terminal voltage tends to *increase* the flux due to the residual magnetism, the emf of the generator will build up of its own accord.   Under the foregoing conditions, this initial field current, due primarily to residual magnetism, increases the flux in the magnetic circuit.   This increase of flux causes an increase in the induced emf.   This, in turn, causes further increase in the field current.

These reactions are obviously cumulative.   That is, an increase in field current produces an increase in emf, and the increase in emf produces a further increase in field current.   The emf or terminal voltage increases to some definite value.

It might seem from the foregoing analysis that the terminal voltage would build up indefinitely.   That it does not do so is due to the saturation of the magnetic circuit.   When the flux reaches the value indicated by *b* (Fig. 210, p. 236), a given increase in field current produces a much smaller proportionate increase in flux.   As the field current increases still further, the rate of increase of flux diminishes, until a point is reached where the increase of flux is extremely small, even with large increase in field current.   For this reason, the voltage to which a generator can built up is *limited by the saturation of the iron*.   If the saturation curve were a straight line, a shunt generator would build up indefinitely.

197. Failure of Generator to Build Up.—A shunt generator may fail to build up for one of the following reasons:

1. *Lack of Residual Magnetism.*   In time, the residual magnetism may diminish to a very small value, particularly if the machine has been idle and has been subjected to vibration.   A reversed field connection, at some time, may have opposed or "bucked" the residual magnetism and caused it to diminish to a value so small that the field current resulting from the initial induced emf is insufficient to produce an appreciable increase in the flux. These conditions are indicated when a voltmeter connected across the armature terminals reads zero practically.

The remedy is to "flash" the generator field.   The shunt field is connected across another source of voltage, to build up the flux due to residual magnetism.   If the dynamo is a compound generator, a convenient method is to connect across the series field a dry cell, in the smaller units, or a low-voltage storage battery, in the larger units, the shunt field being connected across the armature terminals.   If the polarity of the battery is correct, the current which it sends through the series field usually causes the generator to begin to build up.

2. *Too High Shunt-field Resistance.* If the resistance of the shunt field is too great, the generator will not build up. When this condition exists, the voltmeter reading after the field circuit is closed increases but slightly above its initial reading and then ceases to increase. The remedy is to reduce the resistance of the shunt-field circuit by cutting out resistance in the field rheostat.

3. *Reversed Shunt-field Connections.* If the connections of the shunt field are such that the initial field current, primarily due to residual magnetism, opposes or "bucks" the residual magnetism, the generator cannot build up. When this condition exists, the voltmeter reading *decreases* when the shunt-field connection is made. The remedy is to reverse the shunt-field connections. This condition of reversed field connection rarely occurs with a shunt generator which is installed in service and in which the field connection has already been made properly. With compound generators, a flashover, an open field in the generator, or other conditions may cause the line current to reverse and enter the armature at its positive terminal. The current in the series field is reversed, therefore; and as the current under these conditions is usually large, the series-field ampere-turns may exceed the shunt-field ampere-turns and cause a reversal of flux.

4. *Shunt Field Open-circuited.*—Occasionally the shunt-field circuit becomes open-circuited, due to a burnt-out rheostat, poor contact, etc. This condition is indicated by there being no change in the voltmeter reading when the field connection to the armature terminals is made and then broken. The remedy is to test the field circuit and locate the difficulty.

198. Armature Reaction.—Figure 215 (*a*) shows the flux from the field poles and through the armature, when there is no current in the armature conductors. This flux is due entirely to the ampere-turns of the field. The neutral plane, which is the plane perpendicular to the flux, coincides with the geometrical neutral of the system. At the right is shown a vector *F* which represents in magnitude and direction the mmf producing the field flux. At right angles to this vector *F* is the neutral plane.

In Fig. 215 (*b*), there is no current in the field coils, but the armature conductors are shown as carrying current. This current is in the same direction in the armature conductors as it would be if the generator were under load, with north and south

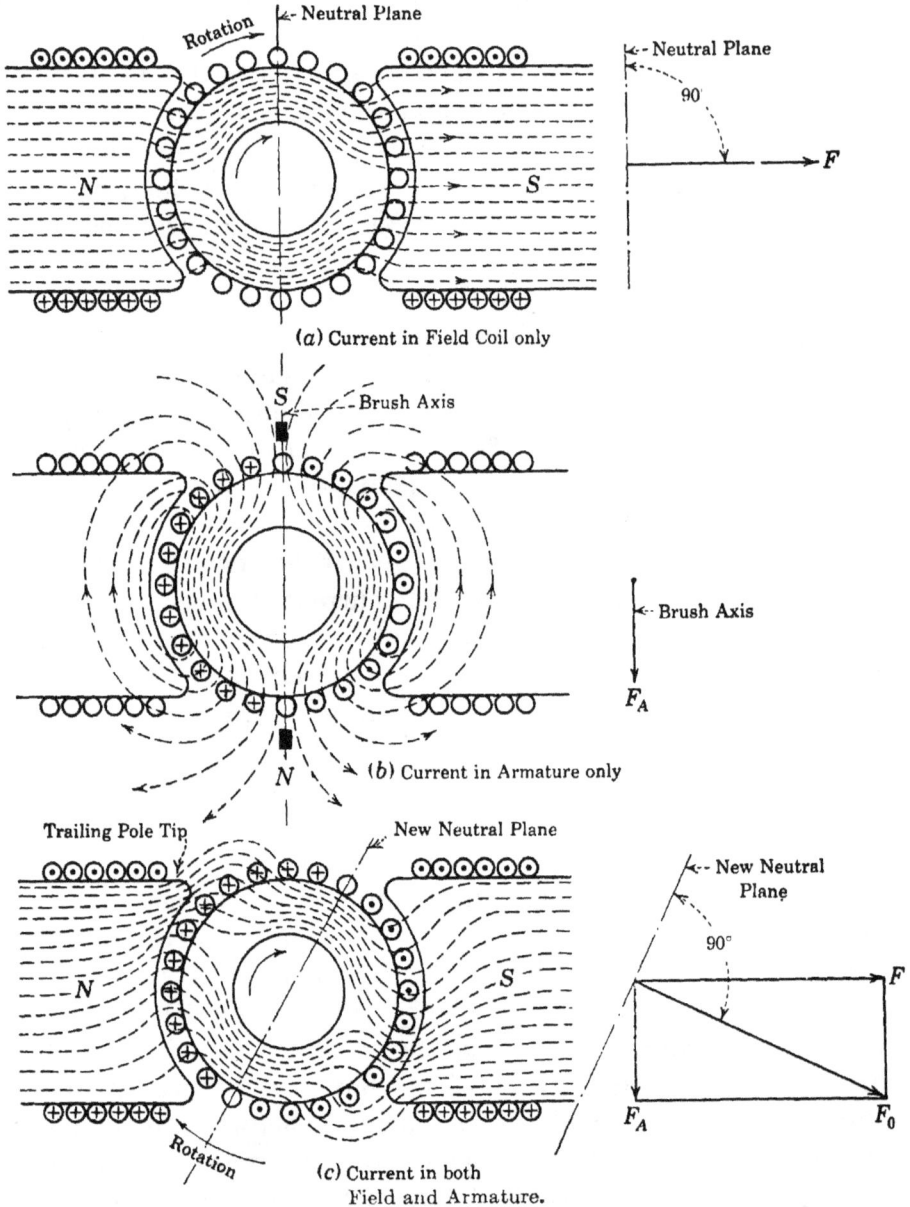

(a) Current in Field Coil only

(b) Current in Armature only

(c) Current in both Field and Armature.

FIG. 215.—Effect of armature reaction on the field of a generator.

poles and direction of rotation as shown in (a) and (c). The current flows in the same direction in all the conductors that lie under one pole. The current is shown as flowing into the paper on the left-hand side of the armature. (This current direction may be checked by Fleming's right-hand rule, p. 211.) The mmfs due to the conductors on the left-hand side of the armature combine to send a flux *downward* through the armature, as shown in the diagram, this direction being determined by the corkscrew rule. The conductors on the right-hand side of the armature are shown as carrying current coming out of the paper and the mmfs due to these conductors combine to send a flux also *downward* through the armature. That is, the mmfs due to the conductors on both sides of the armature combine in such a manner as to send flux downward through the armature. The direction of this flux is perpendicular to the main polar axis. To the right of the figure this armature mmf is represented in magnitude and direction by the vector F_A.

Figure 215 (c) shows the result obtained when the field current and the armature current act simultaneously, which occurs when the generator is under load. The armature mmf crowds the symmetrical field flux shown in (a) into the upper pole tip in the north pole and into the lower pole tip in the south pole. As the generator armature is shown rotating in a clockwise direction, the flux is crowded into the *trailing* pole tip of each pole so that the trailing pole tip is strengthened and the *leading* pole tip is weakened.

The effect of the armature current is to displace the field in the direction of rotation of the generator. It should be kept firmly in mind that the flux is not pulled around by the mechanical rotation of the armature.

To the right of Fig. 215 (c) the effect of armature reaction is shown by vectors. The field vector F and the armature vector F_A combine at right angles to form the resultant field vector F_o. The direction of F_o is downward and to the right, which corresponds to the general direction of the resultant flux. The neutral plane must be at right angles to F_o, provided the direction of the resultant flux is the same as the direction of the resultant mmf.

As the neutral plane is perpendicular to the resultant field, this plane, also, has been advanced. It is shown in Chap. X that the brushes should be set so that they short-circuit the coil undergoing commutation as the coil is passing through the neutral plane. (When the generator delivers current, the brushes should be set a little ahead of this neutral plane, as will be shown later.) If the brushes are advanced to correspond to the advance of the neutral plane (Fig. 215 (c)), all the conductors at the left of the brush axis must carry current into the paper, and the conductors at the right must carry current out of the paper.

Figure 216 (a) shows that portion of the ampere-conductors which are included within an angle β on each side of the geo-

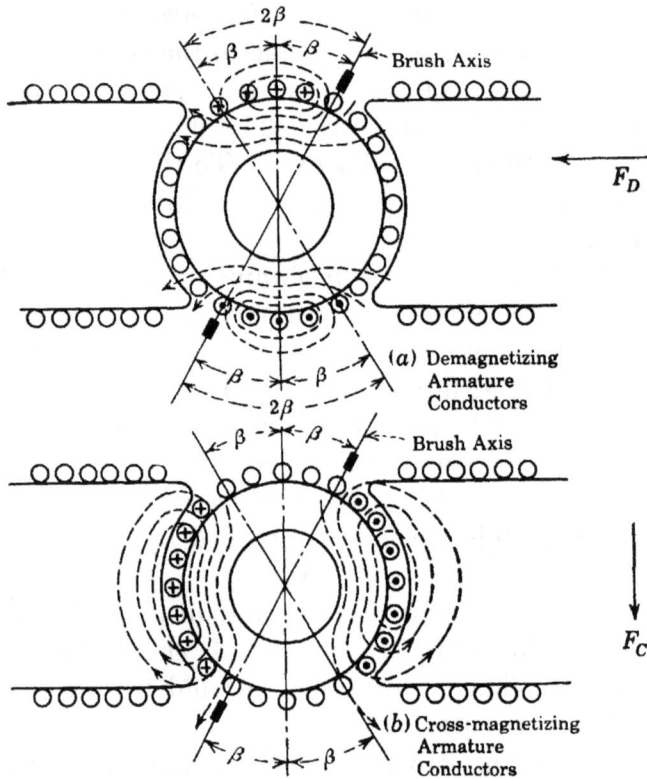

FIG. 216.—Demagnetizing and cross-magnetizing components of armature reaction.

metrical neutral, where β is the angle of brush advance. The conductors at the top of the armature carry current into the paper, since they are at the left of the brush axis (see Fig. 215 (c)). By the corkscrew rule, their mmf through the armature acts from right to left. The conductors at the bottom of the armature carry current out of the paper, since they are at the right of the brush axis. Their mmf through the armature acts also from right to left. Referring to Fig. 215 (c), the mmf of the main poles acts from left to right. Hence, the mmf of the ampere-conductors which are included within twice the brush angle at both top and bottom of the armature *opposes* the mmf of the main field and therefore tends to *reduce* the flux through the armature. These conductors are called *demagnetizing ampere-conductors.*

Figure 216 (*b*) shows the armature ampere-conductors of Fig. 215 (*c*) which are not included in Fig. 216 (*a*). By the corkscrew rule, the mmf of the conductors on both left-hand and right-hand sides of the armature acts downward and hence acts at right angles to the mmf of the main field as in Fig. 215 (*b*). Therefore, these conductors are called *cross-magnetizing ampere-conductors.*

Therefore, if the generator is carrying load and the brushes are advanced, the armature ampere-conductors tend both to *demagnetize* and to *cross-magnetize* the main magnetic field. These demagnetizing and cross-magnetizing mmfs in the armature constitute *armature reaction.* Armature reaction has a very important effect in the operation of all generators and motors.

Example.—In a bipolar generator with a drum winding there are 36 slots on the armature and 12 conductors per slot. The current from the armature is 60 amp. The brushes are advanced 12 space-degrees. Determine: (*a*) the demagnetizing ampere-conductors; (*b*) the demagnetizing ampere-turns; (*c*) the cross-magnetizing ampere-conductors; (*d*) the cross-magnetizing ampere-turns.

Since the armature current is 60 amp and there are two parallel paths, the current per path is 30 amp.

Total armature conductors are $36 \times 12 = 432$.

a. The angle β is 12 deg. The total demagnetizing ampere-conductors are therefore

$$\frac{4 \times 12°}{360°} \times 432 \times 30 = 1{,}728 \text{ ampere-conductors.} \quad Ans.$$

b. Demagnetizing ampere-turns are 1,728/2 or 864. *Ans.*

c. The remaining ampere-conductors must be cross-magnetizing.

$$360° - 48° = 312°.$$

Hence, the cross-magnetizing ampere-conductors are

$$\frac{312°}{360°} \times 432 \times 30 = 11{,}232 \text{ ampere-conductors.} \quad Ans.$$

d. Cross-magnetizing ampere-turns are

$$\frac{11{,}232}{2} = 5{,}616. \quad Ans.$$

The effect of armature reaction may be diminished by increasing the length of the air-gap. This increases the reluctance of the path of the armature flux, and of the field flux as well, and necessitates an increase in the field ampere-turns. Hence, it follows that by increasing the magnetic strength of the field relative to that of the armature the effect of armature reaction is reduced. Operating the tooth tips at high saturation also reduces the effect of armature reaction. (See Fig. 205, p. 230.) Extra windings on the field structure, which are connected in series with the

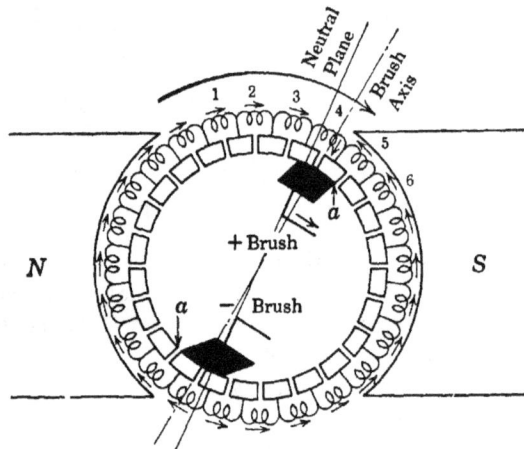

Fig. 217.—Commutation in a generator.

armature and oppose the mmf of the armature, are sometimes used to reduce the effect of armature reaction.

199. Commutation.—It has been shown that the emf induced in any single coil of a d-c generator is alternating and in order that the current may flow always in the same direction to the external circuit a commutator is necessary.

Figure 217 shows diagrammatically a bipolar generator with a Gramme-ring winding. The armature rotates in a clockwise direction. The currents in both sides of the armature flow upward in the diagram and combine to flow out of the positive brush into the external circuit. The figure shows the direction of the currents in all the coils at any instant, and hence the direction of the current in any one coil at successive instants. For example, consider coil 1 as it takes successive positions 2, 3, 4, 5, and 6. For all positions between brushes on the left-hand side of the armature, the current is upward. For all positions between brushes on the right-hand side of the armature, the current is upward. Therefore, in the interval between positions 3 and 5, the current in any coil must have been reversed. This reversal in direction must occur in the very brief interval of time during which the commutator segments to which a coil is connected are passing under the brush. Since the armature coils are embedded in slots and are nearly surrounded by the iron of the armature, they have substantial self-inductance. As is shown in Chap. VIII, self-inductance *opposes* any change of current. When the current in the armature coil changes, an emf of self-induction is produced which tends to *prevent* the change and to prolong the current flow in a given direction. Therefore, if the brushes are in the neutral plane (see Par. 175, p. 211) and the generator is carrying load, sparks will ordinarily appear under the brush and particularly under the toe (*a, a,* Fig. 217) where the commutator segments are breaking contact with the brush. These sparks are due to the fact that self-induction has delayed the complete reversal of the current in the coil up to the instant that the toe of the brush breaks contact with the segment. Hence, at this instant a sudden change in the value of the current must occur as the current assumes its new steady value. This sudden change in current induces an emf of self-induction in the coil, resulting in sparks under the toe of the brush when contact with the segment is broken.

This emf of self-induction may be neutralized in part if the brushes are advanced *ahead* of the neutral plane, as shown in Fig. 217, so that the coil will have induced in it, due to cutting flux, an emf *opposed* to the emf of self-induction. Even then it is usually impossible to neutralize completely this emf of self-induction.

The foregoing, which refers to a single coil only, obviously applies to all the coils of the armature as they undergo reversal of current. Since both armature reaction and this emf of self-induction are proportional to the current, theoretically the brushes should be shifted with every change of load. Practically, and particularly when carbon brushes are used, this is not necessary since the high contact resistance of carbon brushes prevents the short-circuit current becoming excessive.

It is thus evident that commutation consists of two parts: the reversal of the current in the individual armature coils during the time required for a commutator segment to pass under the brush, and the conduction of the useful current from the commutator to the external circuit.

200. Sparking at the Commutator.—The emfs induced in a coil due to the shifting of the neutral plane and also due to its own self-inductance are comparatively low in value, being of the order of magnitude from a few tenths of a volt to perhaps 4 or 5 volts. But they are acting in a circuit of very low resistance. The resistance of the coil itself is extremely low, so that the greater part of the circuit resistance is at the brush contact.

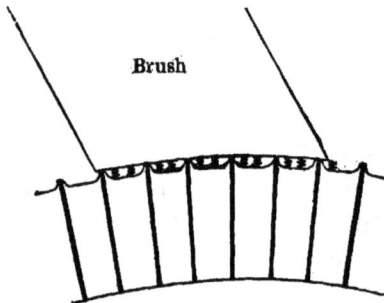

Brush

Fig. 218.—Commutator with high mica.

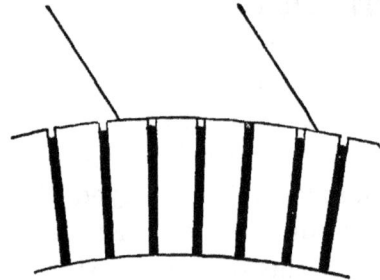

Fig. 219.—Commutator with undercut mica.

If the brush-contact resistance is too low, these short-circuit currents may reach such values as to produce severe sparking at the brushes. Consequently, in most generators and motors, except in very low-voltage, high-current generators, carbon brushes are used on account of their greater contact resistance. Also, they are self-lubricating, owing to their graphitic character, and therefore have but slight tendency to cut the commutator.

The passage of the current from the commutator to the brush is in the nature of an arc phenomenon rather than one of pure conduction. A careful examination will show myriads of minute arcs existing between the brush surface and the commutator. These arcs burn away or volatilize the copper of the segment, leaving "high mica" (Fig. 218). Because of their abrasive action, hard brushes are used to prevent high mica.

If the carbon brushes are too hard, they cut the commutator. Different grades of carbon are required for different operating conditions. In modern practice, the mica is frequently undercut; that is, the top of the mica is lower than the commutator surface (Fig. 219). There is some disadvantage in this, in that small bits of copper, carbon, and dirt collect in the grooves and may ultimately short-circuit the segments. These grooves can be easily cleaned out, however. Undercutting gives better results with

FIG. 220.—Proper method of fitting brushes.

high-speed commutators, since the centrifugal force tends to throw out the particles that lodge between segments.

The result of any arcing under the brush is to pit the commutator. As irregularities and depressions in the commutator surface tend to prevent the brush making intimate contact with the commutator, arcs of increasing magnitude will be formed. Hence, any condition which produces sparking, and so roughens the commutator, increases the sparking and roughening, as these actions are cumulative. If a commutator is sparking badly and the cause of the sparking is not eliminated, the commutator will deteriorate rapidly and soon become inoperative.

The brushes should be fitted very carefully to the commutator surface by grinding with sandpaper in the manner shown in Fig. 220. Carbon on the surface of the commutator should be removed with compressed air. Do not use waste. A slightly roughened commutator may be partially smoothed with fine sandpaper. Do not use emery, as the particles of emery are conducting and may short-circuit the commutator bars. Com-

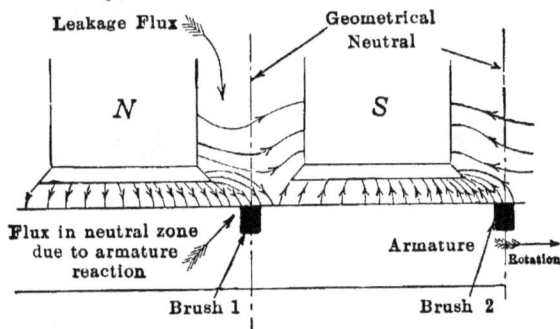

Fig. 221.—Flux distortion due to armature reaction.

mutator compounds of a lubricating character are on the market. They often reduce sparking, particularly if the sparking is due to a slightly roughened commutator. If the commutator is badly grooved by the brushes or is otherwise in poor condition, it may be necessary to turn it down in a lathe.

Other conditions, such as loose mica and loose segments, are more serious in character. It is often possible to remedy these by tightening up the commutator clamp bolts when the commutator is hot.

201. Commutating Poles (Interpoles).—When load is applied to a generator, the flux is crowded into the trailing pole tips by the mmf of the armature ampere-conductors or armature reaction (see Fig. 215 (c), p. 242). If the brushes are allowed to remain in the *geometrical* neutral, sparking will occur, due to the fact that the coils undergoing commutation generate emf and are simultaneously short-circuited by the brushes. This effect of armature reaction is indicated diagrammatically in Fig. 221, where, for simplicity, the armature is shown as a flat surface and the armature conductors are omitted. The flux is shifted in the direction of rotation. The brushes, if allowed to remain in the geometrical neutral, will short-circuit coils whose active conductors are cutting flux, and severe sparking will result. One method of improving commutation, as has been stated already, is to move the brushes ahead, beyond the neutral zone into the fringe of the flux due to the next pole.

The same result may be obtained in a much more satisfactory manner without moving the brushes, by introducing small poles, called *commutating poles* or *interpoles*, between the main poles.

In Fig. 221, some of the flux from the north pole enters the armature midway between the poles or at the geometrical neutral. The armature coils which are cutting this flux are simultaneously short-circuited by the brush. As a result, a large circulatory

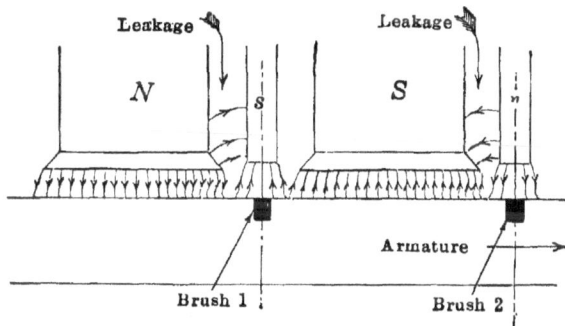

FIG. 222.—Effect of commutating poles on flux distortion.

current flows in the short-circuited coils and causes sparking at the brush. This short-circuit current is further increased by the emf of self-induction (see p. 247).

If a small south interpole of the proper strength be placed in the interpolar region (Fig. 222), not only may the north-pole flux which enters the armature midway between poles be neutralized, but the south pole may be sufficiently strong to cause flux actually to *leave* the armature in this region. The armature conductors cutting this south-pole flux may generate sufficient emf to neutralize the emf of self-induction. Therefore, if this commutating pole is correctly adjusted, good commutation at all loads is obtainable.

Obviously, a commutating pole of north polarity should be placed in the commutating region of brush 2.

Armature reaction is practically proportional to the armature mmf and, hence, to the armature current. The emf of self-induction is proportional to the self-inductance of the armature coils and to the rate of change of current (see Eq. (67), p. 175). The self-inductance of the armature coils is practically constant, and for a given speed the time of complete reversal of current is inversely proportional to the brush width. Therefore, for a given speed, the emf of self-induction is directly proportional to the armature current. Hence, the flux which neutralizes these effects must be proportional to the armature current. The

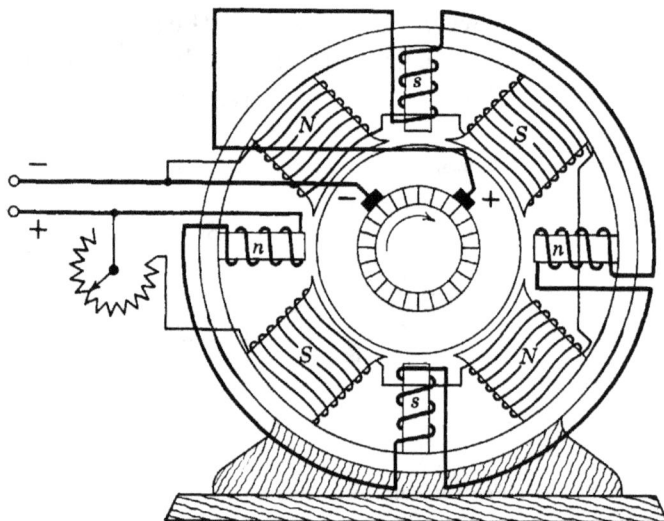

Fig. 223.—Connections of shunt field and commutating poles

commutating poles, therefore, must be excited by a winding connected in *series* with the armature, as shown in Fig. 223. Commutating poles are usually designed with a larger number of turns than is necessary for good commutation. Formerly, the poles were adjusted to the proper strength by means of a shunt or diverter connected across the interpole circuit. (See Fig. 230, p. 258.) Modern practice dispenses with the shunt and adjusts the effect of the interpoles by the use of shims between the interpoles and the yoke. It should be noted that in a generator the sequence of main and commutating poles in the direction of rotation is *Ns Sn*, where the capitals refer to the main poles and the small letters to the commutating poles.

If the commutating poles are properly adjusted, it is not necessary to shift the brushes with change of load.

It should be noted that commutating poles compensate the flux *in the neutral zone only* and do not compensate for the distortion of flux, due to armature reaction, over the entire armature surface.

202. Characteristics of Separately Excited Generator.—If either a separately excited generator or a shunt generator be loaded after building up to voltage, the terminal voltage will drop. This drop in voltage will increase with increase of load. It is important to know the voltage at the terminals of a generator for each value of the load current, because the ability to maintain voltage under load conditions determines in large measure the suitability of a generator for a specific use.

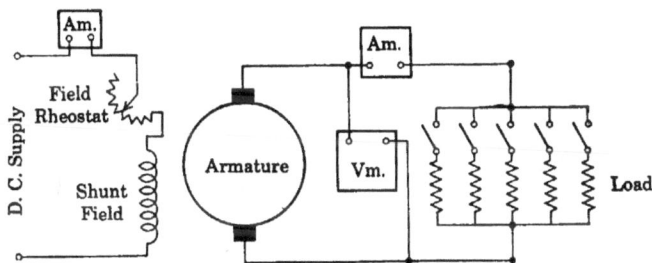

FIG. 224.—Separately excited generator.

Figure 224 shows the connections for determining the characteristics of a separately excited generator. The field, with its rheostat and an ammeter in series, is connected across a constant-potential, d-c supply. An adjustable load is connected across the armature terminals. An ammeter in series with the line measures the armature or load current, and a voltmeter across the armature terminals measures the armature terminal voltage. Throughout the test, the speed is maintained constant at its rated value. The field rheostat is adjusted until rated voltage at rated current is obtained, the load is then removed, and the no-load voltage *oa* (Fig. 225), is obtained. Load is then gradually applied, the speed being maintained constant. At each load, the load ammeter and the voltmeter are read. The test is carried to approximately 125 per cent of rated current.

If the results of this test are plotted, a curve similar to *abf* (Fig. 225) is obtained. The terminal voltage decreases with increase of load, for two reasons. *Armature reaction reduces the flux produced by the main field windings* and therefore reduces the

induced emf, as indicated by the line *ce* (Fig. 225).   (See p. 245.)
*There is a voltage drop in the armature itself due to the current
flowing through the resistance of the armature,* and this drop also
increases with increase of load.   The net voltage at rated load is
bd.   The characteristic of the generator is shown by the line *af.*

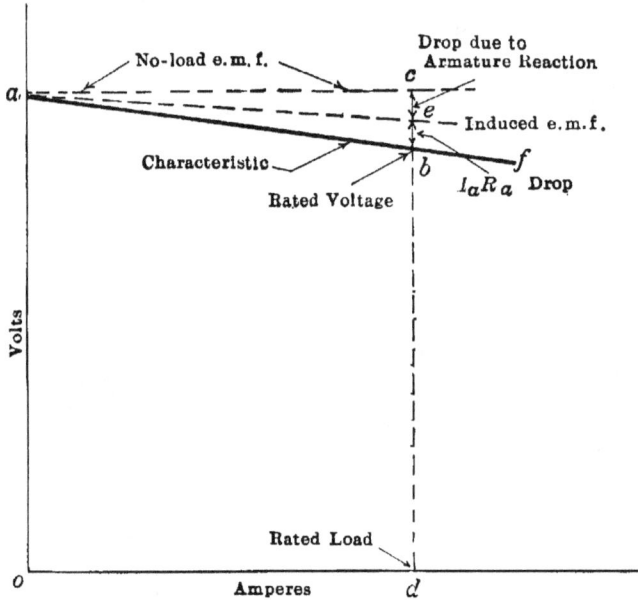

FIG. 225.—Characteristic of separately excited generator.

It is evident that in a generator the *induced* emf exceeds the
terminal voltage by the resistance drop in the armature.

That is, the induced emf

$$E = V + I_a R_a \qquad (90)$$

where V is the terminal voltage, I_a the armature current, and R_a
the armature resistance.

Transposing Eq. (90), the terminal voltage

$$V = E - I_a R_a. \qquad (91)$$

That is, the terminal voltage equals the induced emf minus
the armature-resistance drop.

Equations (90) and (91) should be compared with eqs. (33)
and (34), p. 43.

203. External Characteristic of Shunt Generator.—The operating or external characteristic of the shunt generator is obtained

in a manner similar to that used in obtaining the characteristic of the separately excited generator. The shunt field, in series with its rheostat, is now connected directly across the armature terminals and the generator excites itself. The connections for obtaining the characteristic are shown in Fig. 214, p. 239. An ammeter connected in the field circuit is usually desirable. The procedure is the same as that followed in obtaining the characteristic of the separately excited generator. The speed is held constant, and the volts and amperes are recorded at various loads.

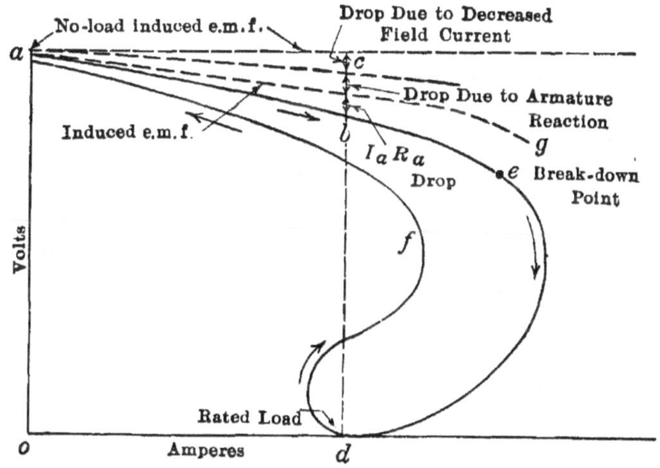

FIG. 226.—External characteristic of shunt generator.

In Fig. 226, the curve *abedfa* shows a typical shunt characteristic. Three causes now contribute to the drop in terminal voltage as load is applied, the voltage drop due to *armature resistance*, the voltage drop due to *armature reaction*, and the voltage drop due to *decrease in field current*. The decrease in field current is due to the fact that the field is now connected across the armature terminals. The voltage drops due to armature reaction and armature resistance result in lessened terminal voltage and, hence, decreased field current. Armature reaction and the lessened field current cause a further decrease in the induced emf. Line *ag* gives the induced emf with change of load.

When the breakdown point *e* on the characteristic is reached, the voltage drops very rapidly almost to zero with but slight increase of current. Little or no additional current can be obtained, even if the generator be short-circuited. In large generators, this breakdown point may occur at three or four times rated-load current. The breakdown is due to the rapid drop in field current as the load is increased. Considerable current *od* flows at zero terminal voltage (short circuit), due to residual magnetism. If the short circuit is gradually removed by increasing the resistance of the load, the voltage returns to the no-load value in the manner given by the part *dfa* of the curve. Because of hysteresis, this no-load value of voltage is slightly less than the initial value, *oa*. Also, on account of the effect of hysteresis, this return curve *dfa* lies below the curve *abe* (see Fig. 212, p. 237). The generator is usually operated on the portion *ab* of the characteristic.

In the shunt generator, the armature current I_a is equal to the sum of the load current I and the field current I_f. That is,

$$I_a = I + I_f. \tag{92}$$

Example.—The no-load voltage of a 50-kw, 220-volt, shunt generator is 232 volts, and the rated-load voltage is 220 volts. The armature resistance, including brushes, is 0.026 ohm, and the resistance of the shunt-field circuit is 52 ohms. Determine at rated load: (*a*) the induced emf; (*b*) the power lost in the field; (*c*) the power lost in the armature; (*d*) the total power generated.

(*a*) The rated current

$$I = \frac{50,000}{220} = 227 \text{ amp.}$$

The field current

$$I_f = \frac{220}{52} = 4.23 \text{ amp.}$$

The armature current

$$I_a = 227 + 4.23 = 231.2 \text{ amp.}$$

The induced emf, Eq. (90), p. 254,

$$E = 220 + (231.2 \times 0.026) = 220 + 6.0 = 226.0 \text{ volts.} \quad Ans.$$

(*b*) $$P_f = \frac{V^2}{R_f} = \frac{(220)^2}{52} = 931 \text{ watts.} \quad Ans.$$

or

$$220 \times 4.23 = 931 \text{ watts (check).}$$

(*c*) $$P_a = I_a^2 R_a = (231.2)^2 \times 0.026 = 1,389 \text{ watts.} \quad Ans.$$

(*d*) The total power generated

$$P = 50,000 + 931 + 1,389 = 52,320 \text{ watts.} \quad Ans.$$

It is to be noted that the total power generated is also equal to the product of the induced emf and the armature current. Thus,

$$P = 226 \times 231.2 = 52,320 \text{ watts}.$$

The induced emf at rated load in (*a*) is less than the no-load voltage because of the effects of armature reaction and of decreased field current.

204. Compound Generator.—The drop in voltage with load, which is characteristic of the shunt generator, makes this type of generator undesirable where constancy of voltage is essential. This applies particularly to lighting circuits, where a very slight change of voltage makes a material change in the candle-

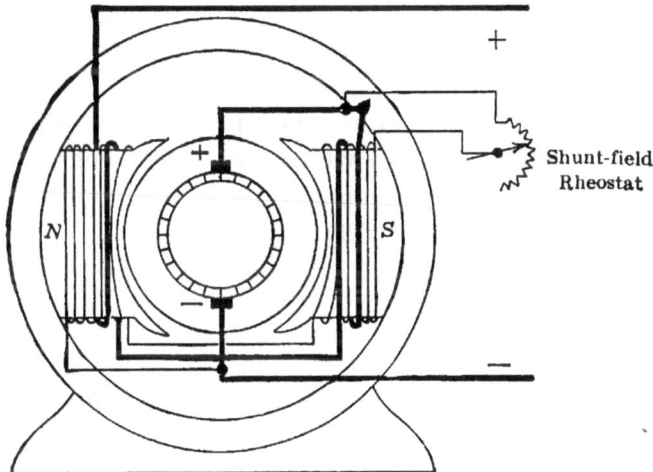

Fig. 227.—Connections of compound generator (short shunt).

power of incandescent lamps. A generator may be made to produce a substantially constant voltage, or even a rise in voltage, as the load increases, by placing on the field core a few turns which are connected *in series* either with the load or with the armature. These series turns are connected so as to *aid* the shunt turns when the generator delivers current (Fig. 227). As the load increases, the current through the series turns also increases and, therefore, the flux through the armature increases. The effect of this increased flux is to increase the induced emf. By proper adjustment of the series ampere-turns, this increase in induced emf or armature voltage may be made to balance the combined drop in voltage due to armature reaction and to the resistance of the armature. If the terminal voltage is maintained substantially constant, the field current will not drop as the load increases.   Therefore, the three causes of voltage drop, namely, armature reaction, I_aR_a drop, and drop in field current (Fig. 226, p. 255), are neutralized more or less completely by the effect of the series ampere-turns.

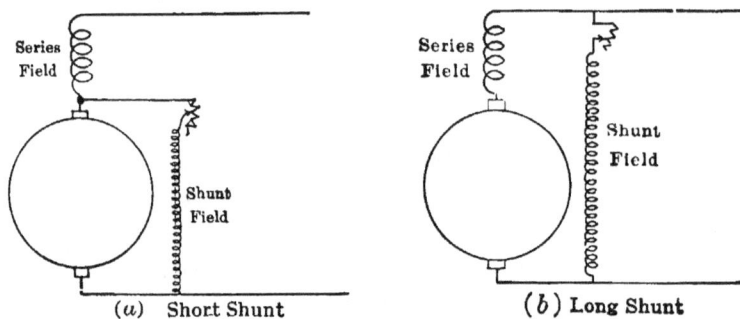

(*a*)   Short Shunt   (*b*) Long Shunt

Fig. 228.—Compound generator connections.

The shunt field may be connected directly across the armature terminals (Fig. 228 (*a*)) in which case the generator is *short* shunt.   If the shunt field be connected across the line terminals outside the series field (Fig. 228 (*b*)), the generator is *long* shunt. The operating characteristic is practically the same in both cases.

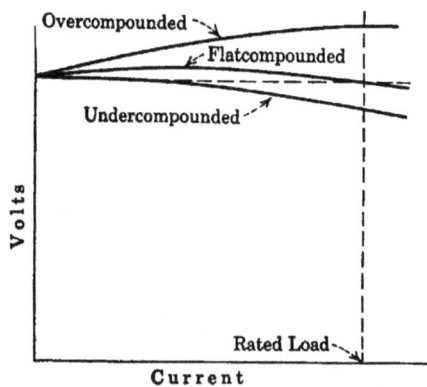

Fig. 229.—Compound generator characteristics.

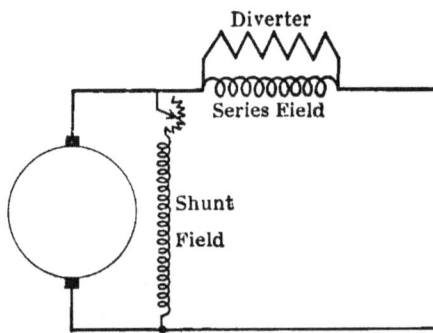

Fig. 230.—Series-field diverter.

If the effect of the series turns is to produce the same terminal voltage at rated load as at no load, the generator is *flat compounded* (see Fig. 229).

When the rated-load voltage is greater than the no-load voltage, the generator is *overcompounded*.   When the rated-load voltage is less than the no-load voltage, the generator is *undercompounded*.   Generators are seldom undercompounded.

All three characteristics tend to droop, due to saturation of the iron. The series ampere-turns do not increase the flux at full load as much proportionately as they do at light load (see Fig. 211, p. 237).

Compound generators are usually wound so as to be somewhat overcompounded. The degree of compounding can then be regulated by shunting more or less current away from the series field. To do this, a low-resistance shunt, called a *diverter*, is used (Fig. 230).

Example.—In a 250-kw, 250-volt, compound generator the resistance of the armature is 0.0090 ohm, the series-field resistance is 0.0015 ohm, and the shunt-field resistance is 17.5 ohms. The generator is connected long shunt (Fig. 228 (*b*)). Determine at rated voltage: (*a*) the rated current of the generator; (*b*) the shunt-field current; (*c*) the armature current; (*d*) the induced emf; (*e*) the power lost in the shunt field, the series field, the armature; (*f*) the total generated power.

a. Rated current

$$I = \frac{250{,}000}{250} = 1{,}000 \text{ amp.} \quad \textbf{\textit{Ans.}}$$

b. Shunt-field current

$$I = \frac{250}{17.5} = 14.3 \text{ amp.} \quad \textbf{\textit{Ans.}}$$

c. Armature current

$$I_a = 1{,}000 + 14.3 = 1{,}014 \text{ amp.} \quad \textbf{\textit{Ans.}}$$

d. Induced emf

$$E = 250 + 1{,}014(0.0015 + 0.0090) = 250 + 10.6 = 260.6 \text{ volts.} \quad \textbf{\textit{Ans.}}$$

e. Power lost in shunt field

$$P_{sh} = 250 \times 14.3 = 3{,}580 \text{ watts.} \quad \textbf{\textit{Ans.}}$$

Power lost in series field

$$P_s = (1{,}014)^2 \times 0.0015 = 1{,}540 \text{ watts.} \quad \textbf{\textit{Ans.}}$$

Power lost in armature

$$P_a = (1{,}014)^2 \times 0.0090 = 9{,}250 \text{ watts.} \quad \textbf{\textit{Ans.}}$$

f. Total generated power

$$P' = 250{,}000 + 3{,}580 + 1{,}540 + 9{,}250$$
$$= 264{,}400 \text{ watts.} \quad \textbf{\textit{Ans.}}$$

Also (see example, p. 256),

$$P' = 260.6 \times 1{,}014 = 264{,}200 \text{ watts } (check).$$

205. Industrial Applications of Compound Generators.—Flat-compounded generators are used principally in isolated plants, such as hotels and office buildings, where the circuit runs are short and where it is desirable that the voltage be maintained constant at all loads without adjustment of the shunt-field rheostat.

Overcompounded generators are used where the load is located at some distance from the generator. As the load increases, the voltage at the load tends to decrease, owing to the voltage drop in the feeder. If, however, the generator voltage rises just

Fig. 231.—Tirrill regulator.

enough to offset this feeder drop, the voltage at the load remains constant. Railway generators are typical of this class of service. They are usually compounded so that the no-load terminal voltage is 600 volts and the rated-load terminal voltage is 650 volts.

206. Series Generator.—In the series generator the field winding is connected in series with the armature and the external circuit. The field winding must consist necessarily of comparatively few turns having a sufficiently large cross-section to carry the rated current of the generator.

Ordinarily, the *series* generator is used for *constant-current* work, in distinction to the *shunt* generator, which maintains substantially *constant voltage*. For a number of years the series generator[1] was used successfully to supply series arc lights for street lighting. At the present time it has been almost entirely replaced by constant-current transformers used in conjunction with mercury-arc rectifiers, where constant-current d-c service is necessary (see Part II, Chap. X.).

207. Tirrill Regulator.—It has been shown that the voltage of a generator varies with the load, speed, etc. By means of a Tirrill regulator the voltage of a generator can be maintained constant even under rapid fluctuations of load. In addition, compensation may be made for line drop. The voltage is controlled by small relay contacts, which short-circuit the shunt-field rheostat, the duration of the short circuit depending on the amount of regulation required. If the generator voltage drops, the time increases during which the shunt-field rheostat is short-circuited; that is, the armature of the relay vibrates more slowly. If the generator voltage rises, the armature of the relay vibrates more rapidly. The field rheostat is set usually so that the generator voltage is 35 per cent below normal when the regulator is disconnected. Figure 231 shows a Tirrill regulator and its mounting.

[1] For a more complete description see C. L. DAWES, "A Course in Electrical Engineering," 3d ed., Vol. I.

CHAPTER XII

THE MOTOR

A *generator* is a machine for converting *mechanical* energy into *electrical* energy.

A *motor* is a machine for converting *electrical* energy into *mechanical* energy. The same machine, however, may be used either as motor or generator.

208. Principle of Motor.—Figure 232 (*a*) shows a magnetic field of uniform intensity, in which a conductor carrying no cur-

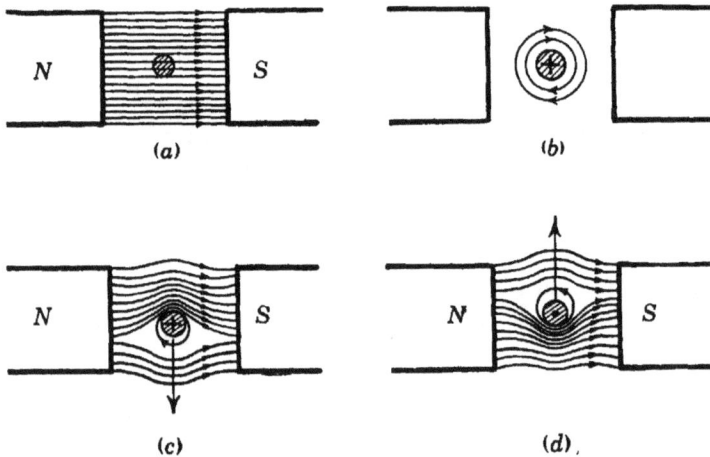

FIG. 232.—Force acting on conductor carrying current in a magnetic field.

rent is placed. In (*b*) the conductor is shown as carrying current into the paper, but the field due to the *N*- and *S*-poles has been removed. A cylindrical magnetic field exists about the conductor, due to the current in it. The direction of this field, by the corkscrew rule, is clockwise.

Figure 232 (*c*) shows the resultant field obtained by combining the main field and that due to the current in the conductor. The field due to the current in the conductor acts in conjunction with the main field above the conductor, whereas it opposes the main field below the conductor. The result is to crowd the flux and

262

increase the flux density *above* the conductor and to reduce the flux density in the region *below* the conductor.

It will be found that a force acts on the conductor, tending to move the conductor *down*, as shown in (c) by the arrow attached to the conductor.

It is convenient to think of this effect as due to the crowding of the lines on one side of the conductor. Magnetic lines of force may be considered as acting like elastic bands under tension. These lines always try to contract and shorten themselves so as to be of minimum length. The tension in these lines on the upper side of the conductor tends to move the conductor down, as shown in (c).

If the direction of the current in the conductor is reversed, the crowding of the lines occurs *below* the conductor, which tends to move *upward*, as shown in Fig. 232 (d).[1]

The operation of the electric motor depends on the principle that *a conductor carrying current in a magnetic field tends to move at right angles to the field.*

209. Force on Conductor Carrying Current.—The force acting on a conductor carrying current in a magnetic field is directly proportional to three quantities: the strength of the field, the magnitude of the current, and the length of the conductor. The force in *dynes* is given by

$$F = \frac{BlI}{10} \text{ dynes} \qquad (93)$$

where B is the flux density in gausses, or lines per square centimeter, l is the active length of the conductor in centimeters, and I is the current in amperes. The directions of the field, of the conductor, and of the force are mutually perpendicular.

Example.—A flat rectangular coil of 20 turns lies with its plane parallel to a magnetic field (see Fig. 236 (a)), the flux density being 3,000 lines per sq cm. The length of the coil transverse to the field is 8 in. (20.32 cm). The current per conductor is 30 amp. Determine the force in pounds which

---

[1] When conductors are embedded in slots, as is customary in the usual dynamo, the actual force is transferred to the armature teeth. However, such force exists only in virtue of the current in the armature conductors, and the armature behaves as if the force acted on the conductors themselves. Therefore, for most purposes, motor phenomena may be treated as if the electromagnetic forces acted on the conductors themselves.

acts on each side of the coil.   (See arrows in Fig. 236 (a), p. 267).

$$B = 3,000.$$
$$l = 8 \times 2.54 = 20.32 \text{ cm.}$$
$$I = 30.$$
$$F_1 = 3,000 \times 20.32 \times \frac{30}{10} = 182,900 \text{ dynes.}$$

As there are 20 turns,

$$F = 20 \times 182,900 = 3,658,000 \text{ dynes.}$$
$$\frac{3,658,000}{981} = 3,730 \text{ grams}$$
$$= 3.73 \text{ kg.}$$
$$3.73 \times 2.20 = 8.21 \text{ lb.}   Ans.$$

210. Fleming's Left-hand Rule.—The relation among the direction of a magnetic field, the direction of motion of a conductor in that field, and the direction of the *induced* emf is given by Fleming's right-hand rule (see p. 211).

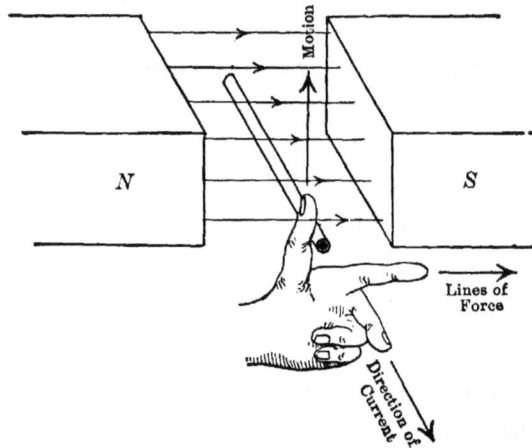

Fig. 233.—Fleming's left-hand rule.

In a similar manner, the relation among the direction of a magnetic field, the direction of a current in that field, and the direction of the resulting force acting on the conductor is given by Fleming's left-hand rule, which is as follows:

Point the forefinger in the direction of the field or flux, the middle finger in the direction of the current in the conductor, and the thumb will point in the direction in which the conductor tends to move.

This is illustrated by Fig. 233.

If Fleming's right-hand rule (for generator action) be applied to Fig. 233, the direction of motion to induce the emf and current shown must be downward or opposite to that for the motor.

Hence, in a generator the conductor moves *against* a force tending to oppose its motion, and so the conductor requires a driving force to keep it in motion. This driving force is supplied by the prime mover which drives the generator.

It will be recognized that the force, due to motor action, which opposes the motion of a conductor in a generator is in accordance

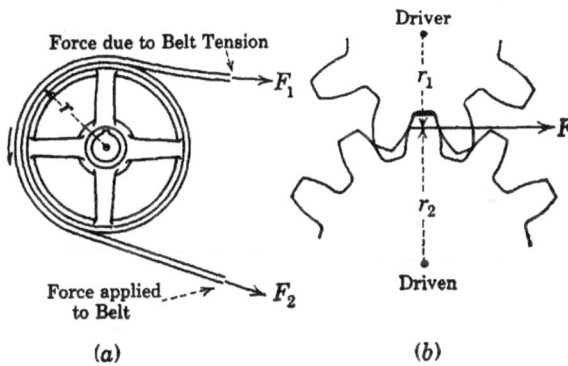

FIG. 234.—Torque developed by belt and by gears.

with the law of the conservation of energy. The electrical energy delivered by the generator results from the action of the prime mover in opposing this force.

Thus, motor action exists in a generator. It will be shown later that generator action exists in a motor.

211. Torque.—When an armature, a flywheel, or any similar device is rotating about an axis, a tangential force is necessary to produce and maintain rotation. This force may be developed within the machine itself, as in a motor or steam engine, or it may be applied to a driven device, such as a pulley, a shaft, a generator, or the driving gears on the wheels of a streetcar (Fig. 234). The total effect of the force is determined not only by its *magnitude* but also by its *arm*, or radial distance from the axis of rotation to the line of action of the force.

The product of the force and its perpendicular distance from the axis of rotation measures the *torque*.

Torque may be considered as a mechanical couple tending to produce rotation. It is expressed in units of force and length.

In the English system, torque is usually expressed in pound-feet. (This distinguishes it from foot-pounds, which represents *work*.)

In the cgs system, the unit of torque is the dyne-centimeter (a very small unit), and in the metric system the unit is the kilogram-meter. A kilogram-meter equals 7.23 lb-ft.

In Fig. 234 (*a*), the tight side of the belt is pulling on the rim at the bottom of the pulley with a tangential force F_2. The loose side of the belt is pulling on the rim at the top of the pulley with a tangential force F_1. The net tangential force acting on the rim of the pulley is F_2-F_1. The radius of the pulley is r, and thus the torque being applied to the pulley is $(F_2-F_1)r$.

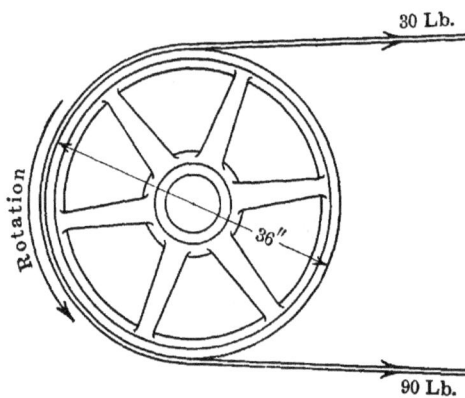

FIG. 235.—Example of torque on pulley by belt.

Figure 234 (*b*) shows a driving gear having a radius r_1 to its line of action, or pitch circle, and a driven gear having a radius r_2 to its pitch circle. Neglecting friction, the torque being applied by the driver is the product of the tangential force F and the radius r_1, or Fr_1. The torque available in the driven gear and shaft is Fr_2.

Since F is the same in both cases and r_2 is greater than r_1, the torque of the driven shaft is greater than that of the driver. The angular speeds of the two shafts are inversely as their radii, and hence as their torques. Therefore, if friction is neglected, the power is the same in the two cases. This must follow from the law of the conservation of energy.

Example.—A belt is driving a 36-in. (91.4 cm) pulley (Fig. 235). The tension in the tight side of the belt is 90 lb (40.8 kg) and that in the loose side is 30 lb (13.6 kg). Determine the torque applied to the pulley.

As the two sides of the belt are acting in opposition, the net tangential pull on the rim of the pulley is

$$90 - 30 = 60 \text{ lb.}$$

This force is acting 18 in. (45.7 cm) or 1.5 ft from the axis of rotation. Therefore, the torque

$$T = 60 \times 1.5 = 90 \text{ lb-ft.} \quad Ans.$$

or

$$T = (40.8 - 13.6)0.457 = 12.43 \text{ kg-m.} \quad Ans.$$

212. Torque Developed by a Motor.—Figure 236 (a) shows a coil of a single turn, whose plane lies parallel to a magnetic field. Current flows into the paper in the left-hand side of the coil and out of the paper in the right-hand side of the coil. Therefore, the left-hand conductor tends to move downward with a force F_1, and the right-hand conductor tends to move upward with a force F_2. As the current in each of the conductors is the same and they lie in the same magnetic field, the force F_1 equals the force F_2. These two forces form a couple tending to rotate the coil about its axis in a counterclockwise direction. Torque is thus developed. In (a), the coil is in the position of maximum

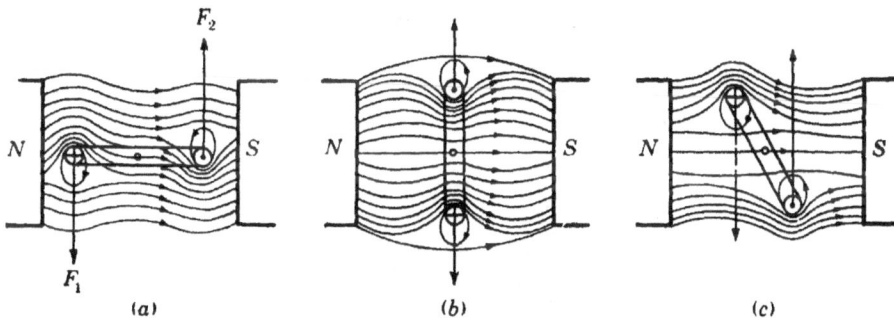

(a)   (b)   (c)

Fig. 236.—Torque developed at different positions of coil.

torque because the perpendicular distance from the coil axis to the lines of action of the forces F_1 and F_2 is a maximum.

When the coil reaches the position (b), neither conductor can move any farther unless the coil spreads. This is a position of zero torque because the perpendicular distance from the coil axis to the lines of action of the forces is zero.

If, however, the direction of the current in the coil be reversed when the coil reaches position (b), and the coil be carried slightly beyond the dead center, as shown in (c), a torque is developed which continues to turn the coil in the counterclockwise direction.

To develop a continuous torque in a motor, the current in each coil of the armature must be reversed and this reversal should occur when the coil is passing through the neutral plane or plane of zero torque. A commutator is therefore necessary. This is analogous to using a commutator in connection with a generator in order that the current delivered to the external circuit shall be unidirectional.

A single-coil motor, shown in Fig. 236, is impracticable, as there are dead centers and the torque developed is pulsating. A two-coil armature would eliminate the dead centers, but the torque developed would still be pulsating.

The best results are obtained with a large number of coils, just as in the armature of a generator. In fact, there is no difference in the construction of a motor armature and a generator armature. In Fig. 237, an armature and a field are shown for a two-pole motor and the force developed by each conductor is indicated by the arrow attached to it.

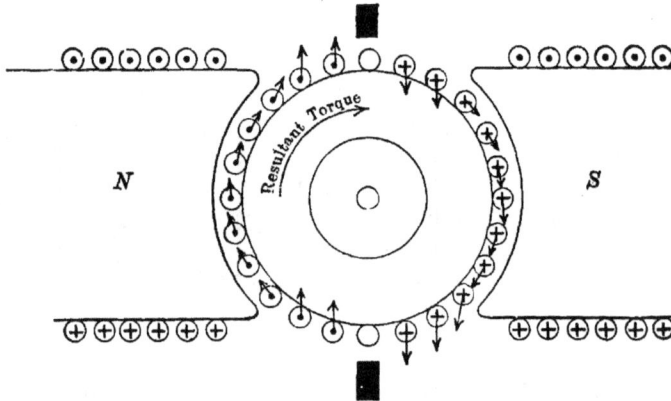

Fig. 237.—Torque developed by conductors in motor armature.

In armatures of this type, only a very small proportion of the total number of coils is undergoing commutation at any instant. Therefore, the variation in the number of active conductors is so slight that the torque developed is substantially constant, if armature current and flux are constant.

Referring to Eq. (93), p. 263, in any given dynamo the flux per pole is proportional to the average flux density B; the radius of the armature and the active length of armature conductor l are fixed. Hence, the torque developed by an armature

$$T = K_t I \Phi \qquad (94)$$

where K_t is a constant of proportionality, involving the dimensions of the armature, the number of conductors, the units, etc.,

I is the armature current in amperes, and Φ is the flux entering the armature from one north pole.

That is, in a given motor,

The torque is proportional to the armature current and to the strength of the magnetic field.

This is a very important relation to keep in mind, for by its use the variation of torque with load can be readily determined for the various types of motors.

Example.—When a motor armature is taking 50 amp from the line, it develops a torque of 60 lb-ft (8.30 kg-m). If the field strength is reduced to 75 per cent of its original value and the current increases to 80 amp, what is the torque developed?

If the armature current remained at 50 amp, the value of torque due to the weakening of the field alone would be

$$0.75 \times 60 = 45 \text{ lb-ft (6.23 kg-m)}.$$

Due to increase in armature current, the final value of torque will be

$$\frac{80}{50}\, 45 = 72 \text{ lb-ft (9.96 kg-m)}. \quad Ans.$$

It must be remembered that the torque expressed by these equations is the entire torque *developed* by the armature or the *electromagnetic torque.* The torque available at the pulley will be slightly less than this, owing to the torque lost in overcoming friction and windage and in supplying the iron losses of the armature.

213. Counter Electromotive Force.—The resistance of the armature of the usual 10-hp., 110-volt motor is about 0.05 ohm. If this armature were connected directly across 110-volt mains, the current, by Ohm's law, would be

$$I = \frac{110}{0.05} = 2{,}200 \text{ amp.}$$

This value of current is not only excessive but unreasonable, especially when one considers that the rated current of such a motor is in the neighborhood of 80 amp. When a motor is in operation, *the current through the armature is evidently not deter-mined by armature ohmic resistance alone.*

The armature of a motor in operation is in every way similar to that of a generator. The conductors on its surface, in addition to carrying current and so developing torque, are *cutting flux* and *must* be generating an emf.

Figure 238 shows a single conductor of a motor armature at the instant when it is directly in front of a north pole. This conductor is free to move. The direction of the applied voltage, hence the direction of the current, is shown by the left-hand arrow. If Fleming's *left-hand rule* be applied, it is found that the relation of direction of current and flux is such that the conductor tends to move downward under the action of the force developed.

As the conductor moves downward, it cuts flux, and an emf is generated in the conductor. If the *right-hand rule* be applied to determine the direction of this *induced* emf, it is found to be acting from *right* to *left* (Fig. 238) and is in *opposition* to the applied voltage and to the current. It follows that the induced emf in a motor armature must always be in opposition to the impressed voltage and to the current. Hence, this *induced* emf is called a *counter emf* or a *back emf*. Obviously, the counter emf *opposes* the current entering the armature. Since the counter emf opposes the line voltage, the net voltage acting in the armature circuit is the *difference* of the line voltage and the counter emf. Let V equal the line voltage and E the back emf. The *net* voltage acting in the armature circuit is

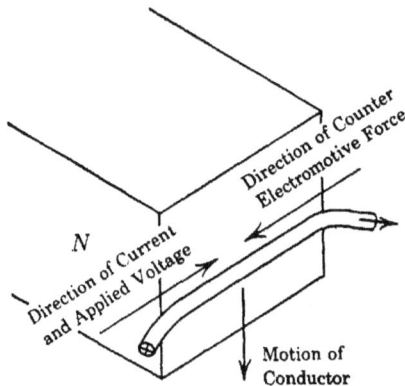

Fig. 238.—Relation of the direction of currents and emfs in motor conductor.

$$V - E \text{ volts.}$$

The armature current, from Ohm's law, is

$$I_a = \frac{V - E}{R_a} \text{ amp} \tag{95}$$

where R_a is the armature resistance.

This equation may be transposed and written

$$E = V - I_a R_a \text{ volts.} \tag{96}$$

This should be compared with Eq. (90), p. 254, which is the corresponding equation for a generator.

In a generator the induced emf is equal to the terminal voltage *plus* the armature-resistance drop. In a motor the induced emf is equal to the terminal voltage *minus* the armature-resistance drop. The counter emf must always be less than the terminal or impressed voltage if current is to flow *into* the armature at the positive terminal.

Example.—Determine the back emf of a 10-hp motor when the terminal voltage is 110 volts and the armature is taking 90 amp. The armature resistance is 0.05 ohm.

$$E = 110 - (90 \times 0.05) = 110 - 4.5 = 105.5 \text{ volts.} \quad Ans.$$

An interesting experiment for demonstrating the existence of counter emf is shown in Fig. 239. A lamp bank is connected

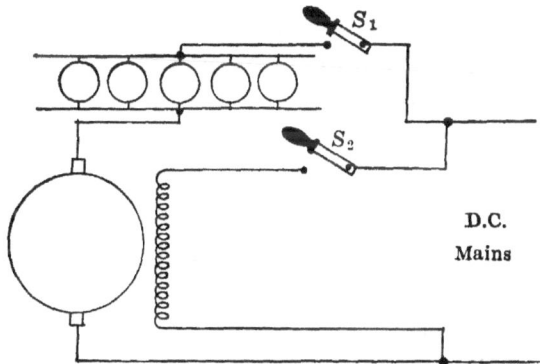

FIG. 239.—Demonstration of counter emf.

in series with the armature of a shunt motor. First, close switch S_2 which closes the field circuit. Then close S_1. At the instant of closing S_1 the lamps will burn brightly, being practically up to candlepower. As the armature speeds up, these lamps will become dimmer and dimmer, showing that the armature is generating a *counter* emf, which opposes the line voltage and so results in less voltage and less current for the lamps. When the armature is up to speed, the lamps will be very dim. If, however, the field switch S_2 now be opened, the flux and, therefore, the counter emf will be reduced to zero practically, which will be shown by the lamps again coming up to candlepower. (In practice, when a motor is in operation, *the field circuit should not be opened under any conditions whatsoever.*)

Equation (89), p. 235, for the induced emf in a generator will apply obviously to a motor. That is, the counter emf

$$E = K\phi S \text{ volts}$$

where K is a constant, ϕ is the total flux entering the armature from *one* north pole, and S is the speed of the armature in revolutions per minute.

214. Internal Power.—On p. 257 it is shown that the *total electrical power* developed in a generator armature is the product of the induced emf and the armature current. Likewise, the *total or internal mechanical power* developed by a motor armature is the product of the induced or counter emf and the armature current. That is,

$$P_m = EI_a. \tag{97}$$

This may be shown as follows: The input to the armature is VI_a where V is the terminal voltage. The armature-resistance loss $I_a^2R_a$ is supplied *electrically*. The only other losses are friction and windage, hysteresis and eddy-current losses in the armature iron. These losses are supplied *mechanically*. (See p. 299.) The only further power to be accounted for is the output at the pulley. It follows, then, that the only portion of the armature input VI_a that does not appear as mechanical power is the armature-resistance loss $I_a^2R_a$. Hence, the *internal* power

$$P_m = VI_a - I_a^2R_a = I_a(V - I_aR_a).$$

But

$$V - I_aR_a = E \text{ (Eq. 96, p. 270).}$$

Hence,

$$P_m = EI_a.$$

The power at the pulley is equal to the internal power minus friction and windage losses and armature-iron losses.

Example.—The armature resistance of a 20-hp, 230-volt, 1,200-rpm, shunt motor is 0.15 ohm. Friction and windage and armature-iron losses are 600 watts. When the armature takes 75 amp at 230 volts, determine: (a) the armature input; (b) the counter emf; (c) the internal power; (d) the power at the pulley in horsepower.

(a) $P_a = 230 \times 75 = 17{,}250$ watts. *Ans.*

(b) $E = 230 - (75 \times 0.15) = 218.7$ volts. *Ans.*

(c) $P_m = 218.7 \times 75 = 16,400$ watts. *Ans.*

Also, $P_m = (230 \times 75) - (75^2 \times 0.15) = 16,400$ watts (*check*).

(d) $P'_m = \dfrac{16,400 - 600}{746} = 21.2$ hp. *Ans.*

215. Speed of Motor.—If Eq. (89), p. 235, be solved for speed,

$$S = K_1 \frac{E}{\phi} \tag{98}$$

where K_1 is a constant equal to $1/K$.

The speed of a motor is directly proportional to the counter emf and inversely proportional to the strength of field.

Substituting for E in (98) its value given in (96), the speed becomes

$$S = K_1 \frac{V - I_a R_a}{\phi}. \tag{99}$$

This is an important equation, for it shows the law of speed variation of a motor with changes in load.

Example.—A 15-hp, 110-volt motor, having an armature resistance of 0.08 ohm, runs at 1,000 rpm when the *armature* current is 50 amp. What is its speed when the *armature* current is 120 amp, if the field strength remains unchanged?

Let S_2 be the speed when the armature current is 120 amp. Applying Eq. (99),

$$\frac{S_2}{1,000} = \frac{K_1 \dfrac{110 - (120 \times 0.08)}{\phi}}{K_1 \dfrac{110 - (50 \times 0.08)}{\phi}}.$$

As ϕ is the same in both cases,

$$S_2 = 1,000 \frac{110 - 9.6}{110 - 4.0} = 1,000 \frac{100.4}{106.0} = 947 \text{ rpm. } \textit{Ans.}$$

To illustrate the effect on speed of changing the field current, the following example is given:

Example.—The field strength of this motor is decreased 10 per cent. What is the speed when the armature current is 120 amp?

Since the armature current does not change, the counter emf E does not change. Hence Eq. (98) may be applied directly.

$$\frac{S_2}{947} = \frac{K_1\dfrac{E}{\phi_2}}{K_1\dfrac{E}{\phi_1}} = \frac{\phi_1}{\phi_2}, \qquad \phi_2 = 0.9\phi_1.$$

Hence,

$$S_2 = 947\frac{\phi_1}{0.9\phi_1} = \frac{947}{0.9} = 1{,}052 \text{ rpm.} \quad Ans.$$

216. Armature Reaction and Brush Position in Motor.—Figure 240 (*a*) shows the currents in the conductors of the armature of a motor, the brushes being in the geometrical neutral. The direction of the current in each half of the armature corresponds to the polarity and direction of rotation shown in (*b*). Due to the armature ampere-turns, a mmf F_A exists in the armature. The direction of the flux produced by this mmf is upward and is at right angles to the polar axis.

Figure 240 (*b*) shows currents in both field and armature.

The combined effect of the ampere-turns of field and armature causes the *resultant* flux to have a direction diagonally upward to the right and to be crowded into the *leading* pole tips. That is, it is distorted in the direction *opposite* to the direction of rotation. Since the neutral plane is perpendicular to the direction of the resultant flux, it also moves backward. Hence, the brushes must be moved *backward* by an angle β.

Therefore, in a *motor* it is necessary to move the brushes *backward* with increase of load (as shown in (*b*)), whereas in a *generator* the brushes are moved *forward* with increase of load. Were it not for the emf of self-induction (see p. 247), the brush axis would coincide with the neutral plane. Owing, however, to the necessity of counteracting this emf, the brushes are set behind this load neutral plane, as shown in Fig. 240 (*b*). That is, in both motor and generator it is necessary to set the brushes beyond the load neutral plane in order to counteract the emf of self-induction. (Fig. 240 should be compared with Fig. 215, p. 242.)

This backward movement of the brushes is accompanied by a *demagnetizing* action of the armature on the field, as may be seen from a study of Fig. 240 (*b*). If β is the angle between the brush axis and the geometrical neutral, the armature ampere-turns included in the angle 2β *oppose* the field ampere-turns, as may be determined by applying the corkscrew rule.

Therefore, as the load is increased, armature reaction tends to increase the speed of a motor. In fact, instances are known

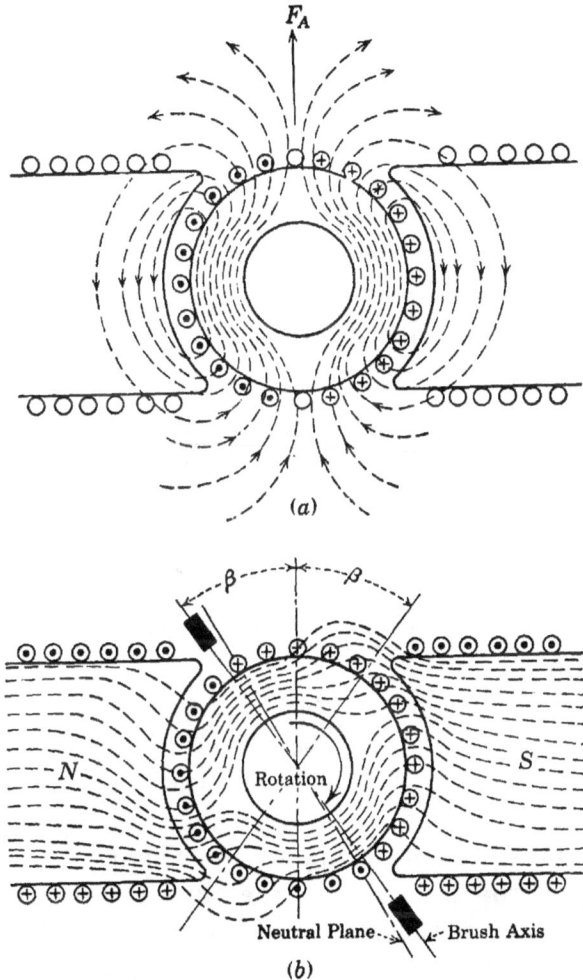

(a)

(b)

Fig. 240.—Armature reaction in motor.

where motors with short air-gaps (thus having high armature reaction) ran away when the load was applied.

217. Commutating Poles.—Commutating poles are used with motors, as well as with generators, for annulling armature reaction *in the neutral plane* and also for inducing an emf in the armature coils undergoing commutation to overcome the emf of self-induction.

Figure 241 shows diagrammatically a bipolar motor. The left-hand pole is north, and the right-hand pole is south. The current flows outward in the conductors on the left-hand side of the armature and inward in the conductors on the right-hand side of the armature. The armature therefore rotates in a clockwise direction (Fleming's left-hand rule). The field mmf F acts from left to right, and the armature mmf F_A acts upwards (see Fig. 240). These directions are determined by the corkscrew rule.

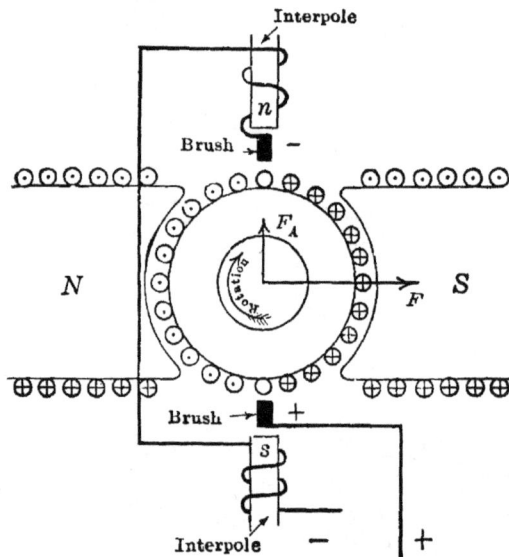

Fig. 241.—Interpoles in motor.

Since the interpoles must oppose the armature reaction in the neutral plane, the interpole at the top of the armature must be north and that at the bottom of the armature must be south, as shown in the figure. Therefore, in motors the relative polarity of the main poles and the interpoles taken in the direction of rotation is $NnSs$, whereas in generators it is $NsSn$ (see Fig. 223, p. 252). The interpoles are connected in series with the armature, as in Fig. 241.

If a motor happens to be sparking badly from some unknown cause, the polarity of the interpoles should be investigated with a compass, as the sparking may be due to their being incorrectly connected.

218. Shunt Motor.—The shunt motor is connected in the same manner as a shunt generator; that is, its field is connected directly across the line in parallel with the armature.

A field rheostat is usually connected in series with the field.

If load is applied to any motor, the motor immediately tends to slow down. Unless the reactions developed within the motor, as a result of this slowing down, are such that the torque increases with the increase of load, the motor stops. With the shunt motor, this decrease of speed lowers the back emf, as the flux remains substantially constant (see Eq. (98), p. 273). If the back emf is decreased, the *difference* between the terminal voltage

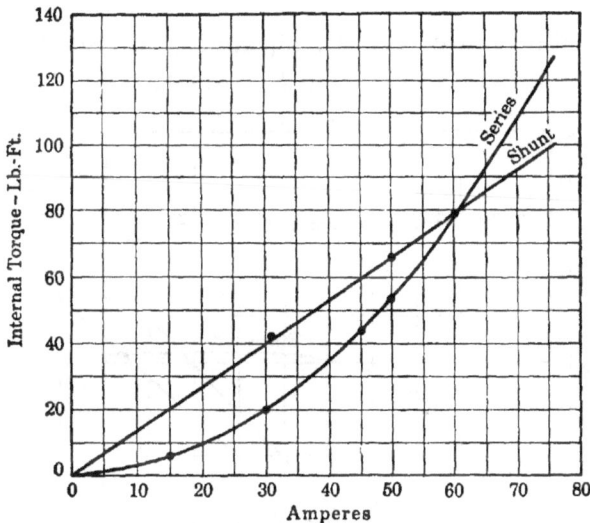

FIG. 242.—Shunt and series motors, torque-current characteristics. (Electromagnetic or internal torque.)

and the back emf is increased and more current flows into the armature (see Eq. (95), p. 270). This continues until the increased armature current produces sufficient torque to meet the demands of the increased load.

The suitability of a motor for any particular duty is determined almost entirely by two factors, the variation of its *torque* with load and the variation of its *speed* with load.

In the shunt motor, the flux is substantially constant. Therefore, from Eq. (94), p. 268, the torque will vary almost directly with the *armature* current. If the torque is plotted as a function of armature current, the resulting graph is practically a straight line (Fig. 242). For example, in Fig. 242, when the current is

30 amp, the armature develops 40 lb-ft (5.53 kg-m) torque, and
when the armature current is 60 amp, the armature develops
80 lb-ft (11.06 kg-m) torque. That is, when the current doubles,
the torque doubles. If the armature rotates, the torque at the
pulley is slightly less than the internal torque by the amount
necessary to overcome friction and windage and to supply the
iron losses.

The speed of a motor varies according to Eq. (99), p. 273,
where

$$S = K_1 \frac{V - I_a R_a}{\phi}.$$

In the shunt motor, K_1, V, R_a, and ϕ are substantially constant, the only variable being I_a. As the load on the motor

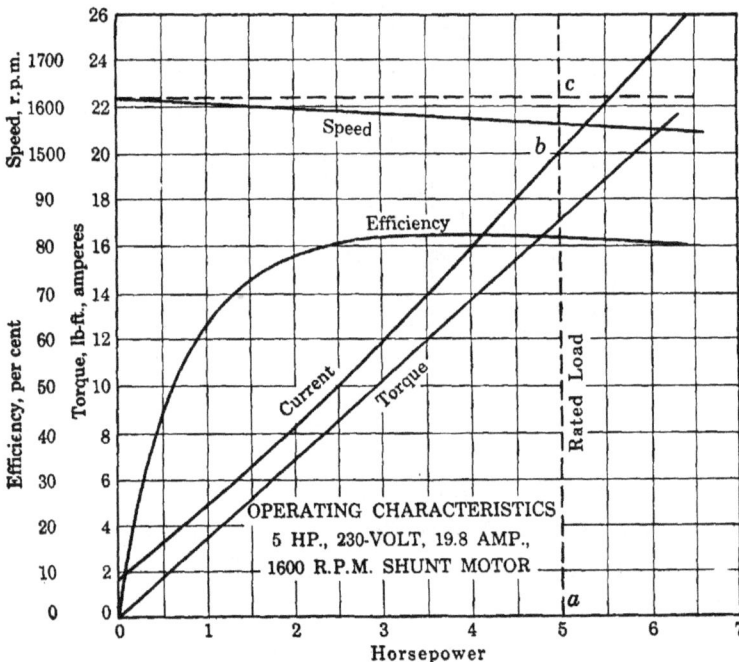

Fig. 243.—Typical shunt-motor characteristics.

increases, I_a increases and the numerator of Eq. (99) decreases.
As a rule, the denominator changes only by a small amount and
the speed of the motor will drop with increase of load, as shown
in Fig. 243. Since $I_a R_a$ is ordinarily from 2 to 6 per cent of V,
the percentage drop in speed of the motor is of the same order of

magnitude. For this reason, the shunt motor is considered a constant-speed motor, even though its speed does decrease slightly with increase of load (see Example, p. 273).

Owing to armature reaction, the resultant field flux ϕ ordinarily decreases slightly with increase of load, and this tends to maintain the speed constant. In modern motors with short air-gaps, the armature reaction may be sufficiently great to give a rising speed characteristic with increase of load. To stabilize the motor, a few series turns are added to the field winding, which act to increase the flux with increase of load (see Par. 220). When series turns are added for this purpose only, the motor is still considered as being a shunt motor.

Figure 243 gives the four essential characteristics of the shunt motor, torque, speed, current, efficiency, each plotted as a function of horsepower. The torque is practically proportional to the horsepower, and the speed drops slightly with load, as has been demonstrated. At light loads the efficiency is low, since the fixed losses, friction, windage, iron losses, field loss, are large in comparison with the output. As the output increases, the efficiency increases. The efficiency reaches a maximum, and then decreases because of the armature loss, which

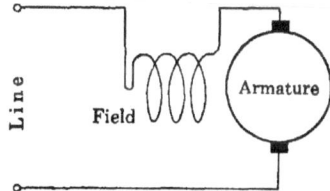

FIG. 244.—Connections of series motor.

increases approximately as the square of the load ($I_a^2 R_a$). The maximum efficiency occurs when the fixed losses and the armature loss are substantially equal.

Shunt motors are used where substantially constant speed is required, as in machine-shop drives, spinning frames, blowers.

The shunt motor can develop full-load torque or even more on starting, provided the starter has the required current-carrying capacity.

219. Series Motor.—In the series motor, the field is connected in series with the armature (Fig. 244). The field has comparatively few turns of wire, and this wire must be of sufficient cross-section to carry the rated armature current of the motor.

In the series motor, the flux ϕ depends entirely on the armature current. If the iron of the motor is operated at moderate saturation, the flux will be nearly proportional to the armature current. Therefore, in the expression for torque,

$$T = K_t I \phi,$$

if ϕ is assumed to be proportional to I, the torque is given by

$$T = K_t' I^2 \qquad (100)$$

when K_t' is a constant.

The torque under these conditions is nearly proportional to the *square* of the armature current (Fig. 242). When the current is 30 amp, the torque is 20 lb-ft (2.77 kg-m); at 60 amp, the torque is 80 lb-ft (11.06 kg-m). That is, doubling the armature current results in quadrupling the torque. This characteristic of the series motor makes its use desirable where large increase of torque is desired with moderate increase in current. In practice, saturation and armature reaction tend to prevent the torque increasing as rapidly as the square of the current.

When Eq. (99), p. 273, is applied to the series motor, the speed

$$S = K_1 \frac{V - I(R_a + R_s)}{\phi} \qquad (101)$$

where K_1 is a constant, V the terminal voltage, I the motor current, R_a the armature resistance including brushes, R_s the series-field resistance, and ϕ the flux entering the armature from a north pole. R_s, the resistance of the series field, is now added to the armature resistance in order to obtain the total motor resistance. Both I and ϕ vary with the load.

As the load increases, the voltage drop due to the resistance of the field and armature increases, because this voltage drop is proportional to the current. Therefore, the back emf becomes less, which causes the motor to run more slowly, although this effect is of the magnitude of only a few per cent. The flux ϕ, however, increases almost directly with the load. Therefore the speed must drop, in order that the back emf be of the proper value, which is usually a few per cent less than the terminal voltage. Both effects tend to slow down the motor. The resistance drop is ordinarily from 2 to 6 per cent of the terminal voltage V, so that its effect on the speed is of this order of magnitude. As the speed is inversely proportional to the flux ϕ, a given percentage change in ϕ produces the same percentage change in speed. Therefore, the speed of the motor is nearly determined by the flux.

When the load is decreased, the flux ϕ decreases correspondingly and the armature must speed up in order to develop the required back emf. If the load be removed altogether, ϕ becomes extremely small, resulting in a very high speed. It is dangerous to remove the load from a series motor, as the armature is almost certain to reach a speed at which centrifugal force will wreck it.

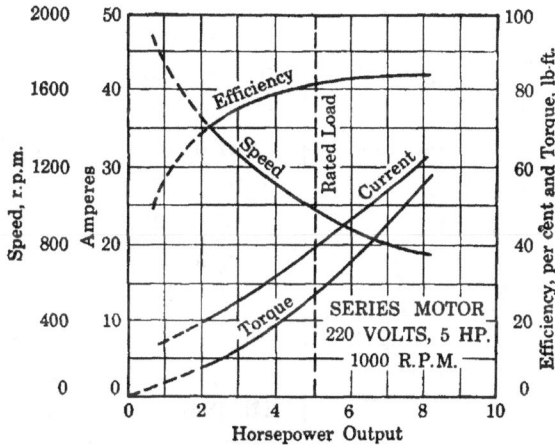

Fig. 245.—Typical series-motor characteristics.

Figure 245 gives the characteristic curves of a 5-hp, 220-volt, series motor, plotted with horsepower as abscissas. The torque curve concaves upward, for the reasons which have been stated. The torque at the pulley increases less rapidly than the square of the current, because of armature reaction, saturation, friction and windage, iron losses. The speed is practically inversely as the current; that is, at large values of current the speed is low and at small values of current the speed is high. The characteristics cannot be determined experimentally for very small values of current, as the speed becomes dangerously high.

Series motors are used for work which demands large starting torque, as for streetcars, locomotives, cranes. Series motors are particularly well adapted to electric railway work, where cars or trains must accelerate rapidly and are often required to start on grades.

220. Compound Motor.—A shunt motor may have an additional series-field winding in the same manner as a shunt generator. This winding may be connected so that it aids the shunt winding, in which case the motor is said to be *cumulative-compound;* or the series winding may oppose the shunt winding, in which case the motor is said to be *differential-compound.*

The characteristics of the *cumulative-compound* motor are a combination of the shunt and series characteristics. As the load is applied, the series turns increase the flux, causing the torque for a given current to be greater than it would be for the simple shunt motor. On the other hand, this increase of flux causes the speed to decrease more rapidly than it does in the simple shunt motor. These characteristics are shown in Fig. 246. The cumulative-compound motor develops a high torque with sudden increase of load. It also has a definite no-load speed and does not run away when the load is removed. Its principal field of application is in driving machines which are subject to sudden heavy loads, such as occur in rolling

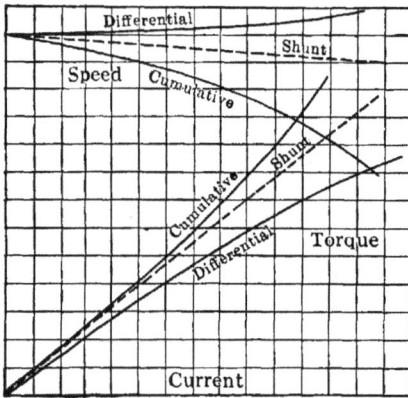

Fig. 246.—Torque and speed characteristics of shunt and compound motors.

mills, shears, punches. This type of motor is used also where a large starting torque and constant running speed are desired, as with elevators.

In the *differential-compound* motor, the series field opposes the shunt field, so the flux is decreased as the load is applied. This results in the speed remaining substantially constant or even increasing with increase of load. This speed characteristic is obtained with a corresponding decrease in the rate at which torque increases with load. Because of its tendency to speed instability and the fact that the speed of the shunt motor is substantially constant, the differential-compound motor is little used. Typical torque and speed curves of the differential-compound motor are shown in Fig. 246.

To reverse the direction of rotation in any motor, the connections to either the armature alone or to the field alone must be reversed. If both are reversed, the direction of rotation remains unchanged. Therefore, insofar as the direction of rotation of the motor is concerned, it is immaterial which line is positive.

221. Motor Starters.—It is shown in Par. 213, p. 269, that if a 10-hp, 110-volt motor were connected directly across 110-volt mains, the resulting current would be 110/0.05, or 2,200 amp. Such a current would not be permissible under any conditions. Hence, resistance should be

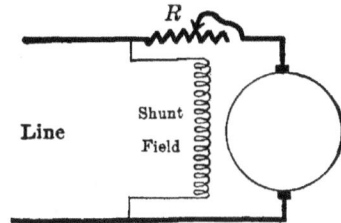

Fig. 247.—Resistance used for starting purposes.

connected in series with the motor *armature* when starting. This resistance may be gradually cut out as the armature comes up to speed and develops back emf.

Figure 247 shows the use of a simple resistance R for starting a motor. It will be noted that this resistance is in the *armature* circuit and that the field is connected directly across the line and outside the resistance. If the field were connected across the armature terminals, putting the resistance R in series with the entire motor, there would be little or no voltage across the field at starting. There would be little torque developed and difficulty in starting would occur.

Figure 248 shows a three-point starter. This does not differ fundamentally from the connections shown in Fig. 247. One line connects directly to an armature and a field terminal connected together. It makes no connection whatever with the starting box. The other line connects to the line terminal of the starting box, which is connected directly to the starting arm. The armature terminal of the starting box, which is connected to the right-hand end of the starting resistance, is connected to the other armature terminal of the motor. The field connection in the starting box is from the first starting contact, through the hold-up magnet, to the field terminal of the box. This field terminal is connected directly to the other terminal of the shunt field.

When the starting arm makes connection with the first contact, the field is connected directly across the line and at the same time the entire starting resistance is in series with the armature. As the arm is moved over the contacts, the starting resistance is gradually cut out.   When the arm reaches the running position,

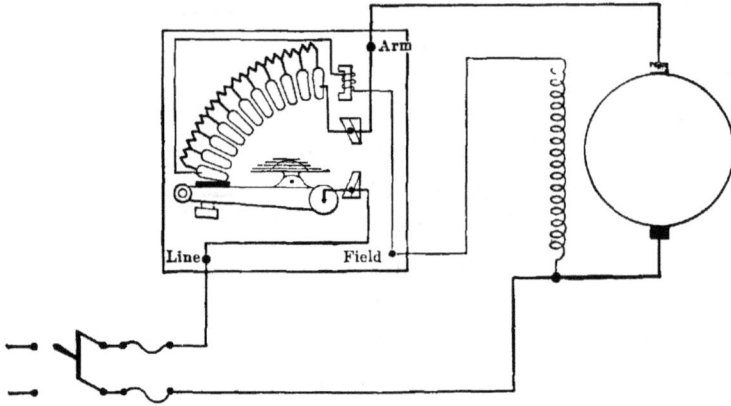

FIG. 248.—Three-point starting box.

the starting resistance is all cut out and, to ensure good contact, the conductors to the line and to the armature frequently are connected directly, by a laminated copper brush, shown in Fig. 248. The field current now feeds back through the starting resistance. This resistance is so low compared with the resistance of the field itself that it has no material effect on the value of the field current. A spring tends to pull the starting arm back to the starting position. When the arm reaches the running position, it is held against the action of this spring by a soft-iron magnet (hold-up magnet) connected in series with the shunt field. If, for any reason, the line is without voltage, the starting arm springs back to the starting position. Otherwise, if the voltage returned to the line after a temporary shutdown, the stationary motor armature would be connected directly across the line and a short circuit would result.

The advantage of connecting the hold-up coil in series with the field is that, should the field circuit become open, the arm springs back to the starting position and so prevents the motor running away.

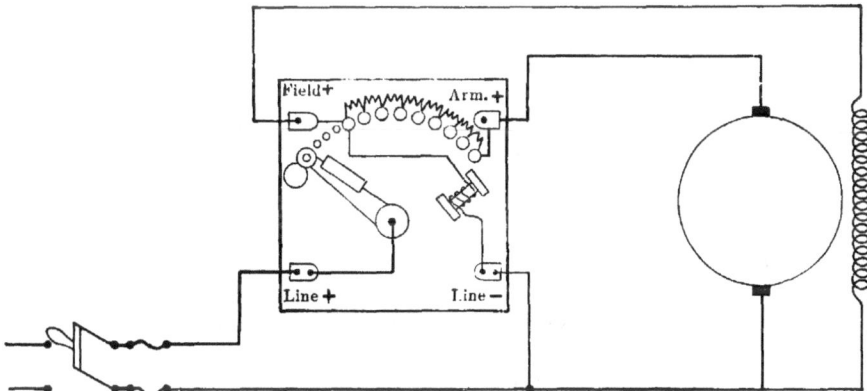

FIG. 249.—Connections of four-point starting box.

The three-point starting box cannot be used to advantage on variable-speed motors having field control. Such motors frequently have a speed variation of 5 to 1. The field current will have approximately this same variation. The hold-up magnet may be too strong, therefore, at the higher values of field current and too weak at the lower values. To obviate this difficulty a four-point box is used (Fig. 249). It is similar to the box shown in Fig. 248, except that the hold-up coil is of high resistance and is connected *directly across the line*. There is now a negative line terminal (Line −) which must be connected to the side of the line which runs directly to the common armature and field terminal. When the voltage goes off the line, the hold-up coil becomes de-energized and allows the arm to spring back to the starting position.

Sometimes the field resistance is contained within the starting box. The box then has two arms of different lengths. The shorter arm is pushed up by the longer arm and cuts out the armature resistance in the ordinary manner. When the starting resistance is entirely cut out, the shorter arm is held by the magnet, and the longer arm, which has no spring, inserts resistance in the field circuit when moved backward.

The *line switch* should always be opened to stop the motor. Do not stop the motor by throwing back the starting arm, because the field discharge burns the contacts. With shunt motors, the line switch can be opened with no appreciable arc, since the motor has a high back emf and the field can discharge gradually through the armature.

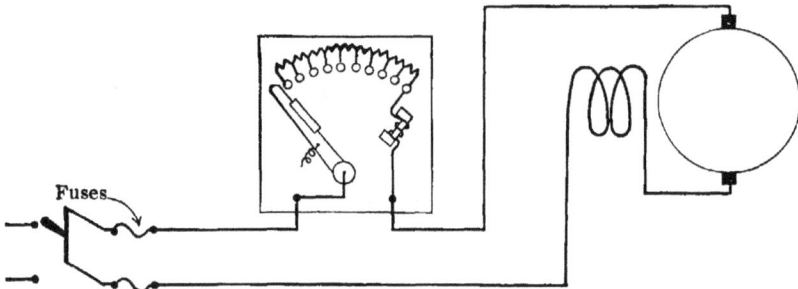

FIG. 250.—Series-motor starter, no-load release.

The *series-motor* starter needs no shunt-field connection. There are two principal types, one having a no-voltage release, and one having a no-load release, shown in Fig. 250. In the no-voltage type, the hold-up coil is connected directly across the line and releases the arm when the voltage goes off the line. In the no-load type, the hold-up coil consists of a few turns in series with the motor. When the motor current falls below the desired value, the starting arm is released. The no-load type is particularly adapted to series motors where there is a possibility of the load dropping to such a low value that the motor speed may become dangerous. Some series-motor starters combine a shunt coil with the series coil.

Controllers are used where the operation of the motor is continually under the direct control of an operator, as in streetcar, crane, and elevator motors. The controller must be more rugged than the starting box, since the controller is in continual use for starting, stopping, and reversing the motor. Such controllers usually have an external resistance, which is cut in and out by fingers in the controller. A shunt-motor field rheostat may also be incorporated within the controller. Controllers are usually fitted with a "reverse," so that the motor may be run in either direction.

Some types of starter are built in the form of controllers. The resistance, usually of the grid type, is designed to carry the rated

Fig. 251.—Grid resistor. (*Cutler Hammer Manufacturing Company.*)

current of the motor continuously so that it may be used to secure speed control also.

222. Resistance Units.—Starting boxes are usually designed for starting duty only. They can carry the starting current of the motor safely for the short period of starting, but they cannot carry such a current continuously. In the smaller types, the resistance is of wire, frequently wound in the form of a helix. It may be self-supporting, or it may be wound on asbestos or porcelain forms. In the larger types, cast-iron grids (Fig. 251) are used. These grids are bolted together. Current lugs are clamped on at suitable points so that the desired range of resistance is readily obtained. Such grids are also used in *controllers* (Pars. 221 and 223).

223. Speed Control.—In the equation for motor speed, $S = K_1\dfrac{E}{\phi}$ (see Eq. (98), p. 273), there are but two factors which can be changed to secure speed control without making changes in the motor construction. These factors are the back emf E and the flux ϕ.

Armature-resistance Control.—In this method, the speed control is obtained by connecting a resistance directly in series with the motor *armature*, keeping the field across the full-line potential (Fig. 252 (a)). A wide range of speed can be obtained by this method, and at the same time the motor will develop any desired

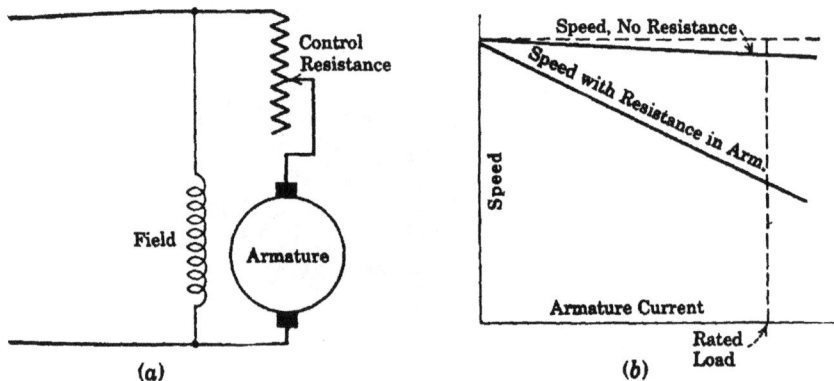

FIG. 252.—Speed control and regulation—armature-resistance method.

torque over its working range, for the torque depends only on the *flux* and armature *current*.

The principal objections to this method of speed control are that an excessive amount of power is lost in the armature series resistance, and the *speed regulation* is very poor. In Fig. 252 (b) are shown for comparison the speed-load curves of a shunt motor with and without resistance in series with the armature. The speed-load curve with series resistance in the armature circuit shows that half speed occurs at rated load. It will be observed that the speed at no load rises to a value which is practically equal to the speed of the motor when there is no resistance in series with the armature. With resistance in the armature circuit, the change of speed with load is excessive. Furthermore, with 50 per cent reduction in speed, about 50 per cent of the power supplied to the armature circuit is lost in the series resistance. Without series resistance, the speed change from no load to full load is the usual 3 or 4 per cent. The control resistance is usually a *controller* having a continuous carrying capacity equal to the full-load armature current.

Example.—The armature resistance of a 20-hp, 230-volt, shunt motor is 0.15 ohm. When the motor operates at no load, the armature current is 3.0 amp and the speed is 1220 rpm. (*a*) What external resistance must be inserted in the armature circuit to reduce this speed to 600 rpm at rated load of 75 amp? (*b*) What is the speed at rated load with no external resistance in the armature circuit? Assume constant field current, and neglect armature reaction.

(*a*) $E_1 = 230 - (3.0 \times 0.15) = 229.5$ volts.
Let E_2 be the induced emf at 600 rpm.
Since the flux does not change,

$$\frac{E_2}{E_1} = \frac{E_2}{229.5} = \frac{600}{1,220} \text{ (see Eq. (98), p. 273).}$$
$$E_2 = 113 \text{ volts.}$$

The total resistance in the armature circuit

$$R' = \frac{230 - 113}{75} = \frac{117}{75} = 1.56 \text{ ohms.}$$

Since $Ra = 0.15$ ohm, there must be inserted $1.56 - 0.15 = 1.41$ ohms.
Ans.

(*b*) At rated load,

$$E_2 = 230 - (75 \times 0.15) = 230 - 11.3 = 218.7 \text{ volts.}$$
$$\frac{S_2}{1,220} = \frac{218.7}{229.5} \text{ (see Eq. (98), p. 273).}$$
$$S_2 = 1,163 \text{ rpm.} \quad Ans.$$

Multivoltage System.—In this system, several different voltages are available at the armature terminals of the motor, while the shunt field of the motor is connected permanently across one of these voltages. Owing to the fact that this system requires several wires for each motor, it is little used in this country.

Field Control.—In the foregoing methods of speed control, the voltage impressed on the armature is varied. Change of speed may be obtained also by varying the flux ϕ by means of a field rheostat. Since the speed varies *inversely* as the flux (Eq. (98), p. 273), the field is *weakened* to *increase* the speed and is *strengthened* to *decrease* the speed. This method is very efficient, since the power lost in the field is small, and for any particular field adjustment the speed variation from no load to full load is also small. With the ordinary motor, the range of speed obtainable by this method is limited by commutation difficulties.

The main field is weakened to increase the speed. For a given power load, the armature ampere-turns remain constant. A study of Fig. 240, p. 275, shows that with constant armature mmf and a decreasing field mmf, the resultant flux tends more and more to take the direction of the armature flux. As a result, the neutral plane shifts further and further backward. In order to commutate properly, the brushes must likewise be shifted backward. This increases the demagnetizing effect of the armature, which results in still further backward shifting of the neutral plane. Shifting the brushes with the neutral plane to improve commutation merely results in increased demagnetization, and ultimately in vicious sparking. There is also a tendency for the motor to become unstable and to run away. Hence, with ordinary motors, field control cannot be used to obtain a wide range of speed, since satisfactory commutation at the higher speeds is practically impossible to secure.

With commutating poles, however, the brushes remain in the geometrical neutral, the demagnetizing action of the armature is small, and good commutation may be obtained over a wide range of speed. A range of 5 to 1 in speed variation is obtainable with properly designed motors having commutating poles.

Because of its efficiency and simplicity, this method of speed control is used ordinarily.

224. Railway-motor Control.—In a two-motor trolley car, two different speeds can be obtained efficiently. The motors are first connected in series through a starting resistance R (Fig. 253 (a)). This resistance is gradually cut out by the controller as the car comes up to speed, and then each motor operates at one-half line voltage. This is the first running position. For any given value of armature current, each motor will run at half its rated speed. Since rated current can flow in both motors, each can develop rated torque. Therefore, under these conditions, each motor develops one-half rated power. As there is no external resistance in the circuit, the motors are operating at an efficiency very nearly equal to that obtainable with full line voltage across the terminals of each motor.

When it is desired to increase the speed of the car, the two motors are connected in parallel and in series with a portion of the resistance R. This resistance is gradually cut out; and when the running position is reached, each motor receives full line voltage (Fig. 253 (*b*)).

Multiple-unit Control.—In the larger electric cars and in electric locomotives, the currents become so great that direct platform control is out of the question from the standpoint of size of controller, safety, and expense. Moreover, when cars are operated in trains, it is necessary that the motors on all the cars shall be under a single control and that each step in the control shall operate simultaneously for all the motors.

In the multiple-unit system, all the heavy-current switching is done usually by pneumatic- or solenoid-operated contactors located beneath the car. These contactors, in turn, are operated by an auxiliary circuit called the *train line*, which receives its power through the master controller operated by the motorman. After the motorman has moved his controller handle to the running position, the closing of the contactors and hence the rate of acceleration are automatic.

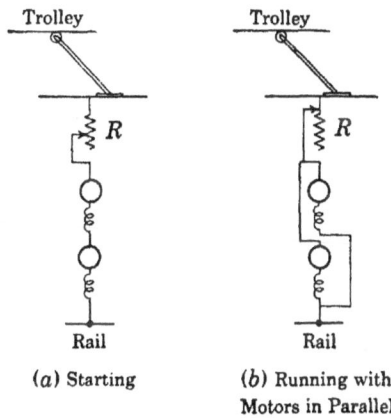

(a) Starting

(b) Running with Motors in Parallel

FIG. 253.—Series-parallel control of series motors.

225. Motor Testing—Prony Brake.—It is often necessary to determine the efficiency of a motor at certain definite loads, and frequently over its entire range of operation. This gives data for obtaining characteristic curves (see Fig. 243, p. 278, and Fig. 245, p. 281). A knowledge of the efficiency may be necessary, as in an acceptance test; further, the motor may be used as a power-measuring device for determining the power taken by some machine, such as a generator, pump, blower. Knowing the motor input, which can be measured with ammeter and voltmeter, and knowing the motor efficiency, the output for any given input can be computed. This output will be the power delivered to the generator, the pump, etc.

The most common method of making direct measurements of
efficiency in motors up to about 50 hp is to use a prony brake.
Such brakes are made in various forms.   One typical form is
shown in Fig. 254.   It consists of a wooden arm of the proper
length, a canvas brake band, and a handwheel for applying tension to the brake band.   By means of this handwheel the motor

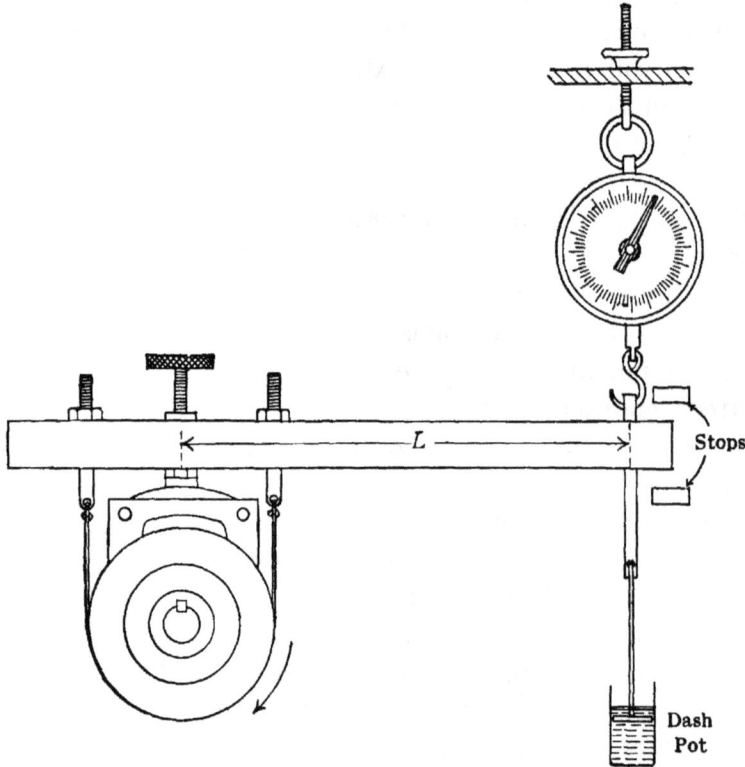

FIG. 254.—Typical prony brake.

load can be controlled.   An oil dashpot is advisable, to prevent
vibrations of the brake arm.

The balance measures the pull on the arm due to the rotation
of the drum, plus the dead weight of the arm.   By multiplying
the net balance reading by the distance L, the torque of the motor
is determined.

There are two simple methods for determining the dead weight,
or tare, of the brake arm.   The brake band is loosened and some
sort of knife-edge, such as a pencil, is placed between the top of
the drum and the brake carriage.   This acts as a substantially

frictionless fulcrum, so that the balance reads the dead weight of the arm alone. Another and easier way is to turn the drum toward the balance by hand, stop, and read the balance. In this case the friction of the brake causes the balance to read too high. If the operation be repeated, rotating the drum in the opposite direction, the balance reading will be too low, owing to the same friction. The average of these two balance readings will give very nearly the correct value for the dead weight of the arm.

Brakes of this type are cooled ordinarily by pouring water into the hollow brake drum. This water prevents the drum from

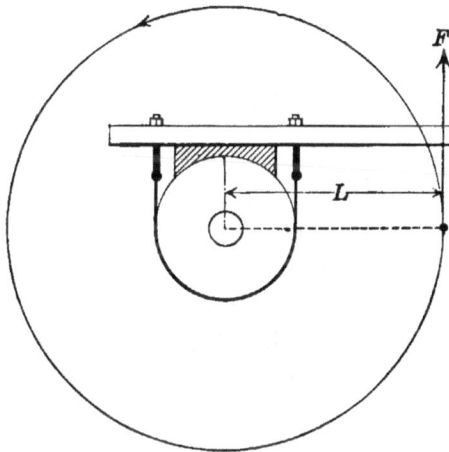

FIG. 255.—Work developed by prony brake.

becoming excessively hot. As the maximum temperature which water can reach in the open air is 100°C, the drum temperature cannot much exceed this. The heat developed in the drum is utilized in converting the water into steam. As a considerable number of heat units is required to convert a small amount of water into steam, a moderate amount of water will keep the drum comparatively cool.

To determine the equation for the horsepower absorbed by such a brake, consider Fig. 255. Let F be the net force in pounds acting at a perpendicular distance L feet from the center of the drum. First assume that the drum is stationary and that the arm is pulled around the drum by the force F. The distance per revolution through which the force F acts is $2\pi L$. The work done in one revolution of the arm around the drum is the force times the distance $= F(2\pi L)$.

The work done in S revolutions $= F(2\pi L)S$.

If S is the rpm, the horsepower

$$\text{Hp} = \frac{2\pi(FL)S}{33,000}.$$

But FL is the torque T; therefore,

$$\text{Hp} = \frac{2\pi TS}{33,000},$$

$$\frac{2\pi}{33,000} = 0.00019.$$

Therefore,

$$\text{Hp} = 0.00019TS. \quad (102)$$

FIG. 256.—Rope brake.

Obviously, the same amount of work is done on the brake surface whether the drum is stationary and the arm rotates or the arm is stationary and the drum rotates. Therefore, Eq. (102) applies to brakes of the type shown in Figs. 254 and 255. It will be noted that in this particular type of brake the horsepower is independent of the diameter of the drum.

Example.—In a brake test of a shunt motor, the ammeter and voltmeter measuring the input read 34 amp, 220 volts. The speed of the motor is found to be 910 rpm and the balance on a 2-ft (0.61 m) brake arm reads 26.2 lb (11.9 kg). The dead weight of the arm is found to be +2.4 lb (1.09 kg). (a) What is the output of the motor? (b) What is its efficiency at this particular load?

(a) Net reading of balance $= 26.2 - 2.4 = 23.8$ lb (10.8 kg).

The torque $T = 23.8 \times 2 = 47.6$ lb-ft (6.58 kg-m).

Hp output $= 0.00019 \times 47.6 \times 910 = 8.23$ hp. *Ans.*

(b) Output $= 8.23 \times 746 = 6,140$ watts.

Input $= 220 \times 34 = 7,480$ watts.

$$\text{Efficiency, } \eta = \frac{6,140}{7,480}\,100 = 82.1 \text{ per cent.} \quad Ans.$$

In brakes of this type, the brake arm should be kept approximately level.

Another simple type of brake is the rope brake shown in Fig. 256. A rope is given a turn and a half around a drum, and the two free ends are held each by a spring balance. The larger balance is on the end of the rope which is being pulled downward by the rotation of the drum. Let F_1 be the reading of the larger balance and F_2 that of the smaller balance. As F_1 and F_2 pull in opposite directions with respect to the rotation of the drum, the net pull at the drum periphery is $F_1 - F_2$.

The torque in pound-feet is

$$T = (F_1 - F_2)R \qquad (103)$$

where R is the radius of the pulley in *feet*.

Example.—In a rope brake of the type shown in Fig. 256, $F_1 = 32.4$ lb (14.7 kg) and $F_2 = 8.2$ lb (3.72 kg). The drum is 10 in. (0.254 m) in diameter. If the motor speed is 1,400 rpm, what horsepower does the motor develop?

The torque

$$T = (32.4 - 8.2)\tfrac{5}{12} = 24.2 \times \tfrac{5}{12} = 10.08 \text{ lb-ft } (1.394 \text{ kg-m}).$$

The horsepower

$$\text{Hp} = 0.00019 \times 10.08 \times 1,400 = 2.68 \text{ hp.} \quad Ans.$$

226. Measurement of Speed.—As a rule, the measurement of the speed of machines is much simpler than the measurement of torque. The most common method is to use a simple revolution counter having a conical rubber tip which fits into the countersink of the shaft.

Tachometers indicate the instantaneous value of speed. There are mechanical tachometers, where the indicator is actuated by centrifugal action. This type should be carefully checked on each occasion of use, as it is subject to error, especially after having been in service for some time.

A simple and convenient type of tachometer is the combination of a d-c magneto and a voltmeter (Fig. 257 (a)). In the magneto, the flux is produced by permanent magnets and so is constant. Therefore, the emf induced in the magneto armature is directly proportional to the speed. If this emf be measured with a voltmeter, the reading multiplied by a constant gives the speed directly. The relation of speed to volts may be plotted, as shown in Fig. 257 (*b*), and the speed read directly from the plot. This plot is a straight line through the origin, so that one point on the plot is accurately known. Manufacturers now supply magnetos with voltmeters calibrated directly in rpm. Although

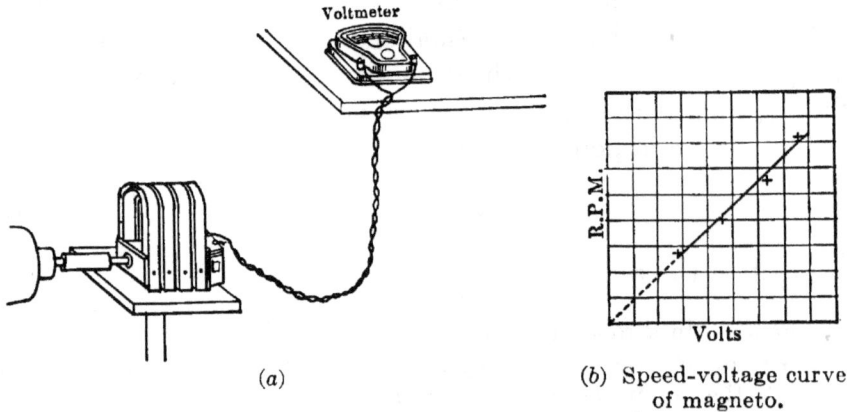

(*a*)

(*b*) Speed-voltage curve of magneto.

FIG. 257.—Speed measurement with magneto and voltmeter.

these do not require calibration ordinarily, they should be checked occasionally. A piece of rubber tubing is conveniently used to attach the magneto to the shaft of the machine whose speed is being measured. It is usually necessary to thread a small stud into the end of the shaft whose speed is to be measured, as shown in Fig. 257 (*a*).

CHAPTER XIII

LOSSES; EFFICIENCY; OPERATION

227. Dynamo Losses.—A part of the energy delivered to any motor or generator is lost within the machine itself, being converted into heat and wasted. Not only is energy lost, but there is the further objection that heating limits the output of a dynamo. If the energy loss in a dynamo becomes excessive, the resulting temperature rise may injure the insulation by carbonizing it.

As motor and generator are similar, they have the same types of losses. Therefore, the following applies to either motor or generator.

COPPER LOSSES

Armature.—The armature windings have resistance, and when current flows through them power must be lost. In addition to the loss in the armature copper, there is an electrical loss in the brushes and in the commutator. Let this total power loss be P_a. Then,

$$P_a = I_a{}^2 R_a \qquad (104)$$

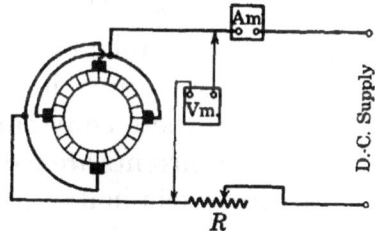

Fig. 258.—Measurement of armature resistance.

where I_a is the armature current and R_a is the armature resistance measured between the terminals of the dynamo and including, therefore, the brushes and their contact resistance. The resistance measurement is often made by the connections shown in Fig. 258, readings being taken with the armature in three or four different positions (also see Figs. 84 and 85, pp. 103 and 104). The resistance R is inserted to limit the current flowing through the stationary armature (see Par. 221).

Shunt Field.—The shunt-field circuit takes a current I_f at the terminal voltage V of the generator or motor. Therefore, the power lost in the shunt field, including the rheostat, is

$$P_f = VI_f. \tag{105}$$

The foregoing losses are all copper losses and can be measured directly or can be calculated with a high degree of precision from instrument readings.

IRON LOSSES

Eddy Currents.—As the armature iron rotates in the same magnetic field as the copper conductors, emfs are induced in this iron. As the iron is a good conductor of electricity and the current paths are short and of large cross-section, large currents will be induced in the armature iron if it is a solid mass as

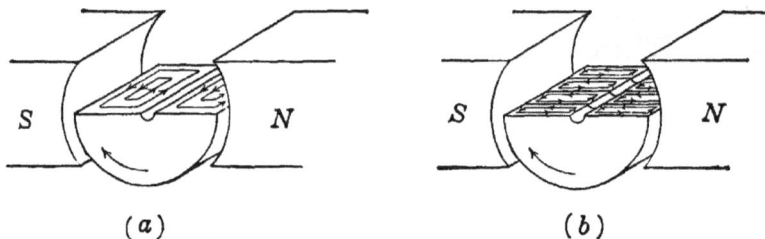

FIG. 259.—Eddy currents in armature iron without and with laminations.

indicated in Fig. 259 (a). These currents produce power loss which cannot be tolerated in commercial machines. By laminating the armature iron in the manner indicated in Fig. 259 (b), the paths of these eddy currents are broken up and the magnitudes of the currents are reduced to low values. Laminating does not entirely eliminate these eddy-current losses, but it does reduce them to small magnitudes. Although the laminations break up the eddy-current paths, they do not interpose reluctance in the magnetic circuit, since they are parallel to the direction of the magnetic flux.

Hysteresis.—It is shown in Chap. VIII that when iron is carried through a cycle of magnetization (p. 165), an energy loss results proportional to the area of the hysteresis loop.

As the armature rotates, all parts of the laminated structure pass alternate north and south poles and the magnetization obviously undergoes reversals. These reversals of magnetization give rise to hysteresis loss. *Laminating the iron does not affect hysteresis loss.*

FRICTION LOSSES

These losses arise from bearing friction, brush friction, and windage.

STRAY POWER

Unlike the copper losses, the iron and friction losses cannot be calculated readily. *Also, they are supplied mechanically and not electrically.* In a generator, they are supplied directly by the driving torque of the prime mover; in a motor, they cause a torque in opposition to the internal torque and thus reduce the torque at the pulley. As these losses depend on either the speed or the flux, or both, and are more or less indeterminate in their nature, they are combined into a single loss, called *stray-power* loss.

228. Efficiency.—The efficiency of a dynamo is the ratio of output to input. Thus,

$$\text{Eff.} = \frac{\text{output}}{\text{input}}.$$

This may also be written in either of the following ways:

$$\text{Eff.} = \frac{\text{output}}{\text{output} + \text{losses}}. \tag{106}$$

$$\text{Eff.} = \frac{\text{input} - \text{losses}}{\text{input}}. \tag{107}$$

Therefore, if the losses in a dynamo are known, the efficiency may be found for any given input or output.

As electrical units rather than mechanical units are used ordinarily in efficiency determinations, Eq. (106) is used for generators (output is electrical) and Eq. (107) for motors (input is electrical).

Example.—A 50-kw, 220-volt generator is delivering its rated load of 50 kw (227 amp at 220 volts). At this load the total generator losses are 4,100 watts. Determine: (a) the generator input in watts; (b) the generator efficiency at rated load; (c) the horsepower required to drive the generator.

(a) Input = 50,000 + 4,100 = 54,100 watts. *Ans.*

(b) Efficiency $\eta = \dfrac{50,000}{50,000 + 4,100} = 0.924$, or 92.4 per cent (Eq. (106)).

Ans.

(c) Hp = 54,100/746 = **72.5 hp.** *Ans.*

Example.—A shunt motor takes 52 amp, at 232 volts. The total motor losses at this load are 2,000 watts. What is the motor efficiency?

Using Eq. (107),

$$\text{Efficiency } \eta = \frac{(232 \times 52) - 2{,}000}{232 \times 52} = \frac{10{,}060}{12{,}060} = 0.835, \text{ or } 83.5 \text{ per cent.}$$

Ans.

229. Efficiencies of Motors and Generators.—The efficiency of electrical apparatus is high as a rule. For example, at rated load a 1-hp motor has an efficiency of about 76 per cent; a 5-hp, 83 per cent; a 10-hp, 86 per cent; a 20-hp, 87 per cent. A 500-kw dynamo may have an efficiency of 94 per cent.[1]

The efficiency of a motor may be determined from simultaneous measurements of input and output, as is shown in Par. 225, where a prony brake is used. With larger motors it becomes difficult to construct a brake capable of absorbing the power.

It is difficult to make direct measurements of generator efficiency. The output is readily measured with ammeter and voltmeter, but a direct measurement of input can be made only by some special device, such as a cradle dynamometer.

In the direct measurement of efficiency, any percentage error in the measurement of either output or input introduces the same percentage error in the efficiency.

In the direct measurement of efficiency, the power necessary for the test must be equal to the rating of the dynamo. In addition to *supplying* this power, there must be means for *absorbing* it. This is not a serious matter with small dynamos; but when large dynamos are tested, supplying and absorbing the necessary power may be difficult, if not impossible.

Because of the foregoing reasons, it is often desirable and even necessary to obtain the efficiency by determining the losses. As is shown in Par. 227, these losses include armature-resistance loss, shunt-field loss, and stray power. With a compound dynamo there will also be the series-field loss.

Example.—A 250-kw, 230-volt, compound generator delivers 800 amp at 230 volts. The field current is 20 amp. The armature resistance is 0.005 ohm and the series-field resistance is 0.002 ohm. The stray power at this load is 2,500 watts. The generator is connected long shunt. What is the generator efficiency at this load?

[1] See DAWES, C. L., "A Course in Electrical Engineering," 3d ed., Vol. I, Direct Currents, p. 533, for approximate efficiencies of motors.

Output = 230 × 800 = 184,000 watts.
Shunt-field loss = 230 × 20 = 4,600 watts
Armature loss = 820² × 0.005 = 3,360 watts
Series-field loss = 820² × 0.002 = 1,340 watts
Stray power = 2,500 watts

Total loss = 11,800 watts

$$\text{Eff.} = \frac{184,000}{184,000 + 11,800} = \frac{184,000}{195,800}, \text{ or 94 per cent.} \quad Ans.$$

230. Measurement of Stray Power.—It is pointed out in Par. 227 that stray power is a function of flux and speed only. If it is desired to determine the stray power of either a motor or a generator, the machine is connected as shown in Fig. 260 and run as a *motor* without load. The field current is adjusted to the value existing under oper-

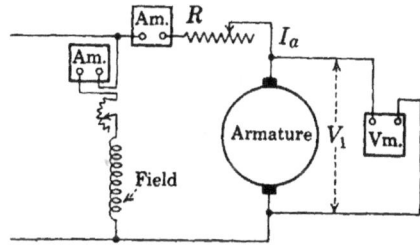

Fig. 260.—Connections for stray-power measurement.

ating conditions. Neglecting armature reaction, this gives the proper value of armature flux.

The resistance R is adjusted until the speed is the same as under operating conditions. Then the stray power

$$S.P. = V_1 I_a - I_a^2 R_a. \tag{108}$$

That is, the stray power is equal to the *armature* input minus the armature-resistance loss, which at no load is nearly negligible.

Since the stray power is but a small percentage of the total power, when the machine is operating under load, a considerable error in the determination of stray power causes only a small error in the efficiency.

Example.—A 50-kw, 230-volt, 800-rpm shunt generator delivers rated load at rated voltage and speed. The armature resistance is 0.032 ohm and the shunt-field current is 4.2 amp. The generator is then run as a motor without load, from 240-volt mains (see Fig. 260). The field current is maintained at 4.2 amp. When the resistance R is adjusted to give a speed of 800 rpm, the voltmeter reads 237 volts (= V_1). The armature current is then 8.2 amp. Determine: (a) the stray power of the generator; (b) the armature-resistance loss at rated load; (c) the field loss at rated load; (d) the total loss at rated load; (e) the generator efficiency at rated load.

(a) From Eq. (108),
$S.P. = 237 × 8.2 - (8.2)^2 0.032 = 1,940 - 2 = 1,940$ watts (nearly). *Ans.*

(*b*) The rated current

$$I = \frac{50,000}{230} = 217 \text{ amp.}$$

The armature current

$$I_a = 217 + 4.2 = 221 \text{ amp (nearly).}$$

The armature-resistance loss

$$P_a = (221)^2 0.032 = 1,560 \text{ watts.} \quad Ans.$$

(*c*) $P_f = 230 \times 4.2 = 966$ watts. *Ans.*
(*d*) $P = 1,940 + 1,560 + 966 = 4,470$ watts (nearly). *Ans.*
(*e*) Using Eq. (106), p. 299, since the output is electrical,

Efficiency $\eta = \dfrac{50,000}{50,000 + 4,470} = \dfrac{50,000}{54,470} = 0.918$, or 91.8 per cent. *Ans.*

The measurements necessary for determining the efficiency of the generator are made *without actually loading it.* Obviously, the same method may be applied to the motor. This method gives a good degree of accuracy, since large errors in the determination of the losses produce but small errors in the resulting efficiency.

231. Ratings and Heating.—*Electrical apparatus* is usually rated at the load which it can safely carry without *overheating.* (Commutation may at times limit the output of d-c machines.)

If the temperature of electrical apparatus becomes too high, the cotton insulation on the armature and field conductors and the insulating varnishes become carbonized and brittle. This may result ultimately in grounds and short circuits within the machine. The A.I.E.E.[1] Standardization Rules recommend 95°C (203°F) as the maximum safe temperature of cotton and other fibrous insulations.

It is important, therefore, to be able to test a dynamo to determine whether it is operating within safe temperature limits. The difficulty in making such tests lies in the fact that the highest temperatures are within the coils, at points not easily accessible. The highest temperature within the machine is called the "hot-spot" temperature.

The temperature at the surface of the winding may be measured by placing a thermometer bulb against the surface and

[1] American Institute of Electrical Engineers.

covering the bulb with a small pad of cotton. It has been found that 15°C added to the thermometer reading will give an approximate value of the hot-spot temperature.

It has been shown already that the resistance of copper conductors changes with the temperature. By utilizing this principle, some idea of the *average* temperature within a winding may be obtained. For copper, the fractional increase of resistance per degree rise of temperature may be obtained from the expression[1] $1/(234.5 + t)$, where t is the surrounding or ambient temperature in degrees C. For example, at an ambient or room temperature of 30°C, the fractional increase of resistance per degree rise is $1/264.5 = 0.00378$.

Example.—With an ambient temperature of 30°C, the resistance of the field of a shunt generator increases from 104 to 112 ohms. What is its temperature rise?

$$\text{The fractional change in resistance is } \frac{112 - 104}{104} = 0.077.$$

$$\text{Temperature rise} = \frac{0.077}{0.00378} = 20°.4 \text{ C.} \quad Ans.$$

In measuring armature resistance to determine temperature rise, it is essential that the resistance of the *copper alone* be

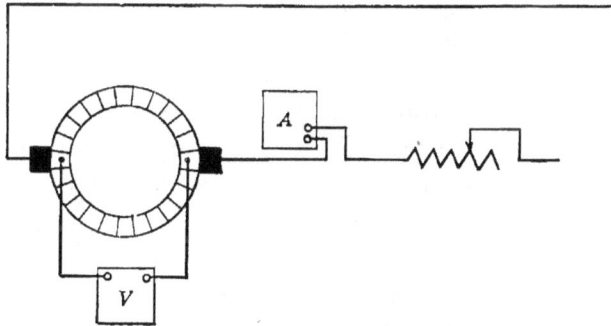

Fig. 261.—Measurement of armature resistance for temperature test.

measured and that the current path through the copper be the same in every measurement. To exclude all resistance except that of the copper, the brush and contact resistances must not be included in the measurement. Therefore, the voltmeter leads must be held on the commutator segments inside the brushes (Fig. 261). Moreover, these segments should be marked, and in

---

[1] See Par. 11, p. 14.

every subsequent measurement they should be directly under the same brushes.

In measuring the shunt-field resistance, the voltmeter should be connected directly across the *winding* so as to exclude the drop in the rheostat.

This resistance method gives an *average* value of the temperature of the windings. To find the hot-spot temperature when the resistance method is used, about 10°C should be added.

232. Parallel Running of Shunt Generators.—In most power plants it is necessary and desirable that the power be supplied by several smaller units rather than by a single large unit. Several smaller units are more reliable than a single large unit.

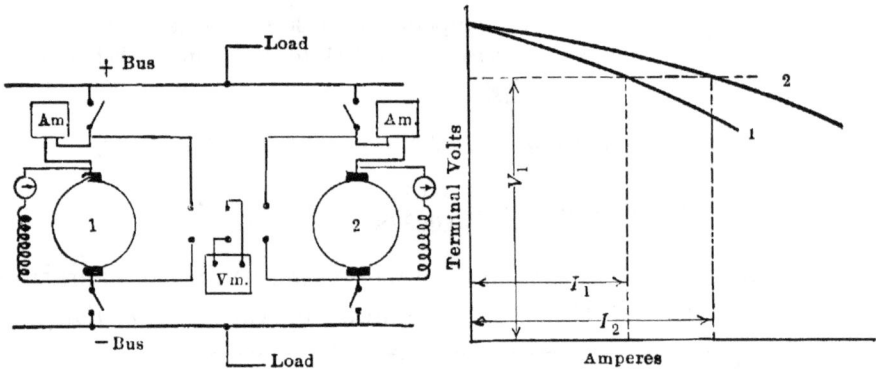

FIG. 262.—Two shunt generators in parallel.

FIG. 263.—Characteristics of shunt generators in parallel.

With several smaller units, if one unit is disabled, the entire power supply is not cut off. The smaller units may be connected in service and taken out of service to correspond with the load on the plant. Additional units may be installed to correspond with the growth of the plant load.

Shunt generators, because of their drooping characteristic, are particularly well suited for parallel operation. In Fig. 262 is shown a diagram of connections of two shunt generators 1 and 2 which may be operated in parallel. The characteristics of these two generators are shown in Fig. 263. It will be noted that generator 1 has the more drooping characteristic.

If the two generators are connected in parallel, their terminal voltages must be the same, neglecting any very small voltage drop in the connecting leads. Therefore, for a common terminal voltage V_1 (Fig. 263), generator 1 delivers I_1 amp and generator 2 delivers I_2 amp. That is, the generator with the more drooping characteristic carries the smaller load.

Hence, two shunt generators of the same rating must have identical characteristics if they are to divide the load equally at all times when operating in parallel. If two shunt generators have different ratings, the voltage drop from no load to full load should be the same.

When operating in parallel, each generator should have its own ammeter. A common voltmeter is sufficient for all generators. The individual generators may be connected to the voltmeter or potential bus through suitable plug connectors or selective switches. A circuit breaker should be connected in the circuit of each generator. The breakers are omitted in Fig. 262.

Assume that 2 is out of service and that 1 is supplying all the load. It is desired to put 2 in service. The prime mover of 2 is started, and 2 is brought up to speed. Its field is then adjusted so that its voltage is just equal to that of the bus-bars, this condition being determined by the voltmeter. The breaker and switch are now closed, and 2 is connected to the system. Under these conditions, however, 2 is not taking any load, as its *induced* emf is just equal to the bus-bar voltage, and no current will flow between points at the same potential. That is, generator 2 is "floating" (p. 46). Its induced emf must be greater than the bus-bar voltage in order that 2 may deliver current to the load. Hence, the field of 2 is strengthened until the generator takes its share of the load. It may be necessary to weaken the field of 1 simultaneously in order to maintain the bus-bar voltage constant.

To take a generator out of service, its field is weakened and that of the other generator is strengthened until the load of the first generator is zero. The breaker and then the switch are opened, clearing the machine. Connecting in and removing a generator from service in this manner will prevent any shocks or disturbance to the prime mover or to the system.

If the field of one generator be weakened too much, the direction of the current is reversed. The generator then operates as a motor and may drive its prime mover.

233. Parallel Running of Compound Generators.—Figure 264 shows two overcompounded generators connected to the bus-bars, positive and negative terminals being properly connected as regards polarity. By proper adjustment of its field, each generator may be made to take its proper share of the load momentarily when the generators are connected in parallel.

Assume that for some reason, such as a slight change in speed, generator 1 takes a slightly increased load. The current in its series-field winding must increase, which strengthens its field and raises its emf, thus causing it to take still more load. On the other hand, as the system load is assumed to be fixed, generator 2 at the same time will drop some of its load, resulting in a weakening of its series field and a consequent further decrease in its load. Almost instantaneously, 1 will be driving 2 as a motor, and ultimately the breaker of at least one of the generators will open.

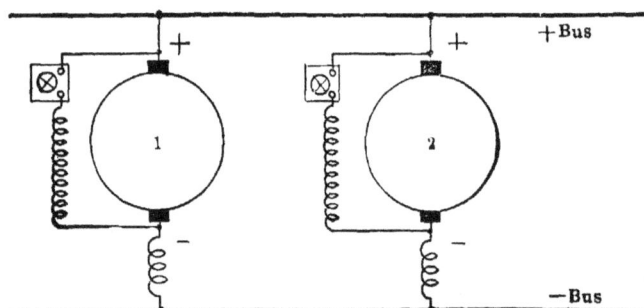

FIG. 264.—Compound generators in parallel.

These two compound generators are in *unstable* equilibrium. That is, any effect tending to throw the generators out of equilibrium is increased by the resulting reactions.

The generators may be made stable by connecting the two series fields in parallel (Fig. 265). This connection, which ties the two negative brushes together, is a conductor of low resistance and is called the *equalizer*. Its operation is as follows: Assume that generator 1 begins to take more than its share of the load. The increase of load current will cause an increase of current not only in the series field of generator 1 but also, by means of the equalizer, in the series field of generator 2. Therefore, both generators are affected in a similar manner, and neither is able to run away with the load.

It should be noted that at any particular load the desired division of load among either shunt or compound generators may be obtained by adjusting their shunt-field rheostats. However, it is usually desirable that the division of load remain constant at all loads, especially if an operator is not in continuous attendance. Therefore, it is desirable that generators operating in parallel should have characteristics such that the voltage drop from no load to full load is the same for all.

The load ammeter in a compound generator should always be connected between the *armature* terminal and the bus-bars. If it is connected in the series-field circuit, the ammeter may not

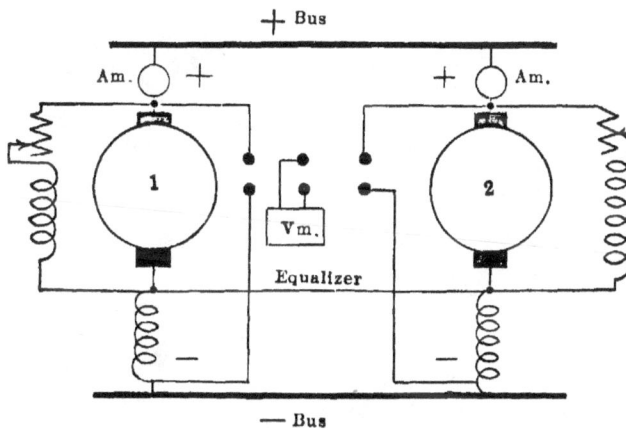

Fig. 265.—Two compound generators in parallel, with equalizer.

indicate the generator current, due to the fact that some of the generator current may be flowing through the equalizer.

Compound generators are put in service and taken out of service in the same manner as shunt generators. The load is adjusted and shifted by means of the shunt-field rheostat.

234. Circuit Breakers.—Generators, motors, and electric circuits in general require protection from short circuits and overloads. The sudden load imposed by a short circuit may injure the generator or its prime mover. Wires may overheat under the short-circuit current, resulting in fire hazard. Two common devices are used for opening both short circuits and overloads, the *fuse* and the *circuit breaker*. The fuse has a much lower first cost and occupies less space. On the other hand, it is worthless after having been blown (unless it is of the refillable type), and considerable inconvenience often results from not having spare fuses at hand. The circuit breaker (Fig. 266) has a higher first cost and requires more space. On the other hand, it operates an indefinitely great number of times without injury and is readily reset. The action of a breaker is more rapid than that of a fuse.

Many times, circuit breakers open on a short circuit or overload which is cleared a moment later by a local breaker or fuse.

FIG. 266.—Automatic reclosing d-c feeder breakers. (*Westinghouse Electric & Manufacturing Company.*)

In order that service shall not continue to be interrupted unnecessarily, automatically reclosing breakers have been developed. After tripping, a mechanism operates to close the breaker. If the short circuit still exists, the breaker cannot reclose; but if the short circuit is cleared, the breaker closes and restores service. The breaker may attempt two or three times to close; and then, if the fault still exists, it becomes permanently locked out and can be closed only by hand. The first and third breakers (Fig. 266) are automatic reclosing breakers for 625-volt, d-c, power feeders. Each feeder breaker is equipped with an automatic resistance-measuring relay (shown on the panel directly beneath the breaker) which controls the reclosing of the breaker.

Circuit breakers should always be mounted at the top of the switchboard. If they are placed at the bottom of the board, the arc which rises may cause personal injury or may damage the switchboard equipment.

AUTOMOBILE STARTING AND LIGHTING SYSTEMS

235. Automobile Electric Systems.—Motor vehicles which are propelled by internal combustion engines are largely dependent on electrical energy for their successful operation. Electrical energy is required for starting, lighting, ignition, and other purposes. A miniature power plant is necessary for supplying this energy. This power plant is ordinarily in the hands of persons not skilled in the operation of electrical machinery; and even if the operator is skilled, it is impossible for him to attend to the power plant while driving. Therefore, this power plant must operate automatically. The source of energy is obviously the engine. The engine drives a generator which, in turn, supplies electrical energy for lights, ignition, and other purposes. The generator also must supply energy for charging the storage battery so that the battery may supply energy for starting and for lights, etc., when the engine is not running.

Present-day automobile systems operate at 6 volts. When the principle of applying an electric cranking motor to an automobile engine was first employed, it was not uncommon to find system voltages of either 18 or 24 volts. As cranking motors were perfected, voltages were lowered and, for some time, 12-volt systems were quite common. The use of 12-volt systems was discontinued in 1926, and since that time American pleasure cars have operated at 6 volts.

Early installations (1912 to 1919) were predominantly of the single-unit type. In a single-unit system the generator and the starting motor were embodied in the same field frame and housing. There was a single armature with two separate windings and two separate commutators. One winding was used for generating and the other for starting. There were also two separate field windings, one for generating and one for starting. A machine of this type is called a *dynamotor*. Because of manufacturing costs, single-unit machines were gradually discontinued, and in 1926 all cars except the Lincoln had changed to the two-unit system. The Lincoln did not make the change until 1931.

In the two-unit system the motor and the generator are entirely separate. The regulation of the generator will be discussed first. As the armature speed increases, the generator must develop a reaction which will counteract the tendency to deliver more current, as otherwise the generator will be overloaded. Until recently, with most cars, regulation has been accomplished entirely by means of "third-brush regulation," which depends on armature reaction for its operation.

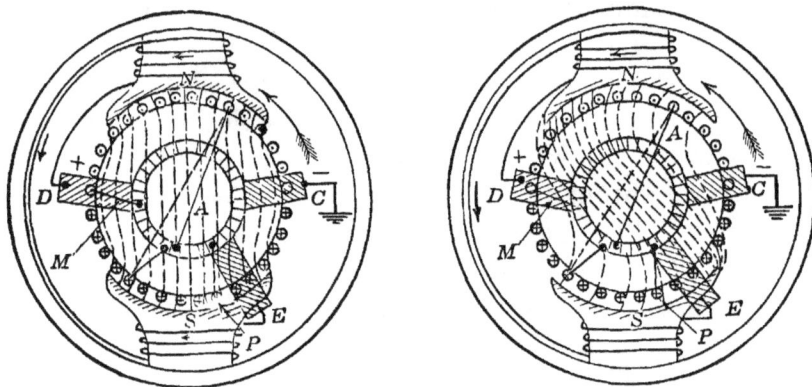

(a) Light load.   (b) Normal load.
Fig. 267.—Third-brush regulation.

236. Third-brush Regulation.—A diagrammatic sectional view, showing the armature, field, commutator, and field connection of a third-brush generator, is given in Fig. 267 (a) and (b). In (a) the condition for low speed and light loading is shown. In (b) the condition for high speed and normal loading is shown. The main positive and negative brushes are shown at D and C, the negative brush being grounded to the frame of the generator. The third or regulating brush E is set along the commutator at a considerable angle from the negative brush C and in a direction opposite to that of rotation. The field connection is made between the positive brush D and the third or regulating brush E.

In both figures a single armature coil A is shown. This coil is connected across two adjacent commutator segments lying between the positive brush D and the regulating brush E.

A study of Fig. 267 (*a*) or (*b*), with particular reference to the connection of coil *A* to the commutator and to the manner in which the other armature coils must be similarly connected to commutator segments, will show that the emf between brushes *D* and *E* is proportional to the number of armature conductors between conductors *M* and *P*, to the flux cut by the conductors between *M* and *P*, and to the speed. With light loads, as in (*a*), the emf between brushes *D* and *E* is approximately 5 volts when the voltage between main brushes *D* and *C* is 7 volts. Under these conditions the field current is approximately 1.25 amp. As the speed increases, the load on the generator tends to increase. The field becomes distorted, owing to armature reaction, as shown in Fig. 267 (*b*). (Compare with Fig. 215, p. 242.) The number of magnetic lines in the armature between conductors *M* and *P* is considerably *decreased*, as a study of Fig. 267 (*b*) shows. This tends to decrease the emf across the field circuit, and hence to decrease the field current, and counteracts in large measure the tendency of the generator output to increase with increase of speed. Overloading the generator is thus prevented.

If the generator is found to be charging the battery at too high a rate, the regulating brush *E* should be moved slightly *against* the direction of rotation; if the charging rate is too low, the brush *E* should be moved slightly *in* the direction of rotation.

Third-brush regulation was quite generally used from about 1918 until 1934, when it became apparent that the generators then in use in American pleasure cars were not able to meet requirements such, for example, as the additional loads placed on them due to radios and other accessories. A third-brush-regulated automobile generator has certain inherent characteristics which make it unsuitable for modern motorcar use. When a battery becomes discharged, its emf drops. This reduces the voltage at the generator terminals, causing the generator output to diminish. The generator output will not come up to its maximum until the line voltage and hence the battery voltage comes up to the maximum value. That is, when the highest charging rate is required, a third-brush-regulated generator develops a lesser output. On the other hand, when a battery becomes fully charged and the line voltage is at its highest value (see Fig. 67,

p. 83), a high charging rate is no longer required, and yet the third-brush generator delivers a large output.

Because of the inability of the third-brush system to keep the battery charged under present-day demands, the system has been largely supplemented by vibrating-point regulation or else vibrating-point regulation has been substituted for third-brush regulation. For example, the Ford car utilizes vibrating-point regulation exclusively, and the Buick utilizes a combination of vibrating-point and third-brush regulation.

237. Starting and Lighting Systems.—The essential parts of the Buick-Eight starting and lighting system are shown in Fig. 268. This system is representative of the general principles now employed in many makes of cars. First, consider the method of generator-voltage regulation, which combines the third-brush and vibrating-point methods. The winding on the S-pole of the generator is shunt, being connected to the positive-brush line at A and to a terminal F which is grounded through resistance R_1 when the armature A' of the voltage regulator causes contacts C to open, and through R_1 and regulator winding W in parallel when the regulator contacts C are closed. The winding on the N-pole of the generator is connected between the third brush E and the terminal F and thence to ground through resistance R_1 alone or through R_1 in parallel with winding W, in the same manner as with the shunt winding. There are two windings on the voltage regulator, a winding W on one core, this winding carrying a substantial part of the current from the two field windings when the contacts C are closed, a small part of the field current being shunted by resistance R_1. The second winding S on the other core of the regulator is connected directly between the live line wire at L and ground.

When the emf of the generator reaches a value of from 6.9 to 7.6 volts, the shunt winding of the cutout relay (Par. 238) causes contacts C' to close, connecting the generator to the line and hence to one terminal of the indicator. If the line voltage becomes too high, the current in the regulator winding S, which is connected between line and ground, causes the regulator contacts C to open. This inserts the comparatively high resistance R_1 in series with the common connection to the two fields at F, thus reducing the current in the two fields and hence reducing the generator emf and the output. The output of the generator is also in part controlled by third-brush regulation, described in Par. 236, such regulation being limited to the current in the winding of the *N*-pole. Thus, with this system, when the emf of the battery is high and the line voltage is also high, the contacts *C'* open, decreasing the currents in the two generator fields and therefore lowering the generator output; when the emf of the battery is low and the line voltage is correspondingly low, the contacts *C*

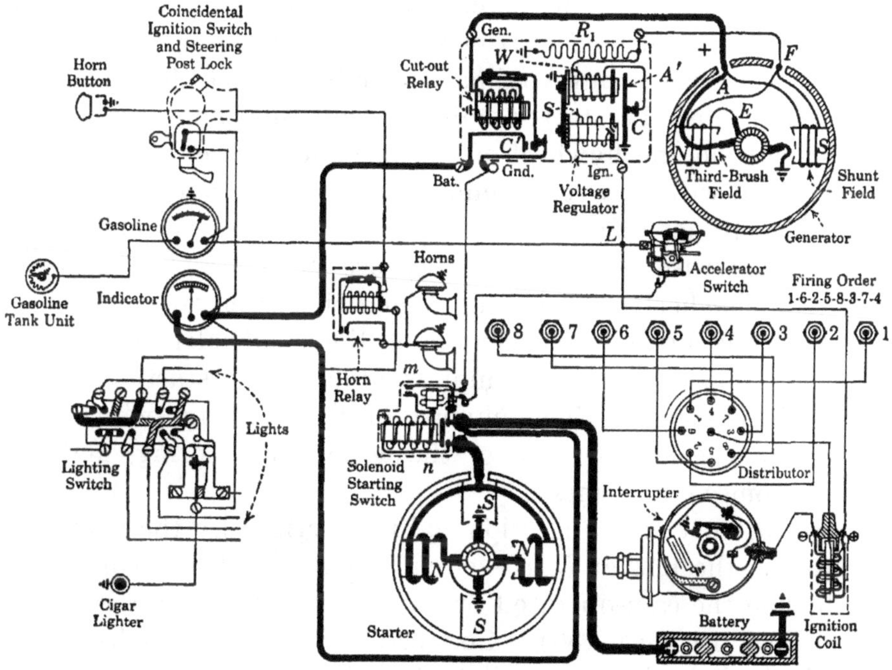

FIG. 268.—Delco-Remy starting and lighting system for Buick "8." (*Standard Engineering and Publishing Company.*)

close, increasing the currents in the two generator fields and raising the generator output. It will be observed that the vibrating system is not unlike the Tirrill regulator (p. 261).

Except during cranking, the indicator is in series with the battery and shows the charge and discharge currents of the battery at all times. For simplicity, in Fig. 268, the different lights, such as head and tail lights, and the wires running to them have been omitted. The operation of the gasoline indicator and of the horns, which operate through a relay, is clear.

A four-pole series motor, with windings on two poles only, is used for cranking. Pressure on the accelerator switch causes the auxiliary contacts *m*, in the small solenoid above the solenoid starting switch, to close. These auxiliary contacts in turn close the circuit of the winding in the solenoid starting switch, which is in series with the starting motor and is shown with a heavier line in Fig. 268. There results a large current in this winding of the solenoid, and the starting plunger is pulled vigorously to the right. This closes the heavy low-resistance contacts *n* which connect the starter motor directly between battery and ground. The closing of the contacts *n* short-circuits the heavy winding on the solenoid starting switch, but the plunger continues to be held by the shunt winding, shown by the light line. As the plunger carries the starter-motor pinion, this pinion will mesh with teeth on the flywheel of the gasoline engine. The pinion is provided with an over-running clutch so that when the engine starts it will not drive the starting motor at the excessive speed which otherwise would result because of the high gear ratio between flywheel and pinion. When the accelerator switch is released, the auxiliary contacts *m* on the small solenoid open by spring action causing the shunt winding of the starting switch to be de-energized. The plunger of the starting switch then moves to the left because of the action of a spring, and the pinion comes out of mesh.

The primary of the ignition coil is energized by the wire connected to the line at *L*, the circuit to ground being completed each time the cam-operated breaker in the interrupter is closed (see Par. 151, p. 180). Connection from the high-tension terminal of the ignition coil to the eight spark plugs is made through the distributor. The firing order is shown on the distributor.

238. Cutout Relay.—In the present-day 6-volt systems, the generator is connected to the engine without the use of an over-running clutch (practically all are now driven by a "Vee" belt) and, therefore, the generator cannot be permitted to "motor" at low engine speeds. The generator must be disconnected from the battery when, because of insufficient speed and emf, it begins to take energy from the battery. When the generator emf builds up to a value which is greater than the emf of the battery, the generator must be connected to the battery. If the generator is not connected to the battery, it will build up to a high emf (see Fig. 211, p. 237) and will overheat owing to excessive field current.

Therefore, the generator and battery are usually connected by a cutout relay (Fig. 269). The relay consists of a soft-iron core C and an armature A pivoted at P. There are two windings on the core, a shunt winding S_h, consisting of a considerable number of turns of fine wire, and a series winding S_e, consisting of comparatively few turns of heavy wire. There are two contacts C_1, one on the armature A connected to the positive terminal of the generator, and a second fixed contact connected through the series winding to the positive terminal of the battery. The contacts C_1 are held open by means of the spring S. The shunt winding is directly across the generator main brushes. When the

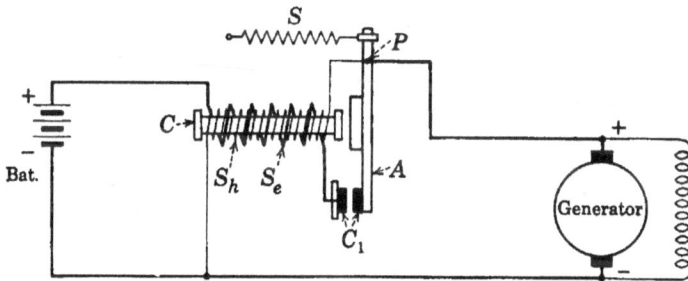

FIG. 269.—Cutout relay.

generator builds up to a sufficiently high emf, the current in the shunt winding causes the core C to attract the armature A, closing the contacts C_1. The battery is then connected across the generator terminals through the series winding S_e. The series winding is connected in such a manner that when current flows from the generator positive terminal to the battery positive terminal the series winding *assists* the shunt winding and increases the pressure between the contacts C_1. When the generator emf drops below that of the battery, the current reverses and flows from the battery positive terminal to the generator positive terminal and supplies energy to the generator. The current in the series winding now being reversed, the series winding *opposes* the shunt winding. When the net flux in the core reaches a sufficiently low value, the spring S causes the relay contacts C_1 to open, disconnecting the generator from the battery. In the usual automobile system, the relay closes with 6.9 to 7.6 volts across the generator and opens with from zero to 3 amp reverse current.

Cutout or "reverse-current" relays are used extensively also in train-lighting systems employing axle drive, and in farm-lighting systems, etc., employing batteries, where automatic connection and disconnection are necessary.

APPENDIX A

Relations of Units

Length

1 inch = 2.54 cm
1 foot = 30.48 cm
1 mile = 1.609 km

Area

1 circular mil = 0.7854 sq mil
1 circular mil = 0.000507 sq mm
1 square inch = 6.452 sq cm
1 square meter = 10.76 sq ft

Volume

1 cubic inch = 16.39 cu cm
1 liter = 1,000 cu cm
      = 1.057 qt
      = 0.2642 gal
1 gallon = 231 cu in.

Weight

1 gram = 981 dynes
1 ounce (av.) = 28.35 grams
1 kilogram = 2.205 lb
1 ton = 2,000 lb
1 long ton = 2,240 lb
1 metric ton = 1,000 kg
          = 2,205 lb

Pressure

1 atmosphere = 14.70 lb on square inch
          = 29.92 in. of mercury at 32°F (0°C)
          = 760.0 mm of mercury at 0°C (32°F)
          = 33.94 ft of water at 60°F (15.6°C)
1 lb on sq in. = 702.9 kg on square meter

Work

1 joule (watt-second) = 10,000,000 ergs
1 gram degree Centigrade (gram calorie) = 4.184 joules
1 pound degree Fahrenheit (Btu) = 252.0 gram degree Centigrade (gram-calorie)
          = 778.5 ft-lb
1 kilogram-meter = 9.81 joules
          = 7.233 ft-lb
1 foot-pound = 1.356 joules
1 horse-power-second = 178.3 gram degree Centigrade (gram-calorie)
          = 0.7074 lb degree Fahrenheit (Btu)
          = 550 ft-lb

Power

1 horsepower = 550 ft-lb per second
          = 33,000 ft-lb per minute
          = 0.7074 Btu per second
          = 746 watts

1 kilowatt = 1,000 watts
          = 1,000 joules per second
          = 1.341 hp
          = 737.6 ft-lb per second
          = 0.239 kg-cal per second

APPENDIX B

Circles and Spheres

Circle

$$A = \pi R^2 = \frac{\pi}{4}D^2 = 0.7854D^2$$

Sphere

$$A = 4\pi R^2 = 12.57R^2 = \pi D^2$$
$$V = \tfrac{4}{3}\pi R^3 = \tfrac{1}{6}\pi D^3$$
$$= 4.189R^3 = 0.5236D^3$$

$\pi = 3.1416$

A = area (surface of sphere)

V = volume

R = radius

D = diameter

APPENDIX C

Specific Gravities

Aluminum	2.67	Mercury	13.60
Copper (drawn)	8.89	Nickel	8.60 to 8.90
Gold	19.26	Platinum	21.37
Iron, bar	7.48	Silver	10.55
Iron, wrought	7.80 to 7.90	Tin (cast)	7.29
Steel	7.60 to 7.85	Zinc (cast)	7.04 to 7.16
Lead	11.45		

1 cu. ft. of water weighs 62.4 lb.

APPENDIX D

Conversion of Thermometer Scales

$$\text{Degrees Fahrenheit} = \tfrac{9}{5}C + 32 = 1.8C + 32$$
$$\text{Degrees Centigrade} = \frac{5}{9}(F - 32) = \frac{F - 32}{1.8}$$

C = degrees Centigrade, F = degrees Fahrenheit

APPENDIX E

Heat Values of Fuels[1]

Btu's per pound

Bituminous coal....	10,000 to 15,000	Dry pine........	9,153
Anthracite coal....	12,500 to 14,000	Gasoline........	19,000 to 21,000
Connellsville coal...	15,000	Kerosene........	20,093
Connellsville coke..	12,600	Fuel oils........	18,500 to 19,000
Dry oak...........	8,316		

Btu's per cubic foot

Water gas................................. 295
Coal gas................................. 500 to 625
Natural gas............................. 1,000 to 1,400

APPENDIX F

Greek Alphabet

Name	Cap.	l.c.	Quantity	Name	Cap.	l.c.	Quantity
			Quantity	Mu	M		
Alpha	A	α				μ	Permeability and "micro"
Beta	B	β					
Gamma	Γ			Nu	N	ν	
		γ	Conductivity (mho per centimeter-cube)	Xi	Ξ	ξ	
				Omicron	O	o	
				Pi	Π	π	(3.1416)
Delta	Δ	δ		Rho	P		
Epsilon	E					ρ	Resistivity
		ϵ	Napierian log. base (2.718)	Sigma	Σ		Summation
						σ	Magnetic or electrostatic surface density of charge
Zeta	Z	ζ					
Eta	H						
		η	Efficiency			ς	
Theta	Θ	θ, ϑ	Phase displacement and power-factor angle.	Tau	T	τ	
				Upsilon	Υ	υ	
				Phi	Φ	φ, ϕ	magnetic flux
Iota	I	ι		Chi	X	χ	
Kappa	K	κ	Dielectric constant	Psi	Ψ	ψ	Electrostatic flux
Lambda	Λ			Omega	Ω		ohms
		λ	Wave length			$\mho$	mhos
						ω	angular velocity

[1] "Mechanical Engineers' Handbook," McGraw-Hill Book Co. Inc.

APPENDIX G
Working Table, Standard Annealed Copper Wire, Solid[1]
American Wire Gage (B. & S.).   English Units

Gage No.	Diameter in mils	Cross-section		Ohms per 1,000 ft.		Ohms per mile	Pounds per 1,000 ft.
		Circular mils	Square inches	25° C. (77° F.)	65° C. (149° F.)	25° C. (77° F.)	
0000	460.0	212,000.0	0.166	0.0500	0.0577	0.264	641.0
000	410.0	168,000.0	0.132	0.0630	0.0727	0.333	508.0
00	365.0	133,000.0	0.105	0.0795	0.0917	0.420	403.0
0	325.0	106,000.0	0.0829	0.100	0.116	0.528	319.0
1	289.0	83,700.0	0.0657	0.126	0.146	0.665	253.0
2	258.0	66,400.0	0.0521	0.159	0 184	0.839	201.0
3	229.0	52,600.0	0.0413	0.201	0.232	1.061	159.0
4	204.0	41,700.0	0.0328	0.253	0.292	1.335	12 .0
5	182.0	33,100.0	0.0260	0.319	0.369	1.685	100.0
6	162.0	26,300.0	0.0206	0.403	0.465	2.13	79.5
7	144.0	20,800.0	0.0164	0.508	0.586	2.68	63.0
8	128.0	16,500.0	0.0130	0.641	0.739	3.38	50.0
9	114.0	13,100.0	0.0103	0.808	0.932	4.27	39.6
10	102.0	10,400.0	0.00815	1.02	1.18	5.38	31.4
11	91.0	8,230.0	0.00647	1.28	1.48	6.75	24.9
12	81.0	6,530.0	0.00513	1.62	1.87	8.55	19.8
13	72.0	5,180.0	0.00407	2.04	2.36	10.77	15.7
14	64.0	4,110.0	0.00323	2.58	2.97	13.62	12.4
15	57.0	3,260.0	0.00256	3.25	3.75	17.16	9.86
16	51.0	2,580.0	0.00203	4.09	4.73	21.6	7.82
17	45.0	2,050.0	0.00161	5.16	5.96	27.2	6.20
18	40.0	1,620.0	0.00128	6.51	7.51	34.4	4.92
19	36.0	1,290.0	0.00101	8.21	9.48	43.3	3.90
20	32.0	1,020.0	0.000802	10.4	11.9	54.9	3.09
21	28.5	810.0	0.000636	13.1	15.1	69.1	2.45
22	25.3	642.0	0.000505	16.5	19.0	87.1	1.94
23	22.6	509.0	0.000400	20.8	24.0	109.8	1.54
24	20.1	404.0	0.000317	26.2	30.2	138.3	1.22
25	17.9	320.0	0.000252	33.0	38.1	174.1	0.970
26	15.9	254.0	0.000200	41.6	48.0	220.0	0.769
27	14.2	202.0	0.000158	52.5	60.6	277.0	0.610
28	12.6	160.0	0.000126	66.2	76.4	350.0	0.484
29	11.3	127.0	0.0000995	83.4	96.3	440.0	0.384
30	10.0	101.0	0.0000789	105.0	121.0	554.0	0.304

[1] Tables of current-carrying capacity are given in Part II

APPENDIX H
Bare Concentric Lay Cables of Standard Annealed Copper
English Units

A.W.G. No.	Circular mils	Ohms per 1,000 ft.		Pounds per 1,000 ft.	Standard concentric stranding		
		25° C. (77° F.)	65° C. (149° F.)		Number of wires	Diameter of wires, in mils	Outside diameter, in mils,
	1,000,000	0.0108	0.0124	3,090	61	128.0	1,152
	900,000	0.0120	0.0138	2,780	61	121.5	1,093
	850,000	0.0127	0.0146	2,620	61	118.0	1,062
	750,000	0.0144	0.0166	2,320	61	110.9	998
	650,000	0.0166	0.0192	2,010	61	103.2	929
	600,000	0.0180	0.0207	1,850	61	99.2	893
	550,000	0.0196	0.0226	1,700	61	95.0	855
	500,000	0.0216	0.0249	1,540	37	116.2	814
	450,000	0.0240	0.0277	1,390	37	110.3	772
	400,000	0.0270	0.0311	1,240	37	104.0	728
	350,000	0.0308	0.0356	1,080	37	97.3	681
	300,000	0.0360	0.0415	926	37	90.0	630
	250,000	0.0431	0.0498	772	37	82.2	575
0000	212,000	0.0509	0.0587	653	19	105.5	528
000	168,000	0.0642	0.0741	518	19	94.0	470
00	133,000	0.0811	0.0936	411	19	83.7	418
0	106,000	0.102	0.117	326	19	74.5	373
1	83,700	0.129	0.149	258	19	66.4	332
2	66,400	0.162	0.187	205	7	97.4	292
3	52,600	0.205	0.237	163	7	86.7	260
4	41,700	0.259	0.299	129	7	77.2	232

QUESTIONS ON CHAPTER I

Resistance

1. Name the properties of electrical energy which make it so useful in modern life.

2. Describe briefly the general nature of the electron as related to the atom. What is the relation of the electron to electric current? What is the essential physical difference between conductors and insulators? Between static and dynamic electricity?

3. When resistance is introduced in a circuit, what two effects are noted? Give a mechanical analogy and a hydraulic analogy which illustrate the effects that occur when current flows through resistance.

4. Into what two general divisions may substances, used for electrical purposes, be classified? Give examples of good conductors; of good insulators.

5. What is the unit of resistance and how is it specifically defined? What is the relation of the ohm to the volt and ampere? What is a megohm? A microhm?

6. In what way does the flow of electricity through conductors resemble the flow of water through pipes? How does an increase of cross-section affect the quantity of water flowing through a pipe if the pressure between its ends is maintained constant? How does an increase of cross-section affect the quantity of electricity flowing through a conductor if the voltage across its ends is maintained constant?

7. At constant temperature, on what three factors does the resistance of a homogeneous conductor of uniform cross-section depend? What is meant by resistivity or specific resistance?

8. At constant temperature, on what three factors does the conductance of a homogeneous substance of uniform cross-section depend? Define conductivity. What relation exists between conductance and resistance? Between conductivity and resistivity?

9. Define a mil; a square mil; a circular mil. Show that the circular mils in a cylindrical conductor are obtained by squaring its diameter, expressed in mils.

10. Define a circular-mil-foot and state its resistance at 20°C for copper.

11. What effect does an increase of temperature have on the resistance of the unalloyed conductors? Give the relation of the resistance at different temperatures to the temperature coefficient of resistance, referred both to 0°C and to any other initial temperature. Show how the relation of resistances at different temperatures may be obtained by reference to an inferred zero resistance.

12. What substances have temperature coefficients that are practically zero and for what are such substances used?

13. What is meant by the resistivity specifications of the A.I.E.E.?

14. How is the equivalent resistance of a number of resistances in series determined? How is the equivalent conductance of a number of conductances in parallel determined? What equation gives the equivalent resistance of a parallel arrangement of resistances?

15. On what simple relation is the A.W.G. based? Give the relation between the cross-sections of conductors differing by three gage numbers; of conductors having successive gage numbers; of conductors differing by two gage numbers.

16. What is the approximate resistance of 1,000 ft of No. 10 copper wire? What is its cross-section in circular mils? Its diameter in mils? Its weight? State the methods by which the properties of a copper conductor of any size may be determined.

17. Why are wires and cables stranded? What is the relation between the number of strands in successive layers of a stranded cable?

18. Why is copper so widely used as a conductor? Under what conditions are silver, aluminum, iron, steel used as conductors? What is "copperweld" steel and where is it used?

PROBLEMS ON CHAPTER I

Resistance

1. The resistivity of copper is 1.724 microhm-cm at 20°C. Determine the resistance, in microhms at 20°C, of a copper bar 6.0 m long and having a cross-section of 4 sq cm.

2. What is the resistance at 20°C of a copper bus-bar, 14 ft long and 0.50 by 4.0 in. cross-section, if the resistivity of the copper is 0.6788 microhm per in.-cube at 20°C?

3. The diameter of No. 4 A.W.G. solid copper wire is 0.204 in. (*a*) Determine the cross-section in sq in. (*b*) Determine the resistance of 600 ft of the wire at 20°C using the data given in Prob. 2.

4. The volume resistivity of aluminum is 2.828 microhms per cm-cube at 20°C (see p. 16). Determine the resistance at 20°C of an aluminum rod 1 cm diameter and 4 m long.

5. The resistance at 20°C of a copper rod 0.25 in. diameter and 15 ft long is 0.0025 ohm. Determine the resistance at 20°C of an aluminum rod of the same diameter and length. (Use data of Probs. 1 and 4.)

6. Determine the resistance at 20°C of a copper rod 0.40 in. diameter and 15 ft long. Repeat for an aluminum rod 0.40 in. diameter and 12 ft long. (Use values of resistivities given in Probs. 1 and 4.)

7. It is desired to wind a resistor of 0.08 ohm with No. 8 A.W.G. iron wire having a diameter of 128 mils (0.325 cm). The resistivity of the iron is 11.8 microhm-cm-cube. What length of wire (in meters) is necessary?

8. The resistance of 500 ft of No. 6 A.W.G. solid copper wire is 0.205 ohm. This wire is drawn down so that the cross-section is halved, the volume and the resistivity remaining unchanged. Determine the resistance of the wire after drawing.

9. In Prob. 8 the diameter of the No. 6 wire is 0.162 in. Determine the resistance if the diameter is drawn down to 0.10 in., the volume and resistivity remaining unchanged.

10. Two cylindrical, copper conductors A and B have the same resistivity and the same volume. The length of conductor A is 24 ft and that of conductor B is 15 ft. The resistance of conductor A is 0.06 ohm. Determine the resistance of conductor B.

11. The conductivity of copper is 580,000 mhos per cm cube. Determine the conductance of 640 ft of 0000 soft-drawn solid copper wire whose diameter is 0.460 in.

12. An aluminum bus-bar 15 ft long has a cross-section 6.0 by 0.50 in. (a) If the conductivity of aluminum is 0.61 that of copper (see Prob. 11), what is the conductance of this bus-bar? (b) Aluminum weighs 0.096 lb per cu in. Determine the weight of the bus-bar.

13. Copper weighs 0.32 lb per cu in. Determine the conductance and weight of a copper bus-bar having the same dimensions as the aluminum bus-bar of Prob. 12.

14. The cross-section of a copper bar is square, 0.60 in. on a side. Determine: (a) the square mils; (b) the circular mils.

15. (a) Determine the cross-section in cir mils of a cylindrical wire 0.5 in. diameter. (b) Determine the cross-section in sq mils.

16. (a) The diameter of 000 A.W.G. wire is 410 mils. Determine its cross-section in cir mils. (b) The cross-section of a copper wire is 26,250 cir mils. Determine its diameter in mils and in inches.

17. (a) In a 19-strand, concentric-lay, copper cable the diameter of the individual strands is 105.5 mils. Determine the cross-section of the cable in circular mils. (b) If this cable were made up of 37 strands, the diameter of the individual strands would be 75.6 mils. Determine the cross-section of this cable in circular mils.

18. A cable having a total cross-section of 450,000 cir mils is made up of 37 strands. Determine the cross-section and the diameter of the individual strands.

19. It is desired to make up a 1,500,000-cir-mil cable with 127 strands. (a) What are the circular mils of each strand? (b) What is the diameter of each strand? (c) Give the gage number (nearest) of the wire to be used in making up the cable. (d) State the number of layers.

20. Determine the resistance of 1,600 ft of 1,000,000-cir-mil cable if the resistance of a circular-mil-foot is taken as 10.5 ohms. Repeat for the same length of 750,000-cir-mil cable.

21. The diameter of a 2-mile length of solid copper conductor is 162 mils. If the resistance of a circular-mil-foot is taken as 10.5 ohms, determine: (a) the circular mils in this conductor; (b) the total resistance.

22. An 800-ft length of copper conductor consists of 61 strands, and the diameter of each strand is 95.0 mils. The resistivity of the copper is 10.6 ohms per cir-mil-ft. Determine: (a) the cross-section of the conductor in cir mils; (b) the total resistance.

23. It is desired to wind an electric furnace element with No. 12 A.W.G. Nichrome resistor wire whose diameter is 80.8 mils. A 120-ft length is required. If the resistivity of Nichrome is 600 ohms-cir-mil-ft (about sixty times that of copper), what is the total resistance of the furnace heating element?

24. It is necessary that the resistance of a resistor-unit be 16 ohms. Constantan wire, having a resistivity of 302 ohms-cir-mil-ft (about thirty times that of copper), is to be used. If No. 16 wire, having a diameter of 50.8 mils, is used, what length is required?

25. The temperature coefficient of resistance of copper at 0°C is 0.00427. The resistance of a 3-mile length of 0000 hard-drawn copper trolley wire at the temperature of 0°C is 0.75 ohm. Determine its resistance: (a) at 20°C; (b) at 50°C.

26. The resistance of a relay magnet coil is measured at 25°C and found to be 240 ohms. Determine its resistance at: (a) 0°C; (b) 20°C.

27. The temperature coefficient of resistance of copper at 30°C is 0.00378 (p. 14). The resistance of a magnet coil at 30°C is 12.5 ohms. Determine its resistance at 55°C.

28. The resistance of the shunt-field winding of a d-c generator is measured and found to be 84.0 ohms after the generator has been standing for a considerable time in a room whose temperature is 24°C. Compute the hot resistance of this same winding after the generator has been in operation and has reached an average temperature of 60°C. (See p. 303.)

29. The resistance between two marked segments of an armature is measured and found to be 0.160 ohm after the dynamo has been standing for a considerable time in a room whose temperature is 22°C. (See p. 303.) After operating for 3 hr, the armature is stopped and the resistance between these same two segments is again measured and found to be 0.181 ohm. Determine the temperature rise of the winding.

30. The resistance of the shunt winding of a generator is 64.5 ohms as measured after the generator has stood for a considerable time in a room whose temperature is 21°C. After the generator has been in operation 4 hr, the resistance of the shunt winding is found to be 72.4 ohms. Determine the temperature rise in the winding over this 4-hr period.

31. The resistance of 40 ft of No. 8 A.W.G. annealed copper wire, whose diameter is 128.5 mils, is measured at 20°C and found to be 0.02544 ohm. Determine whether or not this wire comes within the A.I.E.E. specifications (Par. 16, p. 16).

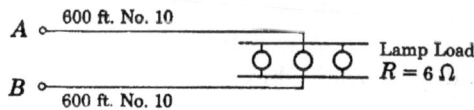

FIG. 32A.

32. In Fig. 32A is shown a lamp load of resistance 6 ohms, fed by a two-wire main, each wire consisting of 600 ft of No. 10 copper wire. The resistance of each wire is 1.11 ohms per 1,000 ft. Determine the resistance of the circuit as measured at the terminals AB.

33. The resistance of the heating element of an electric range is 3.8 ohms. It requires a run of 80 ft of two No. 8 A.W.G. solid copper wires (160 ft of wire) to feed this range from the service entrance. The resistance of No. 8 wire is 0.0642 ohm per 1,000 ft. Determine the total resistance of the circuit.

34. The hot resistance of the field winding of a 250-volt shunt generator is 86.5 ohms. It is desired to reduce the field current to a value requiring a total field-circuit resistance of 168.6 ohms. The present field rheostat has a maximum resistance of 44.5 ohms. How much additional resistance is necessary?

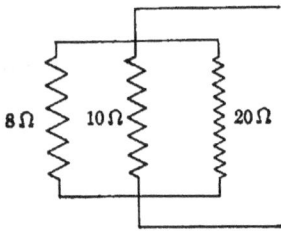

8 Ω   10 Ω   20 Ω

FIG. 35A.

35. A resistance of 8 ohms, a resistance of 10 ohms, and a resistance of 20 ohms are connected in parallel (Fig. 35A). Determine: (a) the conductance of each of the three elements of the circuit; (b) the circuit conductance; (c) the equivalent parallel resistance of the circuit.

36. Three incandescent lamps having hot resistances of 96, 144, 190 ohms are connected in parallel. Determine: (a) the conductance of each lamp; (b) the equivalent conductance of the three; (c) the equivalent resistance of the system.

37. An electric flatiron of 40 ohms resistance is connected in parallel with a toaster of 24 ohms resistance. Determine: (a) the conductance of each; (b) the equivalent parallel conductance of the two; (c) the equivalent parallel resistance of the two.

38. Four resistances of 42, 50, 64, 75 ohms are connected in parallel. Determine the equivalent resistance of the system.

39. Without consulting the wire tables, determine by means of the methods given in Par. 18, p. 19, the circular mils, the diameter, the weight, the resistance of: (a) 1,600 ft of No. 8 A.W.G. copper wire; (b) 800 ft of No. 24 A.W.G. copper wire.

40. Repeat Prob. 39 for: (a) 1,200 ft of No. 30 A.W.G. copper wire; (b) 1,400 ft of No. 27 A.W.G. copper wire; (c) 900 ft of No. 0000 A.W.G. copper wire.

QUESTIONS ON CHAPTER II

Ohm's Law and the Electric Circuit

1. Define the ampere in absolute units. How is it further defined in order that it may be readily determined experimentally? What is meant by the mks system?

2. What is the unit of electrical quantity, and how is it defined? What is the unit of emf, and how is it specifically defined?

3. In what way does the flow of electricity through a wire resemble the flow of water through a pipe? What is meant by pressure drop in each case? Show that a difference of electrical pressure is necessary in order that the current may return from an electrical load to the negative terminal of the generator.

4. Show that voltage or electrical pressure may exist without flow of current. How should an ammeter be connected in circuit? A voltmeter?

5. What is Ohm's law? Express this law algebraically. Transform this algebraic expression into two others, and explain their meaning.

6. How are resistances in series combined to find the equivalent resistance of the combination?

7. How are resistances in parallel combined to find the equivalent resistance of the combination? What is the ratio of the currents flowing in two parallel resistances?

8. What is the practical unit of electrical power, and how is it defined? Give three equations for determining electrical power. Under what conditions is it convenient to use each?

9. In what manner does electrical energy differ from electrical power? What is the practical unit of electrical energy?

10. How is heat energy related to mechanical and electrical energy? State the process by which chemical energy in coal appears as electrical energy and ultimately as mechanical energy. What is the order of efficiency of each conversion process?

11. How is a British thermal unit (Btu) defined? A gram-calorie? What is the value of the gram-calorie in joules?

12. Discuss the method of determining the voltage at a load, provided the current and feeder resistance are known. How may the efficiency of transmission be expressed?

PROBLEMS ON CHAPTER II

Ohm's Law and the Electric Circuit

41. The hot resistance of a 200-watt, gas-filled, incandescent lamp is 72 ohms. What current does it take from 120-volt lighting mains?

42. The resistance of an electric flatiron is 21 ohms. Determine the current and the power which it takes from 115-volt mains.

43. The resistance of an electric-furnace element is 6.5 ohms, and the resistance of each of the two wires connecting it to the power source is 0.04 ohm (Fig. 43A). The voltage at the power source is 234 volts. Determine the current to the furnace and the voltage at the furnace.

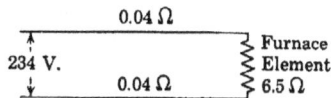

Fig. 43A.

44. The resistance of the shunt field of a 250-volt generator is 22 ohms and the resistance of the field rheostat in series is 7.4 ohms. (See Fig. 214, p. 239.) Determine the current and the power to the field circuit when the voltage across it is 250 volts.

45. The hot resistance of a 6.6-amp, 4,000-lumen, series incandescent lamp is 5.14 ohms. Determine the voltage across the lamp when it is operating at its rated current.

46. Determine the voltage across the field, across the rheostat, and the total voltage, Prob. 44, when the field-circuit current is 8.18 amp.

47. An electromagnet (Fig. 47A) has two pairs of exciting coils. The resistance of each coil of one pair is 0.7 ohm, and the resistance of each coil of the second pair is 1.2 ohms. The coils are all connected in series across 120 volts. Determine: (a) the current; (b) the voltage drop across each 0.7-ohm coil; (c) the voltage drop across each 1.2-ohm coil.

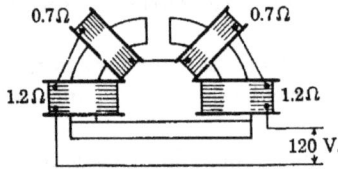

FIG. 47A.

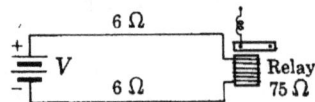

FIG. 48A.

48. A 75-ohm relay (Fig. 48A) requires 160 milliamp for operation. It is connected at the end of a two-wire signal line each conductor of which has a resistance of 6 ohms. At the current necessary to operate the relay, determine: (a) the voltage across the relay; (b) the terminal voltage V of the battery which supplies the power.

49. The field circuit of a shunt motor (Fig. 49A) takes 4.8 amp from 250-volt mains. The voltage drop across the rheostat is 60 volts. Determine: (a) the resistance of the field circuit; (b) the resistance of that portion of the rheostat which is in use; (c) the resistance of the field winding; (d) the power consumed in the rheostat and in the winding.

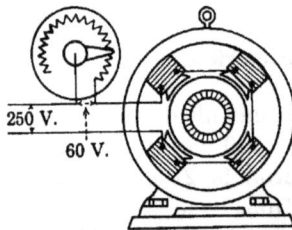

FIG. 49A.

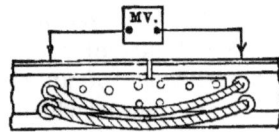

FIG. 50A.

50. A rail bonding is tested for its resistance by causing known current to flow through it and measuring the voltage drop (Fig. 50A). When the current is 160 amp, the millivoltmeter reads 21 millivolts. What is the resistance of the bonding?

51. A two-mile, two-conductor feeder consists of a 1.5-mile length of 300,000-cir-mil cable and a 0.5-mile length of 250,000-cir-mil cable in series. If the resistance of a circular-mil-foot of the copper is 10.6 ohms, determine: (a) the resistance per conductor of the feeder; (b) the total voltage drop due to each length of feeder when the current is 125 amp.

52. Two resistances A and B (Fig. 52A) are connected in parallel across a 50-volt source. The current to resistance A is 4 amp, and the total current to the system is 12 amp. (*a*) Determine the resistance of A and of B. (*b*) By what single resistance can resistances A and B in parallel be replaced?

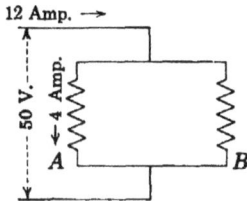

FIG. 52A.

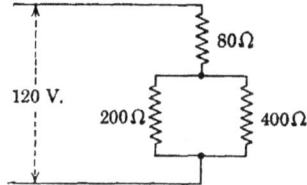

FIG. 53A.

53. In Fig. 53A an 80-ohm resistance is connected in series with two resistances in parallel, one of 200 ohms and the other of 400 ohms. The entire circuit is connected across 120 volts. Determine: (*a*) the value of the single resistance which can replace the parallel circuit; (*b*) the current to the system; (*c*) the current in the 200-ohm and in the 400-ohm resistance; (*d*) the voltage drop across the 80-ohm resistance; (*e*) the voltage drop across the parallel circuit.

54. The resistance of the field winding of a generator is 42 ohms (Fig. 54A). Two field rheostats A and B are connected in parallel, and the two are in series with the field winding. At one particular setting the resistance of A is 12 ohms and that of B is 8 ohms. The entire field circuit is connected across 200 volts at the generator terminals. Determine: (*a*) the equivalent resistance of rheostats A and B in parallel; (*b*) the resistance of the entire field circuit; (*c*) the current in rheostat A and in rheostat B; (*d*) the voltage drop across the field winding; (*e*) the voltage drop across the two rheostats.

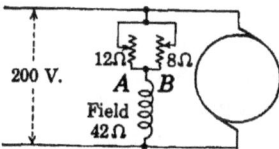

FIG. 54A.

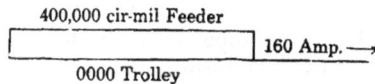

FIG. 55A.

55. A 400,000-cir-mil feeder is connected in parallel with a 0000 trolley wire for 2 miles (Fig. 55A). The resistance of the trolley wire is 0.270 ohm per mile and that of the cable is 0.164 ohm per mile. When the combined load on the trolley system is 160 amp, determine the current in the trolley and in the feeder.

56. Three resistances of 40, 56, 80 ohms are connected in parallel. A current of 4 amp flows in the 40-ohm resistance. Determine: (*a*) the equivalent resistance of the system; (*b*) the voltage across the system; (*c*) the current in each resistance.

57. A resistance *a* of 8.0 ohms, resistances *b* and *c* of 10 and 12 ohms in parallel, and resistances *d*, *e*, and *f* of 15, 20, 30 ohms in parallel are all connected in series across 100 volts (Fig. 57*A*). Determine: (*a*) the equivalent resistance of the system; (*b*) the current in each resistance; (*c*) the voltage across each of the series elements; (*d*) the power loss in each resistance.

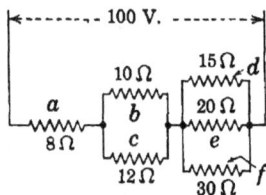

Fig. 57*A*.

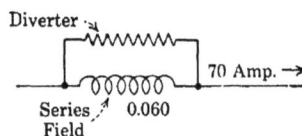

Fig. 58*A*.

58. The resistance of the series field of a compound generator (Fig. 58*A*) is 0.060 ohm. A series-field diverter in parallel with the series field diverts such current from the series field as to give the generator the desired characteristic. The total current delivered by the generator (Fig. 58*A*) is 70 amp. It is desired that, of the total current, 50 amp flow in the series field. Determine: (*a*) the resistance of the diverter; (*b*) the voltage drop across the series field and diverter.

59. Four incandescent lamps are connected in parallel across a 118-volt system. The currents to the lamps are 0.509, 0.848, 1.27, 2.12 amp. Determine the power in watts to each lamp.

60. The resistance of a relay is 184 ohms, and a resistance of 80 ohms is connected in series with it. The relay, in series with the resistance, is operated from a 250-volt circuit. Determine: (*a*) the current; (*b*) the power to the relay circuit; (*c*) the power to the resistance and to the relay itself.

61. In Prob. 58 determine the power lost in the series field and in the diverter: (*a*) when the total current is 75 amp; (*b*) when the total current is 90 amp.

62. Determine the power loss in each of the 0.7-ohm coils and in each of the 1.2-ohm coils (Fig. 47*A*) when the current is 32.5 amp.

63. The resistance between the 300-volt terminals of a d-c voltmeter is 32,500 ohms. Determine the power loss in the voltmeter when the voltage across its terminals is: (*a*) 220 volts; (*b*) 250 volts.

64. A heater resistance of 6.4 ohms is to dissipate 7,500 watts. (*a*) What must be the value of the voltage across the resistance? (*b*) With this same voltage what value of resistance will dissipate 5,000 watts?

65. A small electric resistance furnace requires 2,100 watts for its operation and is supplied from 120-volt mains. (*a*) What should be its resistance? (*b*) In order to reduce the temperature of the furnace, the watts must be reduced to 1,500. What series resistance is necessary, and how many watts are dissipated in this series resistance?

66. A lighting system in a certain building consists of 24 100-watt lamps, each lamp having a hot resistance of 132 ohms. The lamps are all connected in multiple. This load is supplied from 120-volt bus-bars located 150 ft (45.8 m) from the load center over No. 10 (A.W.G.) wire having a cross-section of 10,000 cir mils. (See Fig. 66A.) Assuming a resistance of 10.5 ohms per cirmil-ft, what current and what power are delivered to the load center? What is the voltage at the load center?

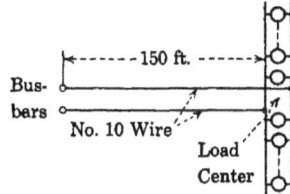

67. A 5-hp, 230-volt, 1,600-rpm, shunt motor takes 20 amp from the line at rated load and voltage. The efficiency at rated load is 0.81. Determine: (a) the watts output; (b) the horsepower output.

Fig. 66A.

68. A compound generator delivers 208 amp at 240 volts, and the efficiency is 0.904. The generator is driven by a synchronous motor. (a) What is the horsepower output of the synchronous motor? (b) If the efficiency of the synchronous motor at this load is 0.91, determine the kw input to the motor.

69. An incandescent lighting installation consists of 40 lamps which take 0.88 amp each at 118 volts and 20 lamps which take 0.60 amp each at 118 volts. On the average, these lights are all in use for $5\frac{1}{2}$ hr a day. At 5 c per kw-hr and on a 30-day basis, what is the monthly lighting bill?

70. An electric furnace requiring 42 amp at 240 volts is in operation, on the average, for $6\frac{3}{4}$ hr a day, 24 days in the month. At an energy rate of 5 c per kw-hr for the first 100 kw-hr, and $2\frac{1}{2}$ c per kw-hr thereafter, what is the monthly cost of operation?

71. A motor delivers an average of 8.8 hp, 8 hr a day, 25 days per month, and the efficiency is 0.80. On a basis of 5 c per kw-hr for the first 200 kw-hr, and 3 c per kw-hr thereafter, what is the monthly energy bill?

72. One gallon of water equals 3.79 liters. At a cost of $2\frac{1}{2}$ c per kw-hr, how much does it cost to heat 20 gal of water from 20° to 100°C? Assume 85 per cent efficiency.

73. A 1-qt (0.946-liter) electric percolator in use 30 days a month raises the temperature of the water from 20° to 100°C, and evaporates 10 per cent of the water once each day. It requires 539 gram-cal to evaporate 1 gram of water at atmospheric pressure (latent heat of vaporization). At 6 c per kw-hr, what is the monthly cost of operation of the percolator? Neglect losses.

74. A motor delivers 10 hp at an efficiency of 0.85 and receives its power from 250-volt bus-bars, over 500 ft (152.5 m) [per wire] of No. 7 A.W.G. copper wire feeder. The cross-section area of No. 7 wire is 20,800 circular mils and the resistance per circular-mil-foot is 10.6 ohms. Determine the voltage at the motor terminals and the efficiency of transmission.

75. A motor similar to that of Prob. 74 is fed at a distance of 700 ft (211 m) from the same bus-bars. It is desired that the voltage at the motor terminals be not less than 230 volts. Determine: (a) the resistance per conductor of the feeder; (b) the total power loss in the feeder; (c) the efficiency of transmission.

76. If in Prob. 75 the distance were 800 ft, what would be: (*a*) the resistance per conductor of the feeder; (*b*) the next largest A.W.G. wire at 25° C. as determined from (*a*)? Determine: (*c*) the actual voltage drop in the feeder; (*d*) the power loss; (*e*) the efficiency of transmission.

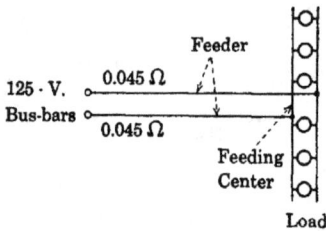

77. A lamp load requiring 80 amp is fed from 125-volt bus-bars over a feeder having a resistance of 0.045 ohm per conductor (Fig. 77*A*). Determine: (*a*) the voltage at the lamp terminals; (*b*) the transmission loss; (*c*) the efficiency of transmission. Show graphically the voltage variation along the feeder. (See Fig. 33, p. 40.) Neglect the voltage drop from the feeding center to the lamps.

Fig. 77*A*.

QUESTIONS ON CHAPTER III

Battery Electromotive Force—Kirchhoff's Laws

1. Distinguish between the internal voltage or emf and the terminal voltage of a battery when the battery is on open circuit; when the battery delivers current.

2. Why does the terminal voltage of a battery drop when the battery delivers current? How may the voltage drop due to the battery resistance be measured? How is the internal resistance determined?

3. What is the direction of current at the positive terminal of a battery when the battery delivers energy? When it receives energy?

4. Before a source of electrical energy, such as a generator, may deliver energy to a battery, how much voltage must be supplied? What is meant when a battery is said to be "floating" on the line?

5. What effect is noted when the line voltage is increased to a value greater than the emf of the battery? Show that the current which enters the positive terminal of a battery obeys Ohm's law.

6. When cells are connected in series, how may the total emf of the combination be determined? How may the total resistance be determined? Discuss the effect of reversing one of the cells in series.

7. What are the equivalent emf and resistance of a number of equal batteries when connected in parallel?

8. State Kirchhoff's first law. Under what conditions is a current preceded by a positive sign? By a negative sign?

9. State Kirchhoff's second law. Under what conditions is a product of current and resistance preceded by a positive sign? By a negative sign? Give a simple hydraulic analogy.

10. In applying Kirchhoff's second law to an electric circuit, what is the sign which should precede a rise in potential? A drop in potential?

11. What sign should precede the emf of a battery when going from negative to positive terminal? From positive to negative terminal?

12. Show that in general if a number of equations involving Kirchhoff's laws be applied to a network, at least one equation must involve the first law.

13. Show that a correct solution of a problem may be obtained, even when the *assumed* direction of the currents is incorrect.

14. Indicate the manner in which Kirchhoff's first law may be applied to the currents of an electric network so as to reduce the number of independent equations.

PROBLEMS ON CHAPTER III

Battery Electromotive Force—Kirchhoff's Laws

78. On open circuit the emf of a dry cell is 1.400 volts. When the cell delivers a current of 1.50 amp, the terminal voltage is 1.235 volts. (*a*) What is the internal resistance of the cell? (*b*) What current would this cell deliver if short-circuited?

79. The emf of a battery consisting of three dry cells in series is 4.26 volts, and the internal resistance of the battery is 0.28 ohm. When the battery delivers a current of 0.6 amp, determine: (*a*) the terminal voltage of the battery; (*b*) the terminal voltage of each cell.

80. The total emf of a 6-volt starting battery, consisting of three lead cells in series, is 6.32 volts, and the internal resistance is 0.024 ohm. What is its terminal voltage when it is doing starting duty delivering a current of 110 amp?

81. When a storage cell, with an internal resistance of 0.007 ohm, is delivering a current of 16 amp, the terminal voltage is 2.00 volts. Determine the internal emf of the cell under these conditions.

82. The emf of a storage cell is 2.12 volts. The terminal voltage becomes 2.00 volts when the cell delivers 22 amp. (*a*) Determine the resistance of the cell. (*b*) Determine the terminal voltage when the current is 40 amp.

83. A storage battery consists of 60 cells in series. On open circuit a voltmeter across the battery indicates 129.0 volts. When the battery delivers 50 amp, the voltmeter indicates 122.0 volts. Determine the internal resistance: (*a*) of the battery; (*b*) of each cell.

84. A resistance of 2.50 ohms is connected across the terminals of a dry cell, of emf 1.40 volts and internal resistance 0.12 ohm. Determine the current delivered by the cell and the terminal voltage of the cell.

85. A storage battery consisting of 60 cells in series has an emf of 130 volts and an internal resistance of 0.05 ohm. The battery supplies current to a lamp load of 4 ohms resistance. Determine: (*a*) the current to the load; (*b*) the terminal voltage of the battery; (*c*) the total power developed by the battery.

86. Determine the current delivered by the storage battery of Prob. 85, when resistances of 8 and 4 ohms in parallel are connected to the battery terminals.

87. Six dry cells having emfs 1.42, 1.40, 1.36, 1.41, 1.38, 1.45 volts and resistances 0.08, 0.09, 0.12, 0.11, 0.12, 0.09 are connected in series with positive to negative terminal.   Determine the current delivered by the battery to an external resistance of 15.0 ohms.

88. Repeat Prob. 87 with the second cell, of emf 1.40 volts and resistance 0.09 ohm, reversed.

89. A house-lighting battery consists of 16 storage cells in series.   The total emf of the battery is 33.6 volts, and the internal resistance is 0.08 ohm. (a) Determine the terminal voltage of the battery when it is being charged at a 15-amp rate.   (b) What is the resistance drop within the battery under these conditions?

90. A generator, with terminal voltage adjusted to 33.6 volts, is connected to the battery of Prob. 89, positive to positive terminal and negative to negative terminal.   (a) Determine the current under these conditions. Determine the current to the battery when the terminal voltage of the generator is: (b) 36 volts; (c) 30 volts.

91. A storage battery having an emf of 132 volts and an internal resistance of 0.05 ohm is connected across 140-volt bus-bars, the terminals of the battery being connected to bus-bars of the same polarity.   (a) Determine the current to the battery.   (b) What extra resistance should be inserted in the battery in order that the charging current shall be 100 amp?

92. A 6-volt lighting and ignition battery consists of three cells in series, each of which has an emf of 2.20 volts and an internal resistance of 0.018 ohm. What voltage must be impressed across the battery terminals in order that the battery may be charged at a 15-amp rate?

93. A storage battery consists of 120 cells each of which has an emf of 2.18 volts and an internal resistance of 0.0005 ohm.   Determine: (a) the emf of the battery; (b) the current which it will deliver to an external resistance of 2.5 ohms.

94. Three batteries having emfs 12.4, 10.2, 8.6 volts and resistances 0.08, 0.10, 0.12 ohm are connected in series aiding.   Determine the current which the system delivers when the external resistance is: (a) 2 ohms; (b) 2.5 ohms.

95. A storage battery installation consists of two batteries A and B, each consisting of 60 cells.   Each battery has an emf of 124 volts and an internal resistance of 0.060 ohm.   The two batteries are connected in parallel with terminals of like polarity connected together.   Determine: (a) the current delivered to an external load of resistance 1.7 ohms; (b) the terminal voltage of the battery.

96. Two equal batteries, each having an emf of 250 volts and an internal resistance of 0.16 ohm, are connected in parallel with terminals of like polarity together.   Determine: (a) the current delivered to an external load of resistance 1.00 ohm; (b) the emf and internal resistance of a single battery equivalent to these two in parallel; (c) the terminal voltage of the equivalent battery.

97. A battery used for marine-engine ignition consists of two groups of dry cells connected in parallel.   Each group consists of six cells.   The cells are all equal, each having an emf of 1.30 volts and an internal resistance of 0.09 ohm.   Determine: (a) the equivalent emf and internal resistance of the battery; (b) the steady current when the ignition circuit, of resistance

9.2 ohms, is connected across its terminals; (c) the terminal voltage under these conditions.

98. Three conductors A, B, C meet at a common junction. The current in A is 15.0 amp flowing toward the junction. The current in B flows away from the junction, and that in C flows toward the junction. The current in B is four times that in C. Determine the currents in B and C.

99. Four conductors A, B, C, D meet at a common junction O. In A the current is 15.0 amp, flowing toward the junction; in B the current is 9.0 amp, flowing away from the junction. The current in C is two-thirds that in D, and both have the same direction with respect to the junction. What are the value and direction of the currents in C and in D?

100. Two batteries (Fig. 100A), one having an emf of 16 volts and resistance of 0.50 ohm, the other an emf of 6.0 volts and resistance of 0.40 ohm, are connected together at their positive terminals by a 1.0-ohm resistance and at their negative terminals by a 3.1-ohm resistance. Assume that the current I flows in a clockwise direction and that point A is at zero potential. Write an equation, involving Kirchhoff's second law, beginning at point A. Solve for the current I. Determine the potential at points B, C, D. Make a graph of the potential variation in the circuit. (See Fig. 41, p. 51.)

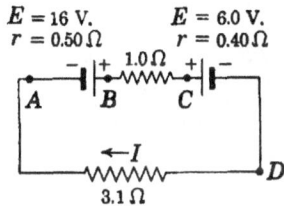

FIG. 100A.

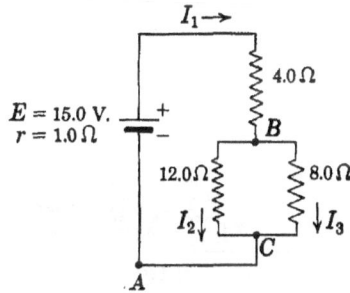

FIG. 101A.

101. A battery, having an emf of 15.0 volts and an internal resistance of 1.0 ohm, is connected to a circuit consisting of 4.0 ohms connected in series with 8.0 ohms and 12.0 ohms in parallel, the 4.0-ohm resistance being connected to the positive terminal of the battery (Fig. 101A). Assuming that the currents I_1, I_2, I_3 flow in the directions indicated in the figure, begin at point A and write two equations involving Kirchhoff's second law. Write a third equation involving the currents at junction B. Solve for I_1, I_2, I_3. What is the terminal voltage of the battery?

102. In Fig. 102A two batteries A and B are connected in parallel with terminals of like polarity together. The emf of A is 12 volts and its resistance is 1.0 ohm; the emf of B is 10 volts and its resistance is 0.5 ohm. A resistance of 4 ohms is connected across the system. By means of Kirchhoff's two laws determine the three currents I_1, I_2, I_3.

103. Repeat Prob. 102 with the 4-ohm resistance replaced by a 3-ohm resistance.

104. Figure 104A shows the network of Fig. 101A except that a battery having an emf of 6.0 volts, an internal resistance of 0.50 ohm, is connected in the 8.0-ohm branch and is acting in opposition to the assumed direction of

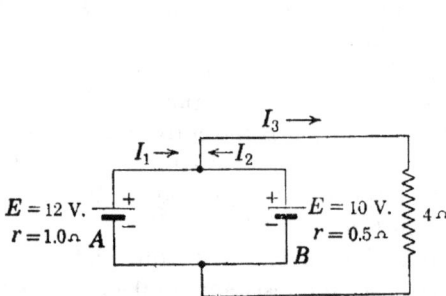

FIG. 102A.

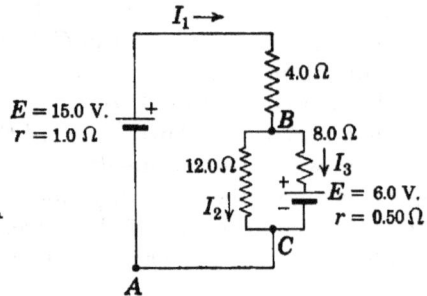

FIG. 104A.

I_3. Find the three currents I_1, I_2, I_3 and the voltage between points B and C.

105. In Fig. 105A is shown a battery, having an emf of 20 volts and internal resistance of 1.0 ohm, connected to an external circuit consisting of resistance and another battery. The assumed directions of the currents I_1, I_2, I_3 are shown. Determine the currents I_1, I_2, I_3, showing their actual directions.

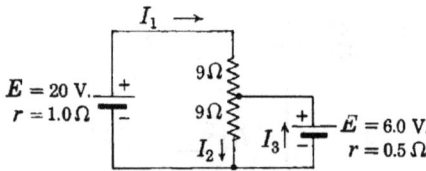

FIG. 105A.

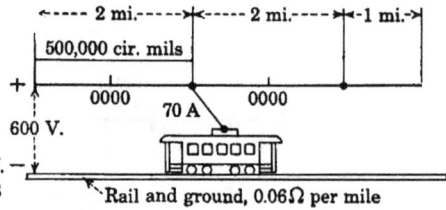

FIG. 106A.

106. A 5-mile trolley system is shown in Fig. 106A. The 5-mile trolley is 0000 hard-drawn copper having a resistance of 0.270 ohm per mile. This trolley is fed at a point 2 miles from the station by a 500,000-cir-mil feeder having the resistance of 0.114 ohm per mile. The combined resistance of rail and ground return is 0.06 ohm per mile. The bus-bar voltage at the station is maintained constant at 600 volts. A car 2 miles from the station is taking 70 amp. Find the voltage at the car and the power taken by the car.

107. Find the voltage at the car and at the end of the line when the car, Prob. 106, has moved a mile farther from the power station and takes 70 amp.

108. Find the voltage at the car, the power taken by the car, and the efficiency of transmission, Prob. 106, when the car takes 60 amp at the end of the line.

109. In the railway system, Prob. 106, a car takes 60 amp at the end of the line and a second car simultaneously takes 70 amp when it is 2 miles from the station. By means of Kirchhoff's laws, find: (a) the voltage at each car; (b) the power taken by each car.

NOTE: Owing to the fact that railway loads are widely fluctuating and are continually changing their position in the system, it is rarely possible, in practice, to formulate definite problems. However, the foregoing problems illustrate the influence of such important factors as the amount of copper which is necessary for maintaining a reasonably constant voltage at the car, the effect of the resistance of the return circuit, etc., on the car voltage. Although railway electrical engineers may not always formulate definite equations, as is done in the foregoing problems, in the design of electrical distribution systems of railways, the general principles illustrated in these problems, together with economic considerations, govern the size of feeders, the nature of rail bonding, etc.

QUESTIONS ON CHAPTER IV

Primary and Secondary Batteries

1. What effect is noted if two copper strips be immersed in dilute sulphuric acid and a low-reading voltmeter be connected to the two strips? What effect is noted if one of the copper strips be replaced by a zinc strip?

2. What two conditions must be fulfilled in order that an emf may exist in an electrolytic cell? When the external circuit is considered, what is the polarity of the copper with respect to the zinc? Why? When the cell itself is considered, what is the polarity of the copper with respect to the zinc? Why? What is meant by one metal being electro-positive to another?

3. Define *galvanic cell, electrolytic cell, electrode, anode, cathode, electrolyte.* In what way is the electrical energy, delivered by an electric cell, supplied?

4. Define a primary cell; a secondary cell. State four qualities which are desirable for commercial primary cells.

5. On what factors does the internal resistance of a cell depend? How may the resistance be decreased? How is the emf affected by the size of the cell?

6. What is meant by *polarization?* In what two ways does polarization reduce cell capacity? How may the effects of polarization be reduced?

7. Of what materials do the cathode and anode of the Daniell cell consist? What electrolytes are used, and how are the electrolytes kept separated?

8. In what ways does the gravity cell resemble the Daniell cell? In what particular does it differ?

9. Of what material does the cathode of the Edison-Lalande cell consist? The anode? What is used for electrolyte? Why is mineral oil necessary? What is the approximate emf of the cell?

10. Of what material does the cathode of a Le Clanché cell consist? The anode? What electrolyte is used? What is the emf of the cell? For what purposes is it used?

11. Of what materials is a dry cell made? What is the usual value of its emf? What is the approximate terminal voltage when it delivers current?

State one common cause of dry cells becoming exhausted.   Give a few of the ordinary uses of dry cells.

12. How is the standard ampere reproduced experimentally?   The standard ohm?   How is the International volt defined?   Of what materials is the cathode of a Weston cell made?   The anode?   What electrolyte is used?   Why should *not* the ordinary voltmeter be used to measure the emf of such a cell?

13. Show that a storage cell on discharge operates on the same principle, electrolytically, as a primary cell.

14. What effect is noted on each of two lead strips immersed in dilute sulphuric acid when a current is sent from one to the other through the acid?   What is the emf between the strips after current has flowed for some time?

15. What changes occur in the strips while they are delivering current, until the cell is exhausted?

16. What changes occur in the specific gravity of a storage cell during charge?   During discharge?   Why?

17. How is the active material formed in the Planté type of plate?   In what manner is a large surface exposed to the action of the acid in the Manchester plate?

18. Describe the construction of a Faure or pasted type of plate.   How is the positive plate formed?   The negative?   What are the advantages of the pasted plate?   Disadvantages?   Where is this type of plate commonly used?

19. Describe briefly the construction of the Exide-Ironclad cell.   What is its principal field of application?

20. Under what conditions are Planté plates used in stationary batteries?   Pasted plates?

21. What materials are used for tanks?   Why is considerable space desirable between the plates and bottom of tanks?   How are lead joints and seams in lead-lined tanks made tight?

22. Name two common types of separator.   What precaution should be taken in the handling of wood separators?

23. Describe the procedure which should be followed when making dilute acid from concentrated acid.

24. How is specific gravity measured?   How may specific gravity be used to determine the condition of charge of a cell?

25. Why is it usually necessary to add only water to make up the loss of electrolyte?   What is the general effect of the specific gravity on the freezing temperature of the electrolyte?

26. State briefly the method of installing a stationary battery.   What is meant by "freshening charge" and "initial charge?"

27. What are the primary requirements of vehicle and automobile batteries?   Why is the specific gravity carried to high values in this type of cell?   What type of plates is used?

28. What is meant by the "normal rating" of a battery?   What is meant by the 8-hr rate?   The 3-hr rate?   Why do the ampere-hours delivered apparently decrease as the rate of discharge increases?

29. Give a rule by which the charging rate for a lead battery may be determined. Why should the rate be adjusted so that violent gassing does not occur? What is meant by "finishing rate?"

30. Describe the constant-current method of charging, and give an example. Describe the constant-potential method and state its advantages. Why is a small series resistance usually desirable?

31. How may the voltage be used to determine the condition of charge of a lead storage cell? Discuss the charging of batteries when only alternating current is available. Describe a simple method of determining the polarity of a d-c supply.

32. What electrolyte is used in the Edison nickel-iron-alkaline battery? Describe the material and the construction of the positive plate or cathode; the negative plate or anode. What method is used to increase the conductance of the cathode?

33. Describe briefly the chemical changes which occur in the positive and negative electrodes of an Edison cell during charge and discharge.

34. What changes, if any, take place in the electrolyte during charge and discharge? Compare with the lead cell.

35. Describe the assembly of the Edison cell. What are used for separators? Of what construction is the tank?

36. On what number of hours discharge is the rated capacity of the Edison cell based? What is the approximate average terminal voltage on discharge?

37. Enumerate some of the advantages of the Edison battery and some of its fields of application.

38. State the fundamental principles of electroplating. What one characteristic is essential in the electrolyte? What material is commonly used for the anode? What is the character of the power supply for electroplating processes?

PROBLEMS ON CHAPTER IV

Primary and Secondary Batteries

110. The internal emf of a Le Clanché cell is 1.40 volts. When a resistance is connected across its terminals, the cell delivers a current of 1.5 amp and the terminal voltage drops immediately to 1.15 volts. Even if the current is maintained constant at 1.5 amp, the terminal voltage continues to fall until it reaches a sensibly constant value of 0.96 volt. Determine: (a) the initial internal resistance of the cell; (b) the *apparent* internal resistance of the cell after its terminal voltage has become sensibly constant. (c) To what is the difference of (a) and (b) due?

111. The emf of a dry cell is 1.42 volts and the internal resistance is 0.10 ohm. After the cell delivers a current of 1.2 amp for some time, the terminal voltage is 1.10 volts. (a) What is the apparent cell resistance with this current flowing? (b) Why does this value of resistance differ from the 0.10 ohm?

112. A battery to supply a current of 0.25 amp to a signal system is to consist of gravity cells in series. The emf of each cell is 1.09 volts and the internal resistance is 0.08 ohm. The resistance of the signal system is 51.4 ohms. How many cells in series are necessary?

113. In Prob. 112, how many dry cells in series would be necessary if the emf of each dry cell is 1.32 volts and the internal resistance is 0.12 ohm?

114. The emf of an unsaturated Weston cell is 1.0182 volts and the internal resistance is 240 ohms. It is attempted to measure the emf of this cell with a 1.5-scale voltmeter whose resistance is 160 ohms. What will be the actual reading of the voltmeter? Why is it impracticable to measure the internal emf of such a cell by this means?

115. In Fig. 115*A* is shown a storage cell whose emf is 2.00 volts. This cell delivers current to the adjustable resistance *R* and the 15-ohm resistance wire *ac* in series. The negative terminal of a Weston cell, whose emf is 1.0184 volts, is connected to the negative terminal *a* of the storage cell. The resistance *R* is adjusted until the storage-cell current is 0.10 amp. A slider *b* is movable along the resistance wire *ac*. (*a*) At what point (ohms from *a*) on the resistance wire *ac* will the voltage from *a* to *b* be 1.0184 volts? (*b*) What will a galvanometer read if connected between points *bd* under these conditions? (This gives a means of determining the emf of a Weston cell without its delivering current and is the underlying principle of the potentiometer. See p. 116.)

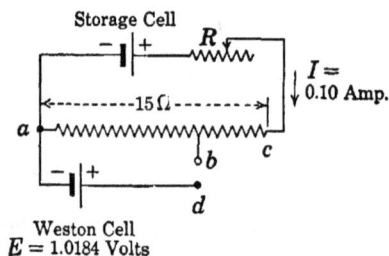

Storage Cell

$I = 0.10$ Amp.

Weston Cell
$E = 1.0184$ Volts

FIG. 115*A*.

116. A dry cell is substituted for the Weston cell in Prob. 115, and a galvanometer is connected between points *bd*. When the slider *b* is 14.30 ohms from *a*, the galvanometer reads zero. (*a*) What is the emf of the dry cell? (*b*) What current does it deliver under these conditions? (*c*) Compare the accuracy of this method of determining the emf of the dry cell with that using a voltmeter.

117. The voltage across the terminals of a storage cell, when charged at the 10.0-amp rate, is 2.40 volts. When the circuit is opened the terminal voltage of the cell is 2.10 volts. (*a*) What is the apparent internal resistance of the cell? (*b*) Approximately how many joules are converted into chemical energy every minute during charge? (*c*) How many joules are converted into heat each minute during the charging period?

118. The apparent internal resistance of the cell of Prob. 117, when on discharge, is 0.02 ohm. (*a*) What is its terminal voltage when discharging at the 12.0-amp. rate? (*b*) How many joules per minute does it develop? (*c*) How many joules per minute are lost within the cell? (*d*) What is the efficiency of the foregoing cell (ratio of energy delivered by cell to energy delivered to cell, Prob. 117)?

119. Plot two curves with specific gravity as abscissas and parts by volume and by weight of water as ordinates, using the data of Par. 63, p. 76.

From these curves determine the number of liters and the number of kilograms of water which it is necessary to add to 4.0 liters of sulphuric acid (sp. gr. = 1.84) to give a solution having a specific gravity of (a) 1.23, (b) 1.26. How should the acid and water be mixed?

120. A battery rated at 160 amp-hr is charged at the 8-hr rate. Its specific gravity at the beginning of charge is 1.160. Assuming that the specific gravity increases according to the curve (Fig. 63, p. 78), how many ampere-hours have been delivered to the battery when a hydrometer indicates 1.190? How many additional ampere-hours will be necessary to charge the battery completely, assuming that the specific gravity will then be 1.200?

121. When fully charged, a battery having Planté plates will discharge 80 amp for 8 hr before it is completely discharged. What is the discharge rate in amp if, after being fully charged, it is completely discharged in 5 hr? (See Par. 67, p. 80.)

122. It is desired to charge a 3-cell, 6-volt battery from 115-volt d-c mains. When charged at the 15-amp rate the terminal voltage is 2.4 volts per cell. (a) What series resistance (ohms and carrying capacity) is necessary? (b) What per cent of the power delivered by the line is used to charge the battery?

123. In Prob. 122 if it requires 8 hr to charge the battery and energy costs 5.0 c per kw-hr, how much does it cost to charge the battery?

124. Repeat Prob. 122 with two batteries connected in series, each the same as the single battery.

125. Repeat Prob. 122 with three batteries connected in series, each the same as the single battery.

126. A 320-amp-hr storage battery consisting of 55 cells is charged at the 8-hr rate. The average terminal voltage of each cell is 2.32 volts. (a) How much power is required to charge the battery? (b) How many kilowatt-hours are required to charge the battery to its full ampere-hour rating? (c) With energy costing 2 c per kw-hr, how much does it cost to charge the battery?

127. Each cell of the battery, Prob. 126, has an open-circuit emf of 2.10 volts and an internal resistance of 0.010 ohm. When the battery is connected across 115.5-volt bus-bars, to what value must the bus-bar voltage drop in order that the battery may deliver 30 amp?

128. An Edison storage battery consisting of five series-connected cells is charged at a constant rate of 18 amp for 5 hr. The initial value of the voltage of each cell is 1.55 volts, and the voltage varies with charge according to the upper curve (Fig. 70, p. 86). What is the total energy required to charge the battery? (Note: Average the ordinates, Fig. 70, to obtain the average voltage during charge.)

129. An electrolytic process requires 140 amp at 32 volts. The generator supplying this energy has an efficiency of 0.80 and is driven by a 220-volt motor whose efficiency is 0.82. (a) How many watts does the motor deliver to the generator? (b) What is the horsepower output of the motor? (c) What is the watt input to the motor? (d) What current does the motor take?   (e) At 2½ c per kw-hr, what is the energy cost per week of operating the electrolytic bath on the basis of 45 hr a week?

QUESTIONS ON CHAPTER V

Electrical Instruments and Electrical Measurements

1. What position does a coil, carrying a current, seek when placed in a magnetic field?   Show: (a) that the poles produced by the coil are attracted or repelled by the poles producing the magnetic field; (b) that the coil seeks such a position that the magnetic flux of the system is a maximum.

2. How is this principle utilized in the D'Arsonval galvanometer?   Describe the construction of the moving coil.   How is it suspended?   How is the current led to the coil?   What is the purpose of the soft-iron core?   How is the moving system damped?   What methods are used for reading the deflection?

3. What is the purpose of galvanometer shunts?   Describe one type, giving the ratios of the shunt resistances to the galvanometer resistance for various decimal shunting ratios.   Sketch the connections of the Ayrton shunt.   What are the advantages of this type of shunt over the first type?

4. Show that the movement of the Weston d-c instrument is an evolution of the D'Arsonval galvanometer.   Describe the changes which are necessary to make the instrument portable.

6. How may the Weston ammeter be made to measure currents of any desired magnitude when its own moving system is designed for very small currents only?   Show that the ammeter may be considered as a voltmeter which measures merely the voltage drop across a fixed resistance.

6. Show that an ammeter may have an indefinite number of ranges. Why do ammeter shunts have four posts, two for the current and two for the connection to the instrument?   Discuss the sources of error and the precautions which must be taken when using an external-shunt ammeter. In what way does the internal-shunt ammeter differ from the external-shunt type?

7. Discuss the similarity between the voltmeter and the ammeter. Sketch the internal connections of a voltmeter having two voltage ranges. Why does a voltmeter have high resistance?

8. How may the range of a voltmeter be increased?   What is a multiplier?

9. On what property of the electric current does the vacuum thermocouple operate?   Describe the construction of the thermocouple and the method of applying it to the measurement of current and voltage.   For what types of measurement is the device particularly well adapted?

10. Describe the manner in which the voltmeter-ammeter method may be used for measuring low resistances.   Why is a rheostat, connected in series with the circuit, often necessary?

11. Why does it become necessary to use special contacts for the voltmeter connection when accurate measurements of very low resistances are desired?

12. Sketch the two connections which are commonly employed when a voltmeter alone is used for measuring resistance.   What are the approximate magnitudes of resistances which can be measured by this method? Why is a voltmeter having very high resistance usually desirable?

13. Make a simple diagram of the Wheatstone bridge, giving the principles upon which it operates. What are the ratio arms? What is the rheostat arm? How may the condition of balance be determined?

14. Describe briefly a systematic procedure for obtaining a balance. Describe the decade bridge.

15. Make a simple sketch of the slide-wire bridge, showing its similarity to the Wheatstone bridge; compare the two types of bridge with regard to accuracy and simplicity.

16. Show the application of the slide-wire-bridge principle to the location of a ground in a cable. Why are the positions of the battery and galvanometer in the Murray-loop test reversed from their positions shown in Fig. 90, p. 110?

17. Show that a sensitive galvanometer, connected in series with very high resistance, for example, the resistance of insulating materials, may be used as an ammeter to measure the current leaking through the insulation. How is the galvanometer calibrated in terms of resistance?

18. Why does the galvanometer deflection, when measuring the leakage current, vary with time? What is standard practice as regards the time of electrification?

19. Make a sketch of a simple potentiometer in which a 2-volt battery supplies current to a resistance-wire while a standard cell, in series with a galvanometer, is connected to the resistance-wire so that the galvanometer may be made to read zero. Also indicate the method by which this resistance-wire, carrying current, may be used to measure an unknown emf.

20. Make a simple diagram of connections of a Leeds and Northrup potentiometer. How is the potentiometer current adjusted to its proper value by means of the standard cell? Show the method used to measure an unknown emf. Name the accessories that are necessary with this type of potentiometer.

21. Sketch the connections of a volt box, and show its use in the measurements of emfs which are in excess of the range of the potentiometer. What is a drop-wire?

22. Show how current may be measured accurately with a potentiometer, which is fundamentally a device for measuring emf. What are standard resistances, and in what values of resistance are they usually made?

23. Sketch the internal connections of a wattmeter. Upon what principle does it operate?

24. In what manner does the watthour meter differ essentially from the other types of instruments which have been described? How is the armature connected to the circuit? The field coils? Why is an auxiliary field coil necessary, and how is it connected? Why are retarding magnets necessary?

25. What relation exists between the revolutions of the disc and the watthours registered? How is the meter calibrated? What adjustments are made if the meter is too fast near full load? Too slow?

26. What adjustments are made if the meter is too fast at light load? Too slow?

27. Sketch the connections of a three-wire watthour meter. Where is it used?

PROBLEMS ON CHAPTER V

Electrical Instruments and Electrical Measurements

130. A D'Arsonval galvanometer whose resistance is 693 ohms deflects 20 cm with a current of 4.80 microamp. What potential difference across its terminals will produce full-scale deflection of 25 cm?

131. The galvanometer of Prob. 130 is shunted with a 99-ohm resistance. (a) What is its deflection when the line current is 22.0 microamp? (b) What is the voltage across its terminals?

132. Compute the values of three resistances to shunt the galvanometer of Prob. 130 so that its current will be one-tenth, one-hundredth, and one-thousandth the line current. (See Fig. 75, p. 92.)

133. Figure 133A shows a galvanometer whose resistance is 1,200 ohms, used in connection with an Ayrton shunt which has a total resistance AB of 10,000 ohms. The resistance Aa is 1.0 ohm. When the line contact C is at a, find: (a) the galvanometer current in microamperes when the line current I equals 6,000 microamp; (b) the current which would flow through the galvanometer if the line contact C were at B. (c) Compare the ratio of resistances Aa and AB with the galvanometer currents. (HINT: Use Eq. 22, p. 31.)

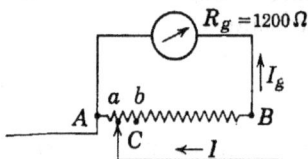

FIG. 133A.

134. Repeat Prob. 133 for a line current of 4,000 microamp with the line contact C connected to b. The resistance Ab is 10.0 ohms.

135. A d-c ammeter without a shunt gives full-scale deflection when the current in its moving system is 0.025 amp. The ammeter resistance is 2.0 ohms. What voltage, in millivolts, across its terminals will cause full-scale deflection?

136. It is desired that the ammeter of Prob. 135 be made to have a full scale of 10 amp. (a) What should be the resistance of its shunt? (b) What must be the resistance of a shunt in Prob. 135 to give a full-scale deflection of 500 amp? (c) How much power is lost in the shunt with full-scale deflection in both (a) and (b)? (NOTE: The current in the instrument is negligible as compared with the current in the shunt.)

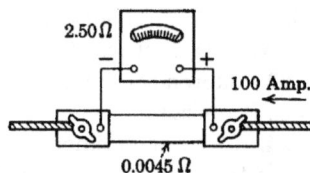

FIG. 137A.

137. The resistance of an ammeter is 2.50 ohms, and the resistance of its shunt is 0.0045 ohm (Fig. 137A). (a) When the external current is 100.0 amp, how much current flows in the ammeter and how much in its shunt? (b) What is the ratio of ammeter current to shunt current?

138. The resistance of a two-scale voltmeter is 16,500 ohms between the terminals marked 0 and 150 volts (Fig. 138A). (a) What is the resistance between the 0 terminal and the 15-volt terminal? (*b*) What voltmeter current gives full-scale deflection in each case?

139. What should be the resistance of a multiplier to give the voltmeter of Prob. 138 a full-scale deflection of 600 volts?

140. Two 300-scale voltmeters *A* and *B* have resistances of 32,000 and 34,000 ohms. What will each read when the two are connected in series across a 500-volt circuit?

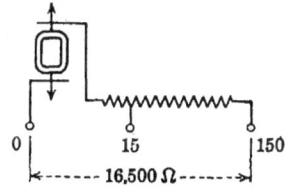

Fig. 138*A*.

141. A 300-scale voltmeter, having a resistance of 32,600 ohms, is connected in series with a 50,000-ohm resistance between a trolley and ground. What is the trolley voltage when the voltmeter reads 215 volts?

142. A low resistance is measured by the voltmeter-ammeter method (Fig. 84, p. 103). The current of 24.0 amp is taken from 110-volt mains through a rheostat. The voltmeter connected directly across the resistance indicates 4.86 volts. Determine: (*a*) the value of the unknown resistance; (*b*) the resistance of the rheostat; (*c*) the power dissipated in the resistance and in the rheostat.

143. The resistance of a copper rod 0.250 in. (0.635 cm) diameter and 48 in. (121.9 cm) long is measured by the voltmeter-ammeter method, shown in Fig. 85, p. 104. The voltmeter contact points are 42 in. (106.7 cm) apart. When the ammeter reads 82.0 amp, the voltmeter reads 47.6 millivolts; the temperature is 20°C. (*a*) What is the resistance of a 1-ft (30.5 cm) length of the rod? (*b*) What is the resistance of a circular-mil-foot of the rod? (*c*) If the resistance of standard copper is 10.37 ohms per cir-mil-ft at 20°C, what is the ratio of the resistivity of the rod to that of the standard?

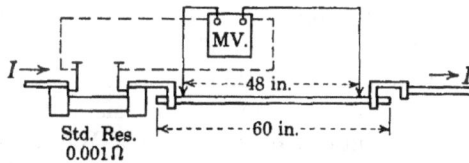

Fig. 144*A*.

144. An aluminum rod 0.375 in. (0.953 cm) diameter and 60 in. (152.4 cm) long is connected in series with a 0.001-ohm standard resistance (Fig. 144*A*). When current flows in the system, a millivoltmeter connected across the standard resistance reads 56.8 millivolts. (See dotted lines.) The millivoltmeter is then connected across the contact points on the aluminum rod 48 in. (121.8 cm) apart and reads 27.6 millivolts. Determine: (*a*) the resistance per foot of the aluminum rod; (*b*) the resistance per circular-mil-foot. (NOTE: In this method of resistance measurement, it is not necessary that the instrument read millivolts. If its scale readings are proportional to the voltage across the instrument terminals, the ratio of the two readings may be used.)

145. A 300-scale, 38,000-ohm voltmeter, when connected across d-c mains (Fig. 145*A*), switch *S* being closed, reads 240.0 volts. When switch *S* is opened, connecting the resistance x in series with the voltmeter, the voltmeter reads 68.0 volts. (*a*) What current flows through the voltmeter and x when S is open? (*b*) What is the voltage across x? (*c*) What is the value of x in ohms?

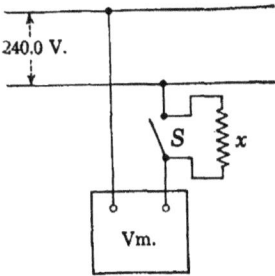

Fig. 145*A*.

146. A voltmeter V_1 (Fig. 146*A*), when connected across d-c mains, reads 600 volts. A voltmeter V_2, whose resistance is 100,000 ohms, is connected between the positive line and the core of a cable. The sheath of the cable is connected to the negative line. V_2 reads 8.0 volts. (*a*) What current flows through V_2? (*b*) What is the voltage across the cable? (*c*) What is the insulation resistance of the cable? (*d*) If the cable is 1,620 ft long, what is the insulation resistance, in megohms, of a mile length?

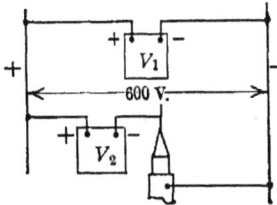

Fig. 146*A*.

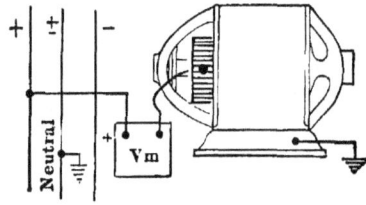

Fig. 147*A*.

147. A 150-scale, 100,000-ohm voltmeter is used to measure the insulation resistance of the armature of a dynamo. The voltmeter is connected first between a positive main and neutral, which is grounded, and reads 126 volts. The negative terminal of the voltmeter is then connected to the commutator and reads 4.0 volts. (See Fig. 147*A*.) The frame of the dynamo is grounded. What is the resistance to ground of the armature?

148. It is desired to measure an unknown resistance X between 6 and 7 ohms to four significant figures, by means of a Wheatstone bridge. (See Fig. 87, p. 105.) What should be the ratio of A to B? If the rheostat arm P reads 6,284 ohms when a balance is obtained, what is the value of the unknown resistance?

149. In the slide-wire bridge (Fig. 90, p. 110), R is a resistance of 10 ohms. A balance is obtained when the slider reads 32.4 cm. What is the value of the unknown resistance X?

150. In Prob. 149, an unknown resistance is substituted for R and a balance is obtained when the slider reads 71.3 cm. What is the value of this unknown resistance?

151. Two single-conductor cables, each 2,500 ft (762 m) long and containing a No. 14 A.W.G. copper conductor, are used for signal purposes. It is desired to locate a fault, which has developed in one cable, by the Murray-loop method. The two cables are looped at the far end, being connected as shown in Fig. 91, p. 111. A balance is obtained when the slider reads 74.4 cm on the 100-cm slide-wire. How far from the home end is the fault?

152. A cable is tested for its insulation resistance, the connections shown in Fig. 92, p. 113, being used. The resistance of the Ayrton shunt is 10,000 ohms, and the galvanometer resistance is 1,200 ohms. When the cable is short-circuited and the Ayrton shunt is set at 0.0001, the galvanometer deflects 22.0 cm. The emf of the battery is 300 volts. Determine: (*a*) the resistance of the circuit; (*b*) the current in the circuit. The short circuit is removed from the cable. The galvanometer deflects 12.2 cm, when the shunt is set at 1.0 after the cable has been electrified 1 min. Determine: (*c*) the current which now flows in the circuit; (*d*) the insulation resistance of the cable in megohms.

153. The insulation of another cable is tested by the same method as that given in Prob. 152. When the cable is short-circuited, the deflection is 22.4 cm, with the shunt set at 0.0001. After the short circuit is removed, the deflection after 1 min is 14.3 cm, with the shunt set at 0.1. Determine: (*a*) the insulation resistance of the cable in megohms; (*b*) the insulation resistance in megohms of a mile length if the cable is 2,400 ft long.

154. A 2-volt storage cell *A* (Fig. 154*A*) delivers current to a 150-ohm resistance-wire *ca* through the rheostat *R*. The rheostat *R* is adjusted until the current in the circuit is 0.01 amp. The negative terminal of a battery *B* is connected to *a*, the negative side of *A*. The positive terminal of *B* is connected through a galvanometer and key to a movable contact *b* on the wire *ca*. If the emf of *B* is 1.0186 volts, what is the resistance *a* to *b* when the galvanometer does not deflect on depressing the key in the galvanometer circuit? (The battery *B* is a standard cell whose emf is known.)

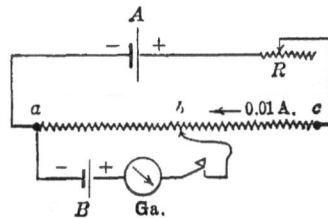

FIG. 154*A*.

155. The battery *B* in Prob. 154 is removed, and a dry cell of unknown emf is substituted. A balance is obtained when the resistance *ab* is 141.2 ohms. What is the emf of the dry cell? What current does it deliver under the condition of balance?

156. In a certain potentiometer, the resistance between each 0.1-volt contact is 10.0 ohms. What is the working current of the potentiometer? If the emf of the standard cell is 1.0182 volts, what is the resistance between the contacts which the standard-cell circuit makes with the potentiometer circuit? (See Fig. 154*A*.)

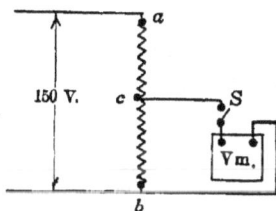

FIG. 157*A*.

157. A drop-wire *ab* (Fig. 157*A*) of 1,000 ohms resistance is connected across 150 volts. At its mid-point *c* a tap is brought out, to which a voltmeter *V*, having a resistance of 4,000 ohms, is connected through switch *S*. (*a*) What is the voltage from *c* to *b* when switch *S* is open? (*b*) What does the voltmeter read when switch *S* is closed?

158. Repeat Prob. 157 (*b*) with a voltmeter whose resistance is 15,000 ohms.

159. It is desired to calibrate a 60-scale ammeter by means of a potentiometer, the connections being as shown in Fig. 97, p. 121. What value of standard resistance should be used?

160. A 15-amp, 115-volt watthour meter is tested for its accuracy by the method described in Par. 100, p. 127. The watthour-meter constant is 0.4. The rotating element makes 50 revolutions in 44.9 sec. The averages of the voltmeter and ammeter readings taken over this interval are 115.0 volts and 14.2 amp. (*a*) What is its percentage of correct registration at this load? (*b*) What adjustment is necessary to bring the registration nearer to its true value?

161. The meter of Prob. 160 is tested at light load. The rotating element makes three revolutions in 46.6 sec, and the averages of the voltmeter and ammeter readings taken over this interval are 115.5 volts and 0.82 amp. (*a*) What is the percentage of correct registration? (*b*) What adjustment is necessary to bring the meter nearer to correct registration?

QUESTIONS ON CHAPTER VI
Magnetism and Permanent Magnets

1. Distinguish between permanent magnets and electromagnets. In general, of what material are permanent magnets made? Electromagnets?

2. What material is used almost exclusively for magnet purposes? Name a few other materials that are used.

3. Describe natural magnet. What is *lodestone?*

4. Describe a magnetic field. Does magnetism actually exist as lines? What is meant by the poles of a magnet? What relation exists between the poles of a magnet and the direction which the magnet assumes when freely suspended in space?

5. What effects are noted when a bar magnet is broken and how may these effects be explained by Weber's theory?

6. State Coulomb's law which gives the magnitude of the force existing between magnetic poles in free space. Define a unit pole. What is pole strength, and how is it measured?

7. What two effects are noticed when a small unit pole is placed in a magnetic field? Describe *lines of force,* distinguishing them from lines of induction.

8. What is the fundamental measure of field intensity, and in what unit is it expressed? What is the relation of field intensity to flux density in air? In magnetic materials?

9. Define flux density. Name the cgs unit.

10. Describe the compass needle. How may it be used to determine the polarity of magnets? In plotting magnetic fields?

11. How are magnetic figures produced, and what do they signify?

12. What is meant by magnetic induction, and what relation exists between the induced and the inducing poles? How does magnetic induction explain the magnetic attraction of iron by magnets?

13. What important law governs the geometrical shape which a magnetic field tends to assume?

14. State the advantages of the horseshoe magnet over the bar magnet. What advantages are gained in making magnets laminated? Give some practical uses of permanent magnets.

15. Describe a magnetic screen, giving its principle of operation.

16. What methods are used to magnetize permanent magnets?

17. What is the nature of the earth's magnetism? Why does the compass needle point true north at only a comparatively few places on the earth's surface? What is meant by the *dip* of the needle?

PROBLEMS ON CHAPTER VI

Magnetism and Permanent Magnets

162. Two north poles, one having a strength of 24 unit poles and the other a strength of 40 unit poles, are placed 8 cm apart in air. Considering these poles as concentrated at points, determine the force in dynes acting between the poles. Is this force attraction or repulsion?

163. A north pole having a strength of 50 unit poles and a south pole having a strength of 60 unit poles are placed 5 in. apart in air. Assuming that these poles are concentrated at points, determine the force in grams acting between the two. Is this force attraction or repulsion?

164. A magnetic pole, when placed in air 12 cm from a north pole having a strength of 240 units, is attracted with a force of 250 dynes. Assuming that both poles are concentrated at points, determine the strength of the attracted pole. Is this pole a north or a south pole?

165. A small north pole of 50 units strength is placed in a magnetic field in air and is acted on by a force of 600 dynes. What is the field intensity in oersteds? What is the flux density in gausses?

166. The pole faces of an electromagnet are 20 cm square and a flux of 200,000 maxwells exists between them. Assuming that the magnetic field is uniformly distributed, determine: (*a*) the field intensity; (*b*) the force in grams exerted on a north pole of 125 units strength placed in this field; (*c*) the direction in which this north pole tends to move.

167. A bar magnet 15 cm long and having poles of 100 units strength at its ends is placed in a uniform field in which the flux density is 6,000 lines per sq in., the axis of the magnet being perpendicular to the direction of the magnetic field. Determine: (*a*) the force in grams acting on each pole; (*b*) the field intensity in oersteds; (*c*) the direction in which the magnet tends to move; (*d*) the turning moment of the magnet in gram-cm. (See p. 136.)

168. A small north pole m of 200 units strength and a small south pole m' of 150 units strength are spaced 7 cm

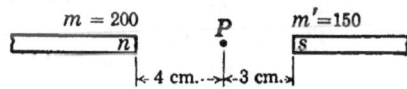

$m = 200$ P $m'=150$

n s

|← 4 cm.→|←3 cm.→|

FIG. 168*A*.

apart in air (Fig. 168*A*). (*a*) Determine the field intensity in dynes per unit pole, or oersteds, at point P, on a straight line joining the poles, 4 cm from m and 3 cm from m'. (*b*) What should be the strength of a small pole placed at P in order that the force acting on this pole may be 384 dynes?

169. A magnetized bar magnet is brought close to the poles of a horseshoe magnet, as shown in Fig. 169*A*. Sketch the resulting magnetic field.

170. Repeat Prob. 169, reversing the bar magnet in position.

171. Repeat Prob. 169, substituting an unmagnetized soft-iron bar for the bar magnet.

172. Sketch the magnetic field produced by the two bar magnets at right angles to each other, shown in Fig. 172*A*.

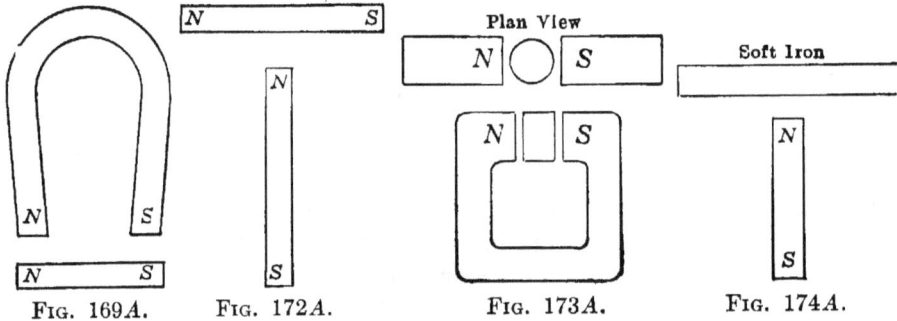

FIG. 169*A*.   FIG. 172*A*.   FIG. 173*A*.   FIG. 174*A*.

173. Figure 173*A* shows an elevation and plan of a horseshoe magnet as used in some types of galvanometers (see Fig. 73, p. 96), the plan view being drawn to an enlarged scale. Between the poles of this magnet is a cylindrical soft-iron core. Sketch the magnetic field in the plan view, and indicate the poles induced on the core.

174. The north end of a permanent bar magnet (Fig. 174*A*) is brought near the middle of a soft-iron bar. Indicate the direction of the lines of induction through the bar and also the polarity of the induced poles.

QUESTIONS ON CHAPTER VII

Electromagnetism

1. How may it be shown that a magnetic field exists about a conductor carrying current? What is the geometrical shape of the field if the conductor is straight and cylindrical and there are no other magnetic fields in the vicinity? In what two ways may it be shown that such a field is cylindrical?

2. What definite relation exists between the direction of the current and the direction of the magnetic field which surrounds the conductor? Give two simple rules by which this relation may be remembered.

3. The current in a horizontal conductor flows from right to left. In what direction will the north end of a compass needle point if held above the wire? Beneath the wire?

4. What is the character of the force existing between two parallel conductors carrying current in the same direction? In opposite directions? Show that the direction of the force may be explained by the simple laws which govern the geometry of the magnetic field.

5. Sketch the direction of the magnetic field produced by a single turn carrying current. Show that a solenoid may be produced by the combination of several such turns connected in series. Give three methods by which the polarity of the ends of such a solenoid can be determined, provided the direction of current in the solenoid be known.

6. Explain, by the simple laws of electromagnetic induction, the reason that a solenoid tends to pull an iron plunger within itself.

7. State the advantages of ironclad solenoids. What characteristic does the "stop" give to the solenoid? Give some commercial applications of the solenoid.

8. Show that the operation of the telegraph relay is an illustration of the law of the magnetic circuit given in Par. 117, p. 141.

9. Make a diagram of connections for a typical electric bell, pointing out its principle of operation.

10. By a simple sketch, indicate the magnetic circuits of a lifting magnet. What are some of its commercial uses, and what are its economic advantages? What is a magnetic separator?

11. Sketch the magnetic circuits and the flux paths for one type of bipolar generator.

12. Sketch the magnetic circuits and the flux paths of a multipolar generator.

13. What is magnetic leakage? Does it represent energy loss? Why should it ordinarily be made a minimum? What general law should be followed in determining the position of generator field coils?

PROBLEMS ON CHAPTER VII

Electromagnetism

175. A portion of a d-c feeder entering the duct of an underground cable system is shown in Fig. 175A. When a compass is held above the feeder, the needle deflects as shown. In what direction does the current in the feeder flow, in or out of the duct?

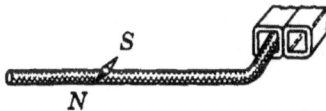

Fig. 175A.   Fig. 176A.

176. Figure 176A shows two positive feeders of a trolley system running on a pole line and carrying current in the same direction. A compass needle is held beneath each conductor. In what direction does the *N*-pole point in each case? In what direction do the conductors tend to move, and what is the direction of the force acting on the insulators? Sketch the resultant field produced by the currents in the conductors, in a plane taken perpendicular to the conductors.

177. Figure 177*A* shows the end view of the two poles of an electromagnet and the direction of the current in the exciting coils.   Indicate the polarity of each of the two poles.

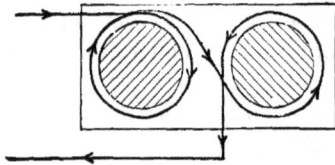

Fig. 177*A*.

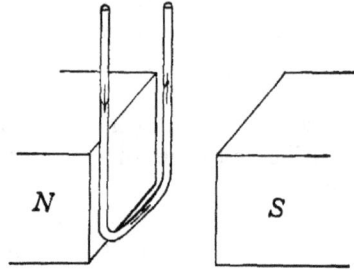

Fig. 178*A*.

178. Figure 178*A* shows a loop of wire in a magnetic field and the direction of the current in the loop.   Does the loop of wire under these conditions strengthen or weaken the existing magnetic field?

179. Figure 179*A* shows two coils on a simple horseshoe magnet.   Connect these coils so that they aid each other.   It is desired that the upper right-hand pole shall be north.   Show the direction of current flow in the excitation circuit.   Sketch the magnetic field between the poles.

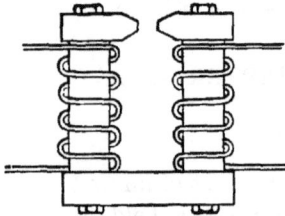

Fig. 179*A*.

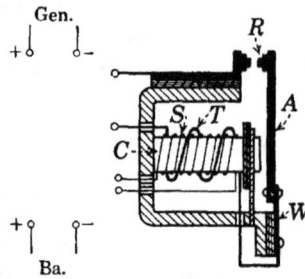

Fig. 180*A*.

180. In Fig. 180*A* is shown an automobile cutout relay.   Its function is to connect the battery to the generator when the generator emf exceeds by a small amount the battery emf and to disconnect the battery when the generator emf drops below that of the battery.   The relay consists of an iron core *C* on which are a shunt winding *S* and a series winding *T*.   There is also an armature *A* to which a contact *R* is attached.   The contact is ordinarily held open by the spring *W*.   When the emf of the generator exceeds slightly that of the battery, the excitation of the shunt coil causes the armature to overcome the force of the spring *W* and the contact *R* closes.   Current then flows from the generator to the battery through the series winding *T* which aids the shunt winding in keeping the contact *R* closed.   When the emf of the generator becomes less than that of the battery, the current in the series winding reverses and opposes the excitation of the shunt coil.   This reduces the magnetism in the core, and the spring *W* causes the contact *R* to open.   (Also see p. 314.)

Connect the shunt and series coils, Fig. 180*A*, to the generator and battery terminals *Gen.* and *Ba.* so that the relay will operate properly.

181. Figure 181*A* gives the diagram of a small buzzer. Connect the terminals 1 to 8 so that the buzzer will operate. Assuming that the terminal 1 is positive, indicate the polarities of the ends of the solenoids *A* and *B*.

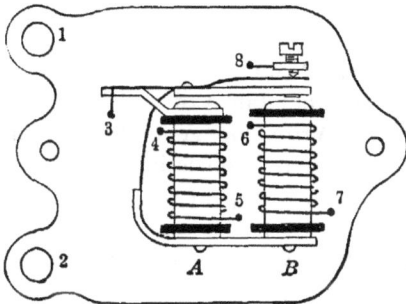

Fig. 181*A*.

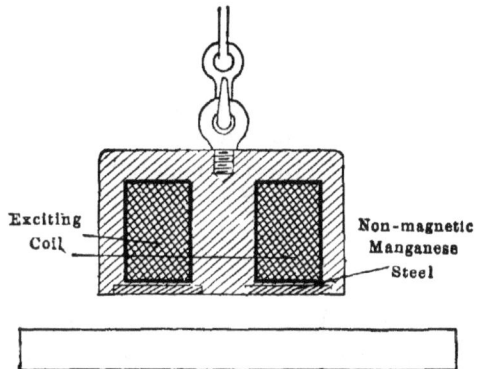

Fig. 182*A*.

182. Figure 182*A* shows a simplified cross-section of a typical lifting magnet, and directly beneath it is a steel bar which the magnet is about to pick up. The current flows through the exciting coil in a counterclockwise direction when viewed from above. (*a*) What is the polarity of the center core? (*b*) Of the outer rim? (*c*) What polarity is induced at the center of the steel bar? (*d*) At its ends?

183. Figure 183*A* shows the cross-section of the magnetic circuit of a four-pole dynamo, with the field windings shown diagrammatically. The positive line terminal *A* is connected to field terminal 1. Complete the

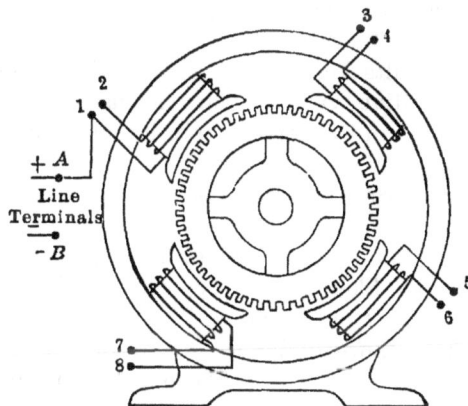

Fig. 183*A*.

field connections. Indicate the polarity of each pole. Indicate the yoke, the field core, the pole shoes, the armature iron, and the armature spider. Sketch the various paths taken by the magnetic flux.

QUESTIONS ON CHAPTER VIII

The Magnetic Circuit

1. In what respect does the magnetic circuit resemble the electric circuit? What three factors, which do not exist to any appreciable extent in the electric circuit, make it difficult to calculate magnetic relations with high precision?

2. Define ampere-turn; mmf; gilbert; reluctance; permeance; permeability; flux; maxwell; gauss.

3. Define unit reluctance. Upon what three factors does reluctance depend? Write the equation which gives the reluctance of any magnetic conductor having a uniform cross-section. Write the general equation which gives the reluctance of any magnetic circuit consisting of several parts in series, each having uniform cross-section. Repeat for permeances in parallel.

4. Sketch the general shape of a curve plotted with flux density as ordinates and mmf per centimeter as abscissas. Why does the curve become concave downward?

5. How is the permeability determined from the magnetization curve? Sketch a typical permeability curve.

6. Write the general equation for determining the flux in a single portion of a magnetic circuit. Write the general equation for the flux in the magnetic circuit when the circuit consists of several parts in series, each of uniform cross-section.

7. Sketch typical magnetization curves for silicon steel, cast steel, cast iron, dynamo steel sheet. Compare their magnetic properties and state their fields of use.

8. Discuss carefully the effect which is noted when a magnetic material forming a closed magnetic circuit is subjected to mmf, which is then reduced to zero. How may the magnetic flux within the specimen be brought to zero? What effect results from carrying the mmf to a negative value equal to the maximum positive value which was used? Describe the effects which result when the mmf is carried from its maximum negative value to its maximum positive value.

9. What is meant by remanence? Coercive force? Why does hysteresis represent energy loss?

10. Define current-flux linkages. Show that linkages are associated with every electric current. Sketch some typical examples. In what manner does "inductance" differ from "linkages"?

11. Define self-inductance under the conditions of constant reluctance. How does self-inductance vary with the number of turns, the reluctance remaining constant?

12. What effect is noted if a galvanometer is connected across the terminals of an insulated coil, and the north pole of a bar magnet is thrust into the coil? What effect is noted when the magnet is stationary? When it is withdrawn?

13. State the relation which exists between the direction of the induced emf and the change of the magnetic linkages.

14. Upon what three factors does the induced emf depend? Why is the equation preceded by a negative sign?

15. State Lenz's law. Upon what fundamental law is Lenz's law based?

16. When a direct current begins to build up in a circuit having self-inductance, what effect is produced by the flux which this current produces? Show that the resulting reactions tend to *prevent* the increase of current.

17. Make a sketch illustrating the rise of current with time.

18. Theoretically, how much time is required for the current to reach its Ohm's law value? Practically, what is the order of magnitude of the time required for the current to reach its Ohm's law value?

19. In what manner does the current in an inductive circuit decrease when the circuit is short-circuited?

20. Upon what three factors does the emf of self-induction depend?

21. Show that an expenditure of energy is not necessary to maintain a steady magnetic field. Write the equation giving the energy of the magnetic field.

22. How may the energy of the magnetic field manifest itself? Show how the energy of the magnetic field is utilized in an ignition system for internal-combustion engines.

23. Show that a change of current in one circuit may induce an emf in another circuit entirely insulated from the first one. What relation does the induced emf in the second circuit bear to the change of current in the first circuit?

24. Define coefficient of coupling. How may this coefficient be increased? Define mutual inductance.

25. Show how mutual inductance is utilized in the operation of the induction coil. Why is a condenser used?

26. Make a wiring diagram of a typical battery-ignition system. State the functions of the cam, the interrupter, the condenser, the ignition coil, the distributor. Compare this system of ignition with induction-coil ignition.

PROBLEMS ON CHAPTER VIII

The Magnetic Circuit

184. In a bipolar generator there are 600 shunt-field turns and 8 series turns per pole. When the shunt-field current is 2.8 amp and the series-field current is zero, determine: (*a*) the ampere-turns per pole; (*b*) the mmf per pole in gilberts. The series-field current is now 16.4 amp, and the series-field ampere-turns aid the shunt-field ampere-turns. Determine: (*c*) the total ampere-turns per pole; (*d*) the total mmf per pole in gilberts.

185. In Prob. 184 the series turns are connected to oppose the shunt ampere-turns. The shunt-field current is now 3.6 amp, and the series-field current is 12.6 amp. Repeat (*a*), (*b*), (*c*), and (*d*).

186. In Fig. 186*A* is shown an electromagnet which is energized by four equal exciting coils *A*, *A'* and *B*, *B'*. There are 600 turns in each exciting coil, and the resistance of each is 5 ohms. Determine the ampere-turns and the mmf in gilberts acting on the magnetic circuit under the following conditions: (*a*) coil *A* alone across 120 volts; (*b*) coils *A* and *A'* in series aiding across 120 volts; (*c*) all four coils in series aiding across 120 volts.

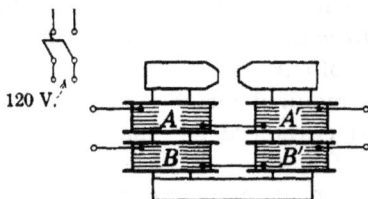

Fig. 186*A*.

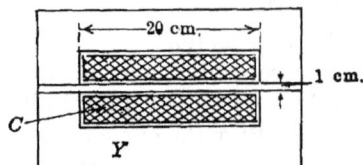

Fig. 187*A*.

187. A steel rod 1 cm diameter fits closely into two holes in a cast-iron yoke *Y* (Fig. 187*A*). The yoke is so massive that its reluctance is negligible as compared with that of the rod. An exciting coil *C* having 200 turns surrounds the steel rod. With the steel rod in place, the flux in the rod is measured and found to be 4,400 maxwells when the current in the coil is 0.416 amp. The rod is then removed, and the flux through the center of the coil is found to be 4.12 maxwells. Determine: (*a*) the flux density *B* in the steel rod in gausses; (*b*) the permeability of the steel at this value of flux density; (*c*) the total mmf in gilberts acting on the rod; (*d*) the mmf per centimeter length of the rod (field intensity *H*, numerically) if the rod is 20 cm in length between the inside faces of the yoke. (*e*) Divide the flux density *B* in (*a*) by the intensity of the magnetic field *H* in (*d*) and compare with (*b*).

188. When the current in the exciting coil (Prob. 187) is increased to twice its value, or 0.832 amp, the flux in the iron increases from 4,400 to 8,300 maxwells. Repeat (*a*) to (*e*), Prob. 187. Explain the change in permeability.

189. In Prob. 187, (*a*) calculate the reluctance of the air-space occupied by the steel rod; (*b*) calculate the reluctance of the steel rod; (*c*) find the ratio of (*a*) to (*b*).

190. In Prob. 189, calculate the permeance in (*a*) and (*b*) and find the ratio of the permeance of the iron to that of the air-space.

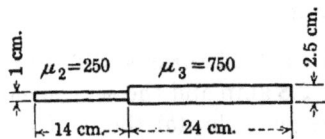

Fig. 191*A*.

191. A portion of a magnetic circuit consists of two cylinders of iron, joined end to end (Fig. 191*A*). The length of one cylinder is 14 cm, the diameter is 1.0 cm, and the permeability is 250; the length of the other cylinder is 24 cm, the diameter is 2.5 cm, and the permeability is 750. Assuming that the flux is distributed uniformly over each cross-section for its entire length, find the reluctance of each cylinder and the combined reluctance of the two. Express results in the proper units.

192. The magnetic circuit of Prob. 191 is clamped between the pole faces of an electromagnet between which an mmf of 1,000 amp-turns exists. Assuming the permeabilities to remain at the values given in Prob. 191, determine: (*a*) the total mmf in gilberts; (*b*) the flux; (*c*) the flux density in each cylinder.

193. In Fig. 193*A* is shown a steel-ring electromagnet in which there is a transverse air-gap 0.1 cm long. The cross-section of the magnet is circular, and the diameter is 1 cm. The mean length of the magnetic path is 18 cm, and the permeability of the steel is 400. The ring is wound with 200 turns, and the current is 2.0 amp.

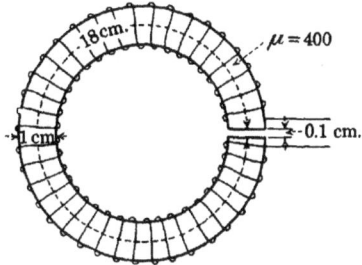

FIG. 193*A*.

Neglecting fringing and leakage, determine: (*a*) the reluctance of the air-gap; (*b*) the reluctance of the steel; (*c*) the mmf in gilberts; (*d*) the flux in maxwells; (*e*) the flux density in gausses.

194. In Prob. 193 determine (*b*), (*c*), (*d*), and (*e*) when the current is 2.5 amp and the permeability of the steel is 500.

195. The cross-section of the cast-steel yoke of a dynamo is 200 sq cm, the mean length is 36 cm, and the flux is 2×10^6 maxwells. Determine: (*a*) the flux density; (*b*) from Fig. 145, p. 162, the permeability of the steel at this flux density; (*c*) the reluctance of the yoke; (*d*) the ampere-turns required to send the flux through the yoke.

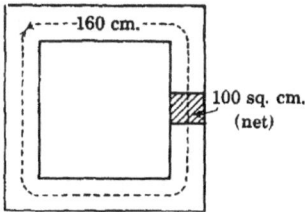

FIG. 196*A*.

196. The mean length of a transformer core of silicon steel (Fig. 196*A*) is 160 cm, and the maximum flux density in this core is 12,000 maxwells per sq cm, or 12,000 gausses. (*a*) From Fig. 147, p. 164, determine the permeability of the steel at this flux density. (*b*) With this flux density, how many gilberts per centimeter length of steel are necessary?

197. In Prob. 196 the net cross-section of the core is 100 sq cm. Determine: (*a*) the reluctance of the magnetic path; (*b*) the total flux in maxwells; (*c*) the total gilberts to produce the flux in (*b*); (*d*) the total ampere-turns.

198. In a transformer core of silicon sheet steel similar to that shown in Fig. 196*A*, the net volume of steel is 12,000 cc and the maximum flux density is 10,000 gausses. From the data of Par. 141, p. 166, determine: (*a*) the ergs hysteresis loss per cubic centimeter per cycle; (*b*) the total hysteresis loss in ergs per cycle; (*c*) the hysteresis loss in watts if the frequency is 60 cycles per sec. (1 watt-sec = 10^7 ergs.)

199. The total reluctance of an anchor-ring solenoid similar to that shown in Fig. 193*A* is 0.15 cgs unit and there are 250 turns. When the exciting current is 2.0 amp, determine: (*a*) the mmf in gilberts; (*b*) the flux in maxwells; (*c*) the ampere-maxwell flux linkages; (*d*) the flux linkages per ampere; (*e*) the self-inductance in henrys.

200. In Prob. 199 the number of turns is doubled and the current is halved. Determine: (*a*), (*b*), (*c*), (*d*), and (*e*). (*f*) How does self-inductance vary with the number of turns, the reluctance of the magnetic circuit remaining constant?

201. In Prob. 193 determine the self-inductance of the ring magnet.

202. Determine the self-inductance of the ring, Fig. 193*A*, under the conditions of Prob. 194.

203. When 5 amp flows in the exciting winding of a solenoid circuit, a flux of 800,000 maxwells is established. There are 600 turns in the winding. The flux is reduced to zero in 0.01 sec by interrupting the current. Determine: (*a*) the average rate of change of flux in maxwells per second; (*b*) the average induced emf, computed from (*a*); (*c*) the self-inductance of the circuit; (*d*) the average induced emf computed from the rate of change of *current*.

204. Determine the stored electromagnetic energy in Prob. 203 when the current is 5 amp; when it is 10 amp.

205. The resistance of the field circuit of a two-pole shunt generator is 240 ohms. When the terminal voltage is 120 volts, the flux is 1,600,000 maxwells. (The shunt-field circuit is connected directly across the generator terminals.) There are 2,500 turns per pole. Determine: (*a*) the self-inductance of the field; (*b*) the energy stored in the magnetic field; (*c*) the average emf induced when the field circuit is interrupted in 0.08 sec.

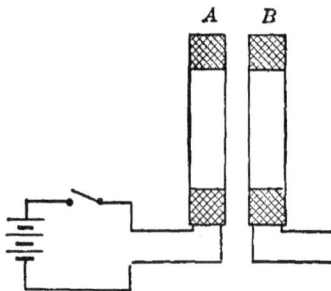

Fig. 206*A*.

206. In Fig. 206*A* are shown two coils *A* and *B* lying in parallel planes and having a common axis. There are 400 turns in coil *A* and 240 turns in coil *B*. When 1.4 amp flows in coil *A* with *B* open-circuited, the effective flux linking *A* is 2,000 maxwells, and of this flux 800 maxwells link *B*. Determine: (*a*) the self-inductance of coil *A*; (*b*) the coefficient of coupling of the two circuits.

207. The current in coil *A*, Prob. 206, is interrupted in 0.02 sec. Using Eq. (66), p. 170, calculate the average induced emf in coils *A* and *B*. Verify by means of Eq. (67), p. 175.

208. At what average rate does the current change, Prob. 207? Knowing the induced emf in *B*, find the mutual inductance, using Eq. (69), p. 179.

209. The self-inductance of a coil *C* is 0.36 henry, that of a coil *D* is 0.80 henry, and the coefficient of coupling of the two coils is 0.9. Determine: (*a*) the mutual inductance of the coils; (*b*) the emf induced in coil *D* when the current in coil *C* changes at the rate of 8 amp per sec.

210. There are 120 turns in the primary of an ignition coil, and when the current is 0.6 amp, the flux in the core is 3,600 maxwells. The primary circuit is interrupted in 0.018 sec. It is necessary to induce 1,000 volts in the secondary in order that the spark may jump the gap. How many secondary turns are necessary?

211. In Prob. 210 determine the mutual inductance between the primary and the secondary. Use Eq. (69), p. 179.

QUESTIONS ON CHAPTER IX

Electrostatics: Capacitance

1. Compare *static* electricity with *dynamic* electricity.

2. Describe an experiment in which two insulated conducting bodies are connected to the two terminals of an influence machine. How do the charges distribute themselves on the bodies when they are brought so that two ends are adjacent but not in contact?

3. If the bodies are well insulated, what effect is noted when the wires connected to the induction machine are disconnected?

4. One end of a positively charged insulated body is brought near one end of an insulated body having no initial charge. What two effects are noted? What effect is noted when the further end of the second body is grounded?

5. Show that the foregoing experiments illustrate that a positive charge induces a negative charge, and *vice-versa;* that unlike charges attract each other and like charges repel each other.

6. From the experiment of Question 4 define "free charge" and "bound charge."

7. Describe a simple method of producing electricity by friction, stating some of the materials which may be used. How may the presence of such electricity be detected?

8. If an electrified rod be held near a suspended pith ball and then be allowed to touch it, describe the phenomena which occur. Account for these phenomena. What is meant by an "induced charge"?

9. What is an electrophorus? What occurs when the metal plate is placed on the charged resinous material? Why is the top of the plate grounded before raising? The energy involved in producing the positive charge is supplied in what way?

10. What is the function of an electroscope? How is it constructed? What occurs when the brass knob is given a positive charge? What occurs, with a totally discharged electroscope, when a positive charge is brought into the vicinity but is not given to the knob?

11. How may a charge of opposite sign to the inducing charge be given to an electroscope? How may the sign of a charge be determined with the electroscope?

12. Describe the construction of a typical influence machine. How are the charges made to build up on the combined carriers and inductors? What is the function of the neutralizing brushes? How is the charge removed from the carriers? What is the object of the Leyden jars?

13. What is meant by an electrostatic field? How may such a field be represented? Give five properties of electrostatic lines.

14. Explain the phenomena of the pith ball first being attracted, then repelled (Fig. 169, p. 169), by means of the laws of induction and also by the laws of the electrostatic field.

15. State Coulomb's law of force between charged bodies.

16. Show that, so far as outside electrostatic effects are concerned, the charge on a sphere acts as if it were concentrated at the center of the sphere.

17. Define a unit electrostatic charge. Compare with the unit magnetic pole.

18. State six fundamental laws of electrostatics.

19. Distinguish carefully between dielectric and insulator. What properties should a good insulator have? A good dielectric?

20. Compare electrostatic lines with magnetic lines of force and with electric current, with particular reference to their effects on their respective media, as their densities increase.

21. What is meant by "dielectric strength"? "Voltage gradient"? Give a simple example of air being ruptured by excessive voltage gradient.

22. Define a condenser. What occurs when the two terminals of a well-insulated condenser are connected across a source of d-c voltage? What occurs when the source of voltage is disconnected? What occurs when the charged condenser is short-circuited? Give a hydraulic analogue of these effects.

23. What relation exists between the charge on the condenser and the voltage across the condenser plates? Define capacitance. What is the unit of capacitance in the practical system? Why is capacitance usually expressed in microfarads?

24. Show that if the air in an air condenser be replaced by other dielectrics, the capacitance usually increases. What is meant by *dielectric constant* or *permittivity?* Of what order of magnitude are the dielectric constants of the ordinary dielectrics?

25. Derive the equation which gives the equivalent capacitance of a number of capacitances in parallel.

26. Repeat Question 25 for a number of capacitances in series. What two fundamental relations are used in deriving this last equation?

27. State three equations which give the energy stored in a condenser. How does the stored energy in a given condenser vary with the voltage?

28. Make a diagram of connections used for measuring the capacitance of a condenser by the ballistic method. How is the ballistic deflection of the galvanometer determined? What does a steady deflection of the galvanometer indicate? How is the constant of the apparatus determined?

29. Make a diagram of connections for measuring the capacitance of a condenser by the bridge method, employing alternating current.

30. Write the equation which gives the relation existing among the ratio arms and capacitances in the other two arms.

31. Describe briefly a method for locating an "open" in a cable by capacitance measurements. Draw a diagram of connections. Derive the equation which gives the distance to the "open."

PROBLEMS ON CHAPTER IX

Electrostatics: Capacitance

212. The distance in air between a positive point charge of 40 esu (electrostatic units or statcoulombs) and a negative point charge of 60 esu is 10 cm. Determine the force in dynes acting between the charges.

213. A force of 20 dynes acts between a positive point charge of 80 esu and a negative point charge of 100 esu in air. Determine the distance in cm between the charges.

214. A sphere of radius 1 cm in air is charged with 150 esu. What is the force in dynes on a unit charge: (a) at the surface of the sphere? (b) at a distance of 10 cm from the center of the sphere?

215. Two spheres in air, 30 cm apart between centers and each of radius 2 cm (Fig. 215A), are con-

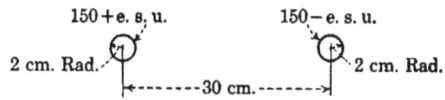

150+e. s. u.   150−e. s. u.

2 cm. Rad.   2 cm. Rad.

←------30 cm.------→

Fig. 215A.

nected across the terminals of an influence machine. The voltage is such that each acquires a charge of 150 statcoulombs. (a) Determine the force in dynes between the spheres. (b) Is this force attraction or repulsion?

216. A unit positive charge is placed midway between the spheres (Prob. 215). Determine: (a) the force in dynes exerted by the positively charged sphere on the unit charge; (b) the direction in which it tends to make the unit charge move; (c) the force exerted by the negatively charged sphere on the unit charge; (d) the direction in which it tends to make the unit charge move; (e) the total force acting on the unit charge.

217. A sample of 15-mil varnished cambric is tested for dielectric strength between flat, parallel electrodes. It punctures at 16,000 volts. At what voltage gradient, in volts per mil, does it puncture?

218. A high-voltage condenser is to be made up of oil-impregnated paper which is guaranteed to withstand 250 volts per mil. The operating voltage of the condenser is 20,000 volts. What thickness of paper should be used?

219. A sample of sheet rubber 40 mils thick punctures at 16,000 volts when tested between flat, parallel electrodes. Determine the dielectric strength in: (a) volts per mil; (b) volts per mm.

220. The capacitance of a condenser is 0.000036 farad. Determine, in proper units, the charge in the condenser when the voltage across the terminals is 300 volts.

221. A 72-μf condenser is connected across a 600-volt source. Determine: (a) the charge in coulombs in the condenser; (b) the time required by a steady current of 0.04 amp to charge this condenser to the coulombs determined in (a).

222. When a condenser is charged to 0.024 coulomb, the voltage becomes 600 volts. Determine the capacitance of the condenser in μf.

223. A steady current of 0.02 amp flows into a condenser for 5 sec, at which time the voltage across the condenser is 500 volts. Determine the capacitance of the condenser in μf.

224. The capacitance of a condenser is 60 μf, and the charge is 0.018 coulomb. What is the voltage across the condenser?

225. A steady current of 0.01 amp charges a 220-μf condenser for 9.68 sec. Determine: (*a*) the voltage across the condenser; (*b*) the energy stored in the condenser in joules.

226. An air condenser, whose capacitance is 0.0072 μf, is connected across 600 volts. Determine: (*a*) the charge in coulombs in the condenser; (*b*) the energy in joules stored in the condenser. This condenser is then immersed in oil, the dielectric constant of which is 2.40, the voltage being maintained at 600 volts. Determine: (*c*) the charge; (*d*) the energy.

227. The charge of an air condenser is 3.6 microcoulombs when the voltage is 300 volts. The condenser is then immersed in transil oil, and the charge becomes 6.0 microcoulombs when the voltage is 200 volts. Determine the dielectric constant of the transil oil.

228. A capacitance of 0.12 μf is connected in parallel with a capacitance of 0.21 μf and the two are then connected across a voltage supply. There is then a charge of 72 microcoulombs on the 0.12-μf capacitance. Determine: (*a*) the resulting capacitance of the system; (*b*) the system voltage.

229. An air capacitor of 0.008 μf is connected in parallel with a second air capacitor of 0.01 μf, and the total charge is 3.6 microcoulombs. The 0.01-μf capacitor is then immersed in oil whose dielectric constant is 2.4. The voltage of the supply remains unchanged. Determine: (*a*) the total capacitance of the system after immersion in the oil; (*b*) the total charge of the system; (*c*) the charge on each capacitor; (*d*) the total stored energy.

230. Three condensers connected in parallel have charges of 2,400, 3,000, 3,600 microcoulombs when the voltage across the combination is 260 volts. Determine: (*a*) the resultant capacitance of the system; (*b*) the total stored energy.

231. Two condensers in which the leakage is negligible have capacitances of 24 and 30 μf and are connected in series across 400 volts. Determine: (*a*) the resultant capacitance of the system; (*b*) the charge on each condenser; (*c*) the voltage across each condenser.

232. The capacitance of two series-connected condensers is measured and found to be 12.85 μf. The capacitance of one of the condensers is 20 μf. Determine: (*a*) the capacitance of the other condenser; (*b*) the voltage across each condenser if 100 volts be impressed across the system; (*c*) the energy stored in each condenser.

233. Three condensers having capacitances of 40, 50, 80 μf are connected in series across 400 volts. Determine: (*a*) the resultant capacitance of the system; (*b*) the voltage across each condenser at the instant of closing the switch.

234. The capacitances of three series-connected condensers are 2.8, 6.0, 8 μf. Determine: (*a*) the resultant capacitance of the system; (*b*) the voltage across each condenser if the voltage across the system is 200 volts; (*c*) the energy stored in each condenser; (*d*) the stored energy of the system. The leakage of the condensers is negligible.

235. A 10-μf and a 20-μf condenser are connected in series across a voltage supply. The voltage across the 10-μf condenser is 80 volts. Determine: (a) the charge on each of the condensers; (b) the voltage across the 20-μf condenser; (c) the energy stored in each condenser.

236. The capacitance of an unknown condenser is measured by the method given in Par. 170, p. 204. When a standard condenser, having a capacitance of 1.0 μf, is in circuit, the maximum ballistic deflection of the galvanometer on charge is 12.6 cm. With the unknown condenser, the maximum ballistic deflection of the galvanometer is 18.8 cm. Determine: (a) the galvanometer constant (Eq. (83), p. 205); (b) the unknown capacitance.

237. The capacitance of an unknown condenser C_x is measured by the bridge method given on p. 205, the connections shown in Fig. 179, p. 206, being used. The capacitance of the standard condenser C_2 is 0.2 μf. When the bridge is in balance, *i.e.*, when no perceptible sound is heard in the telephone receiver, R_1 is 556 ohms and R_2 is 1,000 ohms. What is the value of C_x?

238. If, in Fig. 179, p. 206, $C_x = 4.5$ μf, $C_2 = 0.5$ μf, and $R_2 = 1,000$ ohms, to what value should R_1 be adjusted in order to obtain a balance?

239. In the bridge, Fig. 179, p. 206, $R_1 = 1,000$ ohms, $R_2 = 229$ ohms, $C_2 = 0.4$ μf. Determine the value of the unknown capacitance C_x.

240. A cable is tested for an "open" by the method given in Par. 171, p. 206, the connections given in Fig. 180 being used. The length of the perfect and of the faulty cable is 1,500 ft. The Ayrton shunt is set at 1.0 when the capacitance measurement of the length x of the faulty cable is made, and the ballistic deflection of the galvanometer is 14.2 cm. When the capacitance measurement of the perfect cable plus the looped length $l - x$ is made, the Ayrton shunt is set at 0.1 and the ballistic deflection of the galvanometer is 4.2 cm. Find the distance x.

QUESTIONS ON CHAPTER X

The Generator

1. Define a generator. On what principle does a generator operate?

2. Describe the manner in which the flux linkages of the armature coils of generators are varied, in both bipolar and multipolar generators.

3. Show that the emf in a single coil may be considered also as due to the active conductors of the coil *cutting* flux. When considering the emf induced in a generator armature, why is this point of view the simpler? Show the application of this principle to the generation of emf in a single armature coil.

4. What is the fundamental equation which gives the induced emf in a single conductor cutting a magnetic field? What is the order of magnitude of this emf in the ordinary generator?

5. State Fleming's right-hand rule.

6. Sketch the curve of induced emf for various positions of a flat coil rotating at a uniform speed in a uniform magnetic field. For what positions of the coil is the induced emf zero? A maximum? Why does the emf vary for different positions of the coil? Why does it reverse in sign?

7. What is the character of the emf generated in such a coil? How may current be conducted to the external circuit? Describe slip-rings. What is meant by the "neutral plane"?

8. Why is the current taken from slip-rings not suited for d-c purposes? How may a unidirectional current be obtained from a coil which generates alternating current? Describe a commutator and its operation.

9. Show how the current to the external circuit may be made practically steady.

10. Describe the construction of a Gramme-ring winding. What are its advantages and its disadvantages?

11. What are the advantages of the drum winding over the Gramme-ring winding? Distinguish between end connections and active conductors. What is meant by "winding pitch"? By "fractional pitch"? In what form and in what manner are armature coils placed in the slots?

12. How are the coils connected to commutator segments? In what manner is it desirable to number the coil sides when designing a winding?

13. Define "front pitch," "back pitch," and "average pitch." What two limitations are placed on the pitch? How is the approximate average pitch determined?

14. Define a progressive winding; a retrogressive winding. What determines each?

15. Why is it often necessary to place several coil sides in a single slot? What is meant by a "multiple coil"? What additional restriction is placed on the pitch with this type of winding?

16. How may the number of paths through an armature be determined? What is the effect on the current rating of increasing the number of paths through the armature? The voltage rating? The kilowatt rating? How may the number of paths through a simplex lap winding be readily determined?

17. In what essential manner does the wave winding differ from the lap winding? In a wave winding what is meant by "front pitch," "back pitch," and "average pitch"? In a wave winding what relation must exist among certain winding elements after each passage around the armature?

18. Why is a wave winding much more restricted in the choice of pitch than a lap winding? When is a wave winding retrogressive? Progressive? Why is the use of a "dummy coil" sometimes necessary with wave windings?

19. Why may only two brushes be necessary with wave windings? Why is a greater number of brushes than two ordinarily used?

20. How many paths through the armature does a simplex wave winding give?

21. State two distinct advantages of wave windings. Where are wave windings used? Lap windings?

22. What is the character of the active iron which is used for armatures? How are armature stampings made? What is the armature spider? What are ventilating ducts, and how are they formed? What is the general form of slot insulation, and how are the coils held in place on the armature?

23. State the two functions of the yoke or frame of the dynamo. Of what materials and how is the frame made?

24. Of what materials are the field cores made? Describe two types of construction.

25. What type of insulation is used for the wire of the field coils? How is the series winding often placed on the field structure?

26. Describe the construction of the commutator. How are the segments insulated from one another? How are they held in place?

27. Of what materials are brushes made? Where are carbon and where are copper brushes used? Approximately what brush pressure should be used?

PROBLEMS ON CHAPTER X

The Generator

241. A conductor 30 cm long moves downward at a uniform velocity of 3,500 cm per sec in the magnetic field of Fig. 241A. The field may be considered as having a uniform density of 4,000 gausses. The conductor is perpendicular to the magnetic field, and it moves in a direction perpendicular to itself and to the magnetic field. Determine: (a) the direction of the induced emf in the conductor; (b) the value, in volts, of the induced emf.

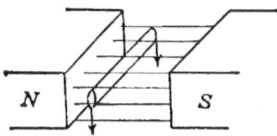

FIG. 241A.

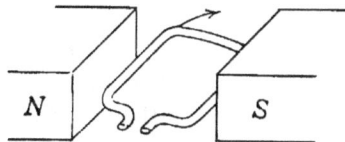

FIG. 242A.

242. A single-turn coil (Fig. 242A) rotates at a uniform velocity of 1,200 rpm, in a uniform magnetic field in which the flux density is 4,000 gausses. The coil has an axial length of 30 cm, and the breadth is 24 cm. Determine: (a) the induced emf in the coil when its plane is perpendicular to the magnetic field; (b) the induced emf when the plane of the coil is parallel to the magnetic field.

243. In Fig. 243A is shown a north pole of a four-pole generator from which flux enters the armature. The uniform flux density under the pole is 40,000 lines per sq in. (6,200 gausses). The combined span of the four poles is 0.7 of the entire armature periphery. The axial length of the active armature copper is 14 in.

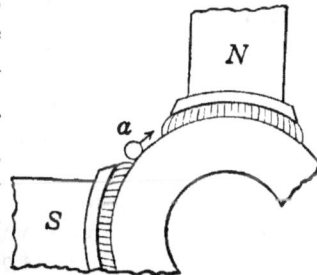

FIG. 243A.

(35.6 cm), and the peripheral velocity of the armature is 100 ft (30.5 m) per sec. Determine: (a) the induced emf per conductor when the conductor is directly under the pole; (*b*) the average induced emf per revolution per conductor. (*c*) How many series-connected conductors must there be between brushes in order that the induced emf between brushes may be 240 volts?

244. The diameter of the armature (Prob. 243) is 22 in. (56.0 cm). Determine its speed in rpm.

245. Repeat Prob. 243 when the average flux density under the poles is 45,000 lines per sq in. (6,980 gausses) and the velocity is 90 ft (27.5 m) per sec.

246. There are 200 surface conductors on the armature of a four-pole generator, and the emf per conductor is 4.6 volts. The armature is simplex lap-wound, giving four parallel paths. The rated current per armature path is 21.75 amp. Determine: (*a*) the emf between terminals; (*b*) the current rating of the generator; (*c*) the kilowatt rating of the generator.

247. In a generator similar to that of Prob. 246, there are 198 conductors on the surface of the armature rather than 200, and a simplex wave winding is used. The emf per conductor and the current rating per armature path are the same. Repeat (*a*), (*b*), and (*c*).

248. There are 62 slots on the armature of a four-pole generator, and the winding is to be of the simplex lap type having two coil sides per slot. Determine: (*a*) the number of coil sides or winding elements; (*b*) the average pitch. (*c*) Find a possible value for the back pitch and for the front pitch.

249. Sketch a portion of the winding, Prob. 248, using the values of back and front pitch determined in that problem. (See Fig. 195, p. 219.)

250. It is desired to wind a 56-slot, four-pole armature with a simplex lap winding, using double coils or four coil sides per slot. (*a*) Determine suitable values of front pitch and back pitch. (*b*) Sketch the winding. (*c*) How many parallel paths are there in this winding?

251. There are 78 slots on the armature of a six-pole generator. Determine suitable values of back pitch and front pitch that will make possible a simplex lap winding employing double coils, that is, four coil sides per slot. (See Fig. 196, p. 220.)

252. Repeat Prob. 251 except that triple coils are used, that is, six coil sides per slot.

253. If possible, design a wave winding with two coil sides per slot for the 62-slot armature of Prob. 248. Select a value of 31 for the average pitch. (*a*) Using 31 for the value of the back pitch and the front pitch and starting with element 1, find the element on which the winding will terminate after one passage around the armature. (*b*) If possible find a pitch that will make a wave winding possible with this armature if all the winding elements are utilized. (*c*) Repeat (*a*) with an armature of 63 slots. Is such a winding progressive or retrogressive? (*d*) Repeat (*a*) for an armature of 61 slots.

254. There are 70 slots on the armature of a six-pole generator, and it is desired to design a wave winding with two winding elements per slot. (*a*) Determine a suitable pitch that will make such a winding possible, and state whether or not the winding is progressive or retrogressive. (*b*) How many parallel paths through this armature? (*c*) Make a sketch similar to Fig.

200, p. 224, showing the passage of the winding about the armature at least twice.

QUESTIONS ON CHAPTER XI

Generator Characteristics

1. From what fundamental relation is Eq. (87), p. 235, derived, giving the induced emf in an armature? What five quantities does this equation involve? How does the induced emf vary with the flux per pole? The speed? The number of armature conductors? The paths through the armature?

2. In a given dynamo, what relation exists among induced emf, flux, and speed? With constant speed, to what factor is induced emf proportional?

3. Sketch a curve giving the relation between flux per pole and field ampere-turns. Discuss the shape of this curve, and compare it with the saturation curves for iron and steel. (See Fig. 147, p. 164.)

4. Show that a curve, obtained at constant speed and plotted with field current as abscissas and induced emf as ordinates, is similar to the curve having field ampere-turns as abscissas and values of flux as ordinates.

5. Sketch a saturation curve taken with increasing and then with decreasing values of field current. Why do not the two curves coincide?

6. Make a diagram of connections such as would be used for determining a saturation curve. Why should the generator be separately excited?

7. Make three sketches showing the connections for three types of generators in common use.

8. Describe in some detail the process by which a shunt generator "builds up" its voltage. What prevents the generator building up indefinitely?

9. Give four reasons why a generator may not build up. How may the probable reason for the generator not building up be determined in each case? What is the remedy in each case?

10. Sketch in cross-section the field and armature cores of a bipolar generator together with their windings. Sketch the resulting flux distribution when there is no armature current.

11. Assume a direction of rotation for the armature, and indicate the direction of the current in the various armature conductors. Show the general direction in the armature of the flux produced by these armature ampere-conductors acting alone.

12. Show the path of the resulting flux when the field is excited and the armature delivers current as in Question 11. What relation does the direction of the resultant flux bear to the direction of rotation?

13. Why is it usually necessary to advance the brushes in the direction of rotation with increase of load? Show that when the brushes are so advanced, certain conductors on the armature *demagnetize* the generator field and others *cross-magnetize* this field.

14. Name three methods by which the effect of armature reaction may be reduced.

15. Of what two parts does commutation consist? What determines the time in which the reversal of current must take place? Why does the reversal of current in the individual coil often give rise to difficulties?

16. Why does it improve commutation to move the brushes ahead of the neutral plane? Why are carbon brushes preferable to metal brushes, in spite of their much higher resistance?

17. Why may the sparking under the brushes be severe, even though the emfs induced in the coils undergoing commutation are comparatively low in value? What is the order of magnitude of these emfs?

18. What is meant by "high mica"? What produces high mica? Describe two methods of reducing or eliminating high mica.

19. Why is it highly desirable that commutators be maintained in excellent condition at all times? How should carbon be removed from the surface of the commutator? Compare the uses of emery paper and sandpaper on commutators. Describe the method of fitting the brushes to the commutator.

20. Why does sparking occur at the brushes of a generator under load conditions if the brushes are allowed to remain in the no-load neutral plane? State a method of neutralizing the flux which exists in the neutral zone because of armature reaction. Why should this flux not only be neutralized but be reversed as well?

21. In a generator, what is the polarity of the interpole following a north main pole in the direction of rotation? Why?

22. How are the interpole windings connected in the circuits of the generator? Give reasons.

23. How are interpoles adjusted to the proper strength?

24. Why does the terminal voltage of a separately excited generator, operated with constant field current and at constant speed, drop as the load increases? Give two reasons. Does the *induced* emf remain constant with increase of load?

25. What relation exists between the induced emf and the terminal voltage?

26. Sketch the connections used in determining the characteristic of a shunt generator. What two factors are held constant during the test?

27. Name the three factors which cause the terminal voltage of a shunt generator to drop as the load increases. What important factor, not present with the separately excited generator, contributes to the drop in terminal voltage?

28. What is meant by a shunt generator "breaking down"? Why can the test for determining the operating characteristic often be carried to short circuit? Sketch the entire characteristic from open circuit to short circuit and thence back to open circuit. Give reasons for the shape of the characteristic.

29. What is a compound generator? Describe the series winding, and show how it is connected. Sketch a "long-shunt" connection and a "short-shunt" connection. Compare the effects of the two connections on the generator operation.

30. Through what agency does the series winding affect the characteristic of the generator?

31. Define and sketch the characteristics of an overcompounded generator, a flat-compounded generator, an undercompounded generator. Where is each used?

32. Sketch the connections of the series generator, and give an example of its use.

33. What is a Tirrill regulator? What principle underlies its operation?

PROBLEMS ON CHAPTER XI

Generator Characteristics

255. There are 65 slots on the armature of a four-pole, 15-kw, 250-volt, 1,500-rpm, shunt generator. There are six series-connected conductors per slot; the armature is simplex lap-wound, giving four parallel paths; and the flux per pole is 2,560,000 maxwells. Determine the induced emf when the generator is driven at its rated speed.

256. There are 64 slots on the armature of a four-pole, 20-kw, 1,200-rpm generator, and there are 12 series-connected conductors per slot. The armature is simplex lap-wound, giving four parallel paths. The pole faces are 10 in. (25.4 cm) square and the average flux density under the pole faces is 40,000 lines per sq in. (6,200 gausses). Determine: (a) the flux per pole in maxwells; (b) the induced emf when the generator runs at rated speed.

257. In a four-pole, 1,800-rpm generator with a wave-wound armature, there are 51 armature slots and eight series-connected conductors per slot. The flux is 740,000 maxwells per pole. Determine the induced emf at 1,800 rpm.

258. On the armature of a 10-pole, 400-kw, 250-volt, 400-rpm generator there are 200 slots, and the flux per pole is 9,300,000 maxwells. The armature is simplex lap-wound. In order that the induced emf at rated speed may be 250 volts, determine: (a) the total number of series conductors on the armature; (b) the number of such conductors per slot.

259. Determine the induced emf in the armature of a 12-pole, 550-kw, 250-rpm generator if there are 15 slots per pole and four conductors per slot. The flux per pole is 8,400,000 maxwells, and the winding is simplex lap-wound.

260. When a six-pole, 50-kw generator is running at 1,500 rpm, the induced emf is 620 volts. There are 72 slots on the armature, with four conductors per slot, and the winding is simplex wave. Determine the flux per pole in maxwells.

261. Find the constant K in Eq. (89), p. 235, for the generator of Prob. 260.

262. The following data were taken during the experimental determination of a saturation curve for a 250-volt, 1,200-rpm, shunt generator:

Field current.....	0	0.5	1.0	1.5	2.0	2.5	3.0	3.5	3.75
Induced volts....	8.0	54	106	158	205	242	269	286	290

The speed was maintained constant at 1,200 rpm.   (a) Plot the saturation curve.   (b) Indicate the point at which saturation begins.   (c) What value of field current gives a no-load emf of 250 volts?   (d) If the generator is self-excited, what value of field resistance gives 250 volts at the terminals at no load?

263. Replot the saturation curve of Prob. 262 for a speed of 1,100 rpm. (a) What field current is now required to produce a no-load voltage of 250 volts?   (b) If the generator is self-excited, what is the corresponding value of field resistance?

264. In Prob. 262, the field rheostat is adjusted so that the resistance of the field circuit is 100 ohms and the field circuit is connected across the armature terminals.   (a) What value of field current flows as a result of residual magnetism?   (b) What emf is induced, due to this field current, provided the initial field current has such a direction as to increase the residual magnetism?   (c) What field current flows as a result of this new induced emf?   (This is the process of building up.)

265. After the field current has reached the value of 3.75 amp, Prob. 262, it is then decreased, the speed being maintained constant at 1,200 rpm. The following data were obtained:

Field current..........	3.5	3.0	2.5	2.0	1.5	1.0	0.5	0
Induced volts..........	288	273	248	213	167	116	60.0	7.0

With these data, plot a curve in conjunction with that of Prob. 262 indicating by arrows the direction of the cycle.   Why do not the two curves coincide?

266. Plot the saturation curves given by the data of Probs. 262 and 265 for a speed of 1,500 rpm, indicating the direction of the cycle.

267. In a bipolar generator there are 36 slots and eight series-connected conductors per slot on the armature.   The generator delivers a current of 80 amp, giving 40 amp per path in the armature.   (a) When the brushes are in the geometrical neutral, determine the total cross-magnetizing ampere-conductors and the total cross-magnetizing ampere-turns.   The brushes are now moved ahead 15 space-degrees.   Determine: (b) the total cross-magnetizing ampere-conductors and ampere-turns; (c) the demagnetizing ampere-conductors and ampere-turns.

268. The resistance of the armature of a 25-kw, 230-volt, shunt generator is 0.107 ohm.   (a) When the armature is delivering its rated current of 112 amp at 230 volts, determine the induced emf in the armature.   (b) The no-load voltage is 250 volts.   Explain why the no-load voltage differs from the induced emf in (a).

269. The resistance of the armature of a 10-kw, 120-volt, shunt generator is 0.11 ohm, and the resistance of the shunt-field circuit including the rheostat is 65 ohms.   The rated current of the generator at 120 volts is 83 amp.   Determine: (a) the shunt-field current; (b) the armature current; (c) the induced emf in the armature.

270. In Prob. 269 determine: (*a*) the power lost in the shunt-field circuit; (*b*) the power lost in the armature; (*c*) the total power generated by the armature.

271. The no-load voltage of a 50-kw, 250-volt, 1,200-rpm shunt generator is 272 volts. The voltage drop at rated load, due to armature reaction, is 4.0 volts, and that due to decreased field current is 5.0 volts. The armature resistance is 0.060 ohm, and the field-circuit resistance is 70 ohms. Determine: (*a*) the rated current of the generator; (*b*) the shunt-field current, assuming as a first approximation that the rated terminal voltage is 250 volts; (*c*) the armature current; (*d*) the voltage drop in the armature due to armature resistance; (*e*) the terminal voltage at rated load.

272. In the generator of Prob. 271 determine: (*a*) the power lost in the shunt-field circuit at no load; (*b*) at rated load; (*c*) the power lost in the armature at rated load; (*d*) the total power generated at rated load.

273. In a 50-kw, 250-volt, 1,200-rpm, flat-compounded generator (Fig. 273*A*), the resistance of the shunt-field circuit, including rheostat, is 72 ohms, the resistance of the series field is 0.010 ohm, and the armature resistance is 0.054 ohm. The no-load and the rated-load voltages are 250 volts at a constant speed of 1,200 rpm. The

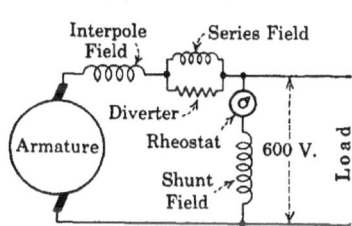

FIG. 273*A*.

generator is connected long shunt. At rated load determine: (*a*) the rated-load current of the generator; (*b*) the shunt-field current; (*c*) the series-field current; (*d*) the armature current; (*e*) the resistance drop in the armature.

274. In Prob. 273 determine: (*a*) the power lost in the shunt-field circuit; (*b*) the power lost in the series field; (*c*) the power lost in the armature resistance; (*d*) the total electric power lost; (*e*) the total power generated.

275. In a 25-kw, 230-volt, 800-rpm, compound generator, the no-load voltage is 220 volts and the rated-load voltage is 230 volts. The resistance of the armature is 0.12 ohm, the resistance of the series field is 0.022 ohm, and the resistance of the shunt-field circuit is 90 ohms. The generator is connected long shunt. Determine: (*a*) the rated-load current; (*b*) the shunt-field current at no load and at rated load; (*c*) the series-field current at rated load; (*d*) the armature current at rated load; (*e*) the voltage across the brushes; (*f*) the induced emf in the armature at rated load.

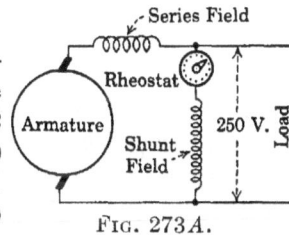

FIG. 278*A*.

276. In Prob. 275 determine at rated load: (*a*) the shunt-field loss; (*b*) the series-field loss; (*c*) the power lost in armature resistance; (*d*) the total power generated.

277. In a 250-kw, 600-volt, 1,200-rpm, compound generator with interpoles, the series field is provided with a diverter and the generator is connected long shunt. (See Fig. 278*A*.) The resistance of the armature is 0.038 ohm; the series-field resistance is 0.008 ohm; the diverter resistance is 0.030 ohm; the interpole-circuit resistance is 0.006 ohm; the shunt-field-circuit resistance is 75 ohms. Determine at rated-load voltage (600 volts): (a) the rated-load current; (b) the shunt-field current; (c) the interpole-circuit current; (d) the series-field current; (e) the diverter current; (f) the armature current; (g) the induced emf.

278. In Prob. 277 determine the power loss in: (a) the shunt-field circuit; (b) the interpole circuit; (c) the series field; (d) the diverter; (e) the armature resistance. (f) Determine the power generated in the armature.

279. In Prob. 277, with 125 per cent rated-load current at 600 volts, determine the power loss in: (a) the shunt-field circuit; (b) the interpole circuit; (c) the series field; (d) the diverter; (e) the armature resistance. (f) Determine the total power generated in the armature.

QUESTIONS ON CHAPTER XII

The Motor

1. Define a motor, and compare the definition with that of a generator.

2. On what fundamental phenomenon does motor operation depend? Show by a sketch the field resulting from placing a single conductor carrying current in a uniform magnetic field, the conductor being perpendicular to the direction of the field.

3. In what manner may the force acting on the conductor be explained by the appearance of the resultant magnetic field?

4. To what three factors is the force acting on a conductor, placed in a magnetic field, proportional? What geometrical relation exists among the conductor, the direction of the field, and the direction of the force?

5. State Fleming's left-hand rule. Describe another simple method by which the relations given by Fleming's left-hand rule and Fleming's right-hand rule may be determined.

6. Define torque. What two quantities determine the magnitude of torque? In what units is torque expressed?

7. Show that if a coil carries current and lies in a magnetic field with its plane parallel to the field, torque is developed. Illustrate by a sketch.

8. Show with sketches two or three positions of such a coil, including the position of maximum torque and the position of zero torque. How may a coil which begins to rotate be made to continue rotating when the coil reaches the position of zero torque?

9. Show with sketches the development of torque in a bipolar and in a multipolar dynamo, having numerous conductors on the armature surface.

10. In any given dynamo, to what two factors is torque proportional?

11. Why is the value of current in a motor armature in operation not equal to the line voltage divided by the armature resistance?

12. Show that a motor armature in operation must have an emf induced within itself. By means of Fleming's left-hand and right-hand rules, or by other methods, show the relation of the direction of the induced emf to that of the line voltage and the armature current.

13. What effect does the counter emf have on the value of current flowing into the armature? What is the *net* voltage acting in the armature circuit? How is the current determined?

14. Why must the line voltage exceed the counter emf if a dynamo is to operate as a motor? What relation exists between the two voltages?

15. Describe a simple experiment which demonstrates clearly the existence of counter emf.

16. In any given motor, what two fundamental quantities determine the speed? Write the equation for speed. How does the speed vary with the counter emf? With the flux?

17. Sketch a bipolar motor showing the direction of the magnetic field due to the poles, the direction of current in the individual armature conductors with the brushes in the geometrical neutral, and the direction of rotation.

18. In Question 17 show the direction of the armature emf and the direction of the resulting field. In what direction must the brushes be moved as the load is applied, in order that excessive sparking may not occur?

19. Show that in a motor a north-commutating pole must follow a north main pole taken in the direction of rotation, and compare with a generator.

20. What immediate reaction results from the application of load to a shunt motor? Show that the reactions which follow cause the motor to develop sufficient torque to carry the increased load.

21. In a shunt motor, how does the internal torque vary with the current? Give reasons.

22. Why does the speed of a shunt motor ordinarily drop slightly with increase of load? What is the effect of armature reaction on the speed of the motor?

23. State some common industrial applications of shunt motors. Discuss the starting torque of such motors.

24. In a series motor, how does the torque vary with current if saturation and armature reaction are neglected? Why? What is the effect of saturation on the torque-current characteristic?

25. Sketch a typical speed characteristic of a series motor. Give two reasons for the decrease in speed with increased load. What are the relative magnitudes of the two factors which cause decrease in speed?

26. Why does a series motor tend to race at light loads? Discuss the efficiency curve of a series motor.

27. Sketch the speed-current curve and the torque-current curve for a cumulative-compound motor. Discuss these curves. Name some industrial applications of this type of motor.

28. Sketch the speed-current curve and the torque-current curve for a differential-compound motor. Discuss these curves. Why is this type of motor little used?

29. Why must resistance be used when starting a motor? Show with a simple sketch the connections which would be used for starting a shunt motor.

30. Sketch the wiring diagram of a three-point starter, giving the reasons for each connection made. What is the object of the hold-up magnet?

31. Sketch the wiring diagram of a four-point starter. Compare its connections and its uses with those of the three-point starter.

32. Sketch the wiring diagram of one type of series-motor starter. What other types are used?

33. Distinguish between a controller and a starter.

34. Describe types of resistance units used in starters and in controllers.

35. With a given motor, what two factors only can be changed in order to change the speed?

36. What factor is changed in the armature-resistance method of speed control? State two important objections to this method.

37. What is "multivoltage" speed control, and what prevents its wide use?

38. With the ordinary motor, why cannot wide ranges of speed control be obtained by variation of field current? What ranges are obtainable with commutating poles?

39. Describe the method commonly used for controlling street-railway motors. How are two efficient operating speeds obtained?

40. What is meant by "multiple-unit control"? Where is it used, and what are its advantages?

41. Derive the equation giving the horsepower absorbed by a prony brake of the arm type.

42. Describe a simple rope brake. Derive the equation for the power absorbed.

43. Why is it possible to measure speed quickly and accurately by means of a combination of d-c magneto and voltmeter? How is the combination calibrated?

PROBLEMS ON CHAPTER XII

The Motor

280. A conductor 30 cm long lies in a uniform magnetic field whose intensity is 1,800 oersteds, the direction of the conductor being perpendicular to the direction of the field. Compute the force in grams acting on the conductor when the current is 50 amp.

281. A rectangular coil consisting of 25 turns of wire is 30 cm long and 20 cm wide and lies in a magnetic field in which the flux density is 1,600 gausses. The plane of the coil is parallel to the direction of the field, and the length of the coil is perpendicular to the direction of the field. (See Fig. 236(a), p. 267.) Determine: (a) the force acting on one of the 30-cm coil sides when the current in each turn of the coil is 20 amp; (b) the turning moment of the coil in gram-cm.

282. In Prob. 281 determine (a) and (b) when the coil is turned 90 degrees in its own plane so that the 20-cm side is perpendicular to the field.

283. A 30-in. (76.2-cm) pulley is driven by a belt. The tension in the tight side of the belt is 220 lb (99.8 kg) and that in the loose side is 70 lb (31.8 kg). Determine: (a) the torque acting on the pulley; (b) the foot-pounds work performed in one revolution of the pulley; (c) the foot-pounds work per minute when the pulley speed is 300 rpm; (d) the horsepower delivered to the pulley.

284. On a gear with a pitch diameter (diameter of the circle of contact) of 12 in. (30.5 cm), there are 36 teeth. This gear is driven by a smaller gear in which there are 12 teeth. The force acting at the points of contact of the teeth is 400 lb (181 kg). Determine: (a) the torque applied to the driven gear; (b) the torque developed by the smaller gear; (c) the work transmitted in foot-pounds per minute, neglecting friction, if the speed of the smaller gear is 600 rpm; (d) the transmitted horsepower.

285. The diameter of a motor armature is 10 in. (25.4 cm), and the axial length is 6 in. (15.2 cm). There are four poles, each pole face being 5.6 in. (14.2 cm) square. The flux density under each pole is 40,000 maxwells per sq in. (6,200 gausses), the flux may be considered as uniformly distributed directly under the pole face, and it may be assumed that there is no fringing at the pole edges. There are 288 conductors on the armature surface, and the current per conductor is 40 amp. Determine: (a) the force in dynes and in grams acting on any one conductor when it is directly under a pole; (b) the total force in kg acting tangentially along the armature surface.

286. In Prob. 285 determine: (a) the total force in lb acting tangentially along the armature surface; (b) the torque in lb-ft developed by the armature.

287. A shunt motor operates with 250 volts at its terminals. The line current is 60 amp, the field resistance (including rheostat) is 140 ohms, and the armature resistance is 0.18 ohm. Determine: (a) the armature current; (b) the counter emf developed by the armature; (c) the power lost in the field circuit; (d) the power lost as heat in the armature resistance.

288. When the motor of Prob. 287 develops a counter emf of 233.8 volts, determine: (a) the net emf acting in the armature circuit; (b) the value of the armature current; (c) the value of the line current.

289. The armature of the motor, Prob. 287, is brought to rest, and by means of a series resistance the line current is made equal to 60 amp. Determine: (a) the value of the field current; (b) the value of the armature current. (c) Compare these results with those of Prob. 287, and explain the difference.

290. The line current of a 50-hp, 230-volt, shunt motor is 182 amp at rated load. The shunt-field current is 3.8 amp, and the armature resistance is 0.051 ohm. Determine: (a) the back emf developed by the motor; (b) the power lost as heat in the armature resistance.

291. In the motor, Prob. 287, determine: (a) the internal power in watts; (b) the internal power in horsepower.

292. In the motor, Prob. 290, determine: (a) the internal power in watts; (b) the internal power in hp. The friction and windage and armature-iron losses are 1,200 watts, and the motor speed is 860 rpm. Determine: (c) the internal torque; (d) the power at the pulley in watts and in hp; (e) the torque at the pulley.

293. The speed of the motor, Prob. 290, is 860 rpm when the line current is 182 amp. The no-load armature current is 5.2 amp. Neglecting armature reaction, determine the speed at no load.

294. While the motor, Prob. 290, is operating with the line current equal to 182 amp and the field current equal to 3.8 amp, the speed is 860 rpm. The field current is then decreased so that the flux is reduced in the ratio of 1 to 0.88. The armature current remains unchanged. (*a*) Determine the new value of speed. (*b*) With the new value of flux determine the no-load speed if the no-load armature current is 5.2 amp.

295. The armature resistance of a 5-hp, 220-volt, 1,200-rpm, shunt motor is 0.62 ohm, and the field resistance is 320 ohms. At no load the motor takes 2.1 amp from the line and runs at 1,280 rpm. The rated-load current is 20.8 amp. Neglecting armature reaction, find the speed at rated load.

296. If, in Prob. 295, armature reaction reduced the flux in the ratio of 1 to 0.92 from the no-load to the rated-load value, what would be the rated-load speed?

297. When a 10-hp, 220-volt, series motor takes 40.0 amp from the 220-volt line, the speed is 820 rpm and the internal torque is 68.3 lb-ft (9.45 kg-m). The armature resistance is 0.33 ohm, and the series-field resistance is 0.20 ohm. (*a*) Neglecting armature reaction and assuming that the motor operates on the straight portion of the saturation curve, determine the speed when the current is 20 amp. (*b*) Determine the internal torque under these conditions.

298. The motor armature of Prob. 297 cannot safely operate at a speed greater than 1,800 rpm. What is the minimum safe current? Assume a straight-line saturation curve, and neglect armature reaction.

299. A 5-hp, 230-volt, shunt motor and a 5-hp, 230-volt, series motor have rated-load speeds of 600 rpm, and both armatures require 20 amp at rated load and develop 43.7 lb-ft (6.04 kg-m) torque. Neglecting all losses and assuming that the motors operate on the straight portion of their saturation curves, determine for armature currents of 10 amp: (*a*) their speeds; (*b*) their internal torques; (*c*) the power developed by each.

300. The speed of a 20-hp, 230-volt, shunt motor is 840 rpm with its rated line current of 75 amp. The field current is 2.1 amp, and the armature resistance is 0.12 ohm. The speed is reduced by inserting 1.0 ohm in the armature circuit. The total motor current does not change. Determine: (*a*) the speed under these conditions; (*b*) the ratio of the torques with and without the added resistance; (*c*) the percentage of the armature input lost in the added resistance.

301. Repeat Prob. 300 with the 1.0-ohm resistance replaced by a 1.5-ohm resistance.

302. A prony brake similar to that shown in Fig. 254, p. 292, is used to test a 10-hp, 250-volt, shunt motor. The perpendicular distance from the center of the shaft to the center line of the balance is 2 ft (0.610 m). The dead weight, or tare, of the arm is +2.2 lb (1.00 kg). When the balance reads 12.9 lb (5.46 kg), the speed is 1,230 rpm, the ammeter giving the current input to the motor reads 18.2 amp, and the line voltmeter reads 250 volts. At this load determine: (*a*) the external torque of the motor; (*b*) the horsepower output; (*c*) the watt output; (*d*) the watt input; (*e*) the efficiency.

303. When the motor of Prob. 302 delivers its rated load of 10 hp, the balance reads 24.1 lb (10.9 kg), the speed is 1,200 rpm, the ammeter reads 35.0 amp, and the voltmeter reads 248 volts.　Repeat (a) to (e), Prob. 302.

304. A rope brake similar to that shown in Fig. 256, p. 294, is used to determine the output of a 2-hp, 115-volt, shunt motor.　The brake drum is 10 in. (0.254 m) in diameter, the speed is 1,750 rpm, one balance reads 11.9 lb (3.90 kg) and the other reads 3.2 lb (1.45 kg).　The ammeter which measures the motor current reads 7.82 amp, and the line voltmeter reads 114.5 volts.　Determine: (a) the torque developed by the motor; (b) the horsepower developed; (c) the efficiency at this load.

305. When the motor, Prob. 304, delivers its rated load of 2 hp, the speed is 1,680 rpm, one balance reads 20.5 lb (9.30 kg) and the other reads 5.4 lb (2.45 kg).　The ammeter reads 16.6 amp, and the voltmeter reads 114.0 volts.　Determine: (a), (b), and (c), Prob. 304, at this load.

QUESTIONS ON CHAPTER XIII

Losses; Efficiency; Operation

1. What two disadvantageous effects on operation do energy losses within dynamos have?　To what is the armature-resistance loss due, and how is its value determined?　What precautions should be taken when measuring the armature resistance?

2. Of what does the loss in the shunt-field circuit consist?　How is it determined?

3. What fundamental laws account for eddy-current losses in the armature iron?　How are these losses reduced to a small value?

4. Why does hysteresis loss occur in the armature iron?

5. On what two factors do the iron losses and the friction losses depend? What are these losses called?

6. Give three equations which may be used for calculating efficiency. Under what conditions is each used?

7. State the approximate rated-load efficiencies of the following: 1-hp motor; 5-hp motor; 10-hp motor; 20-hp motor; 500-kw dynamo.

8. What difficulty is met when it is attempted to measure the efficiency of a large motor by measuring simultaneously the output and input?　How do errors in such measurements affect the results?

9. What difficulty is encountered in attempting to measure the input to a generator?

10. How is a dynamo connected when it is desired to determine its stray power?　Make any distinctions, if such exist, between the methods used for determining the stray power of a motor and those used for determining the stray power of a generator.

11. What effect do errors in the measurement of stray power have on the calculated efficiency?

12. What one factor determines in large measure the rating of electrical apparatus?　Name another factor which may limit the output of d-c dynamos.

13. Give the approximate temperature to which cotton and fibrous insulations may be subjected without rapid deterioration.

14. How may the approximate temperature of windings be determined by thermometer? What is meant by "hot-spot" temperature? How is its approximate value determined?

15. Describe the method, involving change of resistance, which is used for determining the average temperature of windings. What precautions are taken when making such a measurement for an armature winding? For a shunt field?

16. Give four reasons why it is desirable to operate several generating units in parallel rather than to attempt to supply the entire load with one or possibly two units.

17. Show that if two shunt generators operate in parallel, the generator with the more drooping characteristic takes the lesser share of the load.

18. What relation should exist among several shunt generators for the most satisfactory parallel operation? State the procedure to be followed in putting an idle generator in service; in removing a generator from service. How may the load between generators be adjusted?

19. Show that two overcompounded generators, operating in parallel, are in an unstable condition of operation. How may this condition of instability be eliminated?

20. Make a wiring diagram of two compound generators in parallel. Where must the ammeters be connected?

21. Why is overload protection in an electric circuit necessary? Compare circuit breakers and fuses from the point of view of first cost; space required; convenience; operating cost; speed of operation.

22. Describe the operation and state the advantages of automatically reclosing circuit breakers.

23. On what part of the switchboard should circuit breakers be mounted, and why?

24. What factors make the operation of an automobile power plant different from that of a stationary power plant? What is meant by "single-unit" and "two-unit" systems, and which is in general use?

25. Describe the principle underlying "third-brush" regulation. How may the battery charging rate be regulated? Why is this system not adequate to meet present-day demands?

26. Describe a system of generator regulation, involving the third brush, which is used in the Buick car. State the sequence of operations involved in starting the gasoline motor.

27. What is the general function of the cutout relay? Analyze its operation in connecting the generator to the electric system and in disconnecting the generator from the system.

PROBLEMS ON CHAPTER XIII

Losses; Efficiency; Operation

306. The losses in a 10-kw, 230-volt, shunt generator at rated current and voltage are as follows: armature-resistance loss, 520 watts; field loss (including rheostat), 320 watts; stray power, 650 watts. Determine the efficiency of the generator at rated load.

307. The load on the generator, Prob. 306, is reduced so that the armature current is one-half the value in Prob. 306, and the terminal voltage rises to 238 volts. The shunt-field circuit loss is now 340 watts, and the stray-power loss is 660 watts. Determine the efficiency at this load.

308. The generator of Prob. 306 operates as a motor at 230 volts. The input is 10 kw, and the losses are essentially the same as those of Prob. 306. Determine the efficiency under these conditions.

309. A 50-kw, 250-volt, shunt generator is delivering rated-load current at rated terminal voltage. The shunt-field resistance is 60 ohms; the armature resistance is 0.048 ohm; and the stray power is 2,550 watts. Determine: (a) the rated-load current; (b) the field current; (c) the armature current; (d) the field loss; (e) the armature loss; (f) the efficiency.

310. The generator, Prob. 309, operates at half rated load. The field rheostat is adjusted so that the terminal voltage is 250 volts, and the field resistance is now 65 ohms. The stray power is 2,640 watts. Determine: (a) the load current; (b) the armature current; (c) the field loss; (d) the armature loss; (e) the efficiency.

311. A 25-kw, 250-volt, 1,200-rpm, shunt generator delivers rated-load current of 100 amp at 250 volts. The shunt-field resistance is 100 ohms, and the armature resistance is 0.1 ohm. Accordingly, the armature current is 102.5 amp, and the induced emf is 250 + 10.3 or 260.3 volts. In order to measure the stray power, the generator is connected to operate as a motor without load (Fig. 260, p. 301). The field rheostat is adjusted to give 2.5 amp field current, and the rheostat R is adjusted until the speed is 1,200 rpm. The armature ammeter reads 5.3 amp, and the voltmeter reading (V_1) is 262 volts. Determine the stray power of the generator for this condition which corresponds to rated load.

312. From the data of Prob. 311 determine the efficiency of the generator at rated load.

313. A 5-hp, 230-volt, 1,750-rpm, shunt motor takes 18.0 amp at 230 volts. The shunt-field current is 0.70 amp, the armature resistance is 0.75 ohm, and the stray-power loss is 320 watts. Determine: (a) the motor output in watts; (b) in hp; (c) the efficiency; (d) the torque.

314. A 10-hp, 230-volt, 1,500-rpm, shunt motor, when operating near its rated load and at rated speed, takes 36.8 amp at 230 volts from the line. The armature resistance is 0.31 ohm, and the field current is 1.3 amp. The motor is then run light, connections similar to those shown in Fig. 260, p. 301 being used. The field current is maintained at 1.3 amp, and the speed is brought to 1,500 rpm by means of the armature rheostat. The voltmeter across the armature reads 221 volts, and the armature ammeter reads 2.15 amp. Determine: (a) the armature current when the motor is operating near its rated load; (b) the field loss; (c) the stray power; (d) the motor output in watts; (e) the efficiency.

315. In Prob. 314 determine: (a) the motor output in hp; (b) the torque.

316. In Prob. 314 determine the motor torque and efficiency when the line current is 20.0 amp. In the stray-power test the speed is 1,550 rpm, the voltmeter reads 226 volts, and the armature current is 2.20 amp.

317. The armature copper and the field copper of a 10-hp, 220-volt, shunt motor are measured for resistance after the motor has been standing idle for several hours in a room of temperature 20°C. The resistance of the armature between two marked segments is 0.242 ohm, and that of the field excluding the rheostat is 124.0 ohms. The motor is then loaded to rated load. After running 3 hr, the resistances are measured and the armature resistance is found to be 0.268 ohm, and the field resistance 135.4 ohms. Determine: (a) the average temperature rise of the armature during this time; (b) the average temperature rise of the field.

318. After the motor of Prob. 317 operates 6 hr at rated load, it is stopped and the armature resistance is again measured between the two marked segments and found to be 0.275 ohm. The field resistance at this time has increased to 141.0 ohms. Determine (a) and (b), Prob. 317.

319. A 100-hp, 600-volt, series (railway) motor is put under test for temperature rise. The initial series-field resistance is first measured after the motor has been standing for some time in a room of temperature 22°C, and is found to be 0.098 ohm. The armature-copper resistance between marked segments is found to be 0.158 ohm. The motor is then operated at 75 hp continuously for 10 hr. At the end of this period the series-field resistance is measured and found to be 0.109 ohm and the armature resistance between the same marked segments is 0.181 ohm. Determine the measured temperature rise: (a) of the field; (b) of the armature. (c) Are these operating temperatures safe for cotton insulation? (See p. 302.) (NOTE:—Railway motors are rated on intermittent duty and usually are unable to operate continuously at their rated horsepower without overheating.)

320. Two 120-kw, 600-volt, shunt generators A and B are operated in parallel. The characteristic of A is such that when its no-load voltage is 600 volts its rated-load voltage is 550 volts. The characteristic of B is such that when its no-load voltage is 600 volts, its rated-load voltage is 565 volts. For all practical purposes both characteristics may be considered as straight lines. Both generators are connected in parallel at 600 volts with no load on the system. When B is delivering its rated current of 200 amp, determine: (a) the current delivered by A; (b) the voltage of the system; (c) the power delivered by A.

321. Determine the current delivered by B, Prob. 320, when A delivers 120 amp. Find the total power of the system.

322. Two shunt generators A and B are operating in parallel. Their common terminal voltage is 250 volts. The resistance of the armature of generator A is 0.06 ohm, and the induced emf is 258 volts. The resistance of the armature of generator B is 0.08 ohm, and the induced emf is 262 volts. Determine: (a) the current delivered by generator A; (b) the current delivered by generator B; (c) the total power delivered to the bus-bars by the two generators.

INDEX

381

INDUSTRIAL ELECTRICITY

PART II

ELECTRICAL ENGINEERING TEXTS

INDUSTRIAL ELECTRICITY

PART II

BY

CHESTER L. DAWES, S. B., A. M.

Associate Professor of Electrical Engineering, The Graduate School
of Engineering, Harvard University; Fellow, American
Institute of Electrical Engineers; Etc.

SECOND EDITION

PREFACE TO THE SECOND EDITION

The primary object of this text is to develop in a simple manner the principles of alternating currents and alternating-current circuits and to show their applications to such electrical apparatus as machinery, rectifiers, electron tubes, and also to power transmission. Since the first edition of the text, the art has developed so far technically that a more extended knowledge of alternating-current theory, as applied to circuits and apparatus, has become necessary, even for those who do not intend to specialize in the highly technical aspects of electrical engineering. Hence in the revision, the theory has been extended to include such subjects as effective resistance of impedances; polygons of alternating vector quantities; energy and quadrature currents; kilovars and their effect on the operation of power and transmission circuits.

In the chapter on measurements, the many recent improvements in alternating-current measuring instruments and in the electromagnetic oscillograph have been included. The cathode-ray oscillograph has become so important in both research and industry that it also has been added.

The analysis of alternator operation has been extended to include the synchronous-impedance method of determining regulation, and similarly the open-circuit and short-circuit method of determining transformer efficiency and regulation also has been added. Modern practice of fabricating electrical machinery by welding methods is described and illustrated.

In recent years there have been many new developments in rectifiers, in both small and large units, and for power conversion on a large scale it is found advantageous to use the metal-tank rectifier rather than rotating machinery in many of the new installations. Consequently, a new chapter is devoted to rectifiers and in it are included such recent types as the oxide-film rectifier, the thyratron, and the ignitron.

In the chapter on transmission, the recently constructed Boulder Dam–Los Angeles transmission line is briefly described.

The developments in electron tubes and their applications have been rapid during recent years so that a more detailed analysis of tube characteristics has been included; and also the screen grid, the suppressor grid, and superheterodyne reception which is now in general use are described.

In accordance with the requests received from many teachers, the number of problems has been increased substantially over that in the first edition.

In the first edition there were two brief chapters on illumination and interior wiring. Because of the impossibility of covering these subjects adequately in the space available, these chapters have been omitted.

The author is indebted to several persons for their assistance in preparing this volume. The different manufacturers were highly cooperative in supplying the illustrations. R. T. Gibbs, Lecturer in Electrical Engineering at the Graduate School of Engineering, Harvard University, assisted the author and made several excellent suggestions; A. L. Russell, Head of Electrical Department at the Franklin Technical Institute of Boston, rendered valuable assistance, particularly in preparing the problems and their solutions. The chapter on electron tubes was written by R. F. Field, of the General Radio Company, for which the author is most appreciative. The author is particularly indebted to Professor H. E. Clifford, Consulting Editor, who not only made valuable suggestions during the preparation of the text but also reviewed and edited most carefully the completed manuscript.

C. L. Dawes.

Cambridge, Massachusetts,
  Harvard University,
    February, 1942.

PREFACE TO THE FIRST EDITION

This volume is intended as a continuation of "Industrial Electricity," Part I, which is devoted almost entirely to direct-current circuits and direct-current machinery. Since a large proportion of power is now generated as alternating current, and since all communication circuits also utilize alternating currents, the first chapters of the present volume are devoted to the fundamental principles and the simple laws of alternating currents and alternating-current circuits. Only the simplest mathematics is used. Considerable emphasis is placed on the effect of inductance, capacitance, and frequency in the flow of alternating current. A chapter is devoted to the construction and uses of the ordinary alternating-current instruments, particular attention being given to the testing and adjusting of the induction watthour meter. Since, with few exceptions, alternating-current power is generated and distributed through polyphase systems, a chapter is devoted to the study of the relations existing among currents and voltages in such systems.

With the foregoing as a foundation, the construction and the operating characteristics of alternators, polyphase induction motors, single-phase motors, synchronous motors, and converters are discussed and briefly analyzed. The relations of the characteristics of these types of power machinery to their industrial applications are also discussed to a considerable extent.

The last three chapters are devoted to general industrial applications of electricity. In Chap. XII, the more common electrical illuminants, particularly the incandescent lamp, and simple photometric measurements are described. In Chap. XIII an attempt is made to present the underlying principles of electron emission and its applications to the electrical industry. In this chapter the Fleming valve, the kenotron rectifier, the three-electrode vacuum tube, and the applications of the vacuum tube to electrical communication, particularly radio telephony, are discussed in a simple but reasonably comprehensive manner.

Also the principles underlying broadcasting and a brief description of the wiring diagrams of typical receiving sets are included. In the last chapter another important industrial application of electricity is given; that is, the methods and general rules which should be followed in the installation of electric wires in buildings in order that interruptions of service, fire hazard, and personal injury may be minimized.

In the preparation of this volume, it was found advantageous to utilize certain of the material, more particularly figures, from the author's "A Course in Electrical Engineering, Volume II, Alternating Currents." The scope of this volume, however, is intended to be somewhat broader and much less analytical than that of the author's "Alternating Currents."

The author is indebted to Mr. Robert F. Field of the Cruft Laboratory, and Instructor in Physics and Electric Communication Engineering at Harvard University, who is the author of Chap. XIII, "Electron Tubes"; to Mr. Raymond T. Gibbs, Instructor in Electrical Engineering at the Harvard Engineering School, who is author of Chap. XIV, "Interior Wiring"; and, particularly, to Prof. H. E. Clifford, the Consulting Editor of these Electrical Engineering Texts, for his suggestions in the preparation and arrangement of this volume and for his careful review of the manuscript.

C. L. Dawes.

Harvard University,
  Cambridge, Massachusetts,
    October, 1925.

CONTENTS

CHAPTER III

CHAPTER XII

CHAPTER XIII

PAGE

QUESTIONS AND PROBLEMS

INDUSTRIAL ELECTRICITY

PART II

CHAPTER I

ALTERNATING CURRENT AND VOLTAGE

1. General Field of Use of Alternating Current.—At the present time, over 90 per cent of the electrical energy used commercially is generated as alternating current. This is not due primarily to any superiority of alternating over direct current in its applicability to industrial and domestic uses. In fact, there are many circumstances under which d-c energy is absolutely necessary for industrial purposes; for example, charging storage batteries, electrolytic processes, magnetite lamps, and operating street railways and electric locomotives. Where d-c energy is used for these purposes, it is usually generated as alternating current in a large steam or hydroelectric power station, is transmitted some distance, and is converted into direct current. Alternating-current energy may also be converted into d-c energy and used in the immediate vicinity of its place of generation. These conditions exist at Niagara Falls where the energy of the falls is used to generate a-c energy; this is converted to d-c energy for use in electrochemical industries which are situated near by and also is transmitted to other industrial centers at a distance.

The reasons for generating nearly all electrical energy as alternating current are as follows:

1. *Alternators have no commutators. Hence, units having large power ratings may be operated at high speeds.* Not only does this adapt them to the high speeds at which steam turbines operate

1

most economically, but the size and weight of generators decrease as the speed increases. For example, the two 200,000-kva alternators in the Hudson Avenue Station of the Consolidated Edison Company of New York operate at 1,800 rpm, whereas it is practically impossible to operate a 1,000-kw d-c generator at speeds much greater than 1,000 rpm because of commutation difficulties.

The size of the units and the floor space occupied by turbine-driven alternator units are both relatively small when their large

Fig. 1.—208,000-kw, three-unit turbine-generator set; 76,000-kw high-pressure unit; two 62,000-kw low-pressure units, each with 4,000-kw house generator. State Line Station, Hammond, Ind. (*Courtesy of General Electric Company.*)

power rating is considered. For example, Fig. 1 shows the comparatively small size of the 208,000-kw, three-turbine unit at the State Line Station, Hammond, Ind.

2. *Because alternators have no commutators, they can generate energy at comparatively high voltages.* For example, in small plants which generate for local use only, the electrical energy may be generated at 2,300 volts, which is a comparatively low voltage. Where the energy is to be distributed over a moderate area and also where the voltage is to be stepped up for transmission purposes, it is customary to generate at 6,600, 11,000, or 13,200 volts. Although there are alternators for voltages as

high as 20,000 volts, their insulation becomes difficult, and they are justified only when the transmission is at these higher voltages and transformers can be eliminated. Because of sparking and the tendency of commutators to flash over, it is difficult to

FIG. 2.—16,667-kva, 220,000 to 115,500 (Y) to 6,900/13,800 volts, 60-cycle power transformer for Bonneville Dam, Wash. (*Courtesy of General Electric Company*.)

generate d-c energy at voltages even as high as 1,500 volts per commutator.

3. *With alternating current the voltage may be raised and lowered economically by means of transformers.* This permits transmission of power over long distances economically. The weight of copper necessary to transmit a given amount of power a given distance with a fixed loss varies *inversely as the square of the trans-*

mission voltage (see Chap. XII). For example, when the transmission voltage is *doubled*, the weight of copper is *quartered*, other factors being equal. At 100,000 volts, the weight of copper required to transmit a given amount of power a given distance with a fixed loss is *one one-hundredth* that required if 10,000 volts be used.

There is no efficient method of raising and lowering d-c voltages. Therefore, direct current is not suited for power transmission.

Transformers are used also to step down the high transmission voltages so that the voltage is adapted to local distribution and to industrial uses.

Figure 2 shows one of the 16,667-kva, 220,000 to 115,500 to 6,900/13,800-volt transformers used in connection with the Bonneville Dam project on the Columbia River in the state of Washington.

4. *Because of the great distances over which a-c power may be transmitted economically, it is possible to generate electrical energy in large quantities in a single station and to distribute it over a comparatively large territory.* The large boilers, automatic stokers, superheaters, recording instruments, etc., which are used in large stations, result in high boiler-room efficiency. Large turbines have an economy which may be three or four times that of the steam units in a small plant. The generator has an efficiency of 95 to 96 per cent in the larger sizes. Then again, as the boilers and large turbine units require relatively few attendants, the labor and superintendence charges per kilowatt are small.

For these reasons, it is often more economical to generate energy with large units, to transmit it long distances, and even to convert it into d-c energy than to generate the d-c energy at the place where it is to be utilized.

It must be remembered, however, that the reduced generating costs may be offset by the distribution costs resulting from investment in lines, cables, substations, machinery, etc., in addition to the labor and maintenance costs of the distribution system. Also, occasionally, there are industries where a large amount of live steam is required for industrial processes, and a steam plant for the generation of direct current may then be justified.

Figure 3 shows the Boulder Dam hydroelectric power plant, Arizona-Nevada, from which power is transmitted to Pasadena and Los Angeles. (Also see Fig. 295, p. 367.)

5. *For constant-speed work, the a-c induction motor is cheaper than the d-c motor both in first cost and maintenance.* This is due to the fact that this type of motor has no commutator. (See Chap. VIII, p. 263.) On this account, it is frequently desirable to generate a-c energy even in an isolated plant if the power is to be utilized largely for operating motors. The alternating

Fɪɢ. 3.—Hydroelectric power station at Boulder Dam, Arizona-Nevada. (*Courtesy of the Bureau of Power and Light, City of Los Angeles.*)

current, in such cases, has no advantages over direct current except in the more economical operation of the a-c motors.

Alternating-current energy owes its importance to the following features: It can be generated economically, at comparatively high voltages, in units of large power rating operating at turbine speeds. By means of transformers the voltage can be raised and lowered efficiently, so that it is possible to transmit large amounts of energy over long distances and at voltages best adapted to given operating conditions. Generation of large amounts of energy in large central power stations permits substantial generating economies. For constant-speed work the induction motor is more economical than the d-c motor.

2. Sine Curve.—Alternating currents and voltages are not steady like direct currents and voltages but vary with time.

These time variations of current and voltage are usually sine or cosine functions.

Hence, before studying alternating currents and voltages, consideration should be given to some of the properties of sine curves.

If the values of the sine of an angle be plotted as ordinate and degrees as abscissa, a curve similar to that shown in Fig. 4 is obtained. Values of the angle x are plotted as abscissas and the corresponding values of sine x are plotted as ordinates. The values of the sine are found on p. 436 (also see p. 432 for angles greater than 90°). For example, sin 40° is +0.6428. The

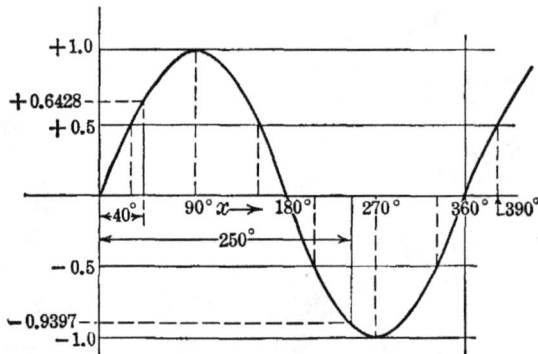

FIG. 4.—Sine curve.

ordinate at 40° is +0.6428; the ordinate at 250° is −0.9397, for sin 250° = sin (180°−250°) = sin (−70°) = −0.9397 (see 31, p. 433). Figure 4 shows that sin 30° = 0.5, sin 150° = 0.5, sin 210° = −0.5, sin 330° = −0.5, etc. The maximum positive ordinate of the curve is +1.0 (sin 90°), and the maximum negative ordinate is −1.0 (sin 270°). The curve is symmetrical about its zero axis; when 360° is reached, all possible values of the sine have been represented. Beyond 360° the curve merely repeats itself. Thus, sin 390° = sin 30°, etc.

A simple method of constructing a sine curve is shown in Fig. 5. A circle is drawn having for its radius the desired maximum value A of the sine curve, and its circumference, starting at one end of the horizontal diameter 6-0, is divided into any number of equal arcs 0-1, 1-2, etc. The horizontal line ab is drawn so that, if extended, it will coincide with the horizontal diameter 6-0. The distance ab is made equal to the desired length of base for the completed curve. The line ab is divided into the same number of equal segments 0-1, 1-2, etc., as the circumference of the circle, and the segments are numbered correspondingly.

The point 0 on the circle is projected horizontally to intersect the ordinate at 0 on line ab; point 1 on the circle is correspondingly projected to intersect at 1' the ordinate at point 1 on line ab; etc. The smooth curve drawn through these intersections gives a sine curve. The circumference of the circle (Fig. 5) is divided into arcs or segments, each arc or segment subtending an angle of 30°. Line ab may be marked correspondingly in degrees, as in Fig. 4.

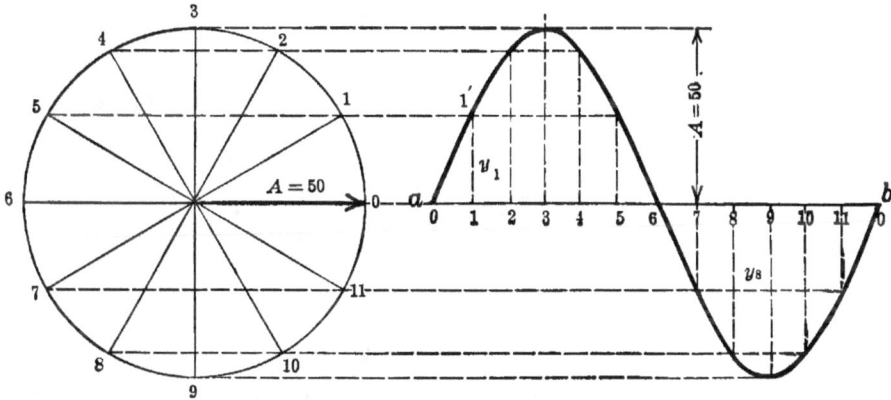

FIG. 5.—Graphical method of constructing sine curve.

Let A be the maximum ordinate of the sine curve. Then, the ordinate y at an angle x,

$$y = A \sin x. \tag{1}$$

For example, if $A = 50$ (Fig. 5), the ordinate y_1 at 30°, or at 1,

$$y_1 = 50 \sin 30° = 25.$$

The ordinate y_8 at 240°, or at 8,

$$y_8 = 50 \sin 240° = 50 \sin (180°{-}240°)$$
$$= 50 \sin (-60°) = -43.30.$$

(See pp. 433 and 434.)

3. Cosine Curve.—The cosine curve has the same form as the sine curve with its position shifted 90°. When the angle x is 0, the cosine curve has its maximum positive ordinate (cosine 0° = 1) whereas the sine curve has its zero ordinate. A cosine curve is shown in Fig. 6. This curve may be plotted using the cosine tables (p. 436). For example, cos 40° = 0.7660, and cos 250° = cos (180°–250°) = –cos 70° = –0.3420 (see 32, p. 433). These values are shown in Fig. 6.

The cosine curve may be plotted in a manner similar to that used for the sine curve (Fig. 5), except that the numbering on the circumference commences at intersection 3, the upper end of the vertical diameter. Thus 3 is changed to 0, 4 to 1, etc., and these points are projected horizontally to intersect the ordinates on line *ab* having corresponding numbers.

The cosine curve is the sine curve moved 90° to the left.

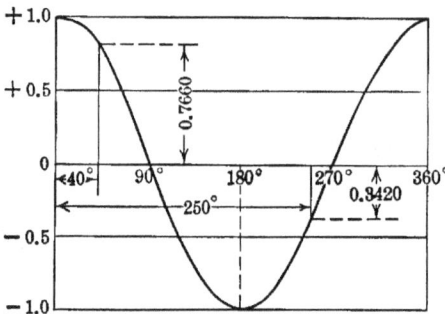

FIG. 6.—Cosine curve. FIG. 7.—Angular measure in radians.

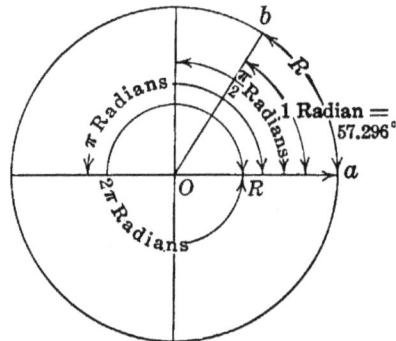

4. Radian.—The common unit of angular measure is the degree. In analytical work it is often desirable to use another unit, the *radian*, as the unit of angular measure.

A radian is an angle at the center of a circle subtended by an arc whose length is equal to the radius.

For example, in Fig. 7, the angle *aob* is subtended by the arc *ab*. The length of the arc *ab* is equal to the radius R or *oa* and angle *aob* is *one radian*, being slightly less than 60°. It is actually equal to 360°/2π, or 57.296° (to five figures).

Because the circumference of a circle is equal to 2π times the radius, the circumference of a circle subtends 2π radians, or, 360° = 2π radians, 180° = π radians, and 90° = $\pi/2$ radians. The sine $\pi/2$ = 1.0; cosine $\pi/2$ = 0; cosine π = –1.0; etc.

Example.—Find the sine of 4π/3 radians.
   2π radians = 360°.
   4π radians = 720°.
   4π/3 radians = 240°.
   sin 240° = –sin 60° = –0.866. (See Fig. 4.) *Ans.*

5. Combining Sine (or Cosine) Curves.—Sine (or cosine) curves may be combined by adding or subtracting their ordinates at each point along the axis of abscissas. If the curves have the same scale of abscissas, the curve resulting from adding or subtracting is also a sine (or cosine) curve, having the same scale of abscissas as the component curves.

For example, in Fig. 8, a sine curve a of maximum value A crosses the zero axis in a positive direction at zero degrees. A second sine curve b of maximum value B crosses the zero axis in a

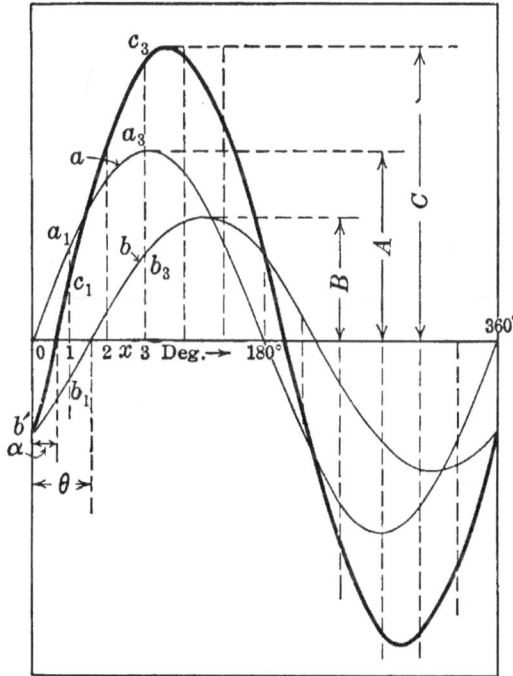

Fig. 8.—Addition of two sine curves.

positive direction at θ degrees ($\theta = 45°$ in Fig. 8) to the right of 0. These curves are said to differ in phase by θ degrees. Let it be required to determine the curve resulting from the addition of these curves. Ordinates 1, 2, 3, etc., are erected along the x-axis, sufficiently close together to give enough points for a smooth resultant curve. At ordinate 0, the ordinate of the a-curve is zero, and the ordinate of the b-curve is $-0b'$. Hence, the algebraic sum of the two curves at this point is $-0b'$, and b' is a point on the resultant curve. At 1 the a-curve has a positive value a_1 and the b-curve a negative value $-b_1$. The resultant ordinate c_1 is found by subtracting b_1 from a_1. At 3, both curves have positive values and the resultant ordinate c_3 is found by adding b_3 to a_3. Thus the resultant curve of maximum value C is found by adding algebraically the ordinates of the two curves a and b for each point on the axis of abscissas.

The resultant curve is a *sine* curve, having the same scale of abscissas as its two component curves. Its maximum value C is less than the arithmetical sum of A and B. It crosses the zero axis in a positive direction α degrees to the right of curve a and $\theta - \alpha$ degrees to the left of curve b.

6. Simple Alternating Electromotive Force.—It is shown in Part I, Chap. X, that when a single coil rotates at constant angular velocity in a uniform field (Fig. 9), an alternating emf

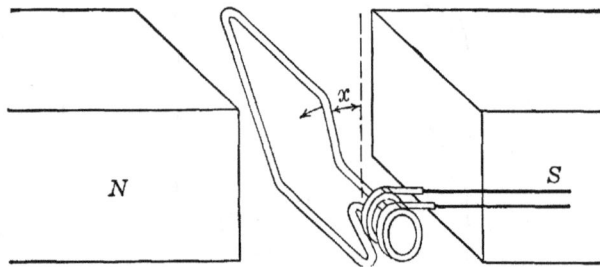

Fig. 9.—Generation of alternating emf.

is generated. This emf is zero when the plane of the coil is perpendicular to the field and reaches its maximum value when the plane of the coil is parallel to the field (see Part I, p. 212, Fig. 186). The value of the emf varies with the value of the angle x (Fig. 9) which the plane of the coil makes with the plane perpendicular to the field. Figure 10 shows in some detail the generation of an alternating emf. In (a) the coil ab is shown rotating in a counterclockwise direction, and at the instant shown it makes an angle x with the neutral plane oo. By Fleming's right-hand rule (see Part I, p. 211), the direction of induced emf in a is outward when it is in this position. The arrow v attached to a is proportional to and gives the instantaneous direction of the constant linear velocity of the conductor.

In order that a conductor cutting magnetic flux may generate an emf, there must be a component of its velocity *perpendicular* to the magnetic flux. That is, the flux, the conductor, and the velocity must be mutually perpendicular (see Part I, p. 209).

For example, in Fig. 10, when the plane of the coil *ab* is perpendicular to the magnetic field, conductor *a* is moving parallel to the magnetic lines. Hence, at this instant, there is no component of velocity perpendicular to the field and no emf is generated at this instant.

In Fig. 10(*a*) the plane of the coil *ab* makes an angle *x* with the neutral plane *oo*. The velocity *v* is oblique to the direction of the magnetic field. This velocity *v* has a component v_1 perpendicular to the magnetic field. The induced emf is propor-

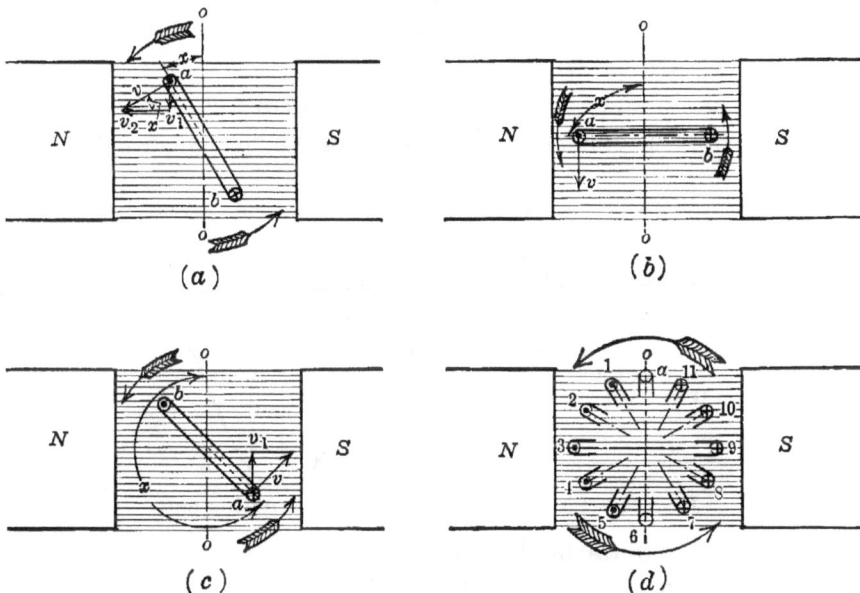

FIG. 10.—Generation of alternating emf by a simple coil rotating in uniform magnetic field.

tional to the perpendicular component v_1. Because $v_1 = v \sin x$, the induced emf is proportional to sin *x*. Hence, the emf induced in coil *ab* is proportional to the sine of the angle *x* and therefore, with constant angular velocity, varies sinusoidally with time (Fig. 11).

In Fig. 10(*b*), the angle *x* is 90°, and sin *x* = 1.0, its maximum positive value. Therefore, at 90° (Fig. 11) the emf has its maximum value, assumed to be 10 volts. In Fig. 10(*c*), the angle *x* is 225°, its sine is negative, and the value of the emf is negative and is equal to 10 sin 225° = −7.07 volts as shown in Fig. 11. Figure 10(*d*) shows the direction of the induced emf in conductors a and b at successive instants during the rotation. It is clear that an alternating emf is induced in coil ab.

Figure 11 shows graphically the value of the induced emf in conductor a, on the basis of a maximum value of 10 volts occurring at Fig. 10(b). The emf in conductor a is positive when it is directed outward. The values of angle x and the numbers corresponding to the positions of a, as shown in Fig. 10(d), are given on the axis of abscissas. For example, when angle x is 30° [1 in Fig. 10(d)], the emf is 10 sin 30°, or 5 volts; when angle x is 150° [5 in Fig. 10(d)], the emf is 10 sin 150°, or 5 volts; when angle x is 210° [7 in Fig. 10(d)], the emf is 10 sin 210°, or −5 volts. These values are shown in Fig. 11.

Thus, when a single coil rotates at uniform angular velocity in a uniform field, it generates an emf which varies as the sine of the

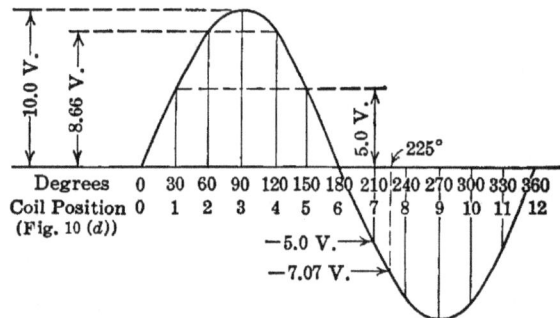

Fig. 11.—Sinusoidal variation of alternating emf.

angle which the plane of the coil makes with a plane perpendicular to the direction of the magnetic field.

7. Time Variation of Alternating Electromotive Force.—Figure 12 shows a sine curve of emf e similar to that of Fig. 11, in which it is assumed that the coil (Figs. 9 and 10) rotates at 60 rps. It therefore completes one revolution in $\frac{1}{60}$ sec.

Also, in $\frac{1}{60}$ sec., it rotates through 2π radians, or 360 space-degrees. Hence, two scales of abscissas may be used for the emf curve, one scale being *space-angles* of the coil, expressed in either degrees or radians, and the other scale being *time*. The scale of abscissas (Fig. 12) is given in both degrees and radians as well as in time. That time as well as space-angle may be used for the scale of abscissas is due to the fact that at constant angular velocity the space-angle through which the coil rotates is proportional to the time.

For convenience the time axis is often given in *time-degrees* rather than in actual time, such as seconds. For example, instead of stating $\frac{1}{240}$ sec. (Fig. 12), 90 electrical time-degrees may be used.

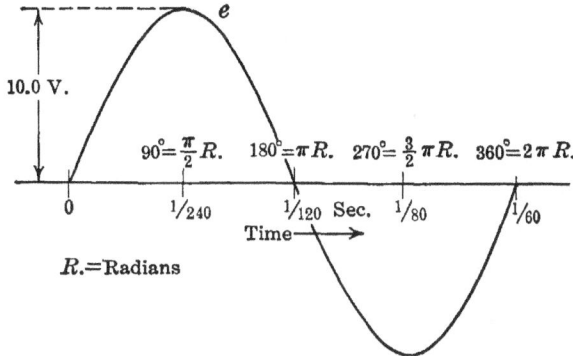

FIG. 12.—Space-degrees and time.

Hence, it may be said that alternating currents and emfs vary sinusoidally with *time*.

8. Cycle; Alternation; Frequency.—After the coil (Fig. 9) has rotated through 360 space-degrees, the emf begins to repeat

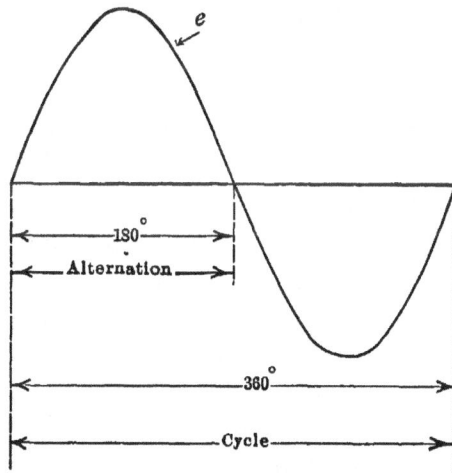

FIG. 13.—Alternation and cycle.

itself. For example when the coil has rotated through 390 space-degrees, its position is the same as when the angle x was equal to 30 space-degrees. Hence, the value and direction of the emf are the same for 390 and 30°. Consequently, when the coil has rotated through 360 space-degrees, the emf has passed through all values of magnitude, direction, and rate of change possible under the given conditions. At subsequent positions of the coil, these values merely repeat themselves at intervals of 360 space-degrees. Hence, the emf (or current) undergoes a complete *cycle* of values for every 360 space-degrees through which the coil rotates. Therefore, one cycle corresponds to 360 electrical time-degrees as shown in Fig. 13. One alternation corresponds to 180 electrical time-degrees. The number of *cycles per second* is the *frequency*.

In a two-pole alternator, the emf goes through one cycle for every revolution of the coil. Hence, the cycles per second, or frequency, is equal to the rps of the coil.

Example.—The armature of a two-pole alternator rotates at 1,500 rpm. What is the frequency?

The rps

$$s = \frac{1,500}{60} = 25.$$

The frequency

$$f = 25 \text{ cycles per sec.} \quad Ans.$$

It is obvious that, with a *two-pole* alternator, one space-degree for the coil corresponds to one time-degree for the emf.

9. Frequency, Poles, and Speed.—Figure 14(*a*) shows a four-pole alternator. For simplicity a single conductor *a* of a coil is shown rotating, rather than the complete coil. As soon as this conductor has passed a north and a south pole, that is, when it has passed from 1 to 5, it has completed one electrical cycle, or 360 electrical time-degrees, as shown in Fig. 14(*b*). Mechanically, it has completed one-half a revolution, or 180 space-degrees, so that in one revolution, or 360 space-degrees, the emf in the conductor will have completed two cycles, and will have gone through 720 electrical time-degrees. Therefore, in a four-pole alternator, 1 *space-degree* corresponds to 2 *electrical time-degrees*. That is, for every space-degree traversed by a conductor, the emf wave completes 2 electrical time-degrees. Such a conductor needs to make only 30 rps, or 1,800 rpm, in order to generate a 60-cycle emf. Likewise, for a 25-cycle emf, such a conductor needs to revolve at only 12.5 rps, or 750 rpm. For a given frequency, as the number of poles increases, the mechani·

cal speed decreases proportionately. The relation between speed, poles, and frequency may be written in the form of an equation

$$f = \frac{P \times S}{2 \times 60} = \frac{P \times S}{120} \quad \text{cycles per sec,} \qquad (2)$$

where P is the number of poles and S is the speed in rpm.

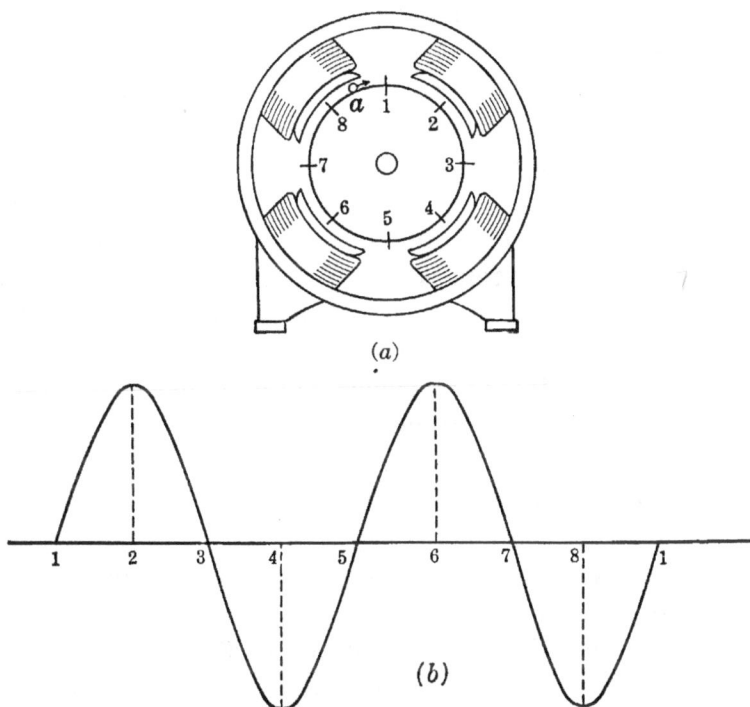

(a)

(b)

Fig. 14.—Two cycles per revolution in four-pole alternator.

The table shows the relation of speed, frequency, and number of poles for a few typical cases.

Poles	Speed, rpm	
	60 cycles	25 cycles
2	3,600	1,500
4	1,800	750
6	1,200	500
8	900	375
40	180	75

Example.—A 60-cycle, engine-driven alternator has a speed of 120 rpm. How many poles has it?

Using Eq. (2) and solving for P,

$$P = \frac{120f}{S} = \frac{120 \times 60}{120} = 60 \text{ poles.} \quad Ans.$$

In practice, nearly all alternators have stationary armatures and rotating field structures, and the above equations apply.

10. Commercial Frequencies.—In the United States, frequencies are standardized at 60 cycles and at 25 cycles per second, although other frequencies are used. In California, for example, 50 cycles is still used although the 60-cycle power which reaches Pasadena from Boulder Dam has necessitated the change-over from 50 to 60 cycles of a considerable amount of apparatus. In the early days of a-c development, 133 cycles was common; but few, if any, plants now generate at this frequency. The principal advantage of higher frequencies is that transformers require less iron and copper and so are lighter and cheaper. The flicker of lamps is not perceptible at 60 cycles, but at 25 cycles it is obvious. On the other hand, the voltage drop in transmission lines and in apparatus varies almost directly as the frequency, so that better voltage regulation throughout the system is obtained with low frequency. Power apparatus, such as induction motors, synchronous converters, and a-c commutator motors, operates better at low than at high frequencies. With one or two exceptions, however, the operation is satisfactory at 60 cycles per sec. A power and lighting company would operate ordinarily at 60 cycles per sec, because the flicker of lamps at 25 cycles per sec is objectionable and the transformers at this lower frequency are heavier and cost more than at the higher frequency. On the other hand, where the connected load is practically all power apparatus, 25 cycles is frequently used. For example, this frequency is used by the New York, New Haven and Hartford R. R. for its electric locomotives; on the Norfolk and Western Ry. for operating electric locomotives; and by the Boston Elevated Ry. Co. for transmitting high-voltage power to its d-c substations. In Europe, frequencies as low as 15 and even 12.5 cycles per sec are common.

11. Electromotive Force Induced in Conductors of Alternator Armature.—Thus far, emfs have been considered as due to the rotation of a single flat coil in a magnetic field of uniform intensity. The emf is sinusoidal under these conditions because the component of conductor velocity perpendicular to the field varies as the sine of the angle through which the coil rotates (Fig. 10). Commercial alternators must be quite different structurally from the simple alternator just described, for alternators built with a single coil and having no iron in the armature would have so small a power output as to be useless. In most commercial alternators the field structure rotates, rather than the armature. This eliminates slip-rings for conducting the power current to the external circuit and simplifies insulation (also see p. 151). The conductors are embedded in slots on the armature surface, and the resulting air-gap is short (see p. 168, Fig. 143). Ordinarily, the winding of any one phase is distributed over a belt consisting of several adjacent slots (see p. 164, Fig. 138).

To illustrate the manner in which an emf varying sinusoidally with time may be induced in the conductors of an alternator armature, a single conductor embedded in a slot is shown in Fig. 15. The armature is stationary, and the field structure is shown as moving from left to right. For simplicity the armature surface is represented as being flat. The pole faces are curved or chamfered in such a manner that the flux density along the air-gap is sinusoidal, the shortest air-gap, hence the greatest flux density, occurring at the center of the poles. The small effect of the slot on the flux is neglected.

The value of *flux density* for each point along the air-gap is given by the ordinates of curve *B* (Fig. 15). For example, if the maximum density, which occurs at the center of the pole, is 7,000 gausses, the density at point *a*, 30 space-degrees to the right of the point where the flux density is zero, is 7,000 sin 30°, or 3,500 gausses. Obviously, the total flux is given by the product of the average height of curve *B* and the area of the pole face.

The equation for induced emf in a conductor is

$$e = Blv\ 10^{-8} \text{ volts,} \tag{3}$$

where *B* is the flux density in gausses, *l* the length of conductor in centimeters, and *v* the velocity of the conductor in centimeters per second. *B, l* and *v* are mutually perpendicular. [See Part I, Eq. (86), p. 210.] Although the conductor itself (Fig. 15) does not move, there need be only *relative* motion between flux and conductor to induce emf. If the conductor is stationary, *v* gives the peripheral velocity of the flux in centimeters per second. If Fleming's right-hand rule be applied, the conductor, although stationary, must be considered as moving from right to left, for it moves in this direction *relative to the flux.* When under the north pole, the direction of the induced emf is found

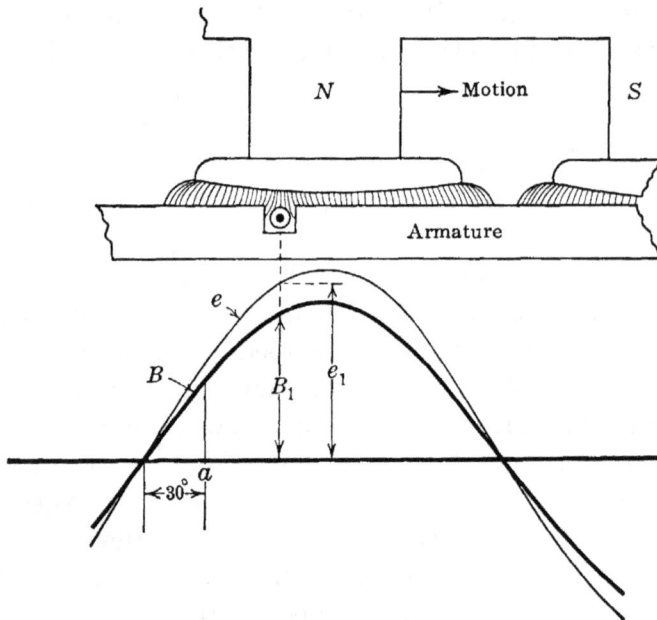

FIG. 15.—Emf induced by rotating field.

to be outward, as is shown by the dot. Because in the foregoing equation *l* and *v* are constant, the induced emf in each conductor *must be proportional to the flux density B of the field in which the conductor finds itself.* Hence, if the flux-density curve is sinusoidal, the emf induced in each conductor of the armature will be sinusoidal. For example, when the conductor is midway between pole centers, the flux density *B* is zero and the emf *e* is zero; when the pole center is directly over the conductor, the flux density *B* is a maximum and the emf *e* is a maximum. At the position shown in Fig. 15 the conductor is in a field whose density is B_1 gausses and the emf induced at that instant is e_1 volts.

It follows that if the flux density at each point along the gap varies as the sine of its electrical space-angle, the emf induced in each armature conductor will vary sinusoidally with time.

Example.—The flux density under the center of the pole of a 60-cycle, four-pole alternator (see Fig. 15) is 7,000 gausses. The diameter of the armature is 1 m, and the active length of armature conductor is 60 cm. The flux density curve along the air-gap is sinusoidal. Determine: (*a*) maximum value of sine wave of emf induced in each conductor; (*b*) effective value.

(*a*) The armature periphery is 100π, or 314 cm.

The rotating field must complete one revolution every two cycles (p. 9), or in $\frac{2}{60}$ or $\frac{1}{30}$ sec. Hence, the velocity at which the flux cuts the armature conductors,

$$v = \frac{314}{\frac{1}{30}} = 9{,}420 \text{ cm. per sec.}$$

Hence, from Eq. (3), the maximum emf

$$e_m = 7{,}000 \times 60 \times 9{,}420 \times 10^{-8} = 39.6 \text{ volts.} \quad \textbf{\textit{Ans.}}$$

(*b*) The effective emf

$$E = \frac{39.6}{\sqrt{2}} = 28.0 \text{ volts.} \quad \textbf{\textit{Ans.}}$$

Usually, with salient-pole alternators such as that shown in Fig. 15, the air-gap is uniform and the flux-density curve is flat rather than sinusoidal. However, the fact that the conductors in any one phase are distributed over a considerable angle on the armature tends to produce a sinusoidal emf wave in the belt as a whole (p. 178). With nonsalient poles which ordinarily have a distributed field winding the flux-density curve along the air-gap is nearly sinusoidal (p. 175).

12. Equations of Alternating Voltages and Currents.—It is shown in Par. 6 that the variation of an alternating emf with time can be expressed as a sine function either of a space-angle or of time. Ordinarily, the voltage is expressed as a function of time, and for mathematical reasons the time-angle is expressed primarily in radians rather than in degrees.

For example, let it be required to express a 60-cycle alternating emf as a sine function of time, the maximum value being 150 volts.

$$\begin{aligned} e &= E_m \sin \omega t \text{ volts} \\ &= 150 \sin \omega t \text{ volts,} \end{aligned} \qquad (4)$$

where e is the value of the emf at any time t, E_m is the maximum value of the emf, and with a given frequency ω is a constant. The time t is given in seconds.

The value of the constant ω must be determined.

If the frequency is 60 cycles per second, the emf completes one cycle, or 2π radians, in $\frac{1}{60}$ sec. Therefore, at the time $t = \frac{1}{60}$ sec. the angle ωt equals 2π radians.

That is,

$$\omega(\tfrac{1}{60}) = 2\pi$$

or

$$\omega = 2\pi 60 \text{ or } 377 \text{ radians per sec.}$$

Therefore, the complete equation of the emf is

$$e = 150 \sin 2\pi 60t = 150 \sin 377t \text{ volts.} \qquad (I)$$

The symbol ω is equal to $2\pi f$ where f is the frequency in cycles per second. Also, ω is the number of radians through which an emf or current passes in one second. For example, the foregoing emf completes 2π radians every cycle. There are 60 cycles per second. Hence, the radians per second are $60 \times 2\pi$, or 377. ω is often called *angular velocity* (in radians per second) (see p. 8).

Example.—Find the value of the foregoing emf wave when the time is equal to 0.003 sec.

Substituting 0.003 in (I),

$$e = 150 \sin (377 \times 0.003) = 150 \sin 1.131 \text{ (radians).}$$
1.13 is obviously the time-angle in radians.

Since 2π or 6.283 radians equal 360°, the angle in degrees,

$$\omega t = \frac{1.13}{6.283}\ 360 = 64.7°.$$
$$\sin 64.7° = 0.904 \text{ (p. 437).}$$

Hence, $e = 150 \times 0.904 = 135.6$ volts. *Ans.*

Example.—A 25-cycle alternating current, having a maximum instantaneous value of 75 amp, varies sinusoidally with time. The value of time may be taken as zero when the current is zero and increasing in a positive direction. Determine: (a) equation of current; (b) time when it first reaches its positive maximum value; (c) value of current when the time is 0.036 sec.

(a) $\omega = 2\pi 25 = 157$ radians per sec.

Hence the equation of the current

$$i = 75 \sin 2\pi 25 = 75 \sin 157t. \quad Ans.$$

(b) This current is plotted in Fig. 16. Obviously, it first reaches its positive maximum value when the time-angle is 90° or $\pi/2$ radians. Hence,

$$2\pi 25t = \frac{\pi}{2}.$$

$$t = \left(\frac{\pi}{2}\right)\frac{1}{2\pi 25} = \frac{1}{100} \text{ or } 0.01 \text{ sec. } Ans.$$

This result is obviously correct since $\frac{1}{25}$ sec is required to complete one cycle. Therefore, $\frac{1}{4} \times \frac{1}{25}$, or 0.01 sec is required to complete one-fourth of a cycle.

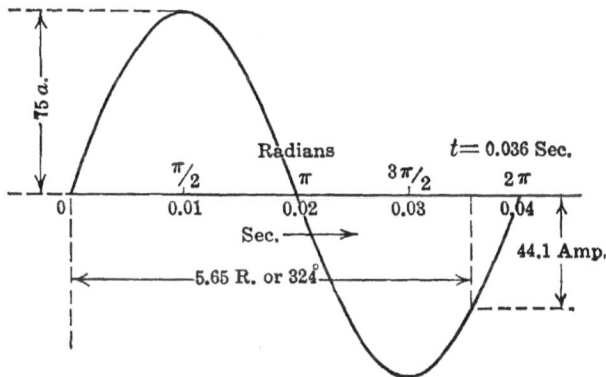

Fɪɢ. 16.—Instantaneous value of a 25-cycle current.

(c) $i = 75 \sin (157 \times 0.036) = 75 \sin 5.65$ (radians).

$$\frac{5.65}{6.283} 360° = 324°.$$

From (29) (p. 433),

$$-\sin 324° = \cos (90° + 324°) = \cos 414°$$
$$= \cos (414° - 360°) = \cos 54°$$
$$= 0.5878$$
$$75 \times (-0.5878) = -44.1 \text{ amp. } Ans.$$

Figure 16 shows this value of current.

The angle could have been found by proportion. It requires 0.04 sec for the current to complete one cycle, or 360°. In 0.036 sec the angle is

$$\frac{0.036}{0.04} 360° = \frac{9}{10} 360° = 324°.$$

13. Alternating-current Ampere.—Figure 17(a) shows an a-c sine wave, having a maximum value of 1.414 amp. At first thought, it might seem that the value in amperes of such a wave should be based on the *average* value. If the wave over one complete cycle is considered, the average value is zero, as there is just as much negative as positive current. A d-c ammeter, if connected to measure this current, would indicate zero, as such an instrument reads *average* value.

The value of an alternating current is based, not on its average value, but on its *heating* effect, and may be defined as follows:

An a-c ampere is that current which, flowing through a given ohmic resistance, will produce heat at the same rate as a d-c ampere.

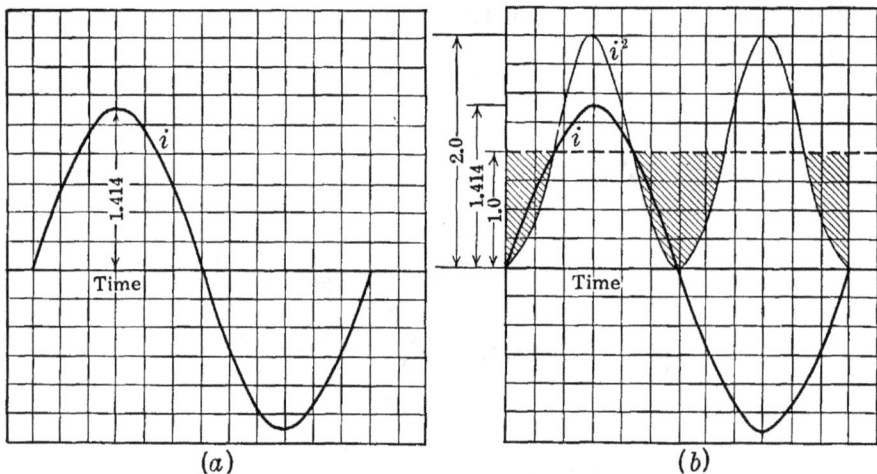

Fig. 17.—Maximum and root-mean-square values of sine-wave alternating current.

Assume that a resistance unit is immersed in a calorimeter and that when a d-c ampere flows through this resistance the temperature of the water is raised 20° in 10 min. An a-c ampere, when flowing through this same resistance unit, will raise the temperature of the water by the same amount in the same time, other conditions such as radiation being the same. That is, both currents produce heat at the same rate.

The heating effect varies as the *square* of the current (i^2R). Therefore, the value in amperes of the wave of current in Fig. 17(*a*) must be based on the *squared* values of its ordinates. Figure 17(*b*) shows the current wave of Fig. 17 (*a*) plotted, together with its squared values. That is, each ordinate of the *i*-wave is squared, and these values are plotted to give the i^2-wave shown. The maximum value of this new wave will be 2.0 ($= \overline{1.414}^2$), since the maximum value of the original current wave is 1.414, or $\sqrt{2}$ amp. The squared wave also lies entirely above the zero axis, for the square of a negative value is positive.

This squared wave has a frequency twice that of the original wave and has its horizontal axis of symmetry at a distance of 1.0 unit above the zero axis, as shown in Fig. 17(b). The average value of this squared wave is 1.0 amp, as shown by the dotted line, because the areas above the dotted line will just fit into the shaded valleys below the dotted line. Therefore, if an equivalent rectangle were made from this wave, its height would be 1.0 unit. This value, 1.0, is the *average of the squares* of the ordinates of the current wave. Average heating varies as the average square of the current so that the heating effect of this current over one cycle is proportional to the area of this rectangle. A d-c ampere flowing through the same resistance would produce the same heating as this current for both have the same average squared value.

The current in Fig. 17(a), therefore, which has a maximum value of 1.414 amp., has a heating value of 1.0 *amp*. This value of the current is called the *root-mean-square* (rms) or *effective* value of the current. The ordinary a-c ammeter indicates this rms, or effective, value of current. Therefore, an a-c ampere, sine wave, which produces heat at the same rate as a d-c ampere, has a *maximum* value of 1.414 ($= \sqrt{2}$) amp. In fact, for any sine-wave current, the ratio of the *maximum* to the *rms* value is equal to the $\sqrt{2}$, or 1.414. The ratio of rms to maximum value is $1/1.414 = 0.707$.

Example.—An alternating current has a maximum instantaneous value of 57 amp. What will an a-c ammeter read when connected in the circuit?

$$\frac{57}{1.414} = 40.3 \text{ amp.} \quad Ans.$$

14. Graphical Determination of Root-mean-square Current.— The method of obtaining the rms value of a current, whether or not it varies sinusoidally with time, is shown in Fig. 18. The curve *oab* is a sine wave of current having a maximum value of 15 amp. Ordinates such as *efg* and *hac* are erected at regular intervals over one-half a cycle. The value of each ordinate, in amperes, of the curve *oab* is squared. For example, the ordinate *ef* to scale is 13 amp; hence, the ordinate *eg* is 169 amp squared. The squared curve *ogcb* is plotted as shown. Its maximum ordinate obviously must be $\overline{15}^2$ or 225 amp squared. The average value of curve *ogcb* is to be found. There are two simple methods of doing this. The values of several equally spaced ordinates are added, and the sum is divided by the number of ordinates. This gives the average value of the squared curve if a sufficient number of ordinates be used.

The area in square inches under curve *ogcbo* may be found with a planimeter. This area divided by the base *ob* in inches gives

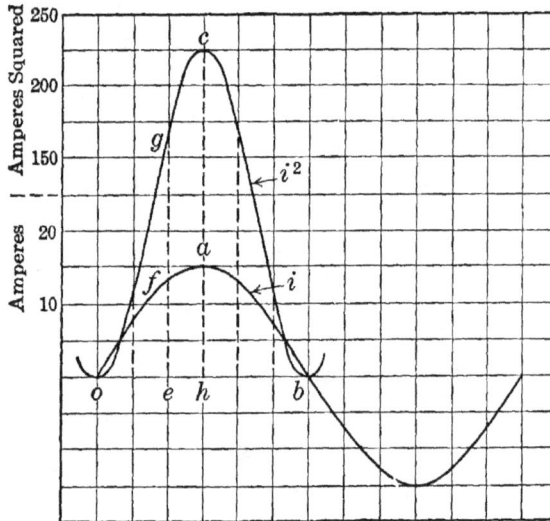

FIG. 18.—Determination of root-mean-square current.

the average height of the curve in *inches*. To find the average value in amperes squared, multiply this average height in inches by the scale of ordinates in amperes squared per inch. For example, assume that area *ogcbo* is 1.68 sq. in., the base is 1.5 in., and the scale of ordinates for the curve is 100 amp squared to the inch. Its average value then becomes

$$\text{Av. } i^2 = \frac{1.68}{1.5} 100 = 112.5 \text{ amp squared.} \qquad (I)$$

Therefore, this average squared current over a half-cycle will produce the same heating in a given resistance as a direct current whose squared value is 112.5 amp². Hence, by definition, the rms alternating current is $\sqrt{112.5} = 10.61$ amp.

The ratio of maximum to rms current is 15/10.61 = 1.414. For simplicity, the foregoing method is given for a sine wave of current, but the method is applicable if the current varies other than sinusoidally with time. Ordinarily, with nonsinusoidal currents, the ratio of the maximum to the effective value is not $\sqrt{2}$.

It will be noted from (I) and also from Fig. 17 that the average squared wave is one-half the square of the maximum value. For example, $112.5 = {}^{225}\!/_2$, or $\overline{15^2}/2$.

15. Average Current or Voltage.—Although for a sine wave the average current or voltage for one cycle is zero, the average for one alternation is obviously not zero. To find the average current for one alternation (Fig. 18), it is only necessary to find the average value of the curve *oab* in amperes. This may be accomplished either by averaging ordinates, or by finding the area, dividing by the base, and multiplying by the ordinate scale. For example, assume the area bounded by *oabo* (Fig. 18) is 0.717 sq in, the base *ob* is 1.5 in., and the scale is 20 amp to the inch. The average current is therefore $(0.717/1.5)20 = 9.56$ amp.

With a sine wave the *average* value is $2/\pi$, or 0.637, times the *maximum* value. Hence, the ratio of the rms value to the average value is $(1/\sqrt{2})/(2/\pi) = 1.11$. The ratio of rms to average value is called the *form factor* of the wave, being 1.11 for a sine wave.

The ratio of average to rms value is 1/1.11, or 0.9.

For example, with the foregoing current, whose maximum value is 15 amp, the average value for a half-cycle is $(2/\pi)15$, or 9.56 amp. The ratio of rms to average current is $10.61/9.56 = 1.11$.

16. Ohm; Volt.—If a resistance of one ohm, as measured with direct current, has no inductance and is so designed that alternating current in flowing through it does not produce any secondary effects, such as eddy currents, or skin effect, it offers a resistance of one *ohm* to alternating current.

When an a-c ampere flows through such a resistance, the drop across its terminals is equal to one a-c *volt*. Hence, the relation between *maximum* and *effective* volts is the same as the relation between *maximum* and *effective* amperes. For a sine wave, the maximum voltage is $\sqrt{2}$, or 1.414, times the rms voltage.

17. Phase Relations.—The current and voltage in the ordinary
a-c system have the same fundamental frequency under normal
operating conditions, although they do not necessarily pass
through their corresponding zero values at the same instant.
Figure 19(*a*) shows two sine-wave currents, one having an rms
value of 8 and the other of 12 amp. Their maximum values
are accordingly 8 $\sqrt{2}$, or 11.3, amp and 12 $\sqrt{2}$, or 17.0, amp,

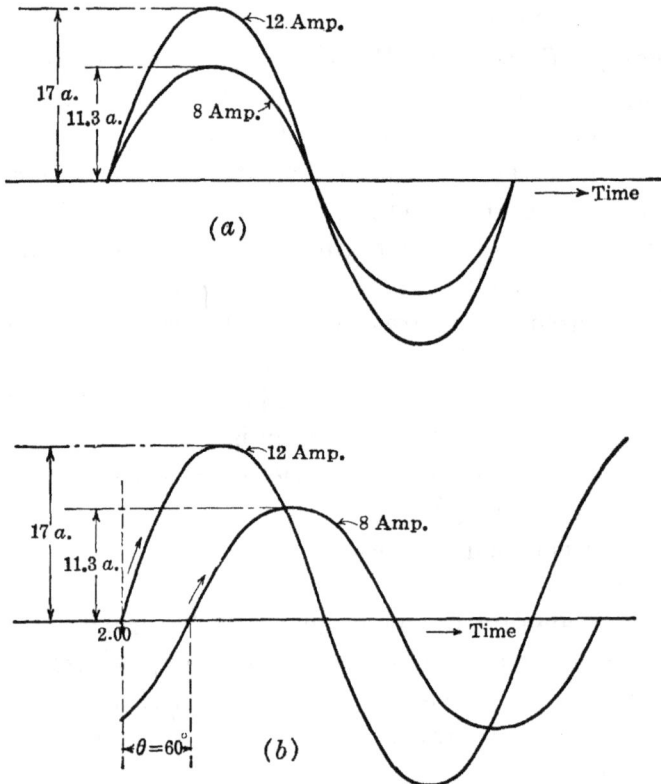

(*a*)

(*b*)

Fig. 19.—Phase relations between alternating currents.

respectively. Both currents pass through zero, increasing positively, at the same instant and are therefore said to be *in phase*
with each other.

Figure 19(*b*) shows two sine-wave currents of 8 and 12 amp
but not passing through zero at the same instant. The 8-amp
current passes through zero, increasing positively, later than
the 12-amp current, for time is increasing from left to right. If
the 12-amp current is passing through its zero value at two

o'clock, the 8-amp current is passing through its corresponding zero value some time later, for any value of time to the right of 2.00 is later than two o'clock. Therefore, the 8-amp current *lags* the 12-amp current.

The time lag shown in Fig. 19(*b*) corresponds to 60° and is represented by the angle θ. Therefore, the 8-amp current lags the 12-amp current by an angle θ, or by 60°. Or the 12-amp current may be said to *lead* the 8-amp current by an angle θ, or by 60°.

In Fig. 19(*a*) the two currents are *in phase* with each other. In Fig. 19(*b*) the two currents have a *phase difference* of 60°.

These phase differences may exist between currents and voltages, between two or more voltages, or between two or more currents.

18. Scalars and Vectors.—Quantities in general are classified as scalars and vectors.

A scalar is a quantity which is completely determined by its magnitude. Examples of scalar quantities are energy, gallons, mass, temperature, and time. Such quantities are added algebraically. For example, 2 gal plus 5 gal equals 7 gal.

A vector has *direction* as well as magnitude. A common example of a vector is force. When a force is under consideration, not only its magnitude but its direction also must be considered. When two or more forces are added, they are not necessarily added algebraically but must be combined in such a way as to take into consideration their directions as well as their magnitudes.

To find the relations existing among currents and voltages in d-c circuits, a knowledge of scalar quantities only is required. To find the relations existing among currents and voltages in a-c circuits a knowledge of vector quantities is required.

Figure 20(*a*) shows two forces acting at the point *O* and represented by the vectors F_1 and F_2. The length of each of these vectors, to scale, is equal to the *magnitude* of the force which it represents. The direction of each of these vectors shows the *direction* in which the force acts. β is the angle between F_1 and F_2. Their sum F_0, or the single force which would have the same effect on their point of application *O* as F_1 and F_2 acting in conjunction, is called their *resultant*. F_0 is one diagonal of the parallelogram having F_1 and F_2 as adjacent sides.

Figure 20(*b*) shows a triangle having F_1 and F_2 as two of its sides, F_1 and F_2 being parallel to and acting in the same directions as F_1 and F_2 of Fig. 20(*a*). The exterior angle between F_1 and F_2 is therefore equal to β. The third side of the triangle F_0 has the same magnitude and direction as F_0 of Fig. 20(*a*). Therefore, the resultant of two vectors may be found by means

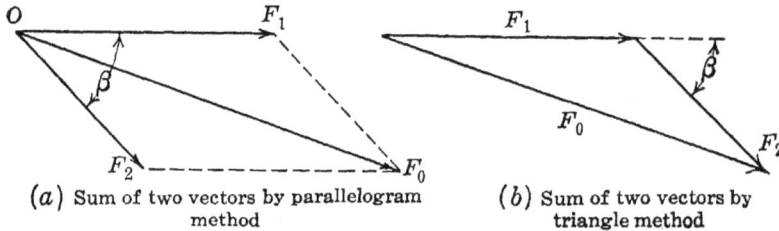

(*a*) Sum of two vectors by parallelogram method

(*b*) Sum of two vectors by triangle method

FIG. 20.—Addition of vector quantities.

of a triangle properly constructed, of which two sides are the two component vectors and the third side is their sum. Such a triangle is called a *triangle of forces.* It is usually simpler to use the triangle of forces than to use the parallelogram of forces.

To subtract one vector from another, reverse the first vector and add it vectorially to the second vector. For example, in

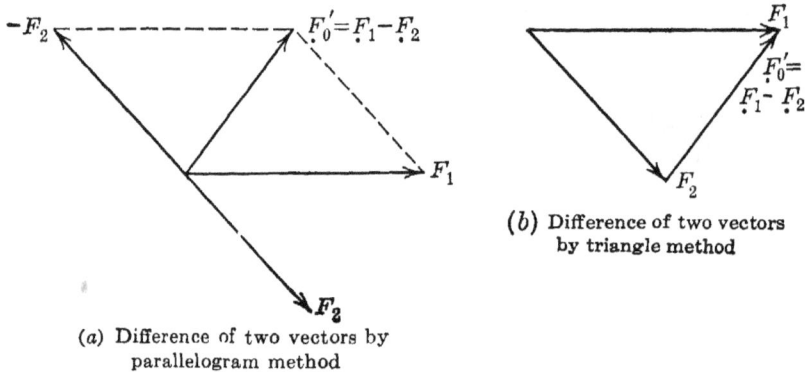

(*a*) Difference of two vectors by parallelogram method

(*b*) Difference of two vectors by triangle method

FIG. 21.—Subtraction of vector quantities.

Fig. 21(*a*), it is desired to subtract F_2 from F_1. F_2 is reversed, giving $-F_2$. F'_0, the vector sum of F_1 and $-F_2$, found by completing the parallelogram, is equal to $F_1 - F_2$. Vectors may be subtracted by the triangle method as shown in Fig. 21(*b*). The vector F'_0, connecting the ends of the two vectors F_1 and F_2 whose difference is desired, is their vector *difference.*

If a parallelogram (Fig. 22), having vectors F_1 and F_2 as adjacent sides, be completed, one diagonal F_0 of the parallelogram is the vector *sum* of F_1 and F_2. The other diagonal F'_0 of the parallelogram is the vector *difference* of F_1 and F_2.

A vector is often indicated by placing a dot under its symbol. For example, in Fig. 22

$$\dot{F}_0 = \dot{F}_1 - \dot{F}_2$$

shows that F_0 is the *vector* sum of F_1 and F_2 and not their algebraic or scalar sum.

When more than two vectors are added, the resultant of two is first found, this resultant is combined with the third vector,

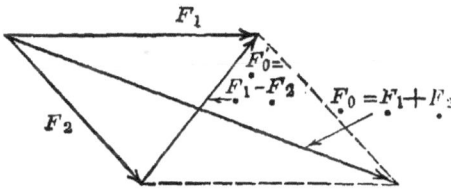

Fig. 22.—Sum and difference of two vectors.

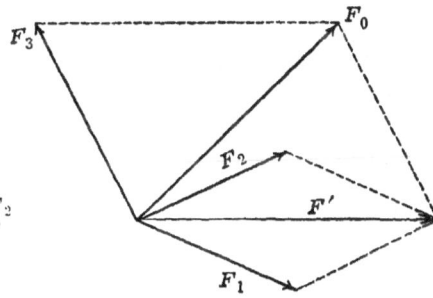

Fig. 23.—Sum of three vectors.

etc. This is illustrated in Fig. 23, in which three vectors F_1, F_2, F_3 are added.

F_1 and F_2 are first combined, and the resultant F' is found. F' is then combined with F_3, giving F_0 as the sum of all three vectors, F_1, F_2, F_3. That is,

$$\dot{F}_0 = \dot{F}_1 + \dot{F}_2 + \dot{F}_3.$$

F' is an intermediate vector and therefore does not appear in the final result.

19. Vector Representation of Alternating Quantities.—It is shown in Fig. 5 (p. 7) that a sine wave can be drawn by projecting from the ends of successive radii to meet corresponding equally spaced ordinates. Thus, the end of radius 1 is projected to meet ordinate 1-1' at 1' to give the ordinate of the sine curve at the angle 01.

It follows that, if the values of current and voltage are represented by sine curves, the value of the current or voltage may be found at any instant by projecting a rotating radius upon a vertical line.

This is illustrated in Fig. 24. A current has a maximum value I_m. This value I_m is laid off as a radius. As I_m rotates in a counterclockwise direction, values of the sine curve are found by projecting I_m in its successive positions corresponding to 1, 2, 3, etc. (Fig. 5). As one cycle is completed for every revolution of the radius I_m, it follows that the radius I_m rotates at a speed in rps equal to the frequency of the current in cycles per

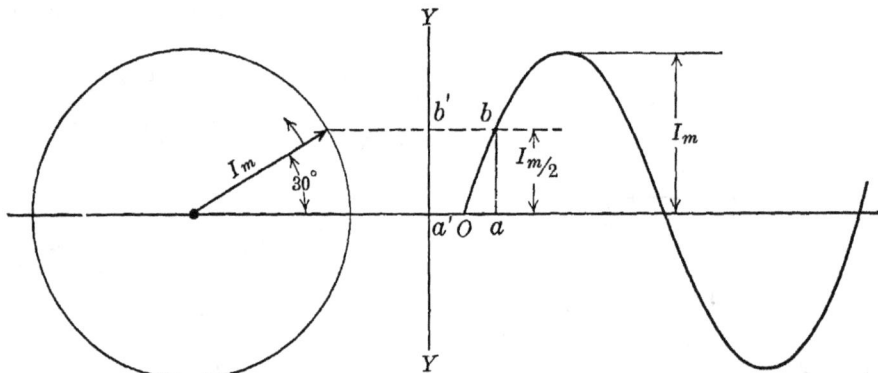

Fig. 24.—Instantaneous values of current from rotating vector.

second. For example, if the current I_m has a frequency of 60 cycles, the radius I_m must rotate at a speed of 60 rps in a counterclockwise direction. Counterclockwise rotation has been adopted internationally as the positive direction of rotation.

When the radius I_m is at the right-hand horizontal position, the value of the current is zero. When I_m has advanced 30°, the point b on the current wave has been reached. The value of the current at this instant is ab. The value of the current is given also by the distance $a'b'$, the projection of I_m upon the *vertical* axis Y-Y. At this particular instant, $ab = a'b' = 0.5I_m$, since $\sin 30° = 0.5$.

The value at any instant of a sine wave of alternating current or voltage is given, therefore, by the projection on a vertical line of a vector equal in length to the maximum value of the current or voltage and rotating at a speed in rps equal to the frequency of the current or voltage. The angular velocity ω of the vector in radians per second is equal to 2π times the frequency.

20. Vector Representation with Phase Difference.—It is required to represent two currents I_1 and I_2, having maximum values I'_1 and I'_2, differing in phase by 60°, I_2 lagging I_1 by 60° as shown by the sine curves (Fig. 25). Assume that I_1 has an *rms* value of 12.0 amp and I_2 an *rms* value of 8.0 amp. The maximum value of I_1 will be $12.0 \sqrt{2} = 17.0$ amp, and the maximum value of I_2 will be $8.0 \sqrt{2} = 11.3$ amp.

Assume that I_1 has its zero value and is increasing in a positive direction when the time is zero. It may be considered as being generated by the radius vector I'_1 rotating in a counterclockwise

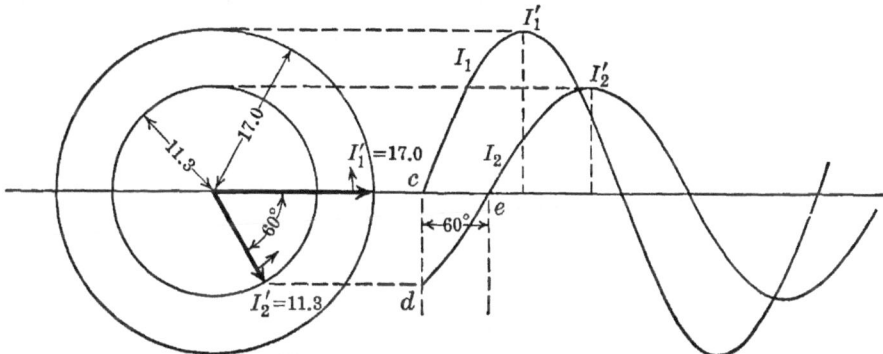

FIG. 25.—Current waves produced by two rotating current vectors differing in phase by 60°.

direction and making an angle of 0° with the horizontal axis when the time is zero.

Since I_2 lags I_1 by 60°, I_2 will not be going through its corresponding zero value until the radius vector I'_1 has advanced 60° in a counterclockwise direction. Therefore, I_2 will be generated by the radius vector I'_2, making an angle of 60° with I'_1 in a clockwise direction.

At the instant when the radius I'_1 is in the horizontal position, the value of I_1 is zero. At this same instant, the radius I'_2 will not have reached its horizontal position, the value of the current being represented by cd (Fig. 25). In fact, the radius I'_2 does not reach its horizontal or zero position until I'_1 has advanced 60° beyond the horizontal. Further, the horizontal distance ce is 60°, the same as the phase angle between the two rotating vectors.

Therefore, these two current curves can be constructed in their proper phase relation by means of two rotating vectors, having lengths of 17.0 and 11.3 amp, having equal angular velocities, and differing in phase by 60° (Fig. 25).

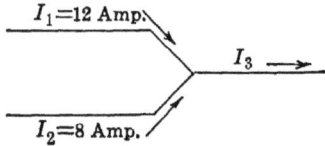

I_1=12 Amp.

$I_3 \longrightarrow$

I_2=8 Amp.

Fig. 26.—Alternating currents combining at junction.

21. Addition of Inphase Currents.—Figure 26 shows two currents I_1 and I_2, having *rms* values of 12 and 8 amp, uniting to flow in a common wire. If these two currents were direct currents, then by Kirchhoff's first law (see Part I, p. 50), their resultant, the current I_3, could have only two possible numerical values, 12 + 8, or 20 amp if the two currents flow in the same direction and 12 − 8, or 4 amp if they flow in opposite directions.

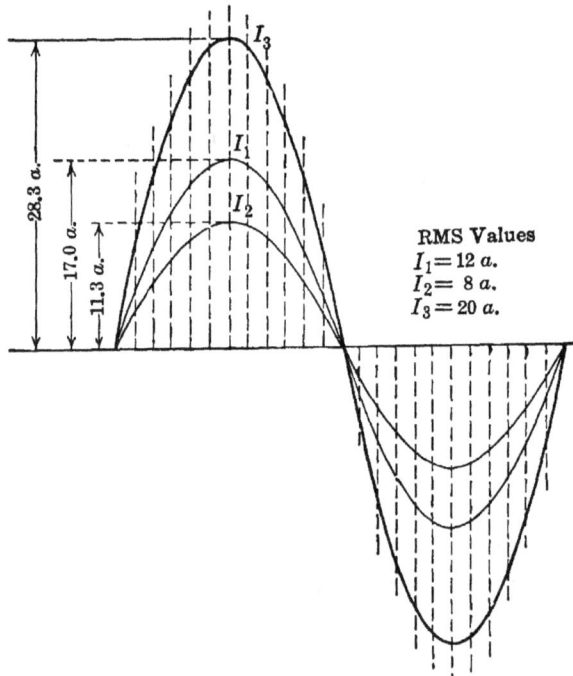

RMS Values
I_1= 12 a.
I_2= 8 a.
I_3= 20 a.

Fig. 27.—Addition of two currents in phase.

If the two currents (Fig. 26) are alternating, their sum I_3 may be equal numerically to *any* value *from* 20 to 4 *amp*, depending on the phase relation existing between I_1 and I_2.

Figure 27 shows the instantaneous values of these two currents plotted over one cycle with the currents in phase. The 8-amp current has a maximum value of 8 $\sqrt{2}$, or 11.3 amp, and the 12-amp current has a maximum value of 12 $\sqrt{2}$, or 17.0 amp. These values are shown in Fig. 27. Since both these currents have the same frequency, hence the same scale of abscissas, the curve found by adding their ordinates at each point will be a sine curve (see Par. 5, p. 9). The resulting current curve I_3 will have a maximum ordinate of 17.0 + 11.3, or 28.3 amp, and its zero points will coincide with those of the two component curves. The rms value of the resultant current I_3 is 28.3/$\sqrt{2}$, or 20.0 amp.

Hence, if two alternating currents of the same frequency are in phase, their sum is merely the algebraic sum of the component currents. This is also true of voltages.

Example.—In Fig. 28 is shown a 5-kva (see p. 252), 2,400/240-volt, 60-cycle transformer. There are two secondary coils, the voltage across each of which is 120 volts. As connected in Fig. 28, these two voltages are in phase with each other. What is the voltage across their open ends *ab*?

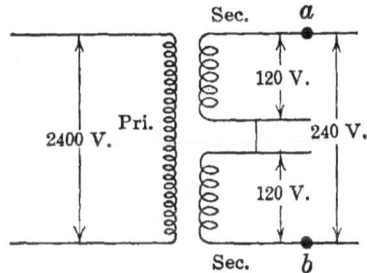

Fig. 28.—5-kva., 2,400/240-volt transformer windings.

$$E_{ab} = 120 + 120 = 240 \text{ volts.} \quad Ans.$$

22. Addition of Currents Differing in Phase.—Let it be required to add the two currents having *rms* values of I_1 and I_2 (Fig. 27) when they differ in phase by 60°, with I_2 lagging (also see Figs. 19 and 25). Let the current I_1 be determined graphically by the rotating vector I'_1 (Fig. 29), whose maximum value is 12.0 $\sqrt{2}$, or 17.0, amp. When time is zero, the current I_1 is assumed to have its zero value and to be increasing positively. Hence, at this instant vector I'_1 is in the horizontal position, as shown. Since I_2 lags I_1 by 60°, it is determined graphically by the rotating vector I'_2, whose maximum value is 8.0 $\sqrt{2}$, or 11.3 amp and which lags I'_1 by 60°.

The sum of the two currents in Fig. 29 is found as for Fig. 27. Ordinates are erected and points on the resultant curve are found by adding algebraically at each abscissa the ordinates of the two currents. This same procedure was also followed in Fig. 8 (p. 9).

As a result of this addition the resultant curve I_3 is found. I_3 lags I_1 by an angle α and leads I_2 by an angle $60°-\alpha$. If the maximum value of I_3 is determined by measurement, it is found to be 24.7 amp.

This corresponds to an rms value of $24.7/\sqrt{2}$, or 17.45, amp. Therefore, the sum of two sine-wave alternating currents, having rms values of 12 and 8 amp and differing in phase by 60°, is 17.45 amp.

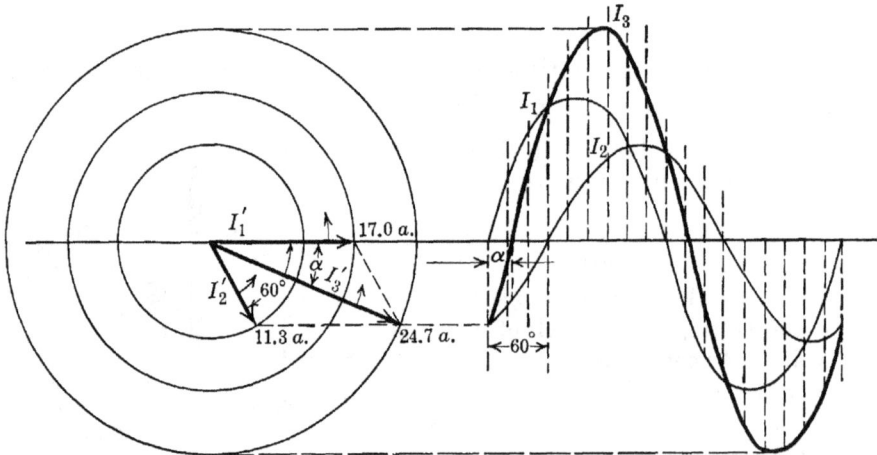

Fig. 29.—Relation of vector addition of vectors to scalar addition of ordinates.

If the rotating vectors I'_1 and I'_2 (Fig. 29) are added vectorially by completing the parallelogram, the resultant vector I'_3 results. This vector I'_3 has a length corresponding to 24.7 amp, the same value of the maximum of the resultant current as found by the addition of ordinates. If a curve is plotted using I'_3 as the rotating vector, projecting as before, it will coincide with I_3 obtained by the addition of ordinates for the 12- and 8-amp waves. The angle α by which the radius vector I'_1 leads I'_3 equals the angle α by which the current curve I_1 leads the current curve I_3.

Hence, the problem can be solved without going through the somewhat tedious process of plotting the curves and adding their ordinates. It is merely necessary to lay off the vectors corresponding to the maximum values of the currents, 60° apart, and add them vectorially, just as forces are combined. The resultant vector will be the maximum value of the curve obtained by adding the curves I_1 and I_2.

In practice, one generally deals with rms values such as are read on ammeters and voltmeters rather than with maximum values. If the rms values of the currents are added in this same manner, their vector sum is the sum of the two alternating currents in rms amperes, since adding rms values merely means adding the vectors of the parallelogram (Fig. 29), reduced to $1/\sqrt{2}$ times the values shown.

This is illustrated in Fig. 30, where the 12- and 8-amp vectors (rms values) are laid off 60° apart, the 12-amp vector leading. The scale (Fig. 30) is considerably larger than that used in Fig. 29. By completing the parallelogram, the resultant current Oc is obtained. This has a value of 17.45 amp. Its value is readily found as follows:

Project ac upon Ocb, where $ac = 8$.

$$ab = ac \cos 60° = 4.00$$
$$bc = ac \sin 60° = 6.93$$
$$Oc = \sqrt{(12 + 4.00)^2 + (6.93)^2} = 17.45 \text{ amp.} \quad \textit{Ans.}$$

The angle α can be readily determined.

$$\tan \alpha = \frac{6.93}{12 + 4} = 0.433.$$
$$\alpha = 23.4°$$

The significance of combining the two currents I_1 and I_2 to give I_3 (Figs. 26, 29, 30) is illustrated in Fig. 31. Two alternators 1 and 2 are connected in parallel to supply current to a load. Alternator 1 supplies an rms current of 12 amp, and alternator 2 supplies an rms current of 8 amp. That is, an ammeter connected in the armature lead from 1 indicates 12 amp, and

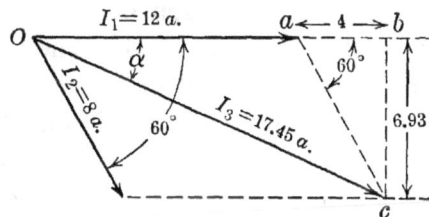

FIG. 30.—Vector addition of currents, using rms values.

an ammeter connected in the armature lead of 2 indicates 8.0 amp. The alternators are so adjusted that the 12-amp current leads the 8-amp current by 60°.

It is required to find the rms current I_3 taken by the load and the values of current taken by the load at various instants.

The rms current I_3 taken by the load is 17.45 amp, lagging I_1 by 23.4°, as just determined.

In Fig. 31(a) the current from either alternator is positive when it flows from the upper terminal. The conditions existing when the time is zero are shown in Fig. 31(b). The value of I_1 is zero at this instant and the value of I_2 is aa', or -9.80 amp (11.3×-0.866). Since 1 delivers no current at this instant, 2 must be supplying all the load current of -9.80 amp. At this instant the load current is negative, or current is returning to the positive terminal of 2. During the time represented by the distance ab (Fig. 31(b)), alternator 2 is supplying negative current

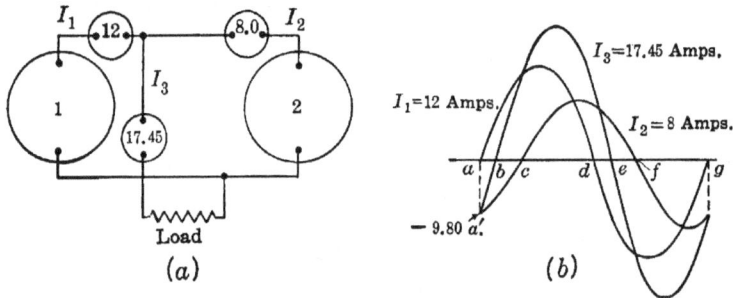

FIG. 31.—Alternators with currents out of phase supplying a common load.

to the load and alternator 1 is supplying positive current. That is, the two alternators are in opposition during this period. More negative current is supplied by 2 than positive current by 1, and the load current in this period is negative. At the instant b, the positive current supplied by 1 is equal to the negative current supplied by 2. Hence, the load current is zero at this instant. Actually, current equal to the value of the ordinate at b is leaving the upper terminal of 1 and is entering the upper terminal of 2 but is not flowing to the load. That is, current merely circulates between the two alternators. In the interval bc the positive current delivered by 1 exceeds the negative current delivered by 2 and positive current flows to the load. In the interval represented by cd, both alternators act in conjunction to deliver positive current to the load. In the interval de, 2 delivers positive current and 1 delivers negative current, but the positive current exceeds the negative current and the net load current I_3 is positive. At the instant *e* the positive current delivered by 2 is just equal to the negative current delivered by 1, so that the net load current is zero. That is, at the instant *e*, current is leaving the upper, or positive, terminal of 2 and entering the upper terminal of 1 but is not flowing to the load. In the interval *ef* the negative current delivered by 1 exceeds the positive current delivered by 2, and the load current I_3 is negative. In the interval *fg*, both alternators act in conjunction to deliver negative current to the load.

Thus, over the cycle the current at some instants is flowing merely between alternators (at *b* and *e*); at other instants, either one alternator or the other is supplying the entire load (at *a, c, d, f, g*); during some periods the two alternators are acting in conjunction to supply the load current (*cd, fg*); at other periods, they are acting in opposition to supply the load current, the net load current being the difference of the alternator currents (*ac, df*). The net effect over the cycle is, however, that an ammeter connected in alternator 1 reads 12.0 amp, an ammeter connected in alternator 2 reads 8.0 amp, and an ammeter connected in the load circuit reads 17.45 amp.

This example illustrates the fact that, with two alternators in parallel, the load current may be less than the algebraic sum of the two alternator currents. The load current is, however, equal to the *vector* sum of the two alternator currents.

From the foregoing, it follows that *the sum of any number of alternating currents depends on their phase relations as well as on their magnitudes.*

The vector addition of currents is further illustrated by the following example:

Example.—A lamp load (Fig. 32(*a*)) takes 9.0 amp (rms) from 110-volt (rms), 60-cycle mains and a single-phase motor takes 12.0 amp (rms). The current taken by the lamps is in phase with the line voltage, and the current taken by the motor lags the line voltage by 45°. Find the resultant current I and the angle θ by which it lags the line voltage.

The instantaneous values of current and voltage are plotted in Fig. 32(*b*), the 9.0-amp current being in phase with the voltage and the 12.0-amp current lagging the voltage by 45°. The currents and voltage are shown vectorially in (*c*), a different scale from that in (*b*) being used. The resultant current I is the diagonal of the parallelogram having I_1 and I_2 as adjacent sides. The value of I is determined as follows: I is projected on I_1 extended.

$$db = dc \cos 45° = 12.0 \times 0.707 = 8.48 \text{ amp.}$$
$$bc = dc \sin 45° = 12.0 \times 0.707 = 8.48 \text{ amp.}$$
$$I = \sqrt{(Od + db)^2 + (bc)^2}$$
$$= \sqrt{(9.0 + 8.48)^2 + (8.48)^2} = 19.43 \text{ amp.} \quad Ans.$$
$$\tan \theta = \frac{bc}{Ob} = \frac{8.48}{17.48} = 0.485.$$
$$\theta = 25.9°. \quad Ans.$$

23. Addition of Voltages.—Since alternating voltages vary sinusoidally with time in the same manner as currents, they are

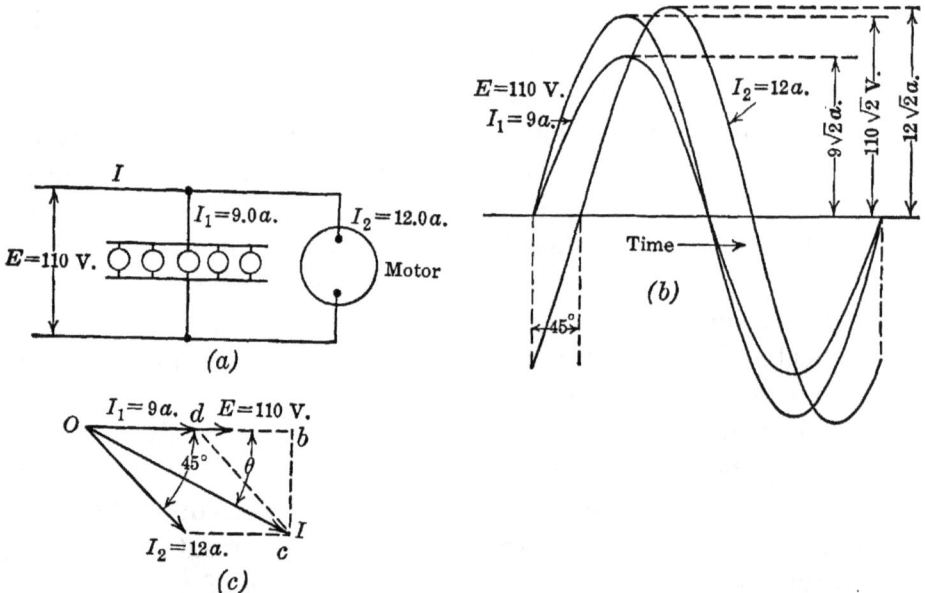

Fig. 32.—Vector addition of currents having 45° phase difference.

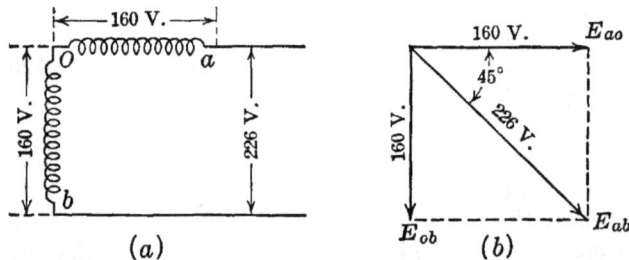

Fig. 33.—Vector addition of two equal voltages having 90° phase difference.

added in exactly the same manner as currents. This is illustrated by the following example:

Example.—Each of two alternator coils Oa and Ob (Fig. 33(a)) is generating an emf of 160 volts. These emfs differ in phase by 90°. Determine the emf across their open ends if they are connected together at O as shown.

Let the vectors E_{ao} and E_{ob}, differing in phase by 90° (Fig. 33(b)), represent the emfs across coils aO and Ob. Combining E_{ao} and E_{ob} vectorially, the emf E_{ab} is obtained. That is, $E_{ab} = E_{ao} + E_{ob}$. As E_{ao} and E_{ob} are at right angles, their resultant is readily found.

$$E_{ab} = \sqrt{E_{ao}^2 + E_{ob}^2} = \sqrt{160^2 + 160^2} = 226 \text{ volts.} \quad Ans.$$

It must be kept constantly in mind that alternating voltages and currents must be combined vectorially.

The only occasion when arithmetical addition is permissible is when voltages or currents are in phase.

CHAPTER II

SINGLE-PHASE ALTERNATING-CURRENT CIRCUITS

24. Circuit Containing Resistance Only.—The simplest a-c circuit is one which contains resistance only. In such a circuit the only voltages involved are the impressed emf and the ohmic drop.

Figure 34 shows a resistance R connected across the terminals of an alternator. An alternating current i, varying sinusoidally with time, flows through this resistance. Let the maximum instantaneous value of the current be I_m amp. The equation of the current is

$$i = I_m \sin \omega t = I_m \sin 2\pi f t, \tag{I}$$

where f is the frequency in cycles per second.

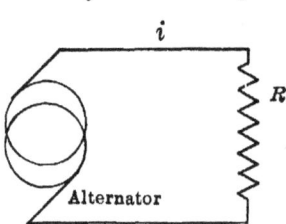

The current, having a maximum value I_m, is shown plotted for one cycle (Fig. 35). From the definition of a-c voltage (see p. 25), the impressed voltage must be utilized wholly in sending the current through the resistance R. Therefore, the impressed voltage must be equal to the resistance drop. The circuit voltage e at *any* instant is equal therefore to the iR drop.

Fig. 34.—Resistance load connected to alternator.

Thus,

$$
\begin{aligned}
e &= iR \\
&= I_m R \sin \omega t \text{ (from (I))}.
\end{aligned}
\tag{5}
$$

It is obvious that the voltage wave e is *in phase* with the current wave i; for when $t = 0$, $\sin \omega t = 0$ and both the current and the voltage are simultaneously going through their zero values and increasing positively. When $\omega t = \pi/2$ or $90°$, $\sin \omega t = 1$, and both the current and the voltage are simultaneously at their maximum positive values. The voltage curve e, in phase with the current curve i, is also shown in Fig. 35.

Since both current and voltage go through their zero, maximum, and other corresponding values together, the radius vectors which determine their curves must be in phase with each other. These radius vectors are equal to the maximum values of the current and voltage (I_m and E_m) as shown in Fig. 35. (Also see Fig. 25, p. 31.) If maximum values are considered, the vector diagram of the circuit is given by the two rotating vectors in

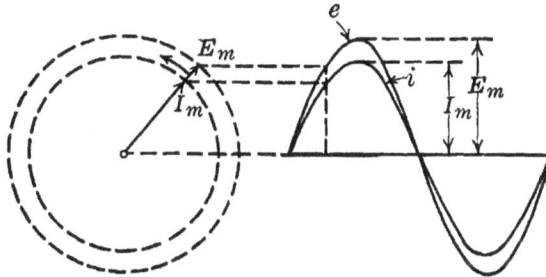

Fig. 35.—Instantaneous and vector values of current and voltage in phase.

Fig. 35. That is, the current and voltage are *in phase* with each other.

It is obvious that, if each of these vectors is reduced in the ratio of 1 to $\sqrt{2}$, it will represent *rms* rather than maximum value. The vector diagram of the circuit when rms values are considered will then be given. This vector diagram is shown in Fig. 36(*a*). The current and voltage vectors are shown arbitrarily as horizontal. They may be shown in any position, as in (*b*), provided they are in phase with each other.

(a)   (b)

Fig. 36.—Vector diagram with current and voltage in phase.

From Eq. (5) the maximum value of the current and of the voltage occurs when $\omega t = 90°$ and $\sin \omega t = 1$. At this instant,

$$E_m = I_m R.$$

Dividing both E_m and I_m by $\sqrt{2}$ gives their *rms* values.

$$\frac{E_m}{\sqrt{2}} = \frac{I_m R}{\sqrt{2}} \quad \text{or} \quad E = IR, \tag{6}$$

where E and I are *rms* values of voltage and current. It also follows from Eq. (6) that

$$I = \frac{E}{R}, \tag{7}$$

which is Ohm's law.

With resistance only in the circuit, the current equals the impressed voltage divided by the resistance. That is, the current obeys Ohm's law just as for a d-c circuit.

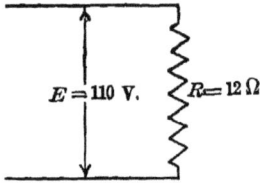

FIG. 37.—Resistance across 110 volts, alternating.

Example.—A 12-ohm, noninductive resistance is connected across a 110-volt (rms), 60-cycle circuit (Fig. 37). Determine: (a) current in the resistance; (b) maximum instantaneous value of voltage and current; (c) equation of impressed voltage; (d) equation of current.

From (6),

(a) $I = {}^{110}\!/_{12} = 9.17$ amp.   *Ans.*

(b) $E_m = 110 \sqrt{2} = 155.5$ volts.   *Ans.*

     $I_m = 9.17 \sqrt{2} = 12.97$ amp.   *Ans.*

(c) $e = 110 \sqrt{2} \sin 2\pi 60t = 155.5 \sin 377t$.   *Ans.*

(d) $i = 9.17 \sqrt{2} \sin 2\pi 60t = 12.97 \sin 377t$.   *Ans.*

25. Effect of Inductance on Alternating-current Flow.—Inductance in a circuit has no effect on the flow of direct current, *so long as the current is steady.* (See Part I, Par. 146, p. 174.) If, however, the current changes, inductance does have an effect. If the current increases, inductance opposes this increase. If the current decreases, inductance opposes this decrease.

Except at the instant when it has either its maximum positive or maximum negative value, an alternating current is changing in magnitude continuously. Hence, except at the instant when the current has its maximum positive or negative value, there must be an emf of self-induction in the a-c circuit. This emf, as well as the impressed voltage, is acting in the circuit and must have an effect on the value of current and accordingly must be taken into consideration. Hence, self-inductance has an effect on the flow of alternating current, even when an ammeter in the circuit shows that the current has a constant rms value.

This fact may be illustrated by a simple experiment. Connect a double-pole, double-throw (D.-P., D.-T.) switch (Fig. 38), so that one set of clips is connected to a 110-volt, d-c supply and the other set to a 110-volt (rms), 60-cycle, a-c supply. Across the blades of the switch, connect a lamp bank in series with an iron-cored inductance coil. This inductance coil may be made by winding 100 to 150 turns of No. 10 or 12 magnet wire over a cardboard or fiber tube, about 2.5 in. (6.35 cm) diameter. The iron core may be made by taping together either iron wire or sheet-steel strips. The core should be of such diameter that it almost fills the tube and slides readily in and out of the tube.

When the switch is thrown to the d-c side, the lamps will come to full brilliancy almost immediately. To be sure, a small

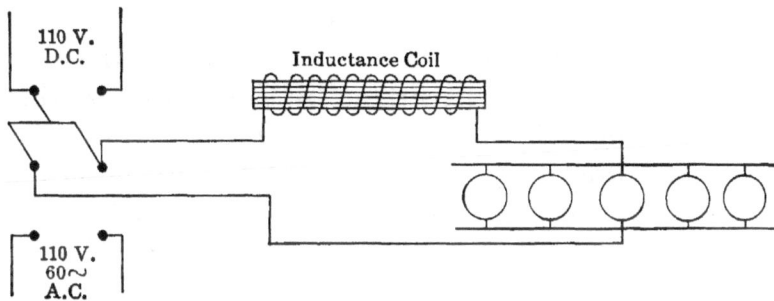

FIG. 38.—Effect of inductance on current flow.

time lag occurs, due to the momentary opposition of the emf of self-induction to the increase of current (see Fig. 39(a)). When the direct current reaches its steady value, aside from the small resistance drop in the winding of the inductance coil, the lamps will burn as brilliantly as if they were connected directly across the 110-volt mains.

A slight flicker may be produced in the lamps, however, by withdrawing suddenly the iron core, either completely or partly out of the coil. The flux linking the turns of the coil is thus suddenly reduced, and an emf is generated in the winding because of the change of flux. The direction of this emf is such as *to oppose* the core being withdrawn; hence, it is in conjunction with the line current which also opposes the core being withdrawn. Therefore, the lamps brighten during the short time that the core is being withdrawn (see Part I, p. 169).

If the core is suddenly thrust into the coil, the lamps are dimmed momentarily. The direction of the induced emf now *opposes* the core entering the coil, hence *opposes* the line current which tends to draw the core into the coil. Hence, the lamps are less bright momentarily.

Aside from these transient effects, however, the lamps burn equally brilliantly for all positions of the core, and they are practically as bright as if they were connected directly across the blades of the switch.

If the core is within the coil and the switch is thrown to the a-c side, the lamps will barely glow or may be dark. The same rms voltage is impressed across the circuit as with the direct current, and the circuit has practically the same *resistance* as before. Therefore, the large reduction in the current must be due to the choking effect of the inductance.

The fact that inductance alone is responsible for the dimming of the lamps may be demonstrated by slowly withdrawing the core from the coil. The lamps increase in brightness the farther the core is withdrawn. This is due to the decrease in the inductance caused by the reduction in the flux linking the turns of the coil. This change in brightness is not a momentary effect as with direct current.

If the frequency be lowered, to 25 cycles, for example, with a given position of the plunger, or core, the lamps burn more brilliantly, owing obviously to the lesser effect of the inductance because of the decreased rate of change of current and hence of flux.

It will be observed from these experiments that if inductance is introduced in a circuit, it reduces the rms value of the alternating current. It therefore has an important effect and must be taken into consideration, as well as resistance, in determining the value of current.

26. Effect of Inductance in Circuit.—From the preceding paragraph, it is clear that inductance must oppose the flow of alternating current. Because of the variation of the alternating current with time an emf is induced. This induced emf *opposes* the change of current. No such emf exists when the circuit contains resistance only. Therefore, in determining the current, this emf of self-induction must be taken into consideration.

Figure 39(*a*) shows the rise of current in a d-c circuit containing resistance and inductance in series, when a steady voltage is impressed (see Part I, p. 172, Fig. 153). The current rises slowly to its ultimate value. On the other hand, when the current begins to decrease in the circuit, the inductance tends to prevent this decrease (Fig. 39(*b*)) (see Part I, p. 173, Fig. 154). In other words, if inductance is present in a circuit, it always *opposes* any *change* in the current. However, with a steady direct current the inductance has no effect. When the current reaches the value *a* (Fig. 39(*a*)) if the circuit voltage is decreased to a small value or even is made zero, the current will no longer continue to increase but will begin to decrease as shown. Under these conditions, there is not sufficient time for the current to

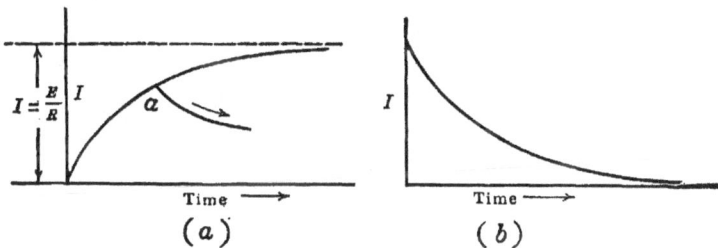

FIG. 39.—Increase and decrease of current in an inductive circuit.

reach its Ohm's law value, and the maximum value *a* which it does reach is accordingly less than the value which would be given by Ohm's law. If the voltage across a circuit increases and decreases with time and the voltage begins to decrease before the current has reached its Ohm's law value, the current is no longer given by the impressed voltage divided by the resistance, as with a steady direct current. This effect occurs in a-c circuits. The impressed voltage is varying sinusoidally with time. With inductance in the circuit, the current does not have time to reach its Ohm's law value either positively or negatively before the voltage is reversed in direction. The current change is opposed by the emf of self-induction, which at any instant is equal in volts to $-L(i/t)$ where L is the inductance in henrys and i/t is the rate in amperes per second at which the current is changing at that instant. The minus sign signifies that this voltage is opposing the change in the current (see Part I, Par. 143, p. 169).

27. Circuit Containing Inductance Only.—Figure 40 shows a sine wave of alternating current i. At a the current is *changing* at its maximum rate in a positive direction. Therefore, at this instant the emf of self-induction must be at its negative maximum value as is shown by the ordinate of the curve $-e$. At b, the current is at its maximum positive value. A tangent, such as b'', drawn to the curve at this instant, is horizontal; therefore, at this instant the current is not changing at all, and hence the emf of self-induction is zero, as is shown by the curve $-e$ crossing the zero axis. At c the current is changing at its maximum rate negatively, and the emf of self-induction must be maximum

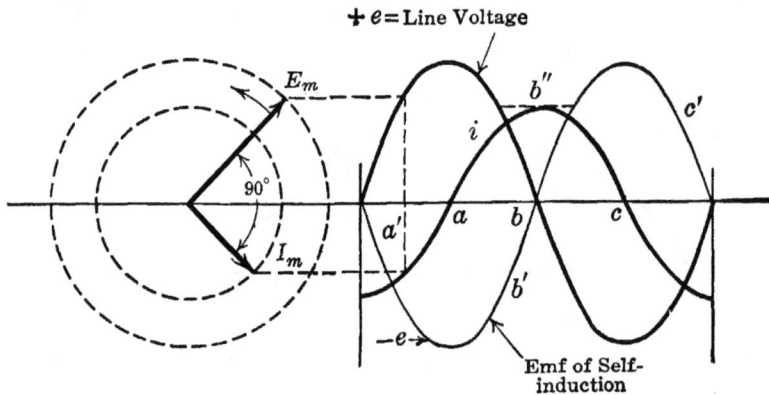

FIG. 40.—Current and emfs existing in an alternating-current circuit containing inductance only.

positive. Continuing in this manner the emf curve $a'b'c'$ is obtained. It will be noted that this is a sine curve and is *lagging* the current by 90°.

This is the only emf in the circuit which *opposes* the change of current. It corresponds to the counter emf of a motor in that the line must supply a voltage opposite and equal to the counter emf before any current can flow into the armature. This same condition exists in the a-c circuit. Before any current can flow into a circuit containing inductance but no resistance, a voltage opposite and equal to the emf of self-induction must be supplied by the line.

Therefore, in Fig. 40 the voltage $+e$, which is the line voltage, is opposite and equal to $-e$, the emf of self-induction.

It will be noted that the *impressed* voltage *leads* the current by 90°, or the current *lags* this voltage by 90°. With inductance only in the circuit, the current *lags* the impressed voltage by 90°. (In practice, it is impossible to obtain a pure inductance, for all inductances must be accompanied by some resistance.)

In Fig. 40 are shown the radius vectors E_m and I_m which determine the sine curves $+e$ and i. Since the impressed voltage e leads the current i by 90°, the radius vector E_m must lead the radius vector I_m by 90° as shown (see Fig. 25, p. 31). Therefore, the radius vectors E_m and I_m together give the vector diagram of this circuit if *maximum* values are under consideration. If a vector diagram involving *rms* values is desired, it is only necessary to reduce E_m and I_m in the ratio of 1 to $\sqrt{2}$. Such a diagram to a larger scale is shown in Fig. 41. In (*a*) the voltage

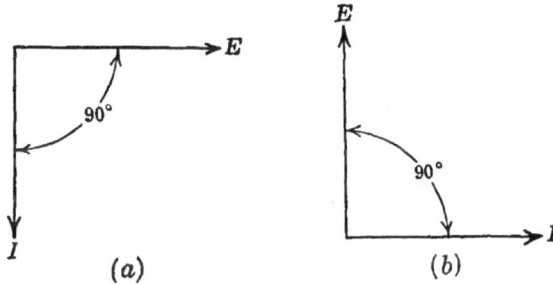

FIG. 41.—Vector diagram with inductance only in circuit.

vector E has been taken arbitrarily along the horizontal axis. The current vector I lags E by 90°. In (*b*) the vector diagram is shown with the vector I taken arbitrarily along the horizontal axis. Both diagrams are correct. In fact the choice of axes is determined by convenience.

It is possible to determine the choking effect of inductance quantitatively. The emf of self-induction is proportional to the *inductance* and to the *rate of change of current*. The rate of change of current is proportional to the *current* and to the *frequency* or to ω which is 2π times the frequency. It may be shown that the rms value of the emf of self-induction $-E$ in an a-c circuit is given by

$$(-E) = -L(2\pi f)I = -L\omega I \text{ volts,} \tag{8}$$

where L is the inductance in henrys and I the current in rms amperes.

With inductance only, the line voltage E must be opposite and equal to this emf of self-induction and is given by

$$E = 2\pi fLI = L\omega I = IX_L \text{ volts,} \qquad (9)$$

where

$$X_L = 2\pi fL \text{ ohms.} \qquad (10)$$

Solving for the current,

$$I = \frac{E}{2\pi fL} \text{ amp.} \qquad (11)$$

Since $2\pi f = \omega$,

$$I = \frac{E}{L\omega} = \frac{E}{X_L} \text{ amp,} \qquad (12)$$

where $2\pi fL = L\omega = X_L$ is the *inductive reactance* of the circuit in ohms. Since the current is equal to the voltage E divided by the inductive reactance X_L, the inductive reactance must be expressed in *ohms*.

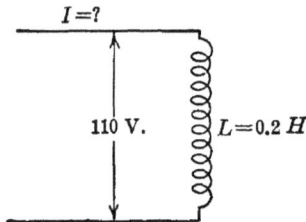

FIG. 42.—Circuit containing inductance only.

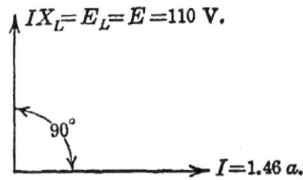

FIG. 43.—Vector diagram for circuit containing inductance only.

Example.—Figure 42 shows a pure inductance of 0.2 henry connected across 110-volt (rms), 60-cycle mains. Determine the current. From Eq. (10),

$$X_L = 2\pi 60 \times 0.2 = 75.4 \text{ ohms.}$$

Using Eq. (12),

$$I = \frac{110}{75.4} = 1.46 \text{ amp (rms).} \quad Ans.$$

Figure 43 shows the vector diagram of this circuit. In this diagram the current vector I is shown arbitrarily along the horizontal axis. The line voltage $E = IX_L = 110$ volts *leads* the current by 90°. Compare this diagram with Fig. 41(b).

Example.—A 25-cycle current of 4.0 amp flows through a pure inductance of 0.05 henry. What is the voltage across the inductance, and what is its phase relation to the current?
From Eq. (10),

$$X_L = 2\pi 25 \times 0.05 = 7.85 \text{ ohms.}$$

Using Eq. (9),

$$E = 4.0 \times 7.85 = 31.4 \text{ volts.} \quad Ans.$$
$$E \text{ leads } I \text{ by } 90°. \quad Ans.$$

In Fig. 44 is shown a laboratory portable-type inductance.

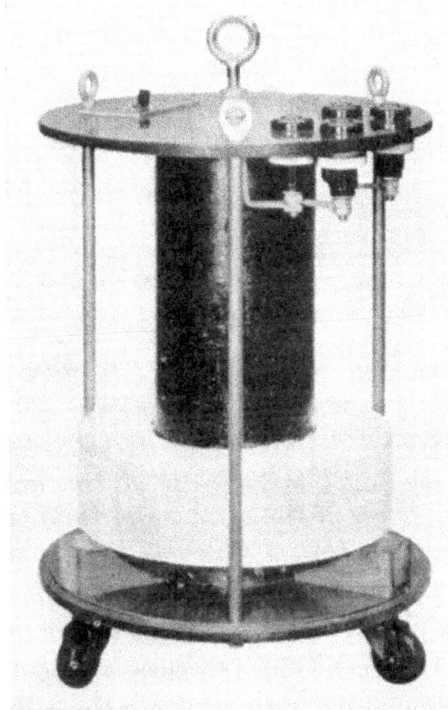

FIG. 44.—Variable inductance without case; laboratory portable type; 110 volts, 60 cycles. (*Courtesy of General Electric Company.*)

28. Mechanical Analogue of Inductance.—It is clear that inductance always opposes *change* of current. It opposes increase in current, and it opposes decrease in current. In Part I (see p. 174), it is stated that inductance in the electrical circuit corresponds to inertia in mechanics. The mechanical analogue of inductance in an alternating-current circuit may be illustrated by a system consisting of a mass M (Fig. 45) on frictionless wheels and rails working in conjunction with a piston and frictionless cylinder. The mass is connected rigidly to the piston rod. The cylinder is filled with a compressible fluid such as a gas and is closed at both ends. When the piston is at the center of the cylinder, the gas pressures on both sides of the piston are made equal. The entire system is frictionless so that no energy is dissipated. (With pure inductance, no energy is dissipated, as will be shown later.)

Electromotive force in the electric circuit corresponds to pressure in mechanics and hydraulics (see Part I, p. 23). Current (rate of flow of quantity) in the electric circuit corresponds to velocity in mechanics. A study of Fig. 45 shows that when the system is set in oscillation, the mass, because of its momentum, will compress the gas in one end of the cylinder and the gas in the other end will be lowered in pressure. When the kinetic

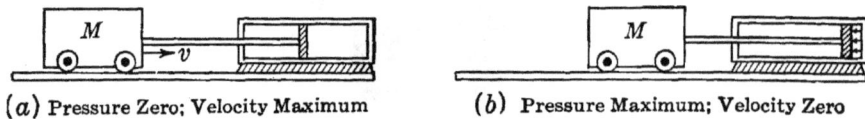

(*a*) Pressure Zero; Velocity Maximum   (*b*) Pressure Maximum; Velocity Zero

Fig. 45.—Mechanical analogue of inductance.

energy of the mass is all utilized in so changing the gas pressure, its velocity becomes zero. The compressed gas which is then at its maximum pressure expands and so accelerates the mass in the reverse direction. The velocity of the mass M reaches a maximum at the center of the stroke and then begins to decrease owing to the compression of the gas in the other end of the cylinder. The maximum velocity v of the mass occurs at the instant when the piston is at the center of the cylinder as in (*a*). At this instant the resultant pressure acting on the piston is zero, for the pressures on both sides are the same. Also, at this instant the kinetic energy of the mass is a maximum and equals $W = \frac{1}{2}Mv^2$ where M is the mass and v its velocity.

At the instant when the moving system reaches the right-hand extremity of its movement (Fig. 45(*b*)), its velocity is zero. The maximum pressure during this half of the stroke occurs at this instant, for the gas between the piston and the right-hand cylinder head is compressed into its minimum volume. The gas on the opposite side of the piston is at its maximum volume and lowest pressure. At this instant the kinetic energy of the mass is zero, for its velocity is zero.

A study of Fig. 40 shows that when the current (electric velocity) is a maximum, the line voltage (electric pressure) is zero. At this instant the energy stored in the magnetic field is a maximum. ($W = \frac{1}{2}Li^2$. See Part I, p. 176.) When the current (electric velocity) is zero (Fig. 40), the line voltage (electric pressure) is a maximum.

The inertia of the mass also opposes change of velocity. When the gas, in expanding, attempts to impart velocity to the mass, inertia *opposes* the increase in velocity. When the gas, by compression, attempts to decrease the velocity of the mass, the inertia of the mass opposes this decrease. In the electric circuit, inductance likewise opposes both increase and decrease of current.

Therefore, an oscillating mechanical system having inertia and an alternating electric system having inductance are similar as regards the relations among velocity and current, pressure and voltage, and energy.

29. Hydraulic Analogue of Capacitance.—In Fig. 46 is shown a cylindrical tank T_1, of considerable volume, which may be supplied with water through a large pipe P_1 connected into its bottom. A smaller cylindrical tank T_2, whose bottom is at the same level as that of T_1, is connected to T_1 by a frictionless pipe P_2.

If water is supplied to T_1 by the pipe P_1, the water level in T_2 will always be the same as the water level in T_1 if the inertia of the water is neglected. That is, the hydrostatic pressure in the two tanks is always the same. If a capacitor is connected across electric mains, it develops a counter emf q/C, where q is its charge in coulombs and C is its capacitance in farads (see Part I, p. 197, Eq. (76)). At every instant, this counter emf of the capacitor must be equal to the line voltage, provided the circuit leads have negligible resistance.

Let the horizontal plane OO be the datum plane, that is, the plane from which the water levels in the two tanks are determined. Assume that by controlling the rate of flow of water through P_1 the water level h in T_1 referred to the datum plane OO is made to vary sinusoidally with time. The *rate of increase* of water level h will be a maximum when h is zero. (Examine any sine curve.) Therefore, the *rate of flow* of water through the pipe P_2 is a maximum at this instant. That is, the hydraulic current is a maximum when the pressure referred to the datum plane is zero. The direction of flow is also positive in that the water flows from tank T_1 to tank T_2.

As the height h in T_1 increases, its *rate of increase* diminishes and the rate of flow of water in pipe P_2 diminishes. The rate of flow of water in pipe P_2 becomes *zero* when the height h of water in tanks T_1 and T_2 reaches its *maximum* value H. At this instant the pressure in tank T_2 is a maximum. Thus, it is seen that the current flow in P_2 reaches its maximum value a quarter-cycle before the pressure in tank T_2 reaches its maximum value. That

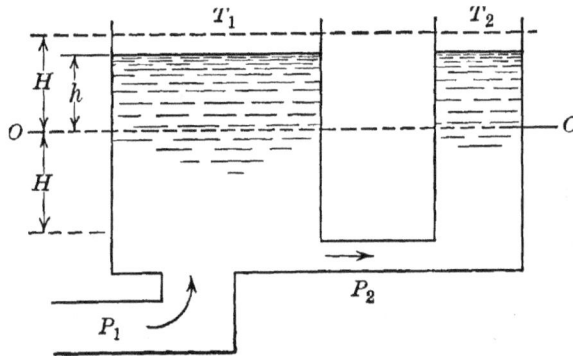

Fig. 46.—Flow of water between tanks.

is, the current or flow of water may be said to *lead* the pressure by 90°.

When the water level in T_1 and T_2 reaches its maximum value, the flow of water in P_2 is zero for an instant. When the water level in tank T_2 begins to diminish, the current in P_2 is reversed in sign and becomes negative. This negative direction of flow continues, not merely until the water level h is zero, but until h is at its maximum negative value ($-H$). The flow then becomes zero for an instant, after which its direction becomes positive.

30. Circuit Containing Capacitance Only.[1]—Electricity is stored in a capacitance or a capacitor in much the same manner that water is stored in the tank T_2. For example, when a d-c voltage is impressed across a capacitance (Part I, p. 196), there is an initial rush of current which charges the capacitance to line potential.

[1] See Part I, Chap. IX, for a discussion of capacitance. Also see Part I, p. 199, for a description of condensers or capacitors.

Consider Fig. 47 in which is shown a capacitance C in series with a galvanometer. By means of a D.-P., D.-T. switch S' the upper wire to the capacitor may be made either positive or negative. A single-pole, double-throw (S.-P., D.-T.) switch S is so arranged that the capacitance may either be connected across the d-c voltage E or be short-circuited. The switch S' is first thrown to the left-hand position, making the upper wire positive. The switch S is then thrown to the left at (a) connecting the upper plate of the capacitance through the galvanometer to the upper wire which is positive. There is a momentary rush of current into the upper plate as is shown by the ballistic deflection of the galvanometer. The galvanometer comes to zero quickly, showing that there is no steady flow of current.

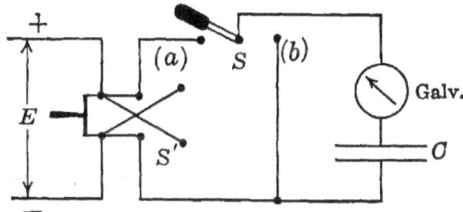

FIG. 47.—Charge and discharge of capacitance with direct current.

Hence, the emf (q/C) of the capacitance must now be equal and opposite to the line emf E. There is now a positive charge on the upper plate of the capacitance and a negative charge on the lower plate. If the switch S is thrown to the right at (b), the positive charge rushes out of the positive plate, as is shown by the ballistic deflection of the galvanometer which is now opposite to the previous deflection. The capacitance being short-circuited, its emf has become zero.

The D.-P., D.-T. switch S' is now thrown to the right, making the upper wire negative. The switch S is then thrown to the left at (a), and the ballistic deflection of the galvanometer continues to be in the same direction as when the capacitance was short-circuited. The capacitance is now charged, with the upper plate negative. If the switch S is now thrown to (b), the ballistic deflection of the galvanometer reverses and the capacitance is again discharged. (Also see Part I, pp. 195 to 197.)

Figure 48(a) shows an alternating emf e impressed across a capacitance C and Fig. 48(b) shows the sinusoidal emf wave e. When the emf e is positive, the upper wire in (a) is positive.

When the emf starts from its zero value at *a* [Fig. 48(*b*)] and increases positively, current flows *into* the upper terminal of the capacitance *C*. Therefore, this current is positive, just as the flow of water was positive (F g. 46) when the pressure in tank T_1 was increasing. As long as the emf across the capacitance continues to *increase*, current must flow *into* it from the positive wire and this current will be positive in sign. When *b* is reached, the increase of emf ceases and the current becomes zero, just as the flow of water ceases (Fig. 46) at the instant of maximum pressure *H*. The capacitance *C* is now charged with the upper plate positive such as occurs when switches *S'* and *S* are both in the left-hand positions (Fig. 47).

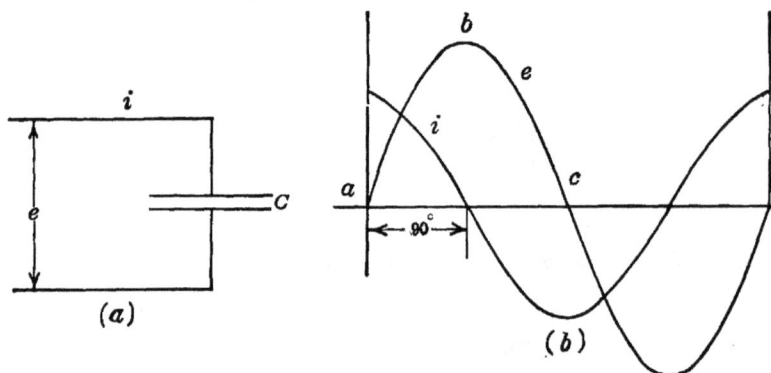

Fig. 48.—Circuit containing capacitance only.

Between *b* and *c* the emf is decreasing so that current is flowing *out of* the capacitance into the positive line; and because the current flow has reversed, the sign of the current is now negative. This corresponds to the reversal of water flow in P_2 (Fig. 46), when the pressure in tank T_1 was decreasing and also to the short circuit of the capacitor *C* in Fig. 47 when switch *S* is thrown to the right. After *e* goes through zero at *c*, the emf is negative and charges the capacitance in the opposite direction; therefore, the current still remains negative. This continues until the emf reaches its negative maximum. This corresponds to the condition in Fig. 47 when switch *S* is in the left-hand position and switch *S'* is thrown to the right. At this point the current reverses and again becomes positive.

An examination of Fig. 48(*b*) shows that when an alternating emf is impressed on a capacitance the current *into* the capacitance *leads* the emf by 90°. It may appear remarkable that current can lead its emf in this manner, but in Fig. 46 the current flow through pipe P_2 reaches its maximum value before the pressure in tanks T_1 and T_2 reach their maximum values.

It will be seen from the foregoing that alternating current does not actually flow *conductively* through the insulation of the capacitor. A perfect capacitor offers an infinite resistance to alternating as well as to direct current. However, with alternating current the capacitor is alternately charged and discharged,

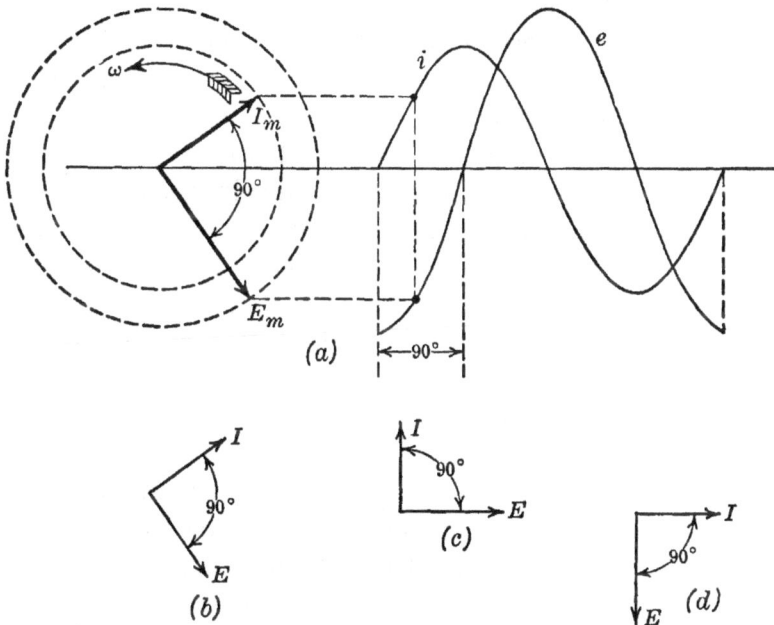

FIG. 49.—Current and voltage and vector diagrams with capacitance only in circuit.

so that a quantity of electricity flows into the positive plate and then out again, etc. It is this quantity of electricity which flows to charge and to discharge the capacitor which constitutes the alternating current. An ammeter placed in the line to such a condenser indicates a current. This is analogous to water flowing in the pipe P_2 (Fig. 46), when the water level in tank T_1 is alternately raised and lowered. A suitable indicating instrument, if such an instrument could be devised, would assume a steady deflection with the alternating water flow in the pipe P_2.

Figure 49(a) shows the vectors E_m and I_m rotating at angular velocity ω, from which the sine curves of emf e and current i

are derived for the circuit containing capacitance only. Since the current leads the emf by 90°, the vector I_m leads E_m by 90°. If a vector diagram involving maximum values of emf and current is desired, it is given by these two rotating vectors E_m and I_m, the current leading the emf by 90°. A vector diagram involving *rms* values of emf and current is ordinarily more useful. To obtain this diagram it is merely necessary to divide E_m and I_m (Fig. 49(a)) by $\sqrt{2}$, and the vector diagram given in Fig. 49(b), involving E and I in rms values, is obtained. Either the current or the emf vector may be horizontal. In Fig. 49(c) the emf E is shown horizontal and the current vector leads it by 90°. It is equally correct to show the current vector horizontal and the emf vector lagging it by 90° (Fig. 49(d)).

31. Quantitative Values of Current in Circuit Containing Capacitance Only.—With inductance only in circuit, the current can be determined quantitatively, provided the voltage, the inductance, and the frequency are known. In a similar manner, the current in a circuit containing capacitance only can be determined, provided the voltage, the capacitance, and the frequency are known.

Obviously, the current is proportional to the voltage, since the quantity which charges the capacitor each half-cycle is proportional to the voltage ($q = Ce$). Also, the current must be proportional to the capacitance, since the quantity which charges the capacitor each half-cycle is also proportional to the capacitance ($q = Ce$). The current must also be proportional to the frequency, since both the positive and the negative quantity per second is proportional to the number of times the capacitor is charged and discharged per second. Actually, the current in rms amperes is equal to the product of the voltage, capacitance, and angular velocity ω. Thus,

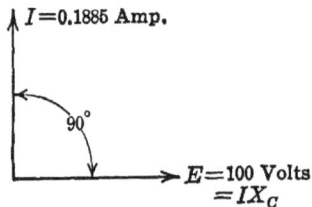

FIG. 50.—Vector diagram of a circuit containing a 5.0-μf capacitance.

$$I = 2\pi fCE = EC\omega \text{ amp,} \qquad (13)$$

where E is in volts (rms), C is in *farads* (not microfarads), and ω is $2\pi f$ where f is the frequency in cycles per second.

Example.—How many amperes will a 5.0-microfarad (μf) capacitance take when connected across 100-volt, 60-cycle mains? Draw a vector diagram for this circuit.

$$C = 0.000005 \text{ farad.}$$
$$\omega = 2\pi 60 = 377.$$
$$I = 100 \times 0.000005 \times 377 = 0.1885 \text{ amp.} \quad Ans.$$

The vector diagram is shown in Fig. 50, where the 0.1885-amp vector is leading the 100-volt vector by 90°.

In Eq. (13), $C\omega$ is in the nature of conductance, or mhos, since it is *multiplied* by the voltage to obtain the current. In order that it may be combined with reactance or ohms, its reciprocal must be used. That is, since $C\omega$ gives mhos, $1/C\omega$ gives ohms.

The current then becomes,

$$I = \frac{E}{\dfrac{1}{C\omega}} \text{ amp.} \tag{14}$$

$1/C\omega$ is the *capacitive reactance* of the circuit, just as with inductance the reactance is called the inductive reactance. Capacitive reactance is expressed by the symbol X_C.

Thus,

$$X_C = \frac{1}{C\omega} \text{ ohms.} \tag{15}$$

Example.—In the foregoing example, find the capacitive reactance of the circuit and also the current.

$$X_C = \frac{1}{0.000005 \times 377} = \frac{1}{0.001885} = 530 \text{ ohms.} \quad Ans.$$
$$I = \frac{100}{530} = 0.1885 \text{ amp.} \quad Ans.$$

Since most commercial capacitors are rated in microfarads, rather than in farads, it may simplify computations to use the following equation:

$$X_C = \frac{1,000,000}{C'\omega} \tag{16}$$

where C' is in microfarads.

Example.—Find the capacitive reactance of a 40-μf capacitor when the frequency is 50 cycles per sec.

Using Eq. (16),

$$X_C = \frac{1,000,000}{40 \times 2\pi 50} = 79.6 \text{ ohms. } Ans.$$

It follows from (14) and (15) that

$$I = \frac{E}{X_C} \text{ amp.} \tag{17}$$

$$E = IX_C \text{ volts.} \tag{18}$$

$$X_C = \frac{E}{I} \text{ ohms.} \tag{19}$$

Example.—In the foregoing example, what 50-cycle voltage across the 40-μf capacitor will maintain a current of 2.0 amp?

From (18), since $X_C = 79.6$ ohms,

$$E = 2.0 \times 79.6 = 159.2 \text{ volts. } Ans.$$

This voltage *lags* the current by 90°.

POWER IN ALTERNATING-CURRENT CIRCUITS

32. Alternating-current Power.—Thus far, the current and voltage relations have been considered in the three fundamental a-c circuits consisting of resistance alone, inductance alone, and capacitance alone. Before combining such circuits by connecting them in series and in parallel, the power relations in each individual circuit will be considered.

Under steady conditions the power in a d-c circuit is always given by the product of the volts across the circuit and the current in amperes flowing in the circuit (see Part I, p. 34). This same rule applies to a-c circuits, provided the *instantaneous* values of amperes and volts are considered. The *average* power, however, is not necessarily the product of the rms volts and rms amperes, the values which are ordinarily measured with instruments. For example, a voltmeter across a circuit reads 120 volts, and an ammeter in the circuit reads 4.0 amp. It is only under special conditions that the power is equal to 120 × 4.0, or 480 watts.

33. Power with Voltage and Current in Phase.—Figure 51 shows a voltage wave *e* and a current wave *i* in phase with each other, which occurs when a circuit contains resistance only.

To obtain the power *at any instant*, the amperes and volts at that instant are multiplied together and a new curve *p* may be plotted, the ordinates being the instantaneous products of *e* and *i*.

Assume that at the instant shown at *a* (Fig. 51) the current represented by the ordinate *ab* is 12 amp; the voltage represented by the ordinate *ac* is 80 volts. The power at that instant is represented by their product, or 960 watts, and is given by the ordinate *ad*. Point *d* is one point on the power curve. Other points are found similarly, and the power curve *p* is plotted. The curve *p* gives the power in the circuit at any instant. The ordinates of this power curve are positive during the first half-cycle, since both voltage and current are positive throughout this

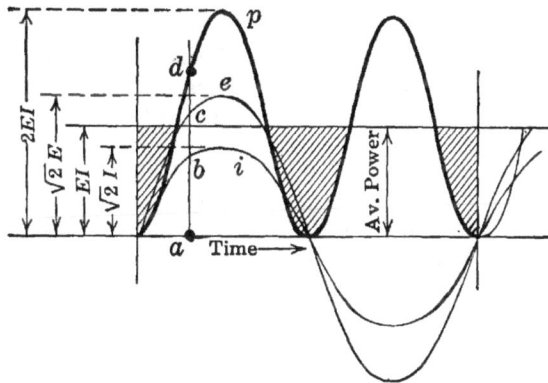

Fig. 51.—Power curve, current and voltage in phase.

period. During the second half-cycle, both voltage and current are negative. The power must still be positive, however, since the voltage and current are still in conjunction. If both voltage and current reverse simultaneously, the power must still be positive. Also, algebraically, the power is positive during the second half-cycle; the power *p* is the product of a negative voltage and a negative current, and the product of two negative quantities is positive. Hence, during both the first and the second half-cycle the voltage and the current act in conjunction and the power is positive. The ordinates of the power curve are positive, therefore, throughout the cycle.

The power curve is a sine wave having double the frequency of the voltage or current. For every cycle of either voltage or current, the power curve touches the zero axis twice, so that in such a circuit the power is zero twice during each cycle. Since the voltage and current waves are in phase, their peaks occur at the same instant; and the corresponding peak of the power curve is, therefore,

$$(\sqrt{2}\,E)(\sqrt{2\,I}) = 2EI,$$

where E and I are the *rms* values of voltage and current.

Accordingly, the power curve has its horizontal axis of symmetry at a distance EI above the zero axis. Consequently, EI must be the *average* value of the power during each cycle, since during the cycle the upper half-wave will just fill the shaded valley below the axis of symmetry of the power curve. When the current and the voltage are *in phase*, the average power is given by their product, as with direct currents.

Example.—An incandescent-lamp load takes 30 amp from 115-volt, 60-cycle mains. (In this type of load the current and voltage are practically in phase.) How much power do the lamps take?

$$P = EI = 115 \times 30 = 3{,}450 \text{ watts. } Ans.$$

34. Positive and Negative Power.—In Fig. 52 is shown a three-cell storage battery whose terminal voltage is 6 volts, as indicated

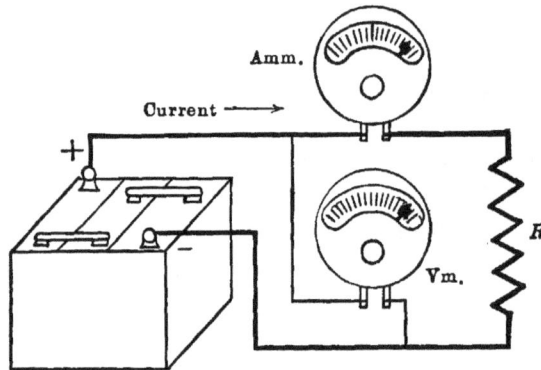

Fig. 52.—Battery delivering energy—its power is positive.

by the voltmeter connected across its terminals. This battery delivers current to a resistance R connected across its terminals. The emf of the battery acts in such a direction as to send current *out* of its positive terminal. This is shown by the zero-center ammeter, connected in the line, the deflection being to the right. The current through the resistance flows from the battery positive terminal to the resistance R connected across the battery terminals. That is, so far as the external circuit is concerned, both voltage and current are acting in conjunction. If the battery is considered a source of energy, the current is *positive* when the current flows out of its positive terminal and the power delivered to the resistance is *positive*.

In Fig. 53, the 6-volt storage battery is connected across the terminals of a generator, the positive terminal of the generator being connected to the positive terminal of the battery and the negative terminal of the generator being connected to the negative terminal of the battery. (Compare this figure with Fig. 37, Part I, p. 46.) If the emf of the generator is just equal to the emf of the battery, no current flows and the battery is merely

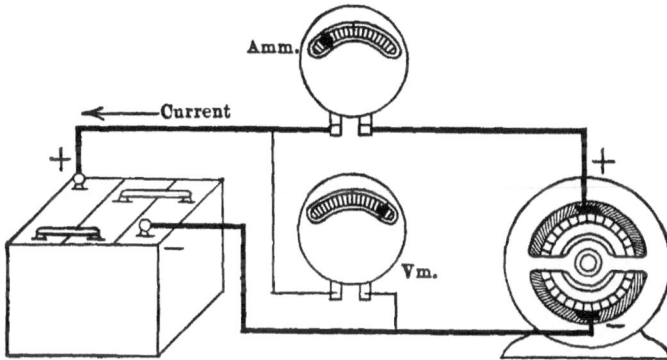

FIG. 53.—Battery receiving energy—its power is negative.

"floating." If the emf of the generator is lowered, current flows from the positive terminal of the battery to the positive terminal of the generator, and the ammeter will deflect to the right. The current is therefore in conjunction with the battery emf, and the power is positive. The battery is delivering energy to the generator and tending to drive it as a motor. That is, so far as the relations of voltage, current, and power are concerned, the conditions are identical with those given in Fig. 52, with resistance only across the battery terminals.

If now, the generator emf is raised so that it exceeds the emf of the battery, current will flow from the generator positive terminal *into* the battery positive terminal. This will be indicated by the zero-center ammeter now deflecting to the left, as shown in Fig. 53. The battery emf has *not* reversed in sign, as is shown by the voltmeter connected across the battery terminals still reading up scale. The direction of flow of the *current*, however, has reversed and is now negative. The battery is now being charged and is receiving energy, therefore. However, the battery is still considered a source of energy, and it must be *delivering negative* power. The current and voltage are now in opposition; that is, they are of opposite sign, and the power is negative.

(If the battery is considered as a translating device or a device for *receiving* energy, the power is positive when the current flows *in* at its positive terminal and negative when it flows *out* of its positive terminal.)

35. Power with Inductance Only.—With inductance only in circuit, the impressed voltage leads the current by 90° (Fig. 40, p. 46). Figure 54 shows sinusoidal waves of voltage and current with inductance only in circuit, the voltage *e* leading the current *i* by 90°. To determine the variation of power with time, the current and voltage at each instant are multiplied together, as was done in Fig. 51.

At points *a, b, c, d, e* (Fig. 54), either current or voltage is zero, and the power must be zero at each of these instants. Between

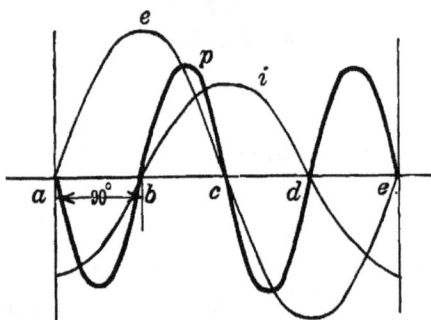

FIG. 54.—Power curve; current and voltage in quadrature, current lagging.

a and *b* the voltage is positive and the current is negative, and they are therefore acting in *opposition*. That is, the current is flowing against the impressed voltage in a manner similar to that shown in Fig. 53. Also, the product of a positive and a negative quantity is negative. Hence, the power between *a* and *b* must be *negative*. This means that the inductance is delivering power to the source, just as the generator (Fig. 53) delivered power to the battery which was considered a source. Also in Fig. 53 the power was negative under these conditions. Between *b* and *c*, both current and voltage are *positive* and are therefore in conjunction. Consequently, the power between these two points must be *positive*. That is, the source is supplying power to the inductance. Between *c* and *d* the current is positive, but the voltage is negative. That is, the current and voltage are again in opposition. Therefore, the power is again negative between these two points. Between d and e both current and voltage are negative and are therefore in conjunction, so that the power is positive. The power curve p, obtained by multiplying the voltage and current at each point, is a sine curve having double the frequency of current or voltage. Its axis of symmetry coincides with the zero axis of current and voltage. Therefore, there must be as much of the power curve above the zero axis as there is below this axis, or the positive area above the axis must be equal to the negative area below the axis for each cycle. That is, all the energy received during a cycle by the inductance from the source is returned by the inductance to the source. The average power over each cycle, therefore, is *zero*. Between b and c and between d and e, when the power is positive, the source of power is storing energy in the inductance ($\frac{1}{2}Li^2$). Between a and b and between c and d, when the power is negative, the inductance is returning energy to the source. (See Fig. 95, p. 117, for actual oscillogram of voltage and current in quadrature, and also power curve.)

The power relations with inductance in the circuit are not unlike a single-cylinder, compressed-air engine having no losses. During part of the cycle, the compressed air entering the cylinder gives energy to the shaft and flywheel. Subsequently, the flywheel and shaft, owing to their inertia, return this energy to the cylinder by compressing the air in the cylinder. It has already been stated that inductance in the electric circuit corresponds to inertia in mechanics.

36. Power with Capacitance Only.—With capacitance only in circuit, the impressed voltage lags the current by 90° as is shown in Fig. 49 (p. 55). Figure 55 shows waves of voltage and current with capacitance only in circuit, the voltage e lagging the current i by 90°. The power curve is determined by multiplying at each instant the voltage ordinate by the current ordinate, at that instant.

As in Fig. 54, either voltage or current is zero at the instants a, b, c, d, e. Hence, at each of these instants the power is zero. Between points a and b, both voltage and current are positive, hence are acting in conjunction, so that the power is positive.

Between points b and c the voltage is positive, but the current is negative. That is, voltage and current are in opposition, so that between these points the power is negative. Between points c and d, voltage and current are both negative, hence are acting in conjunction, and between these points the power is positive. Between points d and e the voltage is negative, the current is positive, and the power is negative.

The power curve p, which is the product of the current and voltage at each instant, is a sine curve having a frequency double that of voltage or current. Its axis of symmetry coincides with the zero axis of voltage and current. Hence, the positive area

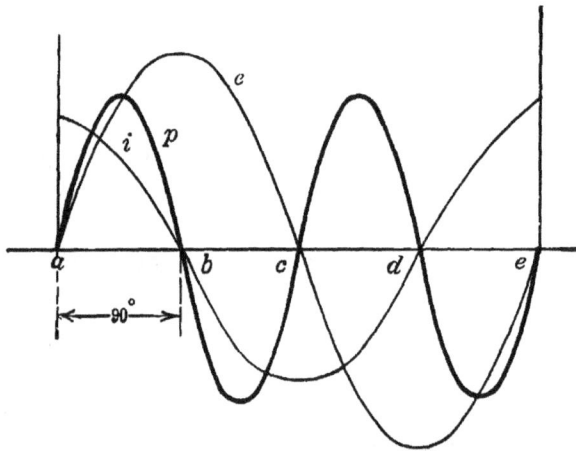

Fig. 55.—Voltage, current, and power curves; circuit containing capacitance only.

of the power curve must be equal to the negative area, so that over a complete cycle the *net or average power is zero*. Between a and b and between c and d, when the power is positive, the source of power is storing energy in the capacitance ($\frac{1}{2}Ce^2$). Between points b and c and between d and e, when the power is negative, the capacitance is returning energy to the source of supply. With a perfect capacitor, such as is assumed, the capacitor returns to the source of power during each cycle all the energy which it receives. Therefore, the average power over a complete cycle must be zero.

These power relations in the electric circuit are not unlike the action of a compressed-air cylinder operating on a spring. During part of the cycle, the expansion of the air in the cylinder compresses the spring. If there are no losses, the spring returns all this energy to the cylinder by compressing the air during the other part of the cycle. Capacitance in the electric circuit corresponds to the reciprocal of elasticity in mechanics.

A comparison of Fig. 55 with Fig. 54 shows that when the power is positive with capacitance, it is negative with inductance, as between points a and b, c and d, etc. That is, during those times when energy is being stored in the capacitance, the inductance is delivering energy to the source. By proper adjustments, the power taken by the inductance may be made equal to the power taken by the capacitance at every instant. Since the power taken by the inductance and capacitance is always of opposite sign, the power supplied by the source at every instant must be zero. As it is impossible to have a perfect inductance coil or a perfect capacitor, this condition of zero power can be realized only approximately.

37. Power with Phase Difference Other Than 90°.—It has been shown that the power is the product of rms voltage and current when voltage and current are in phase and that the power is zero when voltage and current are in quadrature. When voltage and current are neither in phase nor in quadrature but differ in phase by an angle θ, which is greater than zero and less than 90°, the average power is less than the product of volts and amperes but is greater than zero. This is shown in Fig. 56 where the current lags the voltage by an angle θ. (Such a condition occurs when resistance and inductance are connected either in series or in parallel (see Pars. 38 and 48).) The power curve p is the product of the volts and amperes at each instant. As in Figs. 54 and 55, at a, b, c, d, e, either voltage or current is zero and hence the power is zero at each of these instants. Between a and b and between c and d, voltage and current are in opposition, and the power is negative. Between b and c and between d and e, the voltage and current are in conjunction, and the power is positive. During the intervals of time bc and de, the source of power delivers energy to the circuit, this energy being represented by the areas under the positive loops of the power curve. During the intervals of time ab and cd, the circuit is returning energy to the source of power, this energy being represented by the shaded areas. The power relations in such a circuit are not unlike the action of a single-cylinder, internal-combustion engine. During that part of the cycle when the engine is firing, the cylinder gives energy to the flywheel. During the compression part of the cycle, the flywheel returns *some* of this energy to the cylinder. Over the complete cycle, however, the cylinder delivers more energy *to* the flywheel than is returned *by* the flywheel.

The power curve *p* is a double-frequency curve, as before, and its axis of symmetry *xx* is above the zero axis *ae*. A study of Fig. 56 shows that the unshaded areas under the power curve and above the zero axis are greater than the shaded areas below the zero axis. Hence, over a complete cycle there is more posi-

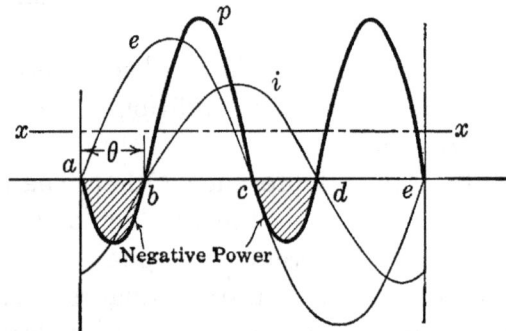

Fɪɢ. 56.—Power curve; current lagging voltage by angle *θ*.

tive energy than negative energy. That is, the average power is not zero, but positive, and is less than the product of *E* and *I*. It will be shown later (Par. 39) that this power

$$P = EI \cos \theta \text{ watts,} \qquad (20)$$

where *θ* is the phase angle between voltage and current. Cos *θ* is called the *power factor* of the circuit. *P* is the true watts and *EI* the apparent watts, or volt-amperes.

The power factor

$$\text{P.F.} = \cos \theta = \frac{\text{true watts}}{\text{apparent watts}} = \frac{P}{EI} \qquad (21)$$

The power factor can never be greater than unity.

Example.—A single-phase motor takes 4.5 amp at 220 volts, the current lagging. The wattmeter indicates 720 watts. (*a*) What is the power factor of the circuit? (*b*) By what angle does the current lag the voltage?

(*a*) From Eq. (21),

$$\text{P.F.} = \frac{720}{220 \times 4.5} = \frac{720}{990} = 0.728. \quad Ans.$$

(*b*) Cos θ = 0.728.   From Appendix E (p. 437),

$$\cos 43.3° = 0.7278.$$
$$\theta = 43.3°. \quad Ans.$$

ALTERNATING-CURRENT CIRCUITS WITH RESISTANCE, INDUCTANCE, AND CAPACITANCE IN SERIES

38. Resistance and Inductance in Series.—Thus far, only special cases of a-c circuits have been considered, such as circuits containing resistance only, inductance only, and capacitance only. It is obvious that these three quantities may be connected in various series and parallel combinations. First, consider a resistance R and an inductance L in series (Fig. 57). Let $X_L = 2\pi f L$ be the inductive reactance of the inductance. (See Eq. (10), p. 48.) A voltage E is impressed across the circuit and as a result a current I flows. Let it be required to determine the relations existing among current, voltage, power, resistance, and inductance. Figure 58(*a*) shows a vector dia-

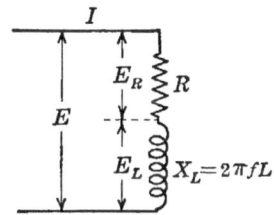

FIG. 57. Circuit containing resistance and inductance in series.

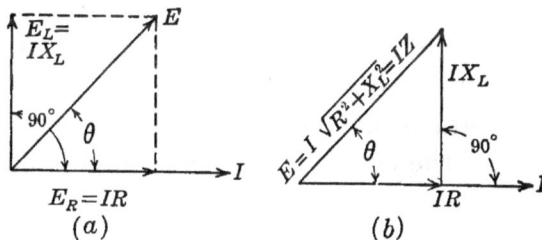

FIG. 58.—Vector diagram for series circuit containing resistance and inductance.

gram for this circuit. Because this is a series circuit the current I is the same in both X_L and R at every instant. Hence, one current vector suffices for the circuit. This current vector I is laid off horizontally to scale. The position of the current vector is arbitrary. It is given the horizontal position merely for convenience. The voltage E_R across the resistance must be equal to IR as has been shown already (see Par. 24, p. 40, and

Fig. 36, p. 41). This voltage $E_R = IR$ is laid off along the current vector I (Fig. 58(a)). The voltage E_L, across the inductance, must be equal to IX_L ($= 2\pi fLI$), and it must also *lead* the current by 90°. This was shown in Par. 27, p. 47, and in Fig. 43, p. 48. Therefore, the voltage $E_L = IX_L$, across the inductance, is laid off at right angles to the current and leading it by 90° (Fig. 58(a)).

The line voltage E must be equal to the sum of its component parts E_R and E_L. Since E_R and E_L are not in phase but are in quadrature, the line voltage E is the *vector* sum of E_R and E_L. The vector addition is shown in Fig. 58(a). The voltage E is the hypotenuse of a right triangle, of which $IR(= E_R)$ and $IX_L(= E_L)$ are sides.

Therefore,

$$E = \sqrt{(IR)^2 + (IX_L)^2}$$
$$= \sqrt{I^2(R^2 + X_L^2)} = I\sqrt{R^2 + X_L^2} \text{ volts.} \qquad (I)$$

Solving equation (I) for the current,

$$I = \frac{E}{\sqrt{R^2 + X_L^2}} = \frac{E}{\sqrt{R^2 + (2\pi fL)^2}} = \frac{E}{Z} \text{ amp.} \qquad (22)$$

$Z = \sqrt{R^2 + X_L^2}$ is the *impedance* of the circuit and is expressed in ohms. It is ordinarily denoted by Z. Equation (22) corresponds to Ohm's law for the d-c circuit. The current in an a-c circuit is directly proportional to the voltage across the circuit and inversely proportional to the impedance of the circuit. That is, if the voltage in volts is divided by the impedance in ohms, the value of the current is obtained in amperes.

Also, the voltage

$$E = IZ \text{ volts,} \qquad (23)$$

and the impedance

$$Z = \frac{E}{I} \text{ ohms.} \qquad (24)$$

An inspection of Fig. 58 shows that the angle θ by which the current lags the voltage may also be determined. Since the tangent of an angle is the ratio of the side opposite to the side adjacent (see p. 431).

$$\tan \theta = \frac{IX_L}{IR} = \frac{X_L}{R} = \frac{2\pi f L}{R}. \tag{25}$$

The cosine of an angle is the ratio of the adjacent side to the hypotenuse. In Fig. 58,

$$\cos \theta = \frac{IR}{\sqrt{(IR)^2 + (IX_L)^2}} = \frac{R}{\sqrt{R^2 + X_L^2}} = \frac{R}{Z} \tag{26}$$

Figure 58(b) shows the vector addition performed by the triangle method, rather than by the parallelogram method (see p. 28). That is, IX_L is added at the end of IR and at right angles to IR, leading. It is obvious that both methods give the same result.

To illustrate the method of computing the current in a circuit having resistance and inductance in series the following example is given:

Example.—A circuit containing 0.1 henry inductance and 20 ohms resistance in series is connected across 100-volt, 25-cycle mains. Determine: (a) impedance of circuit; (b) current; (c) voltage across resistance; (d) voltage across inductance; (e) angle by which voltage leads current.

$X_L = 2\pi 25 \times 0.1 = 157 \times 0.1 = 15.7$ ohms (from Eq. (10), p. 48).

(a) $Z = \sqrt{(20)^2 + (15.7)^2} = \sqrt{646} = 25.4$ ohms.   *Ans.*

(b) $I = \dfrac{E}{Z} = \dfrac{100}{25.4} = 3.94$ amp.   *Ans.*

(c) $E_R = IR = 3.94 \times 20 = 78.8$ volts.   *Ans.*

(d) $E_L = IX_L = 3.94 \times 15.7 = 61.8$ volts.   *Ans.*

As a check, $\sqrt{(78.8)^2 + (61.8)^2} = 100$ volts.

(e) $\tan \theta = \dfrac{X_L}{R} = \dfrac{15.7}{20} = 0.785.$

From p. 438,

$$\theta = 38.1°.\quad Ans.$$

39. Power with Resistance and Inductance in Series.—In Fig. 57, the power dissipated in the resistance must be I^2R. It has been shown already that the average power taken by a pure inductance over a complete cycle is zero. Hence, the inductance (Fig. 57) consumes no power. All the power taken by the circuit must therefore be accounted for in the resistance. The total power

$$P = I^2R = I(IR) \text{ watts.} \tag{I}$$

In the vector diagram (Fig. 58),

$$\cos \theta = \frac{IR}{E},$$

$$IR = E \cos \theta \text{ volts.}$$

Substituting for IR in (I),

$$P = I(E \cos \theta) = EI \cos \theta \text{ watts.} \qquad (27)$$

This equation is the same as Eq. (20) (p. 66). Cos θ is the *power factor* of the circuit. The angle θ is zero when the current and voltage are in phase and the cosine of 0° is unity. The cosine of an angle, and also the power factor, can never exceed unity. When θ is greater than zero, the cosine is less than unity (see Fig. 6, p. 8). Therefore, the power factor is unity when the current and voltage are in phase and is less than unity when they are not in phase.

The power factor is obviously the true power divided by the volt-amperes (EI) or the apparent power (see Eq. (21), p. 66).

Example.—Determine the power factor in the foregoing example. The power

$$P = I^2R = (3.94)^2 \, 20 = 310 \text{ watts.}$$

The volt-amperes

$$EI = 100 \times 3.94 = 394 \text{ watts.}$$

The power factor

$$\text{P.F.} = {}^{310}\!/\!_{394} = 0.787. \quad Ans.$$

Also,

$$\cos \theta = \cos 38.1° = 0.787 \text{ (see p. 437). Check.}$$

40. Resistance and Capacitance in Series.—In Fig. 59 is shown a resistance R connected in series with a capacitance C across a voltage E. It is required to determine the relations existing among current, voltage, power, resistance, and capacitance in this circuit. Since this circuit is a series circuit, the current is the same throughout the circuit at every instant. Hence, one current vector suffices for the circuit. For convenience only, this current vector is laid off horizontally (Fig.

60(a)). The voltage E_R across the resistance must be equal to IR, and it must also be in phase with the current I. Therefore, this voltage is laid off along the current vector as shown in the figure. The voltage E_C across the capacitance is equal to $IX_C(= I/2\pi fC)$ and it *lags* the current by 90° (see Fig. 49, p. 55). The line voltage E must be equal to the sum of its component parts E_R and E_C. Since E_R and E_C are not in phase, the line voltage E is the *vector* sum of E_R and E_C. The parallelogram method of vector

FIG. 59.—Circuit containing resistance and capacitance in series.

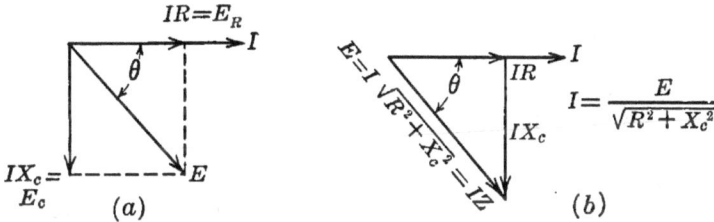

addition is shown in Fig. 60(a) and the triangle method in Fig. 60(b). Since E_R and E_C are in quadrature, the line voltage

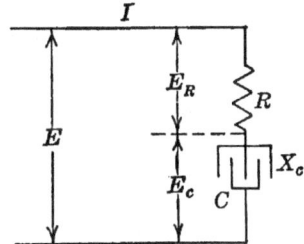

FIG. 60.—Vector diagrams for circuit containing resistance and capacitance in series.

E is the hypotenuse of a right triangle of which $IR(= E_R)$ and $IX_C(= E_C)$ are the other two sides.

Therefore,

$$E = \sqrt{(IR)^2 + (IX_C)^2} = \sqrt{I^2(R^2 + X_C^2)} = I\sqrt{R^2 + X_C^2} \text{ volts.}$$

Solving for the current I,

$$I = \frac{E}{\sqrt{R^2 + X_C^2}} = \frac{E}{\sqrt{R^2 + \left(\dfrac{1}{2\pi fC}\right)^2}} = \frac{E}{Z} \text{ amp.} \quad (28)$$

$Z = \sqrt{R^2 + X_C^2}$ is the *impedance* of the circuit when resistance and capacitance are connected in series. It is denoted by Z and is expressed in ohms (see p. 68). It also follows that

$$E = IZ \text{ volts, and } Z = \frac{E}{I} \text{ ohms.} \quad (29)$$

Since the tangent of an angle is the ratio of the opposite to the adjacent side, an inspection of Fig. 60(*b*) shows that

$$\tan \theta = \frac{IX_c}{IR} = \frac{X_c}{R} = \frac{1}{2\pi fCR}. \tag{30}$$

Since the cosine of an angle is the ratio of the adjacent side to the hypotenuse,

$$\cos \theta = \frac{IR}{\sqrt{(IR)^2 + (IX_c^2)}} = \frac{R}{\sqrt{R^2 + X_c^2}} = \frac{R}{Z}. \tag{31}$$

In all the foregoing formulas, *C* must be expressed in *farads*. The following example is given as illustrative of the method of computing the quantities in a circuit having resistance and capacitance in series.

Example.—A capacitance of 20 μf and a resistance of 100 ohms are connected in series across 120-volt, 60-cycle mains. Determine: (*a*) impedance of circuit; (*b*) current; (*c*) voltage across resistance; (*d*) voltage across capacitance; (*e*) angle between voltage and current.

20 μf = 0.000020 farad.

$$X_c = \frac{1}{2\pi60 \times 0.000020} = 133 \text{ ohms (Eq. (15), p. 57)}.$$

(*a*) $Z = \sqrt{(100)^2 + (133)^2} = \sqrt{27,700} = 166$ ohms.   *Ans.*

(*b*) $I = \frac{120}{166} = 0.723$ amp (Eq. (28)).

(*c*) $E_R = IR = 0.723 \times 100 = 72.3$ volts.   *Ans.*

(*d*) $E_c = IX_c = 0.723 \times 133 = 96.2$ volts.   *Ans.*

$E = \sqrt{(72.3)^2 + (96.2)^2} = 120$ volts (check).

(*e*) $\tan \theta = \frac{X_c}{R} = \frac{133}{100} = 1.33$ (Eq. (30)).

$\theta = 53.1°$.   *Ans.*

41. Power with Resistance and Capacitance in Series.—In Fig. 59, the power dissipated in the resistance is I^2R. It has already been shown that the average power taken by a pure capacitance over a complete cycle is zero. (See Par. 36, p. 63.) Hence, the capacitance (Fig. 59) consumes no power. All the power taken by the circuit must be accounted for in the resistance, therefore. The total power

$$P = I^2R = I(IR) \text{ watts.} \tag{I}$$

In the vector diagram (Fig. 60(*b*)),

$$\cos \theta = \frac{IR}{E},$$

$$IR = E \cos \theta \text{ volts.}$$

Substituting for IR in (I),

$$P = I(E \cos \theta) = EI \cos \theta \text{ (see Eq. (27), p. 70).}$$

As with inductance and resistance in series, $\cos \theta$ is the power factor of the circuit.

Example.—Determine the power and the power factor in the foregoing example.

$$P = I^2R = (0.723)^2 \times 100 = 52.2 \text{ watts.} \quad Ans.$$
$$\text{P.F.} = \cos \theta = \frac{R}{Z} = \frac{100}{166} = 0.602. \quad Ans.$$

Also,

$$\text{P.F.} = \frac{P}{EI} = \frac{52.2}{120 \times 0.723} = 0.602 \text{ (check).}$$

42. Circuit Containing Resistance, Inductance, and Capacitance in Series.—Figure 61 shows a resistance R, an inductive reactance X_L, and a capacitive reactance X_C, connected in series. The voltage across the circuit is E volts, the frequency is f cycles per second, and the current is I amp. Since this is a series circuit, the current I is the same throughout the circuit. In the vector diagram, the current vector I, for convenience, is laid off horizontal (Fig. 62(*a*)). The voltage $IR(= E_R)$ across the resistance is laid off in phase with the current, since the voltage across a resistance must be in phase with the current. The voltage $IX_L(= E_L)$

Fig. 61.—Circuit containing resistance, inductance, and capacitance in series.

across the inductance leads the current by 90° as shown by the vector *Oc*. The voltage $IX_C(= E_C)$ across the capacitance lags the current by 90° as shown by the vector *Od*. In this particular

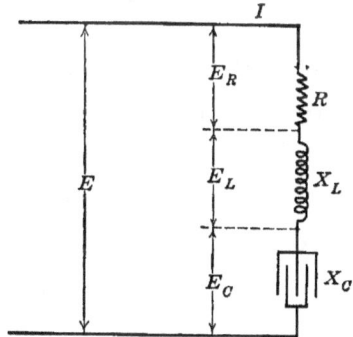

74   *INDUSTRIAL ELECTRICITY*

example the voltage IX_L across the inductance is shown greater than the voltage IX_C across the capacitance.

The line voltage E must be equal to the vector sum of its component voltages. That is, vectorially

$$E = E_R + E_L + E_C \text{ volts.}$$

This vector addition is readily accomplished by first combining IX_L and IX_C (see p. 29). IX_C is less than IX_L and is subtracted from IX_L. This gives the resultant vector

$$Ob = IX_L - IX_C$$

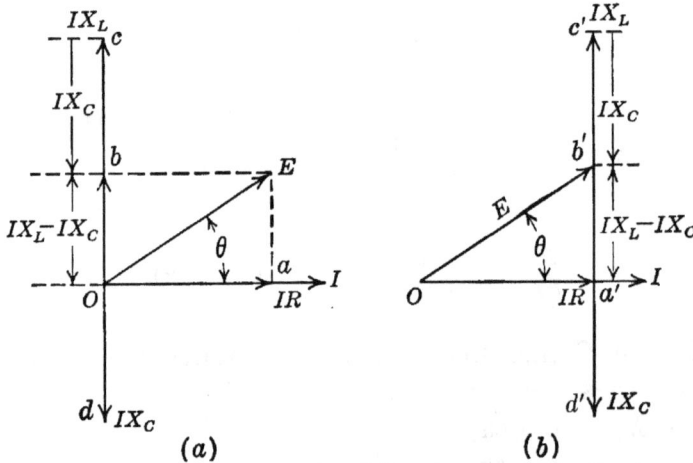

Fig. 62.—Vector diagrams for circuit with resistance, inductance, and capacitance in series.

(Fig. 62(a)). The vector Ob is then added vectorially to IR to give the line voltage E. Since E is the diagonal of a rectangle of which IR and $(IX_L - IX_C)$ are the adjacent sides,

$$E = \sqrt{(IR)^2 + (IX_L - IX_C)^2}$$
$$= \sqrt{I^2R^2 + I^2(X_L - X_C)^2}$$
$$= I\sqrt{R^2 + (X_L - X_C)^2} \text{ volts.}$$
$$I = \frac{E}{\sqrt{R^2 + (X_L - X_C)^2}} \tag{32}$$
$$= \frac{E}{Z} \text{ amp.} \tag{33}$$

The expression $\sqrt{R^2 + (X_L - X_C)^2} = Z$ is the *impedance* of a circuit having resistance, inductance, and capacitance in series.

If X_L is greater than X_C, the quantity within the parenthesis is positive. If X_C is greater than X_L, the quantity within the parenthesis is negative, but since the square of a negative quantity is positive, $(X_L - X_C)^2$ is always positive.

Since $X_L = 2\pi f L$ and $X_C = 1/2\pi f C$, Eq. (32) may be written

$$I = \frac{E}{\sqrt{R^2 + \left(2\pi f L - \dfrac{1}{2\pi f C}\right)^2}} \text{ amp,} \tag{34}$$

where f is the frequency in cycles per second.

The phase angle θ is determined most readily by its tangent.

$$\tan \theta = \frac{IX_L - IX_C}{IR} = \frac{X_L - X_C}{R}. \tag{35}$$

If X_L is greater than X_C, $\tan \theta$ is positive and θ is positive (see p. 432). The voltage then *leads* the current. If X_L is less than X_C, $\tan \theta$ is negative and θ is negative. The voltage then *lags* the current.

Also,

$$\cos \theta = \frac{R}{\sqrt{R^2 + (X_L - X_C)^2}} = \frac{R}{Z}. \tag{36}$$

In Fig. 62(b) the vector addition is accomplished by means of the triangle of vectors (p. 28). At the outer end of IR, the voltage IX_L is added perpendicularly upwards (vector $a'c'$) and the voltage IX_C is added perpendicularly downwards (vector $a'd'$). The resultant $IX_L - IX_C$ of these two vectors is given as $a'b'$ and as in Fig. 62(a) the line voltage $E = \sqrt{(IR)^2 + (IX_L - IX_C)^2}$.

The following examples illustrate the application of the foregoing relations to a specific problem:

Example.—A series circuit consisting of a resistance of 50 ohms, a capacitance of 25 μf, and an inductance of 0.15 henry is connected across 120-volt, 60-cycle mains.

Determine: (a) impedance of circuit; (b) current; (c) voltage across resistance; (d) voltage across inductance; (e) voltage across capacitance; (f) phase angle of circuit.

$$X_L = 2\pi 60 \times 0.15 = 377 \times 0.15 = 56.6 \text{ ohms.}$$

$$X_C = \frac{1}{2\pi 60 \times 0.000025} = 106 \text{ ohms.}$$

(a) $Z = \sqrt{(50)^2 + (56.6 - 106)^2} = \sqrt{(50)^2 + (-49.4)^2}$
$$= \sqrt{2,500 + 2,440} = 70.2 \text{ ohms.} \quad Ans.$$

(b) $I = \dfrac{120}{70.2} = 1.71$ amp (Eq. (33)).   *Ans.*

(c) $E_R = IR = 1.71 \times 50 = 85.5$ volts.   *Ans.*

(d) $E_L = IX_L = 1.71 \times 56.6 = 96.8$ volts.   *Ans.*

(e) $E_C = IX_C = 1.71 \times 106 = 181.1$ volts.   *Ans.*

(f) $\tan\theta = \dfrac{X_L - X_C}{R} = \dfrac{56.6 - 106}{50} = \dfrac{-49.4}{50} = -0.988$ (Eq. (35)).

   $\theta = -44.6°$.

Therefore, the current leads.   *Ans.*

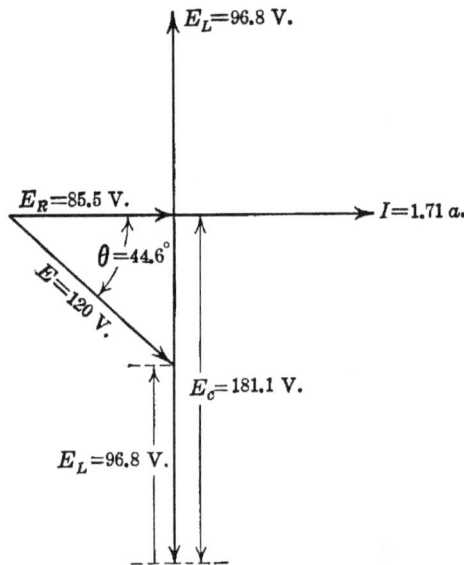

FIG. 63.—Vector diagram for series circuit, giving numerical values.

As a check, the vector sum of the three component voltages may be obtained (Fig. 63).

$$E_L - E_C = 96.8 - 181.1 = -84.3 \text{ volts.}$$
$$E = \sqrt{(85.5)^2 + (-84.3)^2} = \sqrt{7,310 + 7,110} = 120 \text{ volts (check).}$$

The complete vector diagram for this circuit is given in Fig. 63.

Example.—A current of 2.0 amp flows in a circuit consisting of a resistance of 40 ohms, an inductive reactance of 48 ohms, and a capacitive reactance of 20 ohms, connected in series. The frequency is 25 cycles per sec.

Determine: (a) impedance of circuit; (b) circuit voltage; (c) phase angle of circuit; (d) inductance of circuit; (e) capacitance of circuit.

(a) $Z = \sqrt{(40)^2 + (48 - 20)^2} = \sqrt{1,600 + 784} = 48.8$ ohms.   *Ans.*

(b) $E = IZ = 2 \times 48.8 = 97.6$ volts.   *Ans.*

(c) $\tan \theta = \dfrac{48 - 20}{40} = \dfrac{28}{40} = 0.700 \quad \theta = 35.0°. \quad Ans.$

(d) $48 = 2\pi 25 L = 157 L.$

$\quad L = \dfrac{48}{157} = 0.306$ henry. *Ans.*

(e) $20 = \dfrac{1}{2\pi 25 C} = \dfrac{1}{157 C}$

$\quad C = \dfrac{1}{3,140} = 0.000318$ farad, or 318 μf *Ans.*

43. Power in Circuit Containing Resistance, Inductance, and Capacitance in Series.

—In the series circuit (Fig. 61) the power dissipated in the resistance

$$P = I^2 R \text{ watts.} \tag{I}$$

No power is dissipated in either the inductance or the capacitance, as has been shown. Hence, all the power taken by the circuit is accounted for in the resistance. Equation (I) may be written

$$P = I(IR) \text{ watts.} \tag{II}$$

From Fig. 62,

$$IR = E \cos \theta \text{ volts.} \tag{III}$$

Hence, from Eqs. (II) and (III), the power

$$P = EI \cos \theta \text{ watts}$$

as with resistance and inductance in series or with resistance and capacitance in series.

Example.—Find the power and the power factor in the example, p. 75.

$$P = I^2 R = (1.71)^2 \times 50 = 146 \text{ watts.} \quad Ans.$$
$$\cos \theta = \frac{R}{Z} = \frac{50}{70.2} = 0.712 \text{ (Eq. (36))}. \quad Ans.$$

Also,

$$\text{P.F.} = \frac{P}{EI} = \frac{146}{120 \times 1.71} = 0.712 \text{ (check).}$$

44. Resonance in Series Circuit.

—The general equation (34) for the current in a series circuit shows that for fixed values of resistance and impressed voltage the current is a maximum when the denominator is a minimum. The denominator is a minimum when the expression in the parenthesis under the square-root sign is zero.

That is, in Eq. (34)

$$I = \frac{E}{\sqrt{R^2 + \left(2\pi f L - \dfrac{1}{2\pi f C}\right)^2}} \text{ amp,}$$

the current is a maximum when

$$\left(2\pi f L - \frac{1}{2\pi f C}\right) = 0.$$

The current then becomes

$$I = \frac{E}{\sqrt{R^2 + (0)}} = \frac{E}{R} \text{ amp,}$$

its Ohm's law value.

Under these conditions,

$$2\pi f L = \frac{1}{2\pi f C} \text{ ohms,} \tag{37}$$

and

$$2\pi f L I = \frac{I}{2\pi f C} \text{ volts.}$$

$2\pi f L I$ is the voltage across the inductance, and $I/2\pi f C$ is the voltage across the capacitance.

When the current is a maximum under the foregoing conditions, *the voltage across the inductance is equal to the voltage across the capacitance.* As these two voltages are in opposition, they balance each other, so that the *IR* drop is equal to the line voltage. This is illustrated in Fig. 64.

When these conditions exist, the circuit is said to be in *resonance.* The current is then in phase with the line voltage and the power $P =' EI$ watts.

Solving Eq. (37) for the frequency,

$$2\pi f L - \frac{1}{2\pi f C} = 0.$$

$$f = \frac{1}{2\pi \sqrt{LC}} \text{ cycles per sec.} \tag{38}$$

This is the frequency for which a series circuit having inductance L henrys and capacitance C farads is in resonance. It is sometimes called the *natural frequency* of the circuit, because it is the frequency at which the current in the circuit will oscillate, if no external frequency is impressed on the circuit, provided the resistance R is less than $2\sqrt{L/C}$. For example, if in a series circuit a capacitor of C farads is charged to a high voltage and is then discharged into an inductance L henrys of negligible resistance, the frequency of the resulting oscillations is determined by the values of L and C given in Eq. (38). Also, when a radio receiving set is tuned to the incoming signal, inductance and capacitance are so adjusted that the natural or resonant frequency of the set is equal to the frequency of the incoming carrier wave. (See Fig. 65.)

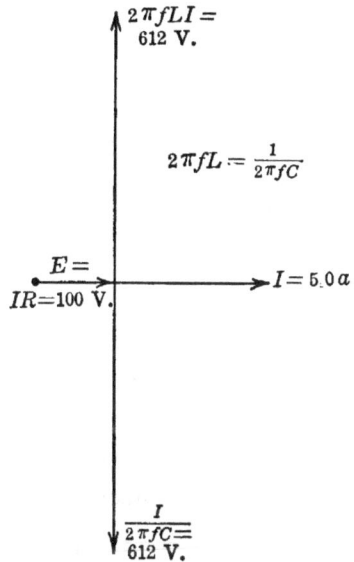

FIG. 64.—Vector diagram for series circuit in resonance.

With the series circuit in resonance, the voltage across the inductance is equal to the voltage across the capacitance. Because the two voltages are in opposition, each may reach a high value, even with moderate line voltage. This is illustrated by the following example:

Example.—The resistance of a circuit is 20 ohms, the inductance 0.3 henry, and the capacitance 20 μf. (*a*) For what value of the frequency will the circuit be in resonance? When the current is 5 amp, determine: (*b*) line voltage; (*c*) voltage across inductance; (*d*) voltage across capacitance; (*e*) power consumed by circuit. (*f*) Draw a vector diagram for the circuit.

(*a*) $f = \dfrac{1}{2\pi \sqrt{0.3 \times 0.000020}} = 65$ cycles. *Ans.*

(*b*) $E = IR = 5 \times 20 = 100$ volts. *Ans.*

(*c*) $E_L = 2\pi fLI = 6.28 \times 65 \times 0.3 \times 5 = 612$ volts. *Ans.*

(*d*) $E_C = \dfrac{I}{(2\pi fC)} = \dfrac{5}{2\pi 65 \times 0.000020} = 612$ volts. *Ans.*

(*e*) $P = EI = 100 \times 5 = 500$ watts. *Ans.*

(*f*) The vector diagram is shown in Fig. 64.

It will be noted that the voltage across the inductance and across the capacitance are equal, each being 612 volts, or more than six times the line voltage.

Example.—A radio receiving set has a fixed inductance of 0.08 millihenry. To what value of capacitance should the capacitor be adjusted in order that the set may be tuned to an incoming signal whose frequency is 1 megacycle (1,000,000 cycles) per sec.

$$\left(\text{Wave length, } \lambda = \frac{(3 \times 10^8)}{1,000,000} = 300 \text{ meters.}\right)$$

Using Eq. (37),

$$2\pi f L = \frac{1}{2\pi f C}.$$

$$C = \frac{1}{4\pi^2 f^2 L} = \frac{1}{4\pi^2 10^{12} 0.00008}$$

$$= 0.000000000317 \text{ farad}$$

$$= 0.000317 \ \mu\text{f}. \quad Ans.$$

It should be noted that when a series circuit is in resonance the current is a *maximum*. This is illustrated in Fig. 65 which

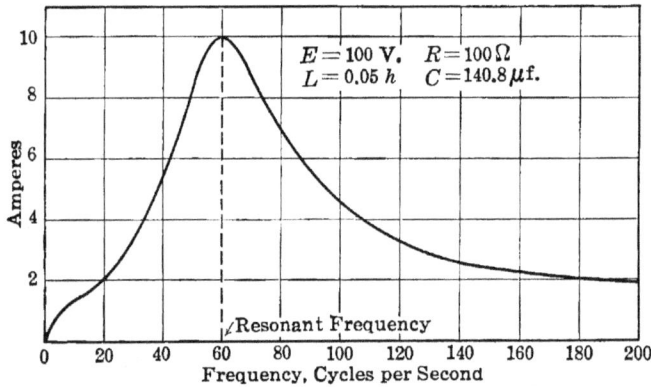

FIG. 65.—Variation of current with frequency in resonant circuit.

shows the variation of current with frequency in a series circuit in which the resistance is 10 ohms, the inductance 0.05 henry, and the capacitance 140.8 μf, the applied voltage being 100 volts. At zero frequency the current is zero (direct current), for the resistance of a pure capacitance is infinite. The current reaches its maximum value at 60 cycles, the frequency to which the circuit is tuned. Thus, substituting in Eq. (38),

$$f = \frac{1}{(6.28 \sqrt{0.05 \times 0.0001408})} = 60 \text{ cycles.}$$

After reaching its maximum value at the resonant frequency of 60 cycles the current diminishes with further increase in frequency and becomes zero at infinite frequency. At infinite frequency, no current can flow through an inductance. (See Eq. (10), p. 48.) In radio the circuit is tuned to the resonant frequency so that the current is a maximum.

If in Eq. (34) (p. 75) the resistance is practically zero and the circuit is in resonance, the current $I = E/0$, or is infinite. That is, a short circuit results. Also, the voltage across the inductance and that across the capacitance theoretically become infinite. Although this may not harm the inductance, it is likely to puncture the capacitor. *Therefore, with a power circuit, do not connect inductance and capacitance in series unless there is sufficient resistance.*

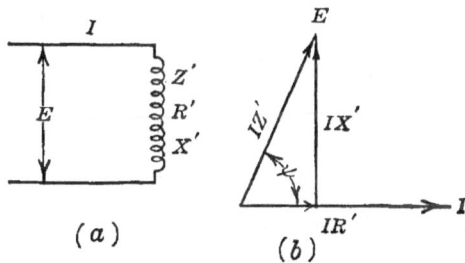

FIG. 66.—Impedance coil.

45. Inductance and Capacitance with Losses.—Thus far, only pure inductance and pure capacitance have been considered. Obviously, it is impossible to obtain either, although pure capacitance may be closely approximated. The wire with which inductance is wound necessarily has resistance. In order to obtain sufficiently high values of inductance, an iron core is usually necessary. Although the core may be laminated, there occur both eddy-current and hysteresis losses in the core. Hence, there is appreciable loss in the circuit and the angle between current and voltage cannot be 90° and is usually considerably less than 90°. Angles as large as 86° are obtainable, but ordinarily the angle lies between 80 and 85°. Hence, coils wound to have inductance are usually called *impedance coils* because of the losses. In making computations of series circuits containing impedance coils the impedances may be considered as consisting of resistance and pure inductance in series. For example, in Fig. 66(*a*) is shown an impedance coil Z' having an iron core. Owing to the fact that this coil has both resistance and iron losses, the angle between the voltage and current is not 90°; but the current lags the voltage by an angle ψ less than 90°, as shown in Fig. 66(*b*). The impedance drop IZ', which is equal to the line voltage E, may be resolved into two components at right angles to each other, IR' along the current vector I, and IX' at right angles to I and leading. R' may then be considered the *effective resistance* of the impedance coil and X' its reactance. A comparison of Fig. 66(*b*) with Fig. 58(*b*) (p. 67), which is the vector diagram for a circuit consisting of resistance and pure inductance in series, shows that they are identical in character.

Because no power is lost in the reactance of the coil, and also from a consideration of Fig. 66(*b*), the total power, supplied to the impedance coil, $P = I^2R'$ watts.

Hence, the effective resistance

$$R' = \frac{P}{I^2} \text{ ohms.} \tag{39}$$

The effective resistance may therefore be determined by measuring the power P to the impedance coil and dividing by the current I squared. (See Fig. 83, p. 104.)

Example.—An impedance coil with an iron core when connected across 110-volt, 60-cycle mains takes 4.0 amp and 30 watts. Determine: (*a*) power factor of impedance coil; (*b*) phase angle; (*c*) effective resistance; (*d*) reactance; (*e*) inductance.

(*a*)   P.F. $= \dfrac{30}{110 \times 4.0} = \dfrac{30}{440} = 0.0682.$ *Ans.*

(*b*) $\cos \psi = 0.0682.$
  $\psi = 86.1°$ (p. 436).   *Ans.*

(*c*) From Eq. (39) the power $P = I^2R'$ where R' is the effective resistance.

$$30 = (4.0)^2R' = 16R'.$$
$$R' = \frac{30}{16} = 1.875 \text{ ohms.}   \textit{Ans.}$$

(*d*) The impedance

$$Z' = \frac{E}{I} = \frac{110}{4.0} = 27.5 \text{ ohms.}$$

The reactance

$$X' = \sqrt{Z'^2 - R'^2} = \sqrt{(27.5)^2 - (1.875)^2}$$
$$= 27.4 \text{ ohms.} \quad Ans.$$

(e) $\quad X' = 2\pi f L.$
$\quad\quad 27.4 = 2\pi 60 L = 377 L.$

$$L = \frac{27.4}{377} = 0.0728 \text{ henry.} \quad Ans.$$

46. Vector Diagram with Resistance and Impedance in Series.

Figure 67(a) shows an impedance coil Z' connected in series with a resistance r. It is required to determine the vector diagram for this circuit. Since this is a series circuit, the current is the same throughout the circuit. The current vector I is laid off

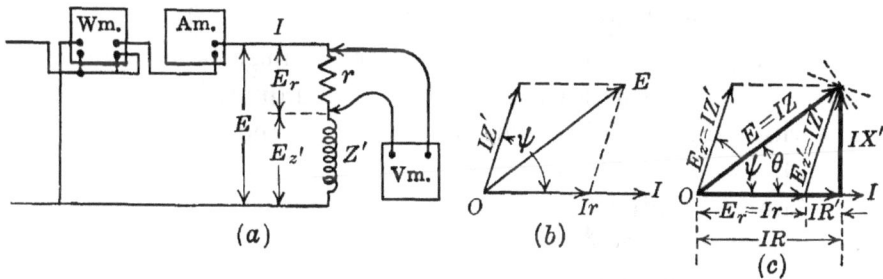

FIG. 67.—Resistance and impedance in series.

horizontally, as shown in (b). The Ir drop is laid off in phase with the current. The voltage $E_{z'} = IZ'$ across the impedance coil leads the current by an angle ψ which is less than 90°. This is also shown in the diagram. The line voltage E is obviously the vector sum of Ir and IZ' as shown.

The vector diagram may be determined experimentally as follows: The current to the circuit is measured with an ammeter (Fig. 67(a)). The voltage E, the voltage E_r across the resistance r, and the voltage $E_{z'}$ across the impedance Z' are measured with a voltmeter. The voltage E_r to scale is laid off in phase with the current vector I (Fig. 67(c)). From the end of E_r an arc equal to $E_{z'}$ to scale is swung in an upward direction. From the origin O, another arc equal to E to scale is swung to intersect the arc swung from the end of E_r. The triangle consisting of the three voltage vectors is then determined. The angle θ may then be determined from the law of cosines (p. 434).

$$E_{z'}^2 = E^2 + E_r^2 - 2EE_r \cos \theta \text{ volts squared.} \qquad (40)$$

$$\cos \theta = \frac{E^2 + E_r^2 - E_{z'}^2}{2EE_r} \qquad (41)$$

The wattmeter may also be connected in circuit to measure the power. (See Fig. 81, p. 102.) If the power to the circuit is P watts, $\cos \theta = P/EI$ which may be used to check Eq. (41).

To determine the effective resistance R' and the reactance X' of the impedance coil, the voltage $E_{z'} = IZ'$ may be resolved into two components, IR' along the current vector and IX' leading the current by 90° as shown in (c). The impedance $Z' = E_{z'}/I$; $IR' = E \cos \theta - Ir$, from which R' may be determined. Then $X' = \sqrt{Z'^2 - R'^2}$. Also, $IX' = IZ' \sin \psi$, and $X' = Z' \sin \psi$.

A vector diagram may be drawn for the circuit in which the total resistance drop is IR, where $R = r + R'$, and the reactance drop is IX'. In other words, this circuit may be considered as having a total resistance R and a pure inductive reactance X' in series. Figure 67(c) should be compared with Fig. 58(b) (p. 67).

A similar vector diagram for resistance and capacitance in series may be drawn except that the voltage across the capacitance *lags* the current. In practice, however, capacitors are very nearly pure capacitances.

The angle by which the phase angle of a capacitor departs from 90° is called the *angle of phase defect of the capacitor*. Mica capacitors often have angles of phase defect which are only 2 or 3′, and it is possible to measure these angles only by the most refined methods. For capacitors in which paraffin paper is used, the angle of phase defect is ordinarily less than 1°, although occasionally the angle may be as large as 2°. If the angle of phase defect of a capacitor departs too far from 0°, the power taken by the capacitor causes its temperature to rise and the capacitor burns out.

Example.—A resistance and an impedance coil are connected in series across a 60-cycle, a-c supply (see Fig. 67(a)). The current is 5.0 amp; and the voltage E across the line, the voltage E_r across the resistance r, and the voltage $E_{z'}$ across the impedance Z' are measured and found to be 120, 90, and 60 volts. Determine: (a) circuit power factor; (b) circuit impedance; (c) resistance r; (d) resistance R'; (e) reactance X'; (f) angle ψ of the impedance Z'.

(a) From Eq. (41),

$$\cos \theta = \frac{\overline{120}^2 + \overline{90}^2 - \overline{60}^2}{2 \times 120 \times 90} = 0.875. \quad Ans.$$

From p. 437, $\theta = 29°$.

(b) $Z = \dfrac{120}{5} = 24$ ohms. *Ans.*

(c) $r = \dfrac{90}{5} = 18$ ohms. *Ans.*

(d) $E \cos \theta - Ir = 120 \times 0.875 - 90 = 105 - 90 = 15$ volts.

$IR' = 15$ volts; $R' = \dfrac{15}{5} = 3$ ohms. *Ans.*

(e) $X' = \sqrt{\overline{Z'^2 - R'^2}}$.

$Z' = \dfrac{60}{5} = 12$ ohms. *Ans.*

$X' = \sqrt{\overline{(12)^2 - \overline{(3)^2}}} = 11.62$ ohms. *Ans.*

(f) $\cos \psi = \dfrac{3}{12} = 0.250$.

From p. 436, $\psi = 75.5°$. *Ans.*

47. Resistance, Impedance, and Capacitance in Series.—In

Fig. 68(a) is shown a series circuit consisting of a resistance R, an impedance Z', and a capacitive reactance X_c; the circuit is connected across an alternating voltage E. It is desired to determine the vector diagram for this circuit by measurements of the voltages and current, as is done in Fig. 67. By means of the voltmeter Vm the four voltages E, E_R, $E_{z'}$, E_c are measured; the current is measured with the ammeter A. In Fig. 67, only three voltages are involved so that the polygon of voltages consists of a triangle (Fig. 67(c)) with the three voltages E_r, $E_{z'}$, E as sides. The polygon is readily constructed; for, with three sides given, a triangle is completely determined. However, in Fig. 68(b) the vector diagram involves a polygon of four voltages E, E_R, $E_{z'}$, E_c. Four sides alone do not determine a polygon; for, with four sides given, an infinite number of different polygons may be constructed. The polygon can be determined only when some additional factor is known. As stated in Par. 46, the phase angle of the usual capacitance is practically 90°; and, except with unusually poor capacitors, in power circuits the phase angle may be taken ordinarily as 90° without error. On this assumption the vector diagram for the circuit of Fig. 68(a) is constructed as follows:

In Fig. 68(b) the current vector I is laid off horizontally. The voltage vector E_R which gives the voltage across the resistance R is laid off in phase with the current vector I, and the vector giving the voltage E_C across X_C is laid off lagging I by an angle α equal to 90°. The voltage E_{R+C}, the resultant of E_R and E_C, is then drawn. Upward from the end of E_{R+C} an arc is swung whose radius to scale is equal to $E_{Z'}$. From the origin O an arc is swung whose radius to scale is equal to E, to intersect the arc swung from the end of E_{R+C}. The positions of E and $E_{Z'}$, also the power-factor angle θ of the circuit and the power-factor angle ψ of the impedance, are now determined. For

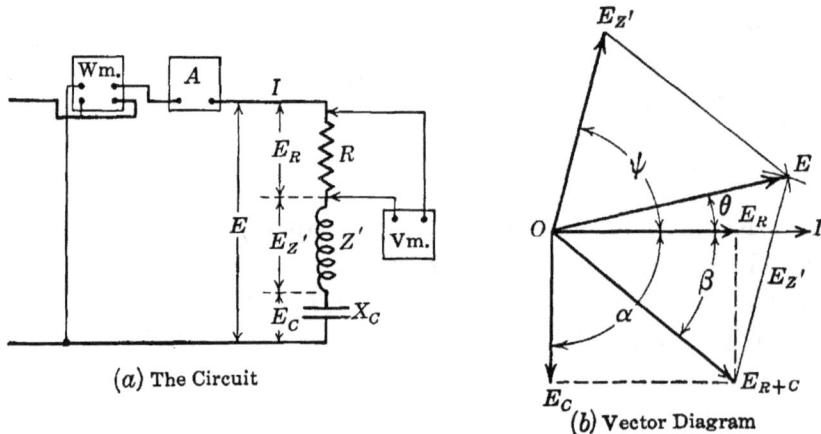

(a) The Circuit

(b) Vector Diagram

FIG. 68.—Resistance, impedance, and capacitance in series.

example, $E_{Z'}$, equal and parallel to $E_{Z'}$ as drawn from the end of E_{R+C}, may be drawn also from the origin O. Figure 68(b) shows that the line voltage E is the vector sum of the three component voltages E_R, $E_{Z'}$, E_C. The angles θ and ψ are now determined and are shown in the diagram. If the phase angle of the capacitor is not 90°, the value of the angle α may be taken accordingly and the diagram constructed. As in Fig. 67(a) a wattmeter Wm may be used to measure the power to the circuit.

Example.—A resistance, an impedance coil, and a capacitor whose phase angle is practically 90° are connected in series (see Fig. 68(a)). The voltages across resistance, impedance, capacitor, and line are measured with a voltmeter as follows: $E_R = 80$ volts; $E_{Z'} = 60$ volts; $E_C = 90$ volts; $E = 115$ volts. The current I is 4 amp. Determine: (a) angle β (Fig. 69); (b) circuit power-factor angle; (c) circuit power factor; (d) circuit power; (e) angle ψ of the impedance coil.

Referring to Fig. 68(b),

(a) $\tan \beta = \dfrac{E_C}{E_R} = \dfrac{90}{80} = 1.125; \qquad \beta = 48.4°. \quad Ans.$

(b) $E_{R+C} = \sqrt{80^2 + 90^2} = 120.4$ volts.

Applying the law of cosines (Eq. (40) and p. 434) to triangle *Oab*,

$$\overline{60}^2 = \overline{120.4}^2 + \overline{115}^2 - 2 \times 120.4 \times 115 \cos (\beta - \theta).$$

$$\cos (\beta - \theta) = \frac{\overline{120.4}^2 + \overline{115}^2 - \overline{60}^2}{2 \times 120.4 \times 115} = \frac{24,130}{27,690} = 0.8713.$$

From p. 437, $\beta - \theta = 29.4°$; $\theta = 48.4° - 29.4° = 19°$. *Ans.*

(c) $\cos \theta = \cos 19° = 0.9455$. *Ans.*

(d) $P = 115 \times 4 \times 0.9455 = 435$ watts. *Ans.*

(e) Power in resistance

$$P_r = 80 \times 4 = 320 \text{ watts.}$$

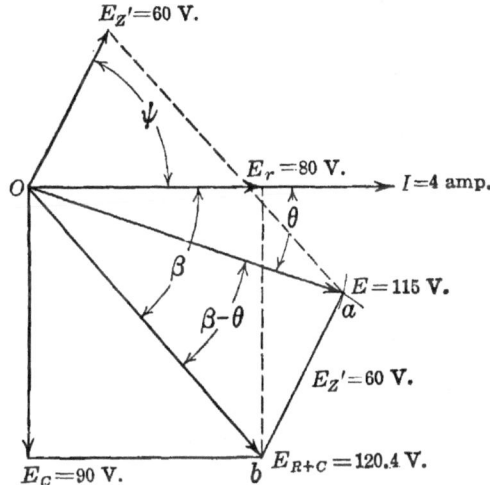

FIG. 69.—Vector diagram with R, Z', and X_c in series.

Since no power is lost in the capacitance, the power in the impedance

$$P' = 435 - 320 = 115 \text{ watts.}$$

$$\cos \psi = \frac{115}{60 \times 4} = 0.479.$$

From p. 436, $\psi = 61.6°$. *Ans.*

PARALLEL CIRCUITS

48. Resistance and Inductance in Parallel.—In practice, parallel circuits are more common than series circuits because of the general use of the parallel or multiple system of transmission and distribution. The determination of the total current when a number of circuits are connected in parallel is not difficult.

In the series circuit the same current flows through each member of the circuit. In the parallel circuit the voltage is the same across each member of the circuit. The method of finding the total current with resistance and inductance in parallel is illustrated by the following example:

Example.—In Fig. 70 a resistance R of 10 ohms, and an inductance L of 0.03 henry are connected in parallel across a 100-volt, 50-cycle circuit. Determine: (*a*) current I_R in resistance; (*b*) current I_L in inductance; (*c*)

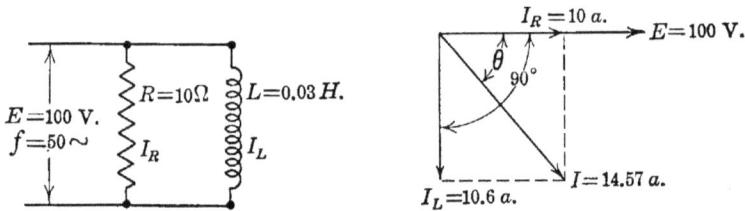

Fig. 70.—Resistance and inductance in parallel.

total current I; (*d*) power taken by circuit; (*e*) power factor of circuit; (*f*) power-factor angle of circuit.

(*a*) Current in resistance

$$I_R = \frac{100}{10} = 10 \text{ amp.} \quad Ans.$$

(*b*) Reactance of inductive branch

$$X_L = 2\pi 50 \times 0.03 = 314 \times 0.03 = 9.42 \text{ ohms.}$$

Current in this branch

$$I_L = \frac{100}{9.42} = 10.6 \text{ amp.} \quad Ans.$$

(*c*) The vector diagram is shown in Fig. 70(*b*). Voltage $E = 100 \ V$ is common to both branches of the circuit and is laid off horizontally for convenience. The current I_R in the resistance (10 amp) is in phase with the voltage. The current I_L in the inductance (10.6 amp) *lags* the voltage by 90°. The total current I is their vector sum. Since I_R and I_L differ in phase by 90°, the resultant current I is the square root of the sum of their squares.

$$I = \sqrt{(10)^2 + (10.6)^2} = \sqrt{212.4} = 14.57 \text{ amp.} \quad Ans.$$

(*d*) The average power taken by the inductance is zero. All the power taken by the circuit must be accounted for in the resistance. Hence,

$$P = EI_R = 100 \times 10 = 1,000 \text{ watts.} \quad Ans.$$

(*e*) P.F. $= \dfrac{P}{EI} = \dfrac{1,000}{100 \times 14.57} = \dfrac{1,000}{1,457} = 0.686. \quad Ans.$

(*f*) An inspection of Fig. 70(*b*) shows that

$$\tan \theta = \frac{10.6}{10} = 1.06.$$

$$\theta = 46.7° \text{ (p. 439). } \textit{Ans.}$$

Also, from (*e*),

$$\cos \theta = 0.686$$
$$\theta = 46.7° \text{ (check).}$$

It follows that, with resistance and inductance in parallel,

$$I = \sqrt{\left(\frac{E}{R}\right)^2 + \left(\frac{E}{X_L}\right)^2} = \frac{E}{RX_L}\sqrt{R^2 + X_L{}^2} \text{ amp.} \qquad (42)$$

49. Resistance and Capacitance in Parallel.—The solution of circuits having resistance and capacitance in parallel is illustrated by the following example:

Example.—A resistance of 40 ohms and a capacitance of 60 µf are connected in parallel across 100-volt, 60-cycle mains (Fig. 71(*a*)).   Determine:

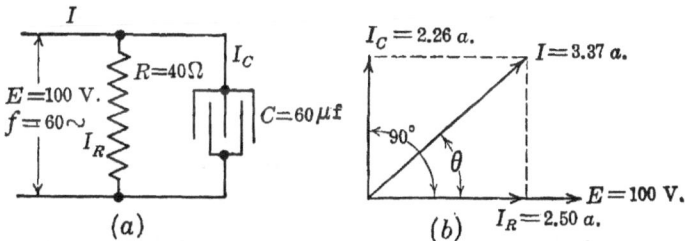

Fig. 71.—Resistance and capacitance in parallel.

(*a*) current in resistance; (*b*) current taken by capacitance; (*c*) total current; (*d*) power taken by circuit; (*e*) power factor of circuit; (*f*) power-factor angle of circuit.

(*a*) Current in resistance

$$I_R = \frac{100}{40} = 2.50 \text{ amp.} \quad \textit{Ans.}$$

The voltage is common to both branches of this circuit and is laid off horizontal (Fig. 71(*b*)).   The current I_R to scale is laid off in phase with the voltage E.

(*b*) Capacitive reactance from Eq. (15) (p. 57)

$$X_C = \frac{10^6}{2\pi 60 \times 60} = 44.2 \text{ ohms.}$$

$$I_C = \frac{100}{44.2} = 2.26 \text{ amp.} \quad \textit{Ans.}$$

This current I_C leads the voltage by 90°, as shown in the vector diagram (Fig. 71(b)).

(c) Since I_R and I_C are in quadrature, the resultant current I is the square root of the sum of their squares.

$$I = \sqrt{(2.50)^2 + (2.26)^2} = \sqrt{6.25 + 5.11} = 3.37 \text{ amp.} \quad Ans.$$

The resultant current I is shown in the vector diagram (Fig. 71 (b)).

(d) The average power taken by the capacitance is zero. Hence, all the power taken by the circuit is accounted for in the resistance.

The total power, therefore,

$$P = 100 \times 2.50 = 250 \text{ watts.} \quad Ans.$$

(e) P.F. $= \dfrac{P}{EI} = \dfrac{250}{100 \times 3.37} = 0.742. \quad Ans.$

(f) From Fig. 71(b),

$$\tan \theta = \frac{I_C}{I_R} = \frac{2.26}{2.50} = 0.904.$$
$$\theta = 42.1° \text{ (p. 438).} \quad Ans.$$

Also, from (e),

$$\cos \theta = 0.742.$$
$$\theta = 42.1° \text{ (check).}$$

From Eq. (42) it follows,

$$I = \frac{E}{RX_c} \sqrt{R^2 + X_c^2} \text{ amp.} \tag{43}$$

50. Resistance, Inductance, and Capacitance in Parallel.— With resistance, inductance, and capacitance in parallel, it is necessary merely to find the current in each branch of the circuit and then to combine these currents vectorially to find the total current. This is illustrated by the following example:

Example.—A resistance of 40 ohms, an inductance of 0.08 henry, and a capacitance of 50 μf are connected in parallel across a 100-volt, 60-cycle supply (Fig. 72(a)). Determine: (a) current in each branch of circuit; (b) total current; (c) power taken by circuit; (d) power factor of circuit; (e) power-factor angle of circuit.

(a)

$$I_R = \frac{100}{40} = 2.50 \text{ amp.} \quad Ans.$$
$$X_L = 2\pi 60 \times 0.08 = 30.2 \text{ ohms.}$$
$$I_L = \frac{100}{30.2} = 3.31 \text{ amp.} \quad Ans.$$

$$X_C = \frac{10^6}{2\pi 60 \times 50} = 53.0 \text{ ohms.}$$

$$I_C = \frac{100}{53.0} = 1.885 \text{ amp.} \quad Ans.$$

The three currents are shown vectorially in Fig. 72(*b*). I_R is in phase with the voltage E, I_L lags the voltage E by 90°, and I_C leads the voltage E by 90°.

(*b*) To find the total current I, it is necessary merely to find the vector sum of I_R, I_L, I_C (Fig. 72(*b*)). Since I_L and I_C are in opposition, their

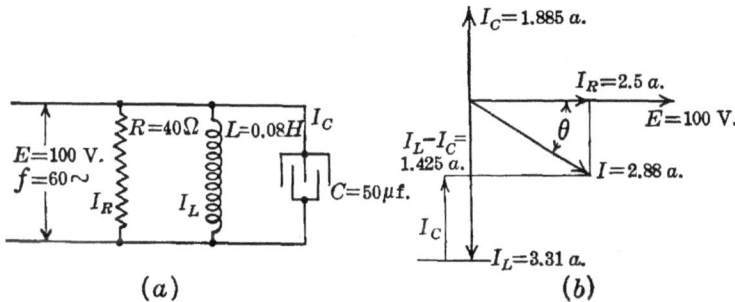

Fig. 72.—Resistance, inductance, and capacitance in parallel.

resultant is $3.31 - 1.885 = 1.425$ amp. This current is then combined in quadrature with I_R to find the total current I.

$$I = \sqrt{(2.50)^2 + (1.425)^2} = \sqrt{6.25 + 2.03} = \sqrt{8.28} = 2.88 \text{ amp.} \quad Ans.$$

(*c*) The inductance and capacitance consume no power. Hence, all the power must be accounted for in the resistance.

$$P = EI_R = 100 \times 2.50 = 250 \text{ watts.} \quad Ans.$$

(*d*) P.F. $= \dfrac{250}{100 \times 2.88} = 0.868.$ *Ans.*

(*e*) From Fig. 72(*b*),

$$\tan \theta = \frac{I_C - I_L}{I_R} = \frac{-1.425}{+2.50} = -0.570.$$
$$\theta = -29.7° \text{ (p. 438).} \quad Ans.$$

Also, from (*d*),

$$\cos \theta = 0.868, \qquad \theta = -29.8° \text{ (check).}$$

51. Antiresonance in Parallel Circuits.—Resonance occurs in series circuits when the voltage across the inductance is equal to the voltage across the capacitance. The current is then in phase with the line voltage; and, for a given value of resistance, the current in the circuit is a *maximum*.

Consider the circuit (Fig. 72) in which resistance, inductance, and capacitance are in parallel. If the current I_L in the inductance is equal to I_C in the capacitance, the two cancel when the vector addition is made (Fig. 73(b)). The current in the resistance is then the resultant current and is in phase with the line voltage E. The instantaneous values of current and voltage are shown in (a). The voltage wave e is sinusoidal. The current i_R in the resistance is in phase with the voltage e. The current i_L in the inductance lags the voltage e by 90°. The current i_C in the capacitance leads the voltage e by 90°. Since the currents i_L and i_C are equal and in opposition, they cancel at every instant. The current i_R in the resistance is therefore the resultant current.

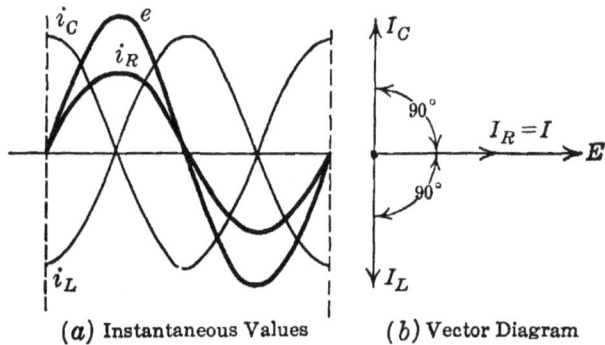

(*a*) Instantaneous Values   (*b*) Vector Diagram
Fɪɢ. 73.—Antiresonance in parallel circuit.

The vector diagram for the circuit is shown in Fig. 73(b). The values are rms, and a larger scale than that of (a) is used. The line voltage E is laid off horizontally, the current I_R in the resistance is in phase with E, the current I_L in the inductance lags E by 90°, and the current I_C in the capacitance leads E by 90°. The current I_L in the inductance is equal numerically to the current I_C in the capacitance. Since these currents are in opposition, the line current I is equal to the current I_R in the resistance.

With a given value of resistance and hence of I_R, the current is a *minimum* under these conditions. The circuit is said, therefore, to be in *antiresonance* in contrast to resonance and maximum current in the series circuit. (Formerly the term "resonance" was applied also to the parallel circuit.)

With pure inductance and pure capacitance only in parallel and the circuit adjusted for antiresonance, the line current is zero. This condition cannot be attained actually, although it may be approximated. That is, the circuit may be so tuned that the current to the inductance and that to the capacitance may be considerable, and yet the current from the power source may be very small.

To summarize: With the series circuit having a fixed resistance the current is a *maximum* when the circuit is tuned for resonance. With the parallel circuit having a fixed resistance the current is a *minimum* when the circuit is tuned for antiresonance.

In both circuits, under these conditions, the resultant current is in phase with the line voltage, the power is the product of the line voltage and resultant current and the power factor is unity.

Example.—A circuit consisting of a resistance of 60 ohms and an inductance of 0.3 henry in parallel is connected across 100-volt, 25-cycle mains. (a) For what value of capacitance in parallel will the circuit be in antiresonance? (b) What is the line current under these conditions?

For antiresonance the current in the capacitance must be equal to the current in the inductance. Hence the capacitive reactance must be equal to the inductive reactance.

The inductive reactance

$$X_L = 2\pi 25 \times 0.3 = 47.1 \text{ ohms.}$$

The capacitive reactance

$$X_C = 47.1 \text{ ohms} = \frac{1}{2\pi 25 C}.$$

$$C = \frac{1}{157 \times 47.1} = 0.000135 \text{ farad}$$

$$= 135 \, \mu\text{f.} \quad Ans.$$

From the foregoing, it is clear that, for antiresonance, $X_C = X_L$.

$$\frac{1}{2\pi f C} = 2\pi f L.$$

$$f = \frac{1}{2\pi \sqrt{LC}}. \tag{44}$$

This equation should be compared with Eq. (38) (p. 78) for the series circuit. Equation (44) holds only with *pure* inductance and *pure* capacitance. With resistance in series with the inductance or capacitance, it does not apply.

52. Resistance and Impedance in Parallel.—When an impedance coil having inductance and losses is connected in parallel with resistance the circuit vector diagram may be determined in a manner similar to that for impedance and resistance in series except that the polygon now consists of currents rather than voltages. (See Par. 46, p. 83.) In Fig. 74(a) an inductive impedance Z' is connected in parallel with the resistance R across the voltage E. The data for the vector diagram may be obtained by measuring the voltage E and the three currents, I_R to the resistance, $I_{Z'}$ to the impedance, and I the total current. A voltmeter and three ammeters for making these measurements are shown.

The vector diagram is shown in (b). The current I_R to scale is laid off in phase with the voltage E. Then an arc with $I_{Z'}$, to scale, as a radius is swung downward from the end of I_R; and another arc with I, to scale, as a radius is swung from the origin

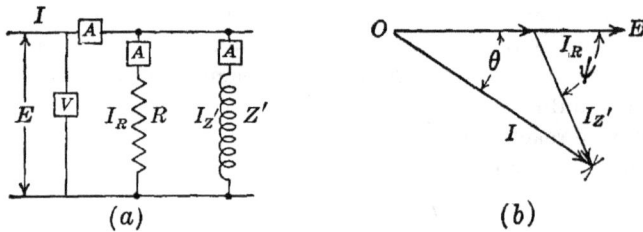

FIG. 74.—Resistance and impedance in parallel.

O to intersect the first arc. When $I_{Z'}$ and I are drawn to the intersection, the diagram is complete. The phase angle θ of the circuit and the phase angle ψ of the impedance are determined. (See example, p. 84, for method of solving.) A wattmeter may also be connected in circuit to measure the power and thus to check the diagram.

53. Resistance, Impedance, and Capacitance in Parallel.—In Fig. 75(a) a resistance R, an impedance Z', and a capacitive reactance X_C are connected in parallel across the voltage E. The vector diagram for this circuit is determined in a manner similar to that for resistance, impedance, and capacitance in series except that the polygon now consists of currents rather than voltages. (See Par. 47, p. 85.) The line voltage E and the four currents, I_R to the resistance, $I_{Z'}$ to the impedance, I_C to the capacitive reactance, and I the total current are measured with the instruments shown.

The vector diagram is shown in (b). The current I_R, to scale, is laid off in phase with the voltage E; the current I_C, to scale, is laid off 90° leading; and the resultant I_{R+C} is found. From the end of I_{R+C} an arc with $I_{Z'}$, to scale, as a radius is swung downward; and from the origin O an arc with I, to scale, as a radius is swung to intersect the first arc. $I_{Z'}$ and I are then drawn to the intersection as shown. $I_{Z'}$ may now be drawn from the origin O, forming one side of a parallelogram. A study of (b)

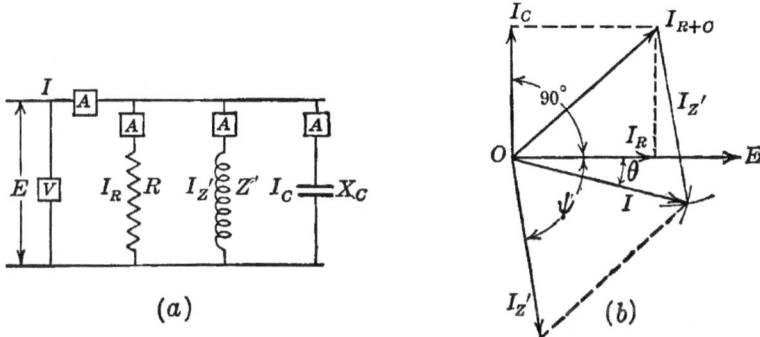

Fig. 75.—Resistance, impedance, and capacitance in parallel.

shows that $I = I_R + I_{Z'} + I_C$. (Also see Figs. 68(b) and 69, pp. 86 and 87.) A wattmeter may be connected to measure the power to the circuit and thus check the diagram.

54. Energy and Quadrature Current; Reactive Power.—In Fig. 76(a) a load takes a current I at voltage E, and the power factor of the load is $\cos \theta$, the current lagging. The vector diagram is shown in (b) in which the current I lags the voltage

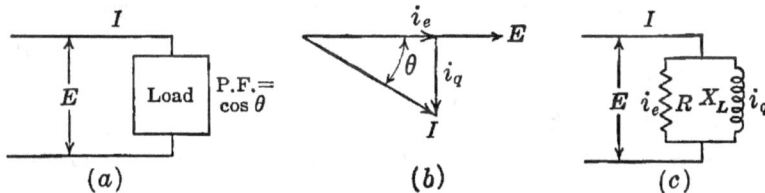

Fig. 76.—Energy and quadrature currents.

E by the angle θ. The current I may be resolved into two components, i_e in phase with E and i_q lagging E by 90°. That is, the actual load current I is the vector sum of i_e and i_q. The power $P = EI \cos \theta$. But $i_e = I \cos \theta$. Hence, the power

$$P = Ei_e \text{ watts.} \tag{45}$$

The component i_e in phase with E is the *energy* current. Hence, *the power is equal to the product of the voltage and the energy current.*

The component i_q is called the *quadrature current* and some-times the wattless current.

$$i_q = I \sin \theta \text{ amp.} \tag{46}$$

i_q contributes no power, since it has a phase difference of 90° with the voltage E.

In Fig. 76(c) is shown the equivalent circuit that may replace the load in (a). A resistance $R(= E/i_e)$ is connected in parallel with a reactance $X_L(= E/i_q)$. As in Eq. (45) the total power, $P = Ei_e$ watts; and as in Eq. (47) the total reactive power, $Q = Ei_q$ vars.

If the current I leads, the quadrature current is positive. The quadrature current contributes no power and so far as power is concerned can be eliminated. However, in practice, quadrature current usually exists, for inductive apparatus such as transformers and induction motors takes lagging current.

Quadrature current becomes disadvantageous when it is necessary for the power to be transmitted. If R is the resistance of the power line, the line loss $P_C = I^2R = (i_e{}^2 + i_q{}^2)R$. If the current I lags the voltage E by 45°, the power factor is 0.707 and $i_e = i_q$. That is, the quadrature current causes the same loss as the energy current. Hence, it is highly advantageous to eliminate the quadrature current where this can be done economically. This is frequently accomplished in practice by the use of capacitors and synchronous condensers. (See p. 321 and Part I, p. 200.)

The product of E and i_q gives the *reactive power* Q. The unit of reactive power is the *var* (volt-ampere reactive). Hence,

$$Q = Ei_q = EI \sin \theta \text{ vars}, \tag{47}$$

and

$$Q = \frac{EI \sin \theta}{1{,}000} \text{ kilovars (kvars)}. \tag{48}$$

Example.—A load is taking 120 kw at 2,300 volts, 60 cycles, and power factor of 0.65 lagging current. Determine: (a) current; (b) energy current; (c) quadrature current; (d) kilovars; (e) rating of a capacitor connected in parallel with the load which will balance the load quadrature current. The vector diagram is shown in Fig. 77.

(a) $I = \dfrac{120,000}{2,300 \times 0.65} = 80.3$ amp. *Ans.*

(b) $i_e = 80.3 \cos \theta = 80.3 \times 0.65 = 52.2$ amp. *Ans.*

Also

$$i_e = \frac{120,000}{2,300} = 52.2 \text{ amp.}$$

(c) $i_q = I \sin \theta$ amp.

$$\cos \theta = 0.65; \qquad \theta = 49.5° \text{ (p. 436)}; \qquad \sin \theta = 0.7604.$$

$$i_q = 80.3 \times 0.7604 = 61.0 \text{ amp.} \quad Ans.$$

$$\sqrt{52.2^2 + 61.0^2} = 80.3 \text{ amp (check).}$$

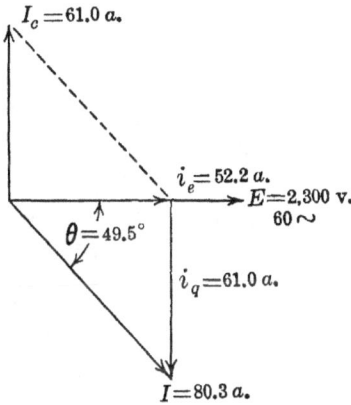

FIG. 77.—Energy and quadrature currents of a power load.

(d) Kvars $= \dfrac{2,300 \times 61.0}{1,000} = 140.3$. *Ans.*

(e) A current of 61.0 amp leading 90° is necessary to counterbalance $i_q = 61.0$ amp which lags 90°.

Using Eq. (13) (p. 56), $I = EC\omega$.

$$C = \frac{i_q}{E\omega} = \frac{61.0}{2,300 \times 377} = 0.0000704 \text{ farad} = 70.4 \ \mu\text{f.} \quad Ans.$$

It will be noted that the magnitude of the quadrature current exceeds that of the energy current. Also, with the capacitor connected in parallel the resultant current is i_e in phase with the voltage E and is a minimum for the given power.

CHAPTER III

ALTERNATING-CURRENT INSTRUMENTS AND MEASUREMENTS

ELECTRODYNAMOMETER-TYPE INSTRUMENTS

55. Siemens Dynamometer.—Several types of a-c instruments operate on the electrodynamometer principle. The Siemens dynamometer (Fig. 78) is an example of this type of instrument in simple form. It consists primarily of two sets of coils. The coil F is fixed; the coil M, whose axis is at right angles to the axis of F, is free to turn through a small angle. M is suspended by a torsionless filament such as a thread, and its turning moment is opposed by a helical spring S. Current is led into the moving coil M through two mercury cups C.

Fig. 78.—Siemens dynamometer.

When the instrument is used as an ammeter, the two coils are wound with a few turns of coarse wire and are connected in series. When current flows through these coils, there is a tendency for the moving coil to swing into the plane of the fixed coil so that their magnetic fields act in conjunction (see Part I, p. 148, Par. 125). When the current reverses, it reverses simultaneously in the two coils so that the torque is always in the same direction. The movable coil is not allowed to deflect, however, but is kept in its zero position by turning the knurled head H at the top of the instrument which acts on the coil through the springs. The angle by which it is necessary to turn this head is proportional to the turning moment of the coil. The turning moment is proportional to the product of the current in the fixed coil and the current in the movable coil. When the two coils are in series the turning moment must be proportional to the current *squared* and hence to the *heating effect* of the current (see p. 21, Par. 13). The instrument is calibrated with a known direct current. An alternating current giving this same deflection has the same heating effect as the direct current. Hence, the rms a-c amperes squared are equal to the d-c amperes squared for equal settings of the torsion head. When direct current is used, it is advisable to reverse the direction of the current and then to average the readings to eliminate the effect of stray fields.

If small wire is substituted for the coarse wire in both fixed and moving coils and a high resistance is connected in series, the instrument can be used as a voltmeter. Likewise, if small wire is substituted for the coarse wire in the moving coil and a high resistance is connected in series with the coil, the instrument may be used as a wattmeter (p. 101).

This type of instrument is difficult to adjust and to manipulate, especially when the current fluctuates. It is not direct-reading and because of its construction is adapted to laboratory work only. At the present time, it is used rarely if at all, but it illustrates the principle on which electrodynamometer instruments operate, namely, that of torque developed between two coils, one surrounding the other, when current flows in the coils.

56. Indicating Electrodynamometer.—As it is neither portable nor direct-reading, the Siemens dynamometer is not adapted either to portable or to switchboard instruments. However, many types of portable and of switchboard instruments operate on the Siemens-dynamometer principle except that the moving coil, instead of being kept at the zero position, is permitted to deflect against the restraining torque of a spring. The general construction of a portable type of electrodynamometer instrument is shown in Fig. 79.

Two fixed coils *FF'* are connected so that their magnetic fields act in conjunction. These coils may be considered as two parts of a single coil, opened to allow space for the spindle of the moving coil.

M is a movable coil mounted on a vertical spindle. There is a hardened steel pivot at each end of the spindle, which turns in jeweled bearings. Two spiral springs similar to those used in d-c instruments (see Part I, p. 96, Fig. 78) oppose the turning of coil M and at the same time carry the current into the coil. As the springs can carry only a small current, the movable coil is wound with fine wire.

Assume that at some instant the direction of the magnetic field ϕ_1, due to the fixed coils, is from left to right. At the same

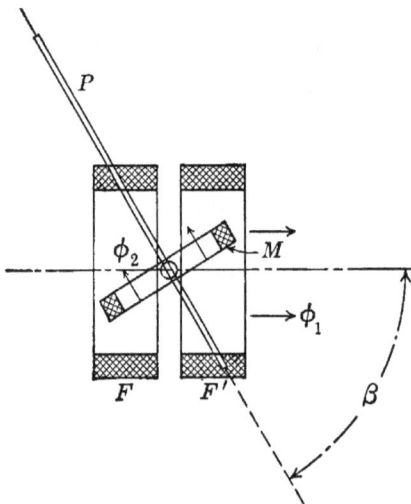

FIG. 79.—Principle of the electro-dynamometer instrument.

instant the current in coil M produces a field ϕ_2 whose direction is along the axis of M. Coils tend to align themselves so that the number of magnetic linkages in the system is a maximum. Therefore, the moving coil M tends to turn in a clockwise direction so that its field ϕ_2 will act in conjunction with ϕ_1. The turning of M is opposed by the control springs.

The torque developed is proportional to ϕ_1, ϕ_2, and $\sin \beta$, where β is the angle between the axis of coil M and the axis of coil FF'. As ϕ_1 and ϕ_2 are proportional to the currents in the coil FF' and M, the torque is proportional to the product of the two currents and $\sin \beta$.

57. Electrodynamometer Voltmeter.—Some types of a-c voltmeter operate on the electrodynamometer principle. The fixed coil FF' and the moving coil M (Fig. 79) are wound with a comparatively large number of turns of small wire and are connected in series. As the instrument itself would take too much current when connected directly across the ordinary circuit, a comparatively high resistance is connected in series.

The current flowing through the dynamometer is therefore proportional to the line voltage and causes coil M to turn, and a pointer attached to M moves over a scale graduated in volts. The scale will not be divided uniformly like that of the d-c voltmeter, since the deflections are very nearly proportional to the *square* of the voltage. The divisions at the lower part of the scale are so small that poor precision is obtained with small **readings.**

This dynamometer type of voltmeter takes about five times as much current as a d-c voltmeter of the same rating and consumes an appreciable amount of power. As the moving coil is in a comparatively weak field, this type of instrument is very susceptible to stray fields and should not be used near inductive apparatus, wires carrying even moderate currents, and stray fields in general.

This instrument may be used for direct as well as for alternating current. Reversed d-c readings should be taken in order to eliminate the effect of the earth's field and of any other stray fields.

Electrodynamometer ammeters are made only for special purposes, since, with alternating current, shunt adjustments are not simple.

58. Wattmeter.—Alternating-current power is equal to the product of rms current and rms voltage only when the power factor is unity. Therefore, the ammeter and voltmeter method, as used with direct currents, can seldom be used to measure a-c power. Consequently, a *wattmeter* is necessary for measuring a-c power.

The wattmeter shown in Fig. 80 operates on the electrodynamometer principle. *M* is a moving coil wound with fine wire and is practically identical with the moving coil of the dynamometer voltmeter (Fig. 79). It is connected across the line in series with a high resistance *R*. The current is led into this coil through the spiral control springs. The fixed coil *FF* is wound with few turns of coarse wire, capable of carrying the load current. As there is no

Fig. 80.—Connections for a wattmeter.

iron, the field due to the current coils *FF* is proportional to the load current at every instant. The current in the moving coil *M* is proportional to the voltage at every instant. Therefore, for any given position of the moving coil, the torque is proportional at every instant to the product of the current and voltage or to the instantaneous power of the circuit. Unless

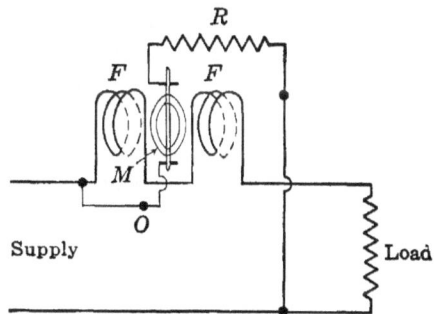

the power factor is unity, there is negative torque for part of the cycle. That is, during the periods when there are negative loops in the power curve (Fig. 56, p. 66) the fields due to the currents in the fixed and moving coils reverse with respect to each other and produce negative torque.

As the torque acting on the moving coil varies from instant to instant, having a frequency twice that of the current and voltage, the coil tends to change its position to correspond with these variations of torque. If the moving system had little inertia, the needle would vibrate so that it would be impossible to obtain a reading. Because of the relatively large moment of inertia of the moving system, the coil and the needle assume a steady deflection for constant values of average power. The moving coil takes a position corresponding to the *average* torque.

It should be noted (Fig. 80) that the voltage terminal O is connected directly to one end of the moving coil. This terminal is ordinarily connected to that side of the line in which the current coil is connected (see also Part I, p. 123, Par. 98).

59. Wattmeter Connections.—In Fig. 81, wattmeter W is shown measuring the power taken by a load. In order to

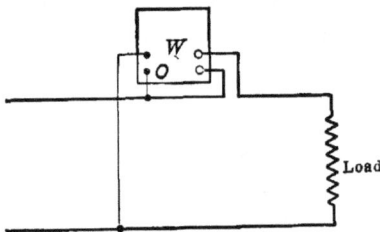

FIG. 81.—Instrument measures power consumed by its own current coil.

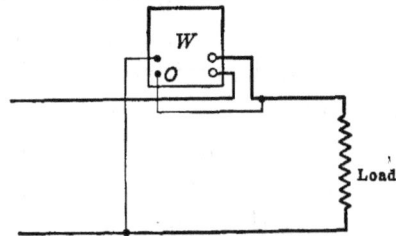

FIG. 82.—Instrument measures power consumed by its own potential circuit.

measure the power correctly, the *load* current should flow through the wattmeter current coil; and the wattmeter voltage coil, in series with its resistance, should be connected directly across the *load*.

In Fig. 81 the load current flows in the wattmeter current coil, but the wattmeter potential circuit is not connected directly across the load and is measuring a voltage in excess of the load voltage by the amount of the impedance drop in the wattmeter current coil. Therefore, the wattmeter reads too high by the amount of power consumed in its current coil. Under these conditions the true power

$$P = P' - I^2 R_c \text{ watts,} \tag{49}$$

where P' is the power indicated by the wattmeter, I is the current in the wattmeter current coil, and R_c is the resistance of this coil. The power consumed in the current coil is ordinarily about 1 or 2 watts at the rated current of the instrument and usually may be neglected.

If the wattmeter is connected as shown in Fig. 82, the wattmeter potential circuit is connected directly across the load but the potential-coil current in addition to the load current flows in the wattmeter current coil. The wattmeter potential circuit may be considered as a small load in parallel with the actual load whose power is to be measured. Therefore, the power taken by this potential circuit must be deducted from the wattmeter reading. The true power taken by the load

$$P = P' - \frac{E^2}{R_p} \text{ watts,} \tag{50}$$

where P' is the wattmeter reading, E the load voltage, and R_p the resistance of the wattmeter potential circuit.

An idea of the magnitude of this correction may be obtained from the following example:

Example.—A wattmeter indicates 157 watts when it is connected as in Fig. 82. The line voltage is 120 volts, and the resistance of the wattmeter potential circuit is 2,000 ohms. What is the true power taken by the load?

$$P = 157 - \frac{\overline{120}^2}{2,000} = 157 - 7.2 = 149.8 \text{ watts.} \quad \textit{Ans.}$$

It will be noted that in this case an error of 5.6 per cent would result if the wattmeter potential-circuit loss were neglected.

The current and potential circuits of a wattmeter must have a rating corresponding to the *current* and *voltage* of the circuit in which the wattmeter is connected. A wattmeter is rated in amperes and volts, rather than in watts, because the indicated watts show neither the amperes in the current coil nor the voltage across the potential circuit.

If the current in an ammeter or the voltage across a voltmeter exceeds the rating of the instrument, the pointer goes off scale and so warns the user. A wattmeter may be considerably overloaded, and yet the load power factor may be so low that the needle is within the limit of the scale. Also, if the rating of the current coil is exceeded and the voltage is low or if the rating of the potential circuit is exceeded and the current is low, the needle may still be on the scale. For these reasons a voltmeter and an ammeter should be used ordinarily in conjunction with a wattmeter so that it may be determined whether either the voltage or the current exceeds the wattmeter rating.

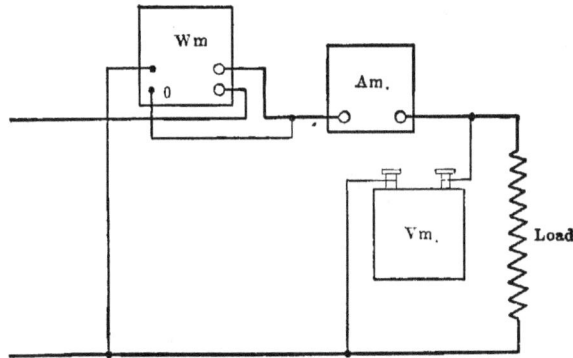

Fig. 83.—Wattmeter, ammeter, and voltmeter connections for measuring power.

Figure 83 shows the connections which may be used to measure the power taken by a single-phase load. In addition to the load power, the wattmeter measures the power taken by its potential circuit, the power taken by the voltmeter and the power taken by the ammeter. Correction for the first two may be necessary, although with steady conditions the voltmeter may be disconnected when the wattmeter and the ammeter are read.

60. Polyphase Wattmeter.—Ordinarily, it requires two or more wattmeters to measure the total power of a two-phase or of a three-phase circuit. (See Pars. 83, 84, and 86, pp. 140, 142, and 148.) If the load fluctuates, it is difficult to obtain accurate simultaneous readings of two or more wattmeters. At a power factor less than 0.5, in a balanced three-phase circuit, the reading of one of the wattmeters is reversed (see p. 143). This necessitates reversing the connections of this instrument, which is often inconvenient. If both wattmeters are combined in a single one, both moving coils being mounted on the same spindle, the turning moments add or subtract automatically, and the total power is read on a single scale.

Figure 84 is an interior view of a Weston polyphase wattmeter in which the two elements are clearly shown. Figure 85 shows one method for connecting a polyphase wattmeter in a three-

FIG. 84.—Interior view, Weston polyphase wattmeter.

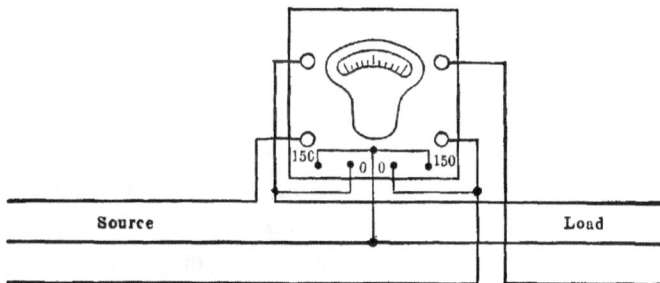

FIG. 85.—Connections for polyphase wattmeter on three-phase circuit.

phase circuit. Note that two wires from the source are connected to the two front current binding posts and two wires to the load are connected to the two rear current binding posts. Each of the O potential-circuit binding posts is connected to the wire corresponding to the current connection on its side of the watt-meter. The other two potential-circuit binding posts (150) are connected to the third wire of the system in which there is no current connection.

IRON-VANE INSTRUMENTS

61. Iron-vane Instruments.—The usual type of a-c voltmeter and ammeter depends for its operation on the magnetization of a light iron vane mounted on a spindle. Such instruments are simple and rugged, with no electrical connections to the moving system, and are of low cost. Although they are not adapted to measurements of highest precision, their precision is satisfactory for most measurements.

One such type of instrument, manufactured by the Weston Electrical Instrument Co., is shown in Fig. 86.

A small strip of soft iron M, bent into cylindrical form, is mounted axially on a spindle which is free to turn. Another similar strip F, which is slightly tapered and with a larger radius than M, is fixed inside a cylindrical coil. In the voltmeter the cylindrical coil is wound with fine wire and is connected in series with a high resistance. When the instrument is connected across the line, the current to the instrument is substantially proportional to the circuit voltage. When this current flows through the exciting coil, both iron vanes become magnetized. The upper edges of the two strips F, M will always have like magnetic polarity, and the lower

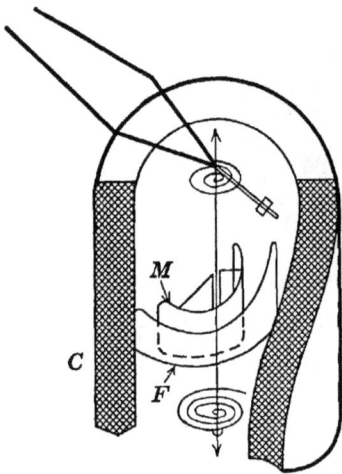

FIG. 86.—Weston iron-vane type of instrument.

edges will always have like magnetic polarity; but when the upper edges are north poles, the lower edges are south poles. Therefore, there will always be a repulsion between the two upper edges, and also between the two lower edges of the strips. This repulsion produces a turning moment in the moving system which is opposed by two flat spiral springs. A pointer mounted on the spindle moves over a graduated scale and indicates the voltage. Air damping is obtained by the use of a light aluminum vane moving in a restricted space.

Another method of using the iron-vane principle, employed by the General Electric Co., is shown in Fig. 87. A fixed coil c

is mounted obliquely to the spindle, and a small iron vane is mounted obliquely on the spindle. When the pointer is at zero, this vane lies at an angle to the coil axis, as at *a* (Fig. 87). When current flows in the coil, the vane attempts to take such a position that the direction of its major axis shall coincide with the direction of the magnetic field produced by the current in the coil, which acts in the direction of the coil axis. This position is shown at *b* (Fig. 87). The vane in seeking this position turns the spindle which carries the pointer. The turning moment is opposed by flat spiral springs. The coil is surrounded by iron laminations which shield the instrument from stray fields. Magnetic damping, such as is used

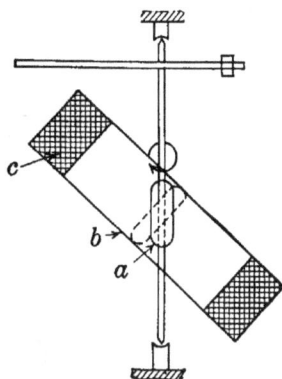

Fig. 87.—Inclined coil, iron-vane type of instrument.

with watthour meters, is employed, a light aluminum vane moving between the poles of permanent magnets.

These two types of iron-vane instrument may be used as ammeters by winding the coils with comparatively few turns of wire of sufficient cross-section to carry the current. In iron-vane types of instrument, no current is conducted to the moving element.

62. Thermal-type Instruments.—Thermal-type instruments depend for their operation on the heating effect rather than on the magnetic effect of current. In the hot-wire type the linear expansion of a wire heated by the current causes the deflection of the pointer. This type has been superseded for the most part by the vacuum thermocouple described in Part I, p. 101, Par. 84. As the deflections in this type of instrument depend on the square of the current (I^2R loss), the instrument reads equally well on both direct and alternating current. Accordingly, it can be used as a transfer from alternating to direct current, and *vice versa*.

63. Induction Watthour Meter.—The d-c watthour meter can be used with alternating current, as the reversal of line voltage reverses both armature and field current simultaneously, and the direction of the torque remains unchanged. At low power factors, however, considerable error may be introduced by the inductance of the armature circuit which causes the armature current to lag the line voltage by a small angle. This error may be compensated by shunting the current coils of the meter with a low noninductive resistance.

In the induction watthour meter, there is no wire-wound armature. The aluminum disc, which rotates between the poles of permanent magnets providing the damping torque, acts also as the armature. Thus the rotating element is simpler and lighter than that of the d-c meter, so that friction and wear on the jewels and pivots are much less. As there are no commutator and

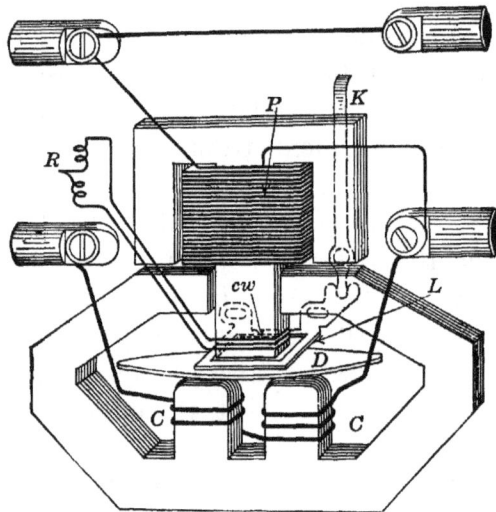

Fig. 88.—Diagram of induction watthour meter.

brushes, maintenance is low. A simple metallic stamping rather than a coil provides for friction compensation. The induction watthour meter is superior to and cheaper than the d-c type so there is little necessity for using the d-c type on a-c circuits.

A rear view of a typical induction watthour meter is shown in Fig. 88. *P* is a potential coil which is highly inductive and is placed on one lug of the laminated magnetic circuit, this lug being over the disc *D*. *CC* are two series, or current, coils placed on two lugs beneath the disc. These coils are so wound that when one tends to send flux upward the other tends to send it downward. *cw* is a small auxiliary or compensating winding placed on the potential lug, and its ends are connected to the resistance R. In order that the meter may register correctly, the potential-coil *flux* must lag the line voltage by 90°. As it is impossible to make the resistance of the potential coil zero, its current will lag by an angle less than 90°. At low power factor, this introduces considerable error in the meter registration. However, by properly adjusting the resistance R, the potential-coil *flux* may be brought into the 90° relation, and the meter will register substantially correctly at all power factors. To adjust the compensation, the meter is made correct at unity power factor; and then the power factor is reduced to some low value, as 0.5. If the registration is now in error, it is due to improper compensation. The meter is again made to register correctly by changing the resistance R, the two small wires of this resistance being either twisted or untwisted and then soldered. If the meter underregisters when the load current lags, the resistance R should be *decreased;* if the meter overregisters with lagging current the resistance R should be *increased.* The reverse is true with leading current.

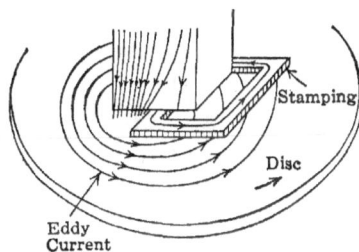

FIG. 89.—Shaded-pole principle of light-load adjustment.

L is a small metallic stamping placed under the potential lug and can be moved laterally by means of the lever K. Its function is to provide the small torque just necessary to overcome the friction of the meter. The operation of this adjustment is as follows:

Figure 89 shows the stamping set off center under the lug. When the flux starts to pass down through the lug, a current is induced in the short-circuited stamping. This current, by Lenz's law (Part I, p. 171), *opposes* the flux entering the stamping so that the flux is crowded to the left-hand side of the lug, as shown. Ultimately, however, the flux through the stamping, and hence in the right-hand side of the lug, reaches its maximum value; but this maximum occurs later than the maximum value of the flux in the left-hand side of the lug. When the flux starts to decrease, the current in the short-circuited stamping tends to *oppose* the decrease in the flux. This again retards the time phase of the flux in the right-hand side of the lug with respect to that in the left-hand side of the lug.   The result is a sweeping of the flux from left to right across the lug.   This sliding flux cuts the disc and induces eddy currents in it.   These currents, reacting with the flux, produce a torque tending to drive the disc in the direction in which the stamping is displaced from its position of symmetry.   This is the "shaded-pole" principle which is used also to start small single-phase induction motors (see p. 308).

The driving torque of the meter is produced by currents induced in the disc, these currents reacting with an alternating magnetic field.   For example, the alternating flux produced by the current lugs CC cuts the disc and induces eddy currents in it.   These currents are in an alternating magnetic field of the same frequency, produced by the potential lug.   Motor action results, and the disc rotates.   Also the currents induced in the disc by the potential lug may be considered as reacting with the magnetic field produced by the current lugs CC.   It can be shown that the combined effect of the flux from the potential lug and that from the current lugs is a sliding field[1] which ordinarily moves from left to right when viewed from the rear of the meter.   By induction-motor action (p. 263) the portion of the disc cut by this sliding field tends to follow the field, and rotation of the disc results.

It may be shown that with proper compensation the driving torque at each instant is proportional to the power at that instant.

In order to register watthours, a retarding torque proportional to the angular velocity of the disc is necessary.   As with the d-c watthour meter, this retarding torque is secured by the disc cutting a field of constant strength produced by *permanent* magnets.   This causes a *retarding* torque which is proportional to the angular velocity of the disc.   Therefore, both the *driving* torque (motor action) and the *retarding* torque (generator action) are produced in the same disc.

64. Checking and Adjustment of Induction Watthour Meter.— The induction watthour meter is calibrated in much the same manner as the d-c watthour meter.   A standard indicating wattmeter is used to measure the average power during a given

[1] See Dawes, C. L., "A Course in Electrical Engineering," Vol. II, Alternating Currents, Chap. IV.

interval, and the revolutions of the disc of the watthour meter are counted, using a stop watch. The average meter *watts* are calculated by means of the equation

$$W = \frac{K \times N \times 3{,}600}{t}, \tag{51}$$

where K is the meter constant, N the revolutions of the disc, and t the time in seconds.

As a rule, an ammeter and a voltmeter are used in connection with such a test (Fig. 90) in order to determine the power factor. Instrument losses should be carefully investigated and corrections made if necessary. In Fig. 90, the wattmeter is connected so

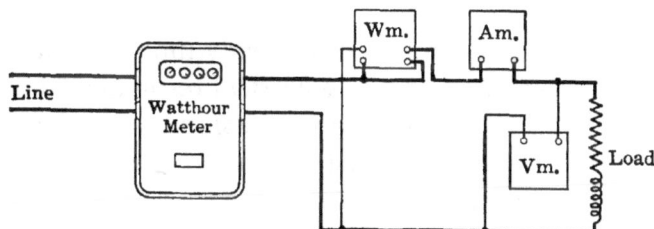

FIG. 90.—Connections for testing alternating-current watthour meter.

that it measures the correct voltage output of the meter; but it does not measure the correct current output, since some of the current coming through the meter flows through the wattmeter potential circuit. Therefore, to obtain the correct output of the watthour meter with this connection of the wattmeter, the loss in the wattmeter potential circuit must be *added* to the wattmeter reading. No correction for the ammeter and voltmeter power is necessary since they form part of the load.

After the meter is adjusted at rated load and unity power factor by means of the retarding magnets, it is adjusted at light load by means of the light-load adjustment, that is, by setting the metallic stamping L (Fig. 88) to the position which gives correct registration. The power factor is then lowered, usually to 0.5, lagging current, and the load is adjusted to rated volt-ampere value. Any error occurring now must be due to improper lagging. The registration is then made correct by adjusting the resistance R (Fig. 88) which is in series with the lagging coil. If the meter registers *low* with lagging current, the resistance R

should be *decreased;* if it registers *high,* the resistance *R* should be *increased.* With leading current, these operations should be reversed.

As has been stated already, the induction watthour meter has several advantages over the d-c meter. As there is no coil-wound armature, the rotating element of the induction meter is much lighter than that of the d-c meter so that the effects of friction are much less. Moreover, the induction watthour meter has no commutator or delicate brushes, which are sources of trouble with the d-c meter.

The induction meter is made also in the polyphase type. Two single-phase elements act on a common spindle, as in the polyphase wattmeter (p. 105). There are two sets of damping magnets.[1]

65. Frequency Indicators.—Frequency indicators are operated on one of two basic principles—electrical resonance and mechanical resonance. The latter type is in common use for measuring frequencies from as low as 15 to as high as 500 cycles per sec, and is simple in operation.

The Frahm vibrating-reed frequency meter (Fig. 91(a)) consists of a set of slender steel reeds *R*, each reed being about ⅛ in. wide and coated with a plastic lacquer. Each reed is tuned to vibrate at a definite frequency. The reeds are fastened in a row to a bridge piece *B* to which is attached the armature *A* of a small electromagnet *M* mounted close to the armature. When the instrument is connected across the circuit, the frequency of which is to be measured, the current flows through the coil of the electromagnet *M* and resistance *G* in series with it. The electromagnet imparts to the armature *A* one impulse for each cycle of current if the core of the electromagnet *M* is a permanent magnet and two impulses if the core is of nonpolarized soft iron. The vibration of the armature *A* is transmitted to the reeds *R*. The particular reed or reeds whose mechanical resonant frequencies are nearest that of the alternating current whose frequency is being measured, come into sympathetic vibration and thus indicate the frequency. The ends of the reeds are bent at right angles and appear against the dark background of the interior of the instru-

---

[1] For a detailed analysis of the watthour meter, see F. A. Laws, "Electrical Measurements."

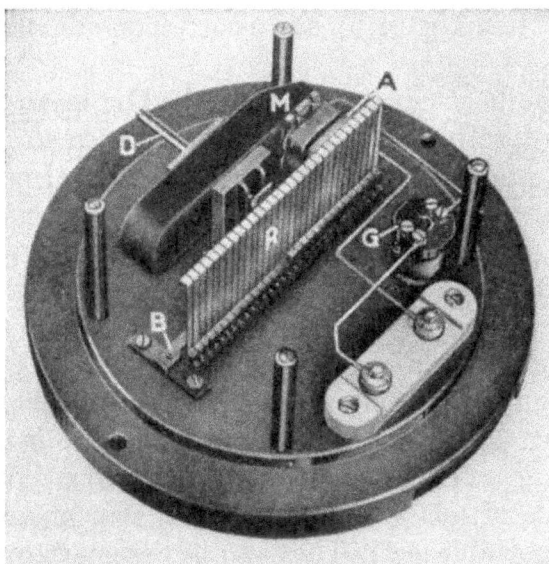

(a)

(b)

FIG. 91.—Frahm vibrating-reed frequency meter.

ᴜ𝑟ent, as shown in Fig. 91(*b*), and a corresponding scale is suitably calibrated.

The amplitude of vibration is controlled by means of the screw *D* [Fig. 91(*a*)] with which the air gap between the armature *A* and the electromagnet *M* can be adjusted. Two frequency scales may be obtained with the same set of reeds. By using an electromagnet having a permanent-magnet core there is one vibration of the reeds for each cycle of current; by using an electromagnet having a nonpolarized soft-iron core, there are two vibrations of the reed for each cycle of current.

This method of frequency measurement is independent of wave form, voltage, and external magnetic fields; also it is practically independent of ordinary temperature variations.

Instruments of this type are made for switchboard mounting and also for portable use. They can be designed to operate with as little as from 2 to 5 milliamperes.

66. Synchroscope.—Before connecting an alternator to the bus-bars and in parallel with other alternators, it is necessary for its voltage to be the same as that of the bus-bars and also for it to be in phase opposition. This corresponds to having d-c generators of the same emf and the same polarity before connecting them in parallel.

Fɪɢ. 92.—Synchroscope. (*Courtesy of General Electric Company.*)

A synchroscope is an instrument for indicating when alternators are in the proper phase relation for connecting in parallel and at the same time showing whether the incoming alternator is running fast or slow. Figure 92 shows such a synchroscope. If the incoming alternator has the same frequency as the bus-bars, the pointer remains stationary. When the alternators are in the correct phase relation for closing the switch, the pointer is over the index at the top center of the dial (Fig. 92). The direction of rotation of the pointer shows whether the incoming alternator is running fast or slow. The alternator switch is usually closed when the pointer is rotating slowly in the "fast" direction and is approaching the index.

67. Oscillograph—Magnetic Type.—It is often desired to investigate transient conditions in electrical circuits, such, for example, as the current and voltage relations during the blowing of a fuse, or during the short circuit of an alternator, or in oscillations produced by switching. Further, it is desirable to have apparatus which will show current and voltage waves in a-c circuits during steady conditions. The oscillograph is an instrument capable of meeting these requirements.

Its principle is quite simple, being that of a D'Arsonval galvanometer (Part I, Chap. V, p. 90), as shown in Fig. 93(a). A small phosphor-bronze strip, or filament, is stretched over two

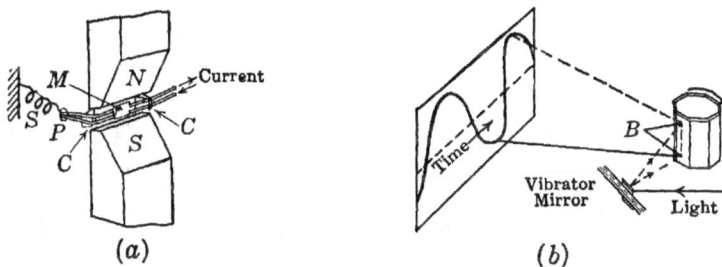

FIG. 93.—(a) Vibrating element of oscillograph. (b) Method of drawing out vibrating beam into a wave.

clefts *CC* around a small pulley *P* and back again. The spring *S* acting on the pulley keeps the two lengths of the strip in tension. This filament is placed between the poles of a strong magnet which in the later models is a permanent magnet. When current flows through the filament, one length of the filament moves outward and the other inward. A very small mirror *M* cemented across the two lengths of the filament near the center of the poles is given a rocking motion by this movement of the filament. A beam of light reflected from this mirror will be drawn out into a straight line by the mirror vibration. If this beam of light is made to strike a rotating mirror, in the manner shown in Fig. 93(b), the rotation of the mirror introduces a time element and the wave is drawn out so that its characteristics are shown.

The instrument is merely a galvanometer having a single turn and a very light moving element. This makes the moment of inertia very small. Also, the filament is under considerable tension, so that its natural frequency of vibration is high, being from 3,000 to 10,000 cycles per sec. These characteristics are necessary in order that the filament may respond accurately to the comparatively high-frequency variations which it is called upon to follow. The moving element is usually immersed in oil so that its movement is properly damped and the filament is kept cool.

Figure 94 shows the general arrangement of a typical magnetic type of oscillograph having two elements.

The light from the source, usually the filament of an automobile headlight, goes first through two focusing lenses and then

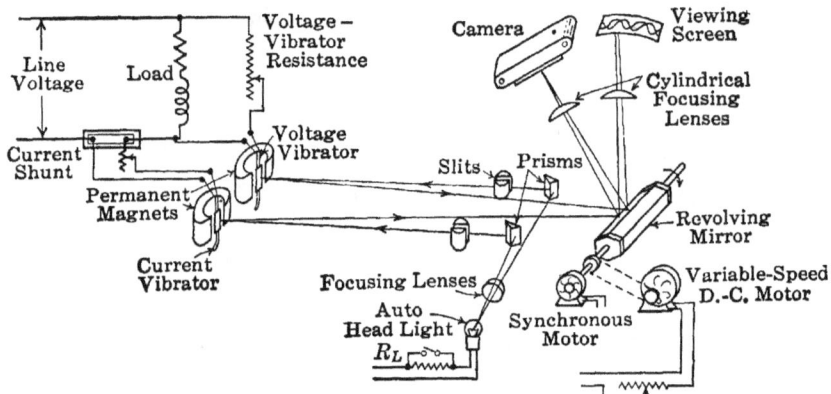

FIG. 94.—Optical system and connections of oscillograph.

strikes each of two total-reflecting prisms. These prisms deflect the light 90°; and it then passes through slits to each of the vibrator mirrors. The slits are adjustable and serve to reduce the section of the beam so that a narrow line may be obtained when the wave is received on the viewing screen or is photographed. The light is reflected back from the vibrator mirrors to a revolving mirror, which in turn reflects it, drawn out as waves, through cylindrical lenses either to the viewing screen or to the camera if a photographic record is desired.

The rotating mirror is usually driven by a small synchronous motor operated from the same circuit as that to which the vibrators are connected, so that waves on the viewing screen remain stationary. A standard roll-film camera may be arranged for photographing (Fig. 94) with the shutter timed to give but a single exposure.

Another method is to wrap the film about a cylindrical drum within a lightproof casing having a narrow transverse opening. The light beams are made to go directly from the vibrator mirrors to the transverse opening through a properly timed shutter, the rotating mirror being moved out of the path of the beams. The drum is driven by a small motor, and its rotation provides the time axis for the film.

In some designs a switch actuated by the shutter mechanism short-circuits a resistance R_L, which is in series with the light, during the brief photographic interval. This produces momentarily an intense beam. Frequently, it is desirable to change the speed of the rotating mirror, usually to obtain the desired time

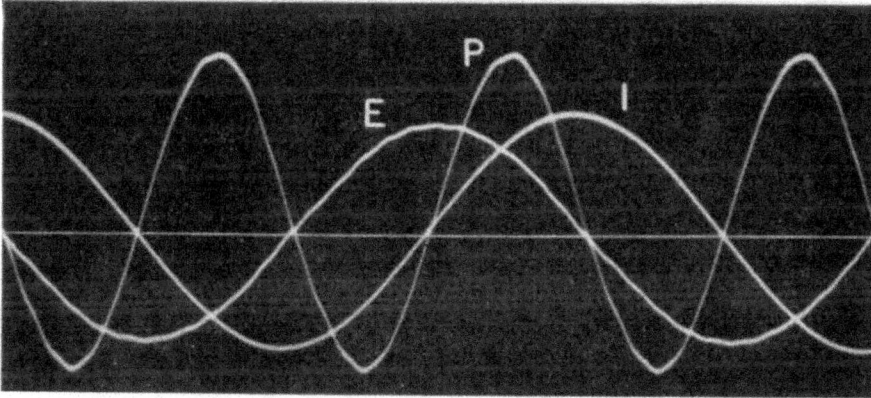

FIG. 95.—Oscillogram of voltage, current, and power waves for air-reactor.

axis in photographing. To do this a small, variable-speed motor, which may be either direct or alternating current, in series with resistance, is also connected to drive the mirror. In some types of oscillograph a single oscillating mirror is used rather than a rotating mirror. On the return movement of the oscillating mirror the light from the source is cut off by a rotating shutter, a double image thus being prevented.

The vibrator which responds to the current is connected across a shunt in series with the circuit (Fig. 94), and the vibrator which responds to the voltage is connected in series with a high resistance across the line. That is, the oscillograph vibrators are connected in circuit in the same manner as the d-c instrument (Part I, pp. 96 to 101). Multivibrator oscillographs with 3, 6, and even 12 vibrators are in use.

Some oscillographs are equipped with a power element also. The magnetic field in which the vibrator operates is energized by the current, and the vibrator is connected across the line in series with resistance. Figure 95[1] shows an oscillogram obtained with a voltage vibrator, a current vibrator, and a power vibrator. The load is an air-core reactor. The power factor is nearly zero, and the power curve goes through zero practically at the points where either the voltage or the current is zero. (Compare with Fig. 54, p. 62.) It should be noted that although there are pronounced ripples in the voltage wave, the current wave is smooth. The reactance has practically eliminated in the current wave the higher harmonics which exist in the voltage wave.

68. Cathode-ray Oscillograph.—Although the magnetic-type oscillograph can respond accurately to frequencies as high as 3,000 cycles per sec, even the small moment of inertia of the vibrating system is too great for accurate response to higher frequencies, particularly those in the higher audio range and in the radio range. Also, the magnetic oscillograph has far too much inertia to follow ultra-high-speed transients such as lightning and lightning-generator discharges. In the cathode-ray oscillograph, however, the deflected element is an electron beam, whose inertia is sensibly zero, so that the oscillograph can respond accurately to transients that occur in a fraction of a microsecond (millionth of a second) even.

A diagram of the oscillograph is shown in Fig. 96. The left-hand end of the tube contains the "electron gun," one element of which consists of an indirectly heated oxide-coated tungsten cathode which emits electrons. These are accelerated by the first and second anodes which are maintained at high positive potentials relative to the cathode, the potential of the second anode being as high as 4,000 volts direct current. The acceleration is due to the positive potential attracting the electrons (negative charges) emitted by the cathode. (See p. 391.) By varying the relatively low potential of the control electrode, the focus of the beam on the screen may be controlled and the beam may be cut off altogether. The beam leaves the electron gun through a very small aperture at the end of the second anode

[1] The author is indebted to Hugo DeFritsch, of the Graduate School of Engineering, Harvard University, for this oscillogram.

and passes through two vertical and two horizontal electrostatic-deflecting plates. By impressing potential on these plates the beam may be deflected both horizontally and vertically. Since the electron beam consists of negative charges in motion, it deflects toward the positive plate. Usually, the wave whose amplitude is to be observed is impressed on the vertical deflecting plates which cause the beam to be deflected in a vertical plane in proportion to the amplitude of the wave. The time axis is produced by a "sweep circuit" which applies a practically uniformly increasing potential to the horizontal deflecting plates and so causes the beam to sweep across the tube at a practically uniform rate and, with alternating current, usually in synchronism

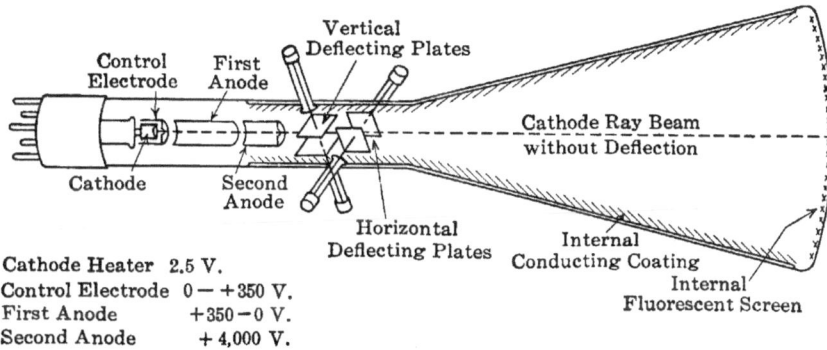

Cathode Heater 2.5 V.
Control Electrode 0 − +350 V.
First Anode +350 − 0 V.
Second Anode + 4,000 V.

FIG. 96.—Cathode-ray oscillograph.

with the frequency of the voltage across the vertical deflecting plates.

The screen material, which is coated on the inside wall at the right-hand end of the tube, fluoresces, usually green, when struck by the beam. The waves, if stationary, may be photographed with an ordinary camera. In order to prevent accumulation of charge on the inner wall of the tube, which would cause the beam to deflect erratically, a conducting coat is applied to the inner wall and the coat is connected to the second anode.

Since the electron beam consists of electric charges in motion and hence constitutes a current, when in a magnetic field it deflects in accordance with Fleming's left-hand rule. (Part I, p. 264.) Hence, the beam may be made to respond to current waves by causing the current to flow in coils placed outside the tube and producing a magnetic field within the tube. At very high frequencies, however, the inductance of the coils may be so great as to disturb the high-frequency circuit.   It then becomes necessary to send the current through a high-resistance shunt connected across the vertical deflecting plates.   To obtain a good deflection with full accelerating potential, a minimum of about 300 volts across the plates is required.   By reducing the accelerating voltage the same deflection may be obtained with values of voltage much less than 300 volts.

This same type of tube is used for television.   (The sending tube is the *iconoscope* and the receiving tube is the *kinetoscope*.) The screen, however, is usually of some color other than green.

CHAPTER IV

POLYPHASE SYSTEMS

With the possible exception of the comparatively few power plants which supply energy to single-phase electric railways, practically all a-c power is generated as polyphase power. Aside from single-phase railway motors, nearly all power apparatus rated at 10 hp and over operates polyphase and takes power, therefore, from polyphase systems. Hence, an understanding of polyphase systems is essential to the study of the industrial applications of alternating currents. It will be shown that polyphase systems are merely the proper combination of two or more single-phase systems; when they are so considered, their analysis is comparatively simple.

69. Reasons for Use of Polyphase Systems.—In many industrial applications of alternating currents, there are objections to the use of single-phase power.

In a single-phase circuit, the power delivered is pulsating. Even when the current and voltage are in phase, the power is zero twice in each cycle (Fig. 51, p. 59). When the power factor is less than unity, not only is the power zero four times in each cycle, but it is *negative* during two periods in each cycle (Fig. 56, p. 66). This means that the circuit returns energy to the generator for a part of the time and is analogous to a single-cylinder gasoline engine in which the flywheel returns energy to the cylinder during the compression part of the cycle. Over the complete cycle, both the single-phase circuit and the flywheel receive an excess of energy over that which they return to the source. The pulsating character of the power in single-phase circuits makes such circuits objectionable in many instances.

A polyphase circuit is somewhat like a multicylinder gasoline engine. With the engine, the power delivered to the flywheel is practically steady, as one or more cylinders are firing when the others are compressing. This same condition exists in polyphase electrical systems. Although the power of any one phase is pulsating and may be negative at times, the *total power* is constant if the loads are balanced. This makes polyphase systems highly desirable for power purposes.

The rating of a given motor, or generator, increases with the number of phases, an important consideration. Below are the approximate ratings of a given dynamo for different numbers of phases, the single-phase rating being assumed as 100.

Single-phase 100
2-phase 140
3-phase 148
6-phase 148
D-c 154

The same dynamo operating three- or six-phase has about 56 per cent greater rating than when operating single-phase. The ratio of polyphase to single-phase output of a synchronous converter is greater even than the foregoing values (p. 333). The much greater output of electrical machinery when operating polyphase is a very important consideration in favor of generating and utilizing polyphase power.

THREE-PHASE SYSTEMS

70. Generation of Three-phase Electromotive Force.—Because the three-phase system is now the most common of the polyphase systems, it will be considered first. Figure 97(*a*) shows three identical simple coils *aa'*, *bb'*, *cc'*. They are fastened rigidly together and are mounted 120° apart on an axis which can be rotated. This combination of coils and axis constitutes an elementary three-phase armature. One side *a*, *b*, *c* of each coil is shown shaded. These shaded sides may be called corresponding sides of the three coils. These corresponding sides are important in that they determine the manner in which the coils are connected electrically. The elementary three-phase armature is shown rotating in a counterclockwise direction in a uniform magnetic field. The current can be conducted from each of the three coils by means of a pair of slip-rings *aa'*, *bb'*, *cc'* (Fig. 97(*b*)). The terminal corresponding to the shaded side *a* of coil *aa'* is connected to slip-ring *a*, the terminal corresponding to the unshaded side *a'* of coil *aa'* is connected to slip-ring *a'*, etc.

When the armature is in the position shown (Fig. 97(a)), the coil aa' is in the magnetic neutral plane and its induced emf is zero (see p. 10) as shown in Fig. 98(a) where the induced emf e_{oa} starts at its zero value and is increasing in a positive direction. When the armature has rotated through 120 space-degrees, the emf in coil bb' will be zero and increasing in a positive direction. This is shown by the emf wave e_{ob} (Fig. 98(a)). That is, emf

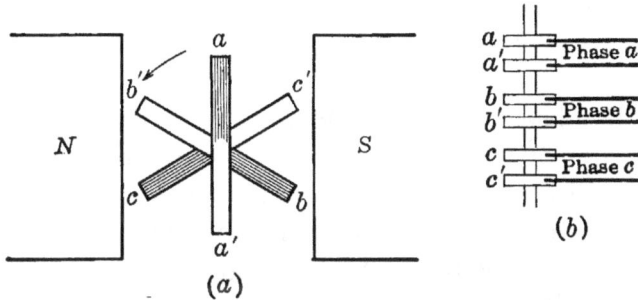

FIG. 97.—Generation of three-phase current.

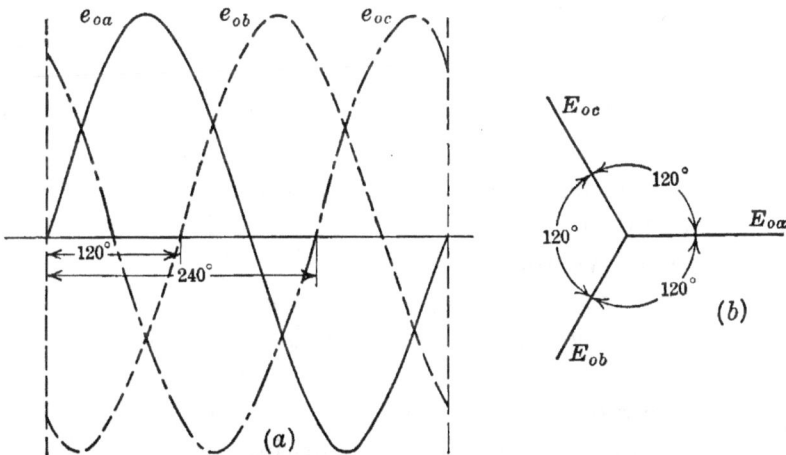

FIG. 98.—Three-phase emf waves and vector diagram.

e_{ob} lags emf e_{oa} by 120 time-degrees, since this is a two-pole alternator.

When the armature has rotated through 240 space-degrees, the emf in coil cc' will be zero and increasing in a positive direction. This is shown by the emf wave e_{oc} (Fig. 98(a)). That is, emf e_{oc} lags e_{oa} by 240 time-degrees and lags e_{ob} by 120 time-degrees. The sequence of phase rotation is e_{oa}, e_{ob}, e_{oc}. These three emfs are equal, since the three coils are identical and rotate at the same angular velocity in the same magnetic field.

Although commercial three-phase alternators differ in construction from the elementary alternator (Fig. 97(a)[1]), the relations existing among the emfs are identical in the two types of machine. That is, the coils on the stationary armature of the commercial alternator are so placed that the emfs generated in the three phases are equal and differ in phase by 120°.

71. Relations among Three-phase Electromotive Forces.—The three emfs e_{oa}, e_{ob}, e_{oc} [Fig. 98(a)] are shown vectorially as E_{oa}, E_{ob}, E_{oc} in Fig. 98(b), the lengths of the vectors representing *rms* values.

It will be observed in Fig. 98(a) that at any one instant the algebraic sum of the emfs is zero. If the three coils were connected in series and in sequence, that is, if the slip-rings were connected a' to b, b' to c, the emf between slip-rings c' and a would be zero. (See p. 134.) For example, when e_{oa} is zero and increasing positively, e_{oc} is positive and equal to 86.6 per cent of its maximum value and e_{ob} is negative and equal to 86.6 per cent of its maximum value. Hence, at this instant, the sum of the three emfs is zero. When e_{oa} is positive maximum, e_{ob} and e_{oc} are negative and each is equal to half its maximum value. Hence, at this instant also, the sum of the three emfs is zero. Similarly, it may be shown that the sum of the three emfs is zero at *every* instant of time. For example, in Fig. 98(b) the sum of the three emf vectors is zero.

Each of the coils of Fig. 97(a) can be connected through its two slip-rings to a single-phase circuit. This gives three independent single-phase circuits, phase a, phase b, and phase c [Fig. 97(b)]. With a rotating field and stationary armature, which is the most common type of alternator, the six slip-rings would not be necessary, but six leads would be taken directly from the armature. In practice, however, an alternator seldom supplies three independent circuits, requiring six wires, but the phases are combined into three-phase, three- or four-wire systems.

72. Symbolic Notation.—The solution of problems involving circuits and systems containing a number of currents and voltages is simplified and is less susceptible to error if the current and voltage vectors are designated by some systematic notation, of which the following is one type. If a voltage is acting to

[1] See Chap. V, p. 166.

send current from point a to point b (Fig. 99(a)), it shall be denoted by E_{ab}. On the other hand, if the voltage tends to send current from b to a, it shall be denoted by E_{ba}. Obviously, $E_{ab} = -E_{ba}$.

It may seem as if alternating currents cannot be considered as having direction, since they are undergoing continual reversal in direction. As will be seen shortly, this consideration is not important. The solution of a-c problems depends primarily on the phase relations existing among the various voltages and currents in the circuit. By using the symbolic-notation method, these phase relations can be expressed correctly, even when the circuit involves more or less complicated networks.

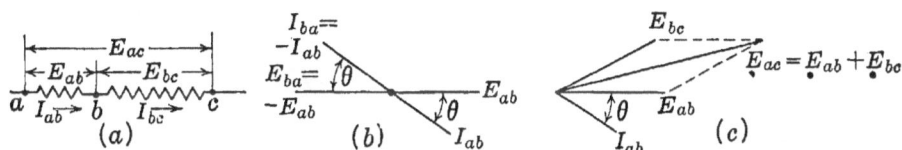

Fic. 99.—Symbolic notation applied to voltage and current vectors.

At times, however, it may be convenient to determine the direction of a voltage or a current vector by the direction of flow of energy.

In Fig. 99(a), corresponding to the voltage E_{ab}, the current I_{ab} can be considered as flowing from a to b. The current from b to a is opposite in direction to that from a to b. Therefore, $I_{ba} = -I_{ab}$. This relation is illustrated in Fig. 99(b), in which I_{ab} differs in phase from I_{ba} by 180°. Likewise, E_{ba} is 180° from E_{ab}.

In Fig. 99(a), the voltage acting from a to c is equal to the voltage acting from a to b plus the voltage acting from b to c, since the whole is equal to the sum of its parts, or $E_{ac} = E_{ab} + E_{bc}$.

The order of the subscripts should be noted. The first and last subscripts on the left-hand side of the equation are the same and in the same order as the first and last on the right-hand side. The adjacent subscripts on the right-hand side must always be the same, as, for example, the b's.

In Fig. 99(c), the voltages E_{ab} and E_{bc} are shown vectorially. Their vector sum E_{ac} is obtained by adding E_{ab} and E_{bc} vectorially.

73. Combining Three-phase Electromotive Forces with 60° Phase Difference.—In Fig. 100(a) are two coils $a'a$ and $b'b$, the emfs in which differ in phase by 120°, the emf $E_{a'a}$ in coil $a'a$ leading the emf $E_{b'b}$ in coil $b'b$. This is shown by the vector diagram in (b). The end a' of coil $a'a$ and the end b' of coil $b'b$ are connected, as in (a). It is required to determine the emf E_{ab} across the free ends ab. As a' and b' are connected, they are the same electrically and the junction is designated by o. Hence, $E_{a'a} = E_{oa}$; $E_{b'b} = E_{ob}$. By the use of symbolic notation,

$$E_{ab} = E_{ao} + E_{ob}. \tag{I}$$

The relation among the subscripts should be noted. The vector E_{ao} was not on diagram (b) originally. However, E_{oa} is given and E_{ao} is found by reversing E_{oa} as shown. E_{ab} is readily found by adding E_{ao} and E_{ob} vectorially. The resultant E_{ab} lags

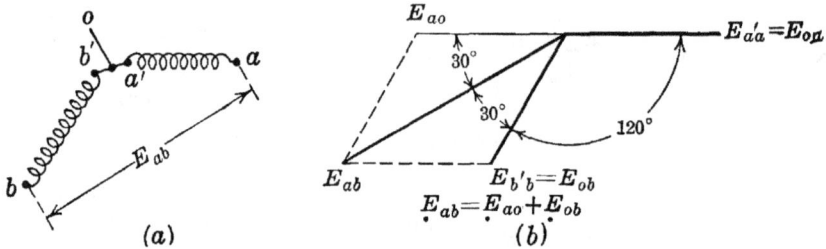

(a) (b)

Fig. 100.—Addition of emfs differing in phase by 60°.

E_{ob} by 30° and leads E_{ao} by 30°. Numerically, E_{ab} is equal to $\sqrt{3}\,E_{ao}$ or $1.732E_{ao}$.

It is to be noted that the symbolic notation of Par. 72 is followed. The first and last subscripts on the left-hand side of Eq. (I) are the same as the first and last subscripts on the right-hand side of the equation. The last subscript o of the first term of the right-hand side of the equation is the same as the first subscript of the second term.

The foregoing method of using subscripts combined with the emf (or current) symbols eliminates the necessity for arrows with the vectors; and, by maintaining the relations of the subscripts in the manner just described, confusion as to the direction of a vector is avoided. It is to be noted that although the vectors E_{oa} and E_{ob} differ in phase by 120°, when ends a' and b' are connected the resultant is found by adding vectors E_{ao} and E_{ob} differing in phase by 60°. As will be shown in Par. 75, this is the method used to determine the emfs in a Y-connected system.

74. Combining Electromotive Forces with 120° Phase Difference.—In Fig. 101(a) the connections of coil $a'a$ [Fig. 100(a)] are reversed, end a being now connected to the end b' of coil $b'b$. It is required to determine the emf $E_{a'b}$ across the two free ends $a'b$ of the two series-connected coils. As ends a and b' are connected, they are the same electrically, and $E_{a'a} = E_{a'b'}$. Applying the subscript notation,

$$E_{a'b} = E_{a'b'} + E_{b'b}.$$

Both $E_{a'b'}$ ($= E_{a'a}$) and $E_{b'b}$ are given in the vector diagram in Fig. 101(b). To find $E_{a'b}$, add $E_{a'b'}$ and $E_{b'b}$ vectorially.

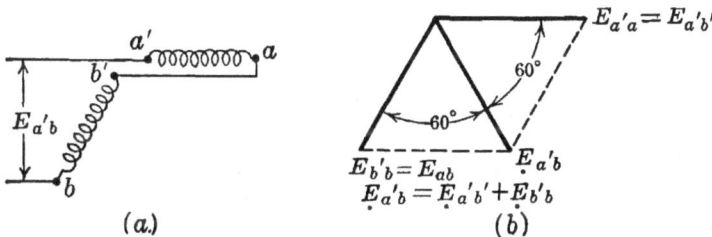

Fig. 101.—Addition of emfs differing in phase by 120°.

$E_{a'b}$ is numerically equal to $E_{a'a}$ and $E_{b'b}$ and makes an angle of 60° with each, lagging $E_{a'a}$ and leading $E_{b'b}$.

As will be shown in Par. 78, this is the method used to determine the emfs in the delta-connected system.

75. Y-connection.—Figure 102 shows the three coils of Fig. 97(a) placed on a cylindrical armature, the shaded and the nonshaded sides being identical in the two figures. For clearness the S-pole is partly cut away. The terminals of the shaded sides of the coils are designated a, b, c, and the terminals of the nonshaded sides a', b', c'. The terminals a', b', c' are brought to a common junction O. The three emfs induced in the coils

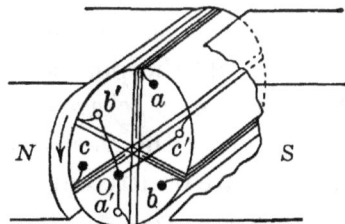

Fig. 102.—Y-connection of armature coils.

are designated $E_{a'a}$, $E_{b'b}$, $E_{c'c}$ and have phase differences of 120° as did the emfs of the three coils of Fig. 97(a). At the instant shown, the emf in coil $a'a$ is zero since the plane of the coil is perpendicular to the flux, and the emf is increasing positively because the shaded side of the coil is about to cut the flux from the N-pole.

When the armature has rotated 120°, the shaded side b of coil $b'b$ will be in the same position as that now occupied by the shaded side of the coil $a'a$; when the armature has rotated 240°, the shaded side c of coil $c'c$ will be in this same position. Hence, the emfs from the common junction O to the terminals a, b, c will have corresponding zero values and maximum values 120° apart (see Fig. 98 (a)). This is the reason for calling the shaded sides of the coils *corresponding sides* (p. 122). If the terminals a, b, c are connected at a common junction, the ends a', b', c' being free, the emfs $E_{aa'}$, $E_{bb'}$, $E_{cc'}$ between the common junction and terminals a', b', c' will differ in phase by 120° but the emfs are the reverse of $E_{a'a}$, $E_{b'b}$, $E_{c'c}$.

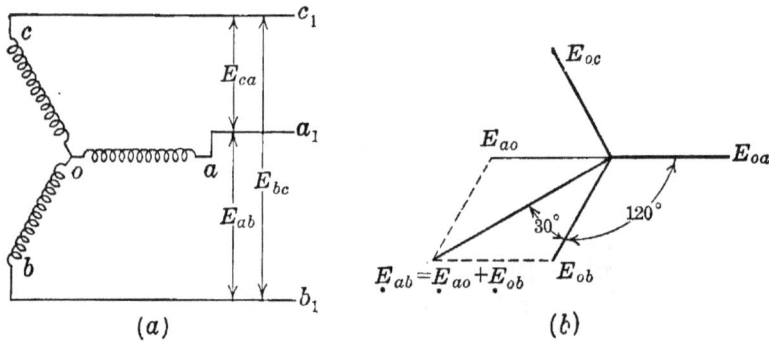

FIG. 103.—Y-connection and corresponding emf vector diagram.

The connection shown in Fig. 102 is the Y-connection. In Fig. 103(a) these same coils are shown in diagrammatic form, but a stationary armature is assumed so that slip-rings for making the connections to the external circuit are not necessary. Since in Fig. 102 the points a', b', c', and o form a common junction, the letters a', b', c' are omitted in Fig. 103(a). Also, in Fig. 103(b) which gives the vector diagram for the three emfs, the letters a', b', c' are omitted and $E_{oa} = E_{a'a}$, $E_{ob} = E_{b'b}$, $E_{oc} = E_{c'c}$. The vectors E_{oa}, E_{ob}, E_{oc} are equal in magnitude and differ in phase by 120°.

The term "Y-connected" comes from the diagrammatic appearance of the coils in (a). In Fig. 103(a), wires aa_1, bb_1, cc_1 from the free ends a, b, c conduct the current to the external circuit. A fourth wire from the neutral o is sometimes employed, giving a three-phase, four-wire system. The three coil-emf vectors are designated by E_{oa}, E_{ob}, E_{oc}. It is required to find the three line emfs E_{ab}, E_{bc}, E_{ca}. From the sequence of the subscript notation and also from consideration of Fig. 100 the emf $E_{ab} = E_{ao} + E_{ob}$. The emf E_{ob} is given in the vector diagram (Fig. 103(b)). The emf E_{ao} was not given originally, but the emf E_{oa} is given. E_{ao} is obtained by reversing E_{oa} as shown. The emf E_{ab} is found by adding E_{ao} and E_{ob} vectorially. As in Fig. 100(b), the emf E_{ab} has a magnitude equal to $\sqrt{3}$, or 1.732, times the coil emf E_{ao}, and it differs in phase from the coil emfs E_{ob} and E_{ao} by 30°, lagging E_{ob} and leading E_{ao}.

In a similar manner the other two line emfs E_{bc} and E_{ca} may be obtained. That is, $E_{bc} = E_{bo} + E_{oc}$ and $E_{ca} = E_{co} + E_{oa}$. The three coil emfs and the three resulting line emfs are shown in Fig. 104. (Notice the sequence of subscripts for the line emfs E_{ab}, E_{bc}, E_{ca}.) *In a balanced three-phase system, the line emf is equal numerically to the coil emf multiplied by $\sqrt{3}$. The three line emfs are equal numerically and differ in phase by 120°. There is a phase difference of 30° between the line emf and the coil emf.*

Example.—Each of the three coil circuits of a three-phase, Y-connected alternator generates an emf of 220 volts. What is the emf across each of the three line terminals?

The three line emfs are

$$E_{line} = \sqrt{3} \times 220 = 381 \text{ volts. } Ans.$$

76. Currents in Y-system.—Since each of the line wires, aa_1, bb_1, cc_1 (Fig. 103(a)) is in series with one of the coil circuits of the Y, the current in each coil circuit must be equal to the current in the line to which it is connected or otherwise current would enter or leave at one of the junctions, as at a. *In a Y-system, therefore, the line currents and their respective coil currents must be equal.*

It follows from Kirchhoff's first law that the vector sum of the three coil currents in a Y-system having no neutral wire must be zero. The fact that this is true for a balanced load may be seen from Fig. 98(a), as the three waves may represent also the three Y-currents, rather than emfs, since the currents also are equal and differ in phase by 120°. It has been shown that at any instant the algebraic sum of the three waves (Fig. 98(a)) is zero. Hence, the algebraic sum of the three currents at every instant is zero.

Figure 104 shows vectorially the three coil currents, each in phase with its coil emf. Since the three currents are equal and differ in phase by 120°, their vector sum is zero.

With unbalanced loads, that is, with unequal currents having phase differences other than 120°, the algebraic sum of the three

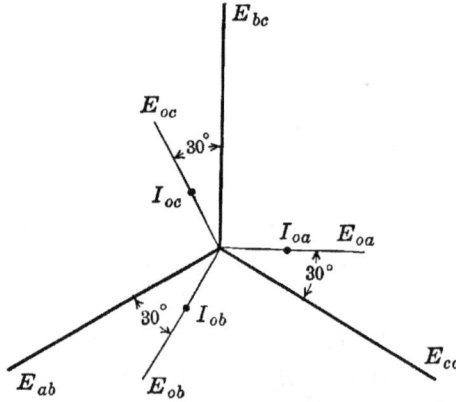

FIG. 104.—Relation of line-to-coil voltage and currents in a Y-system, unity power factor.

currents at any instant and the vector sum of the three currents must also be zero.

With unity power factor, each coil current is in phase with its coil terminal voltage. The currents under these conditions are *not* in phase with the *line* voltages. This is illustrated in Fig. 104. The current I_{oa} is in phase with its coil voltage E_{oa}, the current I_{ob} is in phase with its coil voltage E_{ob}, etc. The current I_{oa} leads the line voltage E_{ca} by 30°. The coil current I_{oa} is the line current $I_{aa'}$. Hence, with *unity power factor* in a three-phase system, the *line current differs in phase by 30° from the line voltage.*

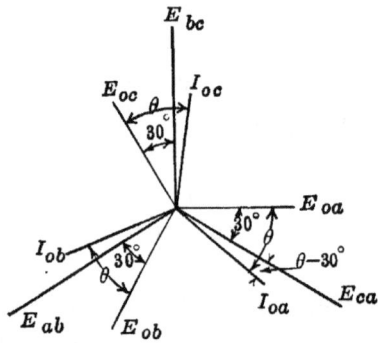

FIG. 105.—Relation of line-to-coil voltages and currents in a Y-system, power factor = cos θ.

If the power factor for a balanced load is other than unity, the coil current lags (or leads) the *coil* voltage by an angle θ, as shown in Fig. 105. For example, the current I_{oa} in coil *oa* lags the *coil* voltage E_{oa} by the angle θ. The current I_{oa} lags the *line* voltage E_{ca} by the angle (θ − 30°).

The current I_{oa} is the line current $I_{aa'}$. Hence, when the coil current differs in phase from the coil voltage by the angle θ, the line current differs in phase from the line voltage by the angle $\theta - 30°$ (or $30° - \theta$).

Figure 105 shows that each of the currents in the other two coils lags its coil voltage by the angle θ but lags the line voltage by the angle $\theta - 30°$. It will be shown that the power factor of such a balanced three-phase system is cos θ, *not* cos $(\theta - 30°)$. In other words the power factor is determined by the phase angle between *coil voltage* and *coil current*.

77. Power in Y-system.—The power developed in each of the three coils (Fig. 103(a)) must be independent of the system to which the coil is connected. If, for example, E_{oa} is the voltage across coil oa and a current I_{oa} flows in coil oa and if the voltage and current are in phase (Fig. 104), the power in coil oa must be equal to $E_{oa}I_{oa}$. That is, the power delivered by each coil is

$$p = E_{oa}I_{oa} \text{ watts (unity P.F.)}.$$

A balanced three-phase system is assumed. Hence, the coil currents are equal numerically, and each is in phase with its coil voltage.

The total power of the system is, therefore,

$$P = 3E_{coil}I_{coil} \text{ watts.} \tag{I}$$

In the Y-system the coil current equals the line current. The coil voltage

$$E_{coil} = \frac{E_{line}}{\sqrt{3}} \text{ volts.}$$

Substituting, in (I), $E_{line}/\sqrt{3}$ for the value of E_{coil},

$$P = \frac{3}{\sqrt{3}} E_{line}I_{coil} = \sqrt{3}\, E_{line}I_{line} \text{ watts.} \tag{52}$$

In a balanced three-phase system, the line power at unity power factor is equal to $\sqrt{3}$ times the line voltage multiplied by the line current.

Example.—A three-phase, Y-connected alternator delivers 80 amp to each line, and the terminal voltage (between lines) is 2,300 volts. The coil voltage and the coil current are in phase with each other, making the power factor of the system unity. Determine: (a) coil voltage; (b) power delivered by each coil; (c) system power.

(a) The coil voltage

$$E_{coil} = \frac{2,300}{\sqrt{3}} = 1,328 \text{ volts.} \quad Ans.$$

(b) $p = 1,328 \times 80 = 106,200$ watts $= 106.2$ kw. *Ans.*
(c) $P = 106.2 \times 3 = 318.6$ kw. *Ans.*
Also, using Eq. (52),
$\quad P = \sqrt{3} \times 2,300 \times 80 = 318.7$ kw (check).

If each coil current lags (or leads) its coil terminal voltage by the angle θ (Fig. 105), the power per coil (from Eq. (27), p. 70) is

$$p = E_{coil}I_{coil} \cos \theta_{coil} \text{ watts.}$$

The total power of the system

$$P = 3E_{coil}I_{coil} \cos \theta_{coil} \text{ watts.}$$

Substituting $E_{coil} = E_{line}/\sqrt{3}$ and $I_{coil} = I_{line}$, the system power becomes

$$P = \sqrt{3} \, E_{line}I_{line} \cos \theta_{coil} \text{ watts,} \qquad (53)$$

and the system power in kilowatts is equal to

$$P' = \frac{\sqrt{3}}{1,000} E_{line}I_{line} \cos \theta_{coil} \text{ kw.} \qquad (54)$$

It is clear from Eq. (53) that $\sqrt{3} \, E_{line}I_{line}$ gives the *volt-amperes* of the system. The power is equal to the product of the volt-amperes and the power factor (see p. 66). The *kilovolt-amperes*

$$\text{Kva} = \frac{\sqrt{3}}{1,000} E_{line}I_{line}. \qquad (55)$$

The system power factor, which is the coil power factor, is, from Eq. (53),

$$\text{P.F.} = \cos \theta_{coil} = \frac{P}{\sqrt{3} \, E_{line}I_{line}}, \qquad (56)$$

where P is the total power of the system in watts.

Therefore, in a balanced three-phase system, the system power factor is the cosine of the angle between the coil current and the coil terminal voltage.

If the system is unbalanced, that is, if the currents or voltages are not equal and do not differ in phase by 120°, there is a question as to what is meant by the system power factor. Where such unbalancing is not very great, Eq. (56) is used, the line currents and line voltages being averaged. The system power factor has little significance when there is substantial unbalance.

Example.—A 3-phase, Y-connected alternator has three coils, each of which delivers 150 amp. at 1,328 volts, and the phase difference between the coil voltage and coil current is 40°, the current lagging. Determine: (a) voltage rating of alternator; (b) kilovolt-ampere rating of alternator; (c) power factor of alternator at this load; (d) power in kilowatts.

(a) $1,328 \sqrt{3} = 2,300$ volts. *Ans.*

(b) From Eq. (55),

$$\text{Kva} = \frac{\sqrt{3}}{1,000} \, 2,300 \times 150 = 598. \quad Ans.$$

(c) Cos $\theta_{coil} = \cos 40° = 0.766$. *Ans.*

(d) From Eq. (54),

$$P' = \frac{\sqrt{3}}{1,000} \, 2,300 \times 150 \times 0.766 = 458 \text{ kw.} \quad Ans.$$

Also, from (b),

$P' = 598 \times 0.766 = 458$ kw (check).

78. Delta-connection.—Figure 106(a) shows the outer end a of one of the three coils or coil circuits of the three-phase alternator (Fig. 103(a)) connected to the inner end o_2 of the next coil o_2b. The outer end b of coil o_2b is connected to the inner end o_3 of coil o_3c; and the outer end c of coil o_3c is connected to the inner end o_1 of coil o_1a, forming a closed circuit.

It is to be noted that points o_2 and a, o_3 and b, o_1 and c are common in pairs. Hence, the o's may be omitted. This is done in Fig. 106(b) which shows the same three coils arranged to form the sides of an equilateral triangle.

The voltage E_{ca} is now the same as E_{o_1a}, E_{ab} the same as E_{o_2b}, and E_{bc} the same as E_{o_3c}. (See Fig. 101.) These three voltages with the new subscripts are shown in Fig. 107.

This method of connecting the three elements or coil circuits of a three-phase system is called the *delta-connection* from the resemblance to the Greek letter delta. (See Part I, p. 319.)

Under balanced conditions and with sinusoidal voltages and currents, no current circulates around the delta, even though it is a closed circuit. The total voltage acting around the delta is $E = E_{ab} + E_{bc} + E_{ca}$. The sum of three vectors equal in magnitude and differing in phase by 120° is zero, as Fig. 107 shows.

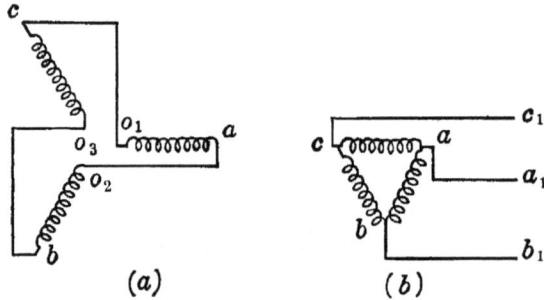

FIG. 106.—Delta connection of alternator coils.

If, for example, E_{ab} and E_{bc} are added vectorially, their sum E_{ac} ($= E_{ab} + E_{bc}$) is equal and opposite to E_{ca}. This sum, when added to E_{ca}, gives a resultant zero. Also, with the three emf waves (Fig. 98(a), p. 123), which represent the three coil emfs, the algebraic sum at every instant is zero, so that the vector sum must be zero.

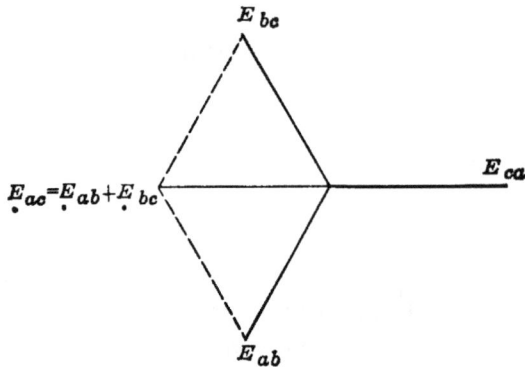

FIG. 107.—Vector sum of three delta voltages equals zero.

The three line connections are made at the junctions of the three coils at a, b, c (Fig. 106(b)). It is clear that each of the three line voltages must be the same as each of the three coil voltages. *Therefore, in a delta-system the line voltages are equal to the coil voltages.*

79. Currents in Delta-system.—Figure 108(a) shows the delta-system of Fig. 106 with the three coil currents I_{ab}, I_{bc}, I_{ca}, flowing from a to b, b to c, c to a. Balanced conditions and unity power factor are assumed. These conditions are shown in the vector diagram (Fig. 108(b)) in which I_{ab} is in phase with E_{ab}, etc.

It is required to find the value of the line current I_{aa_1}. At junction a, I_{ca} flows toward the junction, I_{ab} flows away from the junction, and I_{aa_1} flows away from the junction.

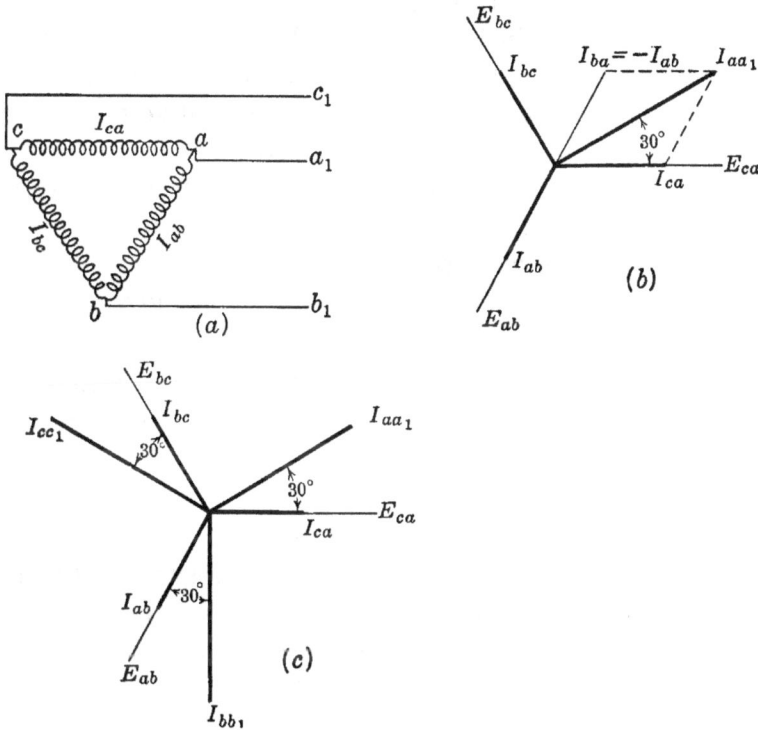

Fig. 108.—Currents and voltages in delta system.

By Kirchhoff's first law, the current

$$I_{aa_1} = I_{ca} - I_{ab} \text{ amp.}$$

To subtract a vector from another, reverse it and add (see p. 28).

In the vector diagram (Fig. 108(b)) the current $(-I_{ab})$ is found by reversing I_{ab}, giving I_{ba}. I_{ba} or $(-I_{ab})$ is then added to I_{ca} to give I_{aa_1}. Also, at junction a,

$$I_{aa_1} = I_{ca} + I_{ba} \text{ amp.}$$

The vector addition is performed in (*b*). Since I_{ba} or $(-I_{ab})$ and I_{ca} are 60° apart and are equal numerically, their vector sum $I_{aa_1} = \sqrt{3}\, I_{ca}$, or $1.732 I_{ca}$, numerically.

If I_{bb_1} is found in a similar manner by adding I_{ab} and I_{cb},

$$I_{bb_1} = I_{ab} + I_{cb} \text{ amp.}$$

Similarly,

$$I_{cc_1} = I_{bc} + I_{ac} \text{ amp.}$$

The three line currents I_{aa_1}, I_{bb_1}, I_{cc_1} will have the magnitudes and directions shown in Fig. 108(*c*). It is to be noted that each line current differs in phase from its coil current by 30°. Also, when the coil currents are in phase with the coil voltages, the line currents differ in phase from the line voltages by 30°. This is also true of the Y-system (see Fig. 104). The foregoing relations between line and coil *currents* hold, whether or not the coil currents are in phase with the coil voltages.

Therefore, in a balanced delta-system, the line currents are equal to $\sqrt{3}$ times the coil currents.

Example.—The three coils of an alternator (p. 129) are connected in delta, and each delivers 90 amp at 220 volts. What is the current in each line?

$$I_{line} = \sqrt{3} \times 90 = 1.732 \times 90 = 156 \text{ amp.} \quad Ans.$$

There is 220 volts across the line.

80. Power in Delta-system.—As in the Y-system, the power developed in any one of the coils of a delta-system must be independent of the system to which the coil is connected.

That is, the power in each coil

$$p = E_{coil} I_{coil} \cos \theta_{coil} \text{ watts,}$$

where θ_{coil} is the phase angle between coil voltage and coil current. The total power developed by the three coils

$$P = 3 E_{coil} I_{coil} \cos \theta_{coil} \text{ watts.}$$

In the delta-system, $E_{coil} = E_{line}$, and $I_{coil} = I_{line}/\sqrt{3}$.
Hence,

$$P = \frac{3}{\sqrt{3}} E_{line} I_{line} \cos \theta_{coil} \text{ watts}$$

$$= \sqrt{3}\, E_{line} I_{line} \cos \theta_{coil} \text{ watts,} \qquad (57)$$

as in the Y-system.

The kilowatts

$$P' = \frac{\sqrt{3}}{1,000} E_{line}I_{line} \cos \theta_{coil} \text{ kw.} \qquad (58)$$

The kilovolt-amperes

$$\text{Kva} = \frac{\sqrt{3}}{1,000} E_{line}I_{line}. \qquad (59)$$

The power factor in the delta-system is the *coil* power factor and is equal to the cosine of the angle between the coil terminal voltage and the coil current. Hence, from Eq. (57),

$$\text{P.F.} = \cos \theta_{coil} = \frac{P}{\sqrt{3}\, E_{line}I_{line}} = \frac{\text{kw}}{\text{kva}}.$$

Example.—Consider the three-phase alternator of p. 133, but now connected in delta, each coil delivering 150 amp at 1,328 volts. The phase difference between coil voltage and coil current is 40°. Determine: (*a*) voltage rating of alternator; (*b*) ampere rating; (*c*) kva rating; (*d*) power factor; (*e*) power output.

(*a*) Since the line voltage and coil voltage are equal, the voltage rating must be 1,328 volts. *Ans.*

(*b*) $I = \sqrt{3} \times 150 = 260$ amp. *Ans.*

(*c*) From Eq. (59),

$$\text{Kva} = \frac{\sqrt{3}}{1,000} 1,328 \times 260 = 598. \quad Ans.$$

(*d*) P.F. $= \cos 40° = 0.766$.

(*e*) $P' = \dfrac{\sqrt{3}}{1,000} 1,328 \times 260 \times 0.766 = 458$ kw. *Ans.*

It is to be noted that the kva rating and the kilowatt rating of an alternator are independent of whether it is connected in Y or in delta.

81. Y- and Delta-connected Loads.—Not only may sources of energy, such as alternator coils, be connected in either Y or delta, but loads, such as resistances, transformer primaries, and coils of three-phase motors, may be connected in either Y or delta. This is illustrated by the following examples:

Example.—Three resistances, each of 10 ohms, are connected in Y across 100-volt, 60-cycle mains (Fig. (109)). Determine: (*a*) current to each resistance; (*b*) current in the line; (*c*) power of the system.

(a) The voltage across each resistance

$$E_R = \frac{100}{\sqrt{3}} = 57.7 \text{ volts.}$$

$$I = \frac{E}{R} = \frac{57.7}{10} = 5.77 \text{ amp.} \quad Ans.$$

(b) Since the resistances are connected in Y, the line current must be the same as the current in the resistances, or 5.77 amp.   *Ans.*

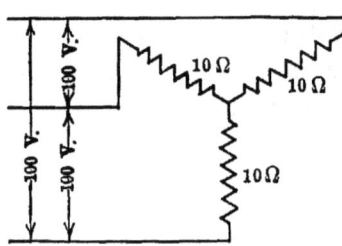

FIG. 109.—Resistances con-   FIG. 110.—Resistances connected in Y.   nected in delta.

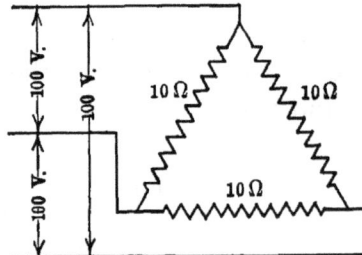

(c) The power taken by each resistance

$$P_R = 57.7 \times 5.77 = 333.3 \text{ watts.}$$

The total power

$$P = 3 \times 333.3 = 1,000 \text{ watts.} \quad Ans.$$

Since the power factor is unity, Eq. (52) (p. 131) may be used.

$$P = \sqrt{3} \times 100 \times 5.77 = 1,000 \text{ watts (check).}$$

Example.—Repeat the foregoing example with the three resistances connected in delta across the same system (Fig. 110).

(a) Since each resistance is across two conductors of the three-phase system, the voltage across each is 100 volts.

The current in each

$$I_R = \frac{100}{10} = 10 \text{ amp.} \quad Ans.$$

(b) The line current

$$I = 10 \sqrt{3} = 17.32 \text{ amp.} \quad Ans.$$

(c) The power taken by each resistance

$$P_R = 100 \times 10 = 1,000 \text{ watts.}$$

The total power

$$P = 3 \times 1,000 = 3,000 \text{ watts.} \quad Ans.$$

Equation (57) may be used, the power factor being unity.

$$P = \sqrt{3} \times 100 \times 17.32 = 3{,}000 \text{ watts.} \quad Ans. \text{ (check)}.$$

It is to be noted that three equal resistances connected in delta across a balanced three-phase system take *three times* the power that they take when connected in Y.

Example.—Each of three lamp banks operates at 115 volts and at this voltage takes 10 amp (Fig. 111). If these lamp banks are connected in Y and to neutral across a three-phase, four-wire system with 115 volts to

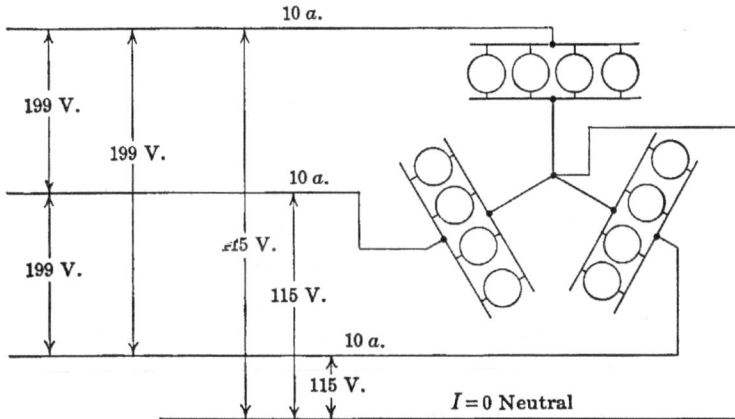

FIG. 111.—115-volt lamp banks connected in Y.

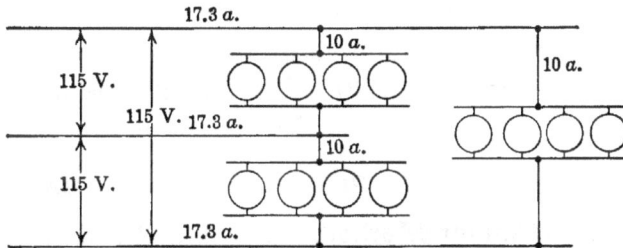

FIG. 112.—115-volt lamp banks connected in delta.

neutral, determine: (a) line current; (b) line voltage; (c) current in neutral; (d) total power.

(a) The line current must equal the current taken by each lamp bank and is, therefore, 10 amp. *Ans.*

(b) The line voltage

$$E = \sqrt{3} \times 115 = 199 \text{ volts.} \quad Ans.$$

(c) Since the three currents are equal and differ in phase by 120°, their vector sum must be zero. Hence, the neutral current is zero.

(d) The power, by Eq. (52) (p. 131),

$$P = \sqrt{3} \times 199 \times 10 = 3{,}450 \text{ watts.} \quad Ans.$$

Example.—If the three lamp-banks of the foregoing problem are connected in delta and operate at the same voltage (Fig. 112), determine: (*a*) current in each line conductor; (*b*) total power.

(*a*) Since the lamp banks are now in delta, the line current

$$I = \sqrt{3} \times 10 = 17.32 \text{ amp.}$$

(*b*) The power

$$P = \sqrt{3} \times 115 \times 17.32 = 3,450 \text{ watts.} \quad \textit{Ans.}$$

In this case the power does not change when the lamp banks are changed from Y- to delta-connection, since the line voltage is simultaneously reduced to $1/\sqrt{3}$ times its former value.[1]

82. Uses of Y- and Delta-connections.

—In many instances, it is immaterial whether the Y- or the delta-connection is used. For example, small alternators operate satisfactorily whether connected in Y or in delta. With units of large output, circulatory currents would probably flow around the delta. These circulatory currents are due to the wave form being nonsinusoidal. With the Y-connection, such circulatory currents cannot exist. At the same time the wave form of the terminal voltage is improved by the use of the Y-connection. The Y-connection gives a neutral with which the system may be grounded. With motors, such as induction motors, both connections are satisfactory. In fact, some manufacturers connect a motor in delta for 220 volts and in Y for 440 volts. Thus, one line of motors can be adapted readily to two different voltages.[2]

METHODS OF MEASURING POWER IN THREE-PHASE SYSTEM

83. Three-wattmeter Method.

—Let (1), (2), (3) (Fig. 113(*a*)) be the three coils of either a Y-connected alternator or a Y-connected load. (The ordinary wattmeter reads upscale when connected as shown, if the coils shown diagrammatically represent a source of energy. If these coils represent loads, the current-coil connections of each wattmeter should be reversed.) If the neutral of the Y is accessible, it is possible to measure the power of each phase by connecting the current coil of a wattmeter in series with the phase and connecting the wattmeter potential

---

[1] See p. 179 for the method of phasing coils in Y and in delta.

[2] For the discussion of the delta- and Y-connection for transformers, see p. 252, Par. 146.

coil across the phase, as in Fig. 113(*a*). Therefore, W_1, W_2, W_3 measure the power in loads 1, 2, 3, regardless of power factor, degree of balance, etc.

The total power

$$P = W_1 + W_2 + W_3.$$

If the loads are balanced,

$$W_1 = W_2 = W_3.$$

If the potential circuits of the three wattmeters have equal resistances, these three potential circuits constitute a balanced Y-load, having a neutral o'. As coils 1, 2, 3 and these three-

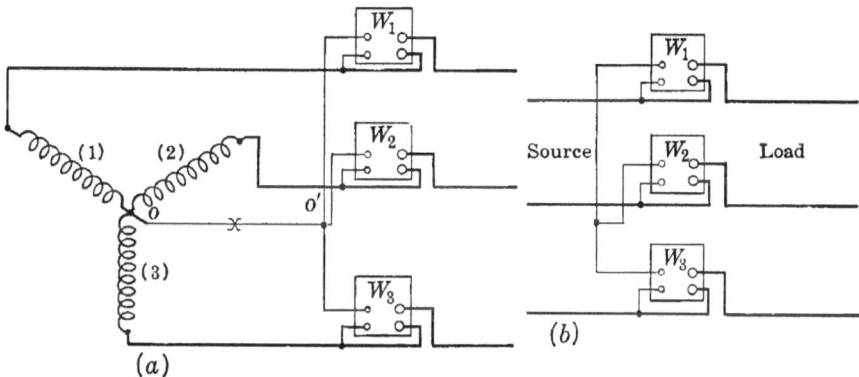

Fig. 113.—Three-wattmeter method of measuring three-phase power.

wattmeter potential circuits are both symmetrical, Y-connected systems, o' must be at the same potential as o. Therefore, no current flows between o and o', and the line can be cut at X without changing existing conditions. Figure 113(*b*) shows the three-wattmeter connection for a three-phase system. It can be shown that the total power is the sum of the wattmeter readings even though the wattmeter potential circuits have different resistances. Under these conditions, however, the wattmeters may not have the same reading, even with balanced loads.

The three-wattmeter method is well adapted to measuring power in a system where the power factor is continually changing, as in obtaining the phase characteristics of a synchronous motor. If the three instruments have equal potential-circuit resistances, and the system is balanced, they read the same regardless of power factor. The three-wattmeter method is necessary in a three-phase, four-wire system, as a system of n wires ordinarily requires at least $n - 1$ wattmeters in order to measure the power.

84. Two-wattmeter Method.—The power in a three-phase, three-wire system can be measured by means of two wattmeters connected as shown in Fig. 114. The current coils of the two wattmeters are connected in series with any two wires such as $a'a$ and $b'b$ (Fig. 114). The potential circuit of each instrument is connected between the wire in which its current coil is connected and the third wire.

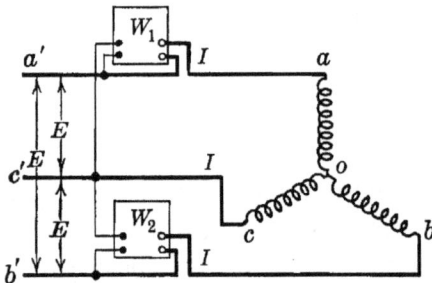

FIG. 114.—Two-wattmeter method of measuring three-phase power.

For example, the potential circuit of W_1 is connected between wire $a'a$ and wire $c'c$; the potential circuit of W_2 is connected between wire $b'b$ and wire $c'c$.

The total power to the load

$$P = W_1 \pm W_2 \text{ watts,} \tag{60}$$

regardless of power factor, unbalancing in the system, etc. The load need not necessarily be Y-connected as shown but may be delta-connected or V-connected or may consist of several loads connected in different combinations.

If the system is balanced, it can be shown that the wattmeter readings are

$$W_1 = EI \cos (30° - \theta) \text{ watts,} \tag{61a}$$
$$W_2 = EI \cos (30° + \theta) \text{ watts,} \tag{61b}$$

where E is the line voltage, I the line current, and θ the power-factor angle of the system, that is, the angle of phase difference between *coil* voltage and *coil* current.

If the power factor is unity, that is, if $\theta = 0$, both wattmeters read the same. Both readings are positive, and the total power is given by the sum of W_1 and W_2. (The wattmeters also read the same when $\theta = 90°$, in which case the reading of W_2 is negative and hence $W_1 - W_2$ is zero. When $\theta = 180°$, both readings are equal and negative.)

If $\theta = 60°$, the power factor is 0.5, since $\cos 60° = 0.5$.   Under these conditions, W_1 still reads positive since

$$\cos(-30°) = \cos(+30°)$$

(see p. 432) and is positive. W_2 must read zero, since $\cos 90°$ is zero.   That is, when the power factor is 0.5, W_1 indicates the entire power.

With power factors less than 0.5, W_1 still reads positive, but W_2 reads negative, since the cosine of an angle between 90 and 120° is negative (see p. 433).   If the wattmeters are connected symmetrically as shown in Fig. 114 and both read upscale when

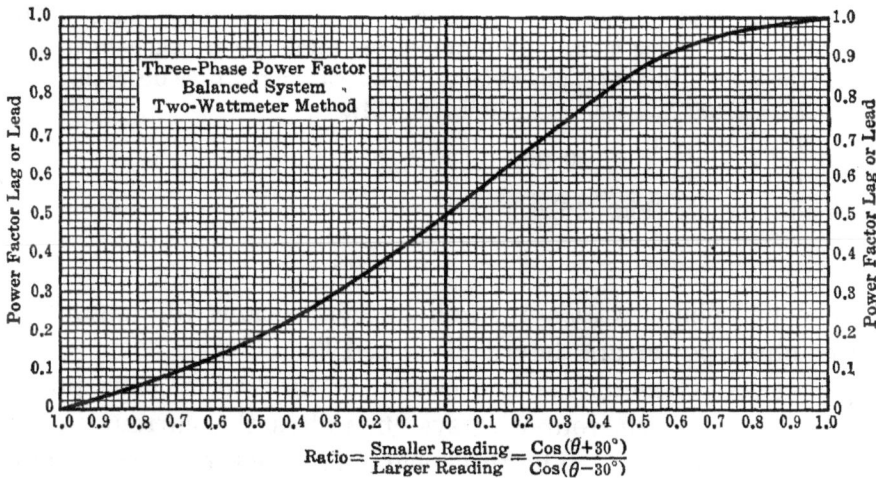

FIG. 115.—Power-factor diagram, two-wattmeter method.

the power factor exceeds 0.5, the deflection of W_2 will be reversed when the power factor is less than 0.5.   The connections of the current coil of W_2 must therefore be reversed in order that the pointer may read upscale.   The reading of W_2 is then *subtracted* from W_1 in order to obtain the total power of the system.   That is, the minus sign in Eq. (60) is now used.

Dividing Eq. (61*b*) by Eq. (61*a*),

$$\frac{W_2}{W_1} = \frac{EI \cos(30° + \theta)}{EI \cos(30° - \theta)} = \frac{\cos(30° + \theta)}{\cos(30° - \theta)} \tag{62}$$

By substituting different values of θ, the curve (Fig. 115) may be obtained.   It shows power factor ($\cos \theta$) plotted as ordinates, and the ratio of the smaller to the larger wattmeter reading

(W_2/W_1) as abscissas.   Having the two wattmeter readings, the system power factor may be obtained without the volt-amperes.

Example.—The power input to an induction motor is measured by the two-wattmeter method.   When the motor is near its rated load, one wattmeter reads $+5,800$ watts and the other reads $+2,500$ watts.   Determine (a) total power taken by motor; (b) power factor at which motor is operating.

(a) $P = 5,800 + 2,500 = 8,300$ watts.   *Ans.*

(b) $\dfrac{W_2}{W_1} = \dfrac{+2,500}{+5,800} = +0.431.$

From the curve (Fig. 115) the power factor is 0.82.   *Ans.*

Example.—When the motor in the foregoing example is operating at light load, the two wattmeters read $+2,400$ watts and $-1,100$ watts.   (It is necessary to reverse the current-coil connections of one wattmeter in order that it may read upscale.   Its reading of 1,100 watts must therefore be preceded by a minus sign.)   Determine: (a) total power taken by motor under these conditions; (b) power factor at which the motor is operating.

(a) $P = 2,400 - 1,100 = 1,300$ watts.   *Ans.*

(b) $\dfrac{W_2}{W_1} = \dfrac{-1,100}{2,400} = -0.458.$

From the curve (Fig. 115),

$$\text{P.F.} = 0.21. \quad Ans.$$

The polyphase wattmeter and its connections for measuring three-phase power are described on p. 105.

The two-wattmeter method cannot be used to measure power with a three-phase, four-wire system unless the current in the neutral wire is zero.   The three-wattmeter method may be used on such a system.

OTHER POLYPHASE SYSTEMS

85. Two-phase Systems.—Although three-phase systems are superseding other polyphase systems, there are many two-phase or quarter-phase systems in use.   The two-phase system is rarely used for transmission but is used for distribution, and in some instances there are special advantages in using two-phase machines.

Two-phase emfs are induced in the elementary generator[1] (Fig. 116(a)) by two coils A and B, 90° apart.   Figure 116(b) shows the emf waves generated by these coils.   The emf of A

---

[1] See Chap. V for the description of two-phase windings as applied to commercial alternators.

leads that of B by 90°.   When one emf is a maximum, the other
is zero.   Figure 116(c) shows these two-phase emfs vectorially.

The two phases, insulated from each other, may be used to
supply two separate single-phase circuits, or they may supply a

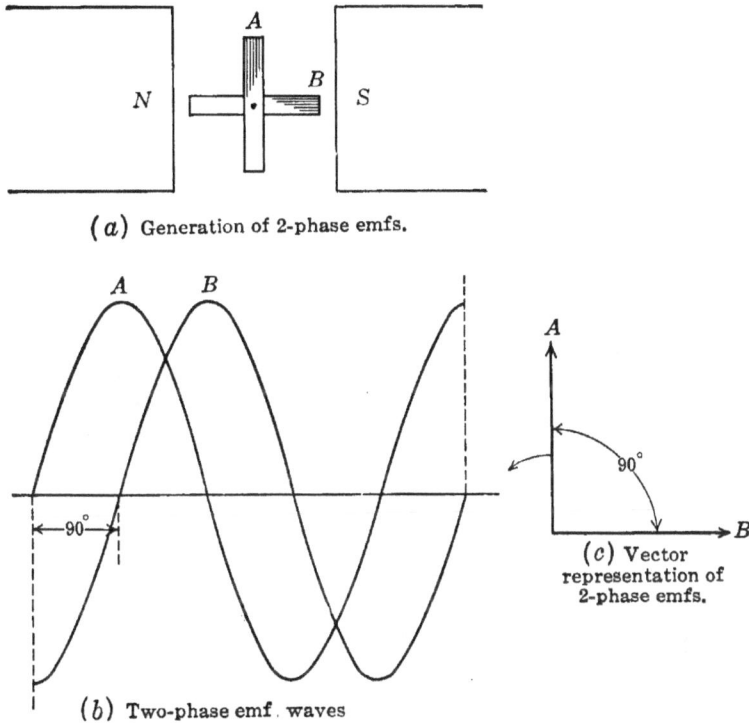

(*a*) Generation of 2-phase emfs.

(*c*) Vector
representation of
2-phase emfs.

(*b*) Two-phase emf. waves
FIG. 116.—Phase relations of two-phase emfs.

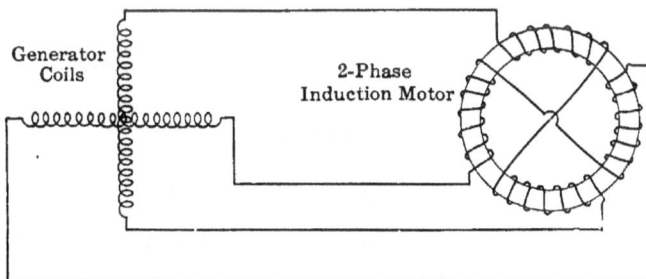

FIG. 117.—Two-phase circuit in which two phases are isolated.

common load such as an induction motor (Fig. 117).   The two
phases are insulated, or isolated, from each other (Fig. 117), and
no one single-phase load can be supplied between the two phases.
Moreover, only one value of voltage is obtainable, for the voltages of the two phases are equal.

If, however, the generator coils are interconnected at their neutral points and a neutral conductor is carried along with the other four conductors, a quarter-phase (or four-phase), five-wire system results (Fig. 118(a)). The coils are then said to be *star-connected*. Moreover, three different voltages are available. For example, if the voltages between the outer wires of each phase are 200 volts, 200 volts is available between wires AA and

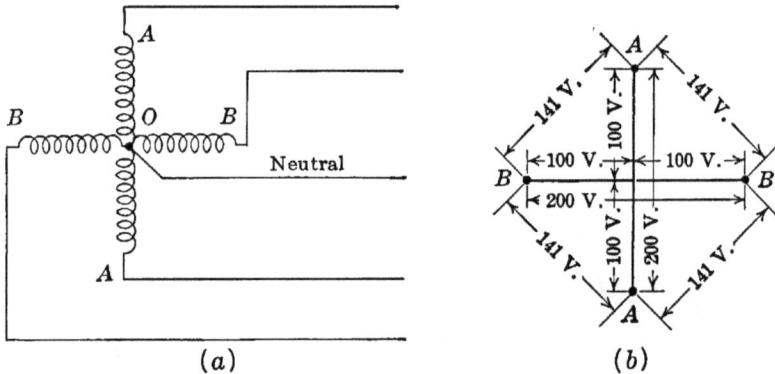

Fig. 118.—Two-phase interconnected system giving four-phase, five-wire system.

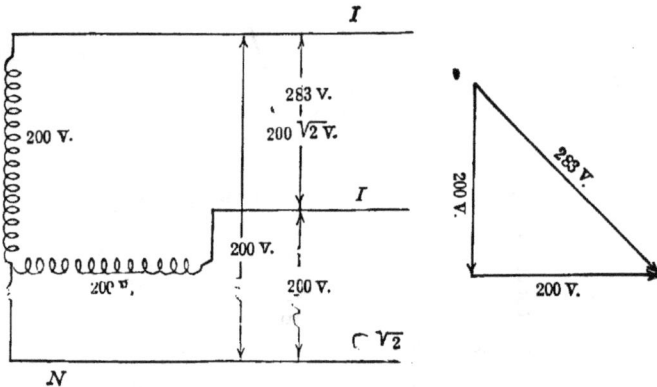

Fig. 119.—Two-phase, three-wire system.

also between wires BB. These two voltages differ in phase by 90°, or are in quadrature with each other. A voltage of 100 volts exists between each outer wire and neutral. Since the voltages from A to O and from B to O are each equal to 100 volts, the voltages between wires A and B are each 100 $\sqrt{2}$, or 141, volts.

This system is not so readily balanced as the three-phase system, which is an objection to its use. Another objection is the greater number of wires.

If one end of the coil A is connected to one end of the coil B, a three-wire, two-phase system results (Fig. 119). This gives an *unsymmetrical* polyphase system. Since the voltage across each coil is 200 volts and these voltages are in quadrature, the voltage across their open ends is $200 \sqrt{2}$ or 283 volts, as shown by the vector diagram.

This system is little used because of its lack of symmetry and the considerable amount of voltage unbalancing which results, even when moderate loads are applied. When currents each of I

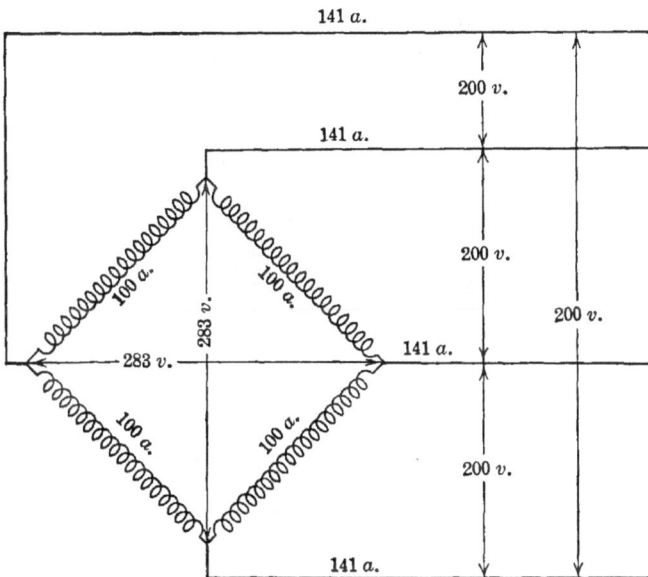

FIG. 120.—Mesh-connected, two-phase winding.

amp and in quadrature flow between the outer conductors and the common return N, the current in the common return is $I \sqrt{2}$ amp.

Therefore, the common return must ordinarily have greater cross-section than that of the two outers.

A two-phase alternator may be mesh-connected in the manner shown in Fig. 120. With a closed two-pole winding, four equidistant taps are brought out as shown. This is called the *mesh-* or *ring-connection*. The coils constituting the winding must be properly connected or the winding will be short-circuited on itself. If each coil generates 200 volts, the voltage across adjacent line wires is 200 volts and that across line wires diametrically spaced is 200 $\sqrt{2}$, or 283, volts, since the voltages across adjacent coils are in quadrature.

If each coil delivers 100 amp, the line currents must be 100 $\sqrt{2}$, or 141, amp, since the currents in two adjacent coils are in quadrature and combine to flow in a common wire. Since each coil delivers 100 amp at 200 volts, the output of the system is

$$\frac{4 \times 200 \times 100}{1,000} = 80 \text{ kva.}$$

The two-phase mesh- or ring-connection corresponds to the three-phase delta-connection.

86. Measurement of Power in Two-phase Systems.—A two-phase isolated system consists merely of two separate single-phase

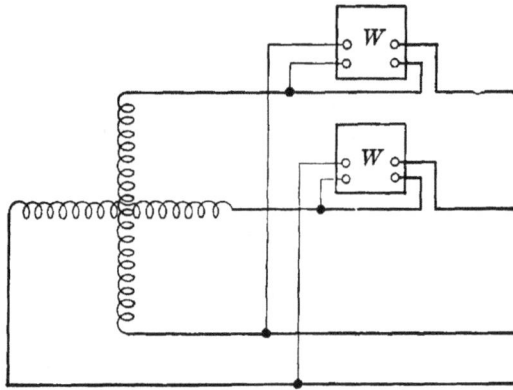

FIG. 121.—Measurement of power in an isolated two-phase, four-wire system.

systems. Hence, a wattmeter is connected in each of the two phases (Fig. 121) and the total power is given by the sum of their readings. If the system is interconnected (Fig. 122), at least three wattmeters are necessary. The total power is the sum of their readings irrespective of power factor and unbalance. If the loads on the two phases happen to be balanced, no current flows between phases at the interconnection point, and two wattmeters connected as in Fig. 121 may be used to measure the total power.

In a four-phase, five-wire system (Fig. 118), four wattmeters are necessary. They are preferably connected with a current coil in each of the outer wires, and each potential circuit is connected between the wire in which its current coil is connected and neutral.

A two-phase, three-wire system may be considered as two single-phase circuits connected at a common point. Hence, two wattmeters connected as in Fig. 123 may be used to measure the total power. It is seen that this connection is identical with the two-wattmeter method (Fig. 114). It is not necessary that

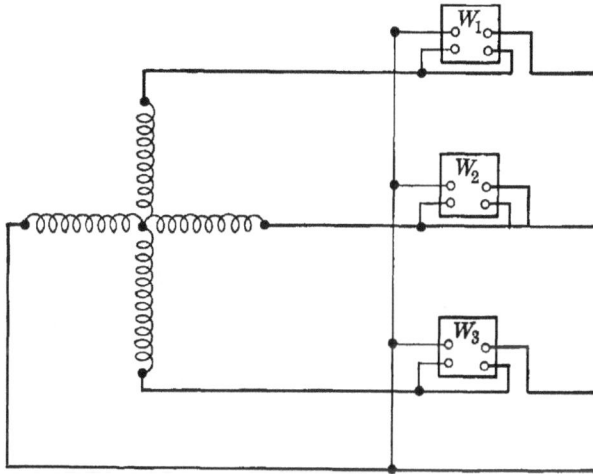

FIG. 122.—Measurement of power in two-phase interconnected system.

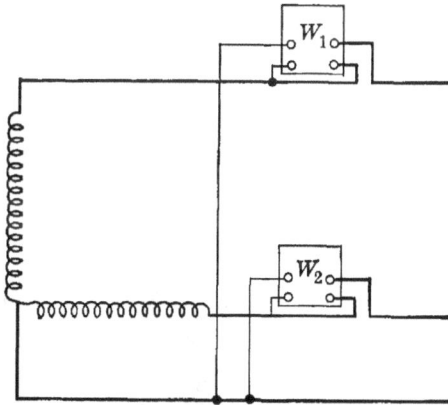

FIG. 123.—Measurement of power in three-wire, two-phase system.

the wattmeters be connected in the two outer lines as shown. One current coil may be connected in the common return. The potential circuit of each wattmeter is connected between the wire in which its current coil is connected and the third wire.

87. Six-phase Systems.—If six similar coils or groups of coils in an alternator winding, 0-1, 0-2, . . . , 0-6 (Fig. 124), are 60 electrical space-degrees apart, a six-phase system may be obtained. If the six coils are connected symmetrically, each with one end to a common point 0, and the proper phase relations are observed, a six-phase star-connection results. If each coil generates 100 volts, the six voltages, 0-1, . . . , 0-6, are each 100 volts. Also, the voltage between each adjacent pair of terminals 1-2, 2-3, etc., is 100 volts. The diametrical voltages, between 1-4, 2-5, etc., are each 200 volts. This six-phase system may be considered as a double Y, of which coils 1, 3, 5 form one Y since their voltages differ in phase by 120° and coils 2, 4, 6 form the second Y, displaced 60° from the first Y.

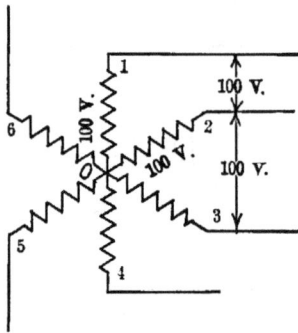

FIG. 124.—Six-phase star connection.

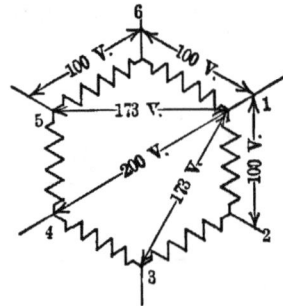

FIG. 125.—Six-phase mesh or ring connection.

Therefore, the voltages across 1-3, 3-5, 5-1, 2-4, 4-6, 6-2 are each 100 $\sqrt{3}$, or 173.2, volts.

The six coils of Fig. 124 may also be connected to form a six-phase mesh- or ring-connection, as shown in Fig. 125. For example, the inner end of coil 2 may be connected to the outer end of coil 1, the inner end of coil 3 to the outer end of coil 2, etc. If the voltage of each coil is 100 volts, the six voltages between adjacent terminals 1-2, 2-3, etc., are each 100 volts. The diametrical voltages 1-4, 2-5, 3-6 are each 200 volts; and the voltages 1-3, 2-4, etc., are each 100 $\sqrt{3}$, or 173.2, volts.

Six-phase connections are generally used on the a-c side of synchronous converters, since the rating of the converter is increased considerably thereby (see pp. 333 and 335).

CHAPTER V

THE ALTERNATOR

It is shown in Chaps. I and III that alternating emfs can be induced by simple coils rotating in magnetic fields. These coils are provided with the necessary slip-rings for conducting the current to an external circuit where it can be utilized. Although these types of alternator demonstrate clearly the fundamental principles underlying the generation of alternating current, they are not practicable for generating electrical energy on a commercial basis, since the small power output for their weight makes them uneconomical. However, the principles involved in the generation of alternating current in these simple alternators and the principles involved in the generation of alternating current in the large commercial types of alternator are *identical*. The commercial types differ from the simple types in that the arrangement of copper and iron is such that much larger outputs are obtained per pound of material. In this chapter the construction and the winding of commercial types of alternator are described.

88. Rotating-field Type.—The generation of an emf in a conductor may take place with the magnetic field stationary and the conductor moving through this field, as in a d-c generator, or with the conductor stationary and the field moving past the conductor. It is merely necessary that there be *relative* motion between conductor and field. In d-c dynamos, the commutator makes it necessary that either the armature be the rotating member or the brushes revolve with the field.

As alternators have no commutator, it is not necessary that the armature be the rotating member. Most commercial alternators have stationary armatures, inside of which the field poles rotate, as shown in Figs. 130, 148 (pp. 157 and 172), etc. This construction has two distinct advantages. A rotating armature requires two or more slip-rings for carrying the current from the armature to the external circuit. Such rings must be more or less exposed and are difficult to insulate, particularly for the higher voltages of 6,600 and 13,200 volts at which alternators commonly operate. These rings may become a source of trouble, due to arc-overs, short circuits, and arcing between brushes and slip-rings. A stationary armature requires no slip-rings, and the insulation of the armature leads can be continuous from the armature coils to bus-bars. It is more difficult to insulate the conductors in a rotating armature than in a stationary one, because of centrifugal force and the vibration resulting from rotation.

When the field is the rotating member, the field current must be conducted to the field winding through slip-rings. As the

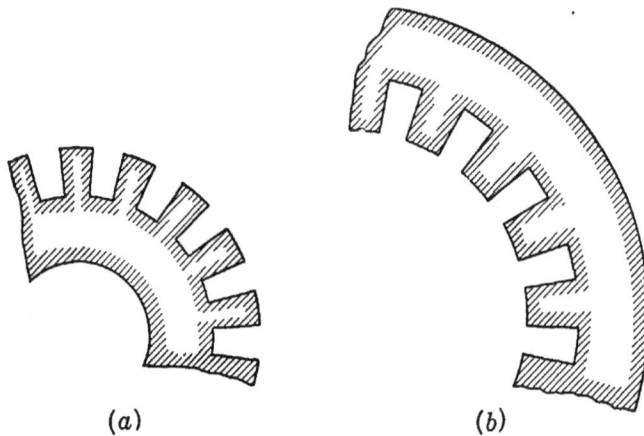

(a)   (b)

FIG. 126.—Effect of slot depth on the width of tooth necks in a rotor and in a stator.

field voltage seldom exceeds 250 volts and the amount of power is small, no difficulty is encountered in the operation of such slip-rings.

The output of a generator is practically proportional to the amount of active copper that can be placed on the armature, that is, in the slots. With a fixed length and diameter of armature the amount of copper can be increased only by increasing the cross-sectional area of the slots. The width of the tooth face should be at least equal to the slot opening in order that there shall be sufficient iron to conduct the air-gap flux into the armature. Hence the slot area can be increased only by deepening the slots. With a rotating armature, the deepening of the slots is limited by the contraction of the tooth necks, as shown in Fig. 126(a). With too small a cross-section at the tooth-necks

the teeth are too weak to withstand the mechanical stresses due
to centrifugal force, and the iron is too highly saturated magnetically. No such difficulty is encountered if the armature is
stationary, for the tooth necks increase in width with the deepening of the slots (Fig. 126(*b*)).

**89. Induced Electromotive Force with Rotating-field Struc-
ture.**—The method by which an alternating emf is induced in
the armature of the rotating-field type of alternator is illustrated
in Fig. 127, which shows two poles of the rotating-field structure
and a single coil *a′abb′* on the stationary armature. For sim-

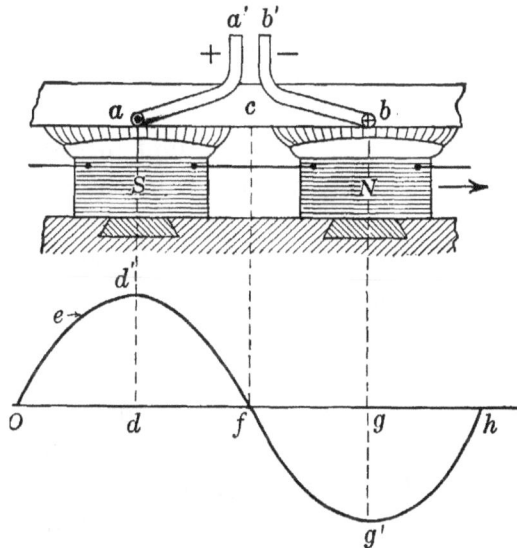

Fig. 127.—Induced emf with rotating-field structure.

plicity the armature surface is shown as a plane. The field
structure is shown as moving from left to right, corresponding
to clockwise rotation. The flux distribution is assumed to be
sinusoidal. At the instant shown, the center of the *N*-pole is
directly beneath coil side *b*, and the center of the *S*-pole is directly
beneath coil side *a*.

The direction of the induced emf may be determined by
Fleming's right-hand rule. Fleming's right-hand rule applies to
a *moving* conductor and a *stationary* field. Hence, assume
(Fig. 127) that the field structure is stationary and that the
armature rotates. Since the field is shown moving from left to
right, the armature must now be assumed to be moving from

right to left in order that the *relative* motion between armature and field shall remain unchanged. Upon applying Fleming's right-hand rule to conductor *a*, the direction of the magnetic-field at the *S*-pole being downward and the relative motion of the conductor being to the left, the direction of the induced emf is outward in conductor *a* and the terminal *a'* is positive. The direction of induced emf is inward in conductor *b*.

Since the direction, as well as the magnitude of an induced emf, depends on the *relative* motion of conductor and flux, the direction of the emf will still be outward in conductor *a* and inward in conductor *b*, if the armature is stationary and the field rotates in the clockwise direction as shown.

At the instant shown, conductors *a* and *b* are directly opposite the pole centers and therefore are being cut by the flux at the maximum rate. Hence, the induced emf in each conductor is a maximum. This is shown by the ordinates *dd'* and *gg'* of the sine curve *e*. As the pole structure moves from left to right, the emf in conductor *a* diminishes and becomes zero at *f* when the center of the *S*-pole is at *c* midway between conductors *a* and *b*. When the center of the *S*-pole is directly under conductor *b*, the emf in conductor *a* is negative, on the assumption, of course, that an *N*-pole follows the *S*-pole. That is, when an *N*-pole moves under conductor *a*, the direction of the induced emf in *a* is inward and the left-hand terminal of the coil is negative. In this manner an alternating emf is induced in each armature conductor as it is cut alternately by *N*- and *S*-pole flux.

The curve *e* then shows the values of emf in conductor *a* as the pole structure rotates. Since the flux density in the air-gap is sinusoidal, the emf wave will be sinusoidal. That is, $e = Blv$ 10^{-8} volts (see Part I, p. 210, Eq. (86)); and as B varies sinusoidally along the gap, l and v being constant, e will be sinusoidal.

The induced emf in coil *ab* may be considered also as being due to change in the flux *linking the coil* (see Part I, page 170, Eq. (66)). At the instant shown in Fig. 127, *no* flux links the coil. When the center of the *S*-pole is directly opposite *c*, *all the S-pole* flux links the coil; when the center of the *S*-pole is directly opposite conductor *b*, again *no* flux links the coil; when the center of an *N*-pole is directly opposite *c*, all the *N*-pole flux links the coil. Hence, the flux *linking* the coil is continually changing from south to zero and from zero to north, thus causing an alternating emf to be induced in the coil.

In the actual alternator, several coils lying in adjacent slots are connected in series to form one phase of the armature circuit. This utilizes more of the winding space on the armature and gives greater output for a given weight of machine. Alternator windings must be designed not only so that emfs are induced in them in the manner just described but so that the desired number of phases is obtained and the output per unit area of armature surface is large.

ALTERNATOR WINDINGS

90. General Principles of Windings.—If the principles underlying the generation of alternating emfs are kept in mind, it is not difficult to design complete windings to give any required number of phases and voltages and at the same time make the winding economical so that the given output is obtained with the minimum amount of copper and iron.

The usual d-c armature generates alternating current; and if it is provided with properly connected slip-rings, alternating current may be obtained from it. On the other hand, only certain types of alternator winding can be used for d-c armatures. The ordinary d-c winding is a *closed winding;* that is, the winding forms a continuously closed, conducting circuit. Alternator windings may be either open or closed. For example, the delta-connected winding is a *closed* winding, whereas the Y-connected winding is an *open* winding.

The general principles which govern d-c windings hold also for windings of alternators. The span of each coil must be approximately one pole pitch; that is, the two sides of any coil must lie under adjacent poles. The coils must be connected so that their emfs add.

There are several types of alternator winding. The more usual types will be described. If the principles involved in the usual windings are known, it is not difficult to apply them to the special windings occasionally met in practice.

91. Single-phase Windings.—At the present time, alternators seldom have single-phase windings. This type of winding is special, and an alternator so wound can seldom be used for purposes other than those for which it is specifically designed. Single-phase windings, however, are used extensively for single-phase motors.

When single-phase power is desired, it is common to utilize a three-phase, Y-connected alternator and to obtain the single

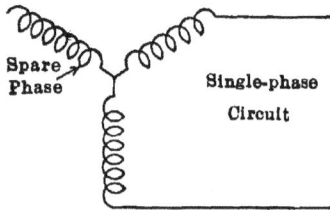

FIG. 128.—Single phase obtained from Y-connected alternator.

phase from two of the Y-connected terminals (Fig. 128). This gives a spare phase, in case of injury to either of the other two phases. A standard three-phase generator is thus utilized and if necessary can be used for three-phase service. The alternators in the Cos Cob Station, which supplies 25-cycle, single-phase current for the New York, New Haven and Hartford R.R. electrification, are Y-connected and operate as shown in Fig. 128.

A knowledge of single-phase windings makes it easier to understand polyphase windings, since polyphase windings are merely two or more single-phase windings having definite geometrical

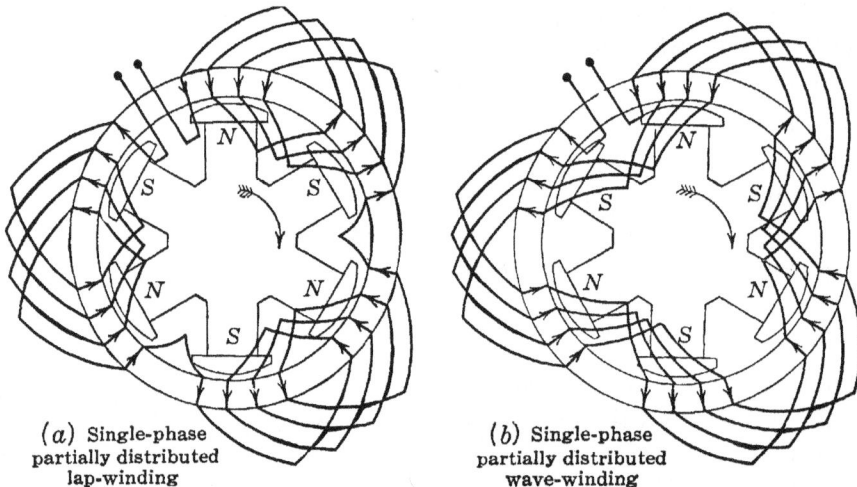

(a) Single-phase partially distributed lap-winding

(b) Single-phase partially distributed wave-winding

FIG. 129.—Single-phase lap and wave windings.

relations to one another. Two simple, single-phase windings for six-pole machines are shown in Fig. 129(a) and (b). Both windings are distributed only partly since all the armature-winding space is not utilized. If conductors were placed in the vacant spaces, their emfs would be so far out of phase with the resultant emfs of the system that they would contribute but little to the total emf. Therefore, it is wasteful of copper to utilize these winding spaces.

In (*a*) a lap winding is shown. All the adjacent *coils* constituting each of the several belts are first connected in series. Then, these *belts* are connected in series. There are three such belts in Fig. 129(*a*). In (*b*) a wave winding is shown. The first coil of a belt is connected in series with the first coil of the next belt, etc., until all the first coils are thus connected. The connection is then continued to the second coil of the first belt, which is connected in series with the second coil of the several belts in

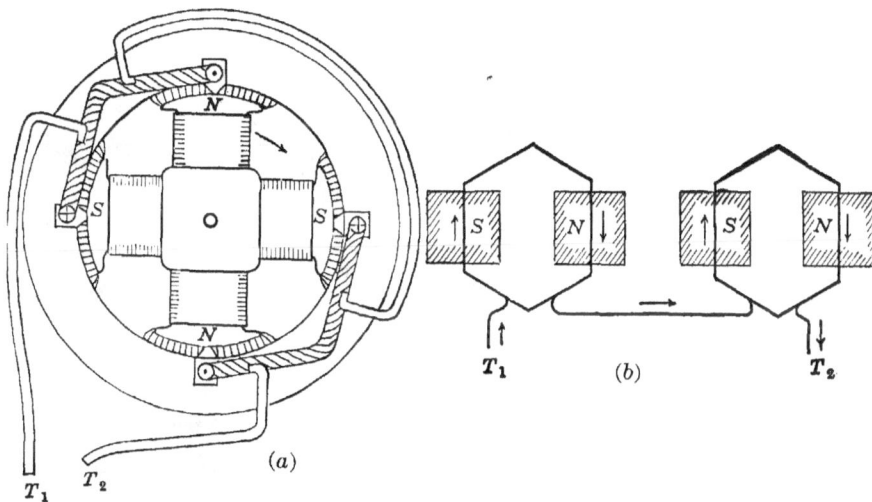

Fig. 130.—Single-layer, half-coil winding, one slot per pole.

order. This method of connection is continued until every coil in every belt is included. It will be noted that in both (*a*) and (*b*) there are the same number of series-connected conductors between the machine terminals. Therefore, both windings give the same emf. This is not true with lap and wave windings in d-c machines (see Part I, Par. 184, p. 226).

A simple type of single-phase winding for a four-pole alternator is shown in Fig. 130. It consists of but two coils placed in four slots and connected in series. Since there is but one coil or one coil group per pair of poles, there is but a half-coil or a half-coil group per pole. The winding consists of but one layer. Hence, this is called a *single-layer, half-coil* winding, having one slot per pole. (The windings of Fig. 129 are also single-layer, half-coil windings having four slots per pole.) The connections of the winding are shown diagrammatically in (*b*).

If two additional coils are placed on the armature of Fig. 130 in such a manner that they span the pairs of poles not spanned by the coils of Fig. 130, a *whole-coil* winding is obtained. That is, there is now one whole coil or one coil group per pole. These additional coils may be placed in the same slots with the original coils, a two-layer winding (Fig. 131) being thus obtained. At the present time, practically all windings are two-layer windings. The whole-coil winding of Fig. 131 may be made a half-coil winding by swinging coil *B* (see (*b*)) so that it spans the same pair

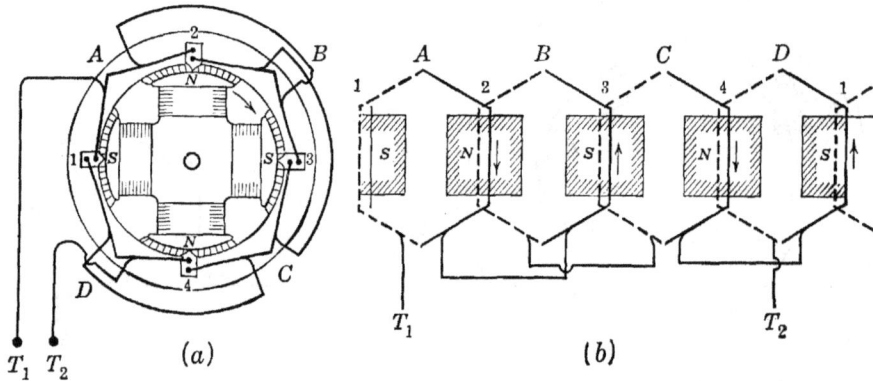

Fig. 131.—Two-layer, whole-coil winding, one slot per pole.

of poles as coil *A* and also by swinging coil *D* so that it spans the same pair of poles as coil *C*. The reversed connection of the alternate coils *B* and *D* should be noted. This is necessary in order that the emfs of all the coils shall be *additive*, as is shown by study of the direction of the arrows. The connections of the winding are shown diagrammatically in (*b*).

Windings utilizing but one slot per pole are not used in practice, ordinarily, since but a fraction of the armature surface is utilized. Figure 132 shows a development of the winding of Fig. 131, but with two slots per pole. The two coils of each group are first connected in series, and the groups are then connected in series as in Fig. 131, reversed connections to alternate coil groups being made. The connections of the winding are shown in (*b*), the coil sides having numbers corresponding to those in (*a*). This type of winding, from the nature of the end connections, is called the *barrel type* and is a lap winding (see Fig. 129(a)).

It is obvious that this type of winding may be extended to three or more slots per pole.

A single-phase *spiral* winding is shown in Fig. 133. Instead of the coils lapping one another, each coil group is made up of separate coils of different breadths and is placed on the armature

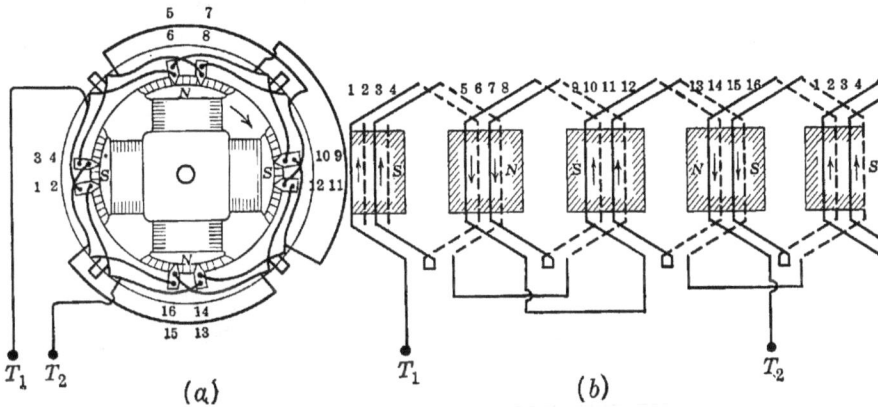

Fig. 132.—Whole-coil, two-layer winding, two slots per pole.

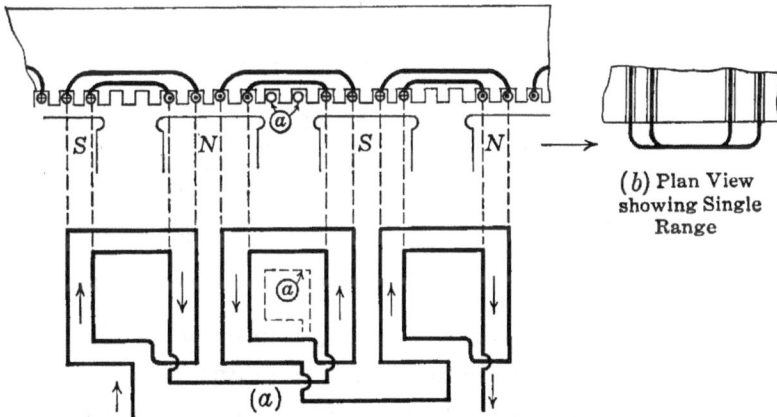

Fig. 133.—Single-phase, single-range, spiral winding.

in the manner shown in Fig. 133(a). The coils of the group are connected in series. The winding derives its name from the spiral manner in which the coils are connected in series as indicated in the wiring diagram (Fig. 133 (a)). A third coil of very small pitch shown dotted could be added to this winding, being placed in the slots marked *a*. This third coil has so small a pitch that the emfs induced in its two sides are practically in phase opposition and therefore contribute very little to the induced emf of the winding as a whole. Hence, the third coil is omitted.

Since the ends of the coils are bent so that they all lie in the same plane, perpendicular to the shaft of the alternator, it is called a *single-range* winding. Two- and three-range windings are used with polyphase machines having spiral windings.

92. Two-phase Windings.—It is shown in Chap. IV (p. 145, Fig. 116) that two-phase emfs are obtained from two simple coils fastened together with their planes at 90° to each other.

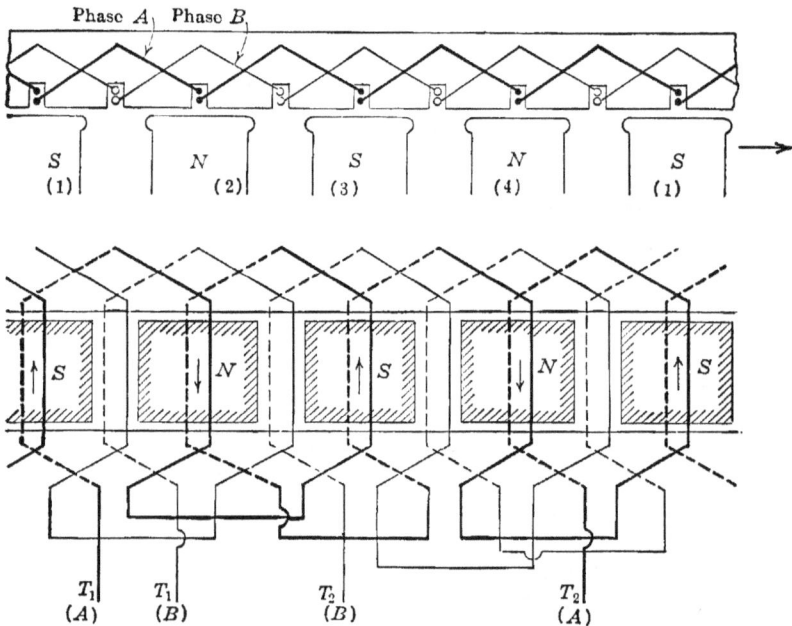

Fig. 134.—Two-phase, two-layer winding, one slot per pole per phase.

The emf induced in one coil lags that induced in the other coil by 90 time-degrees. In commercial dynamos, the same principle of spacing the windings of the two phases 90 electrical space-degrees apart is used, the only difference being that the armature is stationary and the winding must be adapted ordinarily for a considerable number of poles.

Figure 134 shows a simple type of two-phase winding, having but one slot per pole per phase. Each phase of this winding is similar to the single-phase winding of Fig. 131. To obtain this winding, it is necessary merely to add a duplicate winding (Fig. 134) placed in slots halfway between the slots shown in

Fig. 131. In Fig. 134, the *A*-phase is shown with heavy lines and the *B*-phase with light lines. The sides of the coils which lap under are shown dotted. The terminals T_1 and T_2 of each phase are brought out and may be connected in the manner described in Chap. IV (pp. 144 to 148).

Figure 135 shows the spiral winding (Fig. 133) adapted to a two-phase alternator. A second winding, similar to that in Fig. 133 and symmetrically spaced, is added. In order that the coil ends of phase *A* may pass those of phase *B*, the bent-up coil ends must lie in at least two planes perpendicular to the shaft of the alternator. This makes such a winding a *two-range* wind-

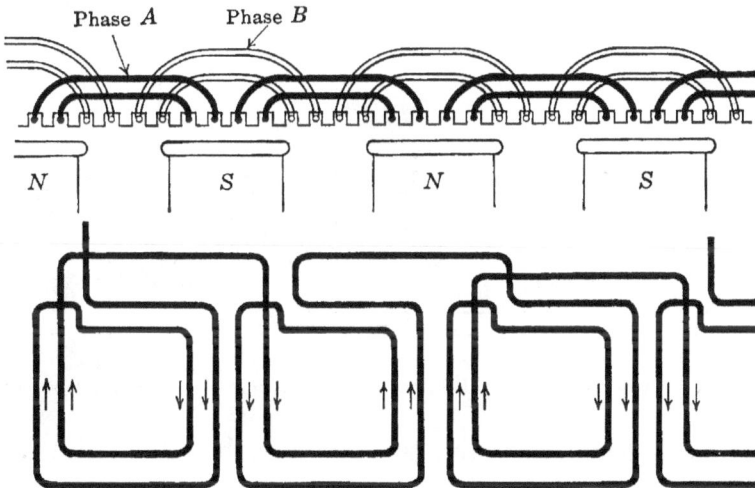

FIG. 135.—Two-phase chain winding, two-range.

ing. The connections of the *A*-phase are shown, those for the *B*-phase being identical. The chain or spiral winding has been superseded practically by the two-layer lap winding.

93. Two-phase Lap Winding.—The lap winding is the usual type of alternator winding. With it, there are few limitations in the choice of number of slots, pitch, etc. All the coils are alike, thus requiring the minimum number of spares, and the winding is flexible in the matter of connections. For example, with a lap winding it is a simple matter by paralleling to change a 440-volt winding to one of 220 volts.

To obtain a two-phase lap winding, it is necessary merely to add slots and coils to the two-phase winding shown in Fig. 134. The connections of the coils of any one phase are almost identical with those in the d-c windings described in Part I, Chap. X. Direct-current lap windings may be used both for single-phase and for polyphase voltages by taps at suitable points connected to slip-rings, as in the synchronous converter (p. 331, Par. 189). Such windings give mesh- or delta-connected windings. The phases of the ordinary lap winding may be connected either in mesh or delta, or in star or Y.

Figure 136 shows a two-phase lap winding, in which there are eight slots per pole, making four slots per pole per phase. This is a full-pitch winding, the coil pitch being eight slots, which is the number of armature slots per pole. The connections of the coils of phase *A* should be noted. All the coils of each coil group

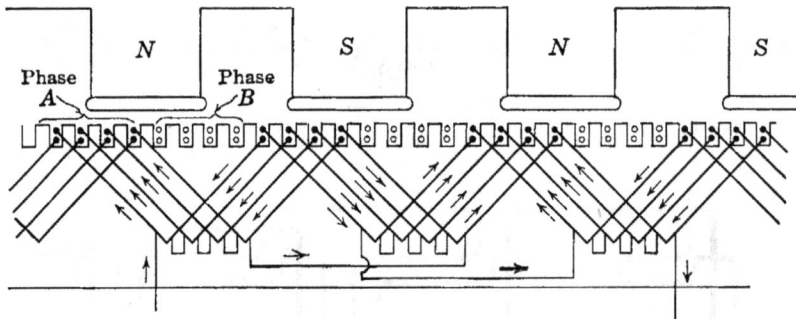

FIG. 136.—Two-phase, full-pitch, lap winding, four slots per pole per phase.

are connected in series, and the group is then treated as a unit. The *A*-phase coil groups then are connected in series as shown. The reversed connection to alternate coil groups should be noted. The arrows show the direction of the induced emf at one particular instant. The connections of phase *B* are omitted for the sake of clearness as they are identical with those of phase *A*. It will be observed that in this full-pitch winding the coil sides in any one slot are both of the same phase. This is not true with fractional-pitch windings.

94. Three-phase Windings.—The difference between two- and three-phase windings is merely in the number of phase belts per pole. Figure 137 shows the simple winding of Fig. 131 adapted for three-phase. For clearness the end connections of phase *A* alone are shown. It is necessary merely to add two more windings, equally spaced, between those of Fig. 131 in order to obtain a three-phase winding having one slot per pole per phase.

The phase rotation given in Fig. 137 is in the sequence *A-B-C*, and not *A-C-B* as might be inferred from the order of the lettering. Note that coil B_1 is 120 electrical space-degrees to the right of A_1, remembering that the distance between the centers of adjacent poles is 180 electrical space-degrees. Coil C_1 is 120 electrical space-degrees to the right of B_1. Therefore, the emfs in these three coils differ in time phase by 120°, and with the rotating field moving from left to right the phase sequence is *A-B-C*. If, for example, the alternator were to be Y-connected,

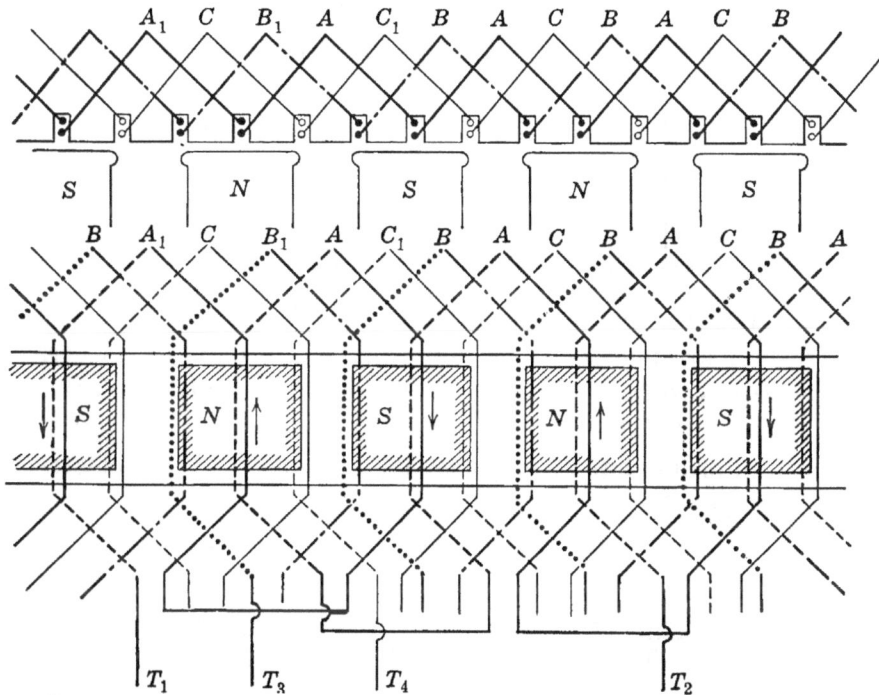

FIG. 137.—Three-phase, two-layer winding, one slot per pole per phase.

the terminal T_1 of the *A*-phase would be connected to terminal T_3 of the *B*-phase and to the terminal T_4 of the *C*-phase to form the common neutral, since the active conductors connected to these terminals are 120 electrical space-degrees apart. The coil *C* between A_1 and B_1 is but 60 electrical space-degrees to the right of A_1; hence, its emf lags that of A_1 by 60 time-degrees. As its emf must be made to lag that of A_1 by 240 time-degrees, its connections must be reversed. If the winding were completed, it would be found that this coil was one of the alternate ones whose terminals are reversed.

95. Three-phase, Full-pitch Lap Winding.—Figure 138 shows a three-phase, full-pitch, two-layer lap winding having 12 slots per pole. This winding differs from that of Fig. 137 in having four slots per pole for each phase rather than one slot per pole per phase. The four coils are connected in series, and this group is connected in the same manner as the single coils of Fig. 137. Note that the phase sequence is *A-B-C* as before, the letters A_1,

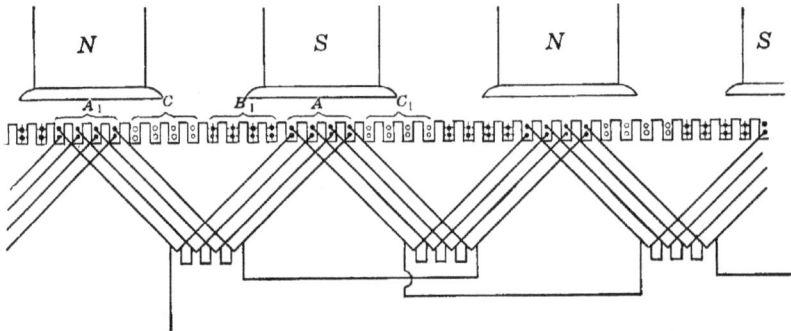

Fig. 138.—Three-phase, full-pitch, two-layer, lap winding.

B_1, C_1 showing the corresponding coil-side belts. B_1 is 120 electrical space-degrees to the right of A_1, and C_1 is 120 electrical space-degrees to the right of B_1. The belt C is only 60 electrical space-degrees to the right of A_1, and its connections must be reversed in order that its emf may lag that of A_1 by 240 electrical time-degrees.

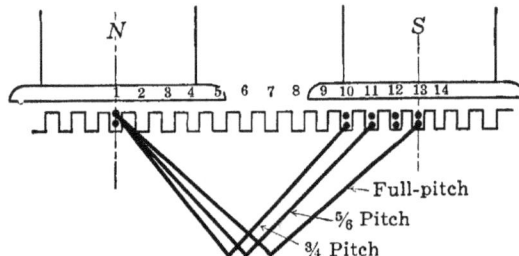

Fig. 139.—Full- and fractional-pitch coils.

96. Three-phase, Fractional-pitch Winding.—Figure 139 shows two poles and a portion of the armature of an alternator having 12 slots per pole. If a *full-pitch* winding is to be placed on this armature, the two sides of every coil must have a span corresponding to one pole pitch, or 12 slots. Therefore, if one side of a coil is in the top of slot 1, the other side must lie in the bottom of slot 13. If, instead of a coil span of 12 slots, the span is 10 slots, the other side of the coil will be in the bottom of slot 11. When the span of the coil is less than a full-pitch value, a *fractional-pitch* winding results. If the coil span is 10 slots, a $^{10}/_{12}$- or $^5/_6$-pitch winding results. If the coil span is 9 slots, a $^9/_{12}$- or $^3/_4$-pitch winding results (Fig. 139).

Fractional-pitch windings are common, since they save copper in the end connections. Also, in alternators the wave form of the induced emf is improved. Owing to the fact that the two coil sides are not simultaneously under corresponding parts of the poles a phase difference exists in their induced emfs and the emf of the coil is less than the full-pitch value. With a $^5/_6$ pitch, the emf is 0.966 of the value for full-pitch, so that the reduction in emf is slight for pitches of the order of $^5/_6$.

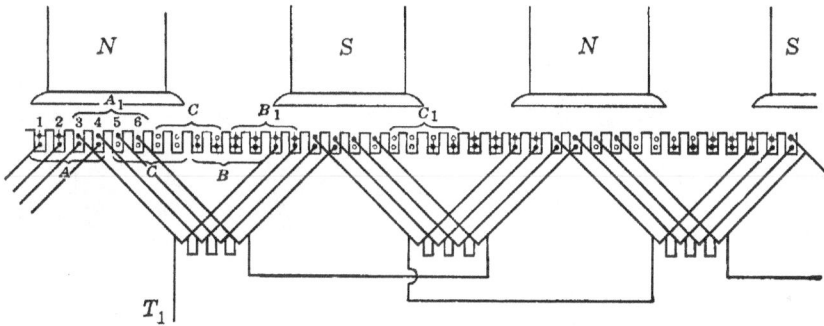

FIG. 140.—Three-phase, $^5/_6$-pitch, two-layer, lap winding, four slots per pole phase.

The factor by which the emf in a full-pitch coil must be multiplied in order to obtain the emf in a fractional-pitch coil is the pitch factor k_p (see Eq. (66), p. 179.)

$$k_p = \cos \frac{180° - \beta}{2}, \tag{63}$$

where β is the angle of the coil span in electrical degrees. For example, with a $^5/_6$ pitch, $\beta = (^5/_6)180° = 150°$. Hence, $k_p = \cos 15° = 0.966$.

Figure 140 shows a three-phase, $^5/_6$-pitch lap winding. It is identical with that of Fig. 138 except that the bottom layer of the winding is moved two slots to the left. The top layer remains unchanged.

It is to be noted that, in the $^5/_6$-pitch winding of Fig. 140, only two of the slots of each phase under any one pole contain conductors of that phase only. The four adjacent slots contain conductors of this phase and also conductors of the other two phases. For example, slots 1 and 2 contain phase-*A* and phase-*B* conductors; slots 3 and 4 contain phase-*A* conductors only; slots 5 and 6 contain phase-*A* and phase-*C* conductors. Of this particular group, slots 3 and 4 contain phase-*A* conductors only. The fact that certain slots contain conductors of different phases reduces slightly the inductance of the winding.

To illustrate the manner in which the armature coils are actually placed on the stator structure, a half of a wound stationary armature for a slow-speed water-wheel alternator is shown in Fig. 141. The core, the ventilating ducts in the core, and the

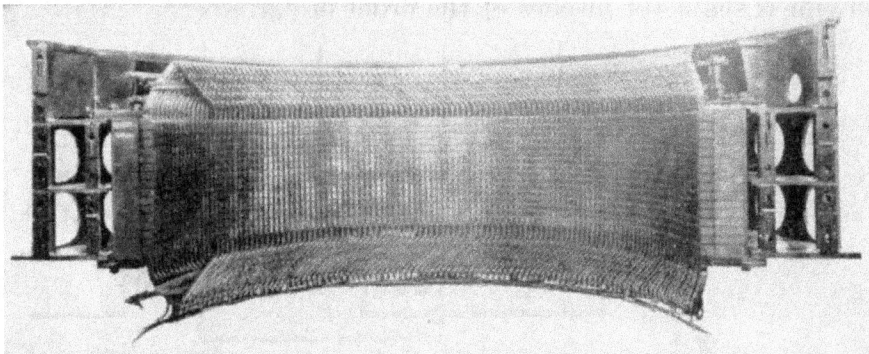

Fig. 141.—Half of wound stationary armature for vertical alternating-current, waterwheel generator, 10,000 kva, 14 pole, 214-$\frac{2}{7}$ rpm, 6,900 volts, 25 cycles. (*Courtesy of General Electric Company.*)

ventilating passages in the box frame are clearly shown. The method of making the end connections and of binding the ends to strengthen them against mechanical stresses should be noted. This armature is part of the alternator now in operation at Madden Dam, Panama.

ALTERNATOR CONSTRUCTION

97. Types of Alternator.—Rotating-field alternators may be divided into two general classes, those with salient-pole rotors and those with cylindrical rotors having the field winding embedded in slots. The salient-pole type is employed at low and moderate speeds such as for engine-driven alternators, for water-wheel-driven alternators, and usually for synchronous motors. The cylindrical, or non-salient-pole, type of rotor is used in high-speed, turbine-driven alternators. Its advantages under high-speed conditions are better mechanical construction to resist centrifugal force, better balance, less noise, and much less windage, an important factor at high speeds.

The engine-driven type must operate necessarily at low speeds and there must be sufficient flywheel effect (WR^2) to insure the

Fig. 142.—Bracing of end connections of turbine-driven alternator to withstand short-circuit stresses. (*Courtesy of Westinghouse Electric and Manufacturing Company.*)

requisite uniformity of speed. (W is the weight of rotor in pounds and R is the radius of gyration.) The speeds of the water-wheel type range from 82 to 500 rpm. The speeds of the cylindrical-rotor type range from 1,200 to 3,600 rpm.

98. Stator or Armature.—The stator, or stationary member, of the alternator is practically always the armature, the field

structure being the rotor or rotating member.   When the alternator is in operation, the armature iron is continuously cut by
the flux of the rotating field and must be laminated in order to
reduce eddy-current losses.   In alternators of small diameter,
each lamination is a single circular punching.

High-speed turbine-driven alternators having armatures of
small diameter are usually built up of single circular stampings.
In the larger sizes, it is necessary to build up the armature of sec-
tor-shaped laminations which alternately butt and lap.   It is difficult to ventilate the armatures of turbine-driven alternators

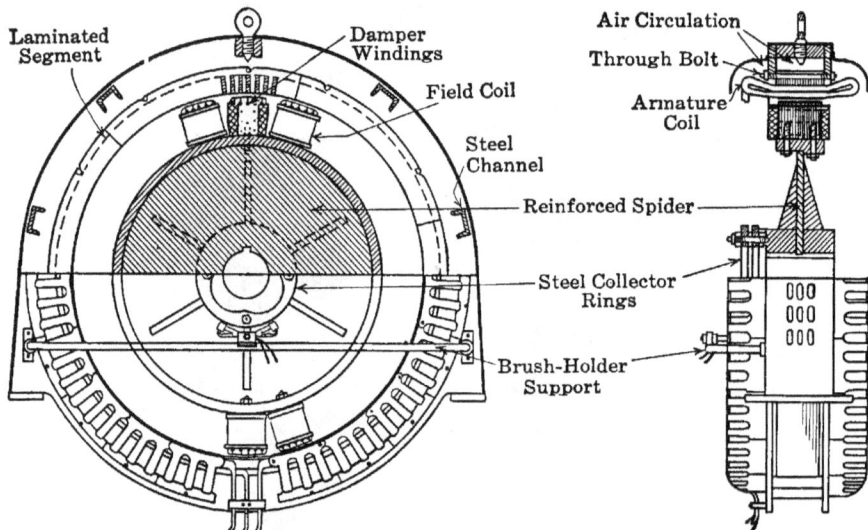

FIG. 143.—Slow-speed, 22-pole alternator for Diesel drive.   (*Courtesy of Westinghouse Electric and Manufacturing Company.*)

owing to their small diameters and the considerable depth of
iron behind the slots.   Hence, there are usually rows of perforations in the stator laminations behind the armature teeth which
form axial ventilating ducts in which the circulation of air or
hydrogen assists in removing the heat developed in the stator
iron and in the armature winding.   (See Fig. 153, p. 176.)   The
end connections of turbine-driven alternator windings protrude
a considerable distance beyond the laminations, particularly in
the two-pole type (Fig. 142).   On short circuit the mechanical
stresses among the end connections caused by the excessive
short-circuit currents would be sufficient to tear the coil ends
from their positions unless special means were taken to brace

them. The bracing of the end connections of a turbine-driven alternator is shown in Fig. 142.

Engine-driven alternators must rotate at comparatively low speeds and hence must have a large number of poles in order to give a sufficiently high frequency and power output. Therefore, the armatures must have a comparatively large diameter. The

FIG. 144.—Fabricated frame of synchronous machine. (*Courtesy of Westinghouse Electric and Manufacturing Company.*)

pole pieces are made of laminations riveted together (Fig. 146) and bolted to the field spider (Figs. 143 and 147). The armature, or stator, is built up of overlapping segments, bolted together and held to the frame by through bolts.

The frame (Figs. 143 and 144) is a hollow-box, fabricated-steel member braced with steel channels. This gives necessary stiffness with minimum weight, and the space within the frame allows a free circulation of air for ventilation. Figure 143 shows the construction of the stator and rotor of a reciprocating-engine-driven alternator, and Fig. 144 shows the completed fabricated stator. The frame is fabricated in accordance with modern practice.

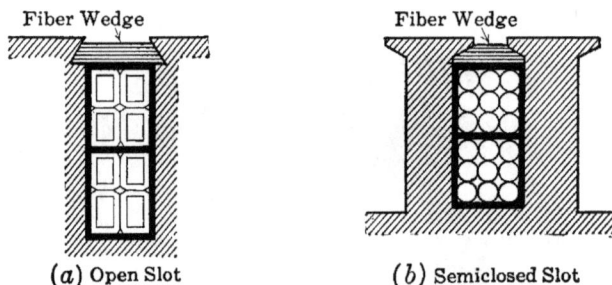

(*a*) Open Slot   (*b*) Semiclosed Slot

Fig. 145.—Open and semiclosed slots.

99. Slots.—Alternator slots are divided into two general classes, the open slot and the semiclosed slot. The open slot, shown in Fig. 145(*a*), is more common because the coils can be form-wound and insulated prior to being placed in the slots, the least expensive and most satisfactory method of winding.

Fig. 146.—Pole piece of a 65-kva, 50-cycle, 377-rpm alternator.

The semiclosed, or overhung, type of slot (Fig. 145(*b*)) is seldom used in alternators but is usually necessary in induction motors in order to reduce the air-gap reluctance. In winding, it is necessary to place the conductors in the slot a few at a time, which increases the expense of winding and is less economical of slot space.

In both types of slot the conductors are usually held in the slot by a fiber wedge, as shown in the figures. The effect of the semiclosed slot may be obtained by the use of open slots and magnetic wedges. These wedges are only partly of iron so that the slot is not entirely closed.

The internal temperatures of modern turbine-driven alternators are so high that built-up mica is found to be the insulation best able to withstand simultaneously the high temperatures and high-voltage stresses. Such mica is pressed around the active part of the conductor, forming a solid, homogeneous mass. Molded forms of this built-up mica line the slots.

Fig. 147.—Diesel-engine-driven alternator rotor with field poles and damper windings. (*Courtesy of Westinghouse Electric and Manufacturing Company.*)

100. Rotating-field Structure.—From the standpoint of their field construction, alternators may be divided into three classes: the very slow-speed, engine-driven alternator (75 to 100 rpm); the medium-speed, such as the water-wheel-driven type (100 to 750 rpm); the high-speed turbine-driven alternator (750 to 3,600 rpm).

The poles of practically all salient-pole generators have cores made up of laminations (Fig. 146) in order to reduce eddy-current losses in the pole faces. In slow-speed alternators the poles are

FIG. 148.—Busch-Sulzer Diesel engine driving a 2,812-kva alternator with exciter. (*Courtesy of Westinghouse Electric and Manufacturing Company.*)

FIG. 149.—Rotor of waterwheel alternator with cast-steel spider and dovetailed salient-pole construction. (*Courtesy of Westinghouse Electric and Manufacturing Company.*)

usually bolted to a fabricated-steel spider (Fig. 147). In Fig. 148 is shown a slow-speed alternator driven by a Diesel engine. At higher speeds the poles are dovetailed to the spider in order

FIG. 150.—Parallel-slot, two-pole rotor for a turbine-driven alternator.

FIG. 151.—Radial-slot type of rotor, having four poles.

to resist the centrifugal forces. This type of construction is illustrated in Fig. 149 in which is shown a 48-pole, water-wheel rotor. The poles are dovetailed to the periphery of the rotor which is formed of thin steel plates held together by through bolts. These plates are supported by the cast-steel spider. The *damper*, or *amortisseur*, winding in the slots of the pole faces should be noted. This winding will be described later in more detail (see p. 340, Fig. 273).

Because of the large centrifugal forces and excessive windage, salient poles cannot be used for high-speed, turbine-driven alternators and the smooth, cylindrical type of rotor is used. There are two common types of such rotors, the parallel-slot type (Fig. 150) and the radial-slot type (Fig. 151).

The winding in the parallel-slot type is of strip copper, wound by hand in the slots. The wires are held in the slots by means of nonmagnetic metallic wedges. As there is not sufficient space to run the shaft through the center of the rotor, it is bolted to the ends by nonmagnetic flanges (Fig. 150). These flanges must be nonmagnetic or they would short-circuit the magnetic poles. This construction gives a smooth rotor with little windage loss and strong mechanically, especially as regards the support of the coil ends. Parallel-slot rotors are seldom used except for two-pole rotors in small alternators. The metal back of the slots tends to become too small in cross-section to withstand the centrifugal forces, if an attempt is made to adapt this type of rotor to more than two poles.

Figure 151 shows a four-pole, radial-slot rotor. Although the coil ends are not held so strongly in this type of rotor as they are in the parallel-slot type, it is better adapted to rotors having more than two poles, because with increase in the number of poles there is not the reduction of iron section back of the slots such as occurs in the parallel-slot type.

The foregoing are of the *non-salient-pole* type of field, and the windings are *distributed-field* windings (see Par. 101).

The field connections are usually carried out through the center of the shaft to slip-rings. The current is conducted to the field winding by means of slip-rings and carbon brushes. The excitation voltage is usually 125 or 250 volts and in the larger stations is supplied by bus-bars devoted wholly to excitation. In smaller installations, the exciter may be mounted directly on the alternator shaft (Fig. 148). When directly connected to the shaft of slow-speed alternators the cost of the exciter is greater than when belt driven, because of the low speed. Large central stations have an exciter bus supplied by separate motor- or steam-driven exciters. To increase reliability, a storage battery is often floated across this bus. Steam-driven exciters are also held in readiness for emergencies.

101. Distributed-field Winding. In salient-pole alternators the turns for exciting each field pole are concentrated in a single coil on the pole. In the non-salient-pole alternator the exciting turns are distributed over the surface of the rotor. Figure 152(a) shows a distributed winding for a single pole. This winding is a spiral winding (see p. 159, Fig. 133) similar to that shown in Fig. 151, although a lap winding may be used. In (b) the curves give the flux density at each point along the air-gap as produced by the inner, the intermediate, and the outer coil, as if each acted alone. The area under each curve gives the total flux produced by each coil. By the corkscrew rule the direction of the flux is upward. Figure 152(c) shows the total flux due to the combined action of the three coils. The flux distribution is more or less sinusoidal, and the emf wave of alternators having this type of field winding is ordinarily more nearly sinusoidal than that of alternators having salient poles.

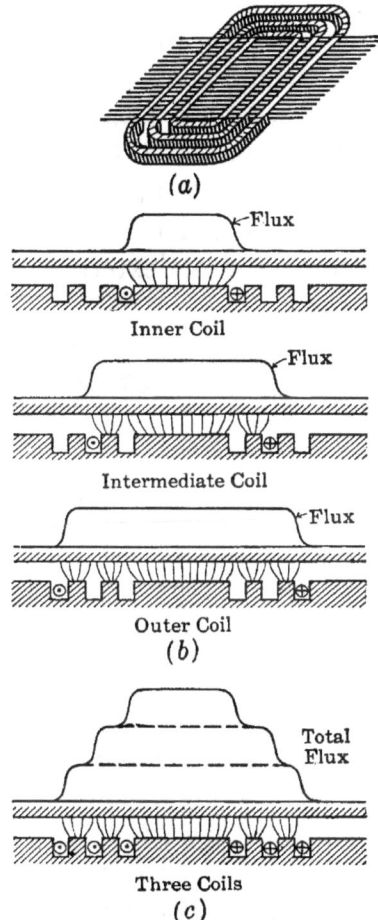

(a)

Inner Coil

Intermediate Coil

Outer Coil

(b)

Three Coils

(c)

Fig. 152.—Distributed field winding and resulting flux.

102. Ventilation.—With the salient-pole type of alternator, there is no particular difficulty in securing adequate ventilation. Large surfaces are exposed to the ambient air, and the poles themselves act as fans to circulate the air. With high-speed, turbine-driven alternators, the surface exposed to the ambient air is small; and since the rotors are smooth, they are not effective in circulating air. Hence the problem of dissipating the heat becomes a more difficult one. For example, a 20,000-kw unit, having an efficiency of 96.5 per cent, requires that 700 kw shall be dissipated. The considerable depth of iron back of the slots and the facts that the armature is long and its diameter is small and that the rotor must ordinarily be a solid-steel forging and thus cannot have air ducts (Figs. 150 and 151) require special means to carry away the heat developed in both armature and field copper and in the iron. Since the rotor is smooth, it produces little fan action as compared with salient-pole rotors (Figs. 147 and 148), and cooling air must be forced through the alternator from external fans.

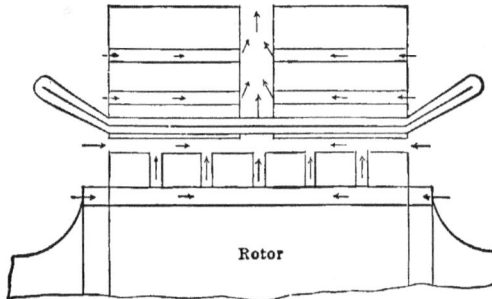

FIG. 153.—Passage of ventilating air through the ducts of a turbine-driven alternator.

The foregoing 20,000-kw unit might require 60,000 to 70,000 cu. ft. of air per minute. The modern method of cooling is to use a totally enclosed system and to employ either air or hydrogen as the cooling medium. Obviously, with hydrogen, a totally enclosed system is necessary.

With respect to air, the advantages of a totally enclosed system are that small particles of dirt contained in the outside air do not accumulate in the air ducts and eventually obstruct them and also that in case of fire in the alternator, oxygen is not being continuously supplied to support combustion.

The advantages of hydrogen are that its density is only 7 per cent that of air and hence the windage losses are reduced by a large amount. This is important with high-speed alternators, since windage constitutes a large part of the losses. Hydrogen has 7.5 times the heat conductivity of air. Also, for a given temperature difference, hydrogen will transfer 30 per cent more heat units from a surface than will air. Because of its explosive nature and the cost of replacement, it must be confined within a gas-tight enclosure. A special sealing joint is needed in the bearing through which the shaft must protrude. In the totally enclosed systems, the air or hydrogen is circulated by blowers and is cooled, after being heated in the alternator ducts, by passing over pipes through which cooling water is circulating.

Figure 153 shows the passage of the air as it is forced through the axial ducts back of the laminations (see Fig. 142) and out through a center radial duct, in a turbine-driven alternator.

ALTERNATOR ELECTROMOTIVE FORCE

103. Induced Electromotive Force.—It will be remembered that the emf induced in a single conductor l cm long, cutting a field whose flux density is B gausses, at a velocity of v cm per sec., is

$$e = Blv10^{-8} \text{ volts,}$$

when B, l, v are mutually perpendicular (see Part I, p. 210, Eq. (86)). The total flux is proportional to B and l. The velocity is proportional to the frequency. It can be shown, therefore, that the rms emf per phase is

$$E = 2.22Z\phi f10^{-8} \text{ rms volts per phase} \qquad (64)$$

where Z is the total number of series-connected conductors per phase, ϕ is the total flux in maxwells (assumed to be sinusoidally distributed) entering the armature from one N-pole, and f is the frequency.

Example.—The armature of a six-pole, three-phase, 60-cycle alternator has 18 slots, and there are 16 conductors per slot. There are 3,000,000 magnetic lines, or maxwells, entering the armature from one N-pole, and this flux is sinusoidally distributed along the air-gap. Determine induced emf per phase.

The total surface conductors

$$= 18 \times 16 = 288.$$

The conductors per phase

$$Z = {}^{288}\!\!\diagup_{3} = 96.$$

Using Eq. (64),

$$E = 2.22 \times 96 \times 3 \times 10^{6} \times 60 \times 10^{-8}$$
$$= 384 \text{ volts.} \quad Ans.$$

If the *turns* per phase N are used rather than the number of *conductors* Z, Eq. (64) becomes

$$E = 4.44N\phi f 10^{-8} \text{ rms volts per phase.} \qquad (65)$$

Equations (64) and (65) assume that there is but 1 slot per pole per phase as in the armature of Fig. 137 (p. 163). Therefore, the emfs induced in all conductors under the same pole are in phase with one another. Ordinarily there is more than 1 slot per pole per phase. For example, in Fig. 138, there are 12 slots per pole and therefore 4 slots per pole per phase. Figure 154(*a*) shows a sketch of such a winding. Since one pole pitch represents 180°

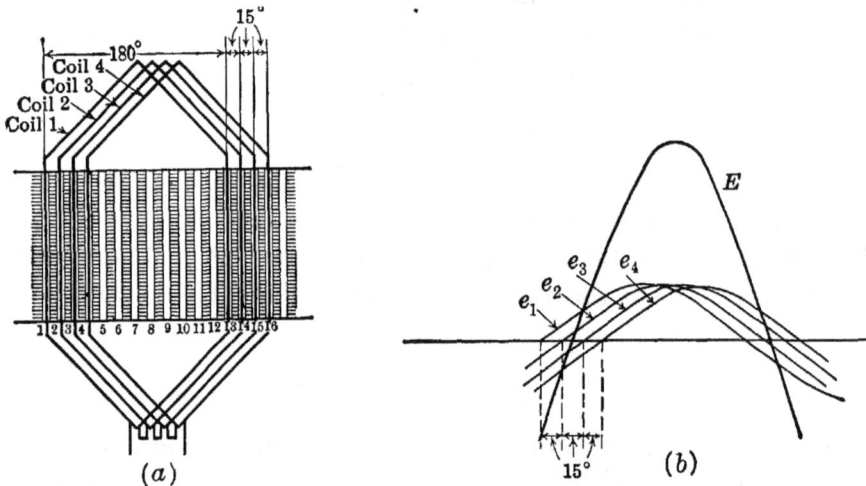

Fig. 154.—Effect of distributed winding on resultant emf.

and there are 12 slots per pole, there must be $^{180}\!/_{12}$, or 15, electrical space-degrees between adjacent slots. If the field pole moves from left to right, the conductors of coil 2 are directly under the center of the pole 15 electrical time-degrees later than those of coil 1. Therefore, the emf e_2 induced in coil 2 will lag the emf e_1 induced in coil 1 by 15 time-degrees, as shown in (*b*). Likewise, the emf e_3 induced in coil 3 will lag that induced in 2 by 15°, etc. Therefore, all four emfs differ in phase by 15°, and their resultant emf E is *less* than it would be if these phase differences did not exist. Hence, a factor k_b called the *breadth factor*, which is less than unity, must be introduced into Eqs. (64) and (65). With a three-phase winding having 4 slots per pole per phase, $k_b = 0.958$.

If a fractional-pitch winding is used, the emf is still further reduced as is stated in Par. 96. That is, Eqs. (64) and (65) must be multiplied by the pitch factor k_p, which is 0.966 for a $\frac{5}{6}$-pitch winding (see Eq. (63), p. 165).

Hence, Eq. (64) becomes

$$E = 2.22\ k_b k_p Z \phi f 10^{-8} \text{ rms volts per phase.} \qquad (66)$$

Example.—(a) Determine the emf in the example, p. 177, assuming $\frac{5}{6}$-pitch and that the winding is distributed in 72 slots rather than in 18 slots. There are now 4 slots per pole per phase and four conductors per slot. The number of conductors Z per phase remains unchanged.

(b) If this alternator is Y-connected, determine its open-circuit terminal emf.

(a) $E = 2.22 \times 0.958 \times 0.966 \times 96 \times 3 \times 10^6 \times 60 \times 10^{-8}$
$= 355$ volts. *Ans.*

(b) $E_{line} = \sqrt{3} \times 355 = 615$ volts. *Ans.*

104. Phasing Alternator Windings.—Three-phase alternator windings may be connected either in Y or in delta. Instances

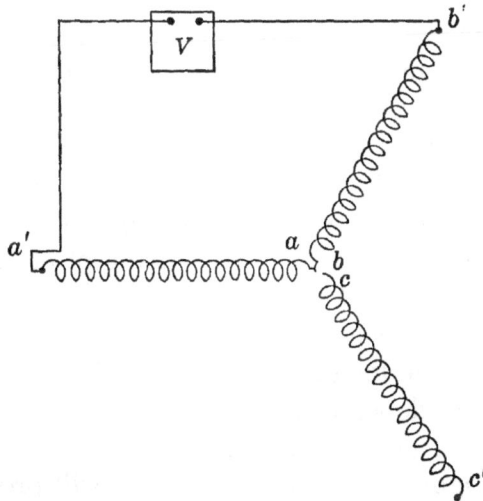

Fig. 155.—Connecting alternator coils in Y.

often occur in practice where six leads come from the alternator, these leads being the three pairs of terminals from the three phases. The proper phase relations must be observed in making the connections, whether they are to be connected in Y or in delta.

Let *aa'*, *bb'*, *cc'* (Fig. 155) be the three coil windings of a three-phase alternator.

Assume that these three windings are to be connected in Y. First, connect ends a and b together. Measure $E_{a'b'}$, the voltage across their open ends. This should equal $\sqrt{3}$ times the coil voltage. It may be equal to the coil voltage, in which case one coil should be reversed. Next, connect the end c of coil cc' to point ab. The voltages $E_{b'c'}$ and $E_{a'c'}$ should each be $\sqrt{3}$ times the coil voltage. If not, the coil cc' should be reversed.

If it is desired to connect the coils in delta, the ends a and b' (Fig. 156) should be connected first. The voltage $E_{a'b}$, across their open ends, should now be equal to the coil voltage. If not (that is, if it is equal to $\sqrt{3}$ times the coil voltage), one of the two coils should be reversed. End c' of coil cc' should then be connected to b. The voltage $E_{ca'}$ across the open ends should be zero, since the vector sum of the three voltages around a delta must be zero (see Par. 78, p. 134). If this voltage is practically zero, the two ends c and a' may be closed. The voltage $E_{ca'}$ may be twice the coil voltage. If this is found to be the case, coil cc' should be reversed. Coil cc' is phased with a voltage $E_{ba'}$ equal to and in phase with its own voltage $E_{c'c}$. Hence the voltage $E_{a'c}$ across the open ends $a'c$ must be either zero or twice the coil voltage.

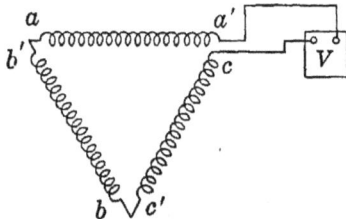

Fig. 156.—Connecting alternator coils in delta.

105. Rating of Alternators.—The rating of electric machinery is determined in general by its temperature rise. This temperature rise is due to the losses in the machine. The I^2R loss in the armature, due to the load current, limits the output of an alternator. This loss depends on the *magnitude* of the armature current and is independent of power factor. For example, 100 amp in a single-phase, 200-volt alternator will produce the same I^2R loss whether the load power factor is unity, 0.4, or any other value. The output in *kilowatts*, however, is proportional to the power factor. If the alternator is limited to 100 amp, its output will be 20 kw at unity power factor but only 8 kw at 0.4 power factor. The rating is 20 kva (kilovolt-amperes) regardless of power factor.

For these reasons, alternators are ordinarily rated in kva. If an alternator is rated in kilowatts, unity power factor is assumed, unless otherwise specified. In stating the output of electric machinery it is always advisable to specify the power factor.

The rating of the prime mover driving an alternator depends on the kilowatt output and is independent of the alternator power factor. The same turbine could be used to drive a 200-kva alternator operating at 0.5 power factor or a 100-kva alternator operating at unity power factor, although the first alternator would have double the kva rating of the second.

Example.—A 3-phase, 60-cycle, 6,600-volt, turbine-driven alternator is rated at 6,250 kw at 0.7 power factor and has a full-load efficiency of 0.95 excluding the field loss. Determine: (a) kva rating of alternator; (b) current rating; (c) horsepower delivered by turbine at full load. (d) If alternator were loaded to its full rating at unity power factor, what horsepower would be required to drive it, on the assumption that the efficiency remains unchanged?

(a) $\text{Kva} = \dfrac{6,250}{0.7} = 8,930.$ *Ans.*

(b) $I = \dfrac{8,930,000}{\sqrt{3} \times 6,600} = 781$ amp.   *Ans.*

(c) $\text{Input} = \dfrac{6,250}{0.95} = 6,580$ kw

$\qquad = \dfrac{6,580}{0.746} = 8,830$ hp.   *Ans.*

(d) $\text{Input} = \dfrac{8,930}{0.95} = 9,400$ kw

$\qquad = \dfrac{9,400}{0.746} = 12,600$ hp.   *Ans.*

CHAPTER VI

ALTERNATOR CHARACTERISTICS AND OPERATION

106. Alternator Regulation.—When load is applied to an alternator, the terminal voltage drops unless the load takes leading current. With leading current the voltage tends to rise when load is applied, as will be demonstrated (see p. 200, Fig. 169). When load is applied to a d-c shunt generator, the terminal voltage drops, as is shown in Part I, Chap. XI. This drop in voltage is due to three causes: the I_aR_a drop in the armature, armature reaction, and the drop in field current which results from the decrease in terminal volts. As commercial alternators are excited from a separate source, there is no decrease of field current due to the drop in the alternator terminal voltage. Therefore, the alternator is more nearly comparable with the separately excited, d-c generator (see Part I, p. 253). Both the I_aR_a drop in the alternator armature and armature reaction ordinarily cause a drop of terminal voltage as load is applied. Another factor which causes the alternator voltage to drop with application of load is the *reactance* of the alternator armature. With leading current, armature reaction and armature reactance usually produce a rise in terminal voltage with increase in load. These several factors will be discussed later in detail.

With fixed conditions of field current and speed the terminal voltage of the separately excited, d-c generator depends on the *magnitude* of the load current. Under similar conditions the terminal voltage of alternators depends not only on the magnitude of the load current but also on the *power factor* of the load. A knowledge of the variation of terminal voltage with current and power factor is usually essential, since the amount by which the terminal voltage varies with different conditions of load has an important effect on the operation of the system to which the alternator is connected. If the alternator supplies incandescent lamps, its voltage must remain very nearly constant or else special regulators are necessary. Moreover, the voltage of an alternator may remain within a few per cent of being constant from no load to full load if the power factor is *unity*, whereas at *low power factor* with lagging current the voltage may drop 30 or 40 per cent from no-load to full-load *current* (not kilowatts). (See Fig. 164, p. 194.)

For alternators, therefore, the variation of voltage with load ordinarily is much larger proportionately than for d-c generators. For example, with shunt generators of commercial size the variation of voltage from no load to full load is small, and it is usually possible to compound a shunt generator so that its terminal voltage is practically constant at all loads. In the alternator, the armature reactance drop, not present in the d-c generator, and the armature reaction produces much larger percentage drops in voltage with increase in load than in the d-c generator. Also, alternators cannot be compounded readily.

In the larger units, the large values of current which occur under short-circuit conditions may cause serious damage to the alternator and to the system. The value of this short-circuit current is closely related to the voltage—load characteristics of the alternators connected to the system, so that a knowledge of the voltage—load characteristics of alternators is helpful in designing accessory apparatus such as exciters, circuit breakers, switches, power-limiting reactances, and relays.

From the operating point of view, therefore, a study of the voltage—load characteristics of alternators is even more essential than the study of similar characteristics of d-c generators. Moreover, it is not only desirable to know the characteristics themselves, but it is even more important to know the factors which determine these characteristics. If these factors and their magnitudes are known, it is easy to understand *why* alternators have characteristics similar to those shown in Figs. 164 and 165 (pp. 194 and 196).

The three factors which cause change of terminal voltage with load are armature *reactance*, armature *resistance,* and armature *reaction.*

107. Armature Reactance.—Since the armature conductors of modern alternators are embedded in slots, hence are nearly surrounded by iron, considerable magnetic flux must link the armature coils when current flows. For example, Fig. 157(*a*) shows four coil sides embedded in a single alternator slot. The insulation and wedge are omitted. It is assumed at the instant shown that the current flows inward in all four conductors. By the corkscrew rule, the flux must have a clockwise direction about the four conductors. It crosses the slot and tooth tips and completes its circuit through the armature iron. It will be noted (Fig. 157(*a*)) that, although all four conductors are linked by some of this flux, more flux links the lower conductors than links the upper conductors.

Figure 157(*b*) shows, to a smaller scale, a complete armature coil embedded in two slots. It will be observed that the flux which links the coil sides in the slots also links the entire coil.

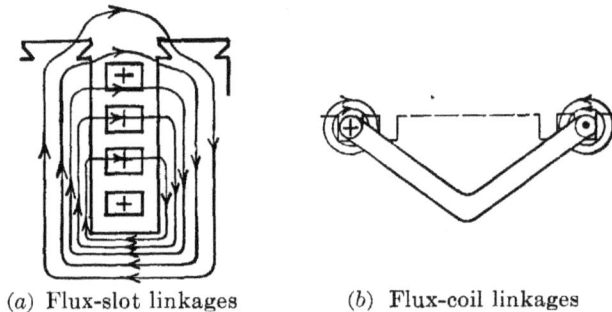

(*a*) Flux-slot linkages   (*b*) Flux-coil linkages
Fig. 157.—Inductance of armature coil.

Some flux links the coil ends; but since the coil ends are not embedded in iron, this flux, although not negligible, is much less than the flux linking the portions of the conductor which lie in the slots. The flux which links the armature coils, but not the field coils, is called the *armature-leakage flux*.

When flux links the current in a circuit, *self-inductance* results (see Part I, p. 168). Since the current in alternator coils is alternating, these coils must have *reactance*. If the self-inductance per *phase* of the armature is L_a, the reactance per phase is

$$X_a = 2\pi f L_a,$$

where f is the frequency.

It is clear that the alternator having the deeper and narrower slots will have the higher self-inductance. Also, the alternator operating at the higher frequency has the greater armature reactance, other factors being equal.

108. Armature Effective Resistance.—The armature iron forms a considerable portion of the path for the flux which links the armature conductors alone (Fig. 157(a)). Since this flux is alternating, it is accompanied by hysteresis and eddy-current losses, occurring in the iron immediately surrounding the slots. The power for these losses must be supplied by the armature current, since this leakage flux linking the armature conductors alone is due to the armature current. The combined hysteresis and eddy-current losses vary nearly as the current *squared*.

As has already been pointed out, more flux links the conductors in the lower part of the slot (Fig. 157(a)) than links those in the upper part of the slot. Hence, the reactance of conductors embedded in a slot increases with their depth in the slot; and if the conductors are in parallel, the current in the conductors in the upper part of the slot is greater than the current in the conductors in the lower part of the slot. Similarly, in any single conductor, the current density will be greater in the upper portion of the conductor.

Also, this slot-leakage flux induces eddy currents in the armature conductors themselves; and if the conductors are of considerable cross-section, this eddy-current loss is not negligible.

The foregoing factors make the loss associated with the armature current considerably greater with alternating current than with an equal direct current. With direct current the armature copper loss is I^2R_o where R_o is the d-c, or ohmic, resistance. To obtain the total loss with alternating current, R_o must be *increased* to a value R, where R is the *effective* resistance of the armature. The ratio of R to R_o varies in different alternators but is of the order of 1.2 to 1.6, the larger values applying to 60-cycle alternators. The effective resistance R ordinarily cannot be measured directly but usually involves a test under short-circuit conditions and also a friction and windage test. In the absence of such data, it is usually sufficiently accurate to measure the d-c resistance R_o and increase this resistance by 50 per cent.

Example.—The resistance between two terminals of a three-phase, Y-connected alternator is measured with direct current and found to be 0.32 ohm. The ratio of effective to ohmic resistance is assumed to be 1.5. What is the effective resistance per phase of this alternator?

Since two phases of the alternator are in series between each pair of terminals, the ohmic resistance of each phase must be 0.32/2, or **0.16, ohm.** The effective resistance per phase

$$R = 1.5 \times 0.16 = 0.24 \text{ ohm.} \quad Ans.$$

109. Armature Reaction.—In d-c dynamos, the armature ampere-turns act on the magnetic circuit in such a way as to

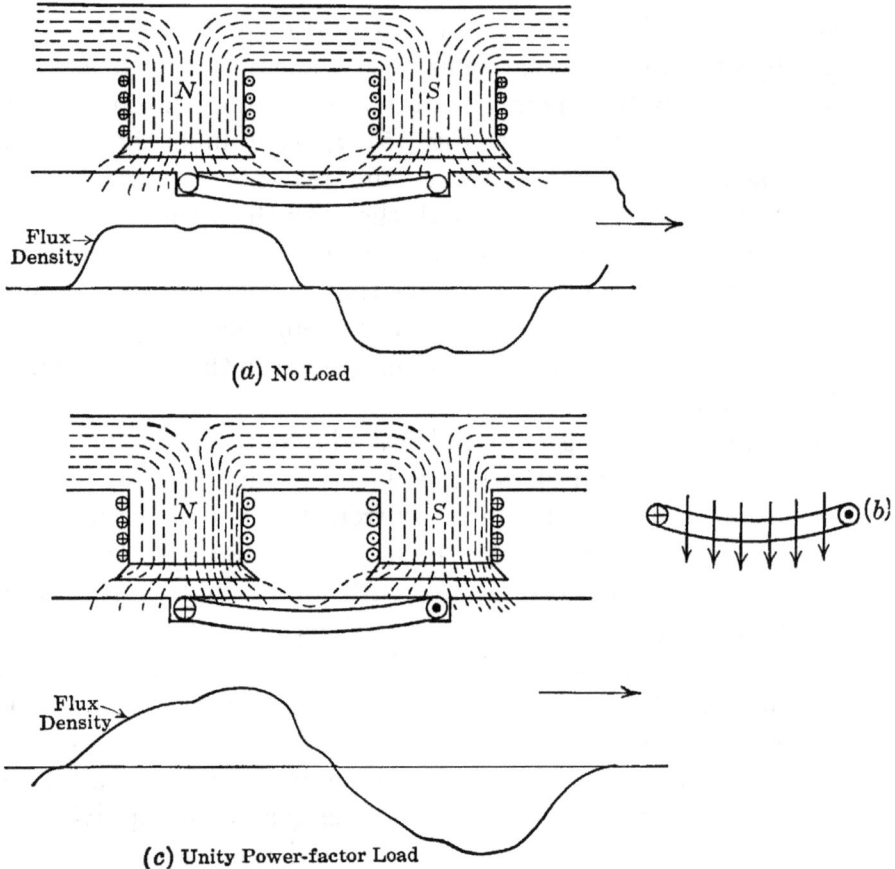

(*a*) No Load

(*c*) Unity Power-factor Load

FIG. 158.—Flux distribution in air-gap of salient-pole alternator.

change both the direction and the magnitude of the flux due to the field poles (see Part I, p. 242). For a given armature current, the direction of the armature reaction depends on the position of the brushes. In an alternator with a given armature current, the direction and magnitude of the air-gap flux obviously cannot depend on brush position but do depend on the phase relation between current and no-load emf and, hence, on the power factor of the load.

Figure 158 shows a portion of the field and armature of an alternator, the direction of rotation of the armature being from left to right. In (a) an armature coil is shown at the instant when its sides are directly under the pole centers and the coil is carrying no current. Since its mmf is zero, it has no effect on the flux distribution. The flux distribution curve is also shown, the ordinates of which are flux density. The curve is practically flat, and the flux under each pole is distributed symmetrically.

a. *Unity Power Factor.*[1]—Figure 158(c) shows the coil with the current in phase with the no-load induced emf. At the position shown the induced emf in the coil has its maximum value. If the current is in phase with this induced emf, corresponding approximately to a load of unity power factor, the current has its maximum value at the same instant. The direction of the induced emf and, hence, of the current in the coil, determined by Fleming's right-hand rule, is inward in the left-hand side of the coil and outward in the right-hand side, as shown in (b) and (c). The coil mmf at this instant, determined by the corkscrew rule, acts downward as shown in (b). The effect of this coil on the magnetic circuit, that is, the effect of armature reaction, is shown in (c). The mmf of this coil being downward strengthens the flux in the right-hand side of the N-pole and weakens it in the left-hand side of the S-pole. This effect is also illustrated by the flux-distribution curve in (c). The curve is peaked on the right-hand side and depressed on the left-hand side. The area under the curve, hence, the total flux, is changed but little.

This same effect occurs in d-c dynamos when the brushes are in the geometrical neutral (see Part I, p. 242).

In an alternator, therefore, it may be said that with unity power-factor loads the effect of armature reaction is to shift the flux in the direction of rotation of the armature,[2] the total magnitude of the flux being changed but little.

b. *Current Lags* 90°.—Figure 159(a) shows an alternator, similar to that shown in Fig. 158, with an armature coil rotating in a clockwise direction. In Fig. 159, however, the current is

---

[1] Unity power factor with respect to *no-load* induced emf and not terminal voltage. This is explained (p. 195). See also Fig. 171(a), p. 202.

[2] With rotating-field poles, the flux is shifted in a direction opposite to that of rotation.

assumed to *lag the no-load induced emf by* 90°.   When the coil is in position (1) its induced emf is a maximum, since the coil sides are directly under the centers of the poles (see p. 153, Fig. 127).   At this instant, however, the current in the coil must be zero.   The current does not reach its maximum value until the coil has traveled 90 electrical space-degrees farther and has reached position (2) where the induced emf is zero.   When in position (1) the induced emf acts *inward* in the left-hand side of the coil. Therefore, the current, which lags 90°, must act inward in the left-hand side of the coil when the coil reaches position (2). When in position (2) the coil center is directly under the center of an *S*-pole as shown in (*b*).   By the corkscrew rule, its mmf is downward and, therefore, acts in direct *opposition* to the *S*-pole.

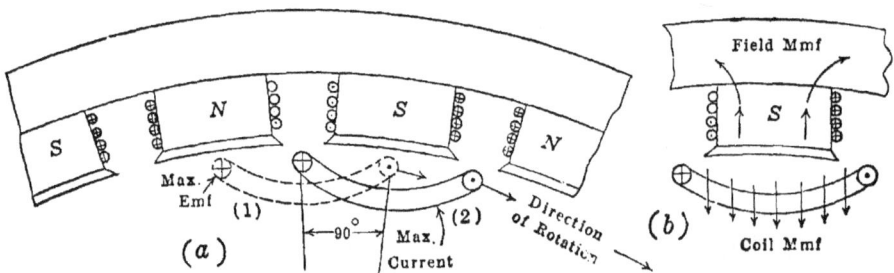

FIG. 159.—Armature reaction due to current lagging 90°.

When the armature current lags the no-load induced emf by 90°, *its mmf acts in direct opposition to the field produced by the field poles.* As a result, the field is materially weakened by a lagging current with a resulting reduction in induced emf.

This result is similar to that of moving the brushes forward 90 electrical degrees in a d-c generator. All the armature ampere-turns are then demagnetizing and weaken the field.

c. Current Leads 90°.—Figure 160 shows the conditions existing when the current leads the no-load induced emf by 90 electrical degrees. The emf reaches its maximum value when the coil sides are directly under the pole centers, position (2) (Fig. 160(*a*)).   The current, however, reaches its maximum value 90 electrical space-degrees *ahead* of this position, or at (1). Since, by Fleming's right-hand rule, the direction of the induced emf is inward in the left-hand side of the coil when it is at (2), the direction of the current, which reaches its maximum value 90 electrical time-degrees earlier, must also act inward in the left-hand side when the coil is at (1). When at (1) the coil is directly beneath an *N*-pole, and by the corkscrew rule its mmf acts downward as shown in (*b*). The mmf or the ampere-turns of the coil, therefore, now *assist* or *strengthen* the main field, as they are acting directly in conjunction with it. *When the armature current leads the no-load induced emf by 90°, its mmf acts in conjunction with the field produced by the field poles.* Accordingly, a leading current *strengthens* the field, with a resulting increase in the induced emf.

If the field current and the armature current in an alternator remain constant, the maximum value of flux and hence of induced emf will occur when the armature current leads the no-load induced emf by 90°; the minimum value of flux and hence of

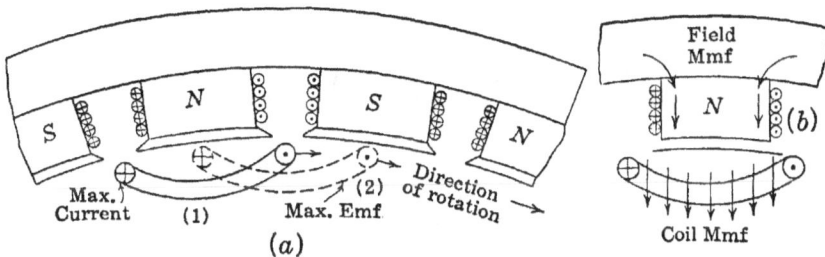

Fig. 160.—Armature reaction due to current leading 90°.

induced emf will occur when the armature current lags the no-load induced emf by 90°. At unity power factor the values of flux and induced emf will be neither maximum nor minimum.

110. Armature Impedance Drop.—In the d-c generator the terminal voltage differs from the induced emf by the resistance drop in the armature and the induced emf may be found by *adding* the armature resistance drop to the terminal voltage (see Part I, p. 254).

In the alternator there is not only a resistance drop in the armature but also a reactance drop due to the self-inductance of the armature. The combined resistance and reactance drop give an *armature impedance drop*. To determine the induced emf in the alternator armature, the *armature impedance drop* must be added *vectorially* to the terminal voltage. This is illustrated by the following examples.

Current in Phase with Terminal Voltage.—Let *V* be the terminal voltage of an alternator (**Fig. 161**(*a*)) when it delivers current *I* at unity power factor.   As the power factor is unity, I is in phase with V.   The armature may be treated as a simple series circuit having resistance and reactance (see Figs. 57 and 58, p. 67).   The resistance drop IR is in phase with the current.   The reactance drop IX leads the current by 90°.   The vector sum of these two voltage drops gives the armature impedance drop IZ.   The emf E_a induced in the armature is the vector sum of the impedance drop IZ and the terminal voltage V.   That is,

$$E_a = V + IZ \qquad\qquad (67)$$

(Compare with Eq. (90), Part I, p. 254.)

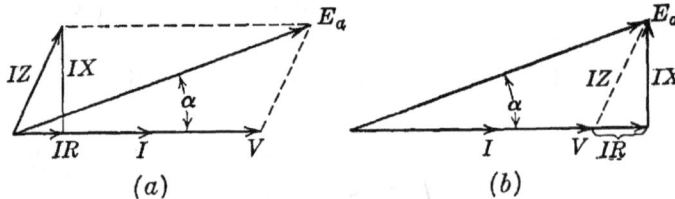

Fɪɢ. 161.—Alternator vector diagram for unity power factor.

The vector addition is accomplished by completing the parallelogram (Fig. 161(a)) in which V and IZ are adjacent sides.   The diagonal E_a, their vector sum, is the emf induced in the armature.

The same result is obtained by adding the IR drop directly to V (Fig. 161(b)) and then adding the IX drop, at right angles to I and leading, at the end of IR.   The vector addition in this case is made by the triangle of vectors described in Chap. I, p. 28. The impedance drop IZ is shown dotted in Fig. 161(b), as it is not used in obtaining E_a in this method.

It is to be noted that with a load of unity power factor the current is in phase with the terminal voltage but lags the *induced* emf E_a by an angle α.

It is a simple matter to find E_a if V, IR, and IX are known. E_a is the hypotenuse of a right triangle of which $(V + IR)$ is one side and IX the other.

$$E_a = \sqrt{(V + IR)^2 + (IX)^2}. \qquad\qquad (68)$$

Example.—A 60-kva, 220-volt, 60-cycle alternator has an effective armature resistance of 0.016 ohm and an armature reactance of 0.070 ohm. What is its induced emf when the alternator is delivering its rated current at a load power factor of unity?

The current

$$I = \frac{60,000}{220} = 273 \text{ amp.}$$

$$IR = 273 \times 0.016 = 4.37 \text{ volts.}$$
$$IX = 273 \times 0.070 = 19.1 \text{ volts.}$$
$$E_a = \sqrt{(220 + 4.4)^2 + (19.1)^2} = 225 \text{ volts.} \quad Ans.$$

Lagging Current.—When the current lags the terminal voltage by the angle θ, the method used at unity power factor is employed to calculate the induced emf. Figure 162(a) shows the current I lagging the terminal voltage V by the angle θ. The IR drop is along the current vector I, and the IX drop is in quadrature with I and leading, as in Fig. 161. The resulting impedance drop IZ is then found, being the resultant of IR and IX. This impedance drop IZ is then added vectorially to V by completing

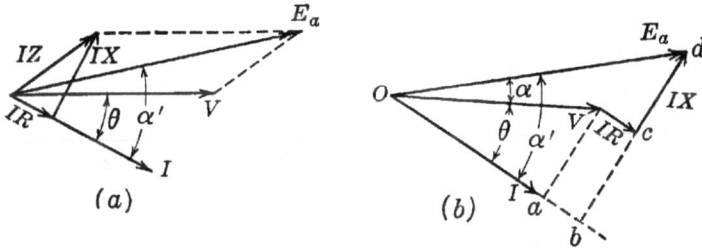

Fig. 162.—Alternator vector diagram for power factor cos θ, current lagging.

the parallelogram having V and IZ as adjacent sides. The diagonal gives the armature induced emf E_a. It will be noted (Figs. 161 and 162) that the position of the armature-impedance triangle is determined by the *current* and not by the generator terminal voltage. Therefore, when the current lags, this impedance triangle swings clockwise with the current.

As before, the impedance drop may be added at the end of V, if the proper phase relations are observed. The most direct method of finding the induced emf E_a is to use the method described under the triangle of vectors (p. 28). IR, which is in phase with the current, is first added vectorially at the end of the terminal voltage V (Fig. 162(b)). Then the reactance drop IX, at right angles to the current and leading, is added at the end of IR in Fig. 162(b), where IR is parallel to I and IX is at right angles to I and leading. The resultant emf found by completing the polygon is the induced emf E_a.

The geometrical solution of the diagram (Fig. 162(b)) is simple. If IR is projected on the current vector I (extended) at ab, a right triangle of voltages, Obd, is formed, of which E_a is the hypotenuse. The values of the two legs of this right triangle may be found as follows:

$$Oa = V \cos \theta.$$
$$ab = IR.$$
$$aV = bc = V \sin \theta.$$
$$cd = IX.$$

$$E_a = \sqrt{\overline{Ob^2} + \overline{bd^2}} = \sqrt{(Oa + ab)^2 + (bc + cd)^2}$$
$$= \sqrt{(V \cos \theta + IR)^2 + (V \sin \theta + IX)^2}. \qquad (69)$$

The current now lags the induced emf E_a by the angle α'.

Example.—In the example on p. 190, determine the induced emf when the alternator is delivering its rated current at rated terminal voltage and 0.7 power factor, lagging current.

The rating of an alternator, as has been pointed out, depends on the current or kva rather than the kilowatts. Therefore, the current rating of the alternator will remain unchanged, although the kilowatts are reduced to 0.7 of their former value.

$$\cos \theta = 0.70, \qquad\qquad IR = 4.37 \text{ volts as before.}$$
$$\theta = 45.6°.$$
$$\sin \theta = 0.715, \qquad\qquad IX = 19.1 \text{ volts as before.}$$
$$E_a = \sqrt{(220 \times 0.70 + 4.4)^2 + (220 \times 0.715 + 19.1)^2} = 237 \text{ volts.} \quad Ans.$$

It is to be noted that the induced emf is now higher than before, although the magnitude of the impedance drop is the same. Therefore, for a fixed value of induced emf, the terminal volts become less with increasing lag of the current, even though the value of the current remains unchanged. This is due to the angle at which the impedance drop subtracts from the induced emf. It would be expected, therefore, that the drop in terminal voltage with increase of load would be greater with lagging current than when the power factor is unity, and such is the fact.

At unity power factor, the armature-resistance drop is the important factor in determining the value of E_a. With lagging current, the resistance drop plays but a small part and the armature reactance drop is the important factor.

Leading Current.—Figure 163(a) shows the alternator vector diagram when the current *leads* the terminal voltage by an angle θ. As before, IR and IX are laid off in phase and at right angles to the current, and their vector sum IZ is found. The parallelogram having IZ and V as adjacent sides is completed and the diagonal, their vector sum, is the emf E_a induced in the armature. This same vector addition is performed in (b) by adding IR, parallel to I, to the end of V, and IX, at right angles to I and leading, to the end of IR. The resultant E_a is the same as in (a).

The value of E_a may be found trigonometrically as in Fig. 162.

The voltage drop IR, parallel to the current, is projected on the current vector I (extended), at ab.

$$Oa = V \cos \theta.$$
$$ab = IR.$$
$$aV = bc = V \sin \theta.$$
$$cd = IX.$$

$$E_a = \sqrt{\overline{Ob}^2 + \overline{bd}^2} = \sqrt{(Oa + ab)^2 + (bc - cd)^2}$$
$$= \sqrt{(V \cos \theta + IR)^2 + (V \sin \theta - IX)^2}. \quad (70)$$

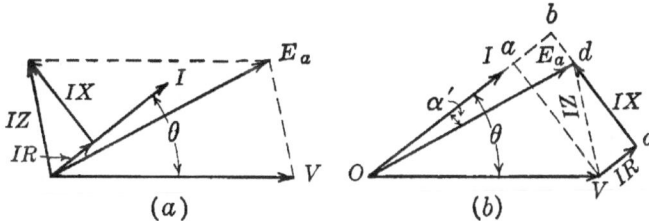

FIG. 163.—Alternator vector diagram for power factor cos θ, leading current.

Equation (70) differs from Eq. (69) only in the sign of IX, which is now negative.

Example.—Repeat the examples, pp. 190 and 192, when the power factor is **0.7**, current leading.

$$\cos \theta = 0.70 \qquad\qquad IR = 4.37 \text{ volts.}$$
$$\sin \theta = 0.715, \qquad\qquad IX = 19.1 \text{ volts.}$$

$$E_a = \sqrt{(220 \times 0.70 + 4.4)^2 + (220 \times 0.715 - 19.1)^2} = 210 \text{ volts.} \quad \textit{Ans.}$$

The induced emf in the armature is now *less* than the terminal voltage. This is a condition which cannot exist in a d-c generator. It results from the phase position of the IZ drop with respect to V. A study of Fig. 163 shows that, as the angle of lead of the current increases, IZ swings counterclockwise and causes the magnitude of E_a to decrease.

111. Alternator Regulation with Lagging Current.—The emf E_a, as determined in the preceding paragraph, is the emf *induced* in the alternator armature under load conditions. In practice, it is a quantity which is not easily measured and which can be calculated only approximately, since it is difficult to determine the armature reactance X. E_a *is not the no-load terminal emf of the alternator.   If there were no armature reaction, E_a would be the no-load emf of the alternator.*

For example, consider a separately excited, d-c generator having an armature resistance of 0.1 ohm and delivering 50 amp at 120 volts at the terminals.   The induced emf is $120 + 50 \times 0.1$, or 125 volts.   When the load is removed, the speed and field current being maintained constant, the no-load emf is measured and found to be 128 volts.   The difference between the 128 volts induced at no load and the 125 volts induced under load is due to the fact that under load the armature reaction has weakened the

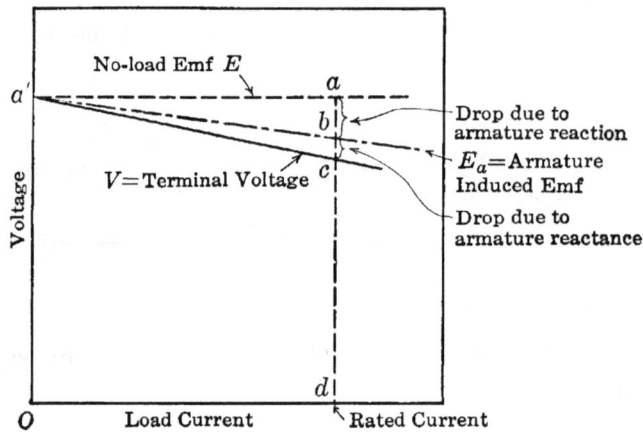

No-load Emf E

a

a'

b

Drop due to armature reaction

E_a=Armature Induced Emf

V=Terminal Voltage c

Drop due to armature reactance

Voltage

d

O Load Current   Rated Current

Fig. 164.—Alternator characteristic with lagging current.

field and reduced the induced emf (see also Part I, p. 254, Fig. 225).

Figure 164 shows the characteristic of an alternator, with lagging current.   At no load the induced emf E is given by Oa' or da.   At rated load the induced emf E_a is db.   The decrease ab from the no-load induced emf E to the rated-load induced emf E_a is due to the reduction in flux brought about by the weakening of the field due to armature reaction with lagging armature current.   The voltage drop bc is due to the armature impedance-drop, although bc is *not equal* numerically to the armature impedance drop since this drop is ordinarily not in phase with the terminal voltage and the induced emf (see Fig. 162).

Thus, with a lagging current, two factors tend to cause the terminal voltage to drop rapidly.   First, the position of the arma-ture coil when its current is a maximum is such that the field is weakened and the induced emf reduced. Second, the phase of the armature impedance drop is such that it subtracts almost directly from the induced emf, and this results in a lowered terminal voltage. Hence, with a given load a lagging current causes greater drop in terminal voltage than a current in phase. To maintain the voltage with lagging current requires an increase in field current.

Even at unity power factor the field is weakened by armature reaction. In Fig. 158, it is shown that, when the current is in phase with the no-load induced emf, armature reaction increases the flux in one side of a pole and decreases it in the other side. However, at unity power factor the current is in phase with the *terminal voltage* and lags the no-load induced emf E by an angle β (Fig. 171(*a*)). Hence, because with a load at unity power factor the current lags the no-load induced emf by some angle β, the field is weakened by armature reaction.

Alternator regulation is defined as the difference between no-load and rated-load voltage, divided by rated-load voltage. For example, in Fig. 164, the regulation of the alternator is $(da - dc)/dc$, or ca/dc. The regulation is a measure of the ability of the alternator to maintain its terminal voltage under load and is not concerned directly with the armature induced emf E_a.

Example.—The terminal voltage of a 400-kva, 60-cycle, three-phase alternator is 2,300 volts when it is delivering its rated load of 100 amp at 0.8 power factor, lagging current. When this load is removed, the no-load voltage becomes 2,760 volts. What is the regulation of the alternator at this power factor?

$$\text{Regulation} = \frac{2,760 - 2,300}{2,300} = \frac{460}{2,300} = 0.20, \text{ or } 20 \text{ per cent.} \quad Ans.$$

112. Alternator Regulation with Leading Current.—A study of Fig. 163 and the example, p. 193, shows that with leading current the terminal voltage may be greater than the induced emf in the armature. With a fixed induced emf, the terminal voltage may *rise* with increase of load. When the current leads, however, it also *strengthens* the field through armature reaction. With a leading current, therefore, the induced emf E_a may actually increase with increase of load, as shown in Fig. 165. Since the induced emf increases with load and the terminal voltage tends to exceed in magnitude the induced emf with increase of load owing to the phase of the armature-reactance drop, the terminal voltage at rated load may be considerably *greater* than its value at no load. In other words, the terminal voltage *drops* when the load is removed from the alternator, the field current remaining constant.

Therefore, an alternator delivering a leading current may have a rising characteristic, similar to that shown in Fig. 165. This characteristic is not unlike that of the overcompounded, d-c generator (see Part I, p. 258, Fig. 229), except that in the over-

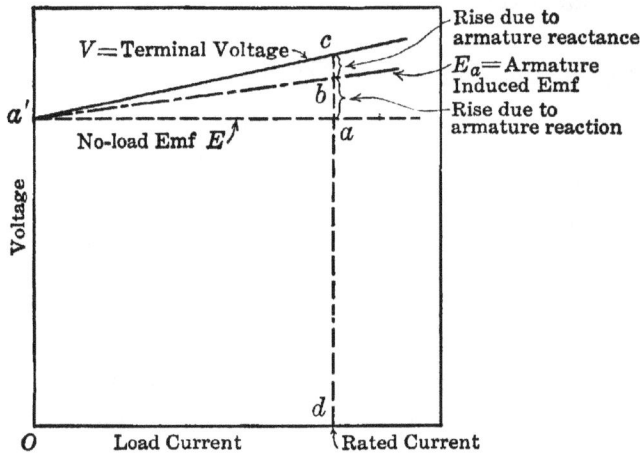

FIG. 165.—Alternator characteristic with leading current.

compounded, d-c generator the *induced* emf must be *greater* than the terminal voltage.

The regulation of the alternator under the conditions shown in Fig. 165 is, by definition,

$$\frac{da - dc}{dc} = -\frac{ac}{dc},$$

and is, therefore, *negative*.

Example.—The terminal voltage of the 400-kva alternator (Par. 111) is 2,300 volts when it is delivering its rated load of 100 amp at 0.8 power factor, leading current. When this load is removed, the terminal voltage drops to 2,100 volts. What is the regulation of the alternator at this power factor?

Regulation $= \dfrac{2,100 - 2,300}{2,300} = \dfrac{-200}{2,300} = -0.087$, or -8.7 per cent. *Ans.*

113. Saturation Curve.—The saturation curve of the alternator is obtained in precisely the same manner as the saturation curve of the d-c generator. (See Part I, p. 238, Par. 194.) The field-winding is connected to a d-c source in series with rheostats

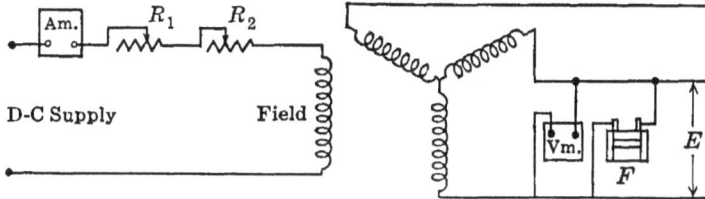

FIG. 166.—Connections for obtaining saturation curve of alternator.

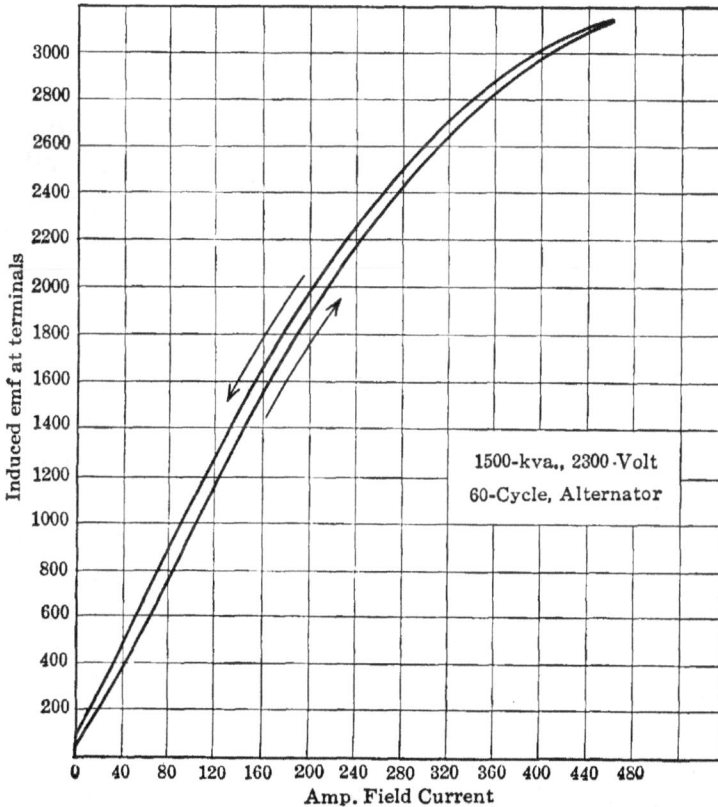

1500-kva., 2300-Volt
60-Cycle, Alternator

FIG. 167.—Saturation curve of 1,500-kva alternator.

and an ammeter (Fig. 166). A voltmeter is connected across one pair of terminals to measure the induced emf. Voltmeters across the other pairs of terminals are not necessary, since the emfs in all three phases should be the same. As a precaution, it is advisable to measure the other emfs at some particular value of field current in order to determine whether or not all three are equal. The speed, which should be maintained constant, is determined by a frequency meter connected across one phase (see p. 112). The saturation curve is obtained by increasing the field current in steps and measuring the induced emf for each value of field current. A saturation curve with decreasing values of field current may also be obtained. Figure 167 gives the saturation curve for a 1,500-kva, 2,300-volt, 60-cycle alternator taken with both increasing and decreasing values of field current.

The alternator in Fig. 166 is Y-connected. If an alternator is delta-connected, the procedure is the same, the induced emf being measured across one pair of terminals.

114. Alternator Load Tests.—It is much more difficult to determine experimentally the characteristics of polyphase alternators than those of d-c generators. If a three-phase alternator is being tested, three loads connected either in Y or in delta are necessary. For each reading, it is necessary to balance all three loads so that the three line currents are equal. If any one load with either the Y- or the delta-connection is changed, at least two currents are affected and it requires time to make adjustments. If the alternator is Y-connected and the neutral is accessible, a Y-connection of the load, with the load neutral connected to the alternator neutral, is most convenient since each line current may be adjusted independently of the other two.

Figure 168 shows the connections for obtaining the load characteristic of an alternator at unity power factor. The alternator is driven by a 230-volt shunt motor M, and an ammeter and voltmeter are connected to measure the motor input, should it be desired to obtain the efficiency of the set. The alternator field is excited from 115-volt, d-c supply, which may be obtained by connecting between neutral and outer of the Edison three-wire system shown. The load is delta-connected, and the power is measured with a polyphase wattmeter.

It is more difficult to obtain loads whose power factors differ from unity, particularly when it is desired that these power factors shall remain constant under varying load. The most practicable method is to use a synchronous motor for a load. Its power load may be changed by applying more load to its shaft mechanically or by loading it with a direct-connected generator. The power factor of the synchronous motor may be varied over a considerable range from lagging to leading current merely by varying the motor field current (see p. 320). Even with a synchronous motor, it requires considerable time to make the necessary frequency, load, and power-factor adjustments at each reading. It usually suffices to obtain the regulation by adjusting the load to rated voltage and current and the desired power factor. The load is then removed, and the terminal voltage is again read, the frequency and field current remaining unchanged. This gives sufficient data for obtaining the regulation.

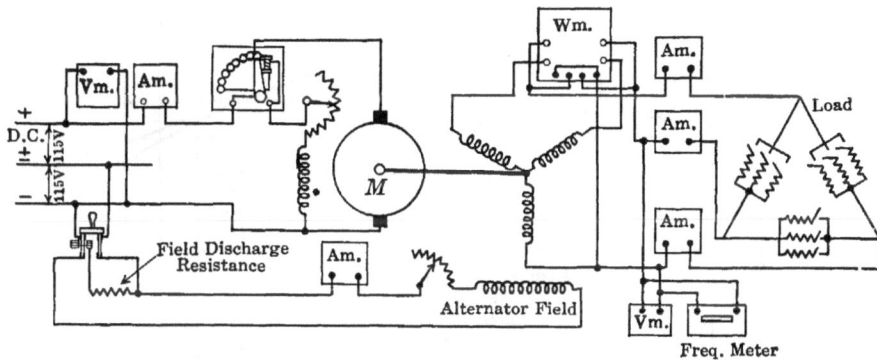

FIG. 168.—Connections for making load test of 3-phase alternator.

The characteristics of a 12-kva, three-phase, 60-cycle, 220-volt alternator at unity power factor and at 0.8 power factor for both lagging and leading current are given in Fig. 169, these characteristics being obtained with the connections shown in Fig. 168. All three characteristics were taken with the alternator field adjusted to give the same *rated* voltage of 220 volts at rated-load current. The regulation at each power factor is

Unity power factor: Regulation $= \dfrac{268 - 220}{220} = 0.218$, or 21.8 per cent.

Power factor $= 0.8$, current lags: Regulation $= \dfrac{286 - 220}{220} = 0.30$, or 30 per cent.

Power factor $= 0.8$, current leads: Regulation $= \dfrac{205 - 220}{220} = -0.068$, or -6.8 per cent.

115. Synchronous-impedance Method.—As stated on p. 183, it is desirable to know the operating characteristics of alternators in order to understand their relations to the system to which they are connected. However, it is not easy to make a direct load test such as is described in Par. 114, even with a unit rated at only 12 kva.

It is difficult to find a balanced three-phase load, particularly one whose power factor can be adjusted. It is likewise difficult to absorb the power output. Hence, some method is desirable which does not involve the actual loading of the alternator. The synchronous-impedance method is one of several such

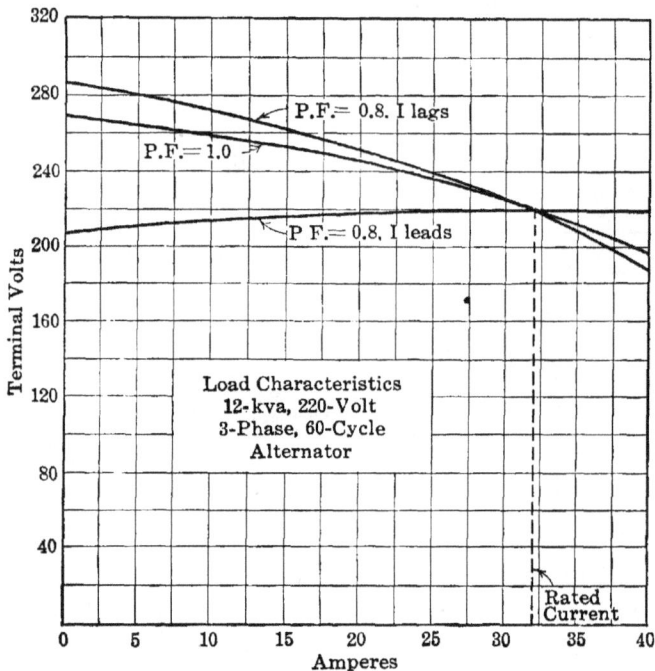

FIG. 169.—Typical load characteristics of 12 kva, 3-phase alternator.

methods used for determining the operating characteristics of an alternator, an open-circuit test and a short-circuit test being the only essential operating tests to be made.

It has been shown that with constant frequency and constant excitation the terminal voltage of an alternator for a given load current and power factor depends on two factors, armature reaction and armature impedance. The armature impedance consists of two components, effective armature resistance and armature reactance. Armature reaction and armature reactance act alike in their effect on the terminal voltage, and accordingly they can be combined into a single quantity called the *synchronous reactance*.

In Fig. 159 (p. 188), it is shown that if the current lags the induced emf by 90°, the armature mmf is in *direct opposition* to the field mmf and hence the flux due to it is *directly opposed* to the flux due to the field poles. It thus acts *directly* to reduce the induced emf. If, in Fig. 162, the current lags the induced emf by 90°, the armature reactance drop acts in direct opposition to E_a and hence to the terminal voltage V if the small effect of armature resistance drop is neglected. This is indicated in Fig. 170(a). Hence, with 90° lag of the current, the armature reaction, by being in direct opposition to the main flux, and the

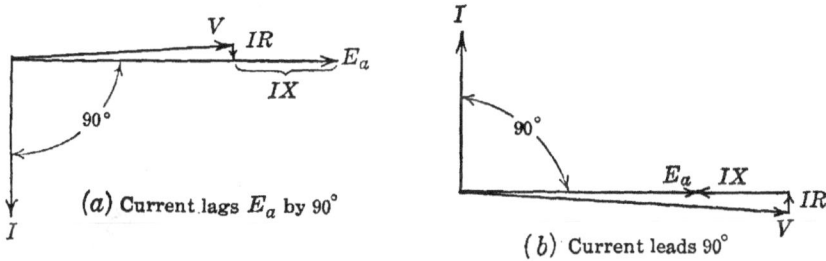

Fig. 170.—Effect of armature-reactance dip on terminal voltage with 90° phase difference.

armature reactance drop are practically in direct opposition to the terminal voltage.

In Fig. 160 (p. 189), it is shown that if the current leads the induced emf by 90° the armature mmf is in conjunction with the field mmf and hence acts to increase directly the flux and hence the induced emf. Also, as indicated in Fig. 170(b), if the current I leads E_a by 90°, the armature reactance drop is in conjunction with E_a and hence with the terminal voltage V, if IR is neglected, and acts to increase both E_a and V.

In Fig. 158 (p. 186), it is shown that if the current is in phase with the induced emf[1] the armature reaction acts at right angles to the flux produced by the field poles. Likewise, in Fig. 161 (p. 190) if the current is in phase with the induced emf E_a the armature reactance drop will be at right angles to the induced emf.

---

[1] In each case the phase angle of the current should be taken with respect to the no-load emf E; but with 90° phase difference, lag or lead, E and E_a are essentially in phase.

Thus, the direction of the armature reaction and the direction of the armature reactance drop are the same in their effect on the terminal voltage, and if the permeance of the magnetic circuit remains constant, each is proportional to the armature current. Hence, the effect of armature reaction and armature reactance can be combined in a single quantity called the *synchronous reactance.*

The application of synchronous reactance is shown in Fig. 171(*a*) and (*b*). Figure 171(*a*) is essentially the same as Fig. 161(*b*), and Fig. 171(*b*) is essentially the same as Fig. 162(*b*), except that in each case *IX* is increased to IX_s where X_s is the *synchronous reactance.* *IX* and E_a are also shown in Fig. 171(*a*) and (*b*). That part *bc* of the synchronous reactance drop *ac* is a

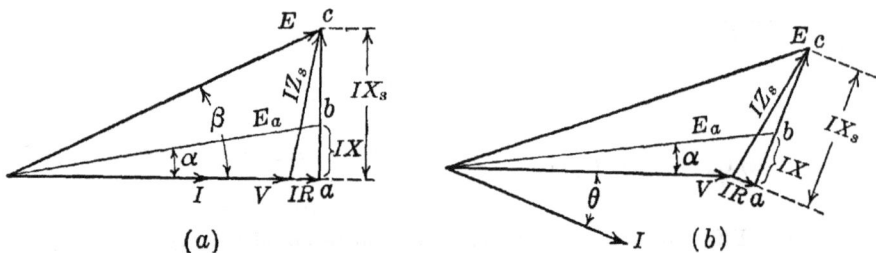

Fig. 171.—Synchronous-impedance diagrams.

fictitious reactance drop which replaces the armature reaction. By using the synchronous reactance the value of the no-load emf *E* is obtained. The vector sum of the armature resistance *R* and the synchronous reactance X_s gives the *synchronous impedance* Z_s. The synchronous impedance drop IZ_s is shown in Fig. 171 (*a*) and (*b*). The method for determining *E* (Fig. 171(*a*) and (*b*)) is the same as for obtaining E_a (Figs. 161 and 162) except that X_s is substituted for *X*. For example, to solve Fig. 171(*b*), X_s is substituted for *X* in Eq. (69) (p. 192).

$$E = \sqrt{(V \cos \theta + IR)^2 + (V \sin \theta + IX_s)^2} \text{ volts.} \quad (71)$$

For unity power factor and also with leading current, substitute X_s for *X* in Eqs. (68) and (70) (pp. 190 and 193).

116. Determination of Synchronous Impedance.—The synchronous impedance is determined by first operating the alternator short-circuited through ammeters as shown in Fig. 172. With the switch *S* closed the field current is increased until the armature short-circuit current is about 200 per cent of rated value. The three ammeters Am are then read. The short circuit is then removed by opening the switch S, and the voltmeter Vm is read. The field current remains constant. Let the average short-circuit current be I_s and the corresponding open-circuit emf be E_s.

The alternator shown in Fig. 172 is Y-connected. Hence, the open-circuit emf per coil is $E_s/\sqrt{3}$. At short circuit the entire emf $E_s/\sqrt{3}$ is utilized in sending the current through the

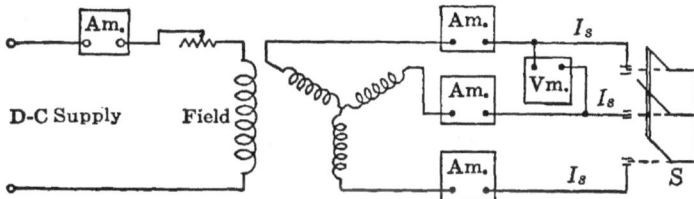

Fig. 172.—Connections to determine synchronous impedance.

synchronous impedance drop of each coil. Hence, in a Y-connected alternator,

$$Z_s = \frac{\left(\dfrac{E_s}{\sqrt{3}}\right)}{I} = \frac{E_s}{\sqrt{3}\,I_s} \text{ ohms.} \tag{72}$$

If the alternator is delta-connected, E_s is equal to the coil emf, and the coil current is $I_s/\sqrt{3}$. Hence, the synchronous impedance per coil for the delta-connected alternator

$$Z_s = \frac{E_s}{\left(\dfrac{I_s}{\sqrt{3}}\right)} = \frac{\sqrt{3}\,E_s}{I_s} \text{ ohms.} \tag{73}$$

It is not necessary to know whether the alternator is Y-connected or delta-connected. Either connection may be assumed; and if the computations are consistent with the assumption, the same value of regulation will be obtained in each case, as will be shown by the example which follows. Upon comparing Eqs. (72) and (73), it is to be noted that if the alternator is assumed to be delta-connected the synchronous impedance per coil is *three times* that for the Y-connection.

If the effective armature resistance R is known (p. 185), the synchronous reactance

$$X_s = \sqrt{Z_s^2 - R^2} \text{ ohms.} \tag{74}$$

However, R is small compared with Z_s and the two are nearly 90° out of phase, so that $X_s = Z_s$ nearly. (See (c) in the following two examples.)

117. Regulation of Y-connected Alternator.—The determination of regulation by the synchronous-impedance method for a Y-connected alternator is given by the following example:

Example.—A 1,500-kva, 6,600-volt, three-phase, Y-connected, 60-cycle alternator is short-circuited through three ammeters (Fig. 172), the field current increased to 150 amp, the average of the three line ammeters then being 260 amp. With this same value of field current, the armature circuit is opened, and the voltmeter reads 5,200 volts.

The ohmic resistance between any two terminals is 0.72 ohm, and the ratio of effective to ohmic resistance is 1.4. Determine: (a) synchronous impedance; (b) effective armature resistance; (c) synchronous reactance; (d) regulation at unity power factor; at 0.8 power factor, current **lagging** and leading.

(a) The synchronous reactance, from Eq. (72),

$$Z_s = \frac{5,200}{\sqrt{3} \times 260} = 11.53 \text{ ohms.} \quad Ans.$$

(b) The ohmic resistance per coil is $0.72/2 = 0.36$ ohm. Hence,

$$R = 0.36 \times 1.4 = 0.504 \text{ ohm.} \quad Ans.$$

(c) From Eq. (74),

$$X_s = \sqrt{(11.53)^2 - (0.504)^2} = 11.53 \text{ ohms.} \quad Ans.$$

(d) Rated current, which is equal to coil current

$$I = \frac{1,500}{\sqrt{3} \times 6,600} = 131 \text{ amp.}$$

Rated coil voltage

$$V = \frac{6,600}{\sqrt{3}} = 3,810 \text{ volts.}$$
$$IR = 131 \times 0.504 = 66 \text{ volts.}$$
$$IX_s = 131 \times 11.53 = 1,511 \text{ volts.}$$

Using Eq. (68) (p. 190) and substituting X_s for X,

$$E = \sqrt{(3,810 + 66)^2 + \overline{1,511}^2} = 4,160 \text{ volts.}$$

$$\text{Regulation} = \frac{4,160 - 3,810}{3,810} = 0.0918, \text{ or } 9.18 \text{ per cent.} \quad Ans.$$

Computed no-load terminal emf

$$E_t = \sqrt{3} \times 4,160 = 7,200 \text{ volts.}$$

The rated current is independent of power factor. At 0.8 power factor, lagging current, the regulation is computed by means of Eq. (71).

$$\cos \theta = 0.80; \quad \theta = 36.9°; \quad \sin \theta = 0.60.$$

$$E = \sqrt{(3,810 \times 0.80 + 66)^2 + (3,810 \times 0.60 + 1,511)^2}$$
$$= \sqrt{(3,114)^2 + (3,797)^2} = \sqrt{24.1 \times 10^6} = 4,910 \text{ volts.}$$

$$\text{Regulation} = \frac{4,910 - 3,810}{3,810} = 0.263, \text{ or } 26.3 \text{ per cent.} \quad Ans.$$

No-load terminal emf

$$E_t = \sqrt{3} \times 4,910 = 8,500 \text{ volts.}$$

At 0.8 power factor, leading current, Eq. (70), (p. 193) is used, with X_s substituted for X.

$$E = \sqrt{(3,810 \times 0.80 + 66)^2 + (3,810 \times 0.60 - 1,511)^2}$$
$$= \sqrt{(3,114)^2 + (775)^2} = \sqrt{10.30 \times 10^5} = 3,210 \text{ volts.}$$

$$\text{Regulation} = \frac{3,210 - 3,810}{3,810} = \frac{-600}{3,810} = -0.158, \text{ or } -15.8 \text{ per cent.} \quad Ans.$$

No-load terminal emf

$$E_t = \sqrt{3} \times 3,210 = 5,560 \text{ volts,}$$

which is *less* than the rated-load terminal voltage of 6,600 volts.

118. Regulation of Delta-connected Alternator.—Assume that the alternator, Par. 117, is delta-connected. Repeat (a), (b), (c), (d).

In the short-circuit test, each coil current will be $260/\sqrt{3} = 150$ amp. When the armature is open-circuited, coil emf is equal to line emf of 5,200 volts.

(a) Hence,

$$Z_s = \frac{5,200}{150} = 34.6 \text{ ohms,} \quad Ans.$$

three times the value for the Y-connection.

The same value is obtained by the use of Eq. (73).

(b) The ohmic resistance between terminals includes that of one coil in parallel with the resistance of the other two in series. Let R_o be the ohmic resistance per coil.

$$\frac{1}{0.72} = \frac{1}{R_0} + \frac{1}{2R_0}.$$
$$R_0 = \tfrac{2}{3} \times 0.72 = 1.08 \text{ ohms.}$$

Effective resistance per coil

$$R = 1.4 \times 1.08 = 1.51 \text{ ohms,} \quad Ans.$$

three times its value with the Y-connection.

(c) $X_s = \sqrt{(34.6)^2 - (1.51)^2} = 34.6$ ohms. *Ans.*

(d) Rated coil current

$$I = \frac{131}{\sqrt{3}} = 75.6 \text{ amp.}$$
$$IR = 75.6 \times 1.51 = 115.3 \text{ volts.}$$
$$IX_s = 75.6 \times 34.6 = 2,620 \text{ volts.}$$

The no-load emf at unity power factor, using Eq. (68) (p. 190) and substituting X_s for X (see Fig. 171(a)),

$$E = \sqrt{(6,600 + 115.3)^2 + (2,620)^2} = 7,200 \text{ volts.} \quad Ans.$$

The no-load emf at 0.8 power factor, lagging current, using Eq. (71),

$$E = \sqrt{(6,600 \times 0.8 + 115.3)^2 + (6,600 \times 0.6 + 2,620)^2}$$
$$= 8,510 \text{ volts.} \quad Ans.$$

The no-load emf at 0.8 power factor, leading current, using Eq. (70) (p. 193) substituting X_s for X,

$$E = \sqrt{(6,600 \times 0.8 + 115.3)^2 + (6,600 \times 0.6 - 2,620)^2} = 5,550 \text{ volts.}$$
$$Ans.$$

Note that the values of no-load terminal emf E are the same whether the alternator is assumed to be connected in Y or in delta.

119. Errors in Synchronous-impedance Method.—The synchronous-impedance method gives too large a value of regulation and hence is called the *pessimistic method*. The value of the synchronous reactance, particularly that part which replaces armature reaction, is too large, for it is determined at short circuit when the iron is not saturated. (See Fig. 167, p. 197.) Also, at short circuit the armature current lags the induced emf by nearly 90°, and the armature mmf is acting directly on the pole axes (see Fig. 159, p. 188) where the permeance of the magnetic circuit is a maximum. In salient-pole alternators under the usual conditions of load the armature mmf acts for the most part on the interpolar space (see Fig. 158, p. 186) where the permeance is low. Hence, the effect of the armature mmf is much greater at short circuit where the value of X_s is determined than under the usual conditions of load. These factors make X_s too large so that the value of E is greater than under normal operating conditions. However, with the synchronous-impedance method there is a factor of safety; and, with some experience, allowance can be made for the amount of error inherent in the method.

120. Field-discharge Switch.—When a highly inductive circuit is opened, a high emf of self-induction may result. Since alternator fields have high self-inductance and considerable energy ($\frac{1}{2}LI^2$) is stored therein, the field winding may be punctured if the field circuit is opened in the ordinary manner. That is, a high emf $e = -L(i/t)$ will be induced (see Part I, Pars. 147 and 148, pp. 174 and 175). This high induced emf is prevented by the use of a *field-discharge switch*. Such a switch has an extra set of clips (Fig. 173) which make contact with one blade only when the switch is being opened. A resistance is connected between these clips and the other side of the line (Fig. 168). At the instant of opening the switch, the field and the line, temporarily, are connected in parallel by the field-discharge resistance. The energy of the field is dissipated partly in this resistance rather than entirely at the switch contacts. Contact with

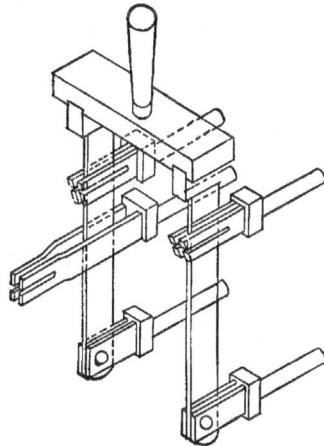

FIG. 173.—Field-discharge switch.

switches opening inductive circuits should be carefully avoided, even in the case of very low voltages. There is danger, not only of being burned by the arc, but also of being injured from the high induced emf.

121. Tirrill Regulator.—An automatic voltage regulator of the Tirrill type for d-c generators is described in Part I, Chap. XI, p. 261. An automatic voltage regulator is more essential in a-c stations, particularly in the smaller ones, than in d-c stations. The voltage changes in the generator and throughout the system are greater with alternating current than with direct current because of the added reactance drop in the generator armature, transformers, feeders, etc. Alternating-current generators cannot be compounded readily to compensate for voltage drop as d-c generators can.

The Tirrill regulator is also designed to be used with a-c generators. As with large d-c generators the regulator acts through the field of an exciter. The underlying principle of the regulator is the same whether used for alternating or for direct current, the voltage being controlled in each case by short-circuiting momentarily the exciter-field rheostat.

When the alternator voltage drops, the contacts which short-circuit the exciter-field rheostat remain closed for longer periods of time, thus increasing the exciter voltage and hence the alternator-field current. If one exciter supplies the fields of all the alternators in the plant, one regulator can readily maintain the bus-bars at the desired voltage.

PARALLEL OPERATION OF ALTERNATORS

122. Division of Load among Alternators in Parallel.—The same reasons which make it necessary to operate d-c generators in parallel (see Part I, Par. 232, p. 304) apply to alternators.

In order to operate satisfactorily in parallel without equalizers, d-c generators must have drooping voltage characteristics.

The speed-load characteristic of the prime mover affects parallel operation only to the extent that it affects the load-voltage characteristic of the generator.

With alternators, load-voltage curves have practically no effect on the division of the kilowatt load. The division of kilowatt load between alternators is determined almost entirely by the speed-load characteristics of their *prime movers*. Furthermore, in order that alternators may operate satisfactorily in parallel, their *prime movers* must have *drooping* speed-load characteristics. Otherwise, instability accompanies their operation in parallel.

The effect of prime-mover speed-load characteristics on the division of kilowatt load is shown in Fig. 174(*b*). In Fig. 174(*a*) are shown diagrammatically two alternators No. 1 and No. 2 connected in parallel.

If they are operating in parallel, they must have the *same frequency* and the *same terminal voltage*. Figure 174(*b*) shows the speed-load characteristic of each of the prime movers driving the alternators. Each prime mover has a drooping speed-load characteristic. (Instead of plotting speed in rpm, the frequency or electrical speed is plotted. For example, a six-pole alternator running at 1,200 rpm would have the same electrical speed as an eight-pole alternator running at 900 rpm.) The speed-load curves of the prime movers are determined by their governors, if such prime movers are steam engines, turbines, water wheels, or internal combustion engines. If motor-driven, the speed-load characteristic of the alternator is that of the motor.

Let *oc* (Fig. 174(*b*)) be the frequency at which the system is operating. By projecting horizontally to intersect the speed-

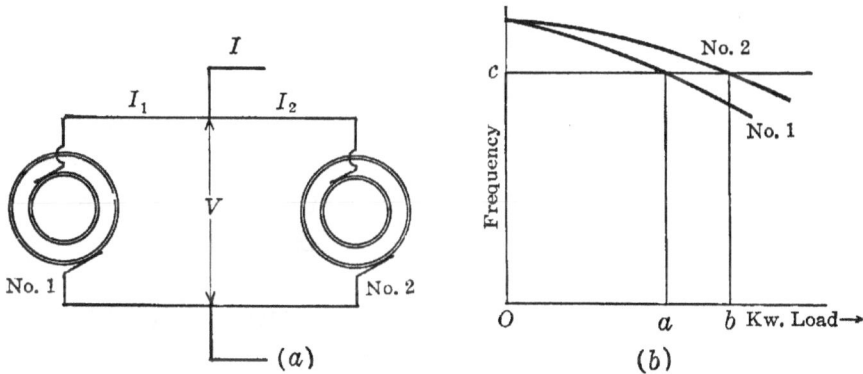

Fig. 174.—Alternators in parallel and speed-load characteristics.

load curves, the load taken by each prime mover at this frequency is obtained. *oa* is the load on prime mover 1, and *ob* is the load on prime mover 2, as both prime movers must operate at the same system frequency. Let the field of alternator 1 be strengthened by means of its field rheostat. At the same time, weaken the field of 2 so that the line voltage does not change. If these were d-c generators, machine 1 would take more load since its induced emf has been increased. But as an alternator, 1 *cannot take more load* because its prime mover can deliver only the load *oa* at this frequency. Alternator 2 cannot drop any load because its prime mover can deliver only the load *ob* at this frequency. Both alternators must always operate at the same frequency or the same electrical speed. With d-c generators, there are no restrictions as to speed. *Therefore, the kilowatt load delivered*

210INDUSTRIAL ELECTRICITY

by alternators in parallel cannot be shifted appreciably by changing the excitation of the generator fields.

To change the kilowatt load of either alternator, the speed-load characteristic of its prime mover must be changed. In engine- or turbine-driven units, this is done by changing the tension in the governor spring or altering the governor device in some manner. Assume, in Fig. 174(*b*), that it is desired to make alternator 1 take the same load as 2. The governor spring of 1 is adjusted so that the characteristic of 1 is raised, as shown in Fig. 175. Both alternators now deliver the same load *oa′* at a frequency *oc′*.

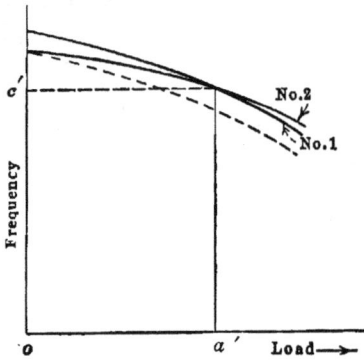

Fig. 175.—Speed-load curves of alternators in parallel—effect of changing governor control.

Under the conditions shown (Fig. 175) the frequency *oc′* is higher than the original frequency *oc* (Fig. 174). If the original frequency is to be maintained, the speed-load characteristic of 2 must be lowered at the same time that the characteristic of 1 is raised. Therefore, to adjust the load between alternators in parallel, the speed-load characteristics of the prime movers must be changed. If the alternators are driven by shunt motors, the speed-load characteristics of the motors may be changed by adjusting the motor-field rheostats. It will be noted, in Fig. 175, that the loads of the two alternators are equal at one frequency only.

If the prime movers had flat speed-load characteristics, the operation of the alternators would be unstable. That is, very small disturbances or changes of frequency would cause very large fluctuations in the kilowatt load delivered by each alternator. This condition would result in serious operating difficulties.

Alternators operating in parallel are in *stable equilibrium.* That is, any circumstance tending to cause the alternators to pull out of synchronism with one another is opposed by reactions which tend to prevent this. For example, assume that alternator 1 (Fig. 174) suddenly loses its driving torque, as might happen if the automatic trip on its turbine governor operated and shut off the steam supply. Alternator 2 automatically would take the entire electrical load, and in addition the electrical reactions developed in the system would be such that motor action would develop in 1, and 2 would drive 1 as a synchronous motor. (In practice alternators in power stations are protected by reverse-energy relays which trip the circuit breakers when electrical energy is supplied to the alternators.)

123. Effect of Field Excitation.—It has been shown that changing the field excitation of alternators in parallel does not change the division of kilowatt load. It does, however, affect the kva load. Assume in Fig. 174(a) that alternators 1 and 2 deliver equal power and equal currents I_1 and I_2 to a load which has unity power factor. Both 1 and 2 are adjusted so that

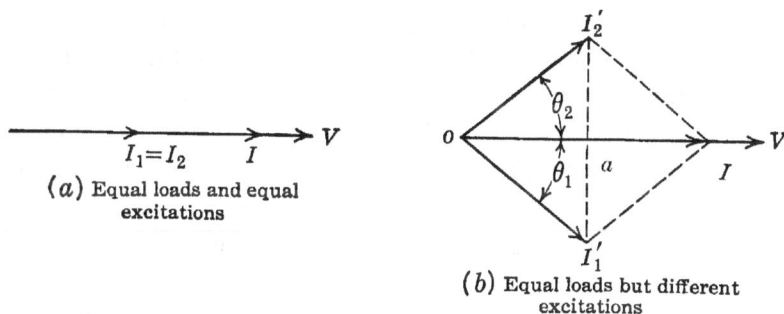

(a) Equal loads and equal excitations

(b) Equal loads but different excitations

FIG. 176.—Effect of changing field excitation of alternators in parallel.

their currents are in phase with the common terminal voltage V (Fig. 176(a)) at which the system operates. The load requires a current I in phase with V, I being the sum of I_1 and I_2.

Now let the field of 1 be strengthened and that of 2 simultaneously weakened, so that the system voltage does not change. If instruments connected to these two alternators are read, the wattmeter readings will not change appreciably. The ammeters, however, will show an increase of current in each alternator. The power-factor indicator in 1 will show that the power factor of 1 has decreased from unity and that the current now lags. The power-factor indicator in 2 will show that the power-factor of 2 has also decreased from unity but that the current now leads.

These conditions are shown in Fig. 176(b). The current I'_1 lags V by θ_1, and I'_2 leads V by θ_2. It is assumed that the load current has not changed. The load current delivered by a central station is determined *entirely by the consumers'* demands, and not by any conditions of operation in the station itself.

Hence, the vector sum of I'_1 and I'_2 must be equal to I. Therefore, if I'_1 lags I, I'_2 must lead I by such an angle that the vector sum of I'_1 and I'_2 is equal to the load current I. Under the foregoing conditions, both I'_1 and I'_2 will terminate on a perpendicular to V at a (Fig. 176(b)). This must be true since the *power* delivered by the two machines has not changed appreciably with the change in excitation. Hence, $VI'_1 \cos \theta_1 = VI_1$, and

$$I'_1 \cos \theta_1 = I_1 = oa.$$

Likewise, $VI'_2 \cos \theta_2 = VI_2$, and

$$I'_2 \cos \theta_2 = I_2 = oa.$$

In order that these conditions may hold, the line $I'_1aI'_2$ must be perpendicular to V at a. (Since the change of I^2R loss in the armature with change of excitation is small, it has been neglected.) Since the current of each alternator has increased, although the power has not changed, the alternator copper losses have increased. Therefore, it is desirable to operate with the currents in phase with each other so as to reduce heating losses. With loads other than unity power factor the same general conditions exist. If the excitation of one alternator is *increased*, its current *lags* further behind the resultant current and the current of the other alternator *leads* the resultant current by a larger angle, the currents adjusting themselves so that their vector sum equals the load current.

The fact that conditions within the alternator adjust themselves to the changes of field excitation is obvious from the following: Through armature reaction a lagging current in an alternator *weakens* the field (p. 188). The alternator with the lagging current operates with the higher induced emf (Fig. 162), resulting from the increased excitation.

Therefore, if the excitation is increased for an alternator in parallel with others, the alternator delivers a more *lagging* current which, through armature reaction, tends to prevent an increase in field strength; the vector giving the impedance drop in the armature increases in magnitude and rotates in a clockwise direction so that it is more nearly in opposition to the induced emf and hence to the terminal voltage (see Fig. 171(b)). Hence, a given increase in the induced emf does not result in the same proportionate increase in the terminal voltage. Thus, the reactions resulting from an attempt to increase the terminal voltage oppose this increase.

Likewise, the alternator with the weaker field delivers a more *leading* current, which strengthens its field; at the same time the vector giving the impedance drop in the armature increases in magnitude and rotates in a counterclockwise direction, tending to increase the magnitude of the terminal voltage (Fig. 163, p. 193).

Thus, a change of field excitation in alternators operating in parallel is accompanied by reactions within the system which oppose the change and all the alternators have the same terminal voltage, thus giving stability to the system.

From the foregoing, it is obvious that a change of excitation in alternators operating in parallel merely changes the power factor for each alternator, and hence the kva load of each. The kilowatt loads are not changed appreciably.

Example.—Two 100-kva, 600-volt, 60-cycle, single-phase alternators, 1 and 2, operate in parallel to supply power to a noninductive load of 165 kw connected across the 600-volt bus-bars. Alternator 1 delivers 90 kw, and 2 delivers 75 kw. Each is operating at unity power factor. (*a*) Determine current delivered by each. The field of 1 is weakened and that of 2 is strengthened so that the bus-bar voltage remains at 600 volts. The governors of the prime movers remain unchanged. The power factor of 1 is now 0.80. Determine: (*b*) current output of 1; (*c*) current output of 2; (*d*) power factor of 2; (*e*) load current.

(*a*) Current

$$I_1 = \frac{90,000}{600} = 150 \text{ amp.} \quad Ans.$$

$$I_2 = \frac{75,000}{600} = 125 \text{ amp.} \quad Ans.$$

See Fig. 177.

(*b*) Since the governors of the prime movers remain unchanged, the power output of each alternator remains the same. Also, the weakening of the field of 1 causes it to deliver *leading* current.

$$I'_1 = \frac{150}{0.80} = 187.5 \text{ amp leading.} \quad Ans.$$

See Fig. 177.

Strengthening the field of 2 causes it to deliver *lagging* current.

(*c*) As the load power factor is still unity, it takes no quadrature current (p. 95). Hence, the sum of the quadrature currents of the two alternators must be zero.

$$\cos \theta_1 = 0.80; \qquad \theta_1 = 36.9°; \qquad \sin \theta_1 = 0.60.$$

Quadrature current of 1 and 2

$$i'_1 = i'_2 = 187.5 \times 0.6 = 112.5 \text{ amp.}$$

Energy current of 2 remains unchanged at 125 amp.

$$\tan \theta_2 = \frac{112.5}{125} = 0.90; \; \theta_2 = 42.0°.$$

$$I'_2 = \frac{I_2}{\cos 42.0°} = \frac{125}{0.743} = 168 \text{ amp.} \quad Ans.$$

(d) Power factor of 2 = $\cos \theta_2 = 0.743$. *Ans.*

(e) Load current is the vector sum of I_1 and I_2 or of I'_1 and I'_2 or 275a and is in phase with V (Fig. 177).

Although the foregoing discussion of parallel operation is given for single-phase alternators, it applies equally well to polyphase alternators. As a matter of fact, with balanced loads, one phase only of the polyphase alternator need be considered and the diagrams would be identical with Figs. 174(b), 175, 176, and 177.

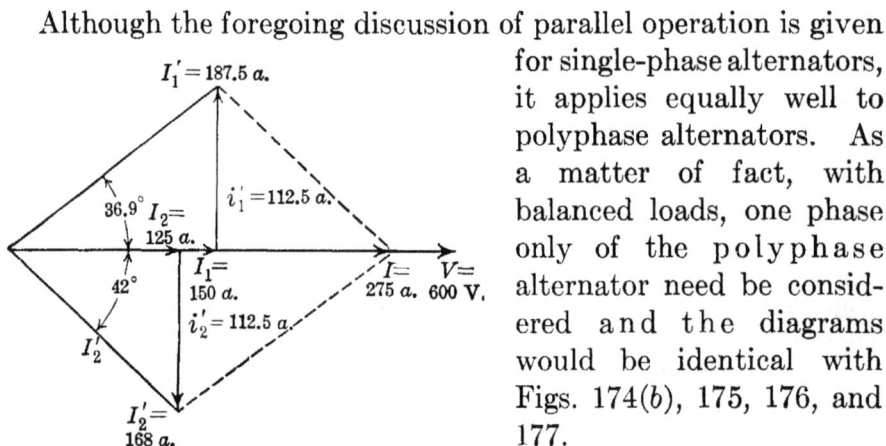

FIG. 177.—Vector diagram for alternators in parallel.

Figure 178 shows the connections for a three-phase alternator ready to be connected across the bus-bars and thus put in parallel with the other alternators of the system.

124. Synchronizing.—Before d-c generators can be put in parallel safely, two conditions must be fulfilled. The two terminal voltages must be equal, or substantially so, and the proper polarity must be observed.

The same two conditions must be fulfilled when alternators are connected in parallel. The equality of voltages can be readily determined by connecting a voltmeter first to one alternator and then to the other. The voltmeter, when so connected, does not give any indication as to polarity, as the indications of an a-c voltmeter are independent of its polarity.

Lamps, however, can be used to determine the correct polarity. Figure 178 shows the connections for phasing a three-phase alter-

nator with the bus-bars. A lamp is connected across each pole
of the three-pole switch which connects the alternator to the line.
The voltage rating of the lamps should be 15 per cent greater
than that of the alternator or line. For example, if the system is
220 volts, two 120-volt lamps in series may be used across each
pole, although these lamps will be subjected to overvoltage during
a part of the synchronizing period. If the alternator is properly

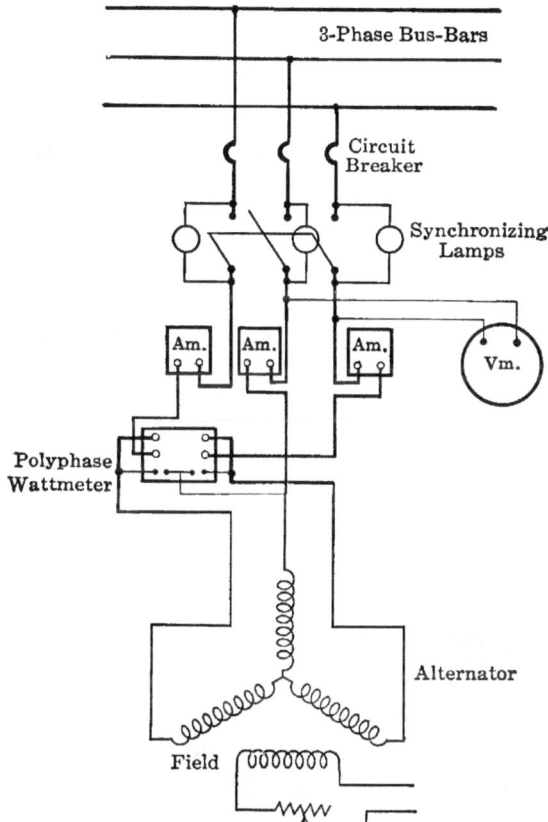

Fig. 178.—Connections for "three-dark" method of synchronizing with lamps.

connected, the three lamps should all become bright and dim
simultaneously. If they brighten and grow dim in sequence, it
means that the phase rotation of the alternator is opposite that
of the bus-bars, so that one phase must be reversed.

The lamps flicker at a frequency equal to the *difference* in the
frequency of the alternator and that of the bus-bars. As the
alternator approaches synchronism with the bus-bars, the flicker
becomes less and less rapid. When all the lamps are dark, the

switch may be closed. The fact that the lamps are dark indicates that the potential difference between each switch blade and its clip is small and that the alternator is practically in phase *opposition* to the bus-bars. Two points across which the potential difference is zero may be connected without any resulting disturbance, so that the switch now may be safely closed, thus putting the alternator in parallel with the bus-bars and thus with the other alternators of the system.

The disadvantage of this method is that lamps are dark even though a considerable voltage may exist across their terminals, and the alternator and bus-bars may be connected in parallel,

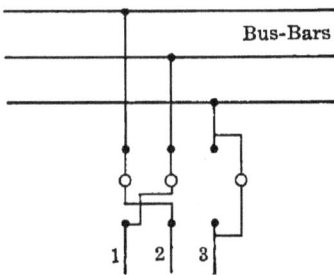

Fig. 179.—Connections for "two-bright-and-one-dark" method of synchronizing with lamps.

therefore, when considerable voltage difference exists between them. This may do no harm with slow-speed or with small units; but with high-speed turbine units, which have little armature reactance and are quite "sensitive," there may be considerable disturbance if there exists a substantial phase difference at the time of connecting in parallel. Another objection to this "three-dark" method is that the lamps do not show whether the incoming machine is running fast or slow.

The foregoing difficulties may be eliminated in part if the connections of two of the lamps, as 1 and 2 (Fig. 179), are crossed. When the alternator is in synchronism with the bus-bars, 1 and 2 are bright and 3 is dark. As one of the bright lamps is increasing and the other is decreasing in brilliancy near the point of synchronism, it is possible to determine accurately the instant at which the switch should be closed. This is called the Siemens-Halske or "two-bright-and-one-dark" method. By noting the sequence of brightness of the lamps, it can be determined whether the incoming alternator is fast or slow. Before crossing the connections, it is desirable to determine the correct phase rotation by means of the "three-dark" method (Fig. 178).

The most satisfactory method is the use of the synchronism indicator, or synchroscope, described in Chap. III (p. 114). Such an instrument shows accurately the position of synchronism. The synchroscope is connected across but one phase. Although the synchroscope may indicate synchronism for this one phase, the other two may be out of phase with the bus-bars owing to incorrect phase rotation.

Initially, the correct phase rotation must be determined by lamps or by other means before depending entirely on the synchroscope. Synchronizing lamps are often used in conjunction with a synchroscope so that the operator has a check on the instrument.

125. Hunting of Alternators.—The driving torque of a reciprocating steam engine, or gas engine, is not uniform during a revolution of the flywheel but varies from zero at the dead centers to a maximum at some intermediate position. Even with a heavy flywheel, this variation of torque may result in impulses in the induced emf, causing it to be ahead of its average position at some instants and behind it at other instants. This causes large currents to flow between alternators in parallel and often causes their rotating members to "oscillate" while rotating. The angular effect of the crank position can be appreciated when it is realized that in a 60-pole alternator a displacement of one mechanical- or space-degree in the rotating member makes a difference of 30 electrical degrees in the phase angle of the emf. The impulses are often communicated to the system, causing synchronous motors and converters to oscillate. These oscillations are called "hunting." Hunting may become serious if the engine governors have a natural frequency of oscillation which is nearly the same as that of the machine rotors. The oscillations may then become cumulative and may even cause the machines to go out of synchronism.

Remedies for hunting are to use heavy flywheels, to put dash-pots on the engine governors, and to use amortisseur or squirrel-cage windings around the field, such as is shown in Fig. 273 (p. 340). Where several engine-driven units are used, they are often paralleled when their cranks occupy different angular positions. This minimizes the effect of the engine impulses on the system, although the effect is increased in the local interchange currents between generators. Because of the uniform driving torques of steam turbines and water wheels, alternators driven by such prime movers rarely "hunt."

CHAPTER VII

THE TRANSFORMER

The static transformer is a device for transferring electrical energy from one a-c circuit to another without a change in frequency. This transference is usually, but not always, accompanied by a change of voltage. A transformer may receive energy at one voltage and deliver it at a *higher* voltage, in which case it is called a *step-up* transformer. A transformer may receive energy at one voltage and deliver it at a *lower* voltage, in which case it is called a *step-down* transformer. A transformer may receive energy at one voltage and deliver it at the *same* voltage, in which case it is called a *one-to-one* transformer.

This energy transfer is accomplished through the medium of an alternating magnetic flux. Hence, a static transformer has no rotating parts and requires little attention, and its maintenance cost is low. The cost per kilowatt for transformers is low as compared with other apparatus, and the efficiency is much higher. As there are no teeth, slots, or rotating parts and the windings can be immersed in oil, it is not difficult to insulate transformers for very high voltages.

Because of its many desirable characteristics, the transformer is a very useful piece of apparatus, and as it can transform economically from low to high voltage and from high to low voltage, it is largely responsible for the extensive use of alternating current. For example, transformers may step up the voltage to high values for transmission purposes and thus effect a saving in copper. Transformers may step down the voltage from its transmission value to values which are suitable for industrial uses, such as for operating motors and supplying lamps (see Fig. 290, p. 360). Hence, the transformer gives a high degree of flexibility in a power system.

126. Transformer Principle.—The transformer is based on the principle that energy may be efficiently transferred by magnetic induction from one set of coils to another set by means of a varying magnetic flux, provided both sets of coils are on the same magnetic circuit.

Electromotive forces are induced by a change in flux linkages. In the generator, the flux is substantially constant in magnitude. The flux linking the armature coils is changed by the *relative mechanical* motion of flux and coils (see Fig. 127, p. 153). In the transformer, the coils and magnetic circuit are all stationary with respect to one another. The emfs are induced by change in *magnitude* of flux with time. This is illustrated in Fig. 180 which shows diagrammatically the parts of a transformer.

The laminated core is made up of rectangular stampings of sheet steel, usually silicon steel, clamped or bolted together.

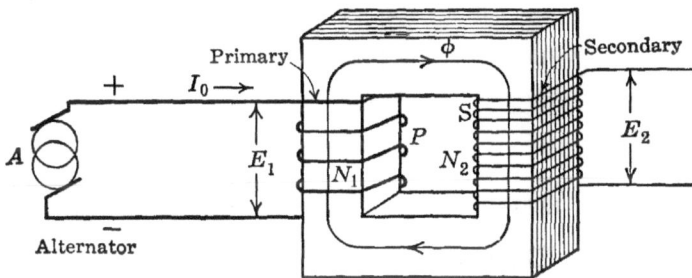

FIG. 180.—Simple transformer, secondary open-circuited.

A continuous winding P is placed on one side, or leg, of the iron core. Another continuous winding S, which may or may not have the same number of turns as P, is placed on the opposite side, or leg. (Actually, P and S would be placed close together on the same leg in order to reduce leakage flux between the windings. For example, see Figs. 191 and 192, pp. 235 and 236. In Figs. 180, 182, 183, the coils are shown on different legs for clearness.)

An alternator A supplies to the primary winding P a current I_0 which varies sinusoidally with time. Since the primary winding links the laminated steel core, its mmf must produce in the core a flux ϕ which varies sinusoidally with time (Fig. 181). This alternating flux links the turns of the secondary winding S. As this flux is alternating, it induces in the winding S an emf of the same frequency as that of the flux. Because of this induced emf, the secondary winding S is capable of *delivering* current and energy.

Therefore, the energy is transferred from P, the primary, to S, the secondary, through the medium of the magnetic flux.

The winding P which *receives* energy is the *primary*. The winding S which *delivers* energy is the *secondary*. In a transformer, either winding may be the primary, the other being the secondary, depending upon which winding receives and which delivers energy.

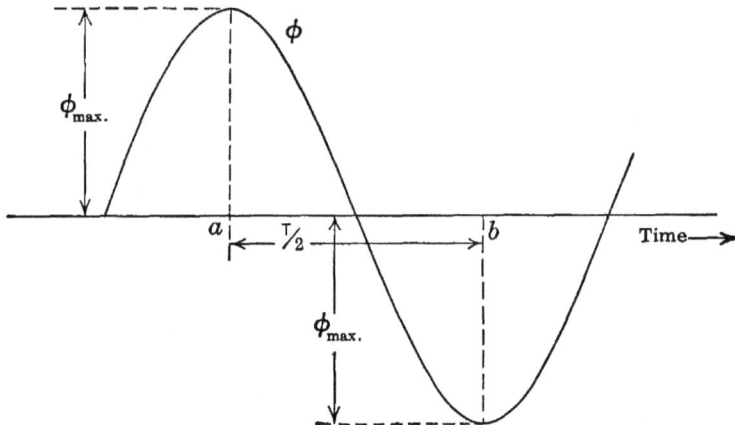

Fig. 181.—Sinusoidal variation of flux with time.

127. Induced Electromotive Force.—The induced emf in a transformer is proportional to three factors: *flux, frequency, number of turns.* The complete equation for the induced emf, assuming a sine wave of flux, is

$$E = 4.44 fN \phi_{max} 10^{-8} \text{ volts,} \tag{75}$$

where f is the frequency in cycles per second, N is the number of turns, and ϕ_{max} is the maximum value of the flux (see Fig. 181).

(Compare this equation with (65) (p. 178) which gives the induced emf in an alternator armature.)

Example.—The secondary of a 20-kva, 60-cycle transformer has 120 turns; and the flux in the core has a maximum value of 720,000 maxwells. What is the voltage E_2 induced in this secondary?

From Eq. (75),

$$E_2 = 4.44 \times 60 \times 120 \times 720,000 \times 10^{-8}$$
$$= 4.44 \times 6 \times 1.2 \times 7.2 = 230 \text{ volts. } Ans.$$

Since the same flux ϕ (Fig. 180) links each of the two windings, it must induce the *same emf per turn* in each winding, and the *total induced emf* in each winding must be proportional to the number of turns in that winding. That is,

$$\frac{E_1}{E_2} = \frac{N_1}{N_2}, \tag{76}$$

where E_1 and E_2 are the primary and secondary *induced* emfs and N_1 and N_2 are the number of turns in primary and secondary.

In the primary, however, the induced emf is a counter emf tending to prevent current entering the primary. In the usual transformer, the terminal voltages of primary and secondary differ from their induced emfs by only a small percentage, ordinarily 1 or 2 per cent, so that for most practical purposes it may be said that the primary and secondary terminal voltages are proportional to the number of turns in primary and secondary.

Example.—In the foregoing example, the primary has 1,200 turns. What is its counter emf?

$$N_1 = 1,200, \qquad N_2 = 120, \qquad E_2 = 230.$$
$$\frac{E_1}{230} = \frac{1,200}{120} = \frac{10}{1}.$$

Therefore

$$E_1 = 2,300 \text{ volts.} \quad Ans.$$

128. Exciting Ampere-turns and Flux.—Figure 180 shows a current I_0 flowing into the primary winding in the upper wire. The directions of flux, voltages and currents, as indicated on the figure, are those existing at the instant when the upper primary line is positive. There is no load on the secondary. Under these conditions the current I_0 flowing into the primary is usually 2 to 5 per cent of the rated current. This no-load current performs two functions. Flowing in the primary turns, it produces the flux in the core and is called the *exciting* or *magnetizing* current. Because of the alternating flux, there are hysteresis and eddy-current losses in the core and the no-load current must also supply this core loss. Since this loss is small, the primary on open circuit is an impedance in which the inductive reactance predominates. Hence, the no-load current lags the terminal voltage by a large angle, of the order of 84 to 87°, and the no-load power factor is of the order of 5 to 10 per cent.

The exciting current produces a flux ϕ whose direction at each instant is determined by the corkscrew rule. At the instant shown in Fig. 180, when the upper primary line is positive the flux must have a clockwise direction through the core. With the frequency and turns fixed, the induced emf in a transformer is *proportional to the flux* (Eq. (75)). Since the counter emf differs in magnitude but slightly from the primary terminal voltage, this flux *must always adjust itself to such a value as to make the counter emf substantially equal to the terminal voltage.* Even at full load and above, the induced or counter emf differs from the terminal voltage by only 1 or 2 per cent. Hence, from Eq. (75), the flux can change by this amount only. Thus, for practical purposes, the flux in a constant-potential transformer is constant under normal operating conditions.

129. Primary and Secondary Ampere-turns with Load.— In Fig. 182 is shown the transformer of Fig. 180 with a load I_2

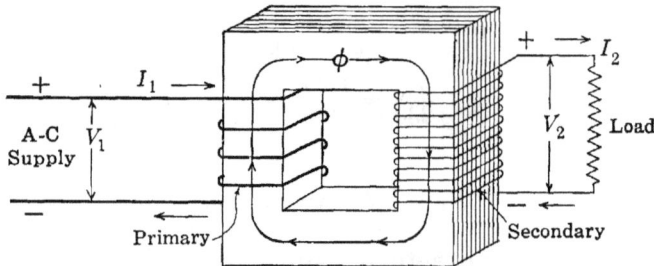

Fig. 182.—Simple transformer, load applied to secondary.

applied to the secondary. At the instant shown, the direction of the current I_2 must be such as to *oppose* the increase in flux ϕ.

This is in accordance with Lenz's law that an induced current always has such a direction as to oppose the effect which produces it (see Part I, p. 171). If the secondary current I_2 were tending to increase the flux ϕ, then by the corkscrew rule the current would flow *in* at the upper terminal (Fig. 182). Since I_2 *opposes* the increase in flux ϕ, the current must actually flow *out* at the upper terminal. The secondary current I_2 then tends to *reduce* the value of the flux in the transformer core. If the flux is reduced, the counter emf of the primary is also reduced. Just as in the d-c motor (see Part I, pp. 269 and 277), a reduction in the counter emf results in less opposition to the flow of current into the primary. Hence, the primary current increases in value and supplies the energy which is being delivered to the load by the secondary. The foregoing is the sequence of reactions which follow the application of load to the secondary and which enable the primary to take from the line the power to supply the increased power demanded by the secondary.

It has been stated already that the mutual flux ϕ remains substantially constant over the working range of the transformer, changing only by the slight amount that is necessary to permit the primary current to adjust itself to the secondary load. If the flux does not change, the *net* ampere-turns acting on the core cannot change. The effect of the secondary ampere-turns is to *reduce* the flux. The effect of the primary ampere-turns is to *increase* the flux. Since the *net* ampere-turns must remain constant, the *increase* in primary ampere-turns over the no-load ampere-turns must be equal and opposite to the secondary ampere-turns. Therefore, any primary ampere-turns in excess of the exciting ampere-turns must be balanced by equal and opposing secondary ampere-turns.

The exciting current is usually of small magnitude and differs considerably in phase from the total primary current. In most constant-potential transformers, this no-load current may be neglected. If the no-load current is neglected in comparison with the total primary current, the *primary* and *secondary* *ampere-turns* are *equal*, and

$$N_1 I_1 = N_2 I_2. \tag{77}$$

Therefore,

$$\frac{I_1}{I_2} = \frac{N_2}{N_1}. \tag{78}$$

That is, *the primary and secondary currents are inversely as the primary and secondary turns.*

The above relation also follows from the law of the conservation of energy. If the transformer losses are neglected and unity power factor is assumed,

$$V_1 I_1 = V_2 I_2.$$
$$\frac{I_1}{I_2} = \frac{V_2}{V_1} = \frac{N_2}{N_1}.$$

Example.—The secondary of the transformer in the example, p. 220, delivers 89 amp at 230 volts and unity power factor. Determine: (a) power delivered by secondary; (b) primary current.

(a) $P = 230 \times 89 = 20{,}470$ watts $= 20.47$ kw. *Ans.*

(b) The secondary ampere-turns (effective)

$$N_2 I_2 = 120 \times 89 = 10{,}680.$$
$$I_1 = \frac{10{,}680}{1{,}200} = 8.90 \text{ amp}, \quad Ans.$$

since there are 1,200 primary turns.

Also, from Eq. (78)

$$\frac{I_1}{I_2} = \frac{I_1}{89} = \frac{120}{1{,}200}.$$

Therefore,

$$I_1 = 89 \frac{120}{1{,}200} = 8.90 \text{ amp}. \quad Ans.$$

130. Leakage Reactance.—In the preceding discussion, it has been assumed that *all* the flux which links the primary links the

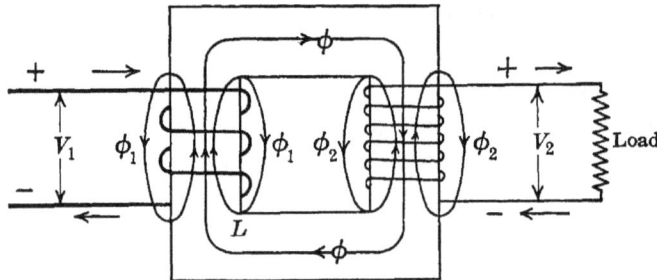

Fig. 183.—Mutual flux, primary leakage flux, and secondary leakage flux in a transformer.

secondary also. In practice, it is impossible to realize this condition. All the flux produced by the primary does not link the secondary, but a part completes its magnetic circuit by passing through the air rather than around through the core, as shown by ϕ_1 (Fig. 183). This flux ϕ_1 is called the *primary leakage flux.*

The mutual flux ϕ is produced by the ampere-turns of both primary and secondary. The primary leakage flux ϕ_1 is produced by the ampere-turns of the primary alone and is, therefore, proportional to the primary current. Since ϕ_1 does not link the secondary, it does not induce any emf in the secondary. ϕ_1 does induce an emf in the primary. This emf is a counter emf tending to prevent the flow of current in the primary. It is proportional to the current and frequency and lags the current by 90°. It is, therefore, a reactance emf and causes a reactance drop I_1X_1 in the primary. X_1 is called the *primary leakage reactance*. It is obvious that a part of the primary impressed voltage is utilized in supplying this reactance drop. This in turn reduces the counter emf, hence the flux and thus the secondary induced emf.

It is stated in Par. 129 that the mmf due to the secondary ampere-turns acts upward (Figs. 182 and 183), opposing the mutual flux ϕ. Further application of the corkscrew rule shows that these ampere-turns act to send flux downward outside the secondary. Some flux actually does complete its circuit in the local path about the secondary, as shown by ϕ_2 (Fig. 183). This flux ϕ_2 links the secondary but not the primary and is, therefore, proportional to the secondary ampere-turns and hence to the secondary current. It is called the *secondary leakage flux*.

The mutual flux ϕ induces an emf which tends to send current out of the secondary at its upper terminal. Although the phase difference usually is not 180° the general direction of flux ϕ_2 is in opposition to ϕ and, hence, tends to prevent current flowing out of the secondary at the upper terminal. Therefore, the flux ϕ_2 tends to reduce the voltage across the secondary. Since the effect of ϕ_2 is proportional to the secondary current and the frequency, this effect is also considered as a reactance voltage I_2X_2. X_2 is called the *secondary leakage reactance*.

The effect of both primary and secondary leakage reactance is to reduce the secondary terminal voltage. If it is desired that the transformer regulate closely, ϕ_1 and ϕ_2 must be reduced to as low a value as possible. Transformers constructed with primary and secondary on separate legs as shown in Figs. 182 and 183, which are diagrammatic, would have too large a leakage reactance for practical purposes. To reduce the magnitude of the leakage fluxes and hence of the leakage reactances, the primary and secondary are split up into sections and closely interwoven (see Figs. 192 and 193). Moreover, the paths of the leakage fluxes are not so simple as indicated in Fig. 183. Some of the flux links a portion of the primary turns, some a portion of the secondary turns, etc.

In transformers of very large rating, a *high* value of leakage reactance is desirable since it limits the short-circuit current.

Large transformers, having low leakage reactance, have been wrecked on short circuit by the enormous stresses between turns caused by the large short-circuit current.

131. Transformer Testing.—Theoretically, it is possible to determine the efficiency of a transformer by measuring simultaneously the output and the input with electrical instruments. The efficiency of small power transformers is 95 to 97 per cent, and the efficiency of large units is 98 to 99.4 per cent. In the larger units there is only a 1 or 2 per cent loss. Unless carefully calibrated, ordinary wattmeters may be in error by 1 per cent. An error of 1 per cent in the measurement of the power may be equal to 50 per cent of the transformer losses, or it may be equal to the total losses in the transformer. Errors greater than 1 per cent may be introduced by the use of instrument transformers (see p. 259). In fact, it is possible for the input and the output wattmeters to read so nearly alike that an almost imperceptible difference exists between their readings. Likewise, the voltage regulation of a transformer is 1 to 3.5 per cent, depending on the power factor (see Fig. 190). Again, poor precision is obtained if the regulation is determined by loading the transformer and measuring the small difference between the no-load and rated-load voltages.

It is far more accurate, in determining the efficiency, to measure the *losses* directly and then from the losses to determine the efficiency. The voltage regulation may also be obtained by measuring the primary and secondary resistances and leakage reactances and then calculating the regulation.

132. Transformer Core Losses.—Three separate losses occur in transformers: the core loss, the primary copper loss, and the secondary copper loss.

Since the magnetic flux in the core is reversed many times a second, there must be hysteresis loss (see Part I, p. 166). Because of its low hysteresis loss, silicon steel is used generally for transformer cores. *Laminating does not reduce hysteresis loss.*

The alternating flux induces emfs in the iron core as well as in the primary and secondary windings. If the core were of solid steel, the accompanying eddy currents would cause prohibitive losses. The dotted line (Fig. 184) shows the paths that the eddy currents tend to take. By laminating the core, the paths of these eddy currents are broken up by the oxide and sometimes varnish on the laminations.

To measure the core losses (hysteresis and eddy current), the open-circuit test is made. The connections are shown in Fig. 185. Rated voltage and frequency are impressed across one of the windings, preferably the winding whose voltage rating most nearly corresponds to a convenient instrument voltage, such as 110 or 220 volts. For example, if the transformer (Fig.

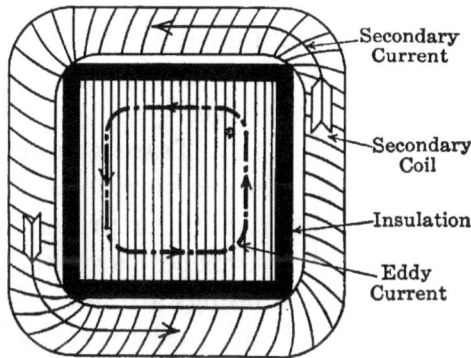

Fig. 184.—Induced current in secondary and in core.

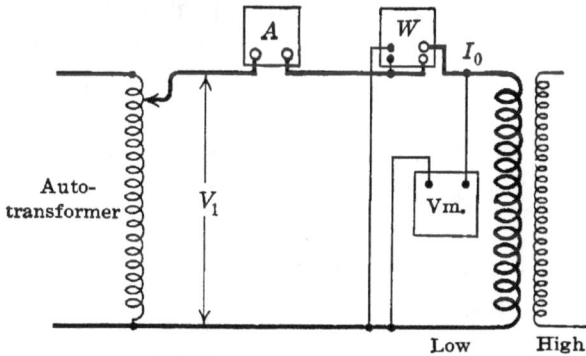

Fig. 185.—Connections for core-loss measurement.

185) were rated at 2,200/220 volts, the measurement would be made on the 220-volt, or low, side. Since the no-load losses are small, instrument losses should be investigated. For example, in Fig. 185 the wattmeter measures the power taken by the voltmeter; hence, either the voltmeter should be open-circuited when the wattmeter is read, or correction made for the power that it consumes.

When rated voltage is impressed across the winding, rated flux exists in the core, and therefore the core loss is at its normal value. Since the flux remains substantially constant at all loads, the core loss also remains substantially constant at all loads. Let the core loss measured at rated voltage be P_0.

It is convenient to be able to control the voltage impressed on the transformer winding, as, for example, by the use of an auto-transformer (Fig. 185). In fact, by varying the voltage from zero to rated value, data for a curve of core loss p_0 as a function of voltage are obtained (Fig. 186). The core loss varies nearly as the square of the voltage.

133. Transformer Copper Losses. The copper losses may be calculated if the primary and secondary resistances are known. Since the windings are not embedded in slots, their resistance with alternating current is but slightly greater than with direct current. Hence, it is sufficiently accurate in most cases to measure the primary and secondary resistance with direct current and to increase these values by about 10 per cent, in order to obtain

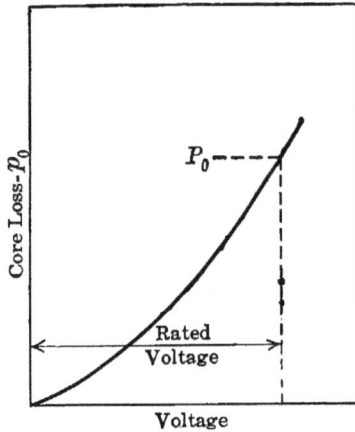

FIG. 186.—Core loss and voltage.

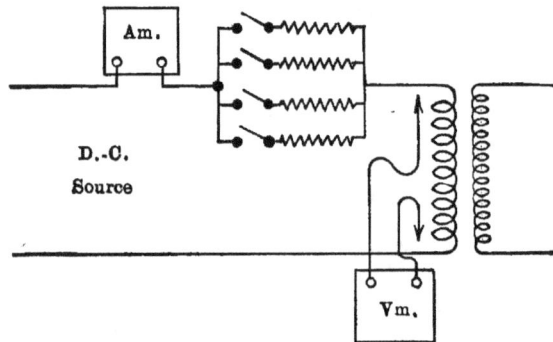

FIG. 187.—Connections for measuring direct-current resistance of transformer windings.

the *effective resistance*. The resistance may be measured by the usual drop-in-potential method (Fig. 187). It is desirable to use small values of current, as large values cause the core to be magnetized so strongly in one direction as to leave it permanently magnetized. Since the windings have high self-inductance, the *voltmeter should be disconnected before opening the circuit.* Otherwise the high emf of self-induction is almost certain to injure the voltmeter.

If R_1 is the primary effective resistance and R_2 is the secondary effective resistance, the total copper loss is

$$P_c = I_1{}^2R_1 + I_2{}^2R_2 \text{ watts.}$$

The effective resistance of the transformer, which includes that of both windings, may be determined by the short-circuit test (Fig. 188). Usually, the low-side winding is short-circuited. At short circuit, about 3 to 5 per cent rated voltage is required to cause rated current to flow in the two windings. Hence, with a

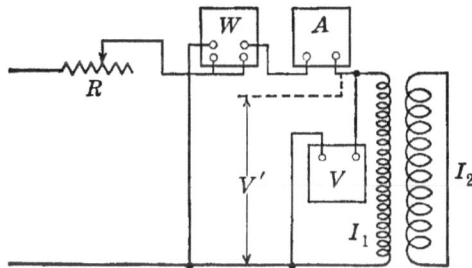

FIG. 188.—Short-circuit test.

winding rated at 2,300 volts, the required voltage is about $0.05 \times 2,300$, or 115, volts, a voltage easy to measure. A series resistance R for controlling the current is usually necessary. Assume that the measurement is made on the primary side. The primary current is I_1; and, from Eq. (78) (p. 223), a current $I_2 = (N_1/N_2)I_1$ must flow in the secondary. Let the power input corrected for instrument losses be P_c.

$$P_c = I_1{}^2R_1 + I_2{}^2R_2 = I_1{}^2R_{01} \text{ watts.}$$
$$R_{01} = \frac{P_c}{I_1{}^2} \text{ ohms.} \tag{79}$$

R_{01} is the equivalent effective resistance referred to the primary. The equivalent effective resistance referred to the secondary

$$R_{02} = R_{01}\left(\frac{N_2}{N_1}\right)^2 \text{ ohms.} \tag{80}$$

Let V' be the voltage across the primary in Fig. 188. The equivalent impedance referred to the primary

$$Z_{01} = \frac{V'}{I_1} \text{ ohms.} \tag{81}$$

The copper loss computed from R_{01} and R_{02} is that due to *effective* resistance and includes the losses associated with alternating current.

134. Transformer Efficiency.—Since the losses in the transformer may be determined experimentally, the efficiency can be calculated readily. Thus, the efficiency

$$\eta = \frac{\text{output}}{\text{output} + \text{losses}}$$

$$= \frac{V_2 I_2 \cos \theta_2}{V_2 I_2 \cos \theta_2 + I_1^2 R_1 + I_2^2 R_2 + P_0} \tag{82}$$

$$= \frac{V_2 I_2 \cos \theta_2}{V_2 I_2 \cos \theta_2 + I_2^2 R_{02} + P_0} \tag{83}$$

$$= \frac{V_2 I_2 \cos \theta_2}{V_2 I_2 \cos \theta_2 + I_1^2 R_{01} + P_0}, \tag{84}$$

where V_2 is the secondary terminal voltage, I_2 the secondary current, I_1 the primary current, $\cos \theta_2$ the secondary power factor, and P_0 the core loss.

Example.—A 100-kva, 60-cycle, 2,300/230-volt transformer is tested for its core loss by impressing 230 volts, 60 cycles, across its low side with its high side open (Fig. 185). The value of core loss is 475 watts.

The primary, or high-side, resistance is measured with direct current and found to be 0.338 ohm; and the secondary, or low-side, resistance is found to be 0.00325 ohm.

In the short-circuit test (Fig. 188), the low side is short-circuited, and measurements are made on the high side. The measurements, corrected for instrument losses are as follows: 1,650 watts; 48.0 amp; 90.0 volts. Determine: (a) equivalent effective resistance referred to primary; (b) equivalent effective resistance referred to secondary; (c) ratio of effective to ohmic resistance; (d) efficiency at rated load and unity power factor; (e) efficiency at half load and unity power factor.

(a) Using Eq. (79),

$$R_{01} = \frac{1,650}{48^2} = 0.717 \text{ ohm.} \quad \textit{Ans.}$$

(b) Using Eq. (80),

$$R_{02} = 0.717 \left(\frac{230}{2,300} \right)^2 = 0.00717 \text{ ohm.} \quad \textit{Ans.}$$

(c) With 48.0 amp in the primary and hence 480 amp in the secondary, copper loss computed from ohmic resistance,

$$P'_c = \overline{48}^2 \times 0.338 + \overline{480}^2 \times 0.00325 = 779 + 749$$
$$= 1{,}528 \text{ watts.}$$

From the short-circuit test the copper loss at 48.0 amp measured with alternating current is 1,650 watts. Hence, the ratio of effective to ohmic resistance is

$$\frac{1{,}650}{1{,}528} = 1.08. \quad Ans.$$

(d) The rated current of the secondary

$$I_2 = \frac{100{,}000}{230} = 435 \text{ amp.}$$

Using Eq. (83),

$$\eta = \frac{230 \times 435}{230 \times 435 + \overline{435}^2 \times 0.00717 + 475}$$
$$= \frac{100{,}000}{100{,}000 + 1{,}355 + 475}$$
$$= \frac{100{,}000}{101{,}830} = 0.982, \text{ or } 98.2 \text{ per cent.} \quad Ans.$$

(e) The core loss remains unchanged. Since the load current is one-half the rated value, the copper loss must be *one-fourth* the value given in (d). Hence,

$$\eta = \frac{230 \times 217.5}{230 \times 217.5 + 339 + 475}$$
$$= \frac{50{,}000}{50{,}814} = 0.984, \text{ or } 98.4 \text{ per cent.} \quad Ans.$$

Example.—Determine the efficiency of the foregoing transformer when the power factor is 0.7, lagging current: (a) at rated load; (b) at one-fourth rated load.

(a) The rating of the transformer, like that of the alternator, is determined by heating, hence by the kva and not by the kilowatt output. The rated current, therefore, is independent of power factor. The efficiency

$$\eta = \frac{230 \times 435 \times 0.7}{230 \times 435 \times 0.7 + \overline{435}^2 \times 0.00717 + 475}$$
$$= \frac{70{,}000}{70{,}000 + 1{,}355 + 475} = \frac{70{,}000}{71{,}830} = 0.975, \text{ or } 97.5 \text{ per cent.} \quad Ans.$$

It will be noted that the losses are the same as in (d) in the previous example.

(b) The secondary current becomes $435 \times 0.25 = 108.8$ amp.

$$\eta = \frac{230 \times 108.8 \times 0.7}{230 \times 108.8 \times 0.7 + \overline{108.8}^2 \times 0.00717 + 475}$$
$$= \frac{17,500}{17,500 + 84.8 + 475} = \frac{17,500}{18,060} = 0.969, \text{ or } 96.9 \text{ per cent.} \quad Ans.$$

It is to be noted that the output is one-fourth that in (a), the copper losses are one-sixteenth, and the core loss remains unchanged.

All-day Efficiency.—The maximum efficiency of a transformer occurs when core loss and copper loss are equal. If a transformer were operating most of the time near rated load, these losses ordinarily would be made equal at that load. However, most transformers, particularly distribution transformers, must be kept energized ready to deliver power 24 hr per day, although for several hours the load may be small if not negligible. Hence, the *all-day* efficiency becomes a criterion of performance; and to keep such efficiency high, the core loss which occurs 24 hr per day is made small relative to the copper loss.

Example.—Determine the all-day efficiency of the foregoing transformer under the following conditions of unity-power-factor load: 1.25 rated load, 1 hr; rated load, 7 hr; 0.5 load, 6 hr; no load, 10 hr.

The all-day efficiency is determined on the basis of energy output and input.

The rated low-side current is 435 amp, and the copper loss at rated current is 1,355 watts. At 1.25 rated current the copper loss is $(1.25)^2 \, 1,355 = 2,120$ watts; at 0.5 load, $(0.5)^2 \, 1,355 = 339$ watts. The core loss is 475 watts.

Twenty-four-hr output

$$W = (100,000 \times 1.25 \times 1) + (100,000 \times 7) + (50,000 \times 6)$$
$$= 1,125,000 \text{ watt-hr.}$$

All-day efficiency

$$\eta = \frac{1,125,000}{1,125,000 + (2,120 \times 1.00) + (1,355 \times 7) + (339 \times 6) + (475 \times 24)}$$
$$= \frac{1,125,000}{1,125,000 + 25,000} = 0.978, \text{ or } 97.8 \text{ per cent.} \quad Ans.$$

135. Transformer Regulation.—By means of the equivalent impedance Z_{01} (Eq. (81)) and the equivalent resistance R_{01} (Eq. (79)) determined from the short-circuit test, it is possible to determine the regulation as with the alternator. The secondary as well as the primary impedance is included in Z_{01}. The equivalent reactance referred to the primary

$$X_{01} = \sqrt{Z_{01}^2 - R_{01}^2}. \quad (85)$$

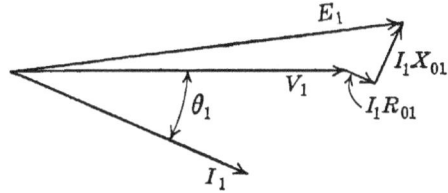

The vector diagram for lagging current is shown in Fig. 189 (compare with Fig. 162, p. 191). The primary terminal volts

FIG. 189.—Transformer vector diagram, lagging current.

$$E_1 = \sqrt{(V_1 \cos \theta_1 + I_1 R_{01})^2 + (V_1 \sin \theta_1 + I_1 X_{01})^2}. \quad (86)$$

(Compare with Eq. (69), p. 192.)

For unity power factor, $\cos \theta_1 = 1$ and $\sin \theta_1 = 0$.

With leading current change $+I_1 X_{01}$ to $-I_1 X_{01}$. (See Eqs. (68) and (70), pp. 190 and 193.)

$$\text{Regulation} = \frac{E_1 - V_1}{V_1}. \quad (87)$$

The same value of regulation is obtained using V_2, I_2, θ_2, R_{02} (Eq. (80)), X_{02}, where

$$X_{02} = X_{01} \left(\frac{N_2}{N_1}\right)^2. \quad (88)$$

Example.—Determine the regulation of the 100-kva transformer (Par. 134) at: (a) *unity* power factor; (b) 0.8 power factor, lagging current; (c) 0.8 power factor, leading current.

From the data of the short-circuit test, using Eq. (81),

$$Z_{01} = \frac{90.0}{48.0} = 1.875 \text{ ohms.}$$

From Eq. (85),

$$X_{01} = \sqrt{(1.875)^2 - (0.717)^2} = 1.733 \text{ ohms.}$$

The regulation will be computed by using secondary values.

$R_{02} = 0.00717$ ohm (p. 230).

$X_{02} = \dfrac{1.733}{100} = 0.01733$ ohm (Eq. (88)).

$I_2 R_{02} = 435 \times 0.00717 = 3.12$ volts; $I_2 X_{02} = 435 \times 0.01733 = 7.53$ volts

(a) $E_2 = \sqrt{(230 + 3.12)^2 + (7.53)^2} = 233$ volts.

$$\text{Regulation} = \frac{233 - 230}{230} = 0.0130, \text{ or } 1.30 \text{ per cent.} \quad \textit{Ans.}$$

$$\cos \theta = 0.80; \ \theta = 36.9°; \ \sin \theta = 0.60.$$

From Eq. (86),

(b) $E_2 = \sqrt{(230 \times 0.8 + 3.12)^2 + (230 \times 0.60 + 7.53)^2}$
    $= \sqrt{(35,000) + (21,100)} = 237$ volts.

$$\text{Regulation} = \frac{237 - 230}{230} = 0.0304, \text{ or } 3.04 \text{ per cent.} \quad \textit{Ans.}$$

(c) $E_2 = \sqrt{(230 \times 0.8 + 3.12)^2 + (230 \times 0.60 - 7.53)^2}$
    $= \sqrt{35,000 + 17,030} = 228.1$ volts

$$\text{Regulation} = \frac{228.1 - 230}{230} = -0.0083, \text{ or } -0.83 \text{ per cent.} \quad \textit{Ans.}$$

The efficiency of the transformer is high even at light load and low power factor, as is shown by the foregoing examples. Figure

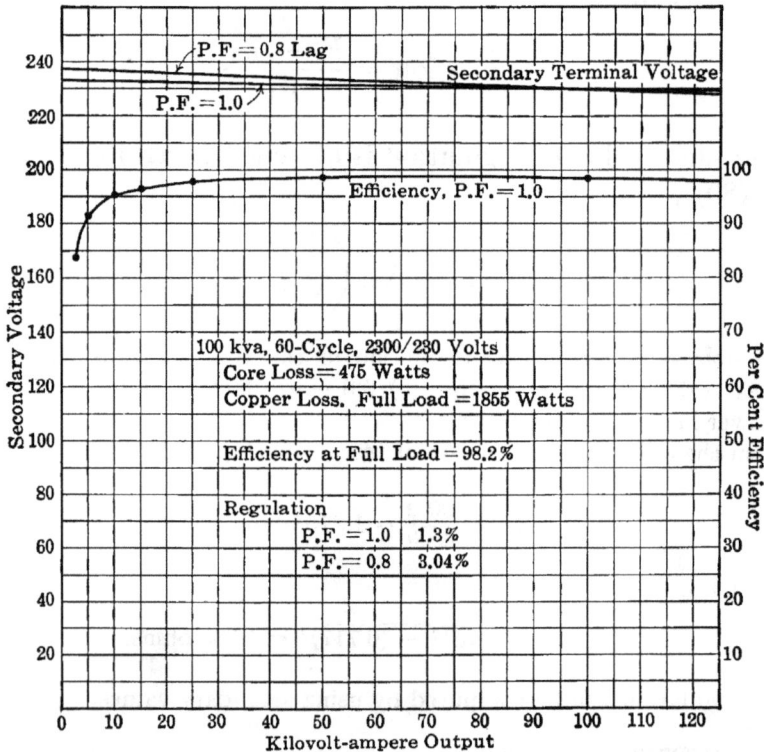

Fig. 190.—Characteristics of a 100-kva, 2,300/230-volt, 60-cycle transformer.

190 gives the efficiency curve of this 100-kva transformer at unity power factor. It is to be noted that the efficiency is high

from 10 per cent rated load to 25 per cent overload and that the curve is nearly flat from 15 per cent rated load to 25 per cent overload.

In Fig. 189, the values of the voltage drops I_1R_{01} and I_1X_{01} are much larger in proportion to V_1 and E_1 than occurs with transformers. It is necessary to exaggerate these drops in order to make the diagram clear.

TRANSFORMER CONSTRUCTION

136. Core- and Shell-type Transformers.—Transformers are divided into two general types, the core type and the shell type. These two types differ in the arrangement of the iron and copper with respect to each other.

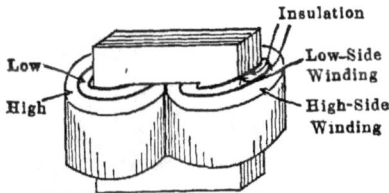

FIG. 191(a).—Arrangement of coils and core in core-type transformer.

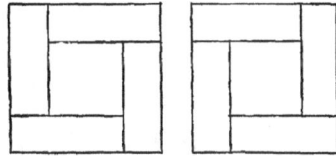

FIG. 191(b).—Arrangement of joints in adjacent lamination layers.

In the core type of transformer the winding, or copper, surrounds the iron core. Although Figs. 180, 182, 183 are diagrammatic merely, they represent core-type transformers. Figure 191(a) shows the general arrangement of the core-type transformer. The core is in the form of a hollow square made up of sheet-steel laminations about 14 mils thick. These laminations are usually built up of rectangular strips, with butt-joints in the individual layers. The joints lap in alternate layers, however, as indicated by Fig. 191(b), which shows the arrangement of joints in two adjacent layers. When a large number of transformers of a single type are being manufactured, the laminations are often made of L-shaped stampings stacked so that the joints alternate.

If a transformer were constructed with the primary and secondary coils on separate legs, as indicated in Figs. 180, 182, 183, an unsatisfactory transformer would result, as the large leakage flux for both primary and secondary would result in poor regulation. By having both a primary and a secondary winding on each leg, as shown in Fig. 191(a), the leakage flux is reduced to a very small value.   If the high-voltage winding were placed next the core, it would be necessary to insulate it from both the core and the low-voltage winding.   Thus, two layers of high-voltage insulation would be necessary.   By placing the high-voltage winding outside, and around the low-voltage winding,

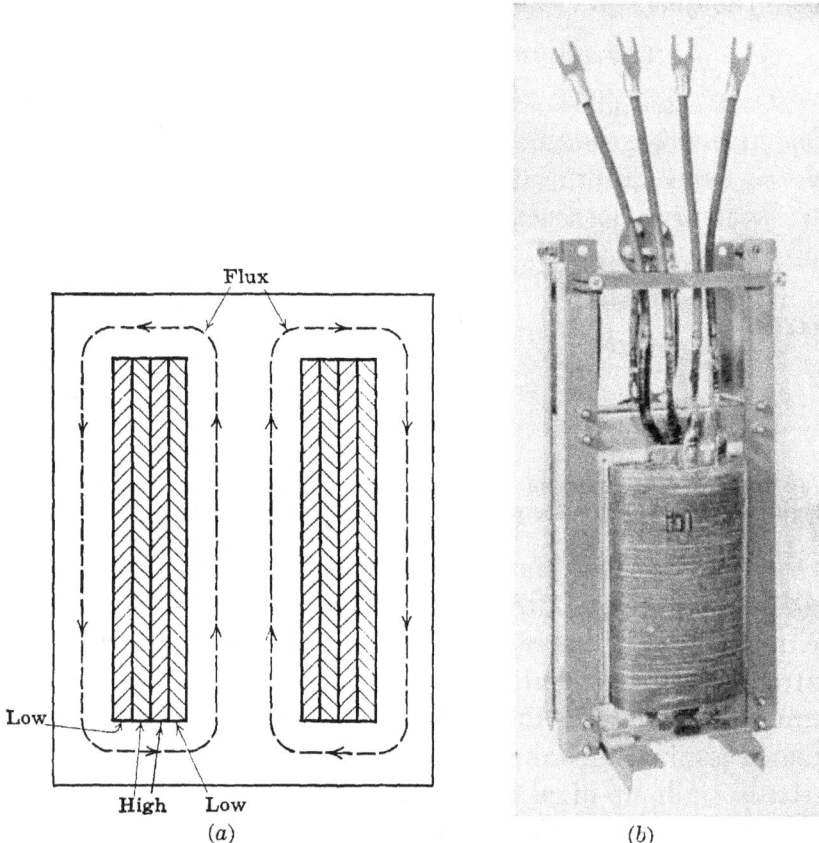

FIG. 192.—Core and windings of a 150-kva, 2,400 to 120/240-volt, 60-cycle shell-type transformer.   (*Courtesy of Wagner Electric Corporation.*)

only the one layer of high-voltage insulation, that between the high- and low-voltage windings, is necessary.

The core type of transformer is well adapted to high voltages, especially in the smaller ratings, because the insulation problem is not difficult.

In the shell type of transformer, the iron surrounds the copper (Fig. 192).   The core has the form of a figure 8.   The entire flux passes through the central part of the core, but outside this central core it divides, half going in each direction as shown in Fig. 192(a). The coils may be pancake form, and are frequently wound with strip copper. The low-side coils are adjacent to the iron, and the high-side coils are between them. This construction minimizes the insulation between the coils and core, and the leakage flux between primary and secondary is minimized. In Fig. 192(a) and (b), cylindrical coils are used. The low-side coils are adjacent to the iron, and the high-side coils are between them. In (b) the coils and core are removed from the case.

137. Type-H Transformer.—In designing a transformer, it is desirable that the mean length of turn shall be as short as possible. This reduces both the weight of copper and the resistance and reactance of the winding. This is accomplished in the type-H transformer of the General Electric Co. by making a shell-type transformer in which the core is cruciform in shape, as shown in Fig. 193. The central core around which the coils are wound is operated at much higher flux density than the four wings. Although the reluctance and losses in this central core are high, the total core loss is not excessive when the entire magnetic circuit is considered. These transformers are used mostly as distribution

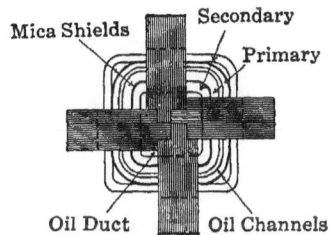

FIG. 193.—Core and windings of type-H transformer.

transformers for stepping down from 2,400 and 1,200 volts to 240 and 120 volts, so that the primary is the high side. Note that the low side, the secondary, is next the iron, one of the two low-side coils being next the central core, and the other being next the iron of the four wings. The two high-side coils lie between the two low-side coils and are not adjacent to the iron. The advantage of this design is that only moderate insulation is required between the low-side coils and the core. As high-voltage insulation, such as mica shields, need be used only between high- and low-voltage coils, a minimum amount of high-voltage insulation is required.

In designing a transformer, provision should be made for keeping it cool. Spaces, or ducts, should be left between coils and between coils and core. Such ducts, or channels, are shown in Fig. 193. The oil in these ducts becomes heated, its specific gravity decreases, and the oil rises. When it comes in contact with the transformer case, it cools, wi.ich in:reases its specific gravity; and it therefore flows downward outside the transformer coils and is subjected to further cooling.   There is a continuous circulation of oil upward through the coils and through the ducts in the core, and this carries away the heat.

The type-H transformer is now made with a wound core (Par. 138). The high-voltage-side view of the assembled core and coils of such a transformer removed from the case is shown in Fig. 194.

138. Wound-core Transformer.—A wound-core design of transformer has been developed by the General Electric Co. In most prior designs the cores are made up of flat laminations stacked and clamped together to form the core, as already described. In the wound-core design the core is formed by winding on a mandrel a long strip of magnetic steel into a tight spiral forming a heavy-walled hollow cylinder. The spiral is wound by machine so that it links the elliptical-shape coils (Fig. 195). This design has several advantages. The direction of the flux is with the grain of the iron which greatly reduces the iron losses. There is but one effective gap in the magnetic circuit, that is, from spiral to spiral, and such a gap is of very short length and of large cross-sectional area. As the cores are wound rapidly by machinery, there is considerable saving of cost in assembling; also there is practically no waste of material.

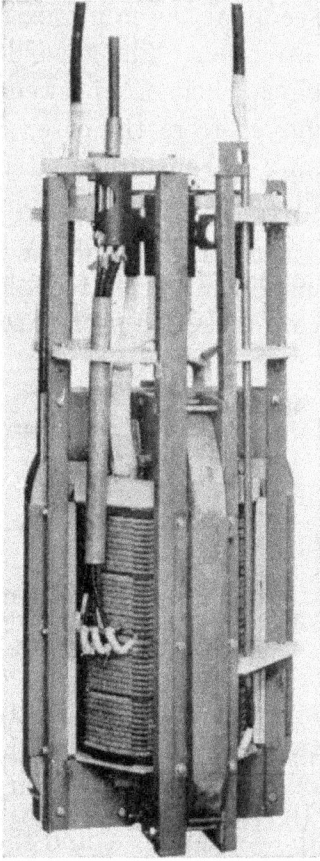

Fig. 194.—Assembled core and coils for single-phase Spirakore transformer, type HS, 333 kva, 60 cycles, 13,200 to 240/480 volts. (*Courtesy of General Electric Company.*)

139. Cooling of Transformers.—All the energy lost in a transformer must be dissipated as heat.   Although this energy is but a small proportion of the total energy undergoing transformation, it becomes quite large in amount in transformers of larger rating. The larger the transformer, the more difficult it becomes to dissipate the heat, for the kilowatt rating of the transformer increases much faster than the area of the heat-dissipating surface.

In general, transformers are cooled by: (*a*) natural circulation of air; (*b*) forced circulation of air; (*c*) natural circulation of oil and air; (*d*) natural circulation of oil and forced circulation of air; (*e*) natural circulation of oil and artificial cooling of oil; (*f*) forced circulation of oil.

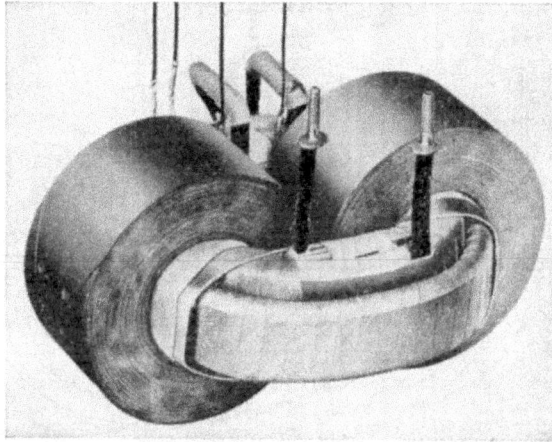

FIG. 195.—Assembled core and coils, without clamps, wound-core distribution transformer, 60 cycles, 1.5 kva, 2,400/416 (*y*) to 120/240 volts. (*Courtesy of General Electric Company.*)

Transformers cooled as described under (*a*) are usually small transformers (also instrument transformers) which are installed in places, such as on the walls or floors of factories, where the use of oil is considered a fire hazard. (To use oil-filled transformers, special fireproof vaults are usually required.)

Transformers cooled as described under (*b*) are for the most part power transformers in substations which are located in thickly settled districts where oil is considered a fire hazard. Common practice is to locate such transformers over an air chamber (Fig. 196) in which the air is maintained under pressure by blowers. The air passes up through the core and windings and is discharged through the top of the case. Since air-cooled transformers lack the dielectric strength which oil adds to the insulation, they are seldom used at voltages exceeding 20,000 to 30,000 volts.

Transformers cooled as described under (c) depend on the natural circulation of the oil through the cooling ducts in the windings and core to carry the heat to the heat-dissipating surfaces where it is carried away by the natural cooling action of the air. The oil in passing through the ducts becomes heated; and,

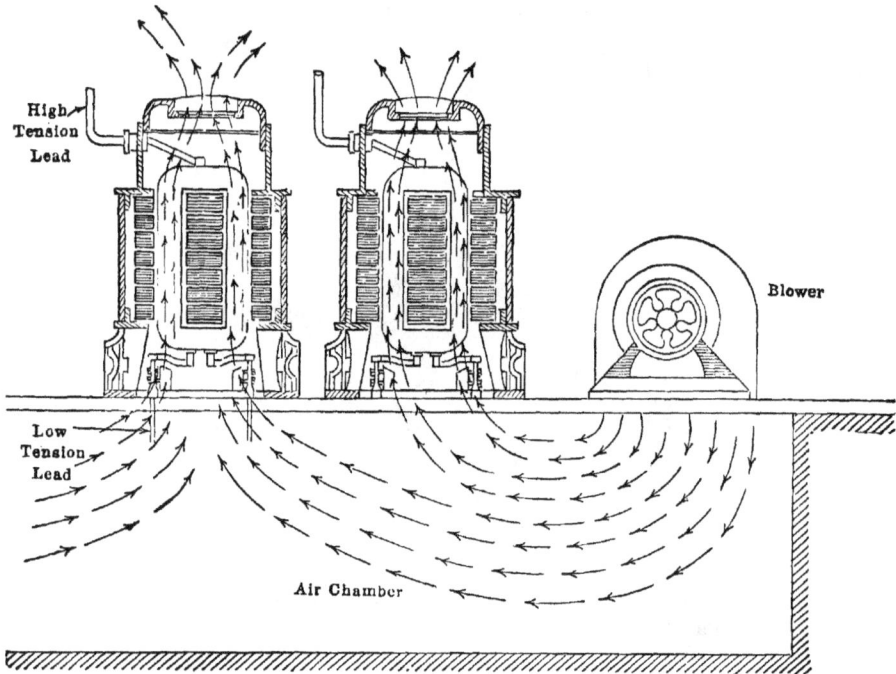

FIG. 196.—Cooling of air-blast transformers.

because of the resulting decrease in its density, it rises and ultimately comes in contact with the upper part of the transformer case and any other heat-dissipating devices which may be used, such as tubes and radiators. On cooling, the oil descends along the cooling surfaces to the bottom of the tank from which it rises again, passing up through the windings.

The General Electric Co. has developed a synthetic dielectric liquid, Pyranol, consisting of chlorinated hydrocarbons, which can also be used as a cooling medium for transformers. It is noninflammable and nonexplosive so that Pyranol-cooled transformers can be located indoors without a fireproof vault being required. Pyranol costs somewhat more than oil. The network transformer (Fig. 301, p. 374) is insulated and cooled with Pyranol.

Since the power rating and the heat losses of transformers increase much more rapidly than the normal surface area, it becomes necessary with the larger units to increase the heat-

Fig. 197.—Single-phase transformer, 60 cycles, 5,000 kva, 112,000 gr (*y*) to 22,000 volts. (*Courtesy of General Electric Company.*)

dissipating surface beyond the normal area of the case. With transformers of moderate rating, the sides of the case are fluted; with larger ratings, return tubes are welded between the top and bottom of the case (Fig. 197). The heated oil is pushed from the top of the tank into the tubes, where it is cooled and descends to the bottom of the tank.

Transformers having this tubular construction are limited in size by considerations of transportation, such as their total weight

and also the clearances of the railroads. The tubular principle
can be utilized, however, by bolting radiators to the casing to
take the place of the tubes (Fig. 198). (See also Fig. 2, p. 3.)

As these radiators are held by bolts, they may be removed
during shipment and bolted in place when the transformer is
installed.

Fig. 198.—40,000-kva, three-phase transformer with regulator providing
both phase angle and a voltage control, for a 69-kv circuit. (*Courtesy of Westinghouse Electric and Manufacturing Company.*)

The natural convection of oil and forced circulation of air (d)
are used most frequently with the radiator-type transformer.
Air is blown through the upper halves of the radiators,
where the temperature is highest, by means of motor-driven
blowers attached to the transformer tank (Fig. 198). The
blowers usually operate automatically, starting only when the
temperature of the transformer has reached a predetermined

value, so that the blowers operate only at the periods of heavier loads.

The transformer (Fig. 198) is also provided with taps in the secondary winding which are arranged so that change from one tap to another can be made when the transformer is under load.

FIG. 199.—Water-cooled, oil-immersed 2,000-kva, 60-cycle, 76,210 to 11,000 volts, single-phase transformer. (*Courtesy of Westinghouse Electric and Manufacturing Company.*)

This permits voltage control. Also the phase of the secondary voltage can be shifted by the introduction of small voltages from auxiliary three-phase windings. This phase control is necessary when the transformer is to be used to tie two systems together. It then becomes necessary to phase the secondary voltages with those of the second system before the connection can be made.

These tap and phase-shifting devices are contained in the steel compartments on the front of the tranformer.

The natural circulation of oil and its artificial cooling (e) are usually accomplished by installing a copper coil in the top of the tank and causing cooling water to circulate through the coil. The coil is naturally in contact with the hottest oil. Careful tests for leaks in the cooling coil are necessary at regular intervals, for even a very small amount of water in the oil greatly impairs its insulating properties. Also, precautions are necessary if the transformer is exposed to freezing temperatures.

Figure 199 shows the exterior view of a 2,000-kva, water-cooled, oil-immersed, single-phase transformer. The piping for the circulating water and the absence of exterior cooling devices, such as radiators, should be noted.

Forced circulation of oil (f) consists in circulating the oil continuously through the transformer and an outside system where it is cooled. This method is rarely used in the United States.

140. Breathing of Transformers.—With large transformers, a layer of gas in contact with the oil is desirable, because of its cushioning effect in case of explosion.

When transformers become warm, the oil and gas expand. The gas at the top of the oil is expelled. When the transformer cools, air is drawn into the transformer. Hence, the transformer "breathes." Unless preventive measures are taken, moisture is drawn in during this process; this moisture is readily absorbed by the oil, and the dielectric properties of the oil are reduced. Moreover, the oxygen in contact with the oil oxidizes it, forming a thick "sludge," which adheres to the windings, clogging the oil ducts and frequently resulting in burnouts. Also, oxygen in contact with the oil gives opportunity for fires and explosions in case of internal flashovers.

The Westinghouse Co. eliminates the deteriorating effects of oxygen and moisture by causing the ingoing and outgoing air to pass through an Inertaire *breather*, mounted on the case. Absorbent materials in the breather remove the oxygen and the moisture from the ingoing air, leaving essentially only dry nitrogen which is chemically inert to the oil.

Another method developed by the General Electric Co. for preventing the deterioration of the oil is to mount an expansion tank, or *conservator*, on top of the transformer case. This tank is connected by pipe to the transformer and is always partly filled with oil. It absorbs the expansion and contraction of the oil so that the main transformer is always full and the surface of the oil is not exposed to oxygen. The conservator is provided with a breather containing a drying agent, such as calcium chloride.

141. Three-phase Transformers.—Three-phase transformers have considerably less weight and occupy much less floor space than three single-phase transformers of the same total rating.

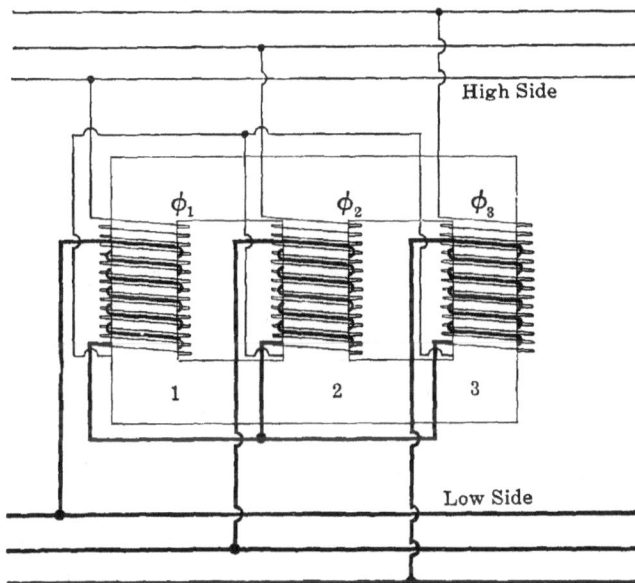

Fig. 200.—Arrangement of windings on a three-phase, core-type transformer connected Y-Y.

Their use is advantageous, therefore, in many instances. Figure 200 shows the construction of the core type of three-phase transformer. The primary and secondary of each phase are wound over one of the three legs of the transformer. Each of the three legs may act in turn as return path for the fluxes from the other two. For example, the three fluxes φ_1, φ_2, φ_3 differ in phase by 120°, as shown in Fig. 201. That is, φ_2 reaches its maximum instantaneous value acting upward, 120° after φ_1, etc. Figure 201 shows that at any instant the sum of the three fluxes is zero. When $t = 0$, $\varphi_1 = 0$, $\varphi_2 = -0.866\ \Phi_{max}$ and therefore acts downward, and $\varphi_3 = +0.866\ \Phi_{max}$ and therefore acts upward. At this instant, the total flux acts upward through 3 and downward through 2, there being no flux through 1. When $\omega t = 30°$, φ_1 and φ_3 are each equal to $\Phi_{max}/2$ and act upward. φ_2 is negative maximum and acts downward. At this instant, therefore, the two fluxes in 1 and 3 act upward and combine to flow downward through 2. Thus, one or more of the legs of the three phases will always act as return path for the fluxes of the others.

As it has a single tank, the three-phase transformer costs less and occupies less floor space than three single-phase transformers

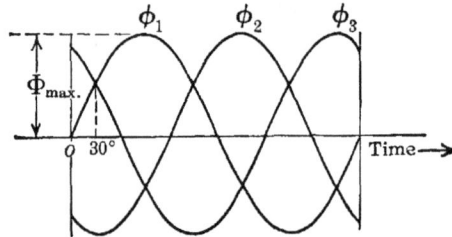

FIG. 201.—Time variation of fluxes in three-phase transformer.

of the same total rating. These advantages are often counterbalanced by the fact that if any one phase becomes disabled the whole transformer ordinarily must be removed from service. If one transformer of a three-phase bank of single-phase transformers becomes disabled, the transformer may be replaced by a single spare which can be readily substituted. The radiator-type transformer (Fig. 198) is a three-phase transformer.

142. Autotransformers, or Compensators.—It is possible to transform a-c energy with a transformer which does not have a separate primary and secondary. Such a transformer, having a 2-to-1 voltage ratio, is shown diagrammatically in Fig. 202. The winding ac may consist of a single winding on one leg of a transformer and having a tap at its mid-point b; or windings ab and bc may each consist of several coils on the same leg or on separate legs, the coils being interspaced to reduce magnetic leakage.

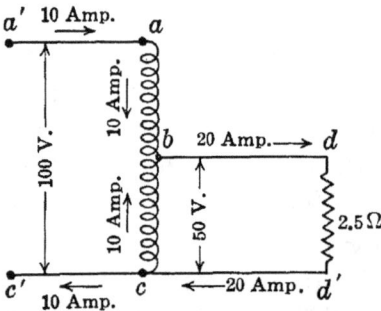

FIG. 202.—Autotransformer giving 2-to-1 voltage ratio.

These coils are connected in series in such a manner that they all aid or tend to send flux in the same direction around the core.   Assume that a voltage of 100 volts is impressed across *ac*.   Since *b* is at the mid-point of the winding, the voltages from *a* to *b* and from *b* to *c* must each be 50 volts.   If a noninductive load of 2.5 ohms is connected across *bc* at *dd'*, the current in *bd* and *d'c* will be 20 amp and the power will be 1,000 watts.   The losses in this type of transformer are small and in this discussion are neglected.   Also, the magnetizing and no-load energy currents as well as the leakage reactances are small and usually may be neglected.   If the losses are neglected, the input to the transformer is 1,000 watts at 100 volts.   The current $I_{a'a}$ entering the winding *ab*, therefore, must be 1,000/100, or 10, amp.   This current is shown flowing downward from *a* to *b* at the instant under consideration.   Since the current flowing from *b* to *d* through *dd'* is 20 amp, then by Kirchhoff's first law the current flowing from *c* to *b* through the winding *cb* must be 10 amp flowing *upward*.   This current combines with the 10 amp flowing *downward* in *ab* and gives the 20 amp required by the load.

This 10 amp flowing upward from *c* to *b* is a *transformed*, or secondary, current.   The 10 amp flowing from *a* to *b* drops 50 volts in potential, representing 500 watts.   By the law of the conservation of energy, this power must either be dissipated or appear elsewhere.   Actually, it is transferred to the flux in the core. This flux, by transformer action, raises the potential of the current flowing from *c* to *b* by 50 volts, since losses are neglected.   Therefore, only 500 watts of the 1,000 watts delivered to the load *dd'* are *transformed*.   The remaining 500

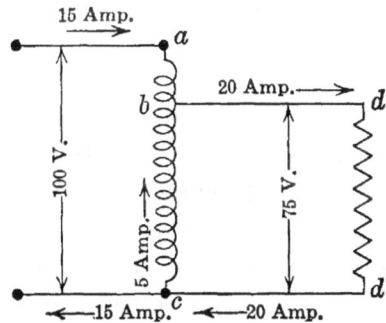

Fig. 203.—Autotransformer giving 4-to-3 voltage transformation.

watts flow *conductively* to *dd'*, in virtue of the current of 10 amp flowing through *ab*, *bd*, and thence to *dd'*.   So far as power is concerned, *ab* may be considered as the primary and *bc* the secondary of the transformer.   It is obvious that the transformer may be reversed; that is, power may be supplied to *bc* at 50 volts and delivered to the line *a'c'* at 100 volts.

Figure 203 shows an autotransformer which transforms from 100 to 75 volts.   The load *dd'* takes 20 amp at 75 volts, or 1,500 watts.   Neglecting losses, the input to the system must therefore be 1,500 watts.   At 100 volts the current input must be 15 amp, as shown.   This current of 15 amp drops 25 volts in potential in flowing from a to b; hence, 25×15, or 375, watts are transferred to the magnetic field.   This power is utilized in raising the potential of 5 amp by 75 volts (375 watts) in winding bc.   Hence, of the 1,500 watts, 375 watts are transformed and 1,125 watts flow conductively to the load through ab and bd.

In the foregoing examples, two facts should be noted.   All the power is not *transformed*, but a certain part flows *conductively* to the load.   Hence, a smaller transformer can be used than would be necessary were all the power transformed as in the usual transformer.   The smaller the ratio of transformation, the smaller the proportion of power *transformed*.   Thus, with a 4-to-3 ratio, only one-fourth the total power is transformed; with a 2-to-1 ratio, one-half the total power is transformed.   Hence, the autotransformer becomes more economical than the usual transformer when the ratio of transformation is small.   With large ratios of transformation there is practically no economy over the usual transformer.   Moreover, since the primary and secondary are connected conductively, then with high potential on one side there is always danger of high potentials to ground on the low side unless special precautions are taken.

143. Industrial Uses of Autotransformers.—It is sometimes necessary to transmit power at voltages slightly greater than it is feasible to generate.   For example, it may be desirable to transmit power at 26,400 volts, particularly if it is necessary to use underground cables through cities.   It is difficult to insulate generators for voltages exceeding 15,000 volts.   However, it is possible to generate at 13,200 volts and transform economically to 26,400 volts by means of autotransformers.   Figure 204 shows the connections which would be employed if a three-phase, Y-connected autotransformer were used for this purpose.   The transformer and generator under these conditions operate as a unit.

Power from Boulder Dam is transmitted to Los Angeles, a distance of 270 miles, at 275,000 volts over a three-phase, 60-cycle transmission line.   At Los Angeles, this voltage is reduced to

132,000 volts, three phase, the voltage of the local system, by means of three-phase, Y-connected autotransformers.

By means of an autotransformer, it is possible to obtain readily and economically a three-wire system from a two-wire supply. Figure 205 shows a 230-volt a-c supply from which it is desired to obtain 115-volt service for lighting. An autotransformer is connected directly across the 230-volt wires, and a wire is carried from its mid-point (Fig. 205), constituting the *neutral*

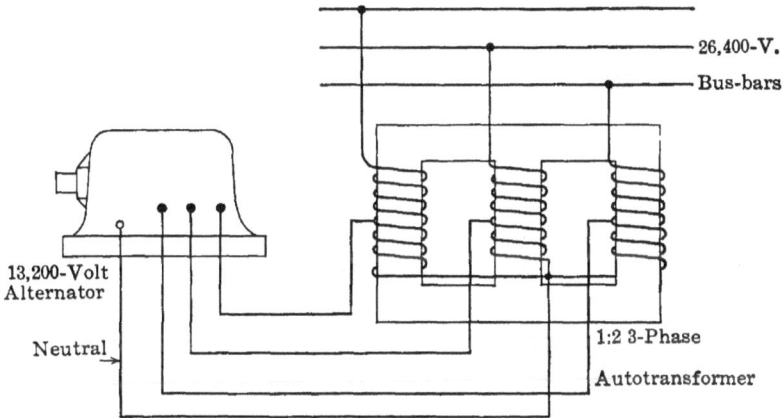

Fig. 204.—Alternator and three-phase autotransformer operating as a unit.

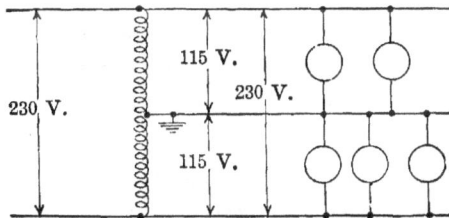

Fig. 205.—Compensator used to obtain three-wire lighting system.

conductor. Obviously the voltage from neutral to either wire is 115 volts, and 115-volt lamps can be connected between either outer wire and the neutral. This gives a 230-to-115-volt, three-wire system (*not* three phase). (For further discussion of the three-wire system, see p. 361.) An autotransformer used in this manner is called a *balance coil*.

Autotransformers are also used extensively as a-c motor starters (see p. 288).

144. Phasing Transformer Windings.—The primaries and secondaries of transformers usually consist of two or more coils

rather than of a single coil. By connecting these coils either
in series or in parallel, several voltage and current ratings may be
obtained with the same transformer. The connections must be
made, however, so that correct phase relations among coil voltages are obtained. In many standard types of transformer the
coil terminals which are brought to a terminal board, or lead
support, are arranged in such a manner that the correct phase
relations are obtained if the well-known conventions are followed
in making the connections (Fig. 207). The wiring diagrams
which usually accompany the transformer may be lost, the
internal connections are often changed during repairs, and it may
be necessary to carry leads from the individual coil terminals
through conduit so that the terminal designations cannot be
readily traced. In such cases, it is necessary to "phase" the
windings.

Figure 206 shows a 2,300 to 1,150/230 to 115-volt distributing
transformer which is the most common type of power transformer. The low-side winding consists of two 115-volt coils ab
and cd. The high-side winding consists of two 1,150-volt coils
$a'b'$ and $c'd'$. Let it first be desired to connect the transformer so
that the low side has a 230-volt rating and the high side a 2,300-
volt rating. Since it is desirable from the point of view of safety
and convenience to use low voltage in testing, a 230-volt source,
if available, is used. The low-side windings are first phased.

If coils ab and cd were connected in series across 230 volts and
happened to act in opposition on their magnetic circuit, a virtual
short circuit would result, the current being limited only by the
small resistance and leakage reactance of these coils. To prevent
short circuit, a resistance, or an impedance, R is inserted in series
with the coils to limit the current (Fig. 206(a)). An ammeter A
may be inserted in series, or a voltmeter V may be connected
across the two coils. If coils ab and bc are in opposition, the
ammeter current will be large since the transformer coils offer
but little impedance, the current being limited almost entirely
by R. The voltmeter will simultaneously read low, since the
impedance drop across ad is small. On the other hand, if the
ammeter reading is small or if the voltmeter reads practically
230 volts, the coils are aiding. (Test lamps may be used in
place of the voltmeter.) If the coils are in opposition, the con-

nections of one should be reversed. On the other hand, if it is desired to operate the coils *ab* and *cd* in parallel at 115 volts, points *a* and *d* may be connected together if the test shows that the coils are in opposition. Also, the coils *ab* and *cd* may be connected in parallel and then put in series with resistance *R* across a 115-volt supply. If the ammeter reading is high and the voltmeter reading is low, the coils are incorrectly connected and one should be reversed.

The foregoing procedure may be pursued safely with the high-side coils, by utilizing the 230-volt supply.

Another method of phasing primaries is shown in Fig. 206(*b*). If a 115-volt supply is available, it may be connected across one

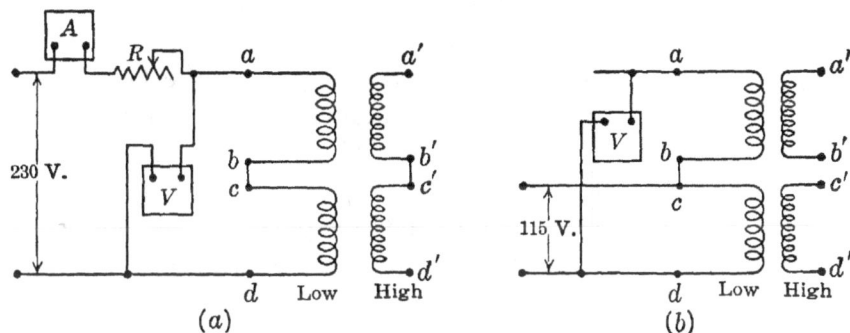

Fig. 206.—Methods of phasing transformer coils.

low-side coil, as *cd*, without the possibility of short circuit. Terminal *b* of coil *ab* is connected to terminal *c*, of coil *cd*, and a voltmeter is connected across the two outer terminals *a* and *d*. If the voltmeter reads approximately 230 volts, the connection is correct for series operation. If it reads zero, points *a* and *d* may be connected together and the coils thus operated in parallel at 115 volts.

The high side may also be phased by tests made on the low side if the connections in Fig. 206(*a*) are used. *With the low side entirely disconnected from the line,* connect high-side coils *a'b'* and *c'd'* in series at *b'c'*. With the resistance or impedance *R* in series with the low side, connect *a'* to *d'*. Then apply voltage to the low-side circuit. If the coils *a'b'* and *c'd'* are aiding, they are short-circuited on themselves. This will be shown by the ammeter in the primary circuit reading high and the voltmeter reading low. If series operation is desired, *the low side is dis-*

connected and the lead short-circuiting $a'd'$ is removed. On the other hand, if it is desired to operate the high side at 1,150 volts, that is, if the coils $a'b'$ and $c'd'$ are to be in parallel, the connections giving the lowest reading of the ammeter and the highest reading of the voltmeter should be retained.

145. Distribution Transformer.—Figure 207 shows the usual method of connecting the coils of a 2,300 to 1,150/230 to 115-volt distribution transformer. The high side consists of two 1,150-volt coils connected to terminals *ad* and *be*, respectively. The line terminals $a'e'$, which enter the transformer through porcelain bushings at the wide side of the case, are permanently connected to terminals *a* and *e*. If it is desired to employ the 2,300-volt connection, the coils *ad* and *be* are connected in series by the two links *bc* and *cd* which connect terminals *b* and *d* together. If it is desired to employ the 1,150-volt connection, the coils are connected in parallel by connecting these links across terminals *ab* and *de*. A study of the arrows shows that in both cases the coils act in conjunction so far as the external circuit is concerned. The leads from the low-side coils 1-3 and 2-4 leave the case through a porcelain bushing. By connecting 2 and 3 together, a 230 to 115-volt, three-wire system is obtained. The neutral wire, connected to 2-3, is grounded. By connecting 1-2 and 3-4, the coils are connected in parallel and a two-wire, 115-volt system is obtained.

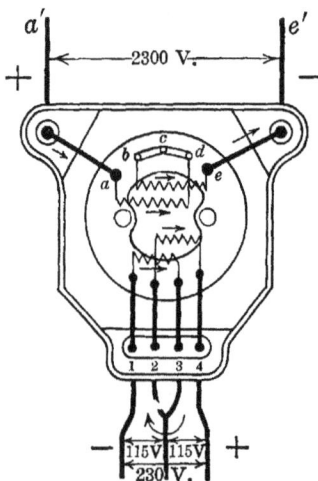

Fig. 207.—Connections of a distribution transformer.

146. Polyphase Transformer Connections.—There are several methods of connecting three-phase transformer banks, such, for example, as Y-Y, Δ-Δ, Δ-Y, Y-Δ, V-V, and T-T.

Figure 208 shows a Y-Y-connected transformer bank, which may be either a step-up or a step-down bank. For simplicity a 1-to-1 ratio and 100 volts between lines on the primary are assumed. With this connection, unbalanced loads to neutral cannot be applied on the secondary without seriously unbalancing the three secondary voltages. This voltage unbalancing can be eliminated by connecting the primary neutral to the generator neutral. If it is desired to obtain a three-phase, four-wire system, the delta-Y system (Fig. 210) may be used, since it is not open to the objection of extreme voltage unbalancing when unbalanced loads are connected between secondary wires and neutral.

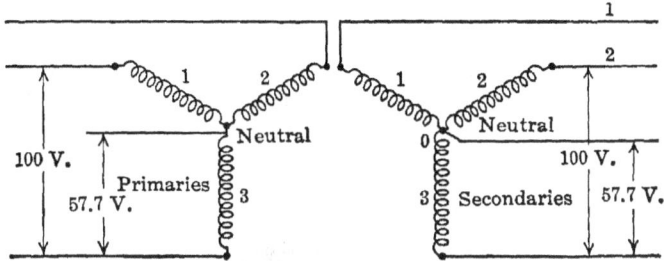

FIG. 208.—Y-Y connection of transformers.

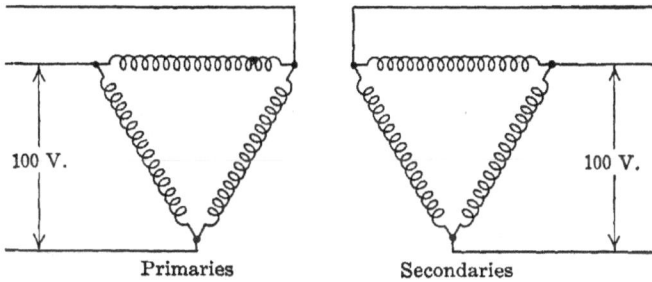

FIG. 209.—Delta-delta connection of transformers.

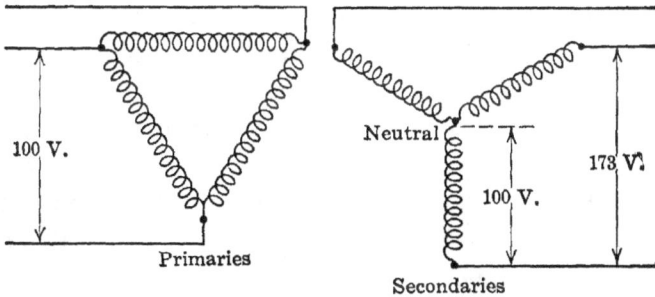

FIG. 210.—Delta-Y connection of transformers.

The delta-delta bank shown in Fig. 209 is often used, especially for moderate voltages. Its chief advantage is that if one transformer becomes disabled, the system may operate in "V," or open delta (Par. 147). In both the Y-Y and the delta-delta connections, the ratio between the primary and secondary line voltages is the same as the individual transformer ratio.

The delta-Y connection shown in Fig. 210 is a useful connection for stepping up the voltage. An advantage of delta-Y connection over the delta-delta connection is that insulation requirements for the secondaries are reduced which is particularly advantageous for high secondary voltages. For example, in a 100,000-volt system, the Y-connected secondaries need be insulated for only 58,000 ($100,000/\sqrt{3}$) volts, whereas delta-connected secondaries must be insulated for 100,000 volts. The Y-delta system is generally used for stepping down from high voltages for the same reason (see p. 360, Fig. 290).

The ratio between line voltages in the delta-Y and Y-delta systems is not the individual transformer ratio, for the line voltage on the Y side is $\sqrt{3}$ times that given by the transformer ratio.

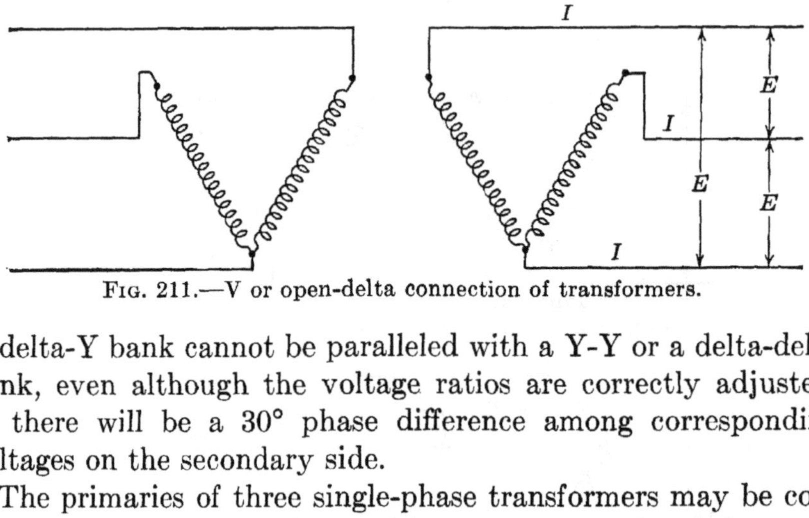

Fig. 211.—V or open-delta connection of transformers.

A delta-Y bank cannot be paralleled with a Y-Y or a delta-delta bank, even although the voltage ratios are correctly adjusted, as there will be a 30° phase difference among corresponding voltages on the secondary side.

The primaries of three single-phase transformers may be connected in either Y or delta without any attention being paid to phase relations. The secondaries must be phased in the same manner as the alternator coils in Par. 104 (p. 179). The primaries of *three-phase transformers*, however, must be connected correctly as regards phase relations. That is, the relations among the three fluxes must be those shown in Fig. 200 (p. 245). The actual phasing is often unnecessary, as the primary and secondary connections are brought out of the case symmetrically.

147. V-connection.—It is pointed out in Par. 104 (p. 180) that coil voltage must exist between the open ends of the two coils of the delta before the third coil is connected. At no load, with only two transformers, three equal three-phase voltages exist around the secondaries and a three-phase transformation is therefore possible with two transformers only. This is called the "V," or open-delta, connection (Fig. 211).

At first thought, it might appear that the V-connection would have two-thirds the rating of the delta-connection. However, both transformers operate at a reduced power factor when connected in V, even though the power factor of the load remains fixed. Therefore, the kva rating of the V-connection is *less* than two-thirds of the kva rating of the delta-connection having individual transformers of equal rating. The ratio of the V-rating to the delta-rating is $1/\sqrt{3}$, or 58 per cent, rather than $66\frac{2}{3}$ per cent. This can be proved as follows:

Let I be the rated current of each transformer and E the line voltage. The secondary output, at unity power factor (Fig. 211), is

$$P_1 = \sqrt{3}\ EI \text{ watts.}$$

Because the transformer rating is determined by the *current*, the secondary output of three of these transformers in delta will be

$$P_2 = 3EI \text{ watts.}$$

Therefore,

$$\frac{P_1}{P_2} = \frac{\sqrt{3}\ EI}{3EI} = \frac{1}{\sqrt{3}}, \text{ or 58 per cent.}$$

Often, in practice, a V-bank of transformers is first installed. The third transformer is added when the increase in load on the system warrants it. The rating of the bank is then increased 73 per cent with an investment increase of 50 per cent.

148. Scott-, or T-, Connection.—By means of the Scott-, or T-, connection, it is possible to transform not only from three-phase to three-phase by means of two transformers, but also from three-phase to two-phase or from two-phase to three-phase. This connection is seldom used, however, for three-phase to three-phase transformation. The method of connecting for three-phase to two-phase transformation is shown in Fig. 212. Two transformers are used, having primaries ad and bc and secondaries $a'd'$ and $b'c'$. The middle point d of the winding bc

must be accessible. One end d of the primary winding ad is connected to the middle point d of the primary bc. The ends of the three primary coils are connected to the three-phase supply ABC as shown in Fig. 212 except that terminal a rather than a_1 is first connected to wire A. The transformer bc is called the *main* transformer and the transformer ad the *teaser* transformer.

The three-phase supply is assumed to have a line voltage of 100 volts, and the transformers have a 1-to-1 ratio. The voltages E_{dc} and E_{db} are each equal to 50 volts and differ in phase by 180°, since coil dc and coil db are both on the same magnetic circuit. Since the total voltage impressed across any system must be equal to the vector sum of the component voltages, the three

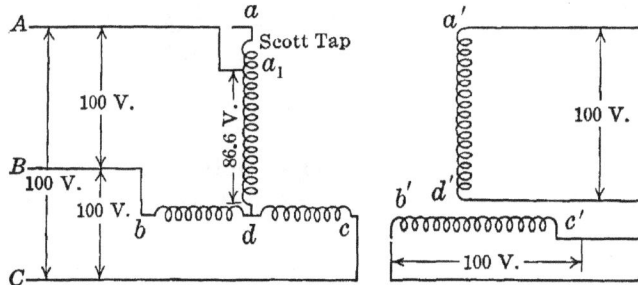

Fig. 212.—Scott or T-connection, three-phase to two-phase.

component voltages E_{da}, E_{db}, E_{dc} must combine in such a way as to give the three three-phase line voltages E_{ab}, E_{bc}, E_{ca}. In order that these component voltages may combine to give the three-phase line voltages, the voltage E_{da} must be equal to $100 \sqrt{3}/2$, or 86.6, volts; and it must differ in phase from E_{db} and E_{dc} by 90°. Obviously, the secondary voltages $E_{a'd'}$ and $E_{b'c'}$ must also differ in phase by 90°. Hence, a two-phase system is obtainable from the secondaries.

Since the voltage across transformer ad is 86.6 volts, its secondary voltage will be only 86.6 volts, whereas the voltage across $b'c'$, the secondary of transformer bc, is 100 volts. This gives two different voltages in the two phases of the two-phase secondary system, which is undesirable. By connecting wire A to a_1, a tap which makes $E_{a_1d} = 0.866E_{ad}$, the volts per turn in transformer ad are increased in the ratio of 1 to 0.866. That is, a_1d is now a step-up transformer having a ratio of 0.866 to 1. The secondary voltage $E_{a'd'}$ will now be 100 volts, which gives a symmetrical two-phase system. The tap a_1 is called the *Scott tap*.

It is obvious that various two-, three-, and four-phase systems may be obtained from these secondaries (see pp. 144 to 149). Furthermore, it is evident that this system may be reversed to transform from two- to three-phase.

149. Constant-current Transformers.—The transformers heretofore considered are constant-potential transformers; that is, both the primary and the secondary voltages remain substantially constant and a change of load is accompanied by a corresponding change of current. Street lamps, however, are ordinarily connected in series and require *constant current*. Constant alternating current is ordinarily obtained from a constant-current transformer.

The construction of the transformer is such that primary and secondary can move with respect to each other. The primary coil may be fixed and the secondary may move, or the secondary coil may be fixed and the primary may move. Both types are found in practice. Figure 213(*a*) shows a recent development in this type of transformer in which the primary is stationary and the movement of the secondary is guided by a hinged joint. The load consists of lamps in series. The current may be brought to the desired value by adjusting the weights on the lever arm at the top of the transformer. A dashpot is provided to prevent rapid changes in the position of the moving coil.

The operation of the transformer is as follows: Assume that the secondary coil is "floating"; that is, it is free to move either up or down and is delivering a definite current to a series load. The currents in the primary and secondary flow in opposite directions. Therefore, there is *repulsion* between the two coils (see Part I, p. 148). Assume that the load changes, that it decreases, for example. This change of load would be produced by *short-circuiting* one or more lamps, causing a *decrease* in the load resistance. Because of the decreased load resistance, the secondary and, as a consequence, the primary current increase. This increases the repelling force between the two coils, and therefore the secondary moves farther away from the primary. As a result a smaller part of the primary flux links the secondary.

and the secondary emf is reduced in consequence.   The secondary
coil will move away from the primary until the secondary current
is again at its normal value.   When the secondary moves away
from the primary, the leakage flux of both secondary and primary

(*a*)

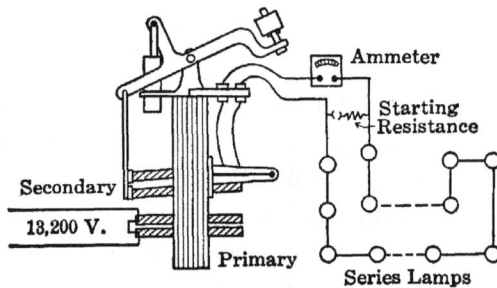

(*b*) Connection

Fig.   213.—Constant-current   transformer.   (*Courtesy   of   General   Electric
Company.*)

must increase (see p. 224).   Because of its large proportionate
leakage flux, this type of transformer has a low power factor
except at or near its maximum load.   This is one objection to its
use.

Since this type of transformer attempts to deliver constant current under all conditions of load, *it should never be open-circuited*, since on opening the circuit, a high resistance in the form of an arc occurs and a very high voltage results. When the transformer is not in service, the secondary is held in its top-most position by a latch device. On starting, a low or starting resistance [Fig. 213(*b*)] is shunted across the secondary. The resulting current repels the coil a sufficient amount to release the latch.

This type of transformer is used alone for supplying incandescent series street lamps. When it is used to supply magnetite lamps, a mercury-arc rectifier (see p. 348) is necessary in addition, since this type of lamp requires unidirectional current.

INSTRUMENT TRANSFORMERS

150. Electrical Measurements at High Voltages.—It is not practicable usually to connect instruments or meters directly to high-voltage circuits. Unless the high-voltage circuit is grounded at the instruments, they may be subjected to high-voltage stresses to ground. This makes it dangerous for anyone to come in contact with the switchboard apparatus. Also, instruments become inaccurate when connected directly to high voltage, because of the electrostatic forces which act on the indicating element.

By means of instrument transformers, instruments may be entirely insulated from the high-voltage circuit and yet indicate accurately the current, voltage, power, and other quantities in the high-voltage circuit. Moreover, low-voltage instruments having standard current and voltage ranges may be used for all high-voltage circuits, irrespective of the voltage and current ratings of the circuits.

151. Potential Transformers.—Except that their power rating is small, potential transformers do not differ materially from the constant-potential transformers already discussed. As instruments only and sometimes pilot lights are connected to their secondaries, such transformers ordinarily have ratings of 40 to 200 watts. The low-voltage side is usually wound for 110 volts, and the ratio is then determined by the voltage of the high-voltage winding. For example, a 13,200-volt potential trans-

former would have a ratio of 13,200/110 = 120/1. Figure
214 shows a simple connection for measuring voltage in a 13,200-
volt circuit by means of a potential transformer. The secondary
should always be grounded at one point to eliminate "static"
from the instrument and further to ensure safety to the operator.

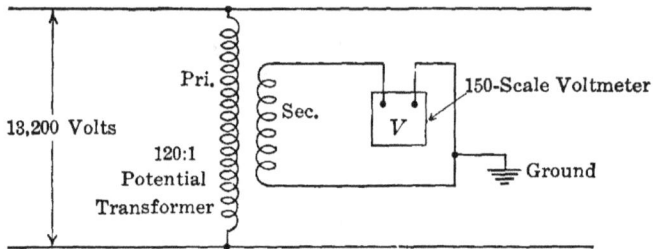

FIG. 214.—Use of potential transformer on a 13,200-volt single-phase circuit.

Figure 217 shows a potential transformer used in conjunction
with a current transformer for measuring power by means of a
wattmeter.

152. Current Transformers.—To avoid connecting a-c ammeters and the current coils of other instruments directly in high-

FIG. 215.—Indoor current transformer, 5 kv, 100 to 5 amp. (*Courtesy of Allis-
Chalmers Manufacturing Company.*)

voltage lines, current transformers are used. In addition to
insulating the instruments from high voltage, they step down
the current in a known ratio. This makes it possible to use a
lower range ammeter than would ordinarily be required if the
instrument were connected directly in the primary line.

The current or series transformer (Fig. 215) has a primary, usually of few turns, wound on a core and connected in series with the line. When the primary has a large current rating, it may consist of a straight conductor passing through the center of a hollow core, as shown in Fig. 216. The secondary, consisting of several turns, is wound around the laminated core. The ratio of current transformation is approximately the inverse ratio of turns. For example, if the primary has 2 turns and the secondary 60 turns, the ratio will be 30 to 1.

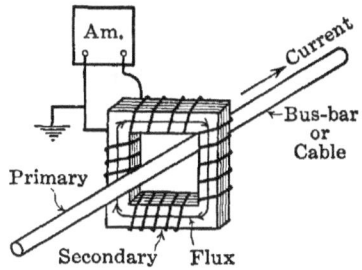

FIG. 216.—Through type of current transformer.

The secondaries of practically all current transformers are rated at 5 amp, regardless of the primary-current rating. For example, a 2,000-amp current transformer has a ratio of 400 to 1, and a 60-amp transformer has a ratio of 12 to 1.

The current transformer differs from the ordinary constant-potential transformer in that its primary current is determined

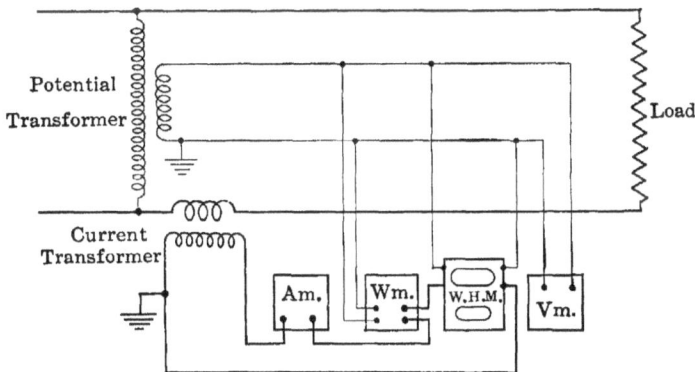

FIG. 217.—Typical connections of instrument transformers and instruments for single-phase measurements.

entirely by the load on the system and not by its own secondary load. If the secondary becomes open-circuited when the transformer is carrying even a moderate current, a high voltage will exist across the secondary because the large ratio of secondary to primary turns causes the transformer to act as a step-up transformer. *This high voltage is dangerous.*

Therefore, a current transformer should always have its secondary short-circuited.

Figure 217 shows the method of connecting a typical instrument load, through instrument transformers, to a high-voltage line. The load on the instrument transformers includes an ammeter, a voltmeter, a wattmeter, and a watthour meter. Each secondary is grounded at one point. Correction for ratio of transformation must be applied to all the instrument readings, the wattmeter and watthour meter involving the ratio of both current and potential transformers. Usually in permanent installations, as on switchboards, the instrument scales themselves are so marked as to take into consideration these ratios, and the primary power may be read directly.

CHAPTER VIII

THE POLYPHASE INDUCTION MOTOR

The induction motor is the most widely used type of a-c motor. This is due to its ruggedness and simplicity, to the absence of a commutator, and to the fact that its operating characteristics are well adapted to constant-speed work.

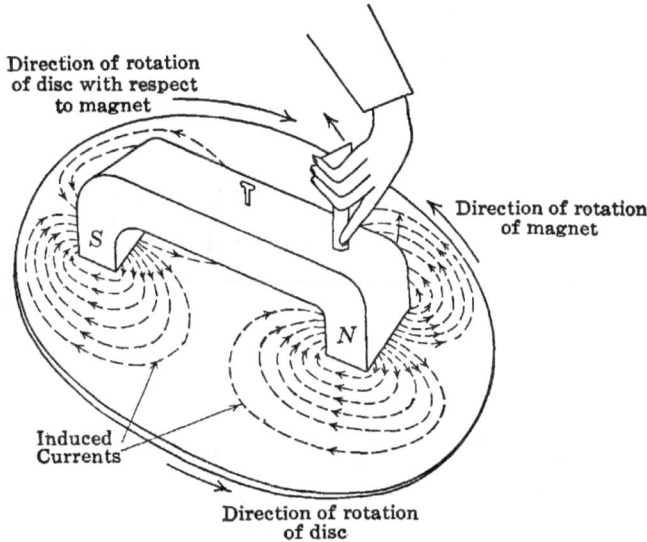

FIG. 218.—Rotation of metal disc produced by rotating magnet.

153. Fundamental Principle of Induction Motor.—The principle of the induction motor may be illustrated as follows: A metal disc (Fig. 218) is free to turn upon a vertical axis. The disc may be of any conducting material, such as iron, copper, or aluminum. A magnet, free to rotate on the same axis as the disc, is placed above the disc; and the ends of the magnet are bent so that its magnetic flux traverses the disc. When this magnet is rotated, the magnetic lines cut the disc and induce currents in it, as shown in the figure. As these currents are in a magnetic field, they tend to move across this field, just as the currents in the conductors of a d-c motor tend to move across the magnetic field of the motor. By Lenz's law, the direction of the force developed among these currents in the disc and the magnetic field producing them will be such that the disc tends to follow the rotating magnet, as shown in the figure. This may be proved by applying Fleming's right-hand rule to determine the direction of the induced currents and Fleming's left-hand rule to determine the direction of motion.

Power is required to overcome the friction of the disc and to supply the I^2R loss due to the induced currents in the disc. This

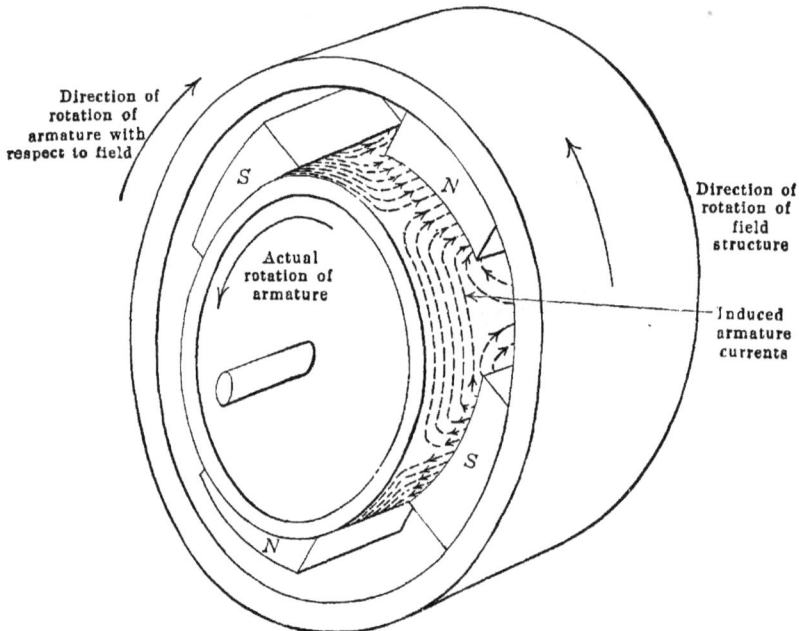

Fig. 219.—Rotation of conducting cylinder due to induced currents.

power must come from the mechanical power required to rotate the magnet. The power transferred to the disc from the magnet can be transferred only by means of a force acting between the two. If force acts between the two, the magnet must pull the disc and the direction of the induced currents in the disc must be such as to cause the disc to *follow* the magnet.

Also, the disc *tends* to attain the same speed of rotation as the magnet. However, the disc cannot attain the speed of the magnet; for were it to reach this speed, there would be no relative motion of disc and magnet and, therefore, no cutting of the disc by the magnetic flux. The currents in the disc would then become zero, no torque would be developed, and the speed of the disc would become less than that of the magnet. Because the disc cannot attain the speed of the magnet, there must always exist a *difference* of speed between the two. This difference of speed is called the revolutions *slip*.

A cylinder may be used instead of the disc, as shown in Fig. 219. In the figure are shown four poles, whose magnetic flux traverses the cylinder. If the frame carrying these poles is rotated by mechanical means, the currents induced in the cylinder will cause the cylinder to rotate in the same direction as that of the rotating frame. This cylinder and the magnetic poles which are associated with it more nearly resemble the commercial induction motor than the disc does, although both the disc and the cylinder operate on the same principle.

It is to be noted that the currents in the disc, or armature, of this type of motor are *induced*, rather than *conducted*, into the armature as in the d-c motor.

The induction motor, therefore, may operate without any conducting connection between the armature and an external circuit.

154. Rotating Field with Two-phase Currents.—The rotating magnetic fields described in the previous paragraph were produced by rotating the magnetic poles mechanically. Such rotating poles are practically the same as the rotating poles of an alternator field. Rotating magnetic fields may be produced, however, by polyphase currents in polyphase windings, such as alternator windings. Such rotating fields are produced entirely by electrical means, there being no mechanical rotation of the pole pieces themselves.

The windings which produce the rotating field are ordinarily on the fixed member, called the *stator*. The stator is made of circular laminations of iron and has slots and a winding identical with that of alternators. In fact, an alternator armature can be used for an induction-motor stator, if the voltage, frequency, and number of poles are the same. The rotating member in which currents are induced by the rotating field is called the *rotor*.

Figure 220(*a*) shows a cross-sectional diagram of a four-pole, two-phase induction motor. The single-layer stator winding, in which there are three slots per pole per phase, is shown diagram-

matically in Fig. 221. For simplicity, the connections of the
A-phase only are shown. The connections of the B-phase are
identical with those of the A-phase. The time variation of the
currents in the two phases is shown in Fig. 220(b). The current

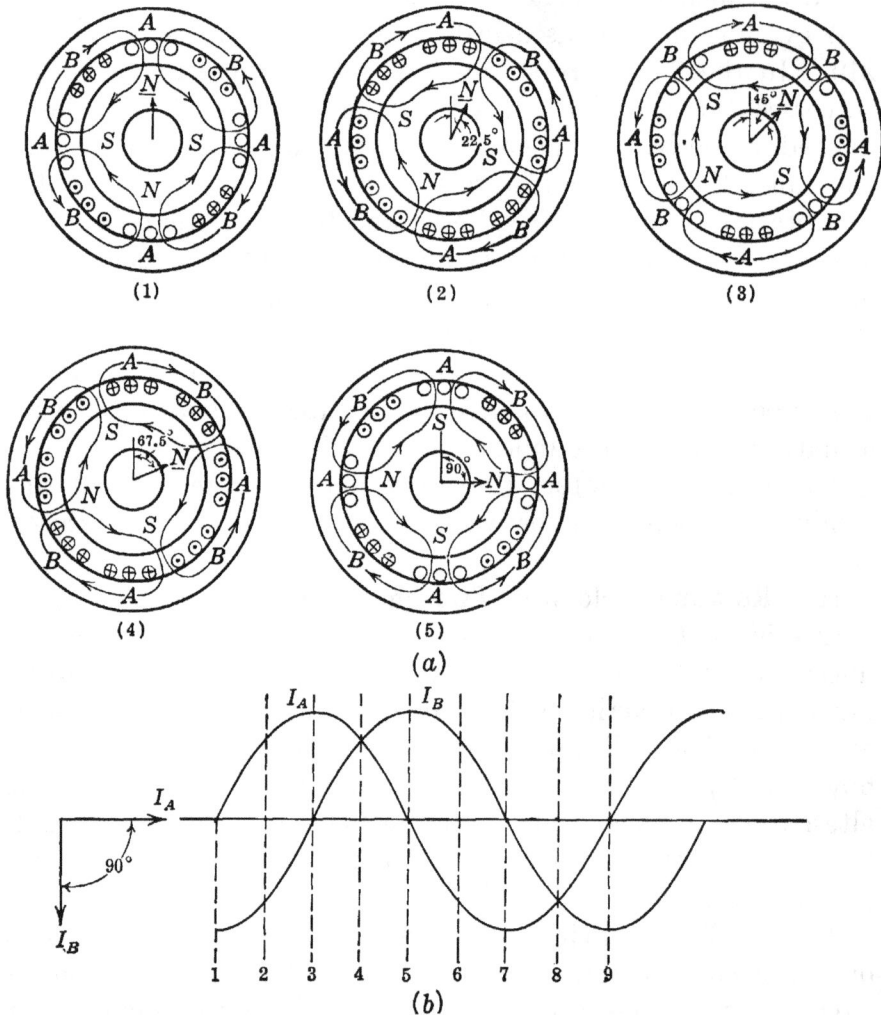

(1)   (2)   (3)

(4)   (5)

(a)

(b)

Fɪɢ. 220.—Production of rotating field by two-phase currents in four-pole
winding.

I_A in the A-phase leads current I_B in the B-phase by 90°. The
radius vectors I_A and I_B associated with these time curves are
also shown in Fig. 220(b).

At the instant 1 (Fig. 220(b)), the current I_A is zero and I_B
is negative maximum. There is, therefore, no current in the

conductors of the *A*-belt in diagram (1) (Fig. 220(*a*)). The armature is wound so that when I_B is negative (Fig. 220(*b*)) the current in the upper left-hand conductor belt of the *B*-phase is inward, that in the adjacent *B*-belts is outward, and that in the opposite *B*-belt is inward. By applying the corkscrew rule, the mmfs of the two upper *B*-belts send flux into the armature in the region between them, forming a north pole $\underline{N}$. Likewise, the mmfs of these same belts when combined with the mmfs of the two remaining conductor belts form alternate south and north poles as shown.

At the instant 2 the current in the *B*-phase is still negative but is 0.707 of its maximum value. The current still flows inward in the same *B*-belts as in (1) but is less in magnitude. The direction of the winding of the *A*-phase is such that when the current in the *A*-phase is positive the current is inward in the top conductor belt as shown. The current in the *A*-belt at instant 2 is 0.707 of its maximum value, and equal currents flow in all the conductors of both phases. Therefore, all adjacent phase-*A* and phase-*B* conductor belts carrying current in the same direction at the given instant act as a single belt. Application of the corkscrew rule

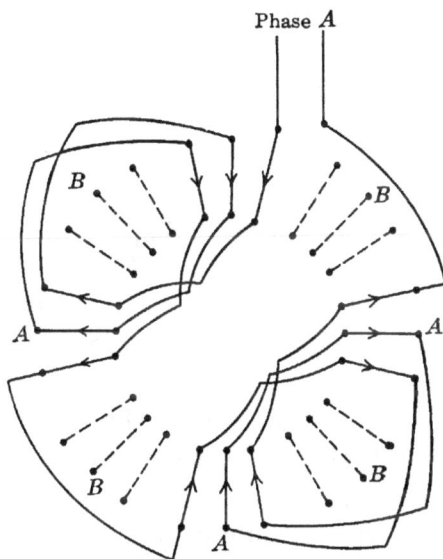

Fig. 221.—Single-layer, four-pole, two-phase induction-motor winding, lap-connected.

shows that the flux produced by the various conductor belts has the direction shown in diagram (2) (Fig. 220(*a*)), *N*- and *S*-poles being formed midway between adjacent belts in which the current flows in opposite directions. It will be observed that in (2) all the *N*- and *S*-poles have advanced 22.5 space-degrees in a clockwise direction from their position in (1).

At instant 3 the current I_B in the *B*-phase is zero. The current I_A in the *A*-phase is positive maximum and has the same direction that it had in (2). Therefore, the direction of the fluxes produced by A-phase conductor belts will be the same as in diagram (2). N- and S-poles will be found midway between adjacent conductor belts, as shown in (3). Note that all N- and S-poles have again advanced 22.5 space-degrees in a clockwise direction from their position in (2).

At instant 4, both I_A and I_B are positive and each is equal to 0.707 of its maximum value. The direction of the currents in the A-phase conductors has not changed from the direction in (2) and (3), but the direction of the currents in the B-phase conductors is opposite to that shown in (1) and (2). Adjacent A-phase and B-phase conductor belts carrying currents in the same direction act as a single belt and produce fluxes as shown in (4). N- and S-poles are formed midway between adjacent belts in which the currents have opposite directions. Again, in (4), all N- and S-poles have advanced 22.5 space-degrees in a clockwise direction from their position in (3).

A study of the currents and the direction of the fluxes which they produce at instant 5 shows that all N- and S-poles have again advanced 22.5 space-degrees in a clockwise direction from their position in (4). Between instants 1 and 5 the currents have gone through 180 time-degrees and in the same interval all N- and S-poles have advanced 90 space-degrees in a clockwise direction.

Further analysis of the direction of the currents and the fluxes which they produce shows that during each time interval 6-7, 7-8, 8-9 the N- and S-poles advance 22.5 space-degrees. At 9, the end of the current cycle, all N- and S-poles have advanced 180 space-degrees. After two complete cycles of current, each N- and S-pole will be in its original position, (1), the rotating field having completed one revolution.

155. Rotating Field with Three-phase Winding.—In Fig. 222(a) is shown an induction motor in section, with a three-phase, four-pole, single-layer winding. In (b) are shown the three-phase current waves I_A, I_B, I_C with the corresponding rotating vectors. These currents flowing in the windings of (a) produce a four-pole rotating field. In the conductor belts marked (+) as $+A$, $+B$, the currents flow inward when the current in (b) is positive. In the conductor belts marked (−) the currents flow outward when the current in (b) is negative.

At instant 1 in (*a*) and (*b*) the current in the *A*-phase is zero; that in the *B*-phase is negative and equal to 0.866 of the maximum value; that in the *C*-phase is positive and is also equal to 0.866 of the maximum value. There are no currents in any of the *A*-belts, and the current is outward in the +*B*-belts and −*C*-belts and is inward in the −*B*-belts and +*C*-belts. By applying the corkscrew rule to (1) in (*a*) the paths of the fluxes are determined as indicated. *N*-poles are formed at top and

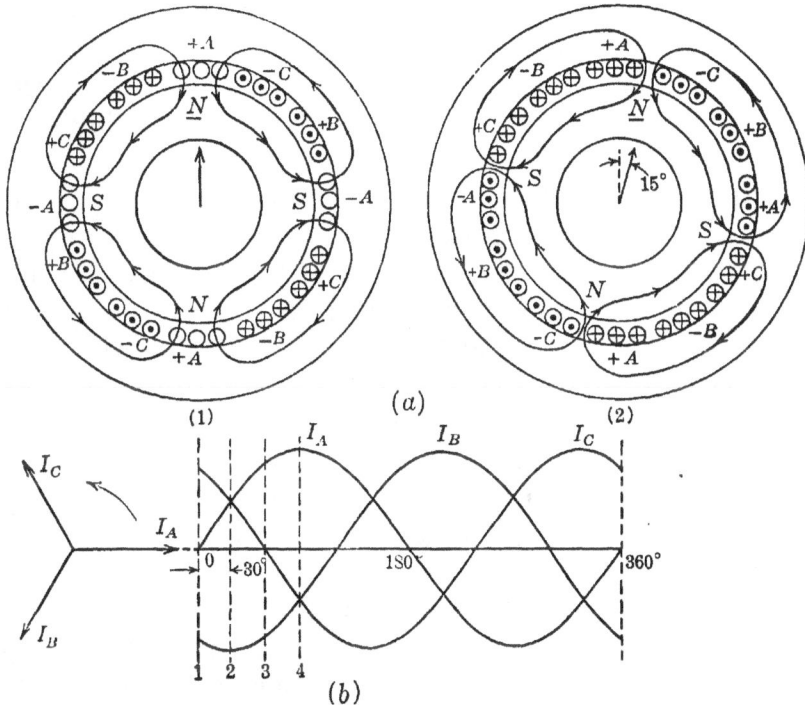

Fig. 222.—Rotating field with three-phase winding.

bottom and *S*-poles at the sides. The arrow points in the direction of the upper *N*-pole.

The instant 2 in (*b*) occurs 30° later in time than 1. The current I_A is positive and equal to 0.5 of its maximum value; the current I_B is negative and equal to its maximum value; the current I_C is positive and equal to 0.5 of its maximum value. At instant 2 in (*a*) the currents in the *B*- and *C*-belts remain unchanged in direction, but that in *B* has increased and that in *C* has decreased. The currents in the +*A*-belts are inward, and those in the −*A*-belts are outward. By applying the corkscrew rule, the direction and path of the fluxes are determined as indicated. It will be noted that four belts are formed, in each of which the current is either inward or outward. Each belt contains all the phase belts; and the current in the center, or *B*-belt, is twice that in the *A*- and *C*- belts.

All four poles have advanced 15° in space in a clockwise direction, this being one-half the angle of advance of the current waves. By correlating the mmfs in (*a*) with the corresponding currents at other instants, such as 3 and 4 in (*b*), the magnetic field will be found to advance 15 space-degrees for each 30 electrical time-degrees of the currents and will have completed one-half a revolution for each cycle of the current as in Fig. 220. The direction of rotation can be reversed by interchanging any two phases, for example, by interchanging the *A*- and *C*-phase belts.

In a full-pitch, two-layer lap or wave winding the arrangement of conductor belts is identical with that of Fig. 222(*a*). (See Fig. 138, p. 164.) With fractional pitch the arrangement is modified only by a slight overlapping of the phase belts. (See Fig. 140, p. 165.) Hence, such three-phase windings will produce a rotating field.

From Par. 154 and the foregoing, it follows that the electrical space-angle between any two windings must be the same as the time-angle between the currents in these same windings.

With a two-pole winding the rotating field completes one revolution for each cycle of the current.

With a six-pole winding, the rotating field completes one revolution for three cycles of the current; with an eight-pole winding, the rotating field completes one revolution for four cycles of the current, etc.

In a two-phase machine the direction of the rotating field may be reversed by reversing the connections to *either* phase; in a three-phase machine the direction of the rotating field may be reversed by interchanging any *two* leads.

156. Synchronous Speed; Slip.—It has been shown in Par. 155 that the angular speed of an a-c rotating field depends on two factors, the frequency of the current and the number of poles for which the machine is wound. The relation among speed, frequency, and poles is given by the following equation:

$$N = \frac{f \times 120}{P},$$ (89)

where N is the speed of the field in rpm, f the frequency in cycles per second, and P the number of poles. (Compare with Eq. (2), p. 15.) The speed N of the rotating field, is the *synchronous speed* of the motor. The usual synchronous speeds for motors at 25 and at 60 cycles per sec. are as follows:

Poles	rpm = N	
	$f = 25$	$f = 60$
2	1,500	3,600
4	750	1,800
6	500	1,200
8	375	900
12	250	600

Slip.—If an armature whose conductors form closed circuits is placed in a rotating field, it will develop torque because the induced currents in the conductors act in conjunction with the rotating magnetic field, as in the disc (Fig. 218) and the cylinder (Fig. 219).

As has been pointed out, the armature cannot attain the speed of the rotating field; for if it did, there would be no cutting of conductors by flux, and there would be no rotor current and no torque.

The difference between the speed of the rotating field and that of the rotor is called the *revolutions slip* of the motor.

Example.—The rotor of a four-pole, 60-cycle motor has a speed of 1,720 rpm. Find its revolutions slip.

The synchronous speed, from Eq. (89),

$$N = \frac{60 \times 120}{4} = 1{,}800 \text{ rpm.}$$

Revolutions slip = 1,800 − 1,720 = 80 rpm. *Ans.*

It is more convenient to express the slip as a fraction of the synchronous speed. Denote the speed of the rotor by N_2 and the synchronous speed by N. Then the slip

$$s = \frac{N - N_2}{N}. \tag{90}$$

Example.—Determine the slip in the foregoing, four-pole motor.

$$s = \frac{1,800 - 1,720}{1,800} = \frac{80}{1,800} = 0.0444, \text{ or } 4.44 \text{ per cent.} \quad Ans.$$

From Eq. (90), the rotor speed

$$N_2 = N(1 - s) \text{ rpm.} \tag{91}$$

From Eq. (89),

$$N = \frac{f \times 120}{P} \text{ rpm.}$$

Hence,

$$N_2 = \frac{f \times 120}{P} (1 - s) \text{ rpm.} \tag{92}$$

The full-load slip in commercial motors varies from 1 to 10 per cent, depending on size and type of motor.

157. Rotor Frequency and Induced Electromotive Force.—If the rotor of a two-pole, 60-cycle polyphase motor is at standstill and voltage is applied to the stator, each rotor conductor will be cut by an N-pole 60 times per second and by an S-pole 60 times per second, as the speed of the rotating field is 60 rps. If the stator is wound for four poles, the speed of the rotating field is halved; but each rotor conductor is then cut by two N- and two S-poles per revolution of the field and, therefore, by 60 N- and 60 S-poles per second, as in the two-pole motor. Consequently, the frequency of the rotor currents at standstill ($s = 1.0$) will be the same as the stator frequency. This holds true for any number of poles. At standstill, the motor is a simple static transformer, the stator being the primary and the rotor being the secondary. (See Par. 168, p. 296.)

If the rotor of a 60-cycle motor revolves at half speed in the direction of the rotating field ($s = 0.5$), the rotor conductors are cut by just one-half as many N- and S-poles per second as when standing still and the frequency of the rotor currents is 30 cycles per sec.

By taking other rotor speeds, it can be seen that the rotor frequency

$$f_2 = sf \text{ cycles per sec,} \tag{93}$$

where f_2 is the rotor frequency, s the slip, and f the stator frequency. *Rotor frequency is equal to stator frequency multiplied by slip.*

Example.—What is the frequency of the currents in the rotor of a 60-cycle, six-pole induction motor, if the rotor speed is 1,164 rpm.

The synchronous speed

$$N = \frac{60 \times 120}{6} = 1,200 \text{ rpm (Eq. (89), p. 270).}$$

The slip

$$s = \frac{1,200 - 1,164}{1,200} = 0.03.$$

$$f_2 = 0.03 \times 60 = 1.8 \text{ cycles per sec.} \quad Ans.$$

From the foregoing, it follows that under ordinary operating conditions the rotor frequency is low. At standstill, however, or in starting, the rotor frequency is that of the stator. The rotor frequency has an important effect on the operating characteristics of the induction motor.

The induction motor can be used as a frequency changer if the rotor is driven mechanically at the correct speed. Current is taken from the rotor, or secondary, through slip-rings. Under these conditions, some of the power to the load is supplied electrically and some mechanically.

158. Torque of the Induction Motor.—With either series or shunt d-c motors operating at constant voltage, the torque will increase with increase of load until the armature is brought to standstill. If the motors are of commercial ratings, the torque at standstill with rated voltage across the armature would be large as compared with rated-load torque. Under these conditions the current would be prohibitively large. The maximum torque that such motors develop is limited by the current which the armature can carry without overheating. Even so, such motors are able to develop several times full-load torque on starting.

With induction motors the torque increases with increase of load only up to a definite *maximum* torque, further increase of load resulting in a *decrease* of torque, the rotor coming to standstill. This maximum torque, called the *breakdown torque*, is from two to four times the rated-load torque. This is illustrated in

Fig. 223, which shows the torque—slip curves for the ordinary squirrel-cage motor. When the slip is nearly zero, the rotor is running at its maximum speed or at nearly synchronous speed. As the load is increased, the torque at first increases almost in proportion to the slip. That is, as the slip increases, the induced rotor currents increase; and since they are reacting with a constant flux, the torque increases ($f = Bli$). At larger values of slip, the torque increases less rapidly than the slip; and ultimately a maximum value of torque, the *breakdown torque*, is reached (Fig. 223). Further increase of load results in *less* torque being developed by the motor, and the rotor comes to standstill.

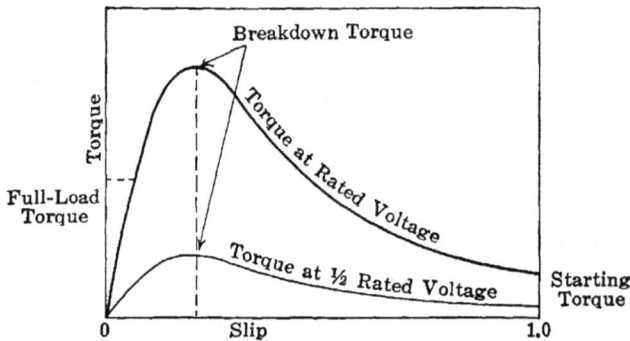

Fig. 223.—Torque—slip curves for squirrel-cage motor.

With the squirrel-cage motor, the starting or standstill torque ($s = 1.0$) is relatively small (Fig. 223).

This ultimate decrease of torque with increase of slip is due primarily to the increase in rotor frequency with slip. The induced emf in each rotor conductor is a maximum when the maximum part of the flux waves in the rotating field is cutting it ($e = Blv$). At low values of slip and hence of rotor frequency the currents in each rotor conductor are substantially in phase with the induced emf and hence are maximum when they are in the maximum part of the flux waves. As the rotor frequency increases, the lag of the induced currents behind their emfs increases ($\tan \theta = 2\pi f L/R$, Eq. (25), p. 69); and at the time when the current in a given conductor reaches its maximum value, the maximum of the flux wave has already passed by it. Hence, the current in such a conductor will be reacting with a lesser flux density; and although the current itself has increased owing to the larger slip, the torque actually will decrease, owing to the increased phase displacement between current and flux.[1]

For a given value of slip, the torque is proportional to the impressed voltage *squared*. For example, if at any given value of slip, the impressed voltage is halved, the air-gap flux is halved and as a result the induced rotor current is halved. The rotor current of half its former magnitude finds itself in a magnetic field of half its former strength; and the torque is, therefore,

FIG. 224.—Squirrel-cage, bar-type rotor. (*Courtesy of Westinghouse Electric and Manufacturing Company.*)

one-fourth its former value. This is illustrated by the torque curve at one-half rated voltage (Fig. 223).

It is important to remember that the torque varies as the *square* of the impressed voltage. A 10 per cent reduction in impressed voltage results in a 19 per cent reduction in breakdown torque.

159. Squirrel-cage Motor.—The squirrel-cage motor is the simplest type of induction motor and is the most generally used. The core of the rotor or armature (Fig. 224), like that of the d-c armature, is usually built up of slotted steel punchings. The

---

[1] For more detailed analyses, see C. L. DAWES, "A Course in Electrical Engineering," Vol. II.

winding, particularly in the larger motors, consists of copper bars placed in the slots, with their ends connected together by conducting rings of copper called *end rings* (Fig. 224). The bars and end rings are either brazed or welded together, or the rotor may be placed in a mold and the ends of the bars cast in a ring of solid copper. Another method is to place the rotors in molds

FIG. 225.—Cast-aluminum, 20-hp, six-pole, 60-cycle, 440-volt induction motor. (*Courtesy of General Electric Company.*)

FIG. 226.—Stator and rotor slots of squirrel-cage induction motor.

and to make the bars and end rings a single integral aluminum casting (Fig. 225).

In nearly all induction motors, except those of large rating, the stator and rotor slots are of the semiclosed type similar to the rotor slots (Fig. 226). In motors of large rating open stator slots such as are shown in Fig. 226 are generally used.

The advantage of the semiclosed slot is that the effective sectional area of the air-gap is increased and the magnetizing current is reduced, therefore. The magnetizing current, whose mmf produces the flux, lags the terminal voltage by approximately 90°. A reduction in magnetizing current increases the power factor, which is a distinct advantage since the power factor of induction motors is low under the best conditions. On the other hand, the semiclosed slot gives a much higher slot inductance than the open slot, and this inductance in the stator and

FIG. 227.—Stator of 50-hp polyphase induction motor. (*Courtesy of Westinghouse Electric and Manufacturing Company.*)

in the rotor tends to lower the power factor and decreases both the starting torque and the breakdown torque of the motor.

Figure 227 shows the stator of a 50-hp induction motor. The fabricated construction, the ventilating ducts, and the open-frame construction should be noted.

160. Operating Characteristics of Squirrel-cage Motor.—The squirrel-cage motor, like the d-c shunt motor, operates at substantially constant speed. As the rotor cannot reach the speed of the rotating magnetic field, it must operate at all times with a certain amount of slip. At no load, the slip is very small. As load is applied to the rotor, more rotor current is required to develop the necessary torque in order to carry the increased load. Consequently, the rotating magnetic field must cut the rotor conductors at an *increased* rate, in order to produce the necessary increase of emf and hence of current. The slip of the rotor must increase, therefore, so that the rotor speed drops. As the

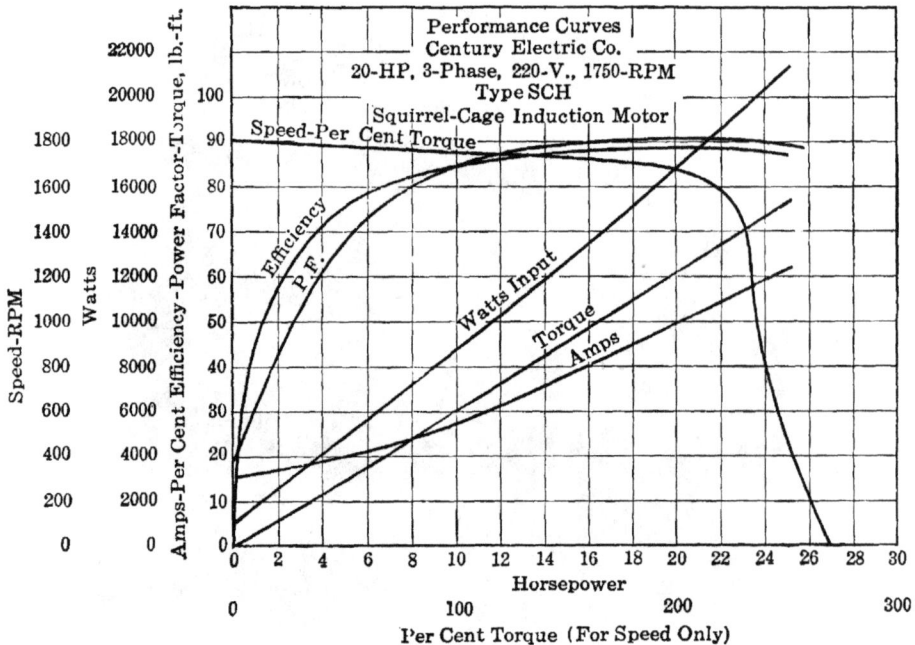

Fig. 228.—Operating characteristics of low-starting current, high-starting torque, three-phase, squirrel-cage induction motor. (*Courtesy of Century Electric Company.*)

resistance of the squirrel cage is low, a small increase in slip produces a large increase in current. The slip, therefore, for ordinary loads is small. In large motors, 100 hp or greater, the slip is of the order of 2 to 4 per cent at rated load. In the smaller sizes of motor, the slip may be as high as 8 to 10 per cent at rated load.

Figure 228 shows the usual characteristic curves of a 20-hp, squirrel-cage motor. It will be noted that the torque, speed, and efficiency curves are similar to those of a shunt motor. Power factor increases with load.

At no load the losses are small, just as with the transformer. Also, the no-load current produces the mmf which sends the flux across the air-gap and through the magnetic circuit. Hence, the no-load current lags the voltage by a large angle, and the no-load power factor is low. Because the no-load or magnetizing current must send the flux across the air-gap, its value is comparatively large, being 20 per cent of rated current in large motors and 40 to 60 per cent of rated current in small motors. As the load on the motor increases, more power and hence more energy current are required and accordingly the power factor increases.

The efficiency curve is similar to that for other types of electrical apparatus (see Part I, p. 278, Fig. 243). At all loads, there are certain fixed losses, such as core loss, friction, and windage. In addition, there are the load losses (I^2R) which increase nearly as the square of the load. At light loads, therefore, the efficiency is low because the fixed losses are large as compared with the input. As the load increases, the efficiency increases to a maximum. Beyond this point the I^2R losses become relatively large, causing the efficiency to decrease.

One disadvantage of the squirrel-cage motor is that, on starting, it takes a large current at low power factor and, notwithstanding this large current, it develops but little torque. When the motor is at standstill, the squirrel cage acts as the short-circuited secondary of a transformer, and if full voltage is applied, the motor takes an excessive current on starting.

Figure 223 shows the variation of torque with slip at line voltage and at one-half line voltage. It will be noted that for small values of slip up to and even beyond full load, which is the ordinary range of operation, the torque is substantially proportional to the slip. At higher values of slip, however, the torque reaches the maximum, or *breakdown torque*, and beyond this maximum the torque decreases as the slip increases until at standstill ($s = 1.0$) the torque may be relatively *small*. This small value of torque has been explained as being due to phase displacement of rotor currents with respect to the flux waves in the air-gap. It may also be explained on the basis of the power transferred across the air-gap. At a given speed, the greater the power transferred across the air-gap, the greater the force acting between stator and rotor and the greater the torque. At standstill, substantially all the power transferred across the gap must be accounted for in rotor I^2R loss.   The resistance of the squirrel-cage rotor is so low that a *large* current is required to produce any considerable I^2R loss and the starting torque must be small even with large rotor currents.   Therefore, even when full line voltage is applied to the stator, the *starting torque* of the squirrel-cage motor is not large.

This low starting torque of the squirrel-cage motor is further reduced when it is necessary to apply reduced voltage to the motor on starting (see p. 290).   If 50 per cent line voltage is applied on starting, only 25 per cent normal starting torque is obtained (Fig. 223).   The simple squirrel-cage motor, therefore, cannot be used where large starting torque is required.

Because of its low rotor resistance, the squirrel-cage motor has excellent operating characteristics for constant-speed work.   The slip is small, and the speed regulation is good.   In addition, the motor is simple and rugged and requires little attention.   Since there are no sliding contacts, it can be used where hazards of an explosive nature exist such as in flour and woodworking mills.   Some of its fields of application are in machine shops, in woodworking shops, in cement mills, and in textile mills; in fact, it is used in most cases where the load requires constant speed with low starting torque.   The speed of the squirrel-cage motor is not adjustable.

161. Wound-rotor Induction Motor.—If resistance be introduced in the rotor circuit of an induction motor, the slip for any given value of torque will increase.

Assume that the motor is developing a definite value of torque at a definite value of rotor current.   At low values of slip the torque is practically proportional to rotor current and air-gap flux.

The air-gap flux of the induction motor is practically constant, since terminal voltage, and hence counter emf, are practically constant.   If resistance is introduced in the rotor circuit, the rotor impedance is increased.   For the low values of slip at which the motor usually operates, armature reactance is small as compared with resistance, and armature impedance is practically all resistance.   If the slip remains constant, the induced emf of the rotor does not change.   The armature current, which is equal to this emf divided by the rotor impedance, decreases owing to the increased rotor resistance. The torque, therefore, decreases.

To bring the torque back to its initial value, the armature current must be increased. To increase the armature current, the armature induced emf must increase. Since the air-gap flux is constant, the increase in the induced emf may be obtained only by flux cutting rotor conductors at a greater rate. For given torque, therefore, slip must *increase* when resistance is inserted in the rotor circuit.

The torque—slip curve will change from curve (1) to curve (2) (Fig. 229). It will be noted that full-load torque is obtained at

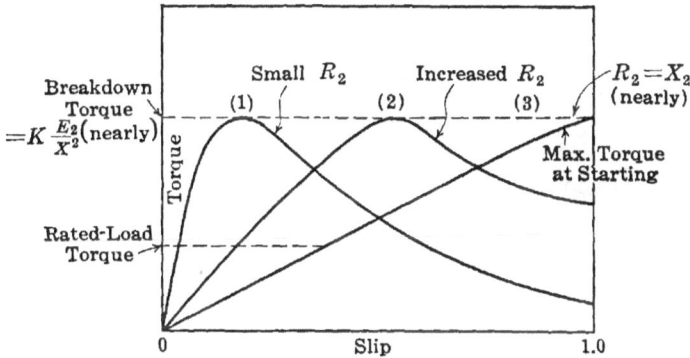

FIG. 229.—Effect on torque—slip curve of inserting resistance in rotor circuit.

increased slip as rotor resistance is increased. The *magnitude* of the maximum, or breakdown, torque will not be affected, but the slip for maximum torque moves toward the point of zero speed ($s = 1.0$). That is, the maximum torque occurs at a greater value of slip. The rotor runs at reduced speed, but the reduced speed is obtained at the expense of efficiency, for the I^2R losses in the rotor circuit are increased.

It follows that speed control may be obtained by the introduction of resistance in the rotor circuit. This method of speed control is similar to the armature-resistance method of speed control in the d-c motor (see Part I, Par. 223, p. 288). The speed-load curve is almost identical with that of the shunt motor with resistance in the armature circuit (Part I, p. 288, Fig. 252(b)). The lowering of speed is accompanied by a material lowering of efficiency and by poor speed regulation. The electrical efficiency of the rotor is equal to the ratio of actual speed to synchronous speed. For example, at 25 per cent slip, the rotor efficiency is 75 per cent. That is, of the power transmitted across the air-gap, 25 per cent is lost as heat in the rotor resistance. The other 75 per cent is converted into mechanical power, although this is not all available at the pulley because of rotor friction and core losses.

If sufficient resistance is introduced in the rotor circuit, maximum torque may be made to occur at standstill, as shown by curve (3) (Fig. 229). That is, breakdown torque is obtained at starting.

This increase of starting torque is due to the increased losses in the rotor circuit, resulting from the rotor current flowing

FIG. 230.—Three-phase wound rotor. (*Courtesy of Century Electric Company.*)

through added rotor resistance. The power required to supply these losses must be transferred across the air-gap, and an increase in the torque between stator and rotor is thus produced.

An adjustable resistance cannot be placed readily in the rotor, so that rotors requiring external resistance are usually wound either two- or three-phase (Fig. 230). The two-phase windings may be connected either star or mesh and the three-phase windings may be connected either Y or delta. Such rotor windings are in every way similar to stator windings. The three terminals of the three-phase winding are connected to three slip-rings (Figs. 230 and 231). Brushes, bearing on each of these three rings (Fig. 231), connect to Y-connected external resistances, usually through a controller. The entire resistance of each phase is in circuit on starting. In addition to producing a good starting torque, the starting current of the motor does not greatly exceed the rated current. As the motor comes up to speed, the external resistance is cut out. The motor then operates on curve (1) (Fig. 229).

Even without the controller, the wound-rotor type of motor is more expensive than the squirrel-cage motor, owing to the greater cost of winding and connecting the rotor coils. The controller and resistors add further to the cost. In the running position, this type of motor has greater slip than the ordinary squirrel-cage motor, because it is not possible to secure the very low resistance obtainable with the squirrel-cage winding. As has been pointed out, such external resistance may be used to obtain

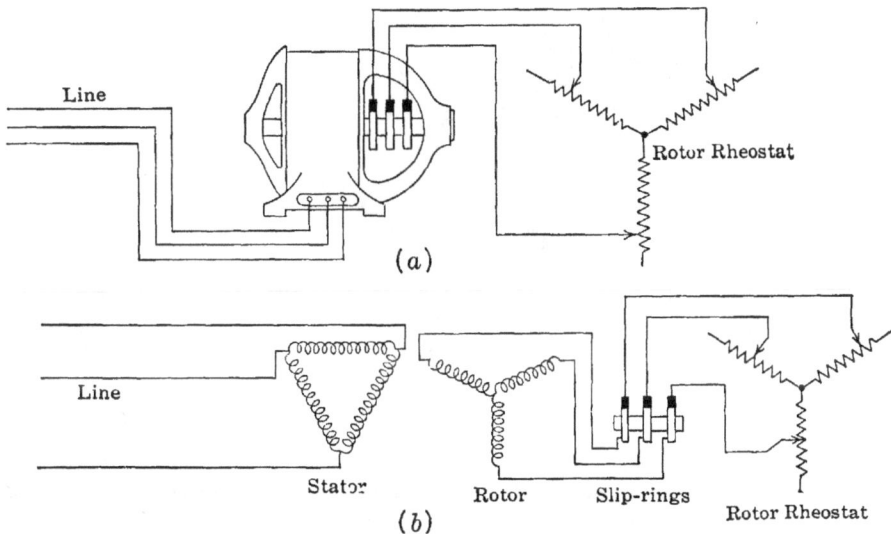

FIG. 231.—Connections for a wound-rotor induction motor.

speed control at reduced efficiency and with poor speed regulation. Hence, the wound-rotor type of induction motor has better starting characteristics, but at operating speeds is less efficient than the squirrel-cage motor.

162. Industrial Applications of Wound-rotor Motor.—Wound-rotor induction motors are used where considerable starting torque is required, and frequently where speed adjustment is desired. Common applications of this type of motor are in cranes, elevators, pumps, hoists, railroads, and calenders.

Where three-phase induction motors are used for railway locomotives, as in the Virginian Ry. and the Norfolk and Western Ry., the wound-rotor type is necessary.

Another use of wound-rotor induction motors is in the electric propulsion of ships. The motors are connected directly to the propeller shafts. Two synchronous speeds are obtained by changing the number of poles. Intermediate speeds are obtained by changing the frequency of the supply.

In Fig. 232 is shown a 3,500-hp, polyphase, wound-rotor induction motor used to drive the rolls in a steel mill.

Fig. 232.—Mill-type, sleeve-bearing, wound-rotor, polyphase induction motor, 3,500 hp, 600 rpm, 2,200 volts, 60 cycles, three-phase. (*Courtesy of General Electric Company.*)

163. Double-squirrel-cage Rotors.—It has been shown that because of its low rotor resistance the simple squirrel-cage motor has excellent constant-speed operating characteristics but low starting torque. On the other hand, high resistance in the rotor circuit gives good starting torque but reduced efficiency and poor speed regulation. By the use of two rotor windings, the double-squirrel-cage motor combines high starting torque and good operating characteristics as well as low starting current.

The rotor slots are in two sections, a top section near the rotor surface of small cross-section and a bottom section of large cross-section (Figs. 233 and 234(c)). The two sections are connected by an elongated contracted section. The top section contains a high-resistance winding; the bottom section, a low-resistance winding, usually of copper bars or aluminum alloy.

The contracted section of the slot which connects the top and bottom sections has a relatively low reluctance so that it is traversed by considerable slot-leakage flux. The leakage flux for one slot is indicated in Fig. 233. The flux linkages per ampere in the low-resistance winding in the bottom of the slot are distinctly greater than those of the high-resistance winding in the top of the slot (Fig. 233). Hence, the inductance L_2 of the low-resistance winding is much greater than the inductance L_1 of the high-resistance winding. On starting, the rotor frequency is the same as the stator frequency, and the reactance $(2\pi f_2 L_2)$ of the low-resistance winding is high. Hence, most of the rotor current flows in the top, or high-resistance, low-reactance, winding, which gives good starting torque and low starting current. As the rotor gains speed, the rotor frequency diminishes and accordingly the reactance of both windings diminishes. This causes a greater proportion of the total rotor current to flow in the low-resistance

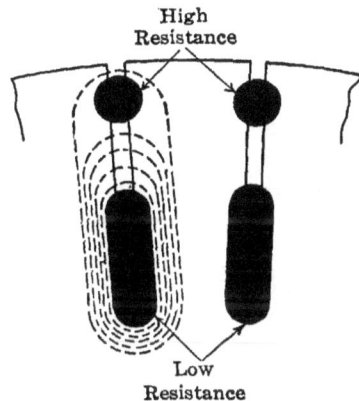

Fig. 233.—Slots in double squirrel-cage winding.

winding. At the operating speed, where the slip is low, the division of current in the two windings is nearly inversely as their resistance, and nearly all the rotor current now flows in the low-resistance winding, giving the motor the running characteristics of a low-resistance squirrel-cage motor. In Fig. 234(c) are indicated three different arrangements of *double-deck*, or double-squirrel-cage, winding as now manufactured.

164. Starting Induction Motors.—A squirrel-cage winding has such low resistance that at standstill it corresponds to the short-circuited secondary of a transformer. Therefore, if the motor is connected directly across the line, the no-load current is large, and with large motors the resulting disturbance to the line voltage may be greater than is permissible. Induction motors up to 7.5 hp may usually be connected directly across the line without undue disturbance to the line voltage, although on starting they may take momentarily as much as six or seven times rated-load current (see Fig. 234(a)).

(a)	(b)	(c)	(d)	(e)
General-purpose. Normal-starting current, 6 to 7 times rated; above 7½ hp reduced voltage start; starting torque about 150 per cent rated; code letters, F to R.[1]	High-reactance. Low-starting current, 4½ to 5 times rated; full-voltage start; normal-starting torque about 150 per cent rated; code letters, B to E.[2]	Double-squirrel-cage. Low-starting current, 4½ to 5 times rated; full-voltage start; high-starting torque, about 225 per cent rated; code letters, B to E.[2]	High-resistance. Low-starting current; full-voltage start; high-starting torque, about 275 per cent rated.	Wound-rotor. Low-starting current, resistance in rotor circuit; high-starting torque.

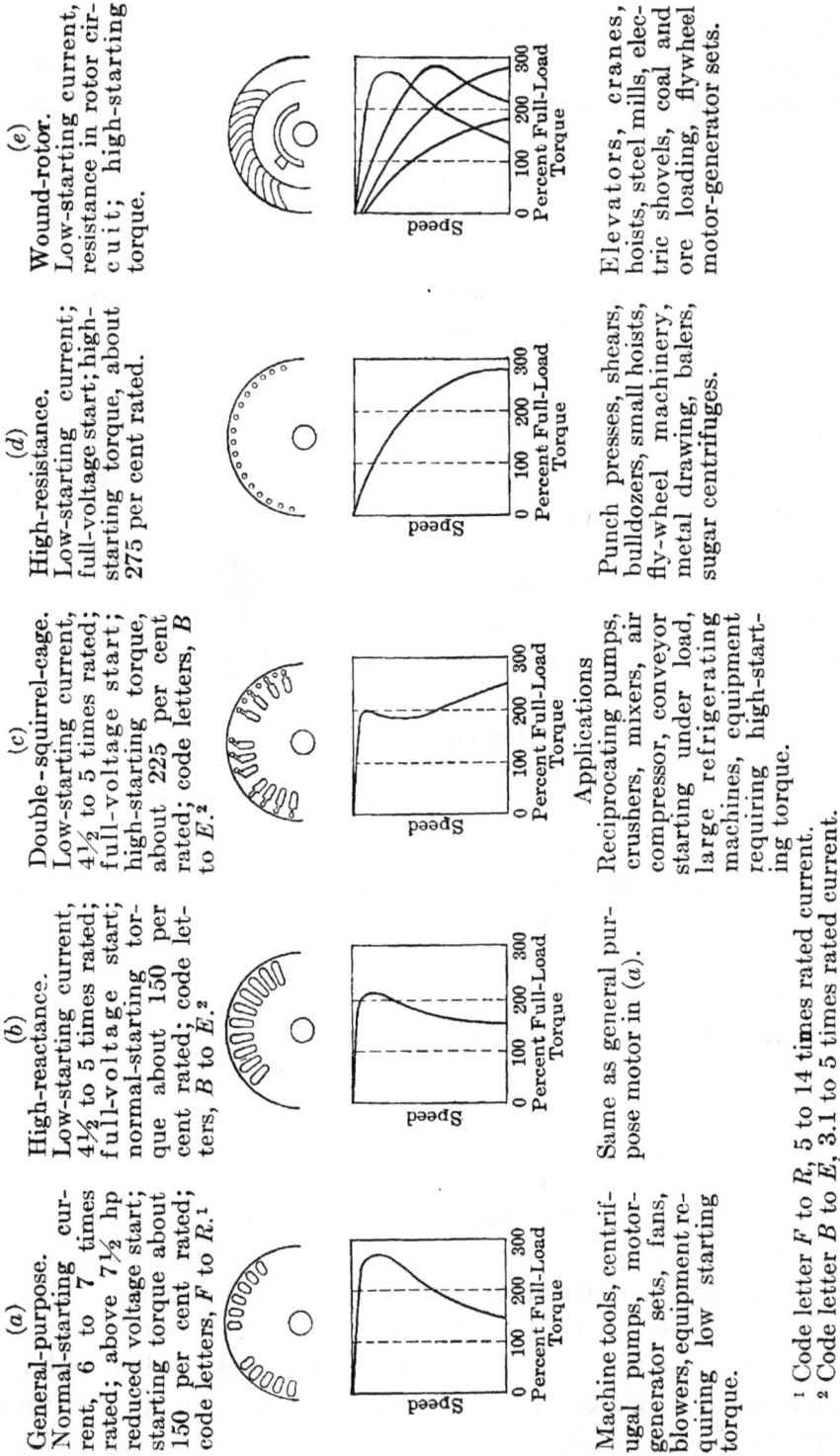

Applications

(a)	(b)	(c)	(d)	(e)
Machine tools, centrifugal pumps, motor-generator sets, fans, blowers, equipment requiring low starting torque.	Same as general purpose motor in (a).	Reciprocating pumps, crushers, mixers, air compressor, conveyor starting under load, large refrigerating machines, equipment requiring high-starting torque.	Punch presses, shears, bulldozers, small hoists, fly-wheel machinery, metal drawing, balers, sugar centrifuges.	Elevators, cranes, hoists, steel mills, electric shovels, coal and ore loading, flywheel motor-generator sets.

[1] Code letter F to R, 5 to 14 times rated current.
[2] Code letter B to E, 3.1 to 5 times rated current.

FIG. 234.—Types of induction motor.

Since starters add considerably to the cost of a motor installation and also may occupy valuable space, manufacturers have directed their efforts to the design of low-starting-current motors for ratings larger than 7.5 hp so that such motors are adapted to across-the-line starting. These motors are designed frequently to give high starting torque. In Fig. 234 are shown the general slot designs and the accompanying speed—torque characteristics for five different types of rotor. In the National Electrical Code, motors are classified by letter according to the ratio of their starting to rated-load current. Such letters should appear on

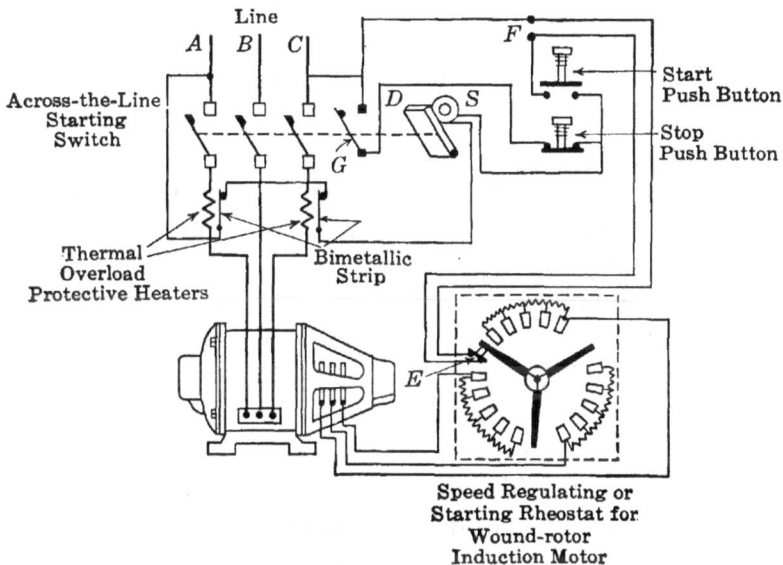

FIG. 235.—Across-the-line starter.

the name plate of the more recent motors. By means of this letter, it is possible to determine the correct ratings of circuit breakers, fuses, and other motor protective devices. In (b), it will be noted that high-reactance, low-starting current is obtained by using totally enclosed, deep, narrow slots. In (c), low-starting current and high-starting torque are obtained by means of the double-squirrel-cage rotor (Par. 163). In (d), low-starting current and high-starting torque are obtained by the use of a high-resistance rotor winding. It is to be noted that this motor is adapted only to intermittent starting and stopping and not to constant-speed drive since its slip is too high and its efficiency too low. The wound rotor in (e) is explained in Par. 161.

In Fig. 235 is shown an across-the-line starter, with its connection to a wound-rotor induction motor. When the "Start" push button is pressed, the solenoid S is energized by the circuit starting at line A, going through the two bimetallic strips in the thermal overload protective heaters, the solenoid S, the "Start" push button, the contacts E, and back to the line wire C. The solenoid S attracts the armature D which is connected mechanically to the three moving-contact members of the starting switch as well as to the auxiliary switch G. As soon as auxiliary switch G is closed, the solenoid circuit becomes energized through the "Stop" push button which is normally closed, so that the "Start" push button may be released. If the arm of the rotor rheostat is not in the starting position, the solenoid circuit is open at the contacts E and the starting switch does not close. If a squirrel-cage motor is used, the terminals F are short-circuited.

To stop the motor the "Stop" push button is pressed. This opens the solenoid circuit, and the switch contacts open. To afford overload protection without the circuit being tripped out by the large momentary values of starting current, thermal overload protective heaters are used. A low resistance is connected in series with two of the three line wires. A bimetallic strip is exposed to the heat developed by each resistance; and when sufficient heat has been developed, the strips deflect to open the solenoid circuit. The strips respond only to the integrated product of rate of heat production and time, so that the solenoid circuit is not opened by large momentary currents such as the starting current. The heaters do protect the motor from continuous overload and from excessive currents caused by an open phase.

From such data as are given in Fig. 234, taken in conjunction with the National Electrical Code, it can be determined whether or not a motor is adapted to across-the-line starting.

Starting Compensator.—When the starting current of a motor is too large for across-the-line starting, a starting compensator is used. The starting compensator is usually a three-phase, Y-connected autotransformer with taps ranging from 25 to 70 per cent. A General Electric starting compensator of this type is shown in Fig. 236. When the switch is in the starting position, the compensator is connected across the line with only the line

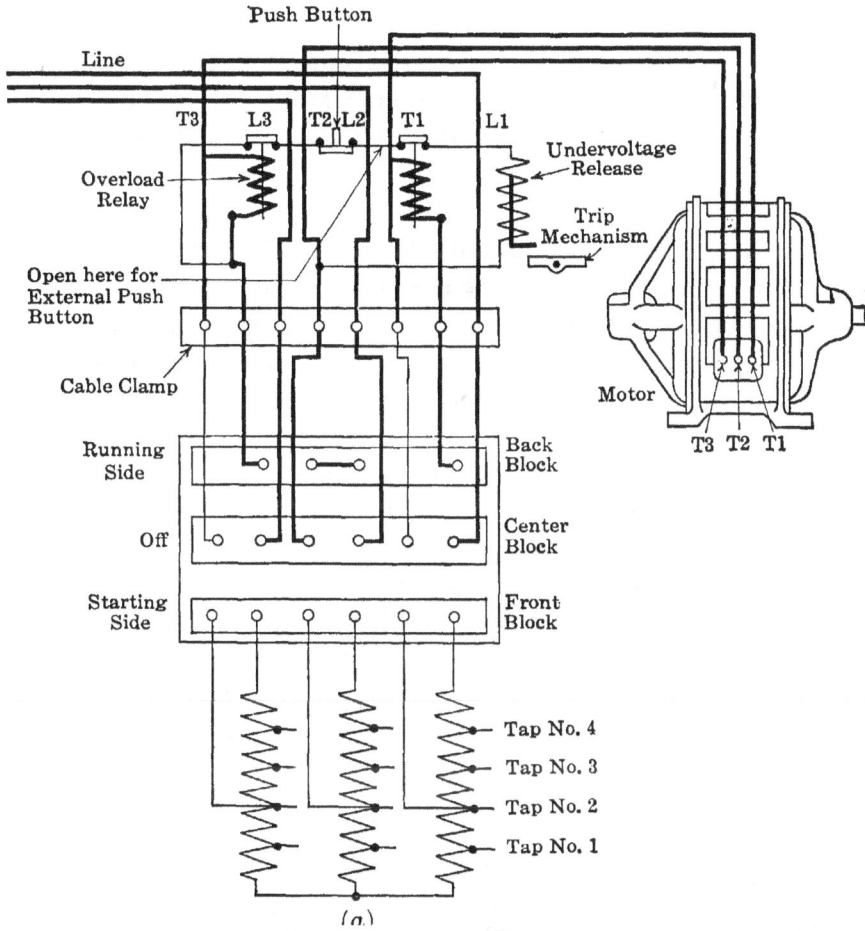

Push Button

Line

T3 L3 T2 L2 T1 L1

Undervoltage Release

Overload Relay

Trip Mechanism

Open here for External Push Button

Cable Clamp

Motor

T3 T2 T1

Running Side

Back Block

Off

Center Block

Starting Side

Front Block

Tap No. 4

Tap No. 3

Tap No. 2

Tap No. 1

(a)

(b)

Fig. 236.—Auto-starter for squirrel-cage induction motor. (*Courtesy of General Electric Company.*)

fuses or circuit breaker for protection. Under these conditions, the three motor lines T_1, T_2, T_3 are connected to three taps, one in each phase of the autotransformer. Hence, the motor voltage is reduced, usually to one-fourth or one-half its rated value. When the switch is in the running position, the compensator is entirely disconnected and the motor is connected directly across the line in series with the overload relays. In Fig. 236 the heavy lines show the path of the current when the switch is in the running position. When the line voltage drops to a value which is too low for the motor to operate satisfactorily, the under-voltage release through its trip mechanism causes the starting arm to spring to the "off" position. The undervoltage release is also operated either by the overload relays or by the push button, each of which opens the undervoltage release circuit.

A compensator supplying a motor with half-voltage reduces the line current to one-fourth its value at full voltage. The motor being at half-voltage takes one-half the current that it would take if directly across the line. As this current is supplied by the secondary of a 2-to-1 autotransformer, the line current is but half the motor current and is, therefore, one-fourth the current that would have been taken had the motor been directly across the line. That is, the line current varies as the tap ratio *squared*. With a 0.6 tap the line current is 0.6^2, or 0.36, of its value when the motor is connected directly across the line.

For a given value of slip the torque of an induction motor varies as the *square* of the line voltage. Hence, the starting torque varies as the square of the tap ratio of the compensator. With a 0.5 tap the starting torque is 0.25 of the starting torque developed when the motor is connected directly across the line.

165. Speed Control of Induction Motors.—The speed of the rotor of an induction motor is given by

$$N_2 = \frac{f \times 120}{P} (1 - s) \text{ rpm, (Eq. (92), p. 272)}$$

where f is the stator frequency in cycles per second, P is the number of stator poles, and s is the slip.

The three factors, frequency, slip, and number of poles, determine the speed of the induction motor. In order to change the speed, it is necessary to change at least one of these factors.

Changing the Slip.—The slip may be changed by introducing resistance in the rotor circuit. This has been discussed in connection with the wound-rotor type of motor. At a given slip, any value of torque up to the breakdown torque may be obtained by this method. Its disadvantages are lowered efficiency and poor speed regulation.

These disadvantages may be avoided by introducing counter emfs instead of resistance in the rotor circuit, either at line frequency, which requires that the rotor have a commutator, or by means of an auxiliary commutating machine which introduces counter emfs at *rotor* frequency through slip-rings. This last method necessitates the use of a commutating type of machine to produce emfs at rotor, or slip, frequency, and ordinarily requires a third machine to drive or be driven by this commutating machine. The investment and complications make this method applicable to the largest units only, a few of which are used for driving rolls in steel mills.

Change of Frequency.—Commercial power systems operate at constant frequency, and it is impossible to control the speed of induction motors by change of frequency when the motors take their power from such systems. In a few special instances, such as in the electric propulsion of battleships, the motors are the only loads connected to the turbine-driven alternators and it is possible to obtain speed control by changing the speed of the turbines themselves. Even here the range of speed variation is limited, because the efficiency of turbines decreases rapidly when their speed departs by any substantial amount from the speed for which they are designed.

Change of Poles.—By means of a suitable switch, the stator connections may be changed in such a manner that the number of poles is changed. This changes the synchronous speed of the motor and, therefore, the speed of the rotor. If the poles are changed in the ratio of 3 to 2, the winding will probably be designed for two-thirds pitch at the higher speed, making it a full-pitch winding for the lower speed. In such a motor the best possible design is not usually obtainable at both speeds. That is, desirable characteristics such as high power factor and high efficiency are sacrificed at one speed in order that a reasonably good motor may be obtained at the other speed. Sometimes the stator connections are changed from delta to Y at the same time that the pole connections are changed. This changes the voltage per phase and makes possible a better motor at each speed. Because of the complications involved in changing the connections, it is not desirable to attempt to obtain more than two speeds by changing the number of poles. To avoid these complicated switching connections, induction motors sometimes have two distinct windings, the two windings being for different numbers of poles.

As a matter of fact the induction motor is inherently a one-speed motor, and speed adjustment and control are not easily obtainable.

166. Induction Generator.—If an induction motor is driven above synchronous speed, the slip becomes negative. The rotor conductors then cut the flux of the rotating field in a direction opposite to that which occurs when the machine operates as a motor. The rotor currents are then reversed with respect to the direction which they have when the machine operates as a motor. By transformer action, these rotor currents induce currents in the stator which are substantially 180° out of phase with the inducing currents. They cause a transfer of energy, therefore, from rotor to stator, and the machine operates as a generator and is called an *induction generator.*

In many respects the operation of the induction generator is materially different from the operation of the synchronous generator or alternator.

The rotor must be driven above synchronous speed in order to obtain generator action, and the power output is practically proportional to the slip, as it is with the induction motor. Hence, the machine does not have a definite speed for a given frequency as the synchronous alternator has, but the speed with constant frequency varies with the load. Because its speed is not in synchronism with line frequency, the machine is often called an *asynchronous* generator. The frequency and voltage of the induction generator are *those of the line to which it is connected,* irrespective of its speed.

With the induction machine the change from motor to generator action is not unlike the change from motor to generator action of a d-c shunt dynamo across constant-voltage bus-bars, when the motor speed is raised. The excitation and the N- and S-poles of the d-c dynamo do not change in the transition from motor to generator action, but the direction of current in the armature does change. Likewise, the excitation and the N- and S-poles of the induction machine do not change in its transition from motor to generator action. The excitation is supplied by the polyphase currents in the stator windings which cause the N- and S-poles to rotate in the air-gap at synchronous speed. This excitation is obtained from the a-c *lines* to which the induction generator is connected. The rotating N- and S-poles of the synchronous alternator obtain their excitation from a d-c source.

This exciting current which comes from the line is a lagging current with respect to the induction-motor current and does not change in the transition from motor to generator, but the energy current of the motor reverses its direction, or changes its phase by 180°. The exciting current, therefore, must *lead* the generator current and the induction generator delivers only *leading* current. As most loads demand *lagging* current, the induction generator cannot supply these loads, whereas the synchronous alternator can deliver current at any power factor, leading or lagging current. The induction generator, therefore, must always operate in parallel with a synchronous machine,[1] the synchronous machine supplying not only the lagging current to the load, which the induction machine cannot supply, but the exciting current required by the induction machine as well. These are serious objections to the use of induction generators.

The advantages of the induction generator are that it does not hunt or drop out of synchronism, it is simple and rugged, and when short-circuited it delivers no sustained short-circuit current since its excitation becomes zero after a few cycles.

Because of the superiority of the synchronous alternator, particularly in its ability to deliver lagging current, the induction generator is seldom used as a power source. However, when induction motors are used in electric-railroad locomotives,

---

[1] If capacitors are connected across its terminals to supply the lagging excitation current, the induction generator may operate independently of a synchronous machine. (Capacitors *take* leading current and hence *supply* lagging current). Such a system requires, however, considerable capacitance and it is not always stable, so that it is not used in practice.

induction-generator action can be used for regenerative braking. If the motors remain connected across the line on downgrade, they automatically become induction generators to brake the train and at the same time pump power back into the line, the transition from motor to generator requiring no auxiliary control apparatus.

167. Measurement of Slip.—There are various methods for measuring slip. The slip may be determined by measuring the rotor speed and subtracting this speed from that of the rotating field as determined by the frequency. Because the slip is but a small percentage of either synchronous speed or rotor speed and is the difference of two nearly equal quantities, it is not possible to determine it accurately by the measurement of each of these quantities and so finding their difference.

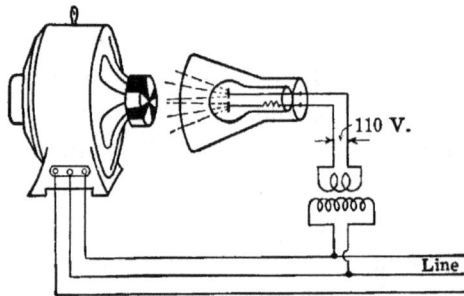

Fig. 237.—Stroboscopic method for measuring slip.

A simple method of measuring slip is shown in Fig. 237. A "target," or disc, is fastened to the end of the shaft or to the pulley of the motor. This disc has the same number of black and the same number of white sectors as the motor has poles. It is illuminated by a neon bulb or some similar type of stroboscopic lamp. The lamp is connected to the same line that supplies the motor, care being taken that the voltage is reduced to the rated value for the lamp. It requires about 50 volts to sustain the neon glow. Hence, the lamp glows only during the time the voltage in each alternation exceeds 50 volts. In the remainder of the time, it emits no light. In one-half cycle the armature of the motor would advance one pole if there were no slip. During this time, each black sector would advance to the position just occupied by the adjacent black sector which preceded it. The same is true of the white sectors. During the period of advancement the sectors are not visible because practically no light is emitted by the lamp. Therefore, each black sector and each white sector is not clearly visible until it has reached the position just occupied by the sector of the same color just preceding it. As the disc is illuminated twice each cycle, while the voltage wave is going through its maximum value, all the sectors are momentarily visible twice each cycle. Therefore, if the disc rotated at synchronous speed, it would *appear* stationary. Owing to the fact that each conductor on the rotor does *not* advance one pole each half-cycle, the sectors will not reach the position of the next adjacent sector of the same color but will fall short of this distance, owing to the slip. The sectors on the disc will then appear not stationary but to be rotating slowly

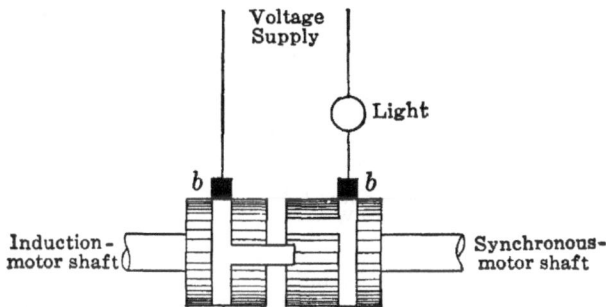

Fig. 238.—Measurement of slip by means of synchronous motor.

backward. The number of rpm that they *appear* to rotate is the revolutions slip of the rotor. Figure 237 shows a stroboscope as applied to a four-pole motor. Occasionally, black and white stripes are painted on the pulley (Fig. 237), to serve the same purpose as the disc.

A mechanical-electrical method of measuring slip is shown in Fig. 238. Two cylinders of insulating material are driven, one by the induction motor shaft and the other by a small synchronous motor having the same number of poles as the induction motor. Each of these cylinders is fitted with a slip-ring, to which a small contact piece is connected. The synchronous motor always runs at the speed of the rotating field. Every time, therefore, that the induction motor slips one revolution, the contact pieces touch each other, closing the circuit between the two sliprings. This is indicated by a flash of the light connected in series with the rings through the brushes *b* (Fig. 238).

In the Electrical Engineering Laboratories at Harvard University, the induction motor and the synchronous motor jointly drive a differential through gears, a method developed in these laboratories. The speed of the differential is the revolutions slip of the induction motor. If desired, the speed of the differential, and hence the slip, may be measured with considerable accuracy with a speed counter. By changing gears, the apparatus is adapted to machines having any number of poles.

168. Induction Regulator.—Without auxiliary apparatus, it is practically impossible to maintain the proper voltage at all the distribution points of a system, because, with a fixed voltage at the station bus-bars, the voltage at the ends of short feeders will ordinarily be greater than the voltage at the ends of long

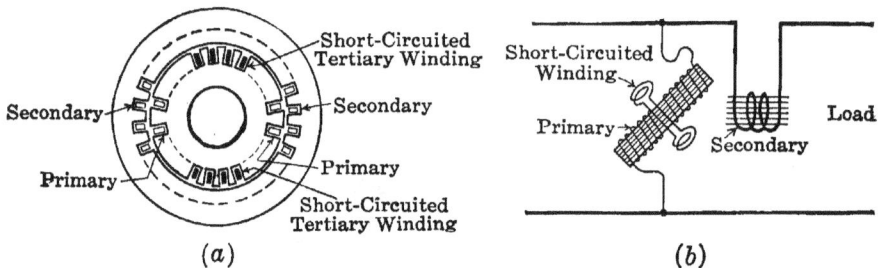

Fig. 239 (*a*).—Single-phase induction regulator.
Fig. 239 (*b*).—Connections of single-phase induction regulator.

feeders. Owing to the ohmic and reactive drops in the lines, the voltage at the load end of the feeder may vary considerably with the load on the feeder. In order to maintain a more constant voltage at the distribution point, without using an excessive amount of copper, an induction regulator is often connected to each feeder. This maintains the voltage at the distribution point practically constant.

The induction regulator is a transformer in which either the primary or the secondary is movable, the one with respect to the other. The principle of the single-phase type is shown in Fig. 239 in which the primary is movable. An ordinary drum type of winding is placed in the slots on the stator and in the slots of the rotor. When the primary is in the plane of the secondary, the maximum emf is induced in the secondary, because the mutual inductance of the windings is a maximum when the primary is in this position. When the primary is at right angles to the secondary, the primary flux does not link the secondary so that the induced emf in the secondary is zero. As the mutual inductance of the windings is zero under these conditions, the secondary acts like a choke coil of very high impedance and it is in series with the line (Fig. 239(*b*)). To prevent this, a short-circuited tertiary winding is placed on the rotor. This

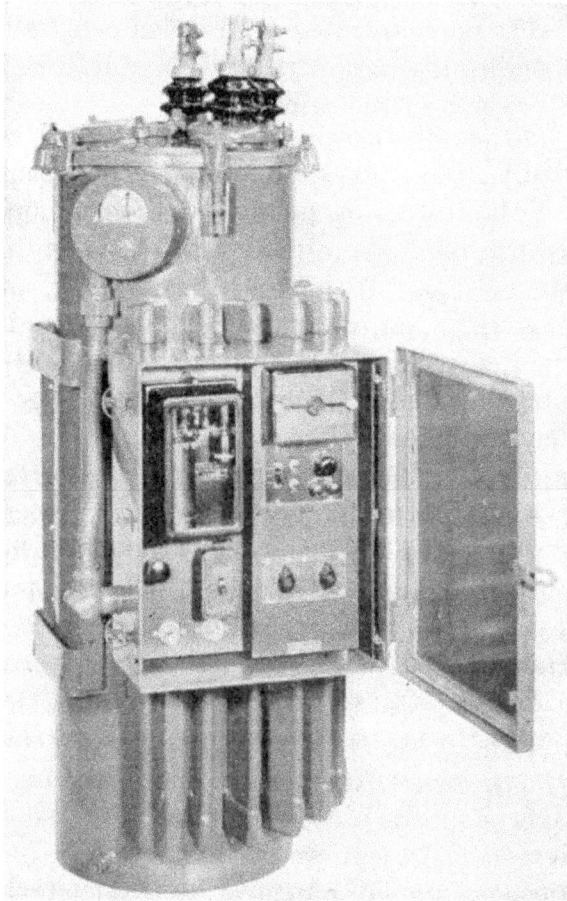

Fig. 240.—Induction voltage regulator, single-phase, 60 cycles, 24 kva, 2,400 volts, 100 amp, at raise-and-lower 10 per cent regulation. (*Courtesy of General Electric Company.*)

acts like a short-circuited transformer secondary and, therefore, reduces the inductance of the regulator secondary to a small value. The primary winding is shunted across the line (Fig. 239(*b*)), and the secondary is connected in series with the line. That is, the regulator operates as an autotransformer. The kva *transformed* is the product of the secondary current and the *regulator* secondary voltage which is a small portion of the total load voltage. Hence the kva transformed is a small proportion of the kva to the load.

When the primary is in the plane of the secondary, in one position, the induced emf in the secondary is a maximum and the regulator acts as a booster. When the primary is turned 180° from this position, the secondary emf is also a maximum, but it now *bucks* the line voltage. Any value of voltage between that corresponding to these two positions may be obtained by varying the position of the primary.

The movable member is turned by a small motor, controlled by relays (Fig. 240). The relays are actuated by a contact-making voltmeter. If the voltage is too high, one set of contacts causes the motor to turn in such a direction as to make the secondary reduce the line voltage. If the voltage is too low, another set of contacts causes the motor to reverse its direction and the secondary boosts the line voltage. Figure 240 shows a 24-kva, induction voltage regulator with the control-cabinet door swung open to show the control panel.

The three-phase induction regulator closely resembles the three-phase, wound-rotor induction motor. The three stator windings, or primaries, are connected across the line in either Y or delta. The three secondaries, which correspond to the three phases of a rotor winding, are insulated from one another and each is connected in series with one of the wires of the three-phase line. As the stator produces a uniform rotating field, the induced emfs in the secondaries are constant and are independent of the position of the rotor. Their boosting and bucking effect, however, depends on the *phase relations* existing between each induced secondary emf and its line voltage.

The three-phase regulator requires no short-circuited tertiary winding.

CHAPTER IX

SINGLE-PHASE MOTORS

169. Alternating-current Series Motor.—The direction of rotation of either the d-c shunt motor or the d-c series motor is the same irrespective of the polarity of the line voltage. If the line terminals are reversed, both the field current and the armature current are reversed, and the direction of rotation remains unchanged. If such motors are supplied with alternating current, therefore, the *net* torque developed acts in one direction only.

With alternating current, the shunt motor develops but little torque. The high inductance of the shunt field causes the field current and, therefore, the main flux to lag nearly 90° in time phase with respect to the line voltage. The armature current cannot lag the line voltage by a large angle if the motor is to develop any substantial power at a reasonably high power factor. Accordingly, with the shunt motor there will be considerable time-phase difference between the flux and the armature current. When the flux is a maximum, the armature current is near its zero value; and when the armature current is a maximum, the flux is near its zero value. Hence, but little torque will be developed under these conditions, and such a motor is therefore impracticable.

In the series motor, however, the same current flows in both the armature and the field. Hence, the flux and the armature current are practically in time phase with each other. The flux is therefore a maximum when the armature current is a maximum, etc. Consequently, the series motor develops approximately the same torque per ampere with alternating current as it does with direct current. With a few changes, the series motor will operate satisfactorily with alternating current.

Since the flux is alternating, the field structure as well as the armature must be laminated to reduce eddy-current losses.

The series-field winding has a comparatively high self-inductance. In order to prevent the voltage drop across the field becoming so high as to make the motor impracticable, as few turns per pole as possible are used. The series-field reactance is further reduced by operating at low frequency, 25 cycles being about the maximum frequency for which a motor can be designed to operate satisfactorily.

Owing to the few turns per pole, it is necessary to make the air-gap of the motor as short as mechanical clearance will permit. Hence, armature reaction is increased (see Part I, Par. 216, p. 274). Moreover, the cross-magnetizing flux resulting from

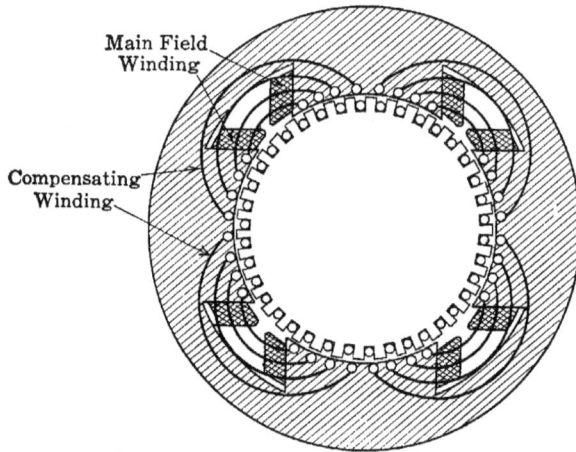

FIG. 241.—Windings of an alternating-current series motor.

armature reaction has the further disadvantage that it increases the *reactance* of the armature, increasing the voltage drop in the armature and thereby lowering the power factor of the motor. The armature reaction is almost entirely suppressed by a compensating winding in the pole faces (Fig. 241). This winding is ordinarily in series with the armature and is connected so that the current in each pole-face slot is opposite in direction and equal in magnitude to the current in the armature slot directly across the air-gap. Hence, the armature mmf is opposed at each point on the armature by a mmf which is opposite in direction and nearly equal in magnitude. The armature reaction is therefore reduced to a minimum. If the compensating winding is short-circuited on itself, it becomes a short-circuited secondary of a transformer, the primary of which is the armature. Hence, its ampere-turns are nearly opposite in phase and are nearly equal in magnitude to those of the primary. The motor is then said to be *inductively* compensated. If the winding is in series with the armature, the motor is said to be *conductively* compensated.

A commutation difficulty occurs with this type of motor which does not occur with the d-c motor. When the armature coils are short-circuited by the brushes during commutation, they are linked by the alternating flux from the field poles. This flux induces a transformer emf in these coils which are short-circuited by the brushes; and unless prevented, large short-circuit currents

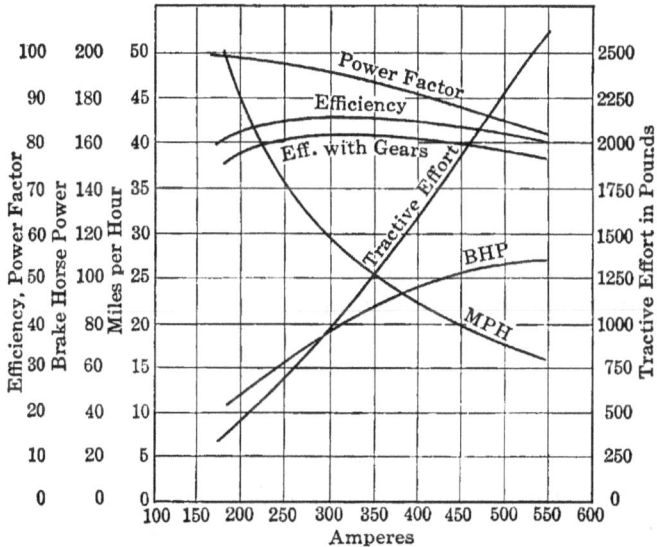

Fig. 242.—Characteristic curves of 430-amp., 235-volt, 25-cycle, single-phase Westinghouse railway motor. Continuous rating, 200 amp., 235 volts.

and severe sparking result. This difficulty is overcome in part by using commutating poles shunted with resistance and by introducing resistance leads in the armature circuit.

Owing to transformer action between field and armature, commutation becomes more difficult than with the d-c motor, and greater refinements in design are necessary.

This type of motor owes its development to the necessity of securing a single-phase motor suitable for railway electrification. Its operating characteristics (Fig. 242) are similar to those of the d-c series motor. The power factor *decreases* with increase of load, whereas the power factor of most other apparatus increases with increase of load (see p. 278, Fig. 228). Initially, this motor was designed and developed by the Westinghouse Electrical and Manufacturing Co. for the New York, New Haven and Hartford R.R. From New Haven to Woodlawn, near Mt. Vernon, the locomotives take power at 11,000 volts, 25 cycles, from an overhead trolley wire, by means of a pantograph trolley. An autotransformer on the locomotive reduces this voltage to 250 volts, the rated voltage of the series motors. The electric locomotives run from Woodlawn into the Grand Central Station, New York City, over the New York Central, 600-volt, d-c system. The same motors are used for both d-c and a-c service, the control devices being switched over when transition is made from one service to the other. The motors which operate at 250 volts on alternating current are connected two in series for d-c operation. This type of motor is also used by the Pennsylvania Railroad.

Owing to the many refinements in design necessary for successful a-c operation and also owing to its greater weight per horsepower, this type of motor has been limited practically to the large units used in electric locomotives.

170. Universal Motors.—Simple series motors of fractional-horsepower rating will operate satisfactorily on a-c circuits at frequencies as high as 60 cycles. Obviously, such motors must have laminated-field structures. The advantages of such motors are that they operate satisfactorily with both alternating and direct current and have high starting torque and that their speed is readily controlled, usually with series resistance. In order to have high power factor and output, they must run at high speeds, from 4,000 to 10,000 rpm in the smaller sizes. Simple motors are available up to ¼ hp, and compensated motors are available up to ½ hp. Typical industrial applications of this type of motor are in sewing machines, vacuum cleaners, and portable electric drills.

The same factors which tend to prevent successful operation of the a-c railroad motor are present in these small motors, but so little power is involved that their effects are not serious.

171. Repulsion Motor.—Figure 243 shows diagrammatically a bipolar motor having a drum-wound armature and a commutator such as would be used for a d-c motor. This armature is placed in a two-pole magnetic field, the field being supplied from a single-phase source. A single-phase flux, varying sinusoidally with time, links the armature winding and induces emfs in it. That is, the armature acts as the secondary of a transformer of which the field winding is the primary.

At the instant shown in Fig. 243(a), the current in the upper line is positive, or is flowing to the motor, and is increasing; the current in the lower line is negative, or is flowing from the motor. The field is wound in such a direction that at this instant the upper pole is north and the lower pole is south. Hence, the direction of the flux is downward through the armature. By Lenz's law (see Part I, p. 171) the direction of the induced emf

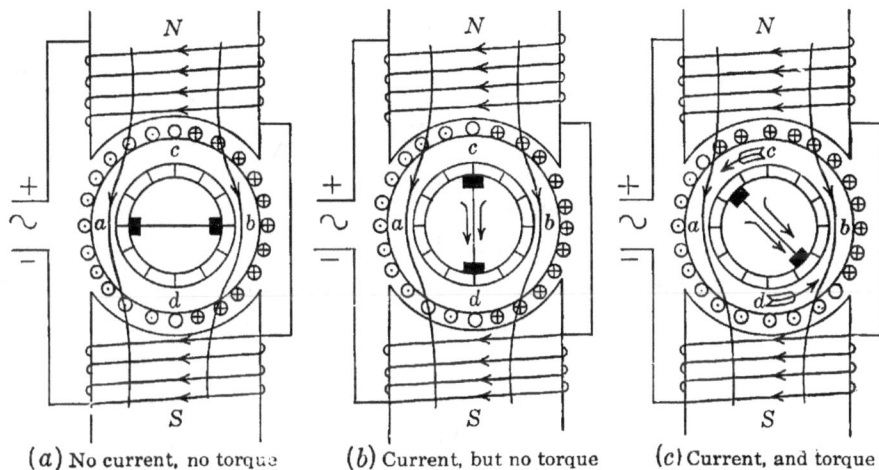

(a) No current, no torque   (b) Current, but no torque   (c) Current, and torque

FIG. 243.—Repulsion motor.

must be such that, if current flows, it *opposes* the increase of flux. Therefore, the direction of the induced emf on the right-hand side of the armature is inward and on the left-hand side is outward. The crosses (+) and dots (.), Fig. 243(a), show the direction of the induced emf, not the direction of current. Points *a* and *b* on opposite sides of the armature lie in the geometrical neutral. Since some coil side at *a* connects to some coil side at *b* to complete a coil, the difference in potential between points *a* and *b* is equal to the induced emf in a single coil. In fact, with a large number of armature slots the potential difference between *a* and *b* is practically zero. For simplicity, assume that each commutator connection is carried radially inward to the commutator rather than spirally, as is usual. If the brushes short-circuit commutator segments in the neutral plane (Fig. 243(a)), no current flows in the armature winding as a whole since the brushes short-circuit points between which no potential difference exists. (A local transformer current may flow in the one or two coils connected between adjacent segments, which the brushes short-circuit, but as this current flows in the neutral plane it can produce no torque.) The brushes are in such a position that torque would be developed if current did flow in the armature winding and brushes, since the currents in all conductors under each pole would flow in the same direction, and the direction of current under the two poles would be opposite.

Owing to the transformer action between field and armature, the maximum potential difference between points in the armature winding occurs along the pole axis or between points c and d. Therefore, if the commutator is short-circuited along the plane cd (Fig. 243(b)), current will flow in the armature winding. The armature winding acts as a short-circuited secondary, the current opposing the change of flux. But no torque will be developed since an equal number of ampere-conductors under each pole carries current in opposite directions as shown in (b) by the crosses and dots. The net torque developed is zero.

In diagram (a), therefore, the brushes are in a position to develop torque, but there is no armature current; in diagram (b), there is a large armature current, but the brushes are in such a position that the net torque is zero.

If the brushes are moved to some position intermediate between (a) and (b), as in (Fig. 243(c)), the brushes short-circuit points of the armature winding between which a difference of potential exists and the direction of the current in the conductors directly under the poles is such that a net torque results, tending to cause rotation, the direction of rotation being counterclockwise. By changing the position of the brushes so that each is moved to the other side of the plane cd, the direction of rotation may be reversed. The brush position for a four-pole repulsion motor is shown in Fig. 244.

As a rule, repulsion motors do not have salient poles such as are shown in Figs. 243 and 244, which are diagrammatic only. As the reluctance of the air-gap must be kept as low as possible, the motor is constructed like the induction motor, having a stator completely surrounding the rotor, a short air-gap, and a distributed stator winding. The characteristics of the repulsion motor are those of the series motor, namely, high starting torque and high speeds at light loads. The simple motor cannot ordinarily be used for continuous duty because of severe sparking, but it can be used merely for starting certain types of motors which have little or no starting torque (see p. 309).

172. Single-phase Induction Motor.—An induction motor may be made to operate with single phase applied to its stator, although single-phase current of itself cannot produce a rotating field. A single-phase motor is shown diagrammatically in Fig. 245. At the instant shown the current enters the lower line

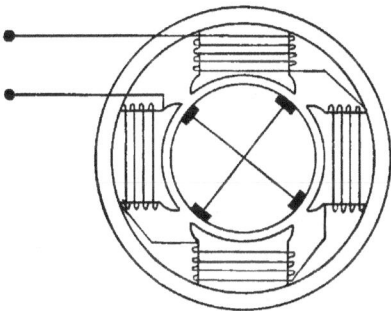

Fig. 244.—Brush position with four-pole repulsion motor.

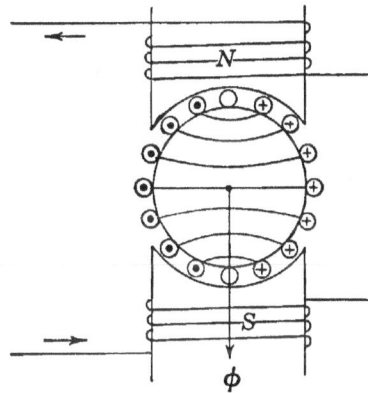

Fig. 245.—Transformer currents in the rotor of a single-phase induction motor.

and is increasing. By the corkscrew rule the upper pole is north and the lower pole is south. The flux ϕ therefore acts downward and is increasing. The effect of the squirrel-cage rotor is as if each conductor on one side of the armature were connected with a conductor on the opposite side of the armature to form a closed turn (Fig. 245). Since the rotor acts as a transformer secondary, the direction of its currents must be such as to *oppose* the increase of flux. As a result, the motor when stationary is a short-circuited static transformer. The magnetic field alternates but always acts along the central axis. As there is no rotational motion of the field, there is no tendency for the rotor to rotate.

If, however, the rotor, by some means, is caused to rotate in either direction, the rotor conductors will *cut* the flux ϕ and accordingly will develop a *rotational*, or *speed*, emf as in a generator. As this emf occurs in conductors which are short-circuited on themselves, comparatively large currents will flow. These currents, acting in conjunction with the stator current, produce a rotating field, whose direction of rotation is the same as that of the armature. Hence, the motor continues to accelerate in this direction until it reaches nearly synchronous speed. Near synchronous speed, it develops considerable torque and will carry load. Like the polyphase motor, its slip increases with load, and it has a definite breakdown torque. For the same speed and weight the single-phase motor has approximately 50 per cent of the output of the polyphase motor. It is slightly less efficient and has lower power factor.

173. Operation of Polyphase Motor as Single-phase Motor.— If one phase of a polyphase induction motor is opened, the motor will continue to operate as a single-phase induction motor, although it will not start under these conditions. The rating and the breakdown torque of a polyphase motor, operating single-phase, are considerably reduced in value; and if rated polyphase load is applied continuously, the motor overheats.

Ordinarily, in starting a polyphase motor, all three lines are closed when the compensator is in the starting position and the motor starts. When the compensator is thrown to the running position, however, a phase may become open through an open fuse or an open circuit in the compensator. The motor then operates single-phase, and the only indication that it may give of this condition is overheating if the load is near the rated polyphase value. It is difficult to locate the open circuit by a voltmeter or by test lamps, since the motor generates a counter emf nearly equal to the line voltage and the voltage difference across the open circuit is small. The best test for an open phase is to insert an ammeter in each line.

174. Starting Single-phase Induction Motors.—As the single-phase induction motor is not self-starting, auxiliary means must be used to supply initial torque. One method is to split the phase by combinations of inductance, resistance, and capacitance.

*Split-phase Methods.—*Figure 246 shows one method of splitting the phase, a two-pole motor being shown. The main winding, which is highly inductive, is connected across the line in the usual manner. Between the main poles are auxiliary poles which have a high-resistance winding, and this winding is also connected across the line, but in series with a switch *S* operated by a centrifugal mechanism in the rotor. As the auxiliary winding has a high resistance, its current will be more nearly in phase with the line voltage than is the current in the main winding. For the best conditions, the two currents should differ in phase by 90°; but this relation is not readily obtainable and, in fact, is not necessary. These two sets of poles produce a sort of rotating field which starts the motor. When the motor comes up to speed, the centrifugal mechanism opens the switch *S* and disconnects the auxiliary winding. The

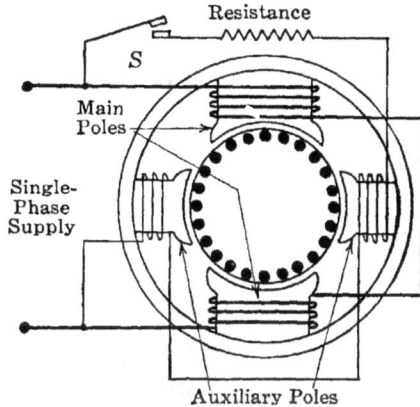

FIG. 246.—Split-phase method of starting single-phase induction motor.

motor then operates with the single-phase pulsating field such as the motor in Fig. 245. Actually, such motors have nonsalient poles. The air-gap is uniform, and the windings are embedded in slots.

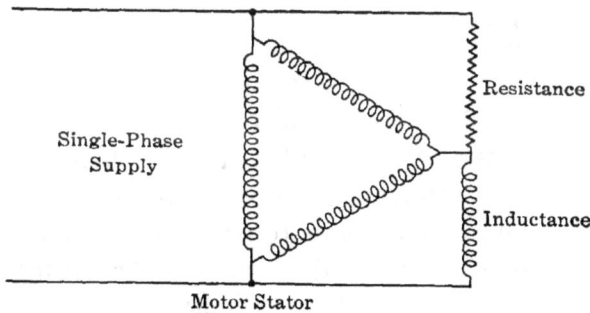

FIG. 247.—Splitting the phase with resistance and inductance.

This same starting principle is applicable to motors having more than two poles.

To reverse the direction of rotation the starting winding is reversed with respect to the main winding.

Another method of splitting the phase is to use a three-phase winding, as shown in Fig. 247, and to connect resistance and inductance as shown. Resistance and capacitance may also be used.   Either a delta- or a Y-connected stator may be used. The resistance and inductance, when connected as shown, displace the phase relations of the currents in the different phases of the stator with respect to one another and so produce a rotating field.

Because of the characteristics of the field combined with the squirrel-cage characteristics of the rotor, the starting torque of split-phase motors is small.

Shaded-pole Method.—The shaded-pole method is shown in Fig. 248.   A short-circuited coil of low resistance is connected around one of the two tips of each pole.   When the flux is increasing, a portion of the flux attempts to go through this shaded tip.

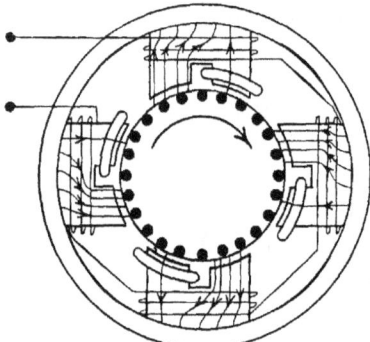

FIG. 248.—Shaded-pole induction motor.

This flux induces a current in the coil which by Lenz's law is in such a direction as to oppose the flux entering the coil.   Hence, at first the greater portion of the total flux is forced to the non-shaded side of the pole, as shown in Fig. 248. Ultimately, however, the main flux reaches its maximum value, where its rate of change is zero.   The opposing emf in the shading coil induced by the main flux becomes zero at this instant and later the opposing mmf of the short-circuited coil ceases, the current in this coil lagging its emf.   Considerable flux then links the short-circuited coil.   After the main flux begins to decrease, the induced current in the shading coil tends to prevent the flux then existing in the shaded pole tip from decreasing.   Accordingly, the flux first reaches its maximum value in the nonshaded side of the pole and later reaches its maximum in the shaded side.   The effect of the shading coil is to retard in time phase a portion of the flux, so that there is a sweeping of the flux across the pole face in the direction of the shading coil, which in Fig. 248 is in a clockwise direction.   This flux cutting the rotor conductors induces currents, which produce a torque sufficient to start the motor.   The shaded pole is not a common method of starting single-phase induction motors and is used only in motors of small size.

It will be remembered that this same shaded-pole principle is used in the light-load adjustment of the induction watthour meter (see p. 109).

Repulsion-motor Starting.—The preceding methods of starting the single-phase induction motor produce weak starting torques which are insufficient to start the motor except under the lightest loads. Where high starting torque is necessary, such as 275 and 300 per cent rated-load torque, repulsion-motor starting is used. The rotor, or armature, winding and commutator are identical with those used in d-c dynamos (Par. 171), except that the brushes usually press axially against the end of the commutator rather than radially against its cylindrical surface. On starting, the brushes are short-circuited on themselves as indicated in Figs. 243(c) and 244. When the rotor reaches about 75 per cent rated speed, a governor weight within the rotor is thrown outward. Through bell-crank levers and rods, this action causes short-circuiting segments within the rotor to be pressed against the inside of the commutator, short-circuiting its segments. At the same time the brushes are forced axially out of contact with the commutator. With the rotor thus short-circuited, the motor now operates as a single-phase induction motor. Because of the development of the capacitor motor, repulsion-motor starting is used only when unusually high starting torque is necessary.

175. Capacitor Motor.—In a capacitor motor the phase is split by connecting capacitance rather than resistance in series with the auxiliary winding of Fig. 246, as shown in Fig. 249. The main winding 1 is connected directly across the single-phase power supply. The auxiliary winding 2 is connected in series with capacitance across the same supply. Unlike the motor of Fig. 246, the auxiliary winding in series with the capacitor is not cut out as the rotor approaches operating

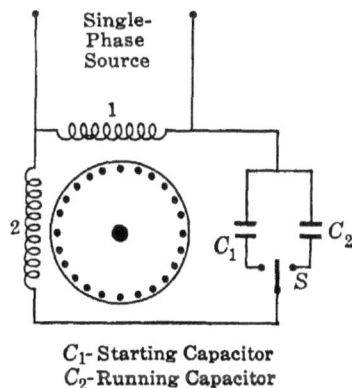

C_1- Starting Capacitor
C_2-Running Capacitor

FIG. 249.—Capacitor motor.

speed. The best starting torque and running conditions are obtained when one capacitance C_1 is used for starting and a smaller capacitance C_2 is used for running. On starting, the capacitance is in series with the standstill impedance of the winding 2. On running the equivalent impedance of the winding is increased so that less series capacitance is required. The capacitor C_1 is connected at starting and the capacitor C_2 is connected under running conditions by means of the switch S, which is controlled by a centrifugal device.

The currents in main and auxiliary windings are practically in quadrature so that a true, two-phase, rotating field is obtained, substantially. When, for economy, a single capacitor is used, its capacitance is a compromise between that for the best starting and that for the best running conditions. Due to the capacitance, the power factor of the motor is high. The capacitor motor became practicable only in recent years when reliable capacitors at low cost became available.

CHAPTER X

SYNCHRONOUS MOTORS AND CONVERTERS

176. Synchronous Motor.—A d-c generator, when supplied with electrical energy at its rated voltage, operates satisfactorily as a motor and at the same electrical rating as when operating as a generator. Likewise an alternator, when supplied with electrical energy at its rated voltage and frequency, operates satisfactorily as a motor and at the same electrical rating as when operating as a generator. When a synchronous machine operates as a motor, it is called a *synchronous motor*.

There is no substantial difference in the design details of the d-c machine, whether it be motor or generator. Although, as a rule, there need not be differences in design details of alternators and synchronous motors, frequently better synchronous motor characteristics are obtained if slight changes are made, such as in the length of air-gap, the field ampere-turns, etc. Alternators can be either of the salient- or non-salient-pole type. In general, synchronous motors are of the salient-pole type, since this type gives greater stability, particularly when the motor loses its excitation.

177. Principles of Operation.—Figure 250 shows a conductor *a* under an *N*-pole and carrying a current flowing toward the observer. By the corkscrew rule, the direction of the mmf about this conductor is counterclockwise. It is, therefore, upward on the right-hand side of the conductor and downward on the left-hand side. The direction of flux from the *N*-pole is downward. Hence, the mmf due to the current in the conductor *a* opposes this flux on its right-hand side and assists it on its left-hand side. The flux density, therefore, will be decreased to the right of *a* and increased to the left of *a*. Conductor *a* will then tend to move from left to right (Fig. 250) (see Part I, p. 267). If the current is alternating, it will reverse its direction for the next half-cycle and the flux density will be increased to the right and

311

decreased to the left of conductor *a*. The torque then acts from right to left. The net torque, therefore, over any number of complete cycles is zero; and no continuous motion of the rotating member can result. This is the condition existing in a synchronous motor at standstill. The armature conductors carry alternating current; and the poles have fixed polarity, being excited with direct current. Therefore, the synchronous motor, as such, develops no starting torque.

If, however, conductor *a* can in some manner be brought under the next pole, which is an *S*-pole, for the half-cycle during which the current in *a* is in the reverse direction, the flux density will continue to be greater to the left of the conductor (Fig. 250) and the resulting torque will continue to be from left to right. Hence, a tendency toward continuous motion will result.

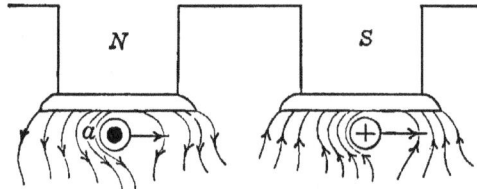

Fig. 250.—Torque developed by synchronous motor.

In a synchronous motor, therefore, a given conductor must move from one pole to the next in each half-cycle, if the motor is to operate. This applies to the rotating-armature type of machine. If the motor is of the rotating-field type, an arc equal to the pole-pitch must pass any given conductor every half-cycle. The synchronous motor then must operate at *constant speed*, if the frequency is constant. During a single revolution there may be slight fluctuations in speed; but if the *average* speed differs by even a small amount from this constant value, the average torque will ultimately become zero and the motor will come to standstill. The relation of speed, number of poles, and frequency for the synchronous motor is the same as for the alternator and for the rotating field of the induction motor. The speed $S = 120f/P$ rpm, where f is the frequency and P the number of poles (see Pars. 9 and 156, pp. 14 and 270).

Example.—What is the rated-load speed and the no-load speed of a 200-kva, 12-pole, 2,300-volt, 60-cycle synchronous motor?

With constant frequency the speed must be the same at all loads. Hence,

$$S = \frac{120 \times 60}{12} = 600 \text{ rpm. } Ans.$$

178. Synchronous Motor as an Elastic Coupling.—The stator of the synchronous motor is wound in the same manner as the stator of the alternator and the induction motor. Hence, if

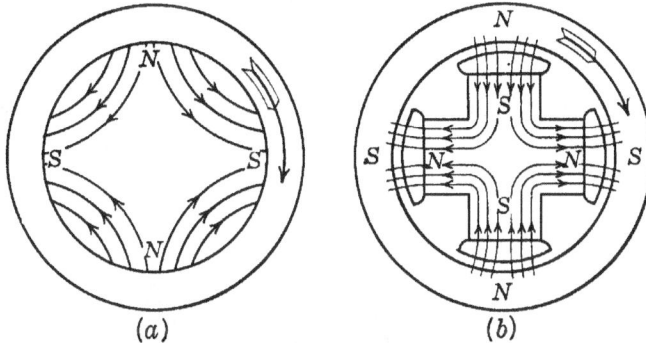

(a)   (b)

FIG. 251.—Interlocking action of salient poles with rotating magnetic field.

polyphase currents are supplied to the stator, a rotating magnetic field must result. Such a rotating field, having four poles and rotating clockwise, is shown in Fig. 251(a). If a salient-pole rotor having two N- and two S-poles is placed in this field (Fig. 251(b)) and is brought near synchronous speed, the S-poles of the rotor will be attracted to the N-poles of the stator and the N-poles of the rotor will be attracted to the S-poles of the stator. The rotating field of the stator, thus being locked with the rotor poles, will drag the rotor around at synchronous speed, or there is magnetic or elastic coupling between stator and rotor not unlike the mechanical coupling (Fig. 252).

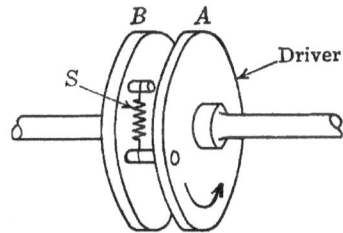

FIG. 252.—Spring-coupling analogue of synchronous motor.

Disc A, rotating in a counterclockwise direction, drives disc B through the spring S which is connected between two pins, one on each disc. So long as the spring coupling remains intact, the disc B must rotate at the same speed as disc A. Likewise, the rotor of the synchronous motor (Fig. 251(b)) must rotate at the speed of the rotating field so long as the magnetic coupling between stator and rotor remains.

179. Effect of Loading Synchronous Motor.—If a load is applied to a d-c shunt motor, the speed is slightly decreased. This reduces the counter emf, permitting more current to enter the armature and enabling the motor to carry the increased load (see Part I, p. 277).

When load is applied to a synchronous motor, its *average speed* cannot decrease since the motor *must* operate at constant average speed. Its field does not change appreciably; hence its counter emf remains substantially constant in magnitude. The synchronous motor, therefore, cannot draw increased current from the line as a result of a reduction in average speed and with the

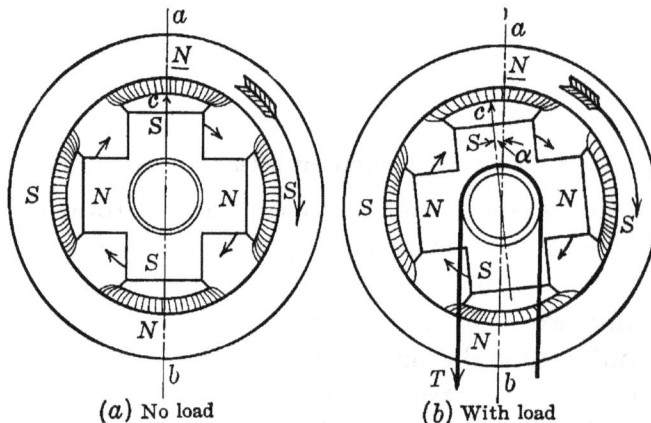

(*a*) No load   (*b*) With load

Fig. 253.—Effect of applying torque to shaft of synchronous motor.

attendant diminution in counter emf, which occurs with the shunt motor.

If increased load torque is applied to the driven disc *B* (Fig. 252), its speed will not change, except momentarily, since it must always run at the speed of *A*. An angular displacement will occur, however, between the discs *B* and *A*, due to the increased elongation of the spring caused by the increase in load torque applied to disc *B*. When load is applied to the rotor of a synchronous motor, this same effect occurs. An angular displacement between the rotor and the rotating field occurs. Figure 253(*a*) shows the motor rotating in a clockwise direction without load. The line *ab* at the center of the N-pole of the stator rotates synchronously with the rotating field being produced by the armature or stator ampere-conductors. The center of the S-pole of the rotor which is locked with $\underline{N}$, shown by the arrow c, coincides with the line ab.

Now apply load torque T to the shaft (Fig. 253(b)). The position of the line ab is not changed appreciably by this load, since its position is determined by the $\underline{N}$-pole produced by the stator ampere-conductors. The rotor, however, is pulled back an angle α from the line ab or from the position that it would occupy if load had not been applied. The stretching of the lines of force in the gap (Fig. 253(b)) tends to pull the rotor into its no-load position. The rotor still continues to operate at synchronous speed as in (a), but the position of the rotor is slightly behind the position which it would occupy were load not applied. That is, there is a flexible magnetic coupling between the rotor and the rotating stator field not unlike the spring coupling

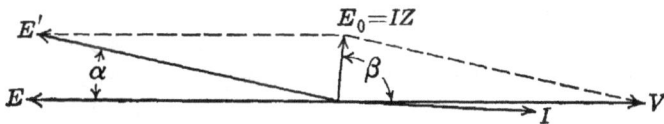

FIG. 254.—Vector diagram of emfs and current in loaded synchronous motor.

between the discs A and B (Fig. 252). Thus, the synchronous motor does not run at reduced speed when load is applied to its shaft, but its rotor shifts its phase backward. When the rotor shifts its phase backward, it changes the phase of the counter emf with respect to the impressed voltage, and this phase shift allows more current to enter the armature and thus to carry the load.

This effect is illustrated in the vector diagram (Fig. 254). The impressed voltage is V and the counter emf is E, practically equal and opposite to V, if the motor is operating at no load, so that the current is small. Assume that mechanical load is applied to the shaft. The torque of the load will cause the rotor to shift backward by an angle α from the position it occupies at no load (Fig. 253(b)). Hence, the induced, or counter, emf E' will now lag its former value E (Fig. 254) by an angle α. The resultant emf E_0, the vector sum of E' and V, acts on the armature impedance Z. Hence, the armature current $I = E/Z$ amp. The ratio of armature reactance to resistance is so large that the current I lags E_0 by an angle β of 90° nearly which causes I to be nearly in phase with V. Hence, I has a large energy component (p. 95) accounting for the increased power that the motor must take from the line in order to supply the power demand caused by the mechanical load applied to the shaft.

Hence, when load torque is applied to the shaft of a synchronous motor, there is a displacement backward in the angular position of the rotor relative to the rotating magnetic field. This produces a lag in the phase of the counter emf, permitting the necessary increase in energy current to flow to the motor from the line. Thus the increased power from the line is determined by a change in the *phase* of the counter emf rather than in its *magnitude*.

This backward angular displacement of the rotor is readily observed by means of a stroboscope (see p. 294).

180. Overexcitation.—When the field of a d-c shunt motor is strengthened, there is a temporary increase in the armature induced, or counter, emf. This decreases the armature current and the torque is lowered, since the decrease in armature current is greater than the corresponding increase in the field. As a result, the motor slows down until its counter emf drops to a value that will allow the flow of sufficient current to carry the load.

When the field of a synchronous motor is increased, the motor cannot slow down, except momentarily, for it must run at constant average speed. Since its speed is constant, its counter emf must increase when the field is strengthened. It might seem that the motor would stop, for its counter emf may become greater than its terminal voltage. In the d-c motor, a counter emf exceeding the terminal voltage would mean generator action, with the result that the machine would cease to operate as a motor.

The synchronous motor, however, may operate as a motor even if its counter emf is greater in magnitude than its terminal voltage. Under these conditions, the motor is said to be *overexcited*. Two reactions occur which enable the motor to operate with an overexcited field.

First, the motor takes a *leading* current. A leading current **in a** synchronous motor *weakens* the field. This is illustrated in Fig. 255, which shows a motor coil moving from left to right. When the axis of the coil is in the position *Y*, coil shown dotted, the coil sides are under the centers of the poles and the induced or counter emf is a maximum. By Fleming's right-hand rule the direction of the *induced* emf is inward in the coil side *a* under the *N*-pole. As the *terminal* voltage is practically in phase opposition to the induced emf, it acts at this instant outward under the *N*-pole or in the left-hand side *a* of the coil and inward in the right-hand side, as shown.

If the current leads this terminal voltage by 90°, it will reach its maximum value one-fourth of a cycle ahead of the voltage, or at a time when the axis of the coil is in position *X*. It will be noted that for this position of the coil the ampere-turns of the coil act upwards in direct opposition to those of the *N*-pole.

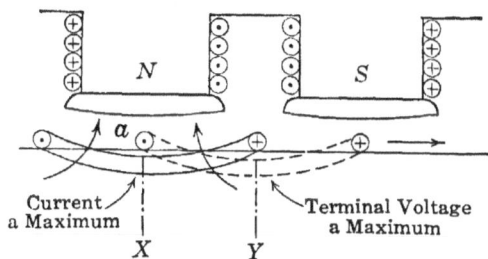

Fig. 255.—Demagnetizing effect of leading current on field of synchronous motor.

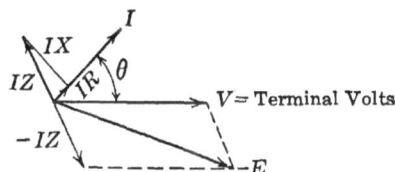

Fig. 256.—Induced armature voltage greater than terminal voltage when synchronous-motor current leads terminal voltage.

Therefore, in the synchronous motor the effect of leading current is to *weaken* the field. In other words, *on overexcitation the armature reaction opposes the effect of increased field current.*

The second effect is illustrated by the vector diagram (Fig. 256). *V* is the terminal voltage, and *I* is the armature current leading *V* by an angle θ. The resistance drop *IR* in the armature is laid off in phase with the current *I*, and the reactance drop *IX* in the armature is laid off at right angles to the current *I* and leading, in the usual manner. The impedance drop *IZ* is the vector sum of *IR* and *IX*. The emf $-E$, necessary to balance the counter emf, is found by subtracting *IZ* vectorially from *V*, just as in the shunt motor the component of terminal voltage which is necessary to balance the counter emf is found by subtracting the *IR* drop from the terminal voltage.

To subtract *IZ* from *V*, $(-IZ)$ is added to *V*. It will be noted that the emf $-E$ is numerically *greater* than the terminal voltage

V. That is, by taking a leading current, the synchronous motor, because of the shift in phase of the armature impedance drop, is able to operate with an induced or counter emf *greater* in magnitude than the terminal voltage, a condition which is impossible with d-c motors.

181. Underexcitation.—When the field of a d-c shunt motor is weakened, the motor speeds up until its counter emf reaches the value which gives the correct value of armature current for the particular load condition.

When the field of a synchronous motor is weakened, the motor cannot speed up permanently, for it must run at a constant average speed. The motor takes a *lagging* current, however. This current has two effects.

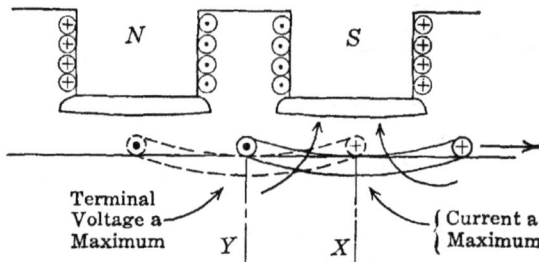

Fig. 257.—Magnetizing effect of lagging current on the poles of synchronous motor.

Figure 257 shows a coil, dotted, whose axis is in position *Y*. In this position the coil sides are opposite the centers of the pole faces; and the counter emf is, therefore, a maximum. The terminal voltage, which is nearly 180° from the counter emf, has its maximum value also for this position of the coil, its direction being indicated in the dotted coil. If the current is lagging the terminal voltage by 90°, it will not reach its maximum value until the coil axis reaches position *X*. The current under these conditions is in such a direction as to *strengthen* the *S*-pole. Therefore, in a synchronous motor, a lagging current *strengthens* the field by armature reaction. Hence, *on underexcitation the motor takes a lagging current which strengthens the field by armature reaction and opposes the effect of decreased field current.*

When the field of the motor is weakened, the motor has not sufficient d-c excitation. It must, therefore, take some of its excitation from the a-c lines by means of a *lagging* current. This is similar to the induction motor which, however, takes *all* its excitation from the a-c line by means of its lagging exciting current.

Figure 258 shows the vector diagram when the motor takes lagging current. The IR and IX drops are laid off with reference to the current in the usual manner, and the IZ drop is obtained. When $(-IZ)$ is added to V, however, $-E$, which is opposite and equal to the counter emf, becomes numerically much less than V. That is, the phase shift of the IZ drop is in such a direction that the motor operates with a counter emf considerably *less* than the terminal voltage.

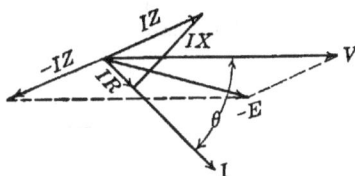

Fig. 258.—Induced armature voltage less than terminal voltage when synchronous motor current lags terminal voltage.

The synchronous motor with salient poles will usually operate even with no field current. If the salient-pole rotor (Fig. 251(*b*), p. 313) is without excitation and is brought near the speed of the rotating stator field, the magnetic lines from the stator will attempt to make the rotor take such a position that the magnetic reluctance is a minimum or the flux is a maximum. In order to accomplish this result, the pole pieces of the rotor when running become locked in with the poles produced by the stator winding (Fig. 251(*b*)). These rotating stator poles pull the salient poles of the rotor around with them and in this manner enable the motor to carry a limited load without d-c excitation. Although the motor may carry a limited load without d-c excitation, its power factor will be low and the current will be lagging, both of which are undesirable.

182. Power Factor of Synchronous Motor.—The power factor of the ordinary *induction motor* for a given load cannot be altered without changing the motor design, and the ordinary induction motor always takes a lagging current. The power factor of the *synchronous motor* at any given load can be altered at will, and the current can be changed from lagging to leading merely by changing the field excitation.

The connections for testing a three-phase synchronous motor are shown in Fig. 259, a polyphase wattmeter being used to measure the input. If the input is kept constant and the field current is varied, it will be found that for small values of field current the motor takes a large armature current and the power factor is low. This is illustrated in Fig. 260 in which the curve P_1 is obtained with constant power input P_1. For field current a

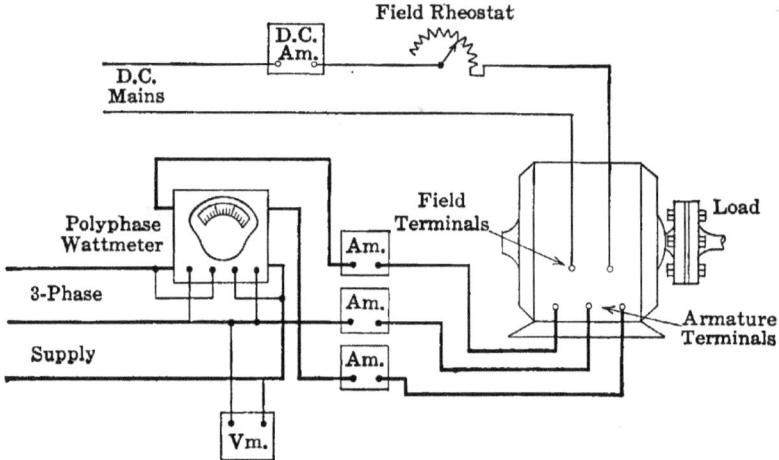

FIG. 259.—Connections for testing synchronous motor.

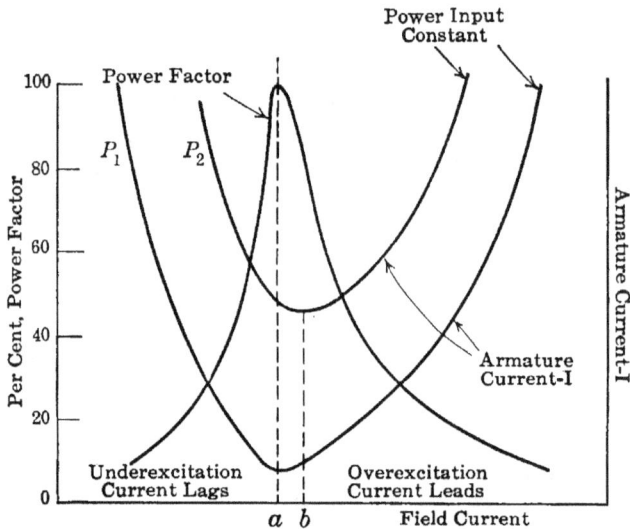

FIG. 260.—Phase characteristics of synchronous motor.

the armature current is a minimum, and the power factor is unity. This value of field current is the *normal* value. For values of field current exceeding a the motor is *overexcited* and takes a *leading* current (Par. 180). As the field is strengthened, beginning with zero field current, the armature current decreases and reaches a minimum at field current a. For values of field current less than the value a the motor is *underexcited* and takes a *lagging* current (Par. 181). The power-factor curve corresponding to P_1 is shown in the figure.

If the power input is increased to P_2 and held constant, the curve P_2 is obtained. The *normal* field current b is slightly greater than it is for P_1. A series of curves similar to P_1 and P_2 for different constant values of power input may be obtained. Such curves are called *V-curves*.

Thus, it is seen that with underexcitation the synchronous motor acts like an inductance in that it takes lagging current. With overexcitation the motor acts like a capacitor in that it takes leading current. The fact that the power factor of the synchronous motor may be varied at will over wide ranges makes it very useful for regulating the power factor of power systems.

The connections of Fig. 259 may be used if a brake test of the motor is to be made. The torque of a synchronous motor is directly proportional to its output since the speed is constant. The current and power factor depend on the field current. Load characteristics are often determined with the field rheostat set to give normal field current or minimum armature current at no load.

183. Synchronous Condenser as Corrector of Power Factor.— The fact that the power factor of the synchronous motor may be varied at will over wide ranges makes it useful on power systems, since by its use the system power factor may be improved. Improved power factor is so important that synchronous motors operating without mechanical load are often connected to power systems for the sole purpose of controlling the system power factor. When used in this manner the motor is called a *synchronous condenser*.

The electric power load in most factories and mills consists chiefly of induction motors. Although the power factor of the induction motor is moderately high when operating near its rated load, the fact that many of the motors in a mill frequently operate at light load may make the system power factor as low as 0.5 or 0.6 (see Fig. 228, p. 278). Such low power factors are undesirable since they cause poor regulation of the generating apparatus, transformers, and lines. Also, since low power factor requires more current for the same power, the generators, transformers, and lines must have greater kva ratings, and the system investment charges are increased.

This low power factor, which is due to the lagging current of the load, may be improved by the use of a synchronous condenser connected in parallel with the load and excited to take leading current. The power factor may be brought to unity, even. A three-phase load (Fig. 261(a)), which may consist of

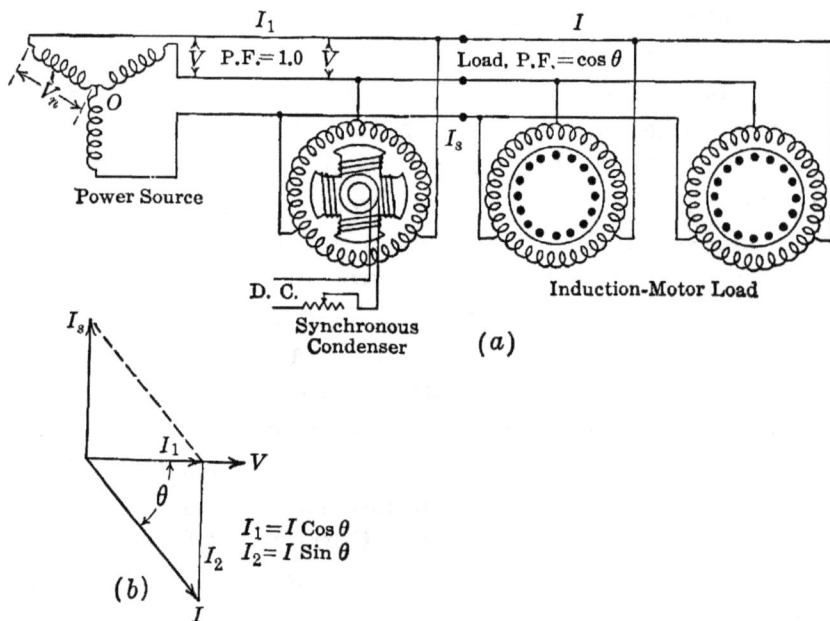

FIG. 261.—Raising power factor to unity by means of synchronous condenser.

induction motors, takes a current I at three-phase line voltage V. The power factor of the load is cos θ, lagging current. This means that the coil current lags the coil voltage by the angle θ (p. 133). In three-phase systems, it is convenient to consider that the system is a Y-system and to work with voltage to neutral. Since the system is balanced, it is necessary to consider only one phase. Let the voltage to neutral be $V_n = V/\sqrt{3}$. Resolve the load current I into two components at right angles to each other. One component I_1, the *energy current*, is in phase with the voltage V_n (p. 95).

The power per phase

$$P = V_n I \cos \theta = V_n I_1 \text{ watts,} \tag{94}$$

since $I_1 = I \cos \theta$.

The power is equal to the product of the voltage and the energy current (p. 95). The component I_2 at right angles to V_n is the *quadrature*, or *wattless*, current and can contribute no power.

It follows that if the quadrature current were eliminated or if it were neutralized by an equal leading quadrature current, the power factor of the system would be raised to unity and the net power of the system would remain unchanged. By means of a synchronous condenser, connected in parallel with the load

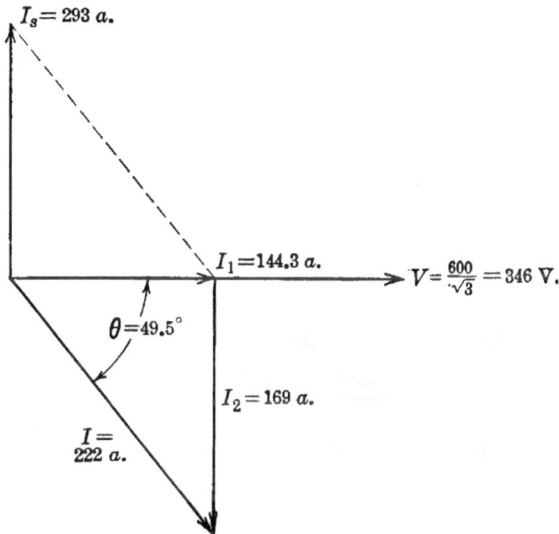

FIG. 262.—Power factor of 150-kw load raised to unity with synchronous condenser.

(Fig. 261(a)), a leading quadrature current I_s is obtained equal to the lagging quadrature current I_2. (Since the synchronous condenser losses are small, the current I_s may be assumed as leading V_n by 90°.) The total current supplied to the load is now I_1, the energy current alone, since I_1 is the vector sum of I and I_s. I_1 is in phase with V_n; therefore, the power factor of the system is now unity.

Example.—The power factor of a 150-kw, 600-volt, 60-cycle, three-phase load is 0.65, current lagging. Determine: (a) energy current; (b) quadrature current; (c) rating of synchronous condenser necessary to raise system power factor to unity.

(a) cos θ = 0.65; θ = 49.5°; sin θ = 0.760. Total load current (Fig. 262)

$$I = \frac{150,000}{\sqrt{3} \times 600 \times 0.65} = 222 \text{ amp.}$$

Energy current $I_1 = I \cos \theta = 222 \times 0.65 = 144.3$ amp. *Ans.*
Also, since the power

$$P = \sqrt{3}\, EI \cos \theta, \text{ and } I_1 = I \cos \theta,$$
$$I_1 = \frac{P}{\sqrt{3}\, E} = \frac{150,000}{\sqrt{3} \times 600} = 144.3 \text{ amp (check)}.$$

(b) Quadrature current

$$I_2 = 222 \sin \theta = 222 \times 0.760 = 169 \text{ amp.} \quad Ans.$$

(c) Synchronous condenser must take leading current of 169 amp at 600 volts. Hence, its rating in

$$\text{Kva} = \frac{\sqrt{3} \times 600 \times 169}{1,000} = 175.8. \quad Ans.$$

The vector diagram is given in Fig. 262. The kva rating of the synchronous condenser here used to obtain unity power factor is greater than the kilowatts required by the load. It is often more economical to use a smaller unit and to obtain a power factor of the order of 0.9 rather than unity.

The conditions for single phase are solved in the same manner (see example, p. 96).

At the present time, capacitors, made of foil and paper immersed in oil, also are used to improve power factor.

Obviously, a synchronous motor may at the same time deliver mechanical load and operate overexcited to improve power factor.

184. Synchronous Condenser as Regulator of Voltage.—If an induction motor or a lamp load (Fig. 263) is supplied from a constant-voltage supply V_s through inductive reactances, the voltage V_r at the load or receiver will be considerably less than V_s owing to the voltage drop in the reactances and to the fact that this drop is nearly in phase with the voltage V_s. If, however, a synchronous motor is connected in parallel with the load, then by overexciting the motor, causing it to take a leading current, the voltage V_r not only may be made to equal V_s but actually may be made to exceed V_s. The synchronous motor, or condenser, therefore, may be utilized to control the voltage at the load or at other points in a power system (see p. 366).

Overhead transmission lines have substantial series inductive reactance (p. 363). Therefore, by connecting synchronous motors (condensers) at the receiving end of the line and varying their field excitation, the voltage at the load may be controlled. The voltage at the load may be made to exceed the voltage at the sending end of the line, a condition which is not possible with

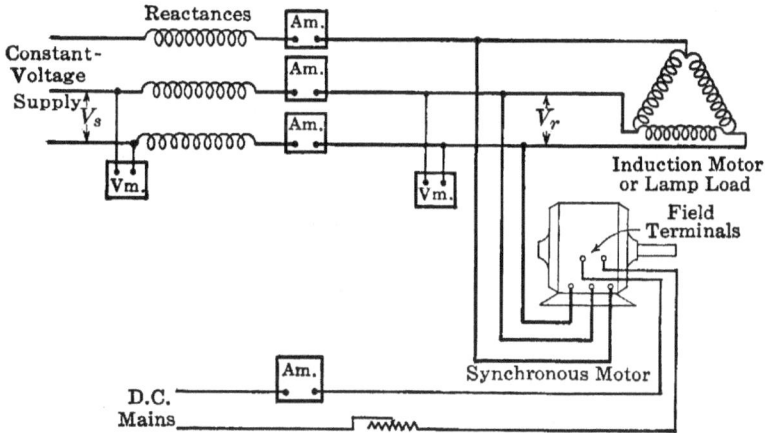

FIG. 263.—Synchronous motor for controlling voltage at end of transmission line.

FIG. 264.—Rotor of 6,600-volt, 60-cycle, 16-pole synchronous motor showing damper winding. (*Courtesy of Westinghouse Electric and Manufacturing Company.*)

direct current unless energy is introduced between generator and load.

185. Amortisseur, or Damper, Windings.—Figure 264 shows the rotating-field structure of a synchronous motor, with a squirrel-cage winding in addition to the d-c salient poles. The conductors of the squirrel cage are embedded in the pole faces of the rotor. Such windings are called *amortisseur* or *damper* windings or simply *dampers*. They assist the motor in starting, and they damp out any tendency of the rotor to oscillate or "hunt." Hunting usually results from pulsations in the power supply which may be caused by the variable torque of reciprocating-engine-driven units or by hunting of turbines themselves due to oscillations of their governors. Moreover, dampers tend to prevent the motor falling out of synchronism on account of line disturbances caused by short circuits, switching, etc.

The action of the damper winding involves the principle of both the induction motor and the induction generator. So long as the rotor is rotating at synchronous speed, the rotating field of the armature or stator does not cut the dampers and they have no effect. Assume that the rotor slows down momentarily. For an instant the rotating field due to the armature mmf is rotating faster than the field structure. This is equivalent to the rotor slipping temporarily, and currents are induced in the dampers. That is, induction-motor action results and the currents in the dampers are in such a direction that they tend to pull the rotor into its normal position.

Again, if the field poles for some reason swing ahead of their normal position, the dampers cut the rotating field in the opposite direction, or the slip becomes negative, temporarily. Induction-generator action follows, putting a load on the rotor and tending to slow it down. These dampers, therefore, always tend to keep the motor in synchronism and thus to prevent hunting. Such windings are often used in alternators, particularly of the reciprocating-engine-driven type, to prevent hunting when operating in parallel.

186. Starting Synchronous Motor.—It has been pointed out that the *synchronous* motor is not self-starting. It must first be brought nearly or actually to synchronous speed before it can operate. There are several methods of accomplishing this.

The motor may be brought up to speed by external means and synchronized like an alternator, after which it operates as a synchronous motor. The d-c exciter is sometimes used to bring the motor up to synchronism if sufficient d-c power is available. Likewise, if the synchronous motor is connected to a d-c generator, this generator may be operated as a motor to bring the synchronous motor up to synchronism. After the motor is synchronized, the field of the d-c machine is strengthened and it then acts as a generator, taking mechanical power from the synchronous motor.

The synchronous motor may start as an induction motor. First, the field circuit is opened. Then a polyphase voltage is impressed on its stator which produces a rotating field about the rotor. As a rule, it is desirable to use a compensator so that reduced voltage is applied to the stator windings. The rotating field induces currents in the pole faces of the rotor and in the amortisseur winding, as well. This is obviously induction-motor action. Because of the comparatively long air-gap and the squirrel-cage characteristics of the damper winding, the usual type of motor develops little starting torque even with large armature current.

However, the starting torque, though small, is usually sufficient to start the motor, which then accelerates until it is at or near synchronism. Before the compensator is thrown into the running position, the field switch is usually closed, so as to minimize disturbances to the system. If the rotor is slipping but slightly, it will usually pull into synchronism when the field switch is closed, the field poles locking in with the poles produced by the armature mmf (Fig. 251, p. 313).

The motor may pull into synchronism before the field circuit is closed. As the motor approaches synchronism, the salient poles lock in with the poles of the rotating stator field (Fig. 251).

When the field circuit is closed, it may excite the motor poles so that their polarity is opposite to that produced by the rotating field, that is, by the armature reaction (Fig. 251). The rotor is then thrown back one pole, or, in other words, it slips a pole. With a large motor, this may cause considerable disturbance to the system; and for this reason the field is usually closed when the compensator is in the starting position. The motor may be made to lock in with the correct polarity by applying a weak d-c field to the motor as it approaches synchronism. This causes the armature reaction to act in conjunction with the d-c excitation, and the poles then come into synchronism with the same polarity as that which will be produced by the d-c excitation

After the motor has pulled into synchronism, it is necessary merely to strengthen the d-c field to the desired value. Then the starting compensator may be thrown quickly into the running position.

When voltage is first applied to the synchronous motor, there may be a high voltage induced in the field winding. The stator acts as the primary of a transformer, a primary having comparatively few turns. The flux produced by the stator, or primary, cuts the field winding at synchronous speed; and as the field has a large number of turns, a high emf is induced in the field. This emf may puncture the field winding. The field winding, therefore, should be insulated for a voltage considerably in excess of that which normal operation requires. When starting, the field is sometimes short-circuited, or is shunted by a resistance in order to decrease this high voltage. The emf induced in the field winding decreases as the rotor comes up to speed, until at synchronism this emf is zero.

187. Industrial Applications of Synchronous Motor.—Large single-phase synchronous motors are rarely used. Like the single-phase induction motor, the direction in which they rotate is determined by the direction in which they are started.

Since the starting torque of the usual synchronous motor is almost entirely due to a squirrel-cage winding, the torque is inherently low. *High-starting-torque* motors may be obtained, however, by the same means as are employed with induction motors. High-resistance dampers may be used. If the slip is so great that the rotor poles cannot lock in with the synchronously rotating stator poles, the field may be short-circuited until lock-in occurs and then may be connected to the d-c excitation source. *Phase-connected dampers* also are used. The dampers, instead of being short-circuited grids, are polyphase windings connected to external resistors through slip-rings. The rotor is then brought up to speed by cutting out the external resistance as with the wound-rotor induction motor.

The inherent disadvantages of the synchronous motor are that it requires direct current for its excitation, its starting torque is small, and the motor is sensitive to system disturbances and may fall out of step when these occur. On the other hand, the ease with which its power factor can be controlled is a distinct advantage, often outweighing all the disadvantages. In the large units, particularly at low synchronous speeds, the synchronous motor is cheaper than the induction motor. The fact that its speed is constant is usually of little importance, since induction motors, especially in the larger sizes, have only 2 or 3 per cent speed regulation.

The synchronous motor is used only in the larger sizes where a source of d-c excitation may be justified. An important field

FIG. 265.—1,200-hp, 720-rpm, 2,200-volt synchronous-motor-driving water pump, Minneapolis, Minn. (*Courtesy of Electric Machinery Manufacturing Company, Minneapolis, Minn.*)

of use is in connection with motor-generator sets for supply to d-c systems. Such motors, situated at different points in a power system, may make it possible to operate the generating station and many of the transmission lines and substations at high power factor, in spite of the low power factor of the consumers' loads. Other important uses of synchronous motors are their application to the driving of ammonia compressors, tube and ball mills, water pumps, and rubber mills. A synchronous motor driving a water pump is shown in Fig. 265.

Synchronous motors have limited application in the propulsion of ships.

188. Synchronous Motors for Timing Purposes.--Because of their absolutely constant speed characteristics, synchronous motors are useful for driving such devices as must be held in synchronism with the supply frequency. Examples are the measurement of slip in the induction motor (p. 295), the driving of oscillograph mirrors (p. 116), stroboscopic devices, electrical clocks, and mechanical rectifiers (p. 342).

As the power required of such motors is extremely small and low power factor is of no consequence, they usually operate without d-c excitation. An example of a motor of this type is the Warren Telechron clock motor (Fig. 266(a)) which is widely

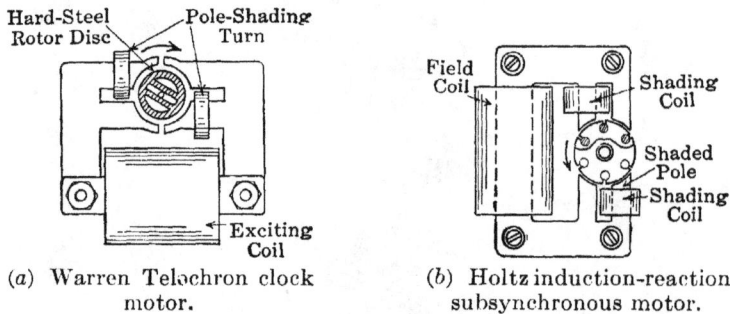

(a) Warren Telechron clock motor.

(b) Holtz induction-reaction subsynchronous motor.

Fig. 266.—Synchronous motors for timing.

used in electric clocks, time switches, demand and recording instruments, and many other timing devices. The motor consists of a laminated stator with an exciting coil designed usually for 110 volts. The two poles are divided, a shading coil of one turn of heavy copper being placed over one-half of each pole, to produce a rotating field (pp. 109 and 308). The rotor consists of two or more thin hardened-steel discs of the shape shown, pressed on a small shaft. There is high hysteresis loss in hardened steel; and when the discs are acted upon by the rotating field, a substantial torque is produced, just as with the induction motor with high resistance loss in the rotor. The rotor accelerates nearly to synchronous speed by this hysteresis action and then locks in, due to the flux of the rotating field seeking the path of minimum reluctance through the two rotor crossbars. As a matter of fact, plain hard-steel circular discs will lock in owing to their becoming permanently magnetized along a diameter. The rotor speed at 60 cycles is 3,600 rpm, which is reduced by a gear train, the gears being enclosed within a case containing lubricating oil.

In Fig. 266(b) is shown a self-starting induction-reaction subsynchronous motor invented by F. C. Holtz of the Sangamo Electric Co.   The stator is the same as that of the Warren motor, consisting of laminations, a field coil, and shaded poles.   In this particular motor, there are six rotor slots containing a squirrel-cage winding.   These rotor slots are so proportioned that six salient poles are formed.   At starting, the torque of the squirrel-cage winding predominates over the lock-in torque of the salient poles with the halves of the field poles (Fig. 266(b)).

The induction-motor torque decreases with increasing speed (curve (3), Fig. 229, p. 281), and at one-third synchronous speed (1,200 rpm at 60 cycles) the salient rotor poles lock with the halves of the field poles each half-cycle to give one-third synchronous speed, the reaction torque predominating over the induction-motor torque at this speed.   This type of motor is also used with timing devices such as demand-meter registers.

SYNCHRONOUS CONVERTER

It has been stated that for economic reasons a large proportion of electrical energy is generated and transmitted as alternating current (see Chap. I).   Direct current, however, is necessary for many purposes, such as for charging batteries, electrolytic work, power and lighting service in large cities, and electric railways.   At present, there are three practicable methods of obtaining direct current from alternating current on a large scale, the motor-generator set, the synchronous converter, and the metal-tank mercury-arc rectifier.   In many applications the use of rotating machinery is desirable, and the choice lies between the synchronous converter and the motor-generator set.   The synchronous converter has some advantages over the motor-generator set in that it is a single machine having one armature, one field, and two bearings.

189. Principle of the Synchronous Converter.—It has been shown already that alternating emf is generated in the individual armature coils of the ordinary d-c generator.   If taps are brought out properly from the armature winding to slip-rings, alternating current may be taken from this same winding and the machine becomes an alternator. If alternating current is supplied to the slip-rings, the machine will operate as a synchronous motor.

With the exception of slip-rings, the construction of the converter is the same as that of the usual d-c dynamo (also see p. 383, Fig. 310), although the relative dimensions may differ. The converter has fixed poles, a rotating armature, a commutator, a shunt field, and usually a series field.

In the synchronous converter, as commonly used, alternating current is supplied to the slip-rings and direct current is taken from the commutator and brushes. Direct current may be supplied to the brushes and commutator and alternating current taken from the slip-rings. The machine is then said to be an *inverted converter.* If, however, the d-c brushes are open-circuited or removed, the machine becomes a synchronous motor of the rotating-armature type. On the other hand, if direct current is supplied to the brushes and commutator and the slip-ring brushes are disconnected, the machine becomes a shunt, or compound, motor.

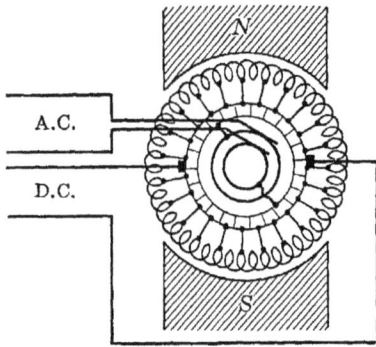

FIG. 267.—Two-pole, single-phase, synchronous converter.

If the machine is driven mechanically and current is taken from the slip-rings only, it becomes an alternator. On the other hand, if current is taken from the commutator only, it becomes a d-c generator. Both alternating and direct current may be taken from it simultaneously, and it then becomes a *double-current* generator.

The connections of a two-pole, single-phase converter are shown in Fig. 267. In a four-pole converter, it is necessary to have two taps from opposite points in the winding to each slip-ring, making four taps in all. A four-pole, three-phase converter must have two taps to each slip-ring, or six taps in all. The taps are not necessarily taken from the commutator segments directly but may be taken from the coils themselves.

The converter owes its economy to the fact that a motor current and a generator current flow in the armature conductors simultaneously. Since the motor current and the generator current flow in opposite directions, the resultant current is small, being considerably less than either the direct or the alternating current. The rating of the armature when operating as a converter is therefore considerably greater than when operating with either direct or alternating current alone. Hence, the commutator is larger in converters than in d-c machines of the same rating.

190. Polyphase Converters.—The rating of a converter increases much more rapidly with increase in the number of phases than is the case with other types of electrical machinery. Also, the rating decreases much more rapidly with decrease in power factor than is true for other types of electrical machinery. These relations are shown by the following table:

EFFECT OF NUMBER OF PHASES AND OF POWER FACTOR ON THE OUTPUT OF SYNCHRONOUS CONVERTER

Number of phases	P.F. = 1.0	P.F. = 0.9
1	0.85	0.74
D-c	1.00	1.00
3	1.33	1.09
4	1.65	1.28
6	1.93	1.45
12	2.18	1.58

It will be noted that the output when operating six-phase is more than double the output when operating single-phase and is 45 per cent greater than when operating three-phase. Converters, therefore, are operated six-phase ordinarily, for it is a simple matter to transform from three- to six-phase (p. 149). Twelve-phase operation is seldom used owing to transformer and wiring complications. Because of the rapid decrease in efficiency and in rating with decrease in power factor (see table above), it is desirable to operate converters at or near unity power factor.

191. Voltage and Current Ratios in Converter.—It is important to know the ratio of voltage and current at the d-c side of the converter to voltage and current at the a-c side.

Both the d-c voltage and the a-c voltage are induced by the same conductors cutting the same flux at the same speed. Hence, *fixed ratios* exist between the *induced* d-c emf and the *induced* alternating emf. The terminal-voltage ratio may differ somewhat from these fixed ratios, owing to the impedance and resistance drops in the armature.

There are the same number of series conductors between the d-c brushes as between single-phase slip-ring taps. The d-c emf is the sum from instant to instant of the emfs induced in all the series-connected conductors between d-c brushes. The alternating emf when it is at its maximum value is equal to the sum of the emfs induced at that instant in all the series conductors between slip-ring taps. Therefore, the maximum value of the alternating emf is equal to the d-c emf. Hence, the d-c emf is equal to $\sqrt{2}$, or 1.41, times the rms single-phase emf. That is, if the single-phase emf is 100 volts (rms), the d-c emf is 141 volts (Fig. 268). If a circle is drawn having 100 volts to scale as a diameter (Fig. 268), the three-phase emf is given by a chord subtending 120°, the four-phase by a chord subtending 90°, and the six-phase by a chord subtending 60°.

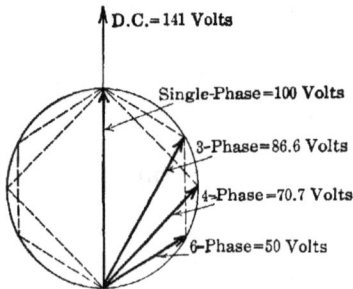

Fig. 268.—Relations existing among voltages in synchronous-converter armature.

On the assumption of unity power factor and no losses, the ratio of alternating voltage and current to d-c voltage and current is given in the following table:

Number of slip-rings	Number of phases	Ratio E_{AC}/E_{DC}	Ratio I_{AC}/I_{DC}
2	1	0.707	1.414
3	3	0.612	0.943
4	4	0.500	0.707
6	6	0.354	0.471

Example.—Find alternating current and voltage at rated load of a 1,000-kw, 500-rpm, 25-cycle, 6-phase, 250-volt synchronous converter. Assume unity power factor, and neglect losses.

The rated direct current

$$I_{DC} = \frac{1,000,000}{250} = 4,000 \text{ amp.}$$

$$I_{AC} = 4,000 \times 0.471 = 1,884 \text{ amp.} \quad \textit{Ans.}$$
$$E_{AC} = 250 \times 0.354 = 88.5 \text{ volts.} \quad \textit{Ans.}$$

These currents and voltages are shown in Fig. 269. The converter is compounded and supplies a 250-volt, three-wire system, the neutral of which is connected to the neutral of the six-phase star-connected transformer secondaries (see p. 150). Any neutral current due to unequal loads on the two sides of the three-wire system returns to or leaves the armature through the secondaries and slip-rings. Two series-field windings are necessary. If but one were used, on the positive side for example, loads from negative to neutral would have no effect on the compounding. This converter is supplied from a 6,600-volt, three-

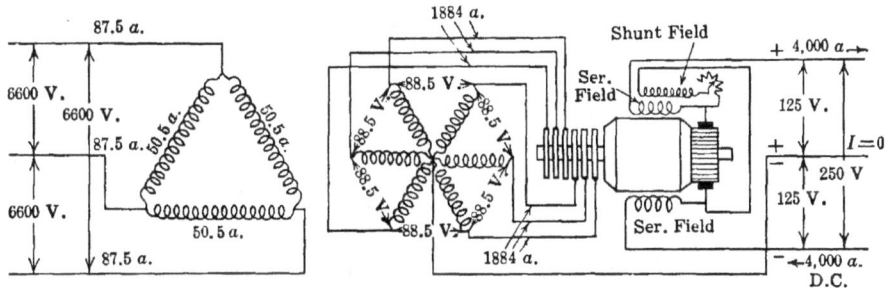

Fig. 269.—Connections of 1,000-kw., 25-cycle, six-phase, 250-volt synchronous converter taking energy from 6,600-volt, three-phase line.

phase circuit. The primaries of the transformers are connected in delta. The currents are determined as follows:

Example.—Determine line currents and primary coil currents in system (Fig. 269). Neglect losses, and assume unity power factor.

Line currents

$$I = \frac{1,000,000}{\sqrt{3} \times 6,600} = 87.5 \text{ amp. } Ans.$$

The transformers with the two split secondaries in series have a 6,600/177.0 ratio of transformation. Hence, the primary coil currents, from Eq. (78), (p. 223),

$$I_C = 1,884 \frac{177.0}{6,600} = 50.5 \text{ amp. } Ans.$$

$$50.5 \sqrt{3} = 87.5 \text{ amp. } (check).$$

192. Voltage Control. *Change of Excitation.*—The *ratio* of the d-c emf to the alternating induced emf in a converter armature is fixed, regardless of field excitation. The ratio of *brush* voltage to *slip-ring* voltage, however, may be changed a limited amount by varying the field excitation. If the field excitation is increased, the converter, since it operates as a synchronous motor, takes a leading current and its induced emf is increased (p. 317, Fig. 256). It is further increased by leading current flowing through the series reactance of the transformers. If the field is weakened, the induced emf is decreased (p. 319, Fig. 258). With sufficient series reactance, the d-c voltage may be varied about 5 per cent above and below its average value. By compounding (Fig. 269) the regulation is accomplished automatically. The principal objection to this method of change of excitation is that regulation is accompanied by a change of power factor

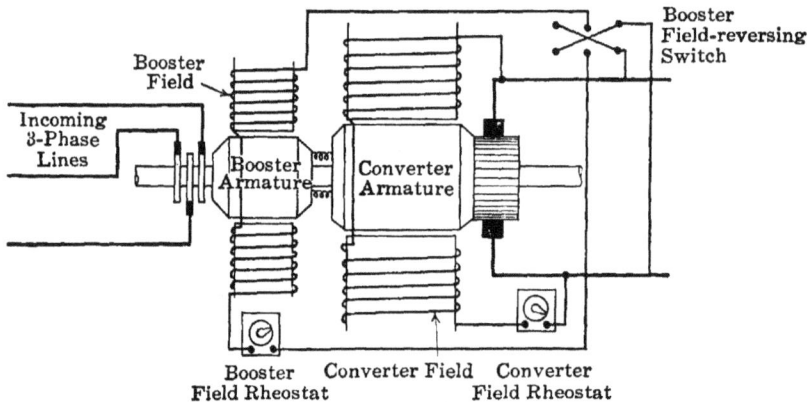

Fig. 270.—Connections of three-phase synchronous converter with series booster.

and converters for best results should operate near unity power factor.

Transformer Taps.—The converter voltage may be adjusted approximately to the desired value by taps on the transformers. Owing to the arcing and burning of sliding contacts, the use of transformer taps for adjustment during operation is not common, though the use of taps for fixed adjustment of voltage is common.

Induction Regulator.—The induction regulator has been described in connection with the induction motor (p. 296). This type of regulator may be connected between the transformers and the converter, and the alternating voltage impressed on the converter terminals may be raised and lowered thereby. This changes the d-c voltage by a corresponding amount. Under these conditions the voltage may be raised independently of power factor, but the extra equipment and the large mechanical forces between stator and rotor are objections to the induction regulator.

Series Booster.—A low-voltage alternator is often connected to the shaft of the converter. This alternator has the same number of poles as the converter. The armature of the alternator is connected in series with the a-c lines supplying the converter (Fig. 270). By increasing the field of the alternator, or booster, the alternating voltage of the converter is increased. The converter

Fig. 271.—2,730/3,705-kw, 300-rpm, 210/285-volt synchronous converter with alternating-current rotating-armature booster. (*Courtesy of General Electric Company.*)

voltage may be decreased, not only by decreasing the booster field, but by reversing it as well. The advantage of this method of control is that the voltage may be varied independently of power factor. Its use is common with large units. The objection to this type of voltage control is the additional machine. Figure 271 shows a converter with booster generator.

193. Inverted Synchronous Converter.—When a converter operates from a d-c source and delivers alternating current, it is an *inverted* synchronous converter. The d-c side has characteristics similar to those of a shunt or compound motor. The a-c side has characteristics similar to those of an alternator. In operating from the a-c supply, the speed of the converter must be in synchronism with the supply, and hence constant. In operating from the d-c supply, the speed is determined by the counter emf and flux, just as in any d-c motor, and the speed may vary. An inductive load on the a-c side weakens the field through armature reaction, in the same manner that the field of an alternator is weakened under similar conditions. The weakening of the field increases the speed of the converter. This increased speed causes the current to lag still more because of the increased frequency. The effect is cumulative and may cause the armature to reach dangerous speeds. Therefore, a centrifugal device is often used which trips the breaker when the speed exceeds the safe value.

194. Starting Synchronous Converter. *Alternating-current Side.*—If polyphase currents are supplied to the armature of a converter, a rotating field is produced about the armature, similar to the rotating field of the induction motor, except that it is produced by a rotating armature about itself. If the armature speed is below synchronism, this field cuts the pole faces and the damper windings (Fig. 273) and induces currents in them. A reaction results between the rotating field and these induced currents, producing torque.

In starting the converter in this manner the armature becomes the primary of a transformer and the shunt and series fields are the secondaries (p. 328). To prevent too high a voltage being induced in the shunt field, it is sectionalized (Fig. 272). The field-splitting switch, therefore, should be opened. To prevent large currents being induced in the series field, the short-circuiting switches and shunts should be opened. The rotating field about the armature cuts coils short-circuited by the brushes, so that excessive sparking occurs under the brushes, particularly if the converter has interpoles. With large converters, therefore, the brushes are raised from the commutator during starting except for one brush in a positive brush arm and one in a negative brush arm, these supplying the field current. The converter is usually started at reduced voltage from taps on the secondaries of its transformers. The field is closed when the converter is near or at synchronism, after which the converter is connected directly across the line. If the converter comes up with the wrong d-c polarity, the field may be opened and then the line switch opened momentarily until the motor slips back one pole. Likewise, before connecting across the line, the armature may be made to slip a pole by throwing over the field-reversing switch (Fig. 272). Before the direct current in the field has had opportunity to reverse owing to the reversed polarity, thus causing the armature to slip another pole, the field switch must be thrown back quickly to its original position. This procedure causes the

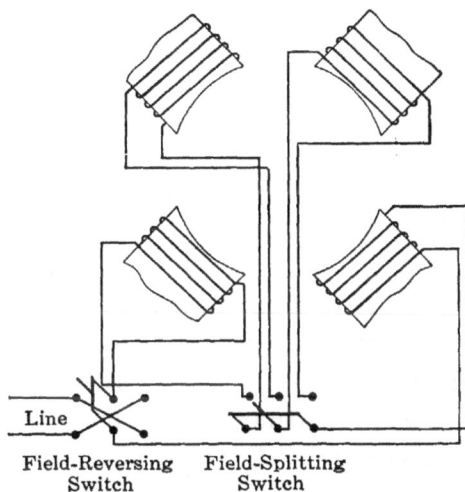

Field-Reversing   Field-Splitting
Switch   Switch

Fig. 272.—Connections of shunt field and shunt-field-splitting switch.

armature to slip back a pole; and, as a result, the d-c polarity reverses.

Direct-current Side.—The converter may be started from the d-c side as a shunt motor. When the converter is started in this manner, the series field should be short-circuited, as it will oppose the shunt field when the converter operates as a motor and will reduce the starting torque. The transformer secondaries are short circuits on the d-c armature at starting, as the frequency is zero and their resistance is very low. The transformers, therefore, should be disconnected. The proper speed is obtained by adjusting the shunt field. As there is practically no voltage control in the simple converter when operating in this manner, it is not always possible to adjust the alternating emf to a value equal to that of the line. To prevent any disturbance which may result from synchronizing at a voltage other than bus-bar voltage, some of the starting resistance is often left in the armature circuit until after the converter has been synchronized.

195. Dampers.—The current which produces torque in the converter armature is the *difference* of the alternating and the direct current. This difference is small. Hence, a small

FIG. 273.—Main pole with damper winding.

momentary change in either direct or alternating current, such as might be caused by switching, by change of load, and by system disturbance, causes a large percentage change in their difference, or the torque current. Converters, therefore, are more sensitive than synchronous motors even and "hunt" readily so that dampers (Fig. 273) are necessary.

196. Experimental Determination of Voltage and Current Relations in Converter.—An instructive laboratory experiment may be performed with a converter connected as shown in Fig. 274. The series reactances may be omitted if the transformers themselves have sufficient leakage reactance. Connect instruments to measure the three-phase input, a voltmeter V_2 to measure the transformer primary voltage, a voltmeter V_3 to measure the slip-ring voltage,

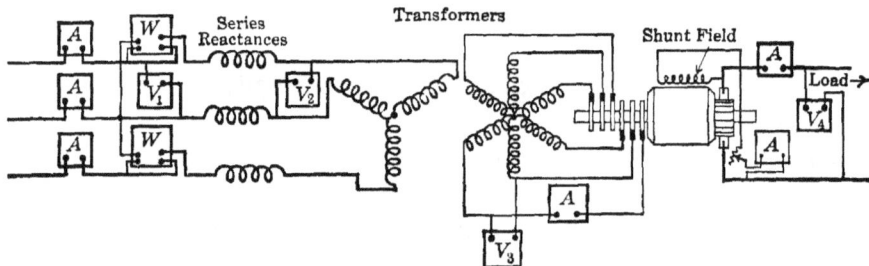

FIG. 274.—Connections for testing synchronous converter.

an ammeter to measure the current between the transformer secondaries and the converter, d-c instruments to measure the converter output, and a d-c ammeter to measure the field current.

Maintain the load on the converter constant at 50 per cent of its rated value. Vary its field over the maximum range of operation, reading all instruments. With field current as abscissas plot as ordinates

1. Voltages V_1, V_2, V_3, V_4.
2. Efficiency of entire unit.
3. Power factor.

Also, check currents by tables of Par. 191 (p. 334). Note effect of power factor on efficiency.

Other experiments may be performed, using the same connections, such as keeping the field current constant at its normal no-load value (P.F. = 1.0) and noting the changes in efficiency and power factor as load is increased. Plot efficiency and power factor as ordinates with output as abscissas.

197. Industrial Applications.—As has been pointed out, the principal industrial application of the synchronous converter is to convert alternating to direct current on a large scale. Units as large as 4,000 kw at 25 cycles and 3,000 kw at 60 cycles are common. The converter has an efficiency of 0.95 to 0.96 at rated load and unity power factor, as compared with less than 0.9 for the motor-generator set. The converter efficiency, however, drops rapidly with decrease of power factor (Par. 190). Being a single machine, the converter occupies less floor space than the motor-generator set. A converter, however, always requires transformers. Up to 15,000 volts, the synchronous motor does not require transformers. Hence, up to 15,000 volts, the cost, floor space, and efficiencies of the transformers must be taken into consideration when the converter is in question. For voltages in excess of this, the motor generator also requires transformers. With a converter alone, the voltage and power factor are not independent of each other; if regulating apparatus, such as a booster generator is used, the cost of the converter unit is increased and its efficiency is decreased.

With the development of the metal-tank mercury-arc rectifier (Chap. XI), new installations of converters for traction and for electrolytic processes have diminished considerably.

CHAPTER XI

RECTIFIERS

Alternating current can be converted into direct current not only by rotating machinery but also by means of *rectifiers*. Either the rectifier suppresses the reversed emf, or the circuit connections are arranged so that both halves of the current wave flow through the receiver circuit in the same direction. Rectifiers of the power type are divided into four general classes: mechanical rectifiers, electrolytic rectifiers, copper-oxide rectifiers, gas discharge and vapor rectifiers. Until the metal-tank, mercury-arc rectifier became available, about 1928, it was possible to convert only relatively small amounts of power by means of rectifiers.

198. Mechanical Rectifiers.—Rectification of alternating current may be accomplished mechanically, one method being by

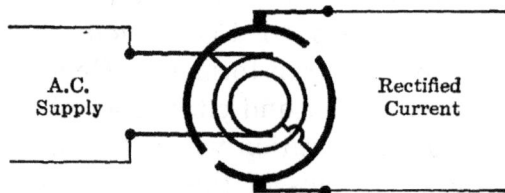

Fig. 275.—Commutating-type rectifier.

means of a commutator driven by a synchronous motor. The segments of the commutator are connected so that when the alternating current reverses, the connections to the d-c circuit are reversed simultaneously (Fig. 275) and a unidirectional current is obtained. As the brushes cannot have zero width, it is difficult to commutate at the point of zero current. As the current and voltage are rarely zero at the same time, such devices spark more or less and are limited to small current and voltage.

Another type of mechanical rectifier is the vibrating-contact type. The contacts are actuated by a solenoid energized by the alternating current and vibrate synchronously with the frequency of the alternating current. In so doing, they reverse the connections to the load circuit, so that the current flows always in the same direction in the load circuit. Because of the greater simplicity, the lesser cost, and the better operating characteristics of the copper-oxide rectifier, mechanical rectifiers are now employed only where special conditions require their use.

199. Half- and Full-wave Rectification.—If a rectifying device permits the passage of current during the positive half-cycles only, *half-wave rectification* is obtained (Fig. 276). Unless there is inductance in the circuit, the current will be zero for half the time. This type of current wave is satisfactory for some purposes, such as battery charging, but is not satisfactory for general power supply.

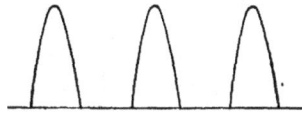

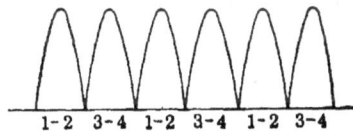

Fig. 276.—Half-wave rectification.

If, however, current flows during both half-cycles, as in Fig. 277(*b*), *full-wave rectification* is obtained. The commutator (Fig. 275) gives full-wave rectification. When a transformer is used with the rectifier, it is a simple matter to obtain full-wave rectification by means of a center-tap connection (Figs. 282 and 287, pp. 348 and 355). When a center tap is not available, the

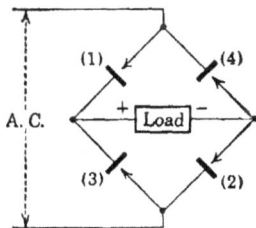

(*a*) Bridge circuit. (*b*) Full-wave rectification.
Fig. 277.

bridge circuit shown in Fig. 277(*a*) may be used. Four half-wave rectifiers connected in a Wheatstone-bridge circuit are used, the direction of positive current being shown by the arrows on the rectifier diagram. When the upper wire is positive, current flows through (1) into the positive terminal of the load and out through (2) to the lower a-c wire, giving alternate waves 1-2 in (*b*). When the lower line is positive, current flows through (3) into the positive terminal of the load and out through (4) to the upper a-c wire, giving alternate waves 3-4 in (*b*). Hence, during every half-cycle, current enters the positive terminal and leaves the negative terminal of the load.

200. Electrolytic Rectifiers.—Electrolytic rectifiers are based on the principle that if a lead plate and an aluminum plate are immersed in a sodium phosphate or ammonium phosphate solution, current can flow only in the direction from solution to aluminum. When the current attempts to reverse and flow from aluminum to solution, a thin insulating film of aluminum oxide is instantly formed over the aluminum plate and acts as an insulator up to about 150 volts. This prevents the current flowing from aluminum to solution. During the reverse cycle the aluminum oxide is dissolved. Such a device may be used, therefore, as a rectifier. Figure 278 shows such a simple rectifier, giving full-wave pulsating current like that shown in Fig. 277(b).

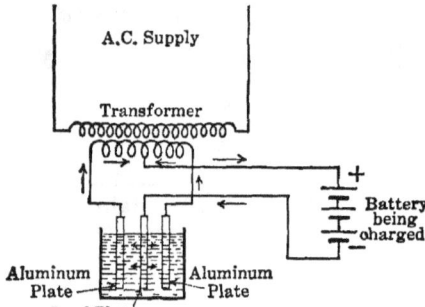

Fig. 278.—Electrolytic rectifier.

Such rectifiers are of low efficiency, 60 per cent or lower, and are of small power capacity. Their advantage lies in their simplicity and low cost.

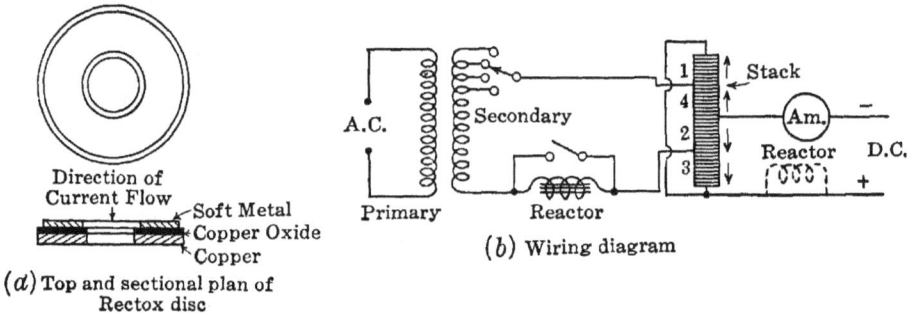

Fig. 279.—Copper-oxide rectifier.

(a) Top and sectional plan of Rectox disc

(b) Wiring diagram

201. Copper-oxide Rectifier.—The copper-oxide rectifier operates on the principle that a layer of cuprous oxide on the surface of copper permits the passage of electrons from the copper to the oxide but prevents their passage in the opposite direction. The conventional direction of current being opposite to the direction of movement of the electrons, the current flows from oxide to copper (Fig. 279(a)). The units consist of washers 1⅛ to 1½ in. in diameter (Fig. 279(a)), mounted on an insulating rod. A soft metal washer, usually of lead, is placed between the copper washers so as to produce more uniform pressure on the oxide. The Rectox rectifier of the Westinghouse Electric and Manufacturing Co. operates on this principle.

The Rectox rectifier is usually employed with the bridge-circuit of Fig. 277(*a*). Each stack is a unit in itself (Fig. 279(*b*)), and several such stacks may operate in parallel to give the desired current rating. The arrows show the direction of rectification in each section of the stack, and the numbers give the corresponding bridge arms of Fig. 277(*a*).

Rectox chargers are adapted for rectifying only small amounts of power, and an important use is for charging batteries. The efficiency of rheostat-regulated chargers is 30 to 40 per cent, and that of step-regulated chargers is 40 to 50 per cent.

Because of its simplicity, low cost, and ruggedness, this type of rectifier is widely used as, for example, in series with d-c instruments to adapt them to the measurement of alternating current. (The instrument will indicate half-wave average values rather than rms values.) Such rectifiers are also used for many other purposes, as in communication circuits for modulating carrier waves.

202. Hot-cathode Rectification.—A considerable number of rectifiers, including vacuum tubes (Chap. XIII), operate on the hot-cathode principle. If a metal cathode is heated to a high temperature, the average velocity of the electrons in the metal increases and a greater number of electrons are able to leave the metal (Par. 231, p. 388). This principle is used in the hot-cathode gaseous rectifiers such as the Tungar, the Rectigon (Par. 203), and the thyratron (Par. 207). This principle may possibly apply in the mercury-arc rectifier.

Figure 280 shows a closed vessel within which is a gas or vapor, such as mercury vapor, at very low pressure. A hot cathode and a relatively cool anode, connected in series with an a-c supply, enter the vessel from opposite ends. Because of its high temperature, the electrons are able to leave the cathode readily. The anode, being relatively cold, emits no electrons, practically. At the instant shown in Fig. 280(*a*), the anode is positive and the cathode is negative. The difference of potential between them creates an electrostatic field, represented by lines, between cathode and anode. The direction of the field is such that electrons travel from cathode to anode. In doing so the electrons collide with molecules of the gas or vapor and thus produce both positive and negative ions by collision. The negative ions travel to the anode, and the positive ions to the cathode, thus establishing a current from anode to cathode. These ions create other ions by collision, so that there is a relatively large number of ions in the space between anode and cathode.

When the potential reverses and the former cathode now becomes positive (Fig. 280(b)), the electrons are withdrawn from the field to the cathode. The former anode becomes negative,

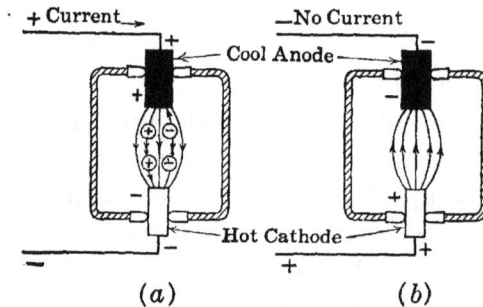

FIG. 280.—Hot-cathode rectification.

and positive ions are withdrawn from the field to this anode. Also, some of the positive ions and electrons recombine. Because of the removal of the electrons and ions from the field, ionization due to collision ceases, and current no longer flows. Hence, such a rectifier, or valve, permits the flow of current from anode to cathode but not from cathode to anode.

203. Tungar and Rectigon.—The Tungar and Rectigon are hot-cathode rectifiers employing an inert gas, such as argon, as the ionized medium. In the bulb (Fig. 281) a tungsten filament is the hot cathode. The filament is heated from a 2.5-volt tap in the transformer secondary. The potential of the filament is practically that of the end *a* of the transformer secondary. When the end *b* of the secondary is positive, the anode is positive, the cathode is negative, and electrons are drawn from the hot cathode. The gas between anode and cathode then becomes ionized in the manner described in Par. 202, and the tube conducts current from anode to cathode. When the end *a* of the secondary is positive, the cathode is positive with respect to the anode, the electrons are withdrawn from the gas to the cathode, the gas between anode and cathode becomes nonconducting, and current ceases to flow. In the figure a battery is being charged, resistance being used to control the current and reactance being used to smooth the current wave. Since conduction occurs only during the positive half-cycles (Fig. 276), this is a *half-wave* rectifier.

204. Mercury Arc.—In the Tungar and Rectigon, the tungsten is the cathode, and the ionized atoms of an inert gas are the current carriers, producing the space charge as well. Since the cathode operates at high temperature, it volatilizes slowly and its life is limited. With mercury-arc rectifiers, the mercury performs the two functions; it supplies the mercury vapor from which the necessary positive and negative ions are produced and also it is the hot cathode. Since the mercury vapor condenses and returns to the cathode pool, there is no deterioration of the cathode with use.

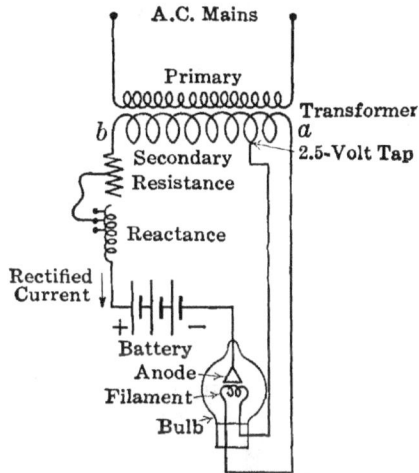

FIG. 281.—Tungar rectifier.

The arc, which is maintained between the anodes and cathode, concentrates on the surface of the mercury pool at the cathode and produces a spot of high electron emission, called the *cathode spot*. In order to maintain the arc continuously, except with the Ignitron (Par. 210, p. 356), mercury-arc rectifiers have two or more anodes. The anodes, usually of iron or graphite, are comparatively cool, their temperature being below that at which electrons are emitted freely. Since one anode is always positive with respect to the cathode, current is always flowing from an anode to cathode, thus maintaining the arc. However, with a pure resistance load, the current becomes zero momentarily twice each cycle. Under these conditions the arc would become extinguished were the current not maintained by the use of a smoothing inductance, as described in Par. 203. The arc drop is from 18 to 30 volts and remains nearly constant, independent of the instantaneous value of the current.

205. Glass-tube Mercury-arc Rectifier.—In Fig. 282 is shown a mercury-arc rectifier of the two-anode, glass-tube type. In order to obtain the correct voltage, the rectifier is supplied ordinarily from the secondary of a transformer. The two terminals A_1 and A_2 are the anodes by which the current enters the tube. The bottom terminal is the cathode, by which the current leaves the tube. A_3 is a starting anode, by means of which the mercury arc is established. The mercury pool at the cathode is also shown. There is also a mercury pool at the starting anode A_3. For starting, the tube is tipped, which causes an arc to form at the break in the mercury path established from A_3 to the cathode. The arc produces mercury vapor, and the main arc is then established.

When the terminal a of the transformer secondary is positive, current enters the tube at anode A_1, is conducted by the vapor to the cathode, and flows through the battery to the center tap N of the autotransformer. When, during the next half-cycle, terminal b is positive, current enters the tube at anode A_2 and likewise flows to the cathode and through the battery to the center tap N. The current waves from anodes A_1 and A_2 are shown as A_1 and A_2 in Fig. 283(a). This is, therefore, a *full-wave rectifier.*

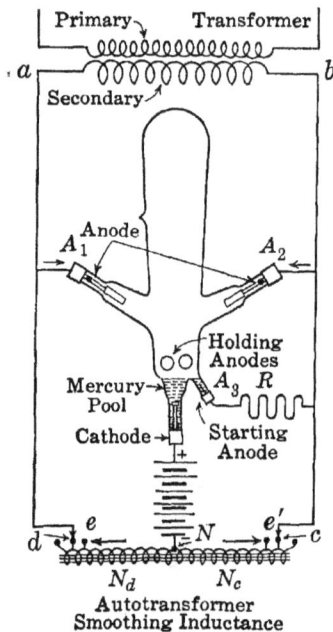

FIG. 282.—Glass-tube mercury-arc rectifier.

If there were no inductance in circuit, the current would go to zero twice each cycle as indicated in Fig. 283(a) which shows the current waves for a resistance, not a battery load. When the current becomes zero, the gas de-ionizes and the arc cannot reestablish itself during the next half-cycle. However, the inductance of the autotransformer prolongs the flow of current (Part I, p. 174), and the current does not go to zero between half-cycles, being maintained as indicated by the solid lines in Fig. 283(b).

When the half-cycle pulse of current from anode A_1 reaches the center tap N, part flows through the winding N_c and completes the circuit to the secondary at b and a. The winding N_c is the primary of an autotransformer and causes the winding N_d to act as secondary, pumping the remainder of the current from N to d, completing the circuit at anode A_1. The current from anode A_2 divides in a similar manner. The output may be controlled by means of the voltage taps c, d, e, e'.

Because of the fragile nature of the glass and the attendant problem of cooling, the power ratings of this particular type of rectifier are limited to a few hundred watts. This type of recti-

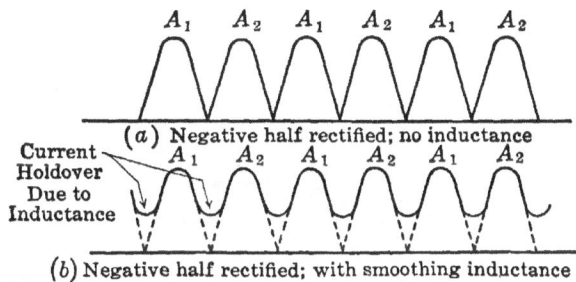

(a) Negative half rectified; no inductance

Current Holdover Due to Inductance

(b) Negative half rectified; with smoothing inductance

FIG. 283.—Rectified alternating-current waves.

fier is used for charging batteries and, in conjunction with constant-current transformers, for obtaining unidirectional current for street lighting.

Glass-tube, mercury-arc rectifiers,[1] made of strong, heat-resisting, special glass are now available up to six anodes. Single bulbs are rated for direct currents up to 400 amp and 1,500 volts and have power ratings as high as 500 kw. They are permanently evacuated, and so require no vacuum pump, and are air-cooled. They are provided with grid control and operate in every way like the metal-tank type (Pars. 206 and 209). They are used to rectify in industrial, mining, and railway service.

206. Metal-tank, Three-phase Rectifiers.—Mercury-arc rectifiers of large power rating were not possible until the metal-tank design was developed. The metal-tank design became practical only when means were found to make gastight seals at the anodes and the cathode. Also, it was necessary to overcome other difficulties, such as impurities on the inner wall of the tank and in the anodes which became sources of electron emission, thus causing

[1] Permavac rectifiers made by the Allis-Chalmers Mfg. Co.

backfiring.　Backfiring, or arc-back, occurs when a cathode spot forms on an anode during the time that the anode is negative. The cause is not thoroughly understood but may be due to the temperature of an impurity in the anode becoming raised above the electron-emission value, or to high concentration of the electrostatic field at an irregularity on the surface of the anode.

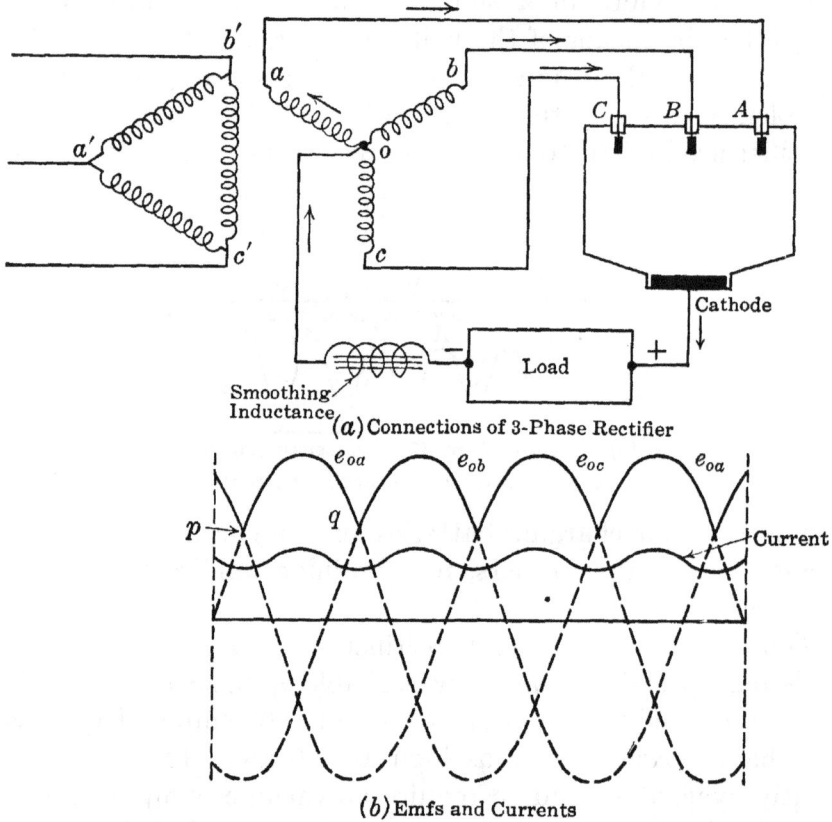

(a) Connections of 3-Phase Rectifier

(b) Emfs and Currents

Fig. 284.—Three-phase steel-tank mercury-arc rectifier.

When an anode thus operates as cathode, current flows to it from an adjacent anode, and a virtual short circuit of the transformer secondaries is the result.

With metal tanks a pumping system operating continuously is necessary to maintain the high vacuum of about 0.0001 mm of mercury.

In single-tank polyphase rectifiers, inductance is not needed to maintain the arc since each anode fires before its predecessor ceases firing so that the mercury vapor is always maintained in an ionized condition. However, in order to improve the current wave form. it is common practice to employ a smoothing inductance (Fig. 284).

A simple, three-phase, metal-tank rectifier is shown in Fig. 284. The three transformer primaries $a'b'$, $b'c'$, $c'a'$ are connected in delta, and the secondaries oa, ob, oc are connected in Y. Terminals a. b. c are connected to the three anodes A, B, C. The load in series with a smoothing inductance is connected between the cathode and the neutral o of the secondary Y.

The three emf waves e_{oa}, e_{ob}, e_{oc} are shown in (b). The arc is transferred from one anode to the next as follows: The emf waves give the anode potentials above the neutral o. Between points p and q the emf e_{oa} of anode A is more positive than the emfs e_{oc} and e_{ob} of anodes B and C, and the electrons are therefore attracted to anode A. Hence, during the interval pq, anode A alone is firing. At the instant q the emf e_{ob} of anode B begins to become more positive than emf e_{oa} of anode A, and the electrons are now attracted to anode B. Accordingly, the arc is transferred from anode A to anode B. In a similar manner, 120° later than instant q, the arc is transferred to anode C. For simplicity the arc drop has been neglected. Because of the smoothing inductance, the ripples in the current are reduced to a small value, as shown by the current wave in (b). The induced emfs due to the changes of current in the smoothing inductance introduce small pulsations in the anode emfs, but for simplicity this effect has been neglected.

GRID-CONTROLLED RECTIFIERS

207. Thyratron.[1]—With vacuum or "hard" tubes such as are used in communication circuits (Chap. XIII), it is possible, by means of potential applied to the grid, to control the current flowing from plate to filament. Also, in a mercury-arc tube or tank, a grid may be inserted between anode and cathode, and by applying potential to this grid the current from anode to cathode may be controlled.

However, in a vapor or gaseous (soft) tube the grid can only *start* the current; it cannot interrupt it no matter how negative the grid is made. This is due to the fact that in the vapor tube there

---

[1] The word *thyratron* is derived from the Greek word *thyra*, a door.

are positive as well as negative ions.   Hence, when the grid is made negative, it attracts positive ions to itself; these surround the grid with a sheath of positive charges, and the positive charges neutralize the negative potential applied to the grid.

In Fig. 285 is shown a thyratron tube employing mercury vapor at very high vacuum.   The hot cathode consists of four metal vanes covered with an activating material (p. 389) from which electron emission can readily occur.   The vanes are surrounded by three concentric nickel cylinders, with spaces between, to

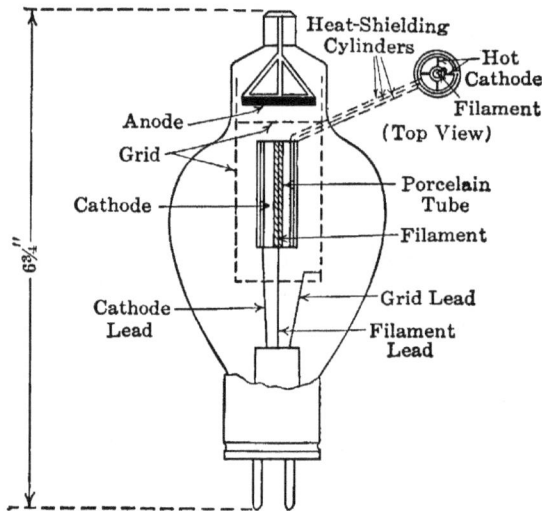

FIG. 285.—Thyratron tube.

heat-shield the vanes and thus reduce the watts per ampere emission.

The cathode is heated by a single tungsten filament passing up through the center of the cathode and insulated from it by a small porcelain tube.   One end of the filament terminates on a cathode vane.   The ions pass from cathode to anode through the open top of the heat-shielding cylinders.   The grid consists of perforated metal between cathode and anode and nearly surrounds both cathode and anode.

Before putting a thyratron into service, the cathode-heater circuit should be operated for at least 5 or 6 min at rated voltage and current.   Otherwise, the tube drop will be excessive and the cathode may be injured owing to bombardment by the positive ions.

208. Thyratron Grid Control.—In gaseous rectifiers the grid can control the arc only to the extent of *starting it*. When the arc once has started, the grid loses control and cannot extinguish the arc. The arc can be extinguished only by interruption of the anode current. This can occur either by an interruption in the external circuit, such as occurs when the alternating current goes through zero, or by making the anode voltage negative.

When once the arc is extinguished, the grid assumes control and can either initiate or prevent further current flow from anode

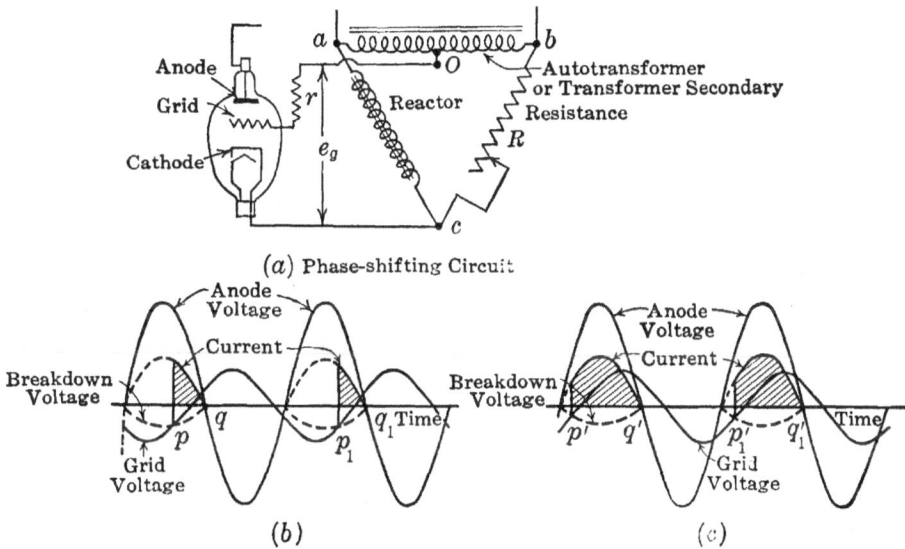

(a) Phase-shifting Circuit

(b)　　　　　　　　　　　　(c)

Fig. 286.—Effect of phase shift of grid potential on thyratron output.

to cathode. This is an important characteristic of all grid-controlled gaseous rectifiers.

Two methods are commonly used to control the firing of the tube. One method is by the magnitude of the voltage applied to the grid. This method is frequently used with grid-controlled power rectifiers (Fig. 288, p. 356). It is necessary to connect considerable resistance in series with the grid. The grid is in reality an anode; and, with constant voltage and no series resistance, the current to the grid will be large when the grid voltage exceeds the arc drop.

The second method is to vary the *phase* of the voltage applied to the grid. This method of grid control is illustrated in Fig. 286. In (a) is shown a convenient phase-shifting circuit. A resistance and a reactor (or capacitor) are connected in series across an autotransformer or transformer secondary *ab*. The grid circuit in series with the grid resistance *r* is connected between *O*, the center tap of *ab*, and *c*, the junction of the resistance and the reactor. By varying the resistance *R* the phase of the grid voltage e_g is varied.

In Fig. 286(*b*) and (*c*) are shown the anode-voltage waves; and, 180° out of phase with them, the breakdown-voltage waves are shown dotted. The breakdown-voltage waves represent the minimum voltage which must be applied to the grid in order that the tube may fire. The grid voltage, the phase of which can be varied, is also shown. Until the time corresponding to point *p* is reached in (*b*), the grid potential is more negative than the value at which the tube can fire, as represented by the dotted wave, and hence up to instant *p* the tube cannot fire. At the instant *p*, the voltage applied to the grid becomes equal to the breakdown voltage, the tube fires, and current flows during the interval represented by the shaded area. *During this interval the grid has no control over the current, since the anode current is not affected by any potential which may be applied to the grid.* At the instant *q*, however, the anode current becomes zero, the gas deionizes, and the grid again assumes control. The tube will then fire again during the next cycle at p_1 where the grid-voltage wave again intersects the breakdown-voltage curve.

In Fig. 286(*c*), the grid-voltage wave has been advanced in phase so that it intersects the breakdown-voltage curve at *p'*. The tube accordingly fires at this instant and continues to deliver current until the anode current again becomes zero at *q'*. During the next cycle the tube fires at p'_1. Hence, the time over which the tube delivers current has been increased, as a comparison of the shaded areas in (*b*) and (*c*) shows. Thus, by varying the phase of the grid voltage the average value of direct current can be easily controlled. The same method of control can be applied to grid-controlled mercury-arc rectifiers (Fig. 288).

In Fig. 287 are shown the connections which are used ordinarily when the thyratron is used as a full-wave rectifier. Since the filament voltage is of the order of 5 volts, a special transformer for the filament supply is desirable. The phase shifter is different from that shown in Fig. 286(*a*), being of the Selsyn type. It is actually a small, three-phase, wound-rotor induction motor with capacitance C and resistance R in series to split the phase in order to produce a rotating field. (See Fig. 247, p. 307.) The grid voltage is obtained from two of the rotor slip-rings. A small autotransformer with a center tap, connected across the slip-ring brushes, makes it possible to supply voltages 180° out of phase with each other to the two grids. By turning the rotor the phase of the grid voltage can be advanced or retarded. The smoothing inductance and the center tap N, by which full-wave rectification is obtained, should be noted.

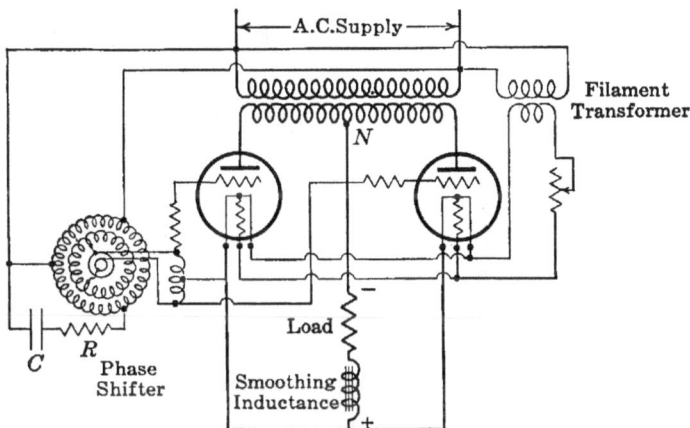

Fig. 287.—Thyratron as rectifier.

Thyratron rectifiers have many industrial applications, such as for regulators for alternator fields, for controlling the current in welding operations, and for lamp dimming in theaters.

209. Metal-tank Multianode Rectifiers.—The metal-tank rectifier of the type shown in Fig. 284 (p. 350) has been developed until it is now available in ratings up to 4,000 kw and 3,200 volts, direct current. Such rectifiers may have as many as 18 anodes in a single tank. In Fig. 288 is shown an 18-anode, 3,000-kw, 630-volt, d-c, metal-tank, grid-controlled rectifier manufactured by the Allis-Chalmers Mfg. Co. It is designed for heavy subway service and is installed in the McKean Street Substation in Philadelphia, Pa.

For the conversion of alternating current to direct current the advantages of rectifiers over motor-generator sets and synchronous converters are: high efficiency over nearly the entire range of load; less weight and space; no special foundations;

low operating and maintenance costs; no falling out of step when system disturbances occur; not affected by dust and moisture; ability to operate at any power frequency; high d-c voltages possible; high instantaneous overload capacity.

Since the arc drop is 25 to 30 volts irrespective of the current, rectifiers do not develop their high efficiencies until about 600 volts is reached. One disadvantage of such rectifiers is the

Fig. 288.—3,000-kw, 630-volt d-c, 18-anode, grid-controlled, mercury-arc rectifier. (*Courtesy of Allis-Chalmers Manufacturing Company.*)

occasional arc-back, or backfiring, described in Par. 206, which produces a momentary short circuit between anodes. One method to minimize this effect is to arrange the grid circuit so that further firing of the anode which backfires is restrained until backfiring has ceased.

Power rectifiers usually operate with grid control actuated either by the magnitude or by the phase of the grid voltage.

210. Ignitron.—The disadvantages of the multianode metal-tank rectifier are: A continuous arc must be maintained by means of an auxiliary anode, with resultant energy loss; diffuse ionization is always present in the tank which is conducive to arc-back; the starting of the arc and its control are obtained by separate means, the control element usually being a grid; precise timing of the firing cannot be obtained with a grid.

The *ignitron* is a single, half-wave, mercury-arc unit in which the ignition of the arc and its control are both accomplished by the same element. A cross-section of a sealed-in unit is shown in Fig. 289.

The anode is of graphite, and the cathode is a mercury pool. The steel tank is water-jacketed. The firing is accomplished by

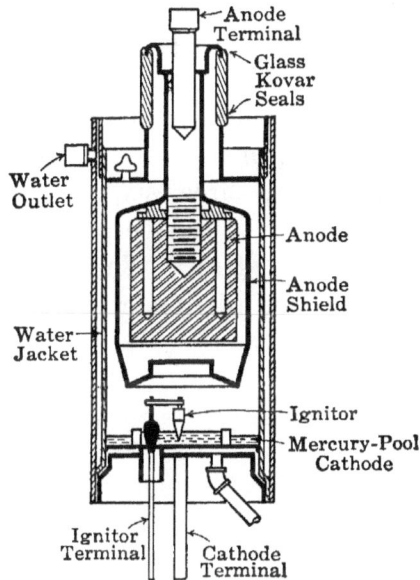

FIG. 289.—Sealed-in metallic-type ignitron. (*Courtesy of Westinghouse Electric and Manufacturing Company.*)

an ignitor invented by Slepian and Ludwig of the Westinghouse Electric and Mfg. Co. The ignitor is a pointed rod of high-resistance refractory material, the pointed end of which dips into the mercury pool. When a positive current impulse is delivered to the ignitor, a spark occurs at the junction of the ignitor and the mercury pool and this spark instantly develops into the mercury arc to the anode if the anode is sufficiently positive. The entire process requires only a few microseconds so that very precise timing is obtained. The arc is extinguished during the following half-cycle when the anode-cathode potential reverses, and the current for the half-cycle goes to zero. The power output can be controlled by timing the firing by means of the ignitor. (See Fig. 286(*b*) and (*c*).)

Thus the ignitron is nonconducting except when actually firing, and accordingly the danger of backfire is minimized. The arc drop is less than with the multianode tank since shields and baffles need not restrict the arc to the degree necessary with the multianode tank; the ignitron has higher overload capacity, longer life, and greater reliability. In the smaller ratings, glass tubes are used; in the intermediate ratings, sealed-in, water-cooled, metal tanks (Fig. 289) are used; in the large power units, all-steel, water-cooled tanks are used.

Ignitrons, on account of their precise timing characteristics, are used extensively in welding; they are also widely used for commercial power purposes and in many installations are preferred to the multianode, metal-tank type. When they are used for commercial power, three, six, and even more ignitron units are combined in order to reduce pulsations in the output-voltage wave. (See Fig. 284(*b*).) The number of such units corresponds to the number of anodes in a multianode metal tank (Fig. 288).

CHAPTER XII

TRANSMISSION AND DISTRIBUTION OF ELECTRICAL ENERGY

211. Typical Transmission and Distribution System.—It is pointed out in Chap. I that for economic reasons electrical energy is ordinarily generated in large amounts in central stations or hydroelectric plants and is *transmitted* to the points where it is utilized. The fact that electrical energy can be transmitted economically over long distances permits the generating station to be located where conditions are most favorable for the generation of energy, for example, near an adequate supply of condensing water and near tidewater so that transportation costs of coal are lowered.

Hydraulic energy, which otherwise would not be available because of its remoteness from centers of population, becomes available owing to the fact that electrical energy can be transmitted economically.

Ordinarily, electrical energy is generated at moderate voltage, is transmitted at high voltage, and is utilized at comparatively low voltage. A typical modern system is shown in Fig. 290, although no attempt is made to show switches, instruments, etc. The energy is generated at 13,200 volts, three phase, in a vertical type water-wheel generator and for transmission is stepped up from 13,200 to 132,000 volts by a delta-Y transformer bank. It is transmitted a distance of approximately 100 miles over a 132,-000-volt line on steel towers. At this voltage a line requires a private right of way, owing to possible dangers from falling wires.

Since it is not possible ordinarily to secure a private right of way through thickly populated districts to the substation, it becomes necessary to step down the voltage to a value that will permit the use of underground cables. Cables cannot be operated at the high voltages which are possible with open-wire lines. Ordinary three-conductor cables do not operate, as a rule,

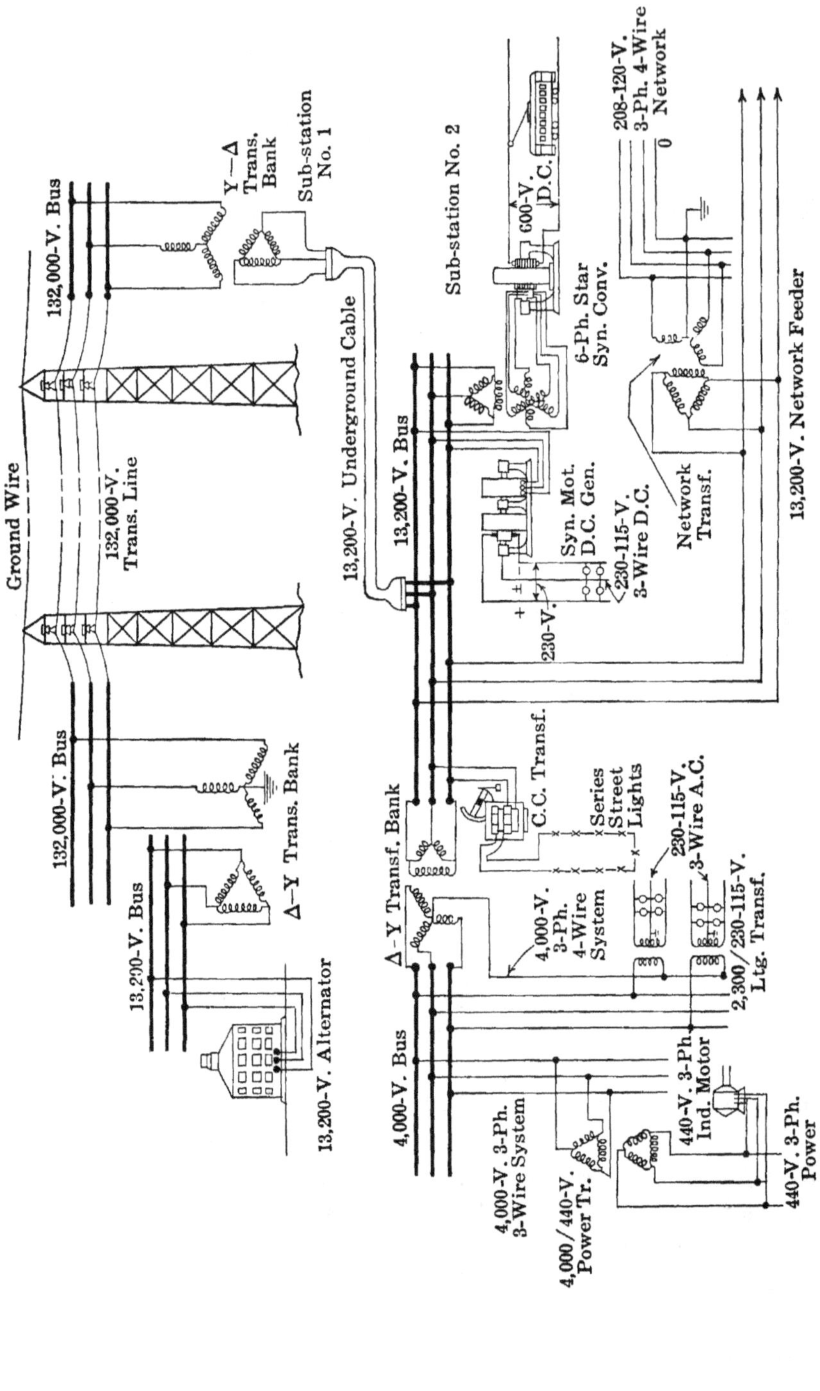

Fig. 290.—Typical transmission and distribution system.

at voltages much higher than 33,000 volts between conductors; and, in the installation shown in Fig. 290, 13,200-volt cables are installed. The energy is stepped down from 132,000 to 13,200 volts by a Y-delta transformer bank in substation No. 1. It is then carried underground to substation No. 2, near the load center, and to other similar substations. Direct-current energy at 600 volts for street railways is obtained from the 13,200-volt bus-bars by means of a synchronous converter taking its energy through a delta-star transformer bank with six-phase secondaries. If 230- to 115-volt, three-wire, d-c service is desired, either another synchronous converter or a synchronous-motor-generator set (Fig. 290) may be used. The synchronous motor operates directly from the 13,200-volt bus-bars.

If series street lights are to be supplied, constant-current transformers may be operated directly from the 13,200-volt bus-bars. For general a-c distribution, a 4,000-volt, three-phase system is used (Fig. 290), the 4,000-volt bus-bars being supplied from the 13,200-volt bus-bars through a delta-Y transformer bank. To eliminate the flicker of lamps due to motors being connected across the line, separate feeders are used for power and for lighting. Lighting loads are supplied from one 4,000-volt, three-phase, four-wire system, which gives $4,000/\sqrt{3}$, or 2,310, volts to neutral. Lighting transformers having a 10-to-1 ratio are connected to neutral, and 230–115-volt, three-wire systems are obtained from their secondaries (see p. 252). The power feeders also run from the 4,000-volt bus-bars. The 440-volt, three-phase service for machine shops, factories, mills, etc., is obtained directly from these 4,000-volt feeders by the use of either V-connected or delta-connected transformers (usually of the three-phase type) located at the premises and stepping down the voltage to 440 volts (Fig. 290).

In downtown districts where the load density is high, a 208–120-volt, three-phase, four-wire network is commonly used. The network is fed directly from high-voltage network feeders by means of network transformers (see p. 374).

The foregoing indicates the manner in which the voltages and currents are obtained for the different classes of power and lighting service required, the source of energy being a power station located over 100 miles away.

212. Voltage and Weight of Conductor.—The factors affecting the transmission of electrical energy are for the most part deter-- mined by the following principle: *The weight of conductor varies inversely as the square of the voltage*, when the power transmitted, the distance, and the loss are fixed.

This relation is based on the fact that the power loss in a given conductor varies as the current *squared*. If the voltage of a system is doubled, the current is halved for the same power. With half the current, the transmission loss is *quartered* if the same size conductors are used. If the transmission loss is to remain unchanged, conductors of one-fourth the original cross-section can be used.

This is illustrated by the following example, which, for sim-plicity, is given for direct current, but which applies equally well to alternating current, since with alternating current the power loss is also equal to I^2R.

Example.—A load of 120 kw at 110 volts is delivered at a distance of 400 ft from the power station over a 1,000,000-cir-mil feeder, each conductor of which has a resistance of 0.0043 ohm. Determine: (*a*) power loss in feeder; (*b*) weight of copper. (*c*) Repeat (*a*) for the same load and distance, but employing 220 volts at the load and using a 250,000-cir-mil feeder, each conductor of which has a resistance of 0.0172 ohm. (*d*) Repeat (*b*) for the 250,000-cir-mil feeder in (*c*). (1,000,000-cir-mil cable weighs 3,090 lb. per 1,000 ft.)

(*a*) The current

$$I_1 = \frac{120,000}{110} = 1,090 \text{ amp.}$$

The total feeder loss

$$P_1 = (1,090)^2 \, 0.0086 = 10,220 \text{ watts. } Ans.$$

(*b*) $\dfrac{800}{1,000} \times 3,090 = 2,472$ lb. *Ans.*

(*c*) $I_2 = \dfrac{120,000}{220} = 545$ amp. $P_2 = (545)^2 \, 0.0344 = 10,220$ watts. *Ans.*

(*d*) Thus the weight of the 250,000-cir-mil cable is one-fourth that of the 1,000,000-cir-mil, or 618 lb. *Ans.*

By doubling the voltage, the weight of copper is reduced from 2,472 to 618 lb. If 550 volts were used, the weight of copper for the same conditions would be only 2,472/25, or 99 lb.

This principle explains why high transmission voltages are used when electrical energy is transmitted over long distances. For example, at 150,000 volts one ten-thousandth (1/10,000) as much copper is required to transmit a given amount of energy, a given distance, with a given loss, as is required at 1,500 volts.

From the standpoint of investment in copper, it is desirable that transmission and distribution voltages be as *high* as possible. On the other hand, other factors, both economic and operating, act to limit the voltage. The cost of insulators increases rapidly with increase of voltage. Higher voltage necessitates greater spacing of conductors, and this requires more costly transmission structures and stations. The cost of transformers, switches, and lightning arresters increases rapidly with increase in voltage. There is thus a voltage which gives the minimum cost of the transmission system. In general, this is found to be approximately 1,000 volts for each transmission mile. For example, a 100-mile line would operate at approximately 100,000 volts.

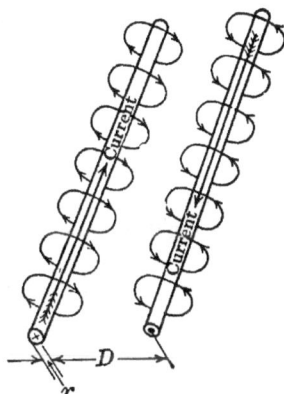

FIG. 291.—Flux linkages in single-phase line.

213. Transmission-line Reactance.—In considering the transmission of power with direct current, the line resistance only need be considered. With alternating current, the line *reactance* ordinarily must be taken into consideration; if the line is long and the voltage is high, the line *capacitance* also must be considered. Cables have large capacitance which cannot be neglected ordinarily.

The reactance of the line is due to the fact that the lines of magnetic induction produced by the current *link* the loop formed by the line conductors. The magnetic flux linking a portion of such a line is shown in Fig. 291. At the instant shown, the current in the left-hand conductor is inward and the current in the right-hand conductor is outward. The flux due to the left-hand conductor alone, the distribution of which is circular about the conductor, is, by the corkscrew rule, clockwise in direction. Since the current in the right-hand conductor is opposite in direction to that in the left-hand conductor, the flux about the right-hand conductor must be counterclockwise in direction. Both fluxes, however, combine to act downward in the single-turn loop formed by the two conductors.   Both these fluxes link the loop, and inductance results.

It can be shown that the inductance of such a loop is

$$L = 2l \left(0.080 + 0.741 \log_{10} \frac{D}{r} \right) \text{ milhenrys,} \qquad (95)$$

where D is the distance between conductor centers and r is the radius of each conductor, both expressed in the same unit. l is the length of the loop in *miles*.   The reactance of the loop is

$$X = 2\pi f L = 4\pi f l \left(0.080 + 0.741 \log_{10} \frac{D}{r} \right) 10^{-3} \text{ ohms,} \quad (96)$$

where f is the frequency.

Example.—Determine the reactance per conductor-mile at 60 cycles of a single-phase line consisting of 0000 copper conductors whose diameters are 0.460 in. and which are spaced 4 ft between centers.

The reactance per *conductor-mile* is one-half that given by Eq. (96).

$$X = 2\pi 60 \left(0.080 + 0.741 \log_{10} \frac{48}{0.230} \right) 10^{-3}$$
$$= 2\pi 60 (0.080 + 0.741 \times 2.318) 10^{-3}$$
$$= 0.677 \text{ ohm.}   \textit{Ans.}$$

(See Appendix I, p. 443.)

214. Transmission-line Capacitance.—If a *d-c* voltage is applied to a transmission line under no-load conditions, no current flows after the first few moments, except the almost negligible leakage current.   If an *alternating* voltage is applied to a transmission line, considerable current may flow, even if there are no appreciable leakage and no connected load.   This current is the *charging current* of the line and leads the line voltage by nearly 90°.   The line acts as a capacitor, the conductors being the electrodes and the air the dielectric.   Each conductor becomes charged, first positively and then negatively, which results in an alternating current.

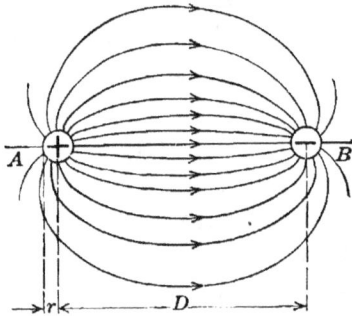
Fig. 292.—Electrostatic flux between line conductors.

This is illustrated by Fig. 292, which shows conductors A and B of a single-phase line.   At the instant shown, conductor A is positive, and conductor B is negative. The electrostatic flux in the field is shown. The capacitance *between conductors* of such a line can be shown to be very nearly

$$C = \frac{0.0194}{\log_{10} \dfrac{D}{r}} \ \mu\text{f per mile,} \tag{97}$$

where D is the distance between conductor centers and r is the radius of each conductor, both expressed in the same unit.

Capacitance has little effect on the normal operation of a transmission line until the voltage becomes about 132 kv and the length of line about 100 miles.

215. Transmission-line Calculations.—Figure 293 (*a*) shows diagrammatically a single-phase transmission line having a resistance of $R/2$ ohms

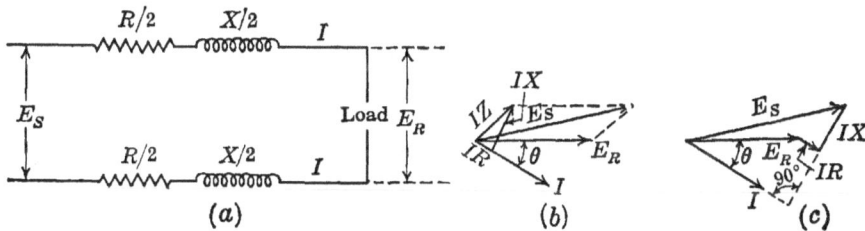

Fig. 293.—Vector diagrams for single-phase transmission line.

per conductor and an inductive reactance of $X/2$ ohms per conductor. The resistance of the entire loop is R ohms and the reactance of the entire loop is X ohms ($X = 2\pi f L$, Eq. (96)). The voltage at the load, or receiver, is E_R; and the voltage at the sending end is E_S. The load takes a current I at the voltage E_R and at power factor cos θ, lagging current. Given the load, or receiver, voltage E_R, the load current I, the load power factor cos θ, it is required to determine the sending-end voltage E_S.

In Fig. 293(*b*) the current I is shown lagging the voltage E_R by the angle θ. The resistance drop IR in the line must be in phase with the current. The reactance drop IX must lead the current by 90°. The vector sum of IR and IX gives the impedance drop IZ. The sending-end voltage E_S must be equal to the receiving-end voltage E_R plus the impedance drop IZ. Hence, IZ is added vectorially to E_R to give E_S (Fig. 293(*b*)). The same result may be obtained by adding IR and IX separately to the vector E_R (Fig. 293(*c*)). This figure is similar to Fig. 162 (p. 191). The solution is also the same as that given for Fig. 162.

$$E_S = \sqrt{(E_R \cos \theta + IR)^2 + (E_R \sin \theta + IX)^2} \ \text{volts.} \tag{98}$$

With leading current,

$$E_S = \sqrt{(E_R \cos \theta + IR)^2 + (E_R \sin \theta - IX)^2} \ \text{volts.} \tag{99}$$

(See Fig. 163(*b*), p. 193.)

Example.—A substation located 6 miles from a central power station requires 1,000 kw at 6,600 volts, 0.8 power factor, lagging current, and 60 cycles. The single-phase line consists of 300,000-cir-mil stranded copper conductors spaced 4 ft apart. Determine: (*a*) voltage at sending end; (*b*) efficiency of transmission.

(*a*) From Appendix H (p. 442) the resistance per mile of 300,000 cir-mil stranded copper conductor is 0.190 ohm. Hence, the total resistance

$$R = 12 \times 0.190 = 2.28 \text{ ohms.}$$

From Appendix I (p. 443) the reactance per mile at 60 cycles of 300,000-cir-mil cable spaced 48 in. is 0.650 ohm. Hence, the total reactance

$$X = 12 \times 0.650 = 7.80 \text{ ohms.}$$

The current

$$I = \frac{1,000,000}{6,600 \times 0.8} = 189.5 \text{ amp.}$$
$$IR = 189.5 \times 2.28 = 432 \text{ volts.}$$
$$IX = 189.5 \times 7.80 = 1,478 \text{ volts.}$$

Using Eq. (98),

$$\cos \theta = 0.8, \qquad \theta = 36.9°, \qquad \sin 36.9° = 0.6.$$
$$E_S = \sqrt{(6,600 \times 0.8 + 432)^2 + (6,600 \times 0.6 + 1,478)^2}$$
$$= \sqrt{32,650,000 + 29,570,000} = \sqrt{62,220,000} = 7,890 \text{ volts.} \quad Ans.$$

(*b*) Line loss $= I^2 R$

$$= (189.5)^2 \times 2.28 = 81,900 \text{ watts, or } 81.9 \text{ kw.}$$

Effi.

$$= \frac{\text{output}}{\text{output} + \text{losses}} = \frac{1,000}{1,000 + 81.9} = \frac{1,000}{1,082} = 0.924, \text{ or } 92.4 \text{ per cent.} \quad Ans.$$

This amount of power would seldom be transmitted single-phase. The same method of calculation would apply, however, to each phase *to neutral* of a three-phase line. That is, the voltages $E_R/\sqrt{3}$ and $E_S/\sqrt{3}$ are used with one-third the power. The reactance per conductor for the same spacing between conductors is the same with three phase as with single phase. The system is then computed on the basis of a single conductor and a neutral return of zero impedance as indicated in Fig. 294. (Also see Par. 116, p. 203.)

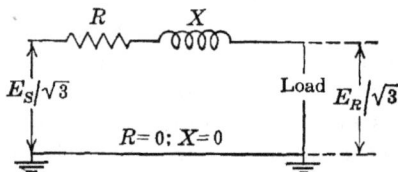

Fig. 294.—Phase to neutral of three-phase line.

It has been stated that if a synchronous motor takes a leading current through a reactance a *rise* of voltage occurs. The reactance of transmission lines is in series with the load. If a load is made to take a leading current by means of synchronous motors or in any other way, this current, because it flows through the line reactance, may cause the load voltage to be higher than the sending-end voltage. In the preceding example, assume that the power factor is 0.8, *leading* current; then, using Eq. (99),

$$E_S = \sqrt{(6,600 \times 0.8 + 432)^2 + (6,600 \times 0.6 - 1,478)^2}$$
$$= \sqrt{32,650,000 + 6,160,000} = 6,230 \text{ volts. } Ans.$$

That is, there is a *rise* of voltage from the sending end to the load without the introduction of energy along the line. This condition is impossible

FIG. 295.—Single-circuit transmission tower, 287,000-volts; Boulder Dam–Los Angeles line. (*Courtesy of the Bureau of Power and Light, City of Los Angeles.*)

with direct current. Hence, within limits, a synchronous motor or condenser may be made to control the voltage of a transmission line.

216. Transmission-line Structures.— For relatively light lines operating at comparatively low voltage, wooden poles are used as line supports. Since the insulators for higher voltage lines are sources of occasional trouble, it is desirable to use as few supports as possible, and longer spans result. Wooden poles have not the requisite length and torsional strength for long spans and the wide spacing which must be used. Steel poles, which ordinarily consist of steel angles and channels held together by latticework, are frequently used where the right of way is narrow, as along a railroad, but such poles are comparatively expensive for their strength. Steel towers of the windmill type (Fig. 295) give the greatest strength for a given amount of material and are the usual supports for high-voltage lines where large amounts of power are involved.

The Boulder Dam–Los Angeles power line operates at 287,000 volts, the highest transmission voltage in the world. Figure 295 shows one of the standard towers used in this line. There are 24 suspension insulators in each string. Note the two ground wires above the power conductors, the function of which is to protect the line from lightning. (See also Fig. 3, p. 5.)

217. Insulators.—The most satisfactory material for high-voltage line insulators is glazed porcelain. This material has high dielectric strength and can withstand simultaneously high

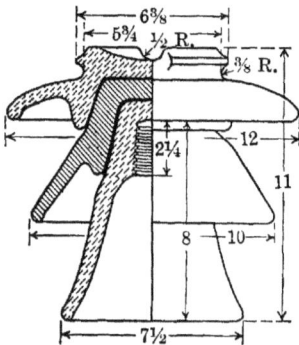

Fig. 296.—60,000-volt pin-type insulator; dry-arcover voltage, 175,000 volts; wet arcover voltage, 120,000 volts; mechanical strength 3,500 lb. Dimensions in inches. (*Courtesy of Locke Insulator Corporation.*)

mechanical stresses, high dielectric stresses, and weathering. For low-voltage lines the pin-type insulator is satisfactory. Pin-type insulators (Fig. 296) can be used for line voltages as high as 77,000 volts, but for such voltages they are bulky and expensive and furthermore produce large torsional stresses in the crossarms.

These disadvantages are eliminated by the use of suspension insulators. The suspension-insulator string is composed of a number of porcelain units suspended in series (Fig. 297). By employing a sufficiently large number of units, a suspension string can be used for the highest voltages. Since tension is the only force that the insulator string can sustain, the horizontal stresses in adjacent spans must be made equal in order that the insulator strings may hang vertically.

Since the number of units in a string is readily adjusted to the system voltage, suspension insulators are used on systems operating from as low as 25,000 volts to the highest voltage systems, such as the Boulder Dam–Los Angeles system which operates at 287,000 volts (Fig. 295).

Fig. 297.—Sixty-cycle flashover of 11-unit suspension string. (*Courtesy of Locke Insulator Corporation.*)

The 11-unit string of Fig. 297 is being subjected to a 60-cycle laboratory flashover test. Note the manner in which the arc "cascades" around the individual units. Under these conditions the heat of the arc usually causes the porcelain to crack and break. "Cascading" is eliminated by the use of arcing rods (at top of strings) and shielding rings (at bottom of strings) (Fig. 295).

Suspension strings are also used for strain insulators, the insulator string then forming part of the span.

218. Lightning Arresters.—Electric power lines are subjected frequently to overvoltage, usually of a transient nature, due to lightning or to disturbances caused by switching, short circuits and arcing grounds. It is desirable to relieve the line of these voltage surges as quickly as possible, since they may flash over insulators (Fig. 297), destroying them and shutting down the line. It is usually not economical to provide protection for each insulator string, although arcing rods and rings which keep the arc away from the string are often used (see Fig. 295).

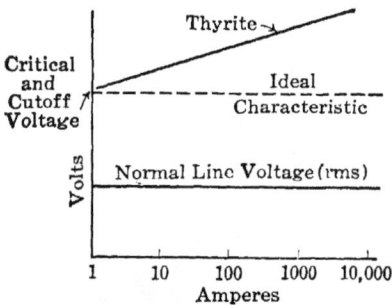

FIG. 298.—Lightning-arrester characteristics.

Also, where conditions justify it, *protector tubes*, fiber tubes with metal ferrules on the ends, are used to limit the voltage across the individual insulator string.

Because of the cost of transformers, generators, and other station apparatus, considerable expense is justified to protect them against overvoltages.

The function of lightning arresters is to relieve the line of overvoltages by supplying a low-resistance path to ground. At the same time the arrester must suppress the dynamic arc which the normal line voltage tends to maintain after the transient discharge has ceased. An ideal lightning-arrester characteristic is shown dotted in Fig. 298, the current being plotted to a logarithmic scale. When the critical voltage is reached, the arrester should discharge up to very large values of current without any voltage rise. Also, the arrester should cut off well above normal line voltage to prevent the line voltage maintaining a dynamic current through the arrester. Lightning arresters are usually rated so that the critical voltage is about $2\frac{1}{2}$ times normal line voltage (rms). Actually, lightning arresters have resistance so that the voltage across the arrester rises with increasing values of current, as shown by the Thyrite-arrester characteristic (Fig. 298).

The Thyrite arrester (Fig. 299), manufactured by the General Electric Company, consists of discs about 6 in. (15.2 cm) diameter and ¾ in. (1.9 cm) thick, made from a nonporous, ceramic material shown in the top of the cutaway (Fig. 299). For electrical contact, the two opposite faces of the disc are sprayed with copper. The discs are insulators until the critical voltage is reached, when they suddenly become good conductors. Thus they allow lightning and other surges to discharge to ground

FIG. 299.—Station-type Thyrite lightning arrester, rated 12 kv maximum line to ground. Thyrite unit cut away to show construction. (*Courtesy of General Electric Company.*)

without undue voltage rise (Fig. 298). After the discharge ceases, the arrester returns to its normal nonconducting condition. Since the critical voltage is well above normal line voltage, dynamic current due to normal line voltage cannot follow the transient discharge. The discs of the arrester are stacked in series (Fig. 299), the number being determined by the voltage level at which the system is to be protected. The stacks are enclosed within a porcelain casing, and there are short sphere gaps in series with each stack, as is shown in the lower part of the cutaway (Fig. 299). Lightning arresters should be connected *near* the apparatus which they are designed to protect.

219. Low-voltage Alternating-current Networks.—One method of distributing a-c energy at low voltage is shown at the left in Fig. 290 (p. 360), in which two 4,000-volt, three-phase distributing systems are shown. Primarily, the four-wire system supplies light, and the three-wire system supplies power. The

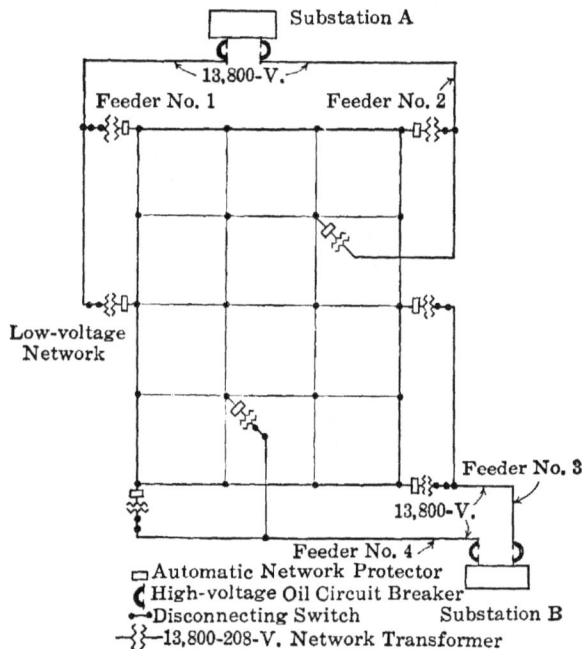

Fig. 300.—Low-voltage a-c network.

connections of a distribution transformer are shown in Fig. 207 (p. 252). In recent years, another type of system has become standard, the 208–120-volt, three-phase, four-wire, low-voltage network, also indicated in Fig. 290. This system is designed to supply districts where the load density is high and where frequent changes in large single loads require a system having a high degree of flexibility. The system is particularly well adapted to downtown districts where changes frequently occur in both lighting loads and miscellaneous industrial applications. Typical loads are stores, theaters, hospitals, elevators, and small industries. The secondary, or low-voltage, system consists of a

208–120-volt, three-phase, four-wire network fed at different points from high-voltage feeders through network transformers. (Some systems operate at 199–115 volts.) A one-line diagram of the system is shown in Fig. 300. In the secondary system the mains are connected together solidly to form a grid, or network, to which the customer loads are directly connected. The secondary system is designed with such size of wire that faults will burn themselves clear. To localize burnouts and thus prevent the destruction of an entire cable by overheating during a short circuit, current "limiters," consisting of a copper conductor of cross-section slightly less than that of the cable, are inserted in series with the cable at the two ends. The limiters are surrounded with cement or some other fire-resisting material to prevent hot metal becoming a fire menace to neighboring cables.

Three-phase, 220-volt induction motors can be operated from the 208-volt, three-phase service, the 220-volt motor design now being such that little difference in operating characteristics occurs with the 5.5 per cent reduction from rated voltage. Lamp loads are connected from the three-phase wire to the grounded neutral and operate at $208/\sqrt{3}$, or 120, volts. Naturally, lamps rated at 120 volts should be used.

The network is supplied by specially designed network transformers (Fig. 301) of about 300- to 500-kva rating, the primaries being connected through disconnecting switches to feeders operating usually at the voltage at which power is delivered to the substation, 13,800 volts or thereabouts (Figs. 300 and 301) being common. Thus the 2,300- and 4,000-volt intermediary system (Fig. 290) is eliminated. The transformers are located usually in manholes, vaults in buildings, and private enclosures. Transformers for subway and manhole installations are of the submersible type and are waterproof (Fig. 301).

Since continuity of service is highly important in downtown districts, particularly in the large cities, means to approximate the quality of service given by direct current with storage battery stand-by are necessary. The network is usually fed from more than one source, two substations being shown in Fig. 300. Also, in order that the failure of a single feeder may not cause interruption of service, there should be at least two feeders from each substation to the network.

FIG. 301(a).—Side view of three-phase Pyranol network unit consisting of transformer, 500 kva, 60 cycles, 12,470 to 208 Y/120 volts, with submersible low-voltage network protector and high-voltage disconnecting and grounding switch. (*Courtesy of General Electric Company.*)

FIG 301(b).—End view with protector door open.   (*Courtesy of General Electric Company*).

To improve further the continuity of service, the automatic network protector (Fig. 300) has been developed. This consists of an air-break switch with closing and tripping mechanism controlled by suitable relays (Fig. 301(*b*)). When a feeder trips out or is removed from service, the network protector automatically disconnects the transformer secondaries from the network, for otherwise the transformer would feed energy from the network back into the feeder which may have become faulty. The relays operate on reverse energy and are so sensitive that they operate to disconnect the transformer secondary when only exciting current flows to it from the network.

When the operator connects the feeder to the substation busbars, the protector will automatically close if the secondary voltage of the transformer is slightly higher than that of the network and if the two voltages are substantially in phase opposition. If the conditions are not correct for closing, the protector merely remains ready to close and operates when the correct conditions do occur.

DIRECT-CURRENT DISTRIBUTION

Because of the difficulty in transforming d-c voltages, there are no commercial d-c transmission systems in the United States. However, there are d-c distribution systems in the centers of some large cities. Such systems were installed initially because the importance of continuity of service made storage-battery reserve almost a necessity. Also, d-c motors are well adapted to elevator service and printing-press drive, the two most important types of load in large cities. Another advantage of d-c distribution is that the capacitive and skin effects in cables which occur with alternating current do not occur with direct current. Also, the system connections are simple. Direct current is also essential for street railways, since no satisfactory a-c street-railway motor has been developed as yet.

However, in new installations of low-voltage distribution systems, a-c networks (Par. 219) are now used. They are more efficient in that the rotating machinery or rectifiers necessary for conversion from alternating current to direct current are eliminated. With several sources of feed and with the present development of protective and relay apparatus, such a-c systems give a high degree of continuity of service.

220. Mains.—Services, or loads, of individual customers are taken directly from *mains*. These loads are ordinarily concentrated at various points along the mains (Fig. 302).

In a city, a d-c distribution system consists ordinarily of a network of connected mains (Fig. 302). The mains are fed at various points, called *centers of distribution*, by *feeders* running directly from substations. The centers of distribution are

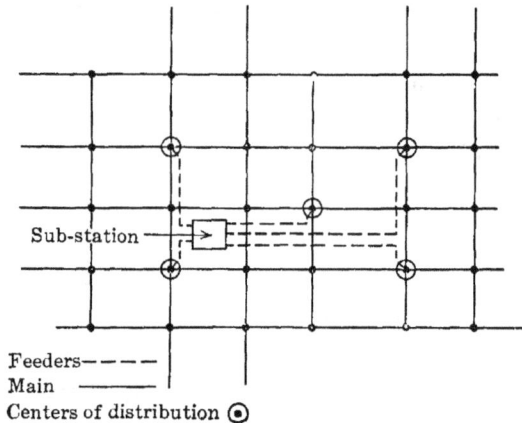

Feeders— — — —
Main   ————
Centers of distribution ⊙
Fig. 302.—Typical feeder and main distribution system.

maintained at approximately the same potential, and the entire system is held within 1 or 2 per cent of the same potential.

221. Feeders.—Feeders are conductors, each feeder consisting usually of a positive and a negative conductor (except trolley feeders), which run directly from the station bus-bars to the centers of distribution (Fig. 302). In distinction to mains, services are not taken from feeders. In most instances the only load on the feeders is the single load at its end or the load taken at the center of distribution. When the ladder system is used, trolley feeders often feed the trolley at regular intervals (see p. 377).

Since there are no intermediate loads on a feeder, the end may be maintained at the potential at which it is desired to hold the center of distribution.

The potential at the center of distribution is determined by means of two *pressure wires*, which are small conductors, about No. 14 A.W.G., in the cable, and are connected from the center of distribution to a station voltmeter. The potential at the centers may be changed by raising and lowering the bus-bar voltage. If a station feeds several centers, three or four sets of bus-bars, maintained at different voltages, are often used. The feeder is connected to the set of bus-bars which give the proper voltage at the center of distribution.

Example.—It is desired to maintain the potential at 234 volts at a center of distribution. A 1,000,000-cir-mil cable, 800 ft long, connects the center with the station bus-bars. If the feeder current is 900 amp, to what voltage must the feeder be connected at the station?

The resistance of 1 cir-mil-ft of copper at 20°C is 10.37 ohms (see Part I, p. 16). The resistance per conductor of the cable is

$$R' = 10.37 \, \frac{800}{1,000,000} = 0.00830 \text{ ohm.}$$

The voltage drop in each conductor

$$IR' = 900 \times 0.00830 = 7.47 \text{ volts.}$$

The total voltage drop

$$IR = 2 \times 7.47 = 14.94 \text{ volts.}$$

It would be necessary to have 234 + 15, or 249, volts across the station end of the feeder. *Ans.*

The feeder would probably be connected across 250-volt bus-bars.

222. Electric-railway Distribution.—Electric-railway generators are generally compounded, the series field being on the negative side. The negative terminal is usually connected directly to the grounded side of the system, or to the rail through a switch. The positive terminal feeds the trolley through an ammeter, a switch, and a circuit breaker (Fig. 303).

On short lines, with light traffic, the trolley alone may suffice to carry the current to the car. Except in small installations, the trolley is of insufficient cross-section to supply the required power. As the size of the trolley wire is limited by the trolley wheel and by mechanical reasons, it cannot be increased conveniently. The same result as from increasing the size of the trolley may be obtained by running a feeder in parallel with the trolley and connecting the feeder to the trolley at short intervals. This is called the *ladder system* of feeding. The trolley and feeder together may be considered as a single conductor.

Where the density of traffic requires several feeders, the best results are obtained by connecting the feeders as shown in Fig. 303(*a*).   Each feeder is protected by a circuit breaker as indicated.

The objections to the preceding methods of feeding are that trouble, due to a ground, for example, at any point on the trolley, involves the entire system.   In cities where traffic is particularly

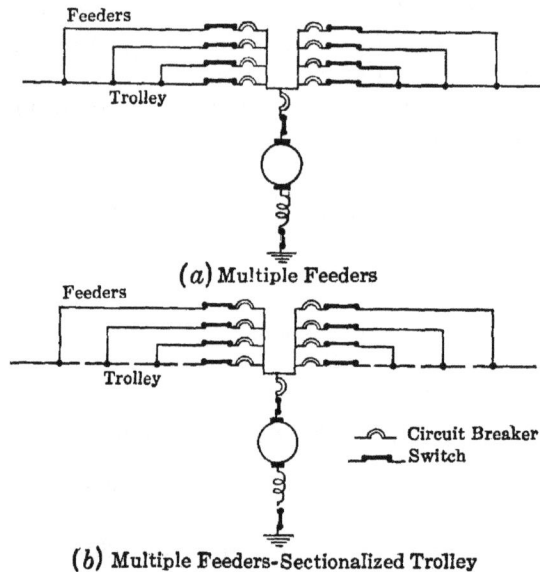

(*a*) Multiple Feeders

_ᴧ_ Circuit Breaker
_ᴧ_ Switch

(*b*) Multiple Feeders-Sectionalized Trolley

Fɪɢ. 303.—Methods of feeding trolley system.

dense, it is not permissible to subject the entire system to the hazard of a shutdown due to a ground at one point only.   The trolley, therefore, is sectionalized (Fig. 303(*b*)).   In this method the trolley is divided into insulated sections, each of which is supplied by a separate feeder.   Trouble in one section is not readily communicated to the other sections.   This increased reliability is obtained at the expense of a less efficient use of the copper, as the feeders are unable to assist one another.

223. Electrolysis.—Most trolley systems use the track as the return conductor for the current taken by the car.   The return currents not only flow through the tracks themselves but seek the paths of least resistance to return to the negative terminal of the station generator.   The return currents, in spreading through the earth, follow such low-resistance conductors as water pipes, gas pipes, and cable sheaths (Fig. 304).   The fact that the current *enters* and flows along these conductors does no harm. When the return currents *leave* the pipes, however, as at *a*, Fig. 304, they tend to carry the metal of the pipe into electrolytic solution, which ultimately results in the pipe being eroded away. To decrease the effects of electrolysis, the following methods are used. (*a*) Provide a return path with as low a resistance as is

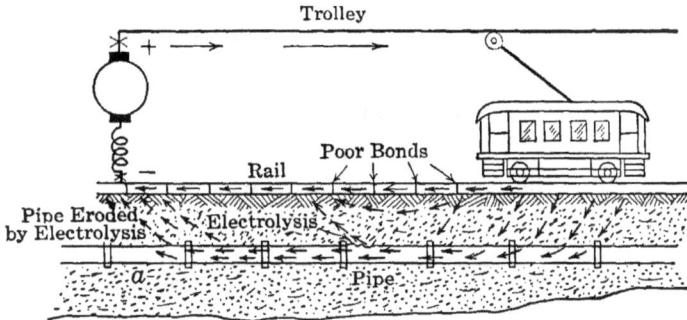

FIG. 304.—Electrolysis by earth currents.

practicable. This is secured by bonding and by negative feeders. The permissible drop in the ground return circuit is often limited to 10 or 15 volts. (*b*) Discourage the entering of the current into the pipes by inserting occasional insulating joints in the pipes.

224. Series-parallel System.—Lighting and household services have become standardized at 115 to 120 volts. (Small isolated plants requiring storage batteries, such as farm-lighting systems, may operate at 32 volts.)

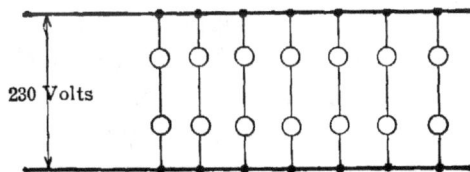

FIG. 305.—Series-parallel system.

From the point of view of the amount of copper, the service voltage should be as high as safety permits, for the amount of copper varies inversely as the *square* of the voltage. From the point of view of incandescent lamps a low voltage is desirable since thicker, shorter filaments are less fragile and the lamps are more efficient. The service voltage must be a compromise between these factors, and 115 to 120 volts has been chosen as the voltage which gives the greatest over-all economy. Moreover, it is not so high as to be dangerous.

If service mains could be operated at 230 volts rather than at 115 volts, the weight of copper would be reduced to one-fourth the value for 115 volts. With lamps, this saving in copper may be effected by connecting the lamps so that two are always in series (Fig. 305).

The disadvantages of this series-parallel system are that lamps can be switched only in groups of two and if one lamp burns out the lamp to which it is connected ceases to operate. Also, both the lamps in series must be of the same current rating.

The series-parallel system is seldom used.

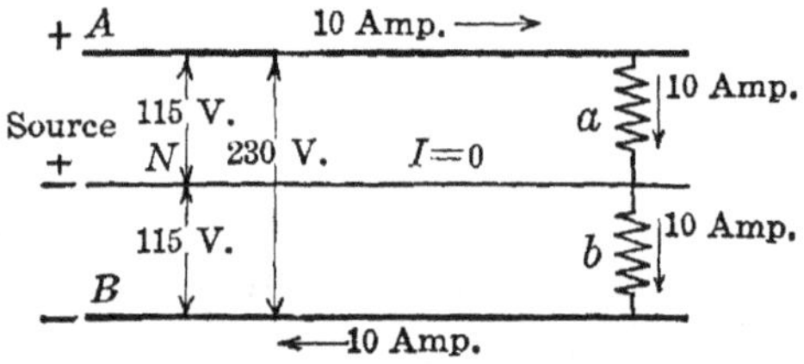

FIG. 306.—Edison three-wire system—balanced loads.

225. Edison Three-wire System.—The objections to the series-parallel system may be eliminated by running a third wire, or *neutral*, between the two outer wires. This neutral maintains all the lamps at approximately 115 volts. The advantage of a higher voltage in reducing the weight of copper is obtained by the use of this system. If there were no neutral, the 230-volt system would require one-fourth the copper of an equivalent 115-volt system. If it is assumed that the neutral of the Edison three-wire system is of the same cross-section as the two outer wires, the total copper for the three-wire system is $\frac{3}{8}$, or $37\frac{1}{2}$ per cent of that for a 115-volt system of the same power rating. The saving in copper, therefore, is $62\frac{1}{2}$ per cent. In practice, the neutral can be made smaller than the two outer wires so that the saving in copper is even greater than $62\frac{1}{2}$ per cent.

The plan of the system is shown in Fig. 306. Two conductors *A* and *B* have 230 volts maintained between them, *A* being the positive and *B* the negative conductor. A third conductor *N* is maintained at a difference of potential of 115 volts from each of the other two conductors. *N*, therefore, must be negative with respect to A and positive with respect to B. That is, current tends to flow from A to N and from N to B.

Figure 306 shows the conditions which exist when the loads on each side of the system are balanced. Each of the loads a and b takes 10 amp. The 10 amp taken by load a flows through to load b and then back through conductor B to the source. This is equivalent to a series-parallel system as both loads are equal and in series. Under these conditions the current in the neutral conductor is zero, and the loads are said to be balanced.

Figure 307(a) shows the conditions existing when the load a on the positive side of the system is 10 amp and the load b on the negative side is 5 amp. Under these conditions the extra 5 amp taken by load a must *flow back* through the neutral to

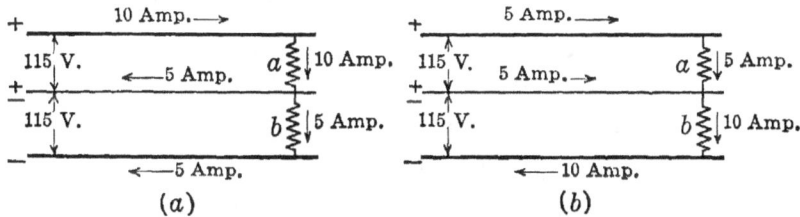

Fig. 307.—Unbalanced three-wire systems.

the generator, or source. There are 5 amp, therefore, in the neutral returning to the generator. In Fig. 307(b), load b is 10 amp, and load a is 5 amp. Under these conditions the extra 5 amp must *flow out* to the load through the neutral. It is to be noted that the current in the neutral may flow in either direction, depending upon which load is the greater. If an ammeter, therefore, is connected in a neutral, it should be of the zero-center type. Moreover, the neutral carries the *difference* of the currents taken by the two loads. In practice the loads are usually disposed so that they are nearly balanced. Twenty-five per cent unbalancing (that is, a neutral current which is 25 per cent of the average of the currents in the two outer conductors) is usually permissible.

If the neutral of the three-wire system is opened and the loads are unbalanced, the voltage across the lesser load, which has the higher resistance, will exceed that across the greater load. This may result in lamps being burnt out.

226. Methods of Obtaining Three-wire Systems. *Two-generator Method.*—Two shunt generators may be connected in series, as shown in Fig. 308. The positive terminal of one is connected to the negative terminal of the other, or the generators are in series between the outer conductors. Both generators may be driven by the same prime mover. Each generator supplies only the load on its own side of the line. The objection to this method is that two separate generators are required.

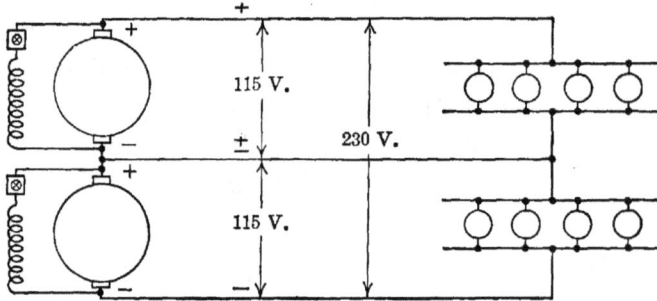

FIG. 308.—Two generators supplying three-wire system.

Balancer Set.—The use of a balancer set is a very common method of obtaining the neutral. This set consists of a motor and a generator mechanically coupled. They are connected in series across the outer wires, and the neutral is brought to their common terminal, as shown in Fig. 309. Either machine may

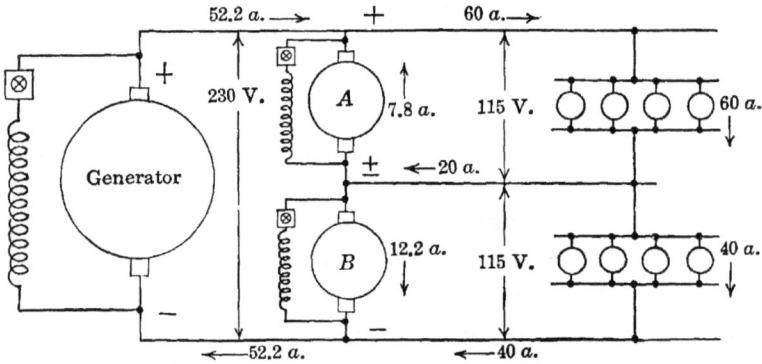

FIG. 309.—Balancer set giving neutral in three-wire system.

act as a motor or generator, depending on which side of the system has the greater load. The action of the set is as follows: If the system is balanced, both machines operate as shunt motors connected in series across the outer conductors. They take just enough current to supply their losses. If the system is unbalanced, as in Fig. 309, where a load of 60 amp is across the positive side of the system and a load of 40 amp is across the negative side of the system, machine *A* across the side having the greater load acts as generator and machine *B* acts as motor. The motor *B* takes *some* of the current returned by the neutral and utilizes the power represented by the product of this current and the voltage across its side of the system to drive the machine *A* as generator. The generator *A* then causes the remainder of the neutral current to be returned to the positive line. In Fig. 309, each machine is assumed to have an efficiency of 80 per cent, giving 64 per cent as the efficiency of the set. The system voltages are also assumed to be balanced. The motor requires 12.2 amp; and the generator delivers only 0.64 of this current, or 7.8 amp, to the positive conductor. If the balancer set were 100 per cent efficient, the motor would take 10 amp and the generator would deliver 10 amp.

With shunt machines, better voltage balance is obtained if the shunt field of the machine on the negative side of the system is connected from neutral to positive and the shunt field of the machine on the positive side of the system is connected from

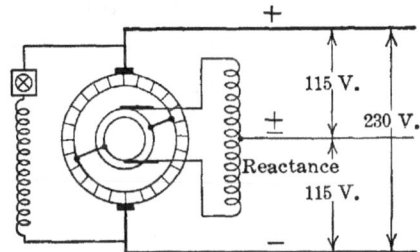

FIG. 310.—Three-wire generator connections (Dobrowolsky method).

neutral to negative. In practice, each machine has a series winding which is cumulative when the machine acts as generator and differential when it acts as motor. This increases both motor and generator action without the voltages of the system becoming too greatly unbalanced.

In practice the rating of the balancer set is of the order of 5 or 10 per cent of the generating capacity, depending on the degree of unbalancing.

Three-wire Generator.—The three-wire generator, or Dobrowolsky method, is a very efficient method used to obtain a neutral. The method was mentioned in connection with the synchronous converter (see p. 335). In its simplest form, the ordinary three-wire shunt or compound generator has two slip-rings in addition to its commutator (Fig. 310) and the emf between these rings must necessarily be alternating. In fact, the generator operates as a double-current generator, delivering practically all its power from the d-c, or commutator, end. The slip-rings deliver only sufficient current to excite the reactance coil which is connected across them. This coil has low resistance but high reactance. Direct current, therefore, can flow through it readily, but it can take only a small alternating current from the slip-rings. The center of the coil is at the center of gravity of the voltages generated in the armature. If the three-wire neutral is connected to the center of this coil, the d-c voltage from either brush to neutral will be the same. Moreover, any direct current returning through the neutral can flow readily into the armature through this reactance, since reactance has no effect on the flow of a steady direct current. The neutral current divides, half flowing in each direction through the reactance.

227. Storage Batteries.—Theoretically, storage batteries can be used to equalize the load on d-c systems, relieving the generating equipment on heavy loads, or "peaks," and being charged during light loads, or "valleys." Practically, because of the high maintenance cost of storage batteries, it is more economical, except under an occasional special condition, to operate extra machinery on the peak loads. In places where a complete shutdown might cause great inconvenience, as in a large city, batteries are frequently installed merely to carry the load in emergencies. They are kept charged, ready to put in service, or are even kept "floating" across the bus-bars, enabling them to take the load automatically in case of shutdown of the generating system.

The most common method of controlling the discharge of such batteries is to have an excess of cells, called *end cells* (Fig. 311), which may be cut in or out, according to the load the battery is called upon to deliver. It is essential to cut cells in and out without opening the circuit.

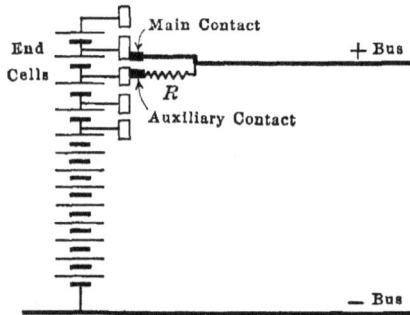

Fig. 311.—End cell control of storage battery (transition period).

For this purpose an end-cell switch is used, similar to that shown in Fig. 311. The main contact is connected to the auxiliary contact through a resistance R. In sliding from one battery contact to the next, the auxiliary contact maintains the circuit connections through the resistance R. Were there zero resistance between the main contact and its auxiliary contact, there would be a dead short-circuit on the individual cells during the transition period. The resistance R is chosen usually so as to allow the normal battery current to flow during the transition period. End-cell switches become rather massive in large battery installations and are often operated by a motor-driven worm. This also permits remote control.

The end cells, not being in continuous service, are discharged to a lesser degree than the others and require individual attention on charging.

228. Series Distribution.—Series distribution is used where the translating devices all take the same current and where the resulting high voltage is not objectionable. The advantages of such a system are the large saving in copper and the simplicity of the circuit. Practically the only systems which fulfil the foregoing requirements are street-lighting systems, in which the lamps all take the same current and operate at the same time and, since the installation is out of doors, the high voltage is not objectionable.

The series system differs from the parallel system in the manner of removing loads. If a load is open-circuited in the parallel system, the other loads are not affected except perhaps by a slight change in voltage. In the series system, the loads are all in series with one another so that the current is the same in each load. If, therefore, the circuit of any one load be opened, the current in all the other loads will be interrupted. As this is not permissible, a load must be *short-circuited* when it is desired to remove it from service.

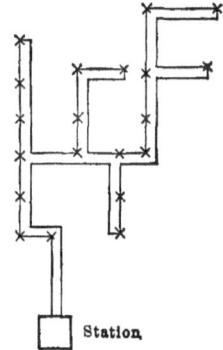

Fig. 312.—Parallel-loop series circuit.

Direct current is required for magnetite arc lamps and is obtained either from a d-c series generator or from a constant-current transformer operating in conjunction with a mercury-arc rectifier (see p. 348). Alternating current for incandescent lamps is also supplied by a constant-current transformer but without the rectifier. All these methods tend to maintain constant current under all conditions of load. If, therefore, the circuit is opened and thus a very high resistance introduced, a constant current is maintained across a high resistance and a high voltage results. For this reason the lamps used on a constant-current system are protected by a thin disc of paper between the lamp terminals (film cut-out). If the lamp burns out, the high voltage punctures the paper and so prevents the circuit being opened.

The parallel-loop system (Fig. 312) is the most common system of series distribution.

CHAPTER XIII

VACUUM TUBES

229. Electrons.—*Vacuum tubes*, electron tubes, or thermionic valves, as they are variously called, depend for their operation on the fact that electricity is atomic. That is, electricity is composed of extremely small *negative* charges called *electrons*. The charge of each electron is 1.6×10^{-19} coulomb, and its mass is 9.04×10^{-28} gram, or $1/1,845$ of that of the hydrogen atom. A *neutral* atom of matter consists of a small, positively charged *nucleus*, with which enough electrons, called *outer electrons*, are associated to give an equal negative charge, so that the resultant charge of the entire atom is zero. The nucleus itself is made up of as many hydrogen nuclei, or *protons*, as there are outer electrons, and an equal or greater number of *neutrons*, which have substantially the same mass as hydrogen nuclei but no electric charge. In nonconductors of electricity, the outer electrons are very closely associated with the nucleus, and it is difficult to remove an electron from the atom. In the metals, which are conductors, a small proportion of the outer electrons appear to be free to pass easily from one atom to the next. But, even so, the density of these free electrons in a metal is very great, being sufficient to give a charge of approximately 16,000 coulombs per cc.

The free electrons in a metal are supposedly in constant motion and are continually colliding with one another and with the atoms of the metal, which are also in motion. Their motion is similar to that of the atoms of a gas in a confined space. As with the atoms of a gas, the velocities of the individual electrons differ widely at any instant, but their velocity as a whole gives an average velocity which is determined by the temperature of the metal.

230. Emission.—The surface of the metal is a boundary surface for the free electrons and is impervious to most of them. A force of repulsion is exerted at this surface on the electrons in contact with it, which turns back all those whose velocities fall below a certain critical value, while allowing those having velocities greater than this critical value to pass through the surface. If the space outside the metal is evacuated, it will gradually become filled with electrons. These electrons collide among themselves, however, and thus some of them are caused to return to the metal through the surface, this process being accelerated by the *space-charge* effect of the electrons. That is, the electrons in the space outside the metal, all being negative charges, mutually repel one another and cause some of their number to be driven back into the metal. A condition of equilibrium is reached when in any given time the number of electrons leaving the metal is equal to the number returning to it.

231. Critical Velocity.—The critical velocity which an electron must have in order to escape from a metal is of the order of 10^8 cm per sec. It varies somewhat for the different metals. It is usually expressed in terms of the difference of potential through which an electron must fall in order to acquire this velocity, because this difference of potential is the quantity actually measured. The energy of an electron carrying a charge e and having a mass m when it has fallen freely through a difference of potential V and acquired a final velocity v is

$$Ve = \tfrac{1}{2}mv^2. \tag{100}$$

The difference of potential corresponding to a velocity of 10^8 cm per sec is about 4 volts.

The emission of electrons from a metal is analogous to evaporation from a liquid. The surface tension of the liquid corresponds to the apparent repulsive force at the surface of the metal. The latent heat of evaporation corresponds to the work done by the electrons against the repulsive force at the surface.

At room temperature, the number of electrons emitted from a metal over any ordinary interval of time is very small, because the average velocity of the electrons within the metal is low and only occasionally does one attain sufficient velocity to escape from the surface.

Raising the temperature of the metal increases the average velocity of the electrons and increases the emission, because more electrons obtain sufficient velocity to escape. The emission increases rapidly with increase in temperature and reaches astonishingly large values at white heat. Emission of electrons may also be produced by the action of ultraviolet light (photoelectric effect), by X-rays, by the various rays from radioactive substances, and by the impinging on the surface of high-speed electrons (secondary emission).

232. Richardson's Law.—Richardson, in 1901, showed that the emission per unit area is an exponential function of the temperature and is also a property of the material.

The current in amperes per square centimeter emitted from a body at temperature T in degrees Kelvin is given by the equation

$$i = AT^2\epsilon^{-\frac{b}{T}}, \tag{101}$$

where ϵ is the natural logarithmic base.

This relation is known as *Richardson's law* of emission. The value of A is the same for most pure metals. Values of A and b for several metals or combinations of metals are given in the following table:

Substance	A	b
Tungsten	60	5.24×10^4
Thorium	60	3.89×10^4
Thorium on tungsten	3.0	3.05×10^4
Barium and strontium oxides on platinum	3.0	2.0×10^4

Of the two parameters, b is the more important at high temperatures, because it enters exponentially. The substances listed in the table are arranged in the order of their emission at a high temperature, tungsten having the smallest emission.

The maximum temperatures at which filaments may be operated depend on their melting points and their physical dimensions. The temperature is usually chosen to give a life of approximately 1,000 hr. This corresponds roughly to current practice with incandescent lamps. The operating temperature for tungsten filaments of 0.01 in. (0.0254 cm) diameter is approximately 2200°C; for thoriated-tungsten filaments, the temperature is about 1600°C; and for platinum filaments coated with barium and strontium oxides, it is about 1300°C. At these temperatures, the emission is approximately the same for all these filaments.

233. Thermionic Efficiency.—In practice, the metal from which emission takes place is in the form of a long filament of small diameter which is heated by an electric current. There is also the separate-heater type in which the emitting cathode is insulated from an internal heater. This is described in Par. 248. The energy necessary to maintain this filament at a given high temperature is determined mainly by the heat radiation from it. The energy radiated varies as the fourth power of the absolute temperature, which is the Stefan-Boltzmann law of thermal radiation. *Thermionic efficiency* is defined as the ratio of the current emitted to the power input, both for the same surface area. For tungsten at 2200°C, this efficiency is approximately 0.01 amp per watt; for thoriated tungsten at 1600°C, it is about 0.05 amp per watt; and for oxide-coated platinum at 1300°C, about 0.06 amp per watt.

Both oxide-coated platinum and thoriated tungsten are less stable in their emission characteristics than pure tungsten and are liable to be affected by traces of gas and by bombardment by positive ions. The surface layer of thorium on a thoriated-tungsten filament is greatly affected by slight traces of gas. Phosphorus or magnesium, used in completing the evacuation of the tube, is deposited on all inner surfaces and is available for absorbing traces of gas formed during operation. When these deposits are completely oxidized, any further formation of gas will attack the thorium surface layer and will end the life of the filament.

234. Space Charge.—It is explained in Par. 230 that when a hot filament which emits electrons is placed in an evacuated chamber the space becomes filled with electrons. This constitutes an *electron gas*, or a *space charge*. The density of the charge is not uniform but obviously is greatest next to the hot filament. Equilibrium is attained when in a given time as many electrons are caused to return to the filament as are emitted by it.

235. Two-electrode Tube (Diode).—If a cold electrode, called the *anode* or *plate* (Fig. 313), is inserted in the evacuated chamber containing a heated filament, some electrons will reach the anode and it will assume the potential of the space which it occupies. The potential will be slightly negative with respect to the filament. Let it be assumed for the present that the filament is heated in such a way that there is no fall of potential along it (see Par. 248). If, now, the anode or plate is connected through a sensitive galvanometer back to the filament, a small current will flow and, in the usual or conventional sense, its direction will be from the filament through the galvanometer to the plate (Fig. 313). Actually, the motion of the electrons, which are negative charges, constitutes the current; and this motion is in the opposite direction, or from the filament to the plate inside the tube.

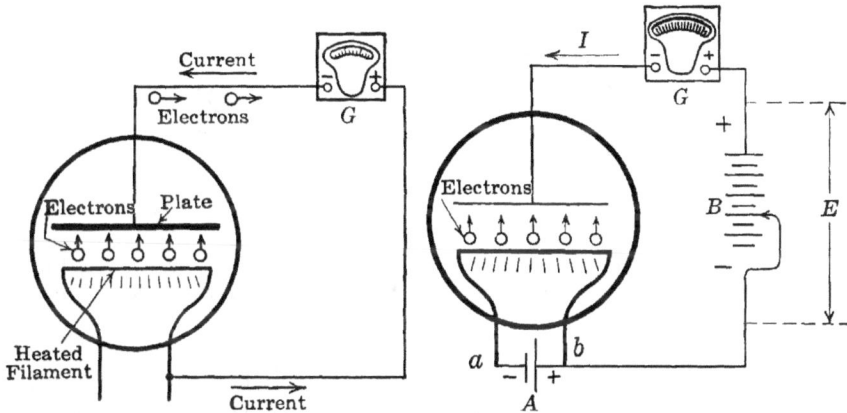

FIG. 313.—Emission of electrons from heated filament with no plate battery.

FIG. 314.—Emission of electrons from heated filament with plate battery.

236. Space-charge Saturation.—When a voltage is applied between plate and filament, making the plate positive with respect to the filament, the positive plate will attract the negative electrons and a much larger current will flow than when the applied voltage is zero. This may be determined experimentally by connecting the tube in the manner shown in Fig. 314. The plate is made positive with respect to the filament by means of the battery B. The voltage E between filament and plate may be varied to E', E'', etc., by the battery tap, as shown. A galvanometer or a microammeter G is connected in circuit to measure the plate current. The manner in which the plate current varies with the temperature of the filament for different applied plate voltages is shown in Fig. 315. For low temperatures and the corresponding small emissions, the small plate voltage E' is sufficiently large so that *all* the emitted electrons are attracted to the plate and the current increases rapidly with the temperature along the curve Oa. As the temperature and, hence, the emission increase, the density of the electron cloud between filament and plate also increases; hence, the repulsive force on the electrons leaving the filament, due to this negative space charge, increases.   Ultimately, this repulsive force becomes equal to the attractive force due to the plate voltage E'.   When

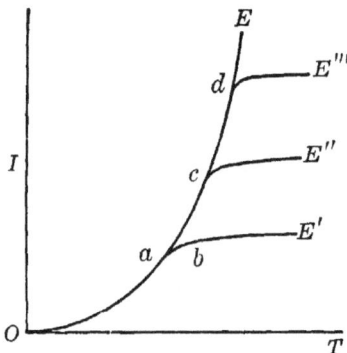

FIG. 315.—Plate current as function of temperature for different voltages.

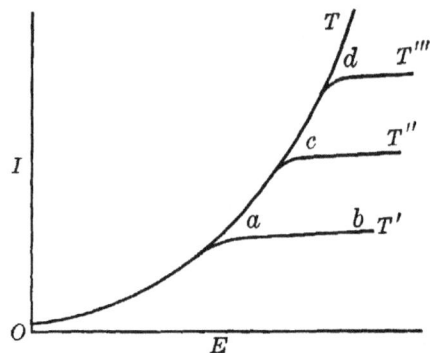

FIG. 316.—Plate current as function of voltage for different temperatures.

this condition is reached, the rate at which the electrons return to the filament is equal to the rate at which they are emitted by the filament, and the plate current becomes substantially constant, as at b (Fig. 315).   This is called *space-charge saturation*. If the plate voltage is raised from E' to E'' (Fig. 315), its attractive effect on the electrons increases.   Obviously, it will require a greater space charge than before to drive the electrons back into the filament at the same rate as the filament is emitting them. Hence, space-charge saturation now occurs at c (Fig. 315), corresponding to a greater value of electron-emission current. The equation of the envelope $OacdE$ is Richardson's law (Eq. (101)).

237. Child's Three-halves Power Law.—The plate current may also be considered as a function of plate voltage for different filament temperatures T', T'', etc., as shown in Fig. 316.   At low plate voltages, the electron emission, or current, is limited by space charge and increases as the voltage to the three-halves power. This is called *Child's three-halves power law.*

$$I = KE^{\frac{3}{2}}. \tag{102}$$

The constant K is a function of the dimensions and relative spacing of filament and plate. At a given filament temperature, such as T', the voltage ultimately reaches such a value that its attractive force at the filament becomes larger than the mutual repulsive forces due to the space charge, and all the electrons emitted are drawn to the plate. The current becomes substantially constant and is independent of the plate voltage. This condition for temperature T' is represented by the part ab of the curve. This is true *filament saturation.* As the filament temperature is increased, the emission increases and filament saturation occurs at higher values of plate current, as at c and d (Fig. 316).

From the foregoing, it is obvious that the plate current is a function of two quantities, or parameters—plate voltage and filament temperature.

238. Edison Effect.—When the filament is heated by a current, there is a fall of potential along the filament because of the resistance drop and the various parts of the filament will have different potentials with respect to the plate.

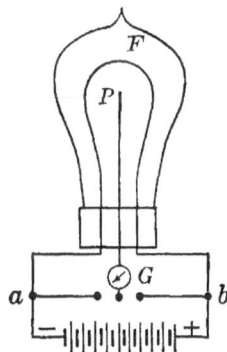

Fig. 317.—Edison effect.

For example, in Fig. 314, the plate is at the potential of the plate battery B above the right end of the filament but is at a higher potential, equal to the sum of the voltages of the B- and A-batteries, above the left end of the filament. If the B-battery (Fig. 314) is removed (Fig. 313) and the plate is connected to the positive end of the filament, its potential above the negative end of the filament will be the voltage of the A-battery. Its potential above the various parts of the filament diminishes toward the positive end of the filament.

A current will flow in the plate circuit, larger than that discussed in Par. 235 but unequally distributed along the filament, being greatest at the negative end of the filament. If the plate be connected to the negative end of the filament, no current flows in the plate circuit, as the plate is negative to all other parts of the filament. Edison first noticed this effect in 1883, when he was developing the incandescent lamp. When he connected a plate P, sealed in the bulb near the filament F, through a sensitive galvanometer G to the negative terminal a of the filament F (Fig. 317), no current flowed through the circuit $PGaF$. When, however, he connected the plate P to the positive terminal b of the filament F, an appreciable current flowed through the circuit $PGbF$. At that time, nothing was known of electrons, and this current was referred to merely as the *Edison effect*. A study of the phenomenon was made by Fleming in 1896, but its true significance was not understood until explained by J. J. Thomson and O. W. Richardson in 1899 and 1901.

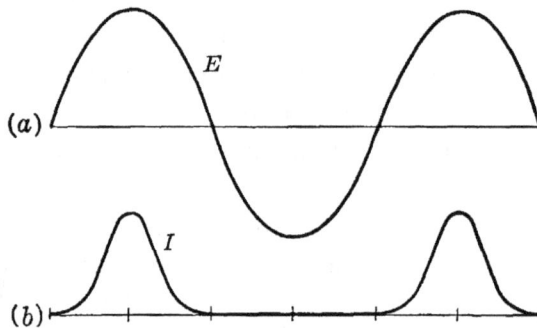

FIG. 318.—Half-wave rectification with Fleming valve.

239. Fleming Valve.—The most valuable property of the two-electrode tube, or *diode,* is its characteristic of unilateral conduction. When the plate is positive with respect to the filament, it draws electrons from the filament, and a current flows from plate to filament which is roughly proportional to the three-halves power of their potential difference. The device then has a finite although variable resistance. When the plate is negative with respect to the filament, the electrons are all driven back into the filament and no current whatever flows. The resistance of the device then becomes infinite. If an alternating emf E is applied to a two-electrode tube (Fig. 318(*a*)), the resultant current is pulsating but unidirectional (Fig. 318(*b*)). The negative loop is entirely suppressed. The positive loop is somewhat distorted because of the variation of resistance of the device with current. The foregoing is called *half-wave* rectification. This rectifying action is identical in principle with that of power rectifiers. (See Fig. 276, p. 343.) Fleming, in 1905, was the first to recognize this rectifying property of a two-electrode tube. He obtained a patent on its use as a detector of high-frequency oscillations, which became one of the fundamental patents in electron-tube development.

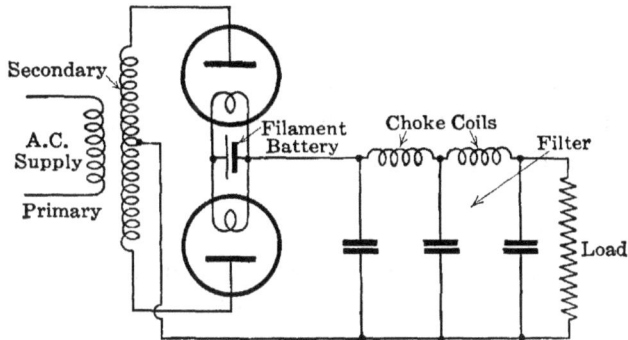

Fig. 319.—Connections which give full-wave rectification.

240. Full-wave Rectification.—Both loops of voltage may be rectified by using two rectifiers, as shown in Fig. 319. The resulting current through the load is shown in Fig. 320. This is called *full-wave* rectification. If the load is of high resistance, the voltage impressed on it may be smoothed out, as shown by

Fig. 320.—Full-wave rectification with two rectifiers.

Fig. 321.—Full-wave voltage rectification with condenser across load.

the full line in Fig. 321, by connecting a large capacitor across the output circuit.

The output voltage across the load is less than the peak voltage applied to the tube by the amount of the voltage drop in the choke coils and in the rectifier tubes themselves. This drop is nearly proportional to the load current in high-vacuum rectifiers and causes the efficiency of the rectification process to be low. Since the efficiency is the ratio of the load resistance to the sum of load resistance and tube resistance, that is, the total circuit resistance, high-vacuum rectifiers should be used only with high-resistance loads. Vapor-type rectifiers (Chap. XI) are used for low-resistance circuits.

The filaments of rectifier tubes are operated usually from a low-voltage winding of the same a-c transformer which supplies the high voltage being rectified. In the tubes of small power, the two units necessary for full-wave rectification are sealed into the same evacuated bulb. These full-wave rectifier tubes are largely used to supply the plate voltage in a-c-operated receiving sets. They are provided with low-pass filters (Fig. 319) for smoothing out the voltage ripple. This filter consists usually of three large capacitors in shunt with the load; between successive capacitors is connected a large choke coil in the high-voltage lead. The cut-off frequency of this filter is placed at about 40 cycles per sec.

241. Rectifier Tubes.—The three quantities which define the operation of a rectifier tube are the maximum inverse-peak voltage which the tube will withstand without breakdown, the maximum peak plate current as determined by filament emission, and the maximum direct current as determined by the heating of the anode due to the voltage drop. The ranges of these three quantities for the smaller rectifier tubes used in radio receivers are

Maximum inverse-peak voltage......... 700 to 1,550 volts
Maximum peak plate current........... 120 to 675 ma
Maximum direct current............... 40 to 225 ma

High-power rectifiers, both the high-vacuum and mercury-vapor types, are used for the production of high-voltage direct current for high-voltage (2,000 to 15,000 volts) oscillators and X-ray tubes and for obtaining high voltages for insulation testing.

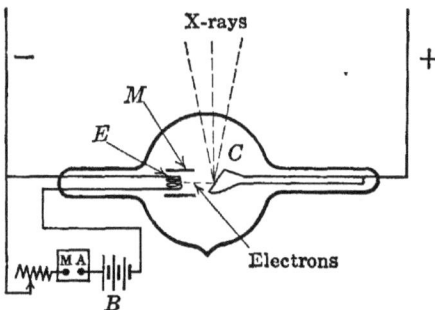

FIG. 322.—Coolidge X-ray tube.

242. X-ray Tubes.—The Coolidge hot-cathode X-ray tube (Fig. 322) is a special two-electrode tube, designed for the production of X-rays. The filament E, heated by the low-voltage battery B, is concentrated and is electrostatically shielded by the molybdenum tube M, so that the electrons are emitted in a fine pencil. The plate, or anode, C is a massive block of tungsten which serves as a target for the electrons. X-rays are given off at the point where the electrons strike the target. The anodes are frequently water-cooled to allow the use of high voltages and relatively large currents, as, for example, 100 kv and 5 ma.

It is interesting to note that, because of the high voltages used, X-ray tubes operate under the conditions for *filament saturation* (Par. 236). All other rectifier tubes operate under the conditions for *space-charge saturation* (Par. 236).

As is well known, X-rays have the property of penetrating substances which are impervious to ordinary light. They are used to a very large extent in medical work in the study of fractured bones, the roots of teeth, etc.; in industry, to locate flaws in castings; in chemistry, to determine the crystal structure of substances; etc.

243. Three-electrode Tube (Triode).—The addition of a third electrode to control the plate current was made by De Forest in 1907, in a tube which he called the *audion*. His patent, issued in that year, became as fundamental as that of Fleming in vacuum-tube development. He placed a lattice, or *grid* as it is now called, between

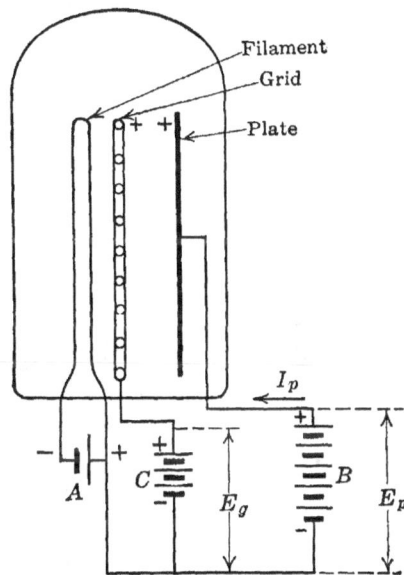

FIG. 323.—Three-electrode vacuum tube, grid positive.

the filament and plate (Fig. 323). The electrons, in passing from filament to plate, must now pass through this grid. If the grid is positive with respect to the filament, it will assist the plate in drawing electrons from the filament and, hence, will increase the plate current. If the grid is negative with respect to the filament, it will act as a space charge acts (Par. 234) and will repel negative charges, or electrons, toward the filament and, hence, will decrease the plate current. Therefore, the plate current is now a function, not only of the filament temperature T and the plate voltage E_p, but also of the grid voltage E_g. In the range in which the current is limited by space charge, the expression for the plate current is

$$I_p = K\left(E_g + \frac{E_p}{\mu}\right)^{3/2}, \qquad (103)$$

where μ is the amplification factor of the tube as defined by Eq. (105). The form of Eq. (103) is similar to that of Eq. (102). The grid voltage E_g is μ times as effective in determining the plate current as the plate voltage E_p because the grid is much nearer to the filament than is the plate. Hence, a very small amount of energy applied to the grid will control a much larger amount of energy passing from filament to plate. It is this characteristic of the tube which permits its use as an amplifier and as an oscillator.

244. Static Characteristics of Three-electrode Tube.—Figure 324 shows the connections which may be used to obtain the

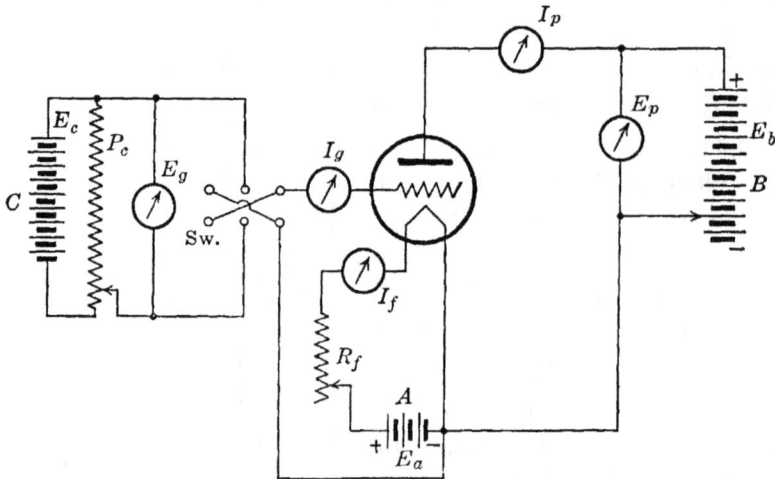

Fig. 324.—Connections for determining static characteristics of three-electrode tube.

static, or steady-current, characteristics of a three-electrode tube, or *triode*. The voltage E_p* applied to the plate by the *B*- or plate battery depends on the type of tube being tested. It is determined by the allowable heating of the plate produced by the d-c power input. The temperature of the plate must not exceed that attained by it during the pumping process. No part

* The symbol E_b is used to denote the steady d-c voltage across the *B*-battery; E_p is used to denote the d-c voltage between plate and filament. These two are seldom equal, though their difference is frequently negligible. A similar distinction is made between the voltage of the *C*-battery E_c and the voltage between grid and filament E_g.

ɔf the tube should become hot enough for the evoluti)n of gas from the heated areas to occur.

The voltage E_g applied to the grid may be obtained readily from dry cells in series (C), a high-resistance drop wire P_c being used to vary the voltage E_g and a reversing switch $Sw.$ to give the desired polarity. The ranges of the operating voltages and currents for receiving tubes are

Grid voltage, negative.................. 0   to   56 volts
Plate voltage........................ 90   to 250 volts
Plate current........................ 0.14 to   60 ma

If the plate voltage is held constant at some value E_1 (Fig. 325) and the grid voltage is varied, a curve abc is obtained. When the grid voltage is zero, current $0b$ flows to the plate,

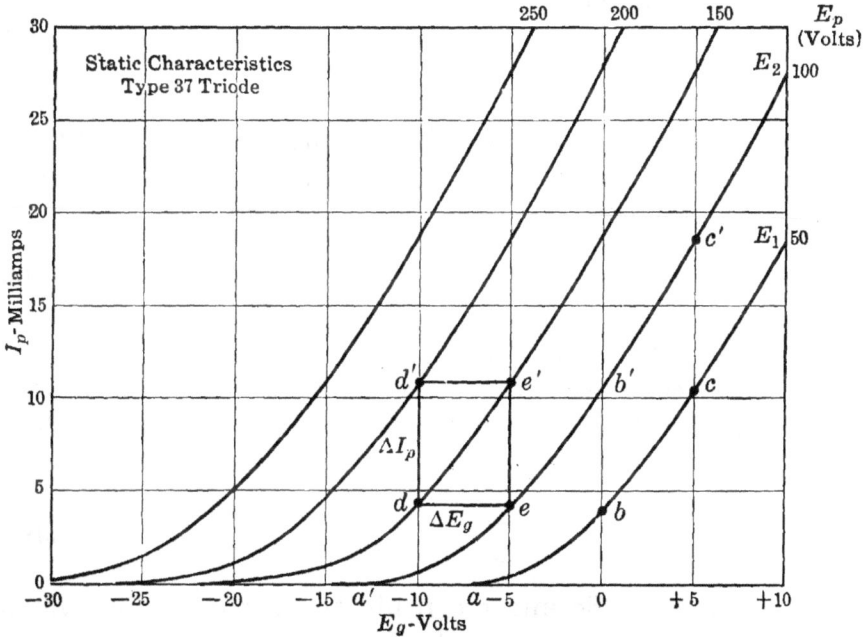

Fɪɢ. 325.—Plate-voltage—grid-voltage characteristics, Type 37 tube.

since the plate itself attracts electrons from the filament. In order to stop the flow of plate current, the grid potential must be negative and of a sufficient value $0a$ to neutralize the effect of the plate voltage. An approximate value may be found by setting I_p in Eq. (103) equal to zero.

$$E_g = -\frac{E_p}{\mu}. \tag{104}$$

The rate of increase of plate current decreases as the grid voltage is increased, owing to filament saturation. (Not shown in Fig. 325.) If the plate voltage is increased to E_2, then, for a given grid voltage, more current will flow in the plate circuit, and curve $a'b'c'$ is obtained. Curves for still greater plate voltages are also shown.

245. Tube Coefficients.—Assume (Fig. 325) that when $E_p = 150$ volts the grid voltage E_g increases by an amount ΔE_g. This increases the plate current by an amount ee' equal to dd' or ΔI_p. If the grid voltage is held constant as at d, the plate voltage must be increased by ΔE_p, or from $E_p = 150$ volts to $E_p = 200$ volts, in order to increase the plate current by ΔI_p. The ratio $\Delta E_p/\Delta E_g$ for constant plate current is the *amplification factor* and is denoted by μ.

$$\mu = \frac{\Delta E_p}{\Delta E_g} \quad \bigg| \, I_p \text{ const.} \tag{105}$$

For three-electrode receiving tubes, μ ranges from 3.5 to 125.

If a change of plate voltage ΔE_p produces a change of plate current ΔI_p, the factor

$$r_p = \frac{\Delta E_p}{\Delta I_p} \quad \bigg| \, E_g \text{ const.} \tag{106}$$

is called the *plate resistance*.

The third combination of these three variables, the ratio of the change of plate current ΔI_p to the change of grid voltage ΔE_g, is called the *transconductance* or *mutual conductance*.

$$g_m = \frac{\Delta I_p}{\Delta E_g} \quad \bigg| \, E_p \text{ const.} \tag{107}$$

These three coefficients are really the partial derivatives of the expression for plate current (Eq. (103)). As such, they are not independent but are related by the equation

$$\mu = g_m r_p \tag{108}$$

or

$$g_m = \frac{\mu}{r_p}. \tag{109}$$

Each of these coefficients is the slope of the family of curves having as coordinates the two variables which form the derivative expressing the coefficient. Each set may be named after the quantity which is held constant. Thus, the family of curves shown in Fig. 325 are curves of equal plate voltage, and their slope is the transconductance g_m of the tube as expressed in Eq. (107). The two other possible plots for these families of curves are discussed in Par. 246.

The amplification factor μ is determined by the relative spacing of grid and plate with respect to the filament, and the degree of completeness with which the grid shields the filament from the electrostatic field of the plate, as determined by the openness of the grid mesh. It is essentially constant over a wide range of grid and plate voltages and filament currents. It follows, then, that transconductance g_m and plate resistance r_g vary reciprocally.

Transconductance increases slowly with plate voltage, with grid voltage in the positive direction, and with filament current. It increases also with the physical dimensions of the tube elements, as to both length and radius. For a given size of tube structure, it is roughly independent of the relative position of the grid with respect to the filament and plate and is independent of the introduction of other electrodes. For receiving tubes, it varies from 275 to 9,000 micromhos.

Plate resistance varies in a manner opposite to that in which transconductance changes, but to a much greater extent, because of the constancy of amplification factor with the various voltages and the constancy of transconductances with physical size. For three-electrode receiving tubes, its limits are 0.8 to 240 kilohms.

246. Other Static Characteristics.—The data represented in Fig. 325 may be plotted with plate voltage as abscissa. The resulting curves will be for constant grid voltage, as shown in Fig. 326. Their slope is the reciprocal of plate resistance r_p (Eq. (106)), that is, it is equal to plate conductance k_p. This type of plot is valuable for studying the operation of tubes as amplifiers (Par. 249).

In Fig. 327 is shown the third way in which the static characteristics of a tube may be plotted. The coordinates are grid voltage and plate voltage. The curves are for constant plate current. Their slope is the amplification factor μ, as expressed in Eq. (105). This type of plot is used in studying the operation of tubes as oscillators (Par. 255, p. 415). The lines of constant

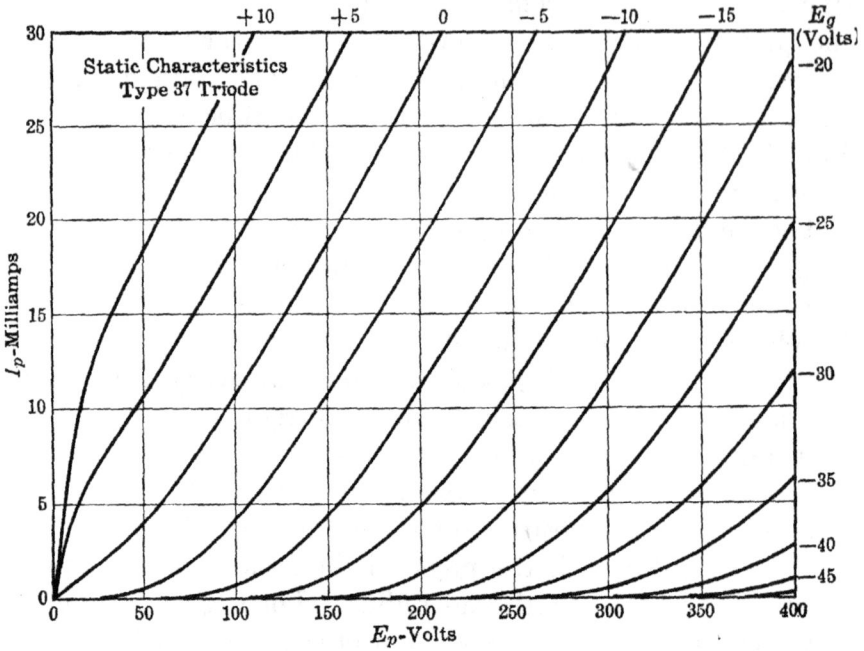

FIG. 326.—Plate-current—plate-voltage characteristics, Type 37 tube.

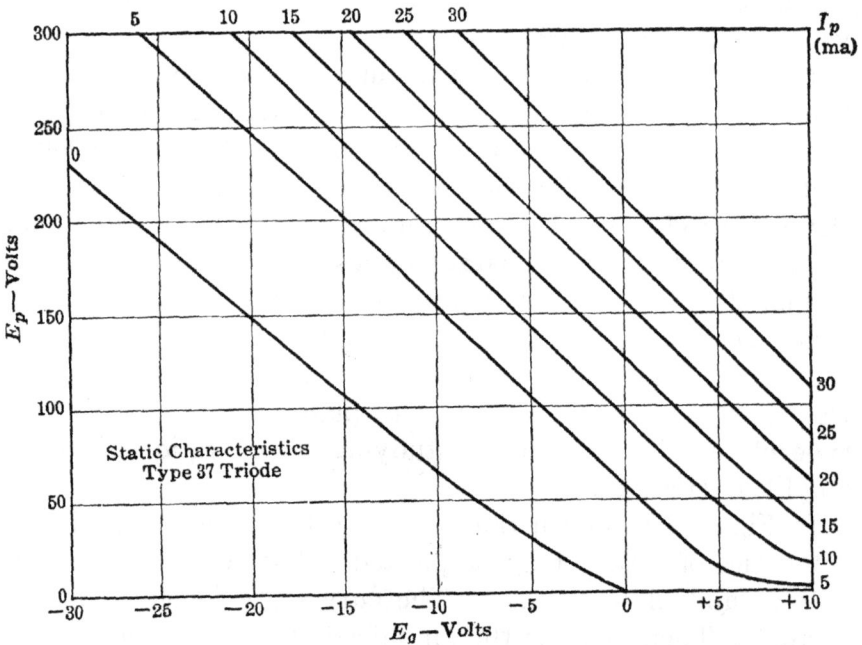

FIG. 327.—Plate-voltage—grid-voltage characteristics, Type 37 tube.

plate current may be thought of as the contour lines of a three-dimensional model in which plate current extends out from the paper at right angles to the plane of the voltage coordinates. The other two types of plotting, Figs. 325 and 326, are cross-sections of this space model, taken parallel to each pair of voltage coordinates in turn.

Grid current may be plotted in a similar manner. Except for tubes used as oscillators and class C amplifiers, the variation of grid current with plate and grid voltage is of little importance.

247. Measurement of Tube Coefficients.—The tube coefficients, amplification factor μ, plate resistance r_p, and transconductance g_m may be measured directly by introducing a small alternating voltage in either the grid or the plate circuit. Because of the use of an alternating voltage, these

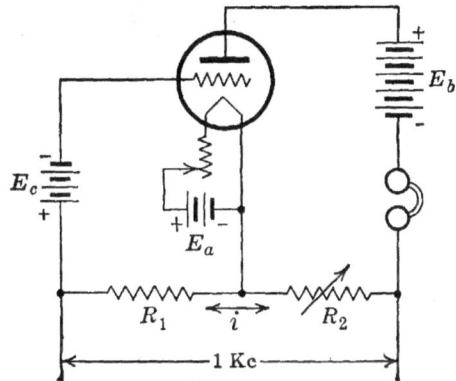

FIG. 328.—Dynamic measurement of amplification factor.

methods are called *dynamic* methods. The results will agree with the values obtained by measuring the slope of the various static characteristics, provided the voltage swings are small.

The connections to be used for measuring the amplification factor μ are shown in Fig. 328. The instruments which must be placed in the various circuits to indicate the steady voltages and currents and the means for varying these voltages, as shown in Fig. 324, have been omitted for the sake of clearness. In addition, because of the use of alternating current, any high-resistance potential divider, used to vary the grid voltage, must be shunted by a capacitor of at least 1 μf capacitance. The telephones should be of low d-c resistance or should be shunted by a low-resistance choke coil so that the d-c voltage across them, due to the steady plate current, is small. When the resistance R_2 is adjusted to produce silence in the telephones, the a-c voltage drop iR_2 across it is equal to that in the plate circuit of the tube due to the voltage iR_1 applied to the grid. Hence, by the equivalent plate-circuit theorem, stated in Par. 249,

$$iR_2 = \mu i R_1$$

or

$$\mu = \frac{R_2}{R_1}. \qquad (110)$$

The resistances R_1 and R_2 should be as small as possible so that the d-c voltage drop across R_2 is small.

Plate resistance may be measured in the manner shown in Fig. 329. This is merely an a-c Wheatstone-bridge circuit, in one arm of which the plate circuit of the tube is connected. As in Fig. 328, resistance R_2 should be as small as possible and the telephones should be shunted by a low-resistance choke coil, in order to keep the d-c voltage across them small. When the bridge is balanced by adjusting R_2,

$$r_p = \frac{R_2 R_3}{R_1}. \qquad (111)$$

FIG. 329.—Dynamic measurement of plate resistance.

In order to cover a wide range of values, resistance R_3 should be adjustable by factors of 10.

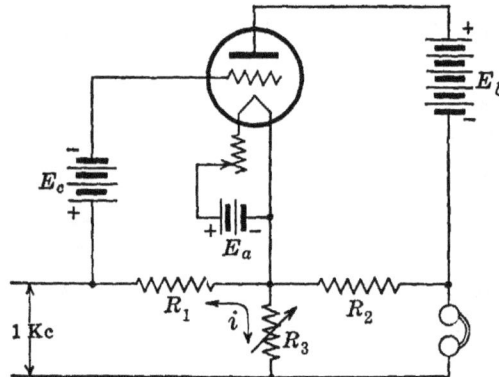

Transconductance is measured by means of the connections shown in Fig. 330. As before, R_2 should be as small as possible

FIG. 330.—Dynamic measurement of mutual conductance.

in order to keep the d-c drop across it small, but the telephones should now have a high resistance. When the resistance R_3 is

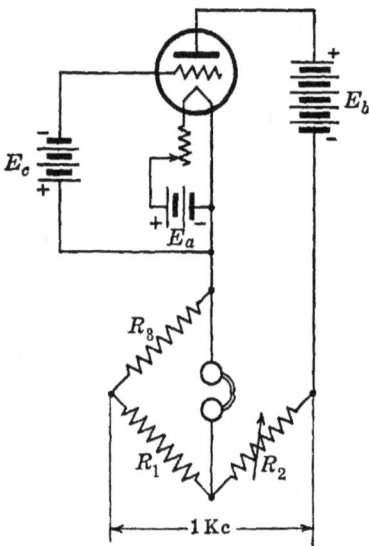

adjusted so that there is silence in the telephones, the a-c voltage drop iR_3 is equal to that across R_2 due to the plate current produced by the grid voltage iR_1.

$$iR_3 = \frac{\mu iR_1}{r_p + R_2} R_2 = \frac{g_m r_p iR_1}{r_p + R_2} R_2,$$

$$g_m = \frac{R_3}{R_1 R_2} \cdot \frac{r_p + R_2}{r_p}. \tag{112}$$

Except for very small values of plate resistance, the values of the various resistances may be chosen so that the error caused by omitting the term $(r_p + R_2)/r_p$ of Eq. (112) will not exceed a few per cent.

In this discussion of the measurement of dynamic characteristics, no consideration has been given to the effect of grid current and of capacitances between the electrodes of the tube. Methods for taking account of these effects have been described by Chaffee.[1]

248. Three-electrode Receiving Tubes.—A large number of types of three-electrode tubes are now in use. They differ according to their use as amplifiers, detectors, and oscillators, at radio- or audio frequencies, and in respect to their power output. In addition, as the art has developed, the type of the power supply for the filament and plate has caused the type of filament to change. The space requirements in present-day compact receivers have reduced the over-all dimensions of the tubes, decreased the spacing of the electrode structures, and caused the development of the multielectrode tubes described in Par. 253, (p. 413). The envelope of the tube may be either glass or metal.[2]

All the recent tubes have oxide-coated filaments, because of the greater efficiency of this type of emitter, as explained in Par. 233. Those intended for operation through a transformer from an a-c supply have usually a separate oxide-coated cathode heated by an internal insulated filament. This type of construction eliminates nearly all the 60-cycle voltage introduced into the plate circuit by the filament and allows tubes having different

---

[1] CHAFFEE, E. L., "Theory of Thermionic Vacuum Tubes," Chap. IX.

[2] A complete description of all tubes with their electrical and constructional characteristics will be found in the tube handbooks of the various tube manufacturers.

grid-bias voltages to be operated from a common filament trans former and plate supply (see Fig. 359, p. 429). The fact that the cathode is an equipotential surface improves somewhat the operating characteristics, and the large surface area of the cathode increases the mutual conductance. Tubes having large output and intended to be used in the last stage of an amplifier for operating a loudspeaker still have the older filamentary cathode. For equal input power, a greater output can be obtained from a directly heated filament than from the heater type, and the resulting amount of a-c hum in the plate circuit is not noticeable. The shape and dimensions of the filament have been chosen so as to minimize the a-c hum.

249. Amplification.—Since the amplification factor μ of a three-electrode tube is considerably greater than unity, the tube

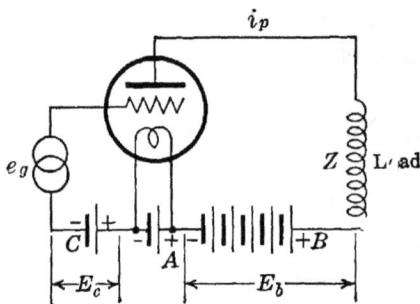

FIG. 331.—Vacuum tube as amplifier.

may be used to amplify small voltages. Figure 331 shows a tube having a steady potential E_b connected in the plate circuit and another steady potential E_c in the grid circuit. E_c gives the grid a negative potential with respect to the filament. These voltages give a definite quiescent point Q on the plate-current–grid-voltage characteristic (Fig. 333). Now, let a small alternating voltage e_g be introduced in the grid circuit. This variation of grid voltage produces a variation of plate current. The actual plate current may be determined by projecting the e_g sinusoid vertically on the I_p–E_g curve. Thus, an alternating current i_p, superposed on the steady plate current, is produced in the plate circuit. This current may be considered as being produced by a voltage μe_g acting through the constant plate resistance r_p and the impedance Z of the load. The equivalence of e_p and μe_g is now referred to as the *equivalent plate-circuit theorem*. Its rigorous proof and that of its counterpart, the equivalent grid-circuit theorem, have been given by Chaffee.[1]

The steady voltages and currents in the plate and grid circuits have no effect in determining the alternating voltages and cur-

[1] *Loc. cit.*, Chap. VIII.

rents, except insofar as they define the portions of the static characteristics over which the tube operates. For distortionless reproduction, the voltages E_c and E_b (Fig. 331) should be chosen so that the alternating emf is operating on the straight part of the characteristic (Fig. 333) and where E_g is negative, so that the grid current I_g is zero.

The grid bias E_c can be obtained without the use of a grid battery by placing a resistor P in the plate circuit as in Fig. 332. The voltage drop E_c produced in this resistor by the steady plate current i_p provides the negative bias E_g for the grid. The effective plate voltage is

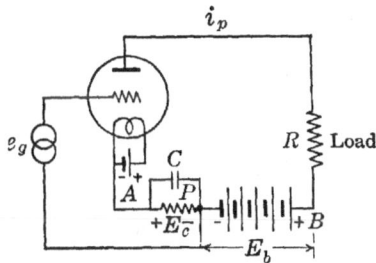

FIG. 332.—Grid bias obtained with resistor.

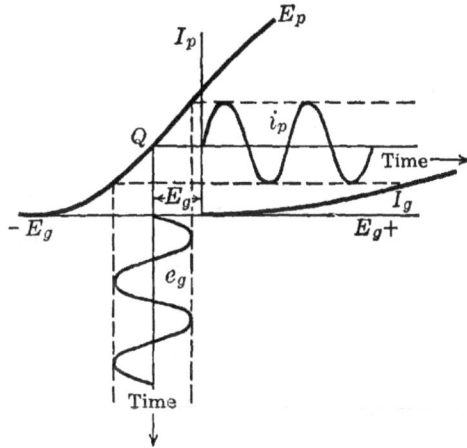

FIG. 333.—Amplification characteristics of vacuum tube.

decreased by the same amount E_c. The capacitance C by-passes the alternating plate current.

When the capacitance C is omitted, an alternating-voltage drop also appears in the grid circuit, opposite in phase to the input voltage e_g when the load R is resistive. This voltage reduces the over-all amplification of the tube, and the effect is called *degeneration*. It makes the remaining gain more nearly independent of variations in the characteristics of the tube and is much used in high-gain multiple-stage amplifiers.

The range of plate voltage and of plate current can be greatly increased by allowing the grid to swing positive until some grid current flows and to swing negative until the plate current is reduced to zero. The plate current will then be distorted, and harmonics introduced. The effect of this on the output voltage may be reduced by the use of a parallel tuned circuit as a load or by using two tubes in a push-pull arrangement, as in Fig. 334. Because of the symmetrical connections of the tubes with respect

to the positive and negative values of the impressed voltage, all
even harmonics are suppressed.

Letters have been assigned to the various types of amplification just described. Class *A* amplification refers to the distortionless operation where the grid never swings positive and the
plate current is never zero. In class *B* amplification, the tube
is biased to cutoff (point *a*, Fig. 325), and the grid is allowed to
swing somewhat positive. In Class *C* amplification, the grid
bias is placed beyond cutoff, and the operating path is as long as

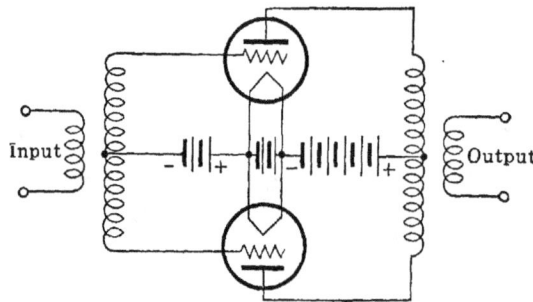

Fig. 334.—Push-pull amplifier.

possible. Intermediate classes are indicated by the symbols *A'*,
AB, and *BC*.

250. Voltage Amplification.—The ratio μ' of the voltage across
the load to that applied to the grid is always less than the amplification factor μ of the tube and approaches that quantity if the
load resistance R (Fig. 332) is made very large compared with
the plate resistance r_p.

$$\mu' = \frac{R}{R + r_p}\mu = \frac{1}{1 + \dfrac{r_p}{R}}\mu. \qquad (113)$$

Making $R = r_p$, however, reduces the gain to only $\mu/2$. The
voltage drop in the load due to the d-c plate current makes it
necessary to provide a large plate-battery voltage, but the
amplification is independent of frequency over a wide range.

A greater voltage amplification than can be obtained with a
single tube is made possible by applying the output of the first
tube to the grid of a second tube. For a resistance load in the
plate circuit of the first tube, the connections shown in Fig. 335
may be used. The capacitors isolate the tubes from the effects

of d-c voltages. The frequency range is determined at the low-frequency end by the ratio of the reactances of these capacitors to the resistances in the grid circuits and at the high-frequency end by the ratio of the reactances of the capacitances of the input circuits of the tubes to the plate-load resistances.

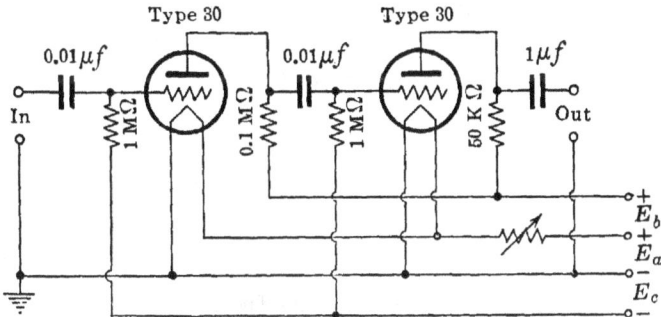

FIG. 335.—Resistance-coupled amplifier.

The voltage amplification of such an amplifier should be about 40 over the frequency range from 25 cycles to 50 kc. By using pentodes (see Par. 252, p. 412) the voltage amplification can be increased to 350.

Because of the very high resistance of the grid circuit of a vacuum tube, a step-up transformer may be used as the plate

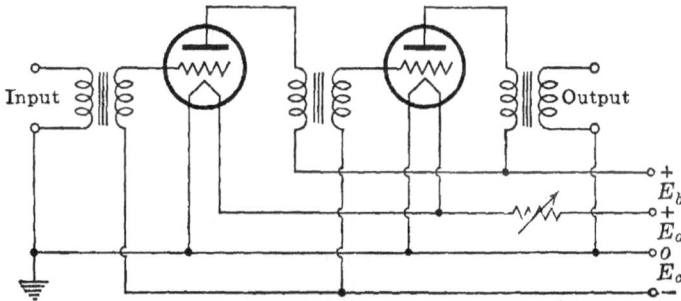

FIG. 336.—Two-stage transformer-coupled amplifier.

load to make connection to another tube. The characteristics of audiofrequency transformers having laminated iron cores are such that, to obtain a reasonably constant voltage step-up ratio over the audiofrequency range from 50 cycles to 10 kc, this ratio should be not greater than 3. It may be increased to 6 or 10 with a corresponding reduction of frequency range. The transformer may be tuned to a single frequency and a voltage gain of at least 50 obtained. The connections for a two-stage transformer-coupled amplifier are shown in Fig. 336. In each amplifier the voltage gain should be about 1,500 over the frequency range from 100 cycles to 7 kc.

The voltage gain which may be obtained in a multistage amplifier is limited by the coupling existing between output and input. The most usual coupling is that due to the impedances in the plate supply which are common to all tubes. The greatly amplified plate current of the last tube produces a voltage drop across these impedances which is thereby introduced into the plate circuit of the first tube. This produces regeneration and self-oscillation, as explained in Par. 254 (p. 414). It occurs most frequently in a-c-operated amplifiers and is reduced by shunting all common impedances with by-pass capacitors and filters.

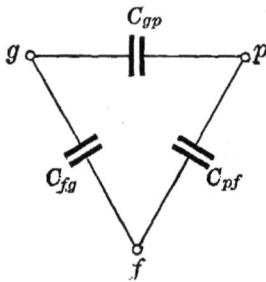

Fig. 337.—Capacitances in triode.

The voltage gain obtainable in a single stage is limited by the magnitude of the capacitances between the electrodes of the tube. There are three of these capacitances, arranged as shown in Fig. 337. Of these, the grid-to-plate capacitance C_{gp} is the only one which transfers energy between the grid and plate circuits and produces regeneration and oscillation. Its value in three-electrode tubes is between 4 and 8 $\mu\mu$f. The existence of a grid-to-plate capacitance of this magnitude limits the voltage gain per stage to 100 at audio frequencies and to 5 to 10 at radio frequencies. In screen-grid tubes the grid-to-plate capacitance C_{gp} is reduced to less than 0.03 $\mu\mu$f, and the regenerative effect is negligible. The other two capacitances, filament to grid (C_{fg}) and plate to filament (C_{pf}), enter mainly into the determination of the input and output impedances of the tube at high frequencies, and their values lie between 2 and 4 $\mu\mu$f.

251. Four-electrode Tubes (Tetrodes).—Four-electrode tubes, or *tetrodes*, generally have two grids in addition to a filament and plate. Of the two grids, one surrounding the other, either may be the control grid; the other may be connected to either filament or plate through a suitable battery and serves merely to modify the characteristics which the tube would have as a triode, if this grid were absent. The two types of connection which are of importance, space-charge grid and screen grid, are shown in Fig. 338.

In the space-charge-grid tetrode (Fig. 338(a)), the two grids are of similar construction, and the outer one is the control grid. The inner grid is made positive with respect to the cathode and serves to decrease the space charge in the neighborhood of the control grid. This increases the transconductance of the

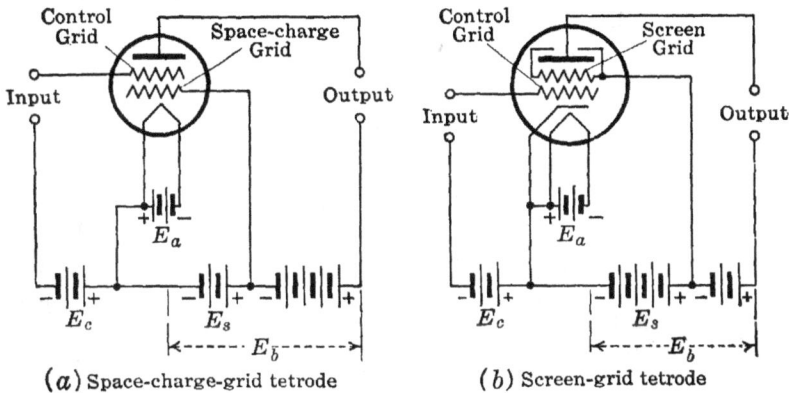

(a) Space-charge-grid tetrode   (b) Screen-grid tetrode

Fig. 338.—Tetrodes.

tube three or four times. This increase is not an important gain, because much the same increase in transconductance can be obtained with closer spacing of the outer electrodes if a large-diameter separate-heater cathode is used. For this reason, this type of connection is used only in low-grid-current tubes, such as the FP-54, where it serves to prevent the positive ions emitted from the cathode from reaching the grid.

The screen-grid tetrode (Fig. 338(b)) was developed to decrease the grid-to-plate capacitance to a negligible amount. The first grid is the control grid, and the second grid is built so that it surrounds the plate and extends outside the plate as well. Because of the nearly complete shielding thus obtained, the capacitance C_{gp} can be made as small as 0.005 $\mu\mu$f. Although the other interelectrode capacitances are considerably increased, the operation of the screen-grid tetrode at radio frequencies permits voltage gains of 25 to 50 per stage.

The usefulness of the screen-grid tetrode is seriously limited by the fact that, if the plate voltage is allowed to become less than the screen-grid voltage, secondary electrons (see Par. 231, p. 389) from the plate, produced by primary electrons from the filament, cause the plate current to decrease rapidly and to have a negative slope over a small range of plate voltage. In this region the plate resistance is negative, as in an arc. When the tube is operated in this manner, it is called a *dynatron*, and the region of negative resistance is referred to as the *dynatron region*. With a parallel tuned circuit connected between plate and filament, oscillations will be produced with no further means for regeneration (see Par. 254), provided the equivalent series resistance of the tuned circuit is greater than the negative resistance of the tube. Secondary emission is affected greatly by the surface conditions of the plate. Hence, the dynatron characteristics will vary greatly for different tubes and with age in the same tube.

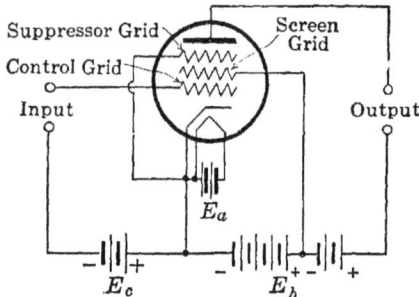

FIG. 339—Five-electrode tube or pentode.

252. Five-electrode Tubes (Pentodes).—The dynatron region of a screen-grid tetrode may be eliminated by introducing a third grid between the screen grid and plate and connected directly to the cathode, as in Fig. 339. This third grid is called a *suppressor grid* because, being always negative with respect to the plate, it causes all secondary electrons emitted by the plate to return to the plate. This tube is called a *pentode*. In a pentode the screen grid cannot extend around the plate to shield it completely from the control grid, as in a screen-grid tetrode, because that would reduce the effectiveness of the suppressor grid. To keep the control-grid-to-plate capacitance small, extra shields are provided either within the glass envelope or external to it. In the latter case the bulb is shaped so as to allow the shield to come very near to the plate.

The plate-current—plate-voltage characteristics are shown in Fig. 340. In pentodes designed for use in intermediate stages of amplifiers, the slope of these curves is small and similar to that in screen-grid tetrodes. Plate resistances are about 1 megohm. Amplification factors range from 600 to 6,000 for values of transconductance of 650 to 9,000 micromhos. Pentodes for use in the output stage of audio amplifiers have plate resistances of

10 to 300 kilohms with amplification factors ranging from 100 to 700.

The operating path extends down to low plate voltages and thus produces a large output, amounting to 10 watts for those tubes having the larger values of transconductance. The amount of distortion existing in these tubes depends largely on the resistance of the plate load and is a minimum when this load

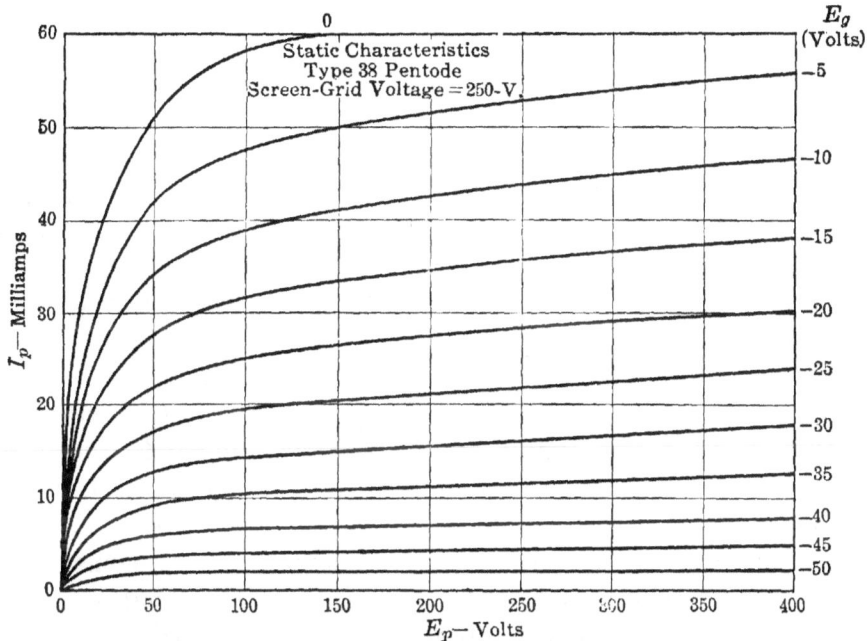

FIG. 340.—Plate-current—plate-voltage characteristics, Type 38 pentode.

resistance is about one-tenth of the plate resistance. At best, the distortion is much larger than in a triode and is due mainly to a third harmonic, which cannot be eliminated by using the push-pull connections described in Par. 249 (p. 408).

Plate-current—plate-voltage characteristics with a much sharper knee are obtained in *beam-power* pentodes by shaping all the electrodes so that the electron stream from cathode to plate is confined to two beams pointed in opposite directions, each covering an angle of about 60°. Beam-forming plates connected to the cathode replace the suppressor grid.

253. Multielectrode Tubes.—Many other combinations of electrodes have been arranged for tubes intended for special uses. Variable amplification tubes, called *supercontrol* amplifiers, are designed to reduce the effect of interfering signals, or cross talk. They depend on the use of a control grid whose mesh is not uniform over the length of the filament. This increases the negative grid bias at which the plate current is reduced to zero, thus decreasing the curvature of the characteristics and reducing distortion. This device is applied only to screen-grid tubes.

Many types of multielectrode tubes have been designed to save space in radio receiving sets. Those which act as both detector and amplifier, detector and oscillator, or detector, oscillator, and amplifier are called *converters*. There are two types of dual amplifiers, one in which the units are connected to give a push-pull amplifier, the other in which the units are connected in cascade to give a two-stage amplifier.

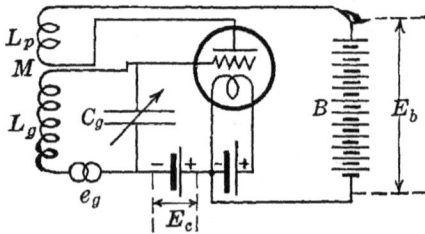

FIG. 341.—Connections for regeneration with tuned grid circuit.

254. Regeneration.—It is shown in Par. 249 (p. 406) that in the amplifier any change of the voltage applied to the grid causes not only a change of current in the plate circuit but also a change of energy in the plate circuit. This change of energy is many times greater than the change of energy applied to the grid. If the proper phase relation between the plate current and the voltage in the grid circuit is obtained, it is possible to feed a portion of this energy of the plate circuit back into the grid circuit and, hence, to reinforce the effect of the grid. The grid, in turn, increases the plate current, which again reacts on the grid. This feedback of energy is called *regeneration*. The connections for one method of regeneration are shown in Fig. 341. The grid is polarized negatively with a grid battery to a potential of E_c volts. An inductor L_g and a variable capacitor C_g are connected in parallel between the grid and the negative terminal of the grid battery. A coil L_p, having mutual inductance M with L_g and connected in the plate circuit, serves to couple inductively the plate and grid circuits.

An alternating voltage e_g introduced into the tuned grid circuit produces a plate current i_p in such a phase that, if the polarity of L_p is correct, the voltage which i_p induces in the tuned grid circuit is in phase with the original voltage e_g. The same effect may be produced by connecting in the plate circuit a similar tuned circuit, consisting of inductance and capacitance in parallel. The capacitance from grid to plate of the tube itself is depended upon for coupling between the plate and grid circuits. If the grid circuit is not tuned, as in an audioamplifier, the voltage induced back into the grid circuit is in quadrature with the impressed voltage e_g, and there is no regeneration.

The effect of regeneration is to introduce a *negative* resistance in the tuned circuit. That is, this tuned circuit may now become a generator of energy, this energy being obtained from the plate, or *B*-battery. If the mutual inductance M between L_p and L_g is made sufficiently large, the total resistance of the tuned circuit may be reduced nearly to zero, where the limit of regeneration is reached. At this limit of regeneration, the a-c plate current is approximately constant and independent of the magnitude of the impressed voltage e_g, as well as that of the initial resistance of the tuned circuit; its value is determined only by the characteristics of the tube. In practice, this maximum theoretical limit cannot be reached, since small disturbances, such as slight mechanical vibrations of the inductors, capacitors, and the tube itself, may cause the total resistance of the plate circuit to become negative momentarily and thus cause the tube to begin oscillating (see Par. 255).

The maximum attainable plate current is therefore not constant but varies directly with the magnitude of the impressed voltage, if the voltage is small, and inversely with the resistance of the tuned circuit, if the voltage is large.

OSCILLATORS

255. Oscillation.—When the mutual inductance between the plate and grid circuits is increased to such a value that the resistance of the tuned circuit becomes zero or negative, sustained oscillations, independent of the voltage e_g (Fig. 341) impressed on the grid, are set up in the system. In fact, the impressed voltage e_g may be removed entirely without affecting the oscillations. These sustained oscillations will start even in the absence of any impressed voltage. Slight mechanical disturbances to parts of the system (see Par. 254) or small electrical disturbances such as occur when the plate circuit is closed are sufficient to start the tube oscillating. Under these conditions, the tube is said to be an *oscillator*. It behaves like an a-c generator, converting the energy of the plate battery into a-c energy in the tuned circuit. The frequency of the alternating current generated is practically equal to the natural frequency of the tuned circuit

$$f = \frac{1}{2\pi \sqrt{LC}} \text{ cycles per sec,} \qquad (114)$$

where L (henrys) and C (farads) are the inductance and capacitance of the tuned circuit (see Eq. (38), p. 78).

The type of oscillating circuit in which the tuned circuit is connected to the grid (Fig. 341) is that used in most receivers in which oscillating tubes are used, such as continuous-wave, carrier-frequency, and superheterodyne receivers.

As far as sustained oscillations are concerned, the tuned circuit, frequently called the *tank circuit*, may be placed equally

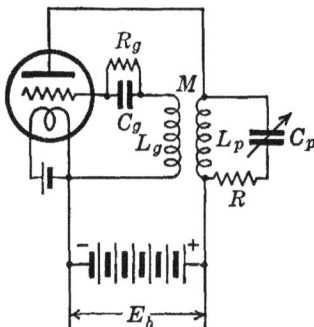

Fig. 342.—Vacuum-tube power oscillator.

well in the plate circuit, as in Fig. 342, the tuned circuit being inductively coupled to the grid by the mutual inductance M. This type of circuit is used in most power oscillators where the tube acts as an a-c generator. The grid bias is obtained from a grid leak and grid capacitor in parallel, as described in Par. 260 (p. 424). The resistance R represents the total resistance in the tuned circuit plus the equivalent resistance of the load on the oscillator, such as the equivalent antenna resistance.

The tuned circuit L_pC_p is at the battery potential E_b above ground. In order to bring it to ground potential for direct connection to an antenna and for safe operation generally, the parallel-feed connection of Fig. 343 is used. C_b is a large blocking capacitor which prevents the tuned circuit from short-circuiting the plate battery E_b. L_b is a radiofrequency choke coil, which keeps radiofrequency current out of the plate battery.

Other circuits frequently used are shown in Fig. 344. The Hartley circuit is simple to construct because the inductance used in the tuned circuit is a single coil with one intermediate tap. The Colpitts circuit is the most difficult to construct because the two capacitors C_p and C_g should be geared together. Its advantage is that a low-impedance path to the filament is provided for the harmonics of both the plate and the grid cur-

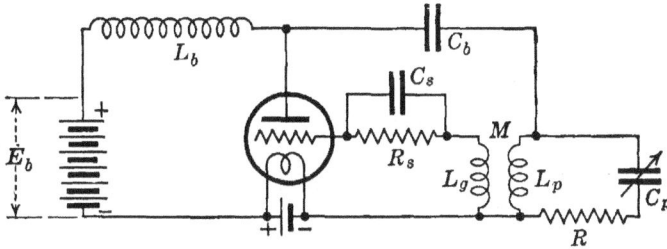

FIG. 343.—Parallel-feed oscillator.

rent. Its output voltage has less distortion than any of the other circuits.

256. Power Tubes.—Most of the tubes designed for use in receiving sets will function as oscillators having power outputs up to several watts. Tubes with greater outputs for use in broadcast transmitters differ from the receiving tubes by being larger in size and by being able to dissipate larger power losses

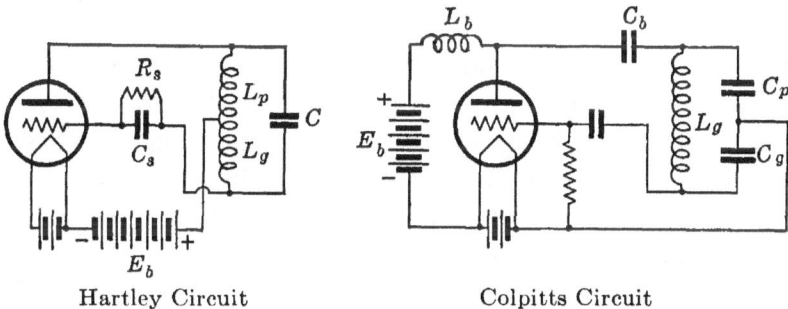

Hartley Circuit   Colpitts Circuit

FIG. 344.—Oscillator circuits.

from their plates. All metal parts are heated during the pumping process, either by electron bombardment or in a high-frequency furnace, to as high a temperature as practicable, to drive out the occluded gases. The tube may then be operated almost up to that temperature without any further evolution of gas. Since most of the power developed in the tube must be dissipated by radiation, the rating of the tube may be increased by blackening the plate and by the addition of cooling fins. When the power loss is greater than 1 kw, the plate is made the outer part of the tube and is water-cooled. The largest tubes now constructed have outputs of 100 kw. The filaments of most power tubes are made of thoriated tungsten because oxide-coated filaments cannot withstand the positive-ion bombardment produced by high plate voltage. For the same reason, pure tungsten is used in the largest tubes.

These power tubes are equally useful as power amplifiers. Screen-grid tetrodes and pentodes are also made with fairly large outputs for use at high frequencies.

MODULATION

257. Modulation.—Electrical communication over wires in its simplest form employs alternating currents of audio frequencies only, either singly or in combination. These currents may be amplified (see Par. 249, p. 406), but only a single communication can be conducted over a single effective circuit at one time. In order to open new channels of communication over any given effective wire circuit, *carrier* wire-telephony and wire-telegraphy are employed. Alternating currents having superaudiofrequencies (3,000 to 33,000 cycles per sec[1]) are used as carriers or vehicles for the audiofrequency currents. In radiotelephony and radiotelegraphy, electromagnetic waves are used as carriers having frequencies of from 10,000 to 60,000,000 cycles per sec. Of themselves, these superaudiofrequencies could not transmit signals, being, for the most part, beyond the range of audibility; and they can transmit only very small amounts of power. By the superposition of audiofrequency currents on these carrier currents, however, it is possible to transmit several messages simultaneously over a given effective communication circuit. The superposition of an audiofrequency current on a carrier-frequency current is called *modulation*.

Two methods of modulation are now in use, *amplitude modulation* (AM) and *frequency modulation* (FM). Amplitude modulation is used in commercial broadcasting in the frequency band of 0.5 to 1.6 megacycles per sec. Frequency modulation is used at higher frequencies, usually in the neighborhood of 45 megacycles per sec. In amplitude modulation, the original constant ampli-

---

[1] The ear may be sensitive to frequencies as high as 15,000 cycles per sec, but most conversational frequencies do not exceed 2,500 cycles per sec.

tude of the carrier-frequency alternating current is made to
vary according to the amplitude of the superposed audiofrequency current. For example, in Fig. 345(*a*) is shown the con-
stant-frequency carrier current having a constant amplitude *A*
and a frequency of *a* cycles per sec. An audiofrequency current, having an amplitude *B* and a frequency of *b* cycles per sec,
is superposed on the carrier frequency. The resulting current is
shown in Fig. 345(*b*). The superposed audiofrequency current

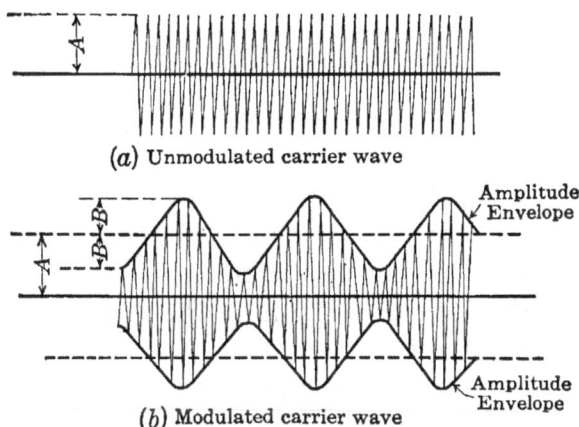

(*a*) Unmodulated carrier wave

(*b*) Modulated carrier wave

Fig. 345.—Modulated carrier current.

is the *envelope* of both the positive and the negative halves of this
modulated carrier current.

The mathematical expression for this modulated current is

$$i = A(1 + m \sin 2\pi bt) \sin 2\pi at, \qquad (115)$$

where *m* is the degree of modulation, being the ratio of amplitudes
of the audio and radio currents

$$m = \frac{B}{A}. \qquad (116)$$

It may be shown that the modulated carrier current (Fig.
345(*b*)) actually consists of three sinusoidal currents, one having
the frequency *a*, the frequency of the original carrier current;
another having a frequency (*a* − *b*), the *difference* between the
carrier frequency and the audio frequency; and a third having a
frequency (*a* + *b*), the *sum* of the carrier frequency and the
audio frequency. The frequencies (*a* + *b*) and (*a* − *b*) are
called *side frequencies*. This is illustrated in Fig. 346, which

shows a portion of the frequency spectrum, the abscissas being frequencies and the ordinates being the amplitudes of the currents.

The amplitudes of these three currents are directly related to the amplitudes of the carrier and audiofrequency currents,

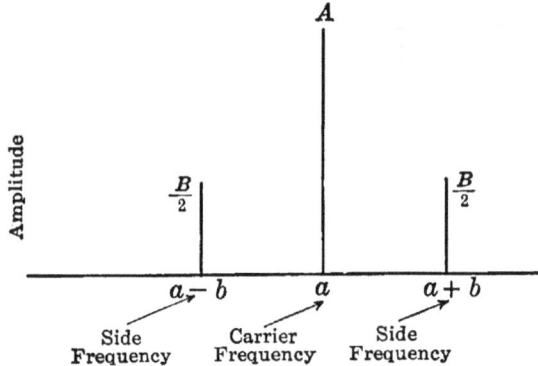

FIG. 346.—Frequency spectrum.

as shown in Fig. 346, but to a different scale. The degree of modulation is the ratio of the sum of the amplitudes of the two side frequencies to the amplitude of the carrier frequency. With the more complex audiofrequency currents, such as would be produced by the voice, the resultant modulated current is quite complex and the side frequencies widen out into *side bands*. The carrier frequency, however, is always sufficiently high so that the side-band frequencies are near it in the frequency spectrum and all are transmitted essentially as a single frequency.

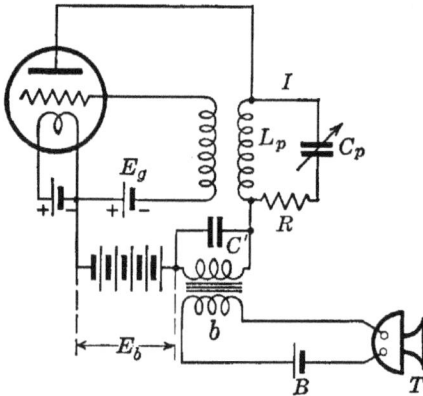

FIG. 347.—Connections for plate-circuit modulation.

In frequency modulation the amplitude of the carrier-frequency alternating current is maintained constant, and the instantaneous frequency is varied at a rate proportional to the audio frequency. The magnitude of the frequency change is proportional to the amplitude of the audiofrequency current.

One method of amplitude modulation is to introduce into the plate circuit of a tube, oscillating at the carrier frequency, an additional voltage of audio frequency whose peak value is somewhat less than the steady plate voltage E_b. This causes the output current I in the tuned circuit to have an amplitude envelope (Fig. 345(b)) proportional to this audiofrequency voltage. The connections for this method of modulation are shown in Fig. 347. The tube oscillates at the carrier frequency due to the tuned circuit L_pC_p, which is inductively coupled to the grid circuit. An audiofrequency transformer is introduced into the plate circuit at b. The primary current of this transformer is shown as coming from a microphone circuit consisting of a battery B in series with a telephone transmitter or microphone T. Ordinarily, there is not sufficient power in the microphone circuit to give sufficient modulation; hence, an amplifier between the microphone circuit and the audiofrequency transformer is necessary (see Figs. 335 and 336, p. 409). The secondary of the transformer introduces the modulation emf into the plate circuit. As the carrier or radiofrequency current is unable to flow through the high inductance of the transformer secondary, a by-pass capacitor C' is necessary. If it were desired to broadcast with this circuit, an antenna would be inductively coupled to L_p, one end of the coupling winding being grounded and the other connected to the antenna.

DETECTION

A modulated high-frequency current, such as is shown in Fig. 345(b), can have no effect on any ordinary sound-producing device, since such a device is unable to respond to such high frequencies. Neither can this high-frequency current produce any effect on the human ear, because its frequency is far beyond audible frequencies. It is, therefore, necessary to demodulate such currents in order that the receiving devices may be actuated by audiofrequency currents similar to those used for modulating. This process of demodulation is called *detection*.

258. Rectification with Two-electrode Tube.—Detection may be accomplished with any rectifying tube, such, for example, as the two-electrode tube (Fig. 348). The tube will eliminate the negative loops (Fig. 345(b), p. 419), leaving a pulsating, unidirectional current (Fig. 349) made up of a unidirectional steady current, an audiofrequency current, and a radiofrequency current. The unidirectional current and the audiofrequency current will flow through the telephones T (Fig. 348), which will reproduce in the sound the initial audiofrequency current. The high-frequency component will be by-passed through the capacitor C.

Although the two-electrode tube is a satisfactory rectifier, it is insensitive

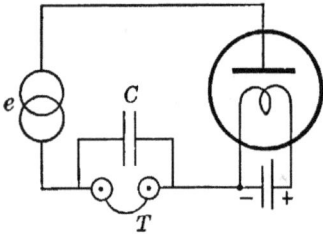

Fig. 348.—Two-electrode tube used as detector.

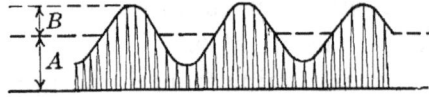

Fig. 349.—Rectified carrier wave.

if used in this manner as a detector because of its excessively high resistance when operated at low voltage. This may be seen in Fig. 351, where the current I for small values of voltage E is extremely small. This difficulty is overcome, in part, by inserting a positive polarizing voltage E_b in series with the tube (Fig. 350). Thus, in Fig. 351, the steady polarizing voltage E_p produces a steady current I_p in the tube circuit. Hence, an alternating emf e impressed on the tube is no longer

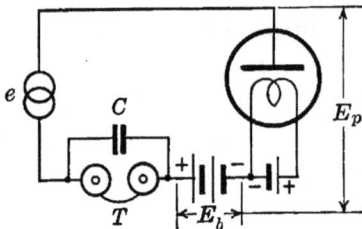

Fig. 350.—Two-electrode tube with polarizing voltage used as detector.

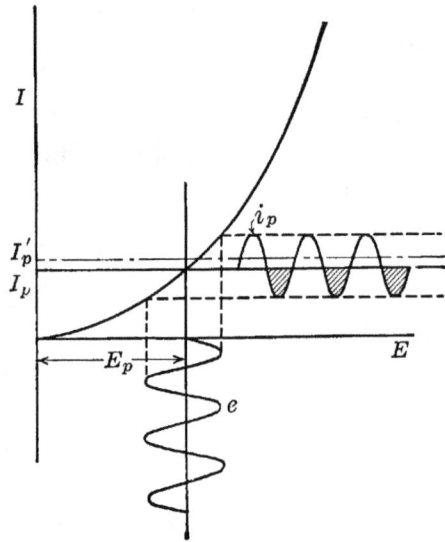

Fig. 351.—Detection with polarized two-electrode tube.

perfectly rectified but produces an alternating current i_p. Owing to the curvature of the characteristic, this current i_p is dissymmetrical, the positive current being larger than the negative current. The negative current is shown shaded. Hence, the average current is increased from I_p to I'_p, and thus the existence of the impressed emf is detected. The *change* in current, $I'_p - I_p$, is greater than the current which would flow for zero polarizing volt·age and has its maximum value when the polarizing voltage corresponds to the point of greatest curvature of the characteristic. When the impressed voltage is modulated, this change in plate current will follow the variations of the modulating current.

259. Detection with Three-electrode Tube with Polarized Grid.—The three-electrode tube will detect in a manner similar to the two-electrode tube with polarizing voltage, rectification depending on operating the tube at a point of curvature on

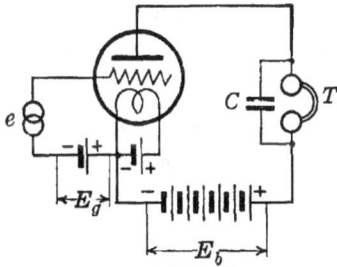

FIG. 352.—Three-electrode tube with polarized grid as detector.

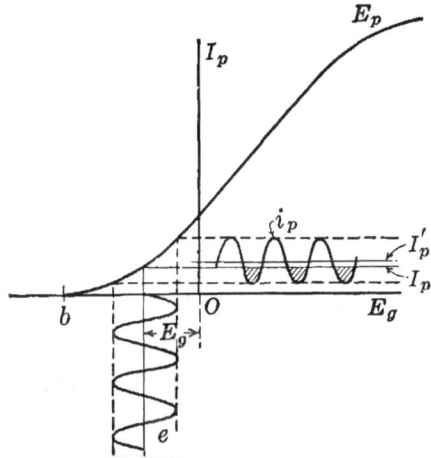

FIG. 353.—Detection with polarized three-electrode tube.

its plate-current—grid-voltage characteristic. The connections for operating a tube as detector are shown in Fig. 352. The grid is polarized negatively with a voltage E_g by the grid or C-battery to such a value as to cause the tube to operate on a point of curvature of the I_p V_g characteristic (Fig. 353). As with the two-electrode tube, a sinusoidal emf e impressed on the grid produces an alternating current i_p in the plate circuit, the reference axis of i_p being I_p. Owing to the curvature of the characteristic, the negative portions of i_p, shown shaded, are less in magnitude than the positive portions, and the average current increases from I_p to I'_p. When the impressed voltage e is modulated, this change in plate current, $I'_p - I_p$, will follow the variations of the modulating current. The radiofrequency plate current i_p is by-passed around the telephones through the capacitor C (Fig. 352). For maximum sensitivity, the polarizing voltage E_g should be such that detection occurs at the point of maximum curvature of the characteristic.

This type of detection is much used in receivers having sufficient amplification ahead of the detector tube to supply a voltage large enough to swing the grid nearly to zero bias. Under these conditions the distortion introduced in the rectified current is a minimum.

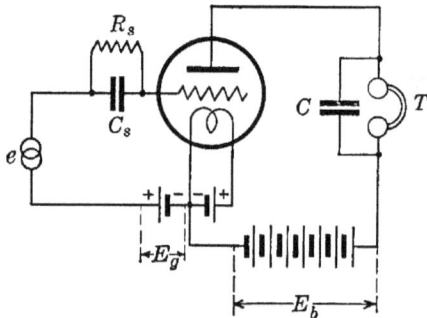

FIG. 354.—Three-electrode tube with grid resistance as detector.

260. Detection with Three-electrode Tube with Grid Resistance.—The three-electrode tube may also detect in a manner which is quite different from the foregoing. The connections are shown in Fig. 354. A high resistance R_s of 1 to 5 megohms is connected in series with and adjacent to the grid. This resistance is shunted by a small capacitor C_s whose capacitance is between 50 and 200 $\mu\mu f$. The grid is polarized positively by the voltage E_g, so that a current I_g flows in the grid circuit, as in

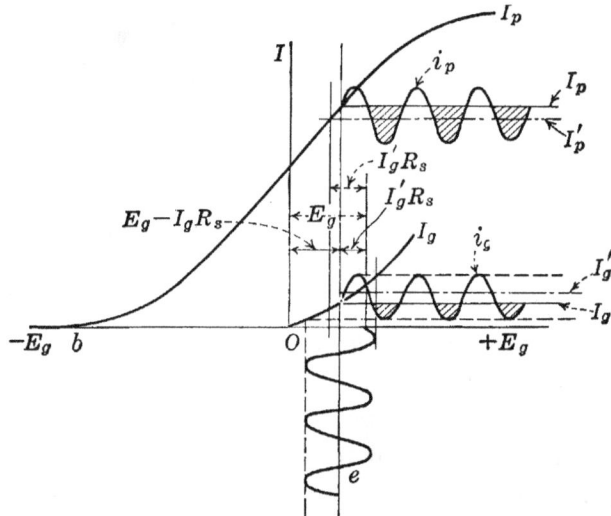

FIG. 355.—Detecting action with grid resistance.

Fig. 355. This current flowing through the high resistance R_s produces in it a voltage drop $I_g R_s$, so that the effective polarization of the grid is $E_g - I_g R_s$. The corresponding plate current is I_p. An alternating voltage e in the grid circuit will produce an alternating current i_g in the grid circuit, whose negative portions, shown shaded, are less in magnitude than its positive portions. Hence, the average grid current is increased from I_g to I'_g. This decreases the polarization of the grid from $E_g - I_g R_s$ to $E_g - I'_g R_s$. The alternating component i_g is by-passed through the capacitor C_s. The average plate current is decreased from I_p to I'_p with a superposed alternating current i_p. With reference to I_p as an axis, the positive portions of i_p are less in magnitude than the negative portions, shown shaded. When the impressed voltage e is modulated, this change in plate current, $I_p - I'_p$, will follow the variations of the modulating current.

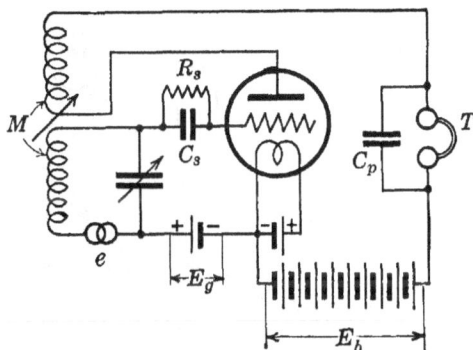

FIG. 356.—Three-electrode tube with grid resistance and regeneration used as detector.

The radiofrequency plate current i_p is by-passed through the capacitor C (Fig. 354).

The large curvature of the grid-current characteristic I_g, the large slope of the plate-current characteristic I_p, and the fact that the high resistance R_s may be made very large all combine to make this type of detection the most sensitive of all the methods thus far discussed.

261. Detection and Regeneration.—The two foregoing types of detection may be operated with a tuned circuit combined with regeneration. (See Par. 254, p. 414.) A very efficient circuit of this character, having a tuned grid circuit, grid resistance, and grid capacitor, is shown in Fig. 356. The incoming signal e is detected, and a portion of the resulting energy of the plate circuit is fed back into the grid circuit through the coupling M. The capacitor C_p shunts the high-frequency currents around the telephone receivers T'.

262. Heterodyne or Beat Reception.—A high-frequency alternating current may have its frequency a lowered by superposing
on it a second current of somewhat lower or higher frequency a'.
The resulting current may be shown to be a modulated current
having a frequency equal to the average of the two frequencies.
Further, the amplitude *envelope* of this resultant frequency has
itself a frequency, $a - a'$ or $a' - a$, that is, the *difference* of the

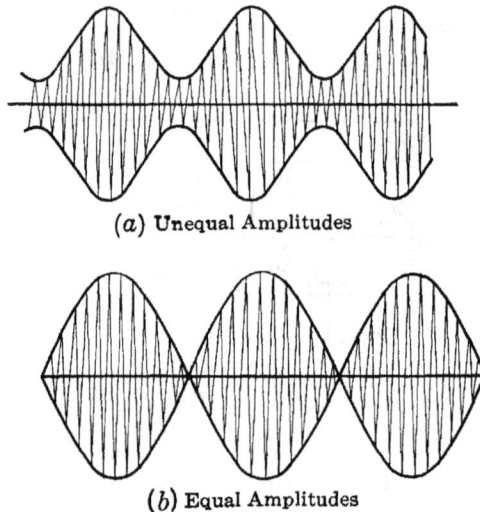

(a) Unequal Amplitudes

(b) Equal Amplitudes

Fig. 357.—Beat-frequency envelopes.

two impressed frequencies.   Figure 357(a) shows the resulting
high-frequency current curve and the resulting amplitude
envelope for the general case, that is, when the amplitudes of the
two currents are unequal.   Figure 357(b) shows the resulting
current curve and the resulting amplitude envelope when the
amplitudes of the two currents are equal.   In neither case is the
envelope sinusoidal, but the divergence is marked only when
the two amplitudes are nearly equal.

A detecting tube will separate the envelope frequency from the
high frequency, thus giving a current having the envelope frequency of $a - a'$ cycles per sec (see Par. 258, p. 421).   This frequency is called the *beat frequency*.

The superposed frequency may be obtained from an oscillating
tube, or oscillator (Fig. 358), whose grid circuit is inductively
coupled to the grid circuit of the detector through the mutual

inductance M'. This method of reception is called *heterodyne reception* or *beat reception*.

In radiotelegraphy, where the high-frequency, or carrier, current is modulated by the dots and dashes of the Morse code, the frequency of the beat note is made so low as to be audible, and the dots and dashes are heard at that frequency. In radiotelephony, or broadcasting, where the high-frequency current is modulated by speech or music, the frequency of the beat note, that is, the amplitude *envelope* of the frequency $a - a'$, is placed between 150 and 450 kc.

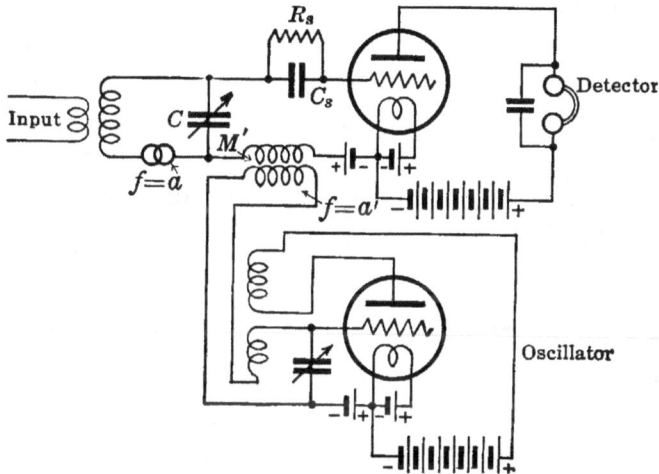

FIG. 358.—Separate heterodyne reception.

For example, if the incoming modulated frequency is 1,000 kc and the superposed frequency is 850 kc, the resulting frequency is

$$\frac{1,000 + 850}{2} = 925 \text{ kc}$$

and the frequency of the amplitude envelope, or the beat frequency, is $1,000 - 850$, or 150 kc.

By means of the detector (Fig. 358), this amplitude envelope of beat frequency is converted into a carrier-frequency current of this same beat frequency and modulated by the original audio-frequency currents, such as those produced by speech and music. This current again must be detected, amplified, etc., in the ordinary manner. This is the principle of the superheterodyne receiver.

The functions of oscillator and detector may be combined in one tube. This method is called *self-heterodyne* or *autodyne* reception. The connections are identical with those of Fig. 356, with the omission of the grid polarizing battery. In the autodyne, the tube is already oscillating, which tends to increase its sensitivity; but this effect is frequently more than offset by the fact that the grid circuit is tuned to the frequency of the oscillating current and not to that of the incoming signal. If these frequencies differ by large amounts, further amplification is necessary to make autodyne reception equal to separate heterodyne reception.

It is possible to detect a speech-modulated, high-frequency current with either the separate-heterodyne or the autodyne method by making the beat note of zero frequency. This greatly increases the detecting action, but serious distortion is likely to be introduced because of the difficulty of maintaining a zero beat-frequency.

RECEIVERS

263. Receiving Circuits.—In modern broadcasting, radiotelephony uses a speech-modulated, high-frequency current such as is produced by the transmitter shown in Fig. 347 (p. 420), except that the tuned circuit L_pC_pR consists of the antenna, so that the output of the oscillator is converted into an outgoing electromagnetic wave. This wave may be received by an antenna or loop, converted into audiofrequency current by a detector and then into sound waves by telephones or a loudspeaker.

Most receivers are designed to operate from an a-c supply, as shown in the wiring diagram of a typical superheterodyne receiver (Fig. 359). A full-wave rectifier and filter (Fig. 319, p. 395) supplies a plate voltage of about 250 volts. The electromagnet of the loud-speaker is frequently used as one of the choke coils of the filter. Lower voltages for the screen grids of the screen-grid tubes and for the plates of any lower voltage tubes are obtained by means of series resistances or a voltage divider. In either case these resistances must be by-passed by large capacitors to provide a low-impedance path for the a-c plate current. Grid-bias voltage is obtained from the voltage drop in a resistance placed in the plate return lead to the cathode. This resistance

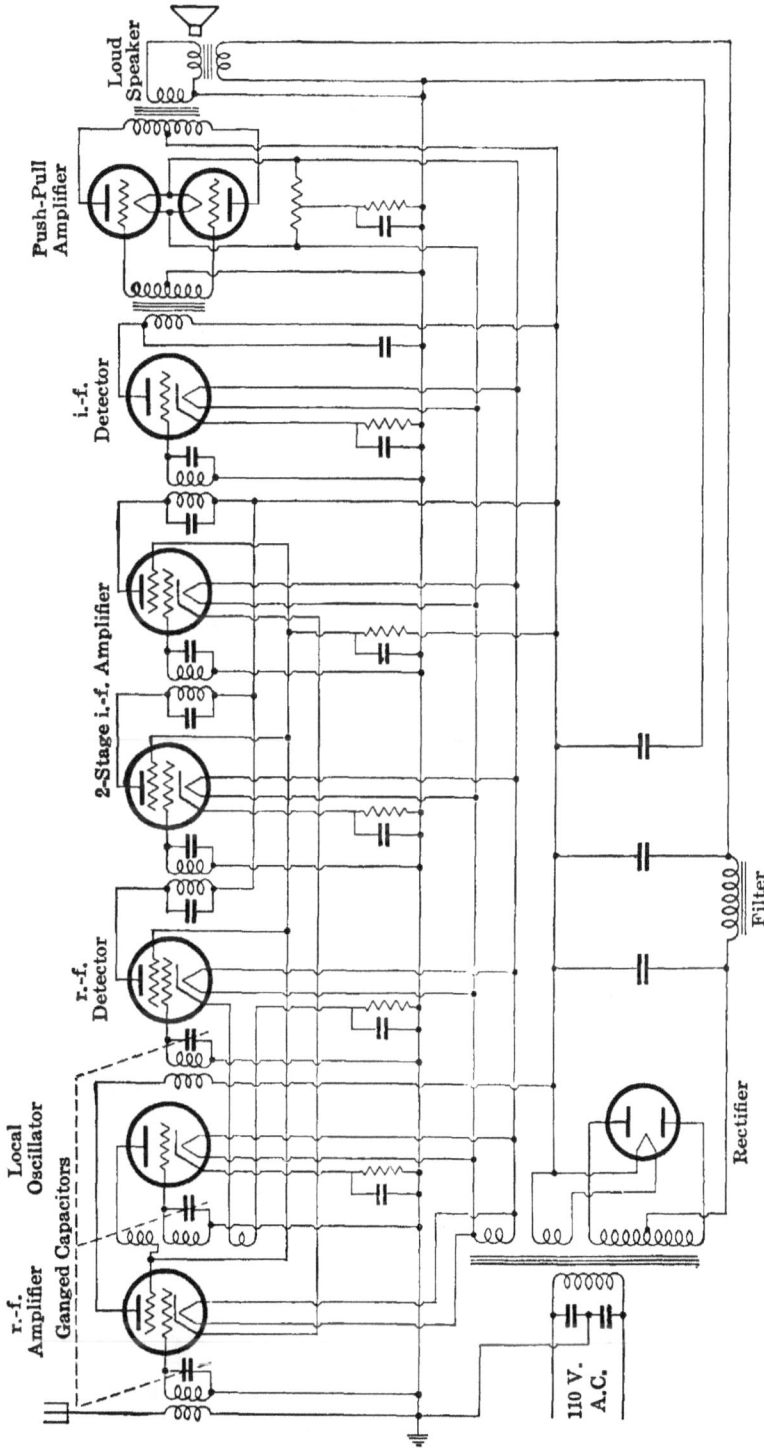

Fig. 359.—Typical superheterodyne receiver.

is by-passed by a suitable capacitor (see Par. 249, p. 407). The filaments of all tubes are heated from one or more separate low-voltage windings on the rectifier transformer. All tubes except the output tubes usually have separate-heater cathodes to reduce the a-c hum in the loudspeaker and to allow all the filaments to be operated in parallel.

The heterodyne method of reception (see Par. 262) is used almost universally because of its great selectivity and sensitivity. This makes possible the use of a power detector which greatly decreases the distortion in the audiofrequency output. The tuned transformers in the intermediate-frequency amplifier are designed so as to give a flat-topped pass band which prevents side-band cutting.

APPENDIX

APPENDIX A

Circular Measure—The Radian

The *radian* is a circular angle subtended by an arc equal in length to the radius of its circle, as shown in the figure. The circle has a radius of r units, and the radian is subtended by an arc whose length is r units.

As the circumference of a circle is $2\pi r$ units, there must be 2π, or 6.283, radians in 360°. Therefore, 1 radian equals $360°/2\pi = 57.30°$. It follows that $180° = \pi$ radians.

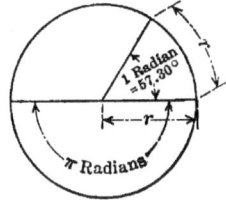

Angular velocity is often expressed in radians per second, and the accepted symbol is ω (omega). In every revolution a rotating quantity completes 2π radians. If the rotating quantity makes n rps, its angular velocity $\omega = 2\pi n$ radians per sec.

APPENDIX B

Trigonometry—Simple Functions

1. The sine (sin) of an angle $= \dfrac{\text{opposite side}}{\text{hypotenuse}}$.

2. The cosine (cos) of an angle $= \dfrac{\text{adjacent side}}{\text{hypotenuse}}$.

3. The tangent (tan) of an angle $= \dfrac{\text{opposite side}}{\text{adjacent side}}$.

4. The cotangent (cot) $= \dfrac{1}{\tan} = \dfrac{\text{adjacent side}}{\text{opposite side}}$.

5. The secant (sec) $= \dfrac{1}{\cos} = \dfrac{\text{hypotenuse}}{\text{adjacent side}}$.

6. The cosecant (cosec) $= \dfrac{1}{\sin} = \dfrac{\text{hypotenuse}}{\text{opposite side}}$.

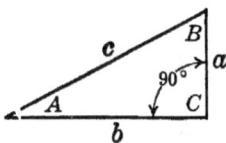

7. $\sin A = \dfrac{a}{c}$.

8. $\cos A = \dfrac{b}{c}$.

9. $\tan A = \dfrac{a}{b}$.

10. $\cot A = \dfrac{b}{a} = \dfrac{1}{\tan A}$.

11. $\sec A = \dfrac{c}{b} = \dfrac{1}{\cos A}$.

12. $\operatorname{cosec} A = \dfrac{c}{a} = \dfrac{1}{\sin A}$.

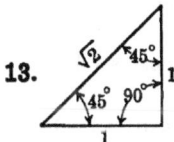

13.

Ratio of sides in a right isosceles triangle.

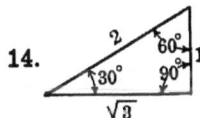

14.

Ratio of sides in a 30–60 right triangle.

431

15. $\sin B = \dfrac{b}{c} = \cos A = \cos (90° - B)$, since $A = 90° - B$.

16. $\cos B = \dfrac{a}{c} = \sin A = \sin (90° - B)$.

17. $\dfrac{\sin A}{\cos A} = \dfrac{a/c}{b/c} = \dfrac{a}{b} = \tan A$.   **21.** $\cos 60° = 0.5$.

18. $\sin 30° = 0.5$.   **22.** $\tan 30° = 1/\sqrt{3} = 0.577$.

19. $\cos 30° = \sqrt{3}/2 = 0.866$.   **23.** $\tan 60° = \sqrt{3} = 1.732$.

20. $\sin 60° = \sqrt{3}/2 = 0.866$.   **24.** $\sin 45° = \cos 45° = 1/\sqrt{2} = 0.707$.

25. $\tan 45° = 1.0$.

APPENDIX C

Functions of Angles Greater than 90°
(*ob* the radius vector is always positive)

FIRST QUADRANT

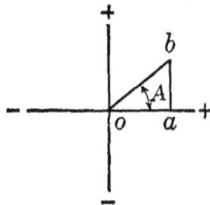

$\sin A = \dfrac{+ ab}{+ ob}$ sin is (+)

$\cos A = \dfrac{+ oa}{+ ob}$ cos is (+)

$\tan A = \dfrac{+ ab}{+ oa}$ tan is (+)

SECOND QUADRANT

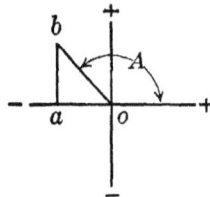

$\sin A = \dfrac{+ ab}{+ ob}$ sin is (+)

$\cos A = \dfrac{- oa}{+ ob}$ cos is (−)

$\tan A = \dfrac{+ ab}{- oa}$ tan is (−)

THIRD QUADRANT

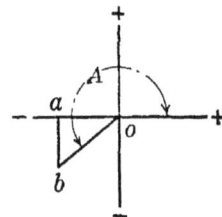

$\sin A = \dfrac{- ab}{+ ob}$ sin is (−)

$\cos A = \dfrac{- oa}{+ ob}$ cos is (−)

$\tan A = \dfrac{- ab}{- oa}$ tan is (+)

FOURTH QUADRANT

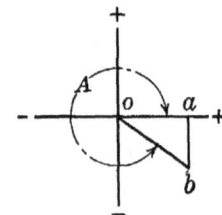

$\sin A = \dfrac{- ab}{+ ob}$ sin is (−)

$\cos A = \dfrac{+ oa}{+ ob}$ cos is (+)

$\tan A = \dfrac{- ab}{+ oa}$ tan is (−)

(Also, see Graphical Representation of Trigonometric Functions on the following page.)

Graphical Representation of Trigonometric Functions

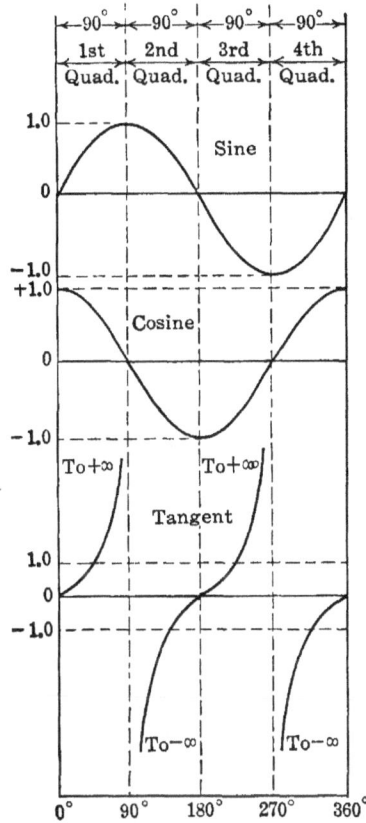

APPENDIX D

Trigonometric Formulas

$$a^2 + b^2 = c^2.$$
$$\frac{a^2}{c^2} + \frac{b^2}{c^2} = \frac{c^2}{c^2} = 1.$$

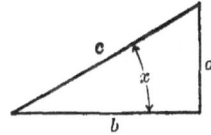

Since $\sin x = \dfrac{a}{c}$; $\cos x = \dfrac{b}{c}$ it follows that

26. $\sin^2 x + \cos^2 x = 1.$
27. $\sec^2 x = 1 + \tan^2 x.$
28. $\sin (90° + x) = \cos x.$
29. $\cos (90° + x) = -\sin x.$
30. $\tan (90° + x) = -\cot x.$
31. $\sin (180° - x) = \sin x.$
32. $\cos (180° - x) = -\cos x.$
33. $\tan (180° - x) = -\tan x.$
34. $\sin (x + y) = \sin x \cos y + \cos x \sin y.$

35. $\sin (x - y) = \sin x \cos y - \cos x \sin y.$

36. $\cos (x + y) = \cos x \cos y - \sin x \sin y.$

37. $\cos (x - y) = \cos x \cos y + \sin x \sin y.$

38. $\tan (x + y) = \dfrac{\tan x + \tan y}{1 - \tan x \tan y}.$

39. $\tan (x - y) = \dfrac{\tan x - \tan y}{1 + \tan x \tan y}.$

40. $\sin 2x = 2 \sin x \cos x.$

41. $\cos 2x = \cos^2 x - \sin^2 x.$

42. $\tan 2x = \dfrac{2 \tan x}{1 - \tan^2 x}.$

Law of Sines.—In *any* triangle,

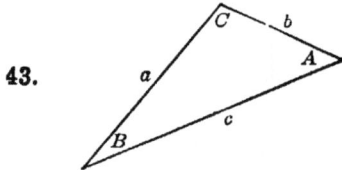

43.
$$\frac{a}{\sin A} = \frac{b}{\sin B} = \frac{c}{\sin C}.$$

Example.—Given $a = 28$, $c = 12$, $C = 20°$.

Find A, B, and b.

$$\frac{28}{\sin A} = \frac{12}{\sin 20°}; \quad \sin A = \sin 20° \frac{28}{12} = 0.342 \frac{28}{12} = 0.798.$$

A is obviously greater than 90°.

$\sin A = \sin (180° - A)$ (see **31**).

$\sin 52.9° = \sin 127.1°.$

Hence, $A = 127.1°$.   *Ans.*

   $B = 180° - 20° - 127.1° = 32.9°.$ *Ans.*

$$\frac{b}{\sin 32.9°} = \frac{12}{\sin 20°}; \quad b = 12 \frac{0.543}{0.342} = 19.05. \quad \textit{Ans.}$$

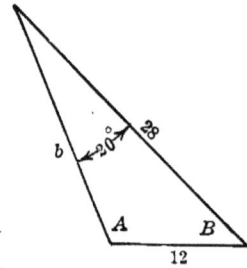

Law of Cosines.—In any triangle the square of any side is equal to the sum of the squares of the other two sides minus twice the product of these two sides and the cosine of their included angle.

That is,

44. $a^2 = b^2 + c^2 - 2bc \cos A$ (see triangle in **43**).

45. $\cos A = \dfrac{b^2 + c^2 - a^2}{2bc}.$

46. $\cos B = \dfrac{c^2 + a^2 - b^2}{2ca}.$

47. $\cos C = \dfrac{a^2 + b^2 - c^2}{2ab}.$

Example.—Given two sides $b = 50$ and $c = 42$ and included angle $A = 40°$. Find a, B, and C.

$a^2 = 50^2 + 42^2 - 2 \times 50 \times 42 \cos 40°.$

$\quad = 2{,}500 + 1{,}764 - 4{,}200 \times 0.766.$

$\quad = 1{,}047.$

$a = 32.36.$ *Ans.*

$$\frac{32.36}{\sin 40°} = \frac{50}{\sin B}.$$

$\sin B = \sin 40 \dfrac{50}{32.36} = 0.6428 \times 1.545 = 0.9931.$

$B = 83.3°.$ *Ans.*

$C = 180° - 83.3° - 40° = 56.7°.$ *Ans*

(See p. 445 for problems in trigonometry.)

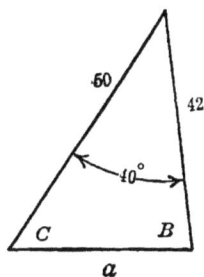

APPENDIX E

MATHEMATICAL TABLES
Natural Sines and Cosines

NOTE.—For cosines use right-hand column of degrees and lower line of tenths

Deg.	°0.0	°0.1	°0.2	°0.3	°0.4	°0.5	°0.6	°0.7	°0.8	°0.9	°1.0	
0°	0.0000	0.0017	0.0035	0.0052	0.0070	0.0087	0.0105	0.0122	0.0140	0.0157	0.0175	89
1	0.0175	0.0192	0.0209	0.0227	0.0244	0.0262	0.0279	0.0297	0.0314	0.0332	0.0349	88
2	0.0349	0.0366	0.0384	0.0401	0.0419	0.0436	0.0454	0.0471	0.0488	0.0506	0.0523	87
3	0.0523	0.0541	0.0558	0.0576	0.0593	0.0610	0.0628	0.0645	0.0663	0.0680	0.0698	86
4	0.0698	0.0715	0.0732	0.0750	0.0767	0.0785	0.0802	0.0819	0.0837	0.0854	0.0872	85
5	0.0872	0.0889	0.0906	0.0924	0.0941	0.0958	0.0976	0.0993	0.1011	0.1028	0.1045	84
6	0.1045	0.1063	0.1080	0.1097	0.1115	0.1132	0.1149	0.1167	0.1184	0.1201	0.1219	83
7	0.1219	0.1236	0.1253	0.1271	0.1288	0.1305	0.1323	0.1340	0.1357	0.1374	0.1392	82
8	0.1392	0.1409	0.1426	0.1444	0.1461	0.1478	0.1495	0.1513	0.1530	0.1547	0.1564	81
9	0.1564	0.1582	0.1599	0.1616	0.1633	0.1650	0.1668	0.1685	0.1702	0.1719	0.1736	80°
10°	0.1736	0.1754	0.1771	0.1788	0.1805	0.1822	0.1840	0.1857	0.1874	0.1891	0.1908	79
11	0.1908	0.1925	0.1942	0.1959	0.1977	0.1994	0.2011	0.2028	0.2045	0.2062	0.2079	78
12	0.2079	0.2096	0.2113	0.2130	0.2147	0.2164	0.2181	0.2198	0.2215	0.2232	0.2250	77
13	0.2250	0.2267	0.2284	0.2300	0.2317	0.2334	0.2351	0.2368	0.2385	0.2402	0.2419	76
14	0.2419	0.2436	0.2453	0.2470	0.2487	0.2504	0.2521	0.2538	0.2554	0.2571	0.2588	75
15	0.2588	0.2605	0.2622	0.2639	0.2656	0.2672	0.2689	0.2706	0.2723	0.2740	0.2756	74
16	0.2756	0.2773	0.2790	0.2807	0.2823	0.2840	0.2857	0.2874	0.2890	0.2907	0.2924	73
17	0.2924	0.2940	0.2957	0.2974	0.2990	0.3007	0.3024	0.3040	0.3057	0.3074	0.3090	72
18	0.3090	0.3107	0.3123	0.3140	0.3156	0.3173	0.3190	0.3206	0.3223	0.3239	0.3256	71
19	0.3256	0.3272	0.3289	0.3305	0.3322	0.3338	0.3355	0.3371	0.3387	0.3404	0.3420	70°
20°	0.3420	0.3437	0.3453	0.3469	0.3486	0.3502	0.3518	0.3535	0.3551	0.3567	0.3584	69
21	0.3584	0.3600	0.3616	0.3633	0.3649	0.3665	0.3681	0.3697	0.3714	0.3730	0.3746	68
22	0.3746	0.3762	0.3778	0.3795	0.3811	0.3827	0.3843	0.3859	0.3875	0.3891	0.3907	67
23	0.3907	0.3923	0.3939	0.3955	0.3971	0.3987	0.4003	0.4019	0.4035	0.4051	0.4067	66
24	0.4067	0.4083	0.4099	0.4115	0.4131	0.4147	0.4163	0.4179	0.4195	0.4210	0.4226	65
25	0.4226	0.4242	0.4258	0.4274	0.4289	0.4305	0.4321	0.4337	0.4352	0.4368	0.4384	64
26	0.4384	0.4399	0.4415	0.4431	0.4446	0.4462	0.4478	0.4493	0.4509	0.4524	0.4540	63
27	0.4540	0.4555	0.4571	0.4586	0.4602	0.4617	0.4633	0.4648	0.4664	0.4679	0.4695	62
28	0.4695	0.4710	0.4726	0.4741	0.4756	0.4772	0.4787	0.4802	0.4818	0.4833	0.4848	61
29	0.4848	0.4863	0.4879	0.4894	0.4909	0.4924	0.4939	0.4955	0.4970	0.4985	0.5000	60°
30°	0.5000	0.5015	0.5030	0.5045	0.5060	0.5075	0.5090	0.5105	0.5120	0.5135	0.5150	59
31	0.5150	0.5165	0.5180	0.5195	0.5210	0.5225	0.5240	0.5255	0.5270	0.5284	0.5299	58
32	0.5299	0.5314	0.5329	0.5344	0.5358	0.5373	0.5388	0.5402	0.5417	0.5432	0.5446	57
33	0.5446	0.5461	0.5476	0.5490	0.5505	0.5519	0.5534	0.5548	0.5563	0.5577	0.5592	56
34	0.5592	0.5606	0.5621	0.5635	0.5650	0.5664	0.5678	0.5693	0.5707	0.5721	0.5736	55
35	0.5736	0.5750	0.5764	0.5779	0.5793	0.5807	0.5821	0.5835	0.5850	0.5864	0.5878	54
36	0.5878	0.5892	0.5906	0.5920	0.5934	0.5948	0.5962	0.5976	0.5990	0.6004	0.6018	53
37	0.6018	0.6032	0.6046	0.6060	0.6074	0.6088	0.6101	0.6115	0.6129	0.6143	0.6157	52
38	0.6157	0.6170	0.6184	0.6198	0.6211	0.6225	0.6239	0.6252	0.6266	0.6280	0.6293	51
39	0.6293	0.6307	0.6320	0.6334	0.6347	0.6361	0.6374	0.6388	0.6401	0.6414	0.6428	50°
40°	0.6428	0.6441	0.6455	0.6468	0.6481	0.6494	0.6508	0.6521	0.6534	0.6547	0.6561	49
41	0.6561	0.6574	0.6587	0.6600	0.6613	0.6626	0.6639	0.6652	0.6665	0.6678	0.6691	48
42	0.6691	0.6704	0.6717	0.6730	0.6743	0.6756	0.6769	0.6782	0.6794	0.6807	0.6820	47
43	0.6820	0.6833	0.6845	0.6858	0.6871	0.6884	0.6896	0.6909	0.6921	0.6934	0.6947	46
44	0.6947	0.6959	0.6972	0.6984	0.6997	0.7009	0.7022	0.7034	0.7046	0.7059	0.7071	45
	°1.0	°0.9	°0.8	°0.7	°0.6	°0.5	°0.4	°0.3	°0.2	°0.1	°0.0	Deg.

Natural Sines and Cosines.—(*Concluded*)

Deg.	°0.0	°0.1	°0.2	°0.3	°0.4	°0.5	°0.6	°0.7	°0.8	°0.9	°1.0	
45	0.7071	0.7083	0.7096	0.7108	0.7120	0.7133	0.7145	0.7157	0.7169	0.7181	0.7193	44
46	0.7193	0.7206	0.7218	0.7230	0.7242	0.7254	0.7266	0.7278	0.7290	0.7302	0.7314	43
47	0.7314	0.7325	0.7337	0.7349	0.7361	0.7373	0.7385	0.7396	0.7408	0.7420	0.7431	42
48	0.7431	0.7443	0.7455	0.7466	0.7478	0.7490	0.7501	0.7513	0.7524	0.7536	0.7547	41
49	0.7547	0.7559	0.7570	0.7581	0.7593	0.7604	0.7615	0.7627	0.7638	0.7649	0.7660	40°
50°	0.7660	0.7672	0.7683	0.7694	0.7705	0.7716	9.7727	0.7738	0.7749	0.7760	0.7771	39
51	0.7771	0.7782	0.7793	0.7804	0.7815	0.7826	0.7837	0.7848	0.7859	0.7869	0.7880	38
52	0.7880	0.7891	0.7902	0.7912	0.7923	0.7934	0.7944	0.7955	0.7965	0.7976	0.7986	37
53	0.7986	0.7997	0.8007	0.8018	0.8028	0.8039	0.8049	0.8059	0.8070	0.8080	0.8090	36
54	0.8090	0.8100	0.8111	0.8121	0.8131	0.8141	0.8151	0.8161	0.8171	0.8181	0.8192	35
55	0.8192	0.8202	0.8211	0.8221	0.8231	0.8241	0.8251	0.8261	0.8271	0.8281	0.8290	34
56	0.8290	0.8300	0.8310	0.8320	0.8329	0.8339	0.8348	0.8358	0.8368	0.8377	0.8387	33
57	0.8387	0.8396	0.8406	0.8415	0.8425	0.8434	0.8443	0.8453	0.8462	0.8471	0.8480	32
58	0.8480	0.8490	0.8499	0.8508	0.8517	0.8526	0.8536	0.8545	0.8554	0.8563	0.8572	31
59	0.8572	0.8581	0.3590	0.8599	0.8607	0.8616	0.8625	0.8634	0.8643	0.8652	0.8660	30°
60°	0.8660	0.8669	0.8678	0.8686	0.8695	0.8704	0.8712	0.8721	0.8729	0.8738	0.8746	29
61	0.8746	0.8755	0.8763	0.8771	0.8780	0.8788	0.8796	0.8805	0.8813	0.8821	0.8829	28
62	0.8829	0.8838	0.8846	0.8854	0.8862	0.8870	0.8878	0.8886	0.8894	0.8902	0.8910	27
63	0.8910	0.8918	0.8926	0.8934	0.8942	0.8949	0.8957	0.8965	0.8973	0.8980	0.8988	26
64	0.8988	0.8996	0.9003	0.9011	0.9018	0.9026	0.9033	0.9041	0.9048	0.9056	0.9063	25
65	0.9063	0.9070	0.9078	0.9085	0.9092	0.9100	0.9107	0.9114	0.9121	0.9128	0.9135	24
66	0.9135	0.9143	0.9150	0.9157	0.9164	0.9171	0.9178	0.9184	0.9191	0.9198	0.9205	23
67	0.9205	0.9212	0.9219	0.9225	0.9232	0.9239	0.9245	0.9252	0.9259	0.9265	0.9272	22
68	0.9272	0.9278	0.9285	0.9291	0.9298	0.9304	0.9311	0.9317	0.9323	0.9330	0.9336	21
69	0.9336	0.9342	0.9348	0.9354	0.9361	0.9367	0.9373	0.9379	0.9385	0.9391	0.9397	20°
70°	0.9397	0.9403	0.9409	0.9415	0.9421	0.9426	0.9432	0.9438	0.9444	0.9449	0.9455	19
71	0.9455	0.9461	0.9466	0.9472	0.9478	0.9483	0.9489	0.9494	0.9500	0.9505	0.9511	18
72	0.9511	0.9516	0.9521	0.9527	0.9532	0.9537	0.9542	0.9548	0.9553	0.9558	0.9563	17
73	0.9563	0.9568	0.9573	0.9578	0.9583	0.9588	0.9593	0.9598	0.9603	0.9608	0.9613	16
74	0.9613	0.9617	0.9622	0.9627	0.9632	0.9636	0.9641	0.9646	0.9650	0.9655	0.9659	15
75	0.9659	0.9664	0.9668	0.9673	0.9677	0.9681	0.9686	0.9690	0.9694	0.9699	0.9703	14
76	0.9703	0.9707	0.9711	0.9715	0.9720	0.9724	0.9728	0.9732	0.9736	0.9740	0.9744	13
77	0.9744	0.9748	0.9751	0.9755	0.9759	0.9763	0.9767	0.9770	0.9774	0.9778	0.9781	12
78	0.9781	0.9785	0.9789	0.9792	0.9796	0.9799	0.9803	0.9806	0.9810	0.9813	0.9816	11
79	0.9816	0.9820	0.9823	0.9826	0.9829	0.9833	0.9836	0.9839	0.9842	0.9845	0.9848	10°
80°	0.9848	0.9851	0.9854	0.9857	0.9860	0.9863	0.9866	0.9869	0.9871	0.9874	0.9877	9
81	0.9877	0.9880	0.9882	0.9885	0.9888	0.9890	0.9893	0.9895	0.9898	0.9900	0.9903	8
82	0.9903	0.9905	0.9907	0.9910	0.9912	0.9914	0.9917	0.9919	0.9921	0.9923	0.9925	7
83	0.9925	0.9928	0.9930	0.9932	0.9934	0.9936	0.9938	0.9940	0.9942	0.9943	0.9945	6
84	0.9945	0.9947	0.9949	0.9951	0.9952	0.9954	0.9956	0.9957	0.9959	0.9960	0.9962	5
85	0.9962	0.9963	0.9965	0.9966	0.9968	0.9969	0.9971	0.9972	0.9973	0.9974	0.9976	4
86	0.9976	0.9977	0.9978	0.9979	0.9980	0.9981	0.9982	0.9983	0.9984	0.9985	0.9986	3
87	0.9986	0.9987	0.9988	0.9989	0.9990	0.9990	0.9991	0.9992	0.9993	0.9993	0.9994	2
88	0.9994	0.9995	0.9995	0.9996	0.9996	0.9997	0.9997	0.9997	0.9998	0.9998	0.9998	1
89	0.9998	0.9999	0.9999	0.9999	0.9999	1.0000	1.0000	1.0000	1.0000	1.0000	1.0000	0°
	°1.0	°0.9	°0.8	°0.7	°0.6	°0.5	°0.4	°0.3	°0.2	°0.1	°0.0	Deg.

APPENDIX F

Natural Tangents and Cotangents

Note.—For cotangents use right-hand column of degrees and lower line of tenths

Deg.	°0.0	°0.1	°0.2	°0.3	°0.4	°0.5	°0.6	°0.7	°0.8	°0.9	°1.0	
0°	0.0000	0.0017	0.0035	0.0052	0.0070	0.0087	0.0105	0.0122	0.0140	0.0157	0.0175	89
1	0.0175	0.0192	0.0209	0.0227	0.0244	0.0262	0.0279	0.0297	0.0314	0.0332	0.0349	88
2	0.0349	0.0367	0.0384	0.0402	0.0419	0.0437	0.0454	0.0472	0.0489	0.0507	0.0524	87
3	0.0524	0.0542	0.0559	0.0577	0.0594	0.0612	0.0629	0.0647	0.0664	0.0682	0.0699	86
4	0.0699	0.0717	0.0734	0.0752	0.0769	0.0787	0.0805	0.0822	0.0840	0.0857	0.0875	85
5	0.0875	0.0892	0.0910	0.0928	0.0945	0.0963	0.0981	0.0998	0.1016	0.1033	0.1051	84
6	0.1051	0.1069	0.1086	0.1104	0.1122	0.1139	0.1157	0.1175	0.1192	0.1210	0.1228	83
7	0.1228	0.1246	0.1263	0.1281	0.1299	0.1317	0.1334	0.1352	0.1370	0.1388	0.1405	82
8	0.1405	0.1423	0.1441	0.1459	0.1477	0.1495	0.1512	0.1530	0.1548	0.1566	0.1584	81
9	0.1584	0.1602	0.1620	0.1638	0.1655	0.1673	0.1691	0.1709	0.1727	0.1745	0.1763	80°
10°	0.1763	0.1781	0.1799	0.1817	0.1835	0.1853	0.1871	0.1890	0.1908	0.1926	0.1944	79
11	0.1944	0.1962	0.1980	0.1998	0.2016	0.2035	0.2053	0.2071	0.2089	0.2107	0.2126	78
12	0.2126	0.2144	0.2162	0.2180	0.2199	0.2217	0.2235	0.2254	0.2272	0.2290	0.2309	77
13	0.2309	0.2327	0.2345	0.2364	0.2382	0.2401	0.2419	0.2438	0.2456	0.2475	0.2493	76
14	0.2493	0.2512	0.2530	0.2549	0.2568	0.2586	0.2605	0.2623	0.2642	0.2661	0.2679	75
15	0.2679	0.2698	0.2717	0.2736	0.2754	0.2773	0.2792	0.2811	0.2830	0.2849	0.2867	74
16	0.2867	0.2886	0.2905	0.2924	0.2943	0.2962	0.2981	0.3000	0.3019	0.3038	0.3057	73
17	0.3057	0.3076	0.3096	0.3115	0.3134	0.3153	0.3172	0.3191	0.3211	0.3230	0.3249	72
18	0.3249	0.3269	0.3288	0.3307	0.3327	0.3346	0.3365	0.3385	0.3404	0.3424	0.3443	71
19	0.3443	0.3463	0.3482	0.3502	0.3522	0.3541	0.3561	0.3581	0.3600	0.3620	0.3640	70°
20°	0.3640	0.3659	0.3679	0.3699	0.3719	0.3739	0.3759	0.3779	0.3799	0.3819	0.3839	69
21	0.3839	0.3859	0.3879	0.3899	0.3919	0.3939	0.3959	0.3979	0.4000	0.4020	0.4040	68
22	0.4040	0.4061	0.4081	0.4101	0.4122	0.4142	0.4163	0.4183	0.4204	0.4224	0.4245	67
23	0.4245	0.4265	0.4286	0.4307	0.4327	0.4348	0.4369	0.4390	0.4411	0.4431	0.4452	66
24	0.4452	0.4473	0.4494	0.4515	0.4536	0.4557	0.4578	0.4599	0.4621	0.4642	0.4663	65
25	0.4663	0.4684	0.4706	0.4727	0.4748	0.4770	0.4791	0.4813	0.4834	0.4856	0.4877	64
26	0.4877	0.4899	0.4921	0.4942	0.4964	0.4986	0.5008	0.5029	0.5051	0.5073	0.5095	63
27	0.5095	0.5117	0.5139	0.5161	0.5184	0.5206	0.5228	0.5250	0.5272	0.5295	0.5317	62
28	0.5317	0.5340	0.5362	0.5384	0.5407	0.5430	0.5452	0.5475	0.5498	0.5520	0.5543	61
29	0.5543	0.5566	0.5589	0.5612	0.5635	0.5658	0.5681	0.5704	0.5727	0.5750	0.5774	60°
30°	0.5774	0.5797	0.5820	0.5844	0.5867	0.5890	0.5914	0.5938	0.5961	0.5985	0.6009	59
31	0.6009	0.6032	0.6056	0.6080	0.6104	0.6128	0.6152	0.6176	0.6200	0.6224	0.6249	58
32	0.6249	0.6273	0.6297	0.6322	0.6346	0.6371	0.6395	0.6420	0.6445	0.6469	0.6494	57
33	0.6494	0.6519	0.6544	0.6569	0.6594	0.6619	0.6644	0.6669	0.6694	0.6720	0.6745	56
34	0.6745	0.6771	0.6796	0.6822	0.6847	0.6873	0.6899	0.6924	0.6950	0.6976	0.7002	55
35	0.7002	0.7028	0.7054	0.7080	0.7107	0.7133	0.7159	0.7186	0.7212	0.7239	0.7265	54
36	0.7265	0.7292	0.7319	0.7346	0.7373	0.7400	0.7427	0.7454	0.7481	0.7508	0.7536	53
37	0.7536	0.7563	0.7590	0.7618	0.7646	0.7673	0.7701	0.7729	0.7757	0.7785	0.7813	52
38	0.7813	0.7841	0.7869	0.7898	0.7926	0.7954	0.7983	0.8012	0.8040	0.8069	0.8098	51
39	0.8098	0.8127	0.8156	0.8185	0.8214	0.8243	0.8273	0.8302	0.8332	0.8361	0.8391	50°
40°	0.8391	0.8421	0.8451	0.8481	0.8511	0.8541	0.8571	0.8601	0.8632	0.8662	0.8693	49
41	0.8693	0.8724	0.8754	0.8785	0.8816	0.8847	0.8878	0.8910	0.8941	0.8972	0.9004	48
42	0.9004	0.9036	0.9067	0.9099	0.9131	0.9163	0.9195	0.9228	0.9260	0.9293	0.9325	47
43	0.9325	0.9358	0.9391	0.9424	0.9457	0.9490	0.9523	0.9556	0.9590	0.9623	0.9657	46
44	0.9657	0.9691	0.9725	0.9759	0.9793	0.9827	0.9861	0.9896	0.9930	0.9965	1.0000	45
	°1.0	°0.9	°0.8	°0.7	°0.6	°0.5	°0.4	°0.3	°0.2	°0.1	°0.0	Deg.

Natural Tangents and Cotangents.—(*Concluded*)

Deg.	°0.0	°0.1	°0.2	°0.3	°0.4	°0.5	°0.6	°0.7	°0.8	°0.9	°1.0	
45	1.0000	1.0035	1.0070	1.0105	1.0141	1.0176	1.0212	1.0247	1.0283	1.0319	1.0355	44
46	1.0355	1.0392	1.0428	1.0464	1.0501	1.0538	1.0575	1.0612	1.0649	1.0686	1.0724	43
47	1.0724	1.0761	1.0799	1.0837	1.0875	1.0913	1.0951	1.0990	1.1028	1.1067	1.1106	42
48	1.1106	1.1145	1.1184	1.1224	1.1263	1.1303	1.1343	1.1383	1.1423	1.1463	1.1504	41
49	1.1504	1.1544	1.1585	1.1626	1.1667	1.1708	1.1750	1.1792	1.1833	1.1875	1.1918	40°
50°	1.1918	1.1960	1.2002	1.2045	1.2088	1.2131	1.2174	1.2218	1.2261	1.2305	1.2349	39
51	1.2349	1.2393	1.2437	1.2482	1.2527	1.2572	1.2617	1.2662	1.2708	1.2753	1.2799	38
52	1.2799	1.2846	1.2892	1.2938	1.2985	1.3032	1.3079	1.3127	1.3175	1.3222	1.3270	37
53	1.3270	1.3319	1.3367	1.3416	1.3465	1.3514	1.3564	1.3613	1.3663	1.3713	1.3764	36
54	1.3764	1.3814	1.3865	1.3916	1.3968	1.4019	1.4071	1.4124	1.4176	1.4229	1.4281	35
55	1.4281	1.4335	1.4388	1.4442	1.4496	1.4550	1.4605	1.4659	1.4715	1.4770	1.4826	34
56	1.4826	1.4882	1.4938	1.4994	1.5051	1.5108	1.5166	1.5224	1.5282	1.5340	1.5399	33
57	1.5399	1.5458	1.5517	1.5577	1.5637	1.5697	1.5757	1.5818	1.5880	1.5941	1.6003	32
58	1.6003	1.6066	1.6128	1.6191	1.6255	1.6319	1.6383	1.6447	1.6512	1.6577	1.6643	31
59	1.6643	1.6709	1.6775	1.6842	1.6909	1.6977	1.7045	1.7113	1.7182	1.7251	1.7321	30°
60°	1.7321	1.7391	1.7461	1.7532	1.7603	1.7675	1.7747	1.7820	1.7893	1.7966	1.8040	29
61	1.8040	1.8115	1.8190	1.8265	1.8341	1.8418	1.8495	1.8572	1.8650	1.8728	1.8807	28
62	1.8807	1.8887	1.8967	1.9047	1.9128	1.9210	1.9292	1.9375	1.9458	1.9542	1.9626	27
63	1.9626	1.9711	1.9797	1.9883	1.9970	2.0057	2.0145	2.0233	2.0323	2.0413	2.0503	26
64	2.0503	2.0594	2.0686	2.0778	2.0872	2.0965	2.1060	2.1155	2.1251	2.1348	2.1445	25
65	2.1445	2.1543	2.1642	2.1742	2.1842	2.1943	2.2045	2.2148	2.2251	2.2355	2.2460	24
66	2.2460	2.2566	2.2673	2.2781	2.2889	2.2998	2.3109	2.3220	2.3332	2.3445	2.3559	23
67	2.3559	2.3673	2.3789	2.3906	2.4023	2.4142	2.4262	2.4383	2.4504	2.4627	2.4751	22
68	2.4751	2.4876	2.5002	2.5129	2.5257	2.5386	2.5517	2.5649	2.5782	2.5916	2.6051	21
69	2.6051	2.6187	2.6325	2.6464	2.6605	2.6746	2.6889	2.7034	2.7179	2.7326	2.7475	20°
70°	2.7475	2.7625	2.7776	2.7929	2.8083	2.8239	2.8397	2.8556	2.8716	2.8878	2.9042	19
71	2.9042	2.9208	2.9375	2.9544	2.9714	2.9887	3.0061	3.0237	3.0415	3.0595	3.0777	18
72	3.0777	3.0961	3.1146	3.1334	3.1524	3.1716	3.1910	3.2106	3.2305	3.2506	3.2709	17
73	3.2709	3.2914	3.3122	3.3332	3.3544	3.3759	3.3977	3.4197	3.4420	3.4646	3.4874	16
74	3.4874	3.5105	3.5339	3.5576	3.5816	3.6059	3.6305	3.6554	3.6806	3.7062	3.7321	15
75	3.7321	3.7583	3.7848	3.8118	3.8391	3.8667	3.8947	3.9232	3.9520	3.9812	4.0108	14
76	4.0108	4.0408	4.0713	4.1022	4.1335	4.1653	4.1976	4.2303	4.2635	4.2972	4.3315	13
77	4.3315	4.3662	4.4015	4.4374	4.4737	4.5107	4.5483	4.5864	4.6252	4.6646	4.7046	12
78	4.7046	4.7453	4.7867	4.8288	4.8716	4.9152	4.9594	5.0045	5.0504	5.0970	5.1446	11
79	5.1446	5.1929	5.2422	5.2924	5.3435	5.3955	5.4486	5.5026	5.5578	5.6140	5.6713	10°
80°	5.6713	5.7297	5.7894	5.8502	5.9124	5.9758	6.0405	6.1066	6.1742	6.2432	6.3138	9
81	6.3138	6.3859	6.4596	6.5350	6.6122	6.6912	6.7720	6.8548	6.9395	7.0264	7.1154	8
82	7.1154	7.2066	7.3002	7.3962	7.4947	7.5958	7.6996	7.8062	7.9158	8.0285	8.1443	7
83	8.1443	8.2636	8.3863	8.5126	8.6427	8.7769	8.9152	9.0579	9.2052	9.3572	9.5144	6
84	9.5144	9.677	9.845	10.02	10.20	10.39	10.58	10.78	10.99	11.20	11.43	5
85	11.43	11.66	11.91	12.16	12.43	12.71	13.00	13.30	13.62	13.95	14.30	4
86	14.30	14.67	15.06	15.46	15.89	16.35	16.83	17.34	17.89	18.46	19.08	3
87	19.08	19.74	20.45	21.20	22.02	22.90	23.86	24.90	26.03	27.27	28.64	2
88	28.64	30.14	31.82	33.69	35.80	38.19	40.92	44.07	47.74	52.08	57.29	1
89	57.29	63.66	71.62	81.85	95.49	114.6	143.2	191.0	286.5	573.0	∞	0°
	°1.0	°0.9	°0.8	°0.7	°0.6	°0.5	°0.4	°0.3	°0.2	°0.1	°0.0	Deg.

APPENDIX G

Logarithms of Numbers

N	0	1	2	3	4	5	6	7	8	9
10	0000	0043	0086	0128	0170	0212	0253	0294	0334	0374
11	0414	0453	0492	0531	0569	0607	0645	0682	0719	0755
12	0792	0828	0864	0899	0934	0969	1004	1038	1072	1106
13	1139	1173	1206	1239	1271	1303	1335	1367	1399	1430
14	1461	1492	1523	1553	1584	1614	1644	1673	1703	1732
15	1761	1790	1818	1847	1875	1903	1931	1959	1987	2014
16	2041	2068	2095	2122	2148	2175	2201	2227	2253	2279
17	2304	2330	2355	2380	2405	2430	2455	2480	2504	2529
18	2553	2577	2601	2625	2648	2672	2695	2718	2742	2765
19	2788	2810	2833	2856	2878	2900	2923	2945	2967	2989
20	3010	3032	3054	3075	3096	3118	3139	3160	3181	3201
21	3222	3243	3263	3284	3304	3324	3345	3365	3385	3404
22	3424	3444	3464	3483	3502	3522	3541	3560	3579	3598
23	3617	3636	3655	3674	3692	3711	3729	3747	3766	3784
24	3802	3820	3838	3856	3874	3892	3909	3927	3945	3962
25	3979	3997	4014	4031	4048	4065	4082	4099	4116	4133
26	4150	4166	4183	4200	4216	4232	4249	4265	4281	4298
27	4314	4330	4346	4362	4378	4393	4409	4425	4440	4456
28	4472	4487	4502	4518	4533	4548	4564	4579	4594	4609
29	4624	4639	4654	4669	4683	4698	4713	4728	4742	4757
30	4771	4786	4800	4814	4829	4843	4857	4871	4886	4900
31	4914	4928	4942	4955	4969	4983	4997	5011	5024	5038
32	5051	5065	5079	5092	5105	5119	5132	5145	5159	5172
33	5185	5198	5211	5224	5237	5250	5263	5276	5289	5302
34	5315	5328	5340	5353	5366	5378	5391	5403	5416	5428
35	5441	5453	5465	5478	5490	5502	5514	5527	5539	5551
36	5563	5575	5587	5599	5611	5623	5635	5647	5658	5670
37	5682	5694	5705	5717	5729	5740	5752	5763	5775	5786
38	5798	5809	5821	5832	5843	5855	5866	5877	5888	5899
39	5911	5922	5933	5944	5955	5966	5977	5988	5999	6010
40	6021	6031	6042	6053	6064	6075	6085	6096	6107	6117
41	6128	6138	6149	6160	6170	6180	6191	6201	6212	6222
42	6232	6243	6253	6263	6274	6284	6294	6304	6314	6325
43	6335	6345	6355	6365	6375	6385	6395	6405	6415	6425
44	6435	6444	6454	6464	6474	6484	6493	6503	6513	6522
45	6532	6542	6551	6561	6571	6580	6590	6599	6609	6618
4f	6628	6637	6646	6656	6665	6675	6684	6693	6702	6712
47	6721	6730	6739	6749	6758	6767	6776	6785	6794	6803
48	6812	6821	6830	6839	6848	6857	6866	6875	6884	6893
49	6902	6911	6920	6928	6937	6946	6955	6964	6972	6981
50	6990	6998	7007	7016	7024	7033	7042	7050	7059	7067
51	7076	7084	7093	7101	7110	7118	7126	7135	7143	7152
52	7160	7168	7177	7185	7193	7202	7210	7218	7226	7235
53	7243	7251	7259	7267	7275	7284	7292	7300	7308	7316
54	7324	7332	7340	7348	7356	7364	7372	7380	7388	7396

Logarithms of Numbers.—(*Concluded*)

N	0	1	2	3	4	5	6	7	8	9
55	7404	7412	7419	7427	7435	7443	7451	7459	7466	7474
56	7482	7490	7497	7505	7513	7520	7528	7536	7543	7551
57	7559	7566	7574	7582	7589	7597	7604	7612	7619	7627
58	7634	7642	7649	7657	7664	7672	7679	7686	7694	7701
59	7709	7716	7723	7731	7738	7745	7752	7760	7767	7774
60	7782	7789	7796	7803	7810	7818	7825	7832	7839	7846
61	7853	7860	7868	7875	7882	7889	7896	7903	7910	7917
62	7924	7931	7938	7945	7952	7959	7966	7973	7980	7987
63	7993	8000	8007	8014	8021	8028	8035	8041	8048	8055
64	8062	8069	8075	8082	8089	8096	8102	8109	8116	8122
65	8129	8136	8142	8149	8156	8162	8169	8176	8182	8189
66	8195	8202	8209	8215	8222	8228	8235	8241	8248	8254
67	8261	8267	8274	8280	8287	8293	8299	8306	8312	8319
68	8325	8331	8338	8344	8351	8357	8363	8370	8376	8382
69	8388	8395	8401	8407	8414	8420	8426	8432	8439	8445
70	8451	8457	8463	8470	8476	8482	8488	8494	8500	8506
71	8513	8519	8525	8531	8537	8543	8549	8555	8561	8567
72	8573	8579	8585	8591	8597	8603	8609	8615	8621	8627
73	8633	8639	8645	8651	8657	8663	8669	8675	8681	8686
74	8692	8698	8704	8710	8716	8722	8727	8733	8739	8745
75	8751	8756	8762	8768	8774	8779	8785	8791	8797	8802
76	8808	8814	8820	8825	8831	8837	8842	8848	8854	8859
77	8865	8871	8876	8882	8887	8893	8899	8904	8910	8915
78	8921	8927	8932	8938	8943	8949	8954	8960	8965	8971
79	8976	8982	8987	8993	8998	9004	9009	9015	9020	9025
80	9031	9036	9042	9047	9053	9058	9063	9069	9074	9079
81	9085	9090	9096	9101	9106	9112	9117	9122	9128	9133
82	9138	9143	9149	9154	9159	9165	9170	9175	9180	9186
83	9191	9196	9201	9206	9212	9217	9222	9227	9232	9238
84	9243	9248	9253	9258	9263	9269	9274	9279	9284	9289
85	9294	9299	9304	9309	9315	9320	9325	9330	9335	9340
86	9345	9350	9355	9360	9365	9370	9375	9380	9385	9390
87	9395	9400	9405	9410	9415	9420	9425	9430	9435	9440
88	9445	9450	9455	9460	9465	9469	9474	9479	9484	9489
89	9494	9499	9504	9509	9513	9518	9523	9528	9533	9538
90	9542	9547	9552	9557	9562	9566	9571	9576	9581	9586
91	9590	9595	9600	9605	9609	9614	9619	9624	9628	9633
92	9638	9643	9647	9652	9657	9661	9666	9671	9675	9680
93	9685	9689	9694	9699	9703	9708	9713	9717	9722	9727
94	9731	9736	9741	9745	9750	9754	9759	9763	9768	9773
95	9777	9782	9786	9791	9795	9800	9805	9809	9814	9818
96	9823	9827	9832	9836	9841	9845	9850	9854	9859	9863
97	9868	9872	9877	9881	9886	9890	9894	9899	9903	9908
98	9912	9917	9921	9926	9930	9934	9939	9943	9948	9952
99	9956	9961	9965	9969	9974	9978	9983	9987	9991	9996

APPENDIX H

Resistance of Copper Wire, Ohms per Mile 25°C (77°F)

Size, cir. mils, A.W.G.	Number of wires	Outside diam., mils	Ohms per mile
STRANDED			
500,000	37	814	0.1130
450,000	37	772	0.1267
400,000	37	728	0.1426
350,000	37	681	0.1626
300,000	37	630	0.1900
250,000	37	575	0.2278
0000	19	528	0.2690
000	19	· 470	0.339
00	19	418	0.428
0	19	373	0.538
1	19	332	0.681
2	7	292	0.856
3	7	260	1.083
4	7	232	1.367
SOLID			
0000		460	0.264
000		410	0.333
00		365	0.420
0		325	0.528
1		289	0.665
2		258	0.839
3		229	1.061
4		204	1.335

For more detailed tables, see Part I, pp. 320 and 321.

APPENDIX I
Inductive Reactance per Single Conductor, Ohms per Mile*

Size cir. mils, A.W.G.	60 cycles per sec. Spacing, in.												
	12	24	36	48	60	72	84	96	108	120	132	144	156
STRANDED													
500,000	0.451	0.535	0.584	0.619	0.647	0.669	0.688	0.703	0.718	0.730	0.742	0.752	0.762
450,000	0.458	0.541	0.591	0.625	0.653	0.675	0.693	0.709	0.724	0.736	0.748	0.758	0.767
400,000	0.464	0.548	0.598	0.632	0.660	0.682	0.700	0.716	0.731	0.743	0.755	0.765	0.775
350,000	0.472	0.556	0.606	0.640	0.668	0.690	0.708	0.724	0.739	0.751	0.763	0.774	0.783
300,000	0.482	0.566	0.615	0.650	0.677	0.699	0.718	0.734	0.748	0.760	0.772	0.783	0.792
250,000	0.493	0.577	0.626	0.661	0.688	0.711	0.729	0.745	0.759	0.772	0.783	0.794	0.804
0000	0.503	0.587	0.636	0.672	0.698	0.722	0.739	0.755	0.770	0.782	0.793	0.804	0.814
000	0.517	0.601	0.650	0.685	0.713	0.735	0.754	0.769	0.784	0.796	0.808	0.818	0.828
00	0.531	0.615	0.664	0.699	0.726	0.748	0.767	0.782	0.798	0.810	0.822	0.832	0.842
0	0.546	0.629	0.678	0.714	0.740	0.762	0.781	0.797	0.812	0.824	0.836	0.846	0.856
SOLID													
0000	0.510	0.594	0.642	0.677	0.704	0.726	0.746	0.762	0.776	0.788	0.800	0.810	0.820
000	0.524	0.608	0.656	0.692	0.718	0.740	0.760	0.776	0.790	0.802	0.814	0.824	0.834
00	0.538	0.622	0.670	0.706	0.732	0.754	0.774	0.790	0.804	0.816	0.828	0.838	0.848
0	0.552	0.636	0.684	0.720	0.746	0.768	0.788	0.804	0.818	0.830	0.842	0.852	0.862
1	0.566	0.649	0.698	0.734	0.760	0.782	0.802	0.818	0.832	0.844	0.856	0.866	0.876
2	0.580	0.664	0.712	0.748	0.774	0.796	0.816	0.832	0.846	0.858	0.870	0.880	0.890
3	0.594	0.678	0.726	0.762	0.788	0.810	0.829	0.846	0.860	0.872	0.884	0.894	0.904
4	0.608	0.692	0.740	0.776	0.803	0.824	0.843	0.860	0.874	0.886	0.898	0.908	0.918
5	0.622	0.706	0.754	0.790	0.817	0.838	0.858	0.874	0.888	0.900	0.912	0.922	0.932
6	0.636	0.720	0.768	0.804	0.831	0.853	0.872	0.888	0.902	0.915	0.926	0.936	0.946

* From formula $x = 2\pi f \left(80 + 741.1 \log \dfrac{D}{r} \right) 10^{-6}$ ohms.

APPENDIX J

Allowable Current-carrying Capacities of Conductors in Amperes*

(Based on room temperature of 30°C, 86°F, and number of conductors in raceway or cable)

Size, A.W.G.	Cross-section, cir mils	Types R and RW			Types RP, RPT, RU, SN			Types RH and RHT		
		Conductors in raceway			Conductors in raceway			Conductors in raceway		
		1 to 3 100 %	4 to 6 80 %	7 to 9 70 %	1 to 3 100 %	4 to 6 80 %	7 to 9 70 %	1 to 3 100 %	4 to 6 80 %	7 to 9 70 %
14	4,107	15	12	11	18	14	13	22	18	15
12	6,530	20	16	14	23	18	16	27	22	19
10	10,380	25	20	18	31	25	22	37	30	26
8	16,510	35	28	25	41	33	29	49	39	34
6	26,250	45	36	32	54	43	38	65	52	46
4	41,740	60	48	42	72	58	50	86	69	60
2	66,370	80	64	56	96	77	67	115	92	81
1	83,690	91	73	64	110	88	77	131	105	92
0	105,500	105	84	74	127	102	89	151	121	106
00	133,100	120	96	84	145	116	102	173	138	121
000	167,800	138	110	97	166	133	116	199	159	139
0000	211,600	160	128	112	193	154	135	230	184	161
	250,000	177	142	124	213	170	149	255	204	179
	300,000	198	158	139	238	190	167	285	228	200
	350,000	216	173	151	260	208	182	311	249	218
	400,000	233	186	163	281	225	197	336	269	235
	500,000	265	212	186	319	255	223	382	306	267
	600,000	293	234	. . .	353	282	. . .	422	338	
	700,000	320	256	. . .	385	308	. . .	461	369	
	750,000	330	264	. . .	398	318	. . .	475	380	
	800,000	340	272	. . .	410	328	. . .	490	392	
	900,000	360	288	. . .	434	347	. . .	519	415	
	1,000,000	377	302	. . .	455	364	. . .	543	434	
	1,250,000	409	327	. . .	493	394	. . .	589	471	
	1,500,000	434	347	. . .	522	418	. . .	625	500	
	1,750,000	451	361	. . .	544	435	. . .	650	520	
	2,000,000	463	370	. . .	558	446	. . .	666	533	

CORRECTION FACTOR FOR ROOM TEMPERATURES OVER 30°C

Deg C	Deg F			
40	104	0.71	0.82	0.88
45	113	0.50	0.71	0.82
50	122	0.00	0.58	0.75
55	131	. . . .	0.41	0.67
60	140	. . . .	0.00	0.58
70	158	. . . .	. . . .	0.35

Type Letters: R—code grade rubber; RW—moisture-resistant rubber; RP—performance grade rubber; RPT—performance grade rubber (small-diameter building wire); RU—90 per cent unmilled grainless rubber; SN—solid, flame-retardant, moisture-resistant, synthetic compound; RH—heat-resistant grade rubber; RHT—heat-resistant grade rubber (small-diameter building wire).

Maximum Operating Temperatures: R—50°C (122°F); RW—50°C (122°F); RP—60°C (140°F); RPT—60°C (140°F); RU—60°C (140°F); SN—60°C (140°F); RH—75°C (167°F); RHT—75°C (167°F).

1. Nos. 18 and 16. The allowable current-carrying capacity of No. 18 is 5 amp, except that in heater cords of types AFS, AFSJ, HC, HPD, and HSJ it is 10 amp. The allowable current-carrying capacity of No. 16 is 7 amp, except that in heater cords of types AFS, AFSJ, HC, HPD, and HSJ it is 15 amp.

2. Aluminum Conductors. For aluminum conductors, the allowable current-carrying capacity shall be taken as 84 per cent of those given in the table for the respective sizes of copper conductor with the same kind of insulation.

* From the "National Electrical Code," 1940 ed., with raceway percentage computations taken from "The New Code," General Cable Corporation.

QUESTIONS AND PROBLEMS

PROBLEMS IN TRIGONOMETRY

1. The hypotenuse of a right triangle (Fig. 1(T)) is 26 in., and one leg is 12 in. (*a*) What is the sine of angle A? (*b*) From the tables, find the value of angle A in degrees. (*c*) Determine the cosine of B and its value in degrees. (*d*) From the sine of B, determine side b of the triangle.

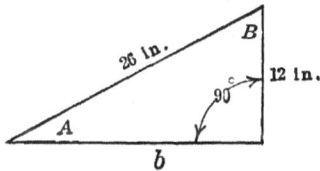

FIG. 1 (*T*).

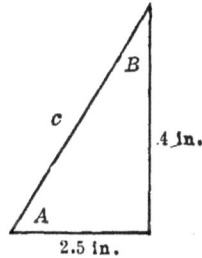

FIG. 2 (*T*).

2. The two legs of a right triangle (Fig. 2(T)) are 4 and 2.5 in. (*a*) Determine angle A. (*b*) Determine angle B. (*c*) Using sine or cosine functions, determine hypotenuse c.

3. Show that $\sin 30° = 0.5$; $\sin 60° = \sqrt{3}/2 = 0.866$; $\tan 60° = \sqrt{3} = 1.732$; $\sin 45° = 1/\sqrt{2} = 0.707$.

4. Given the isosceles triangle abc (Fig. 4(T)), with the side ab equal to 24 in. and the angles at a and b each equal to 68°, find: (*a*) perpendicular dropped from c to ab; (*b*) sides ac and bc; (*c*) angle at c.

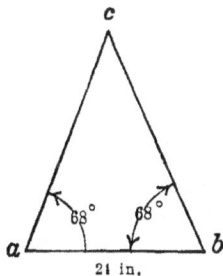

FIG. 4 (*T*).

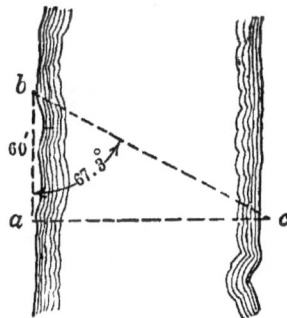

FIG. 5 (*T*).

5. It is desired to measure the width of a river (Fig. 5(T)). From (*a*) on one bank a sight is taken on a small sapling c on the opposite bank, the line

445

ac being practically perpendicular to the river bank. The transit telescope is turned 90°; and a point *b*, 60 ft from *a*, is located. The transit is then set up at *b*, and the sapling at *c* is again sighted. The angle *abc* is found to be 67.3°. Find width *ac* of river.

6. It is desired to determine the height of a precipice *bc* (Fig. 6(*T*)), the face *bc* being practically vertical. From point *a*, 400 ft horizontally from *b* at the base of the precipice, a sight is taken on *c*, and the angle *bac* is found to be 54°. Find height *bc* of precipice.

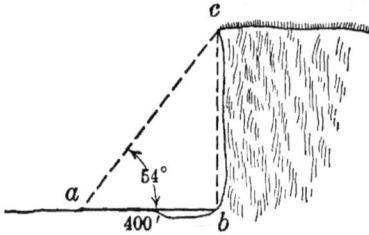

FIG. 6 (*T*).

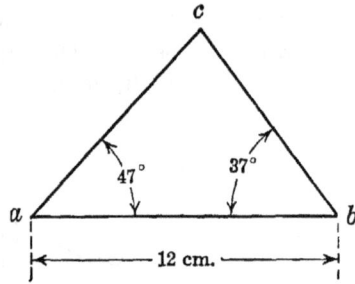

FIG. 7 (*T*).

7. In triangle *abc* (Fig. 7(*T*)), side *ab* is 12 cm, angle *cab* is 47°, and angle *abc* is 37°. Find: (*a*) angle *acb*; (*b*) side *bc*; (*c*) side *ac*.

8. The side *ab* of the triangle *abc* (Fig. 8(*T*)) is 36 cm, side *bc* is 75 cm, and angle *bca* is 20°. Find: (*a*) angle *bac*; (*b*) angle *abc*; (*c*) side *ac*.

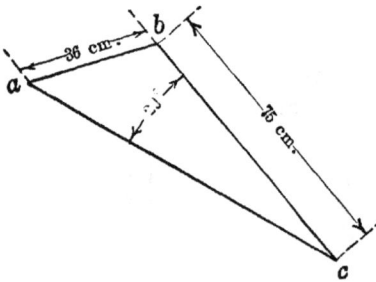

FIG. 8 (*T*).

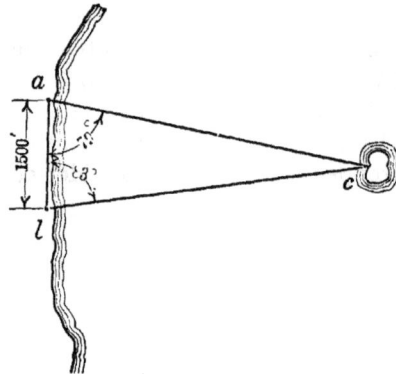

FIG. 9 (*T*).

9. It is desired to determine the distance from the shore line *ab* (Fig. 9(*T*)) to an offshore island *c*. A base line *ab* 1,500 ft long is established along the shore. When *c* is sighted from *a*, angle *bac* is found to be 79°. When *c* is sighted from *b*, angle *abc* is found to be 83°. Find distance of island *c* from both *a* and *b*.

10. In order to measure the width of a river (Fig. 10(*T*)), a base line *ab*, 400 ft long, is established on one bank. When a sapling at *c* on the farther bank is sighted from *a*, angle *bac* is found to be 68°; when this is sighted from b, angle abc is found to be 59°. Find width of river, that is, perpendicular distance from c to ab.

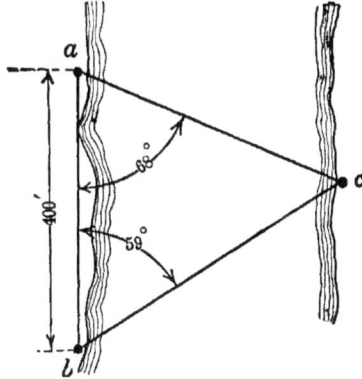

FIG. 10 (T).

11. In order to measure the height of a mountain peak c (Fig. 11(T)) above the surrounding plain, a horizontal base line ab, 2,000 ft long, is established, the base line and the peak lying in the same vertical plane. When peak is sighted from a, angle dac is found to be 21°; and when peak is sighted from b, angle dbc is found to be 34°. Find height cd of peak above surrounding plain.

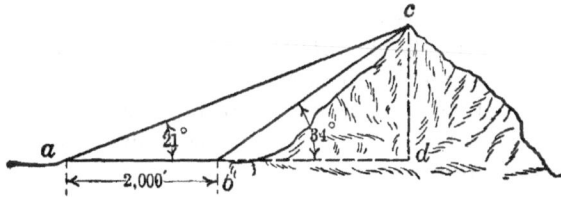

FIG. 11 (T).

12. The sides ac and bc of a scalene triangle are equal, respectively, to 8 and 18.5 in. (Fig. 12(T)), and the included angle is equal to 43°. Find: (a) side ab; (b) angle cab; (c) angle abc.

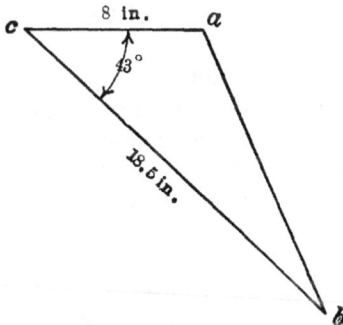

FIG. 12 (T).

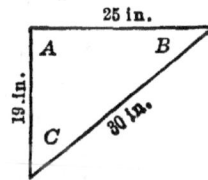

FIG. 13 (T).

13. The three sides of a triangle are equal, respectively, to 19, 25, and 30 in. (Fig. 13(T)). Find the three angles A, B, and C.

14. It is desired to determine the distance *ac* across an inlet (Fig. 14(*T*)). The lengths of two intersecting lines *ab* and *bc* are known to be **2,200** and

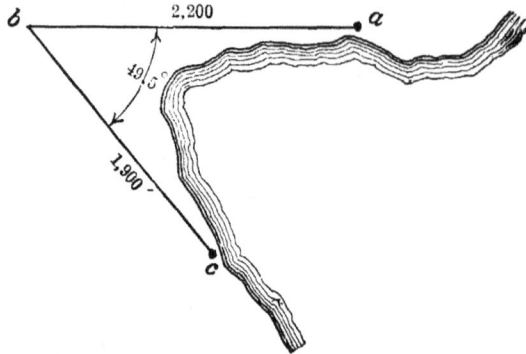

FIG. 14 (*T*).

1900 ft, respectively, and the angle *abc* between *ab* and *bc* is found to be 49.5°. Find distance *ac*.

QUESTIONS ON CHAPTER I

Alternating Current and Voltage

1. Give five reasons for generating electrical energy as alternating current, even although some of this energy must ultimately be converted to direct current. Explain carefully the influence of the commutator and the transformer on the choice of direct or alternating current.

2. What is meant by a sine curve? Show how a sine curve may be plotted with the aid of a table of sines. Show how a sine curve may be plotted using a method of projection.

3. Describe a cosine curve. How may such a curve be determined graphically? By means of cosine tables?

4. Define a radian. How is the value of the radian determined in degrees?

5. What result is obtained if sine or cosine curves, having the same scale of abscissas, are combined by adding their ordinates? If two such curves differ in phase, what is the relation of the amplitude of the resultant curve to the amplitudes of the component curves?

6. Why does a simple plane coil rotating at constant speed in a uniform magnetic field generate an emf that varies as the sine of its angular displacement from the neutral plane? Show that such an emf must also vary sinusoidally with time. What relation exists between degrees and time?

7. Define "cycle," "alternation." How many cycles does a single coil generate per revolution in a two-pole field? In a four-pole field? Derive the equation giving the relation among poles, speed, and frequency.

8. What three frequencies are used in the United States? State the objections to higher frequencies; to lower frequencies. For what class of service are the higher frequencies used; the lower?

9. Describe briefly the construction of commercial alternators. What is meant by "sinusoidal flux distribution"? Show with a simple fundamental relation that with sinusoidal flux distribution a sine emf is generated in the conductors of an alternator.

10. Define "angular velocity." What is its relation to the frequency?

11. Define an a-c ampere. On what basis is it determined? What is meant by the rms, or effective, value of the current? What is the relation of effective to maximum value if the current is sinusoidal?

12. Show how to determine graphically the rms value of any current curve.

13. How is the average current over a half-cycle determined? What is the ratio of average to maximum value for a sine wave of current? What is the ratio of rms to average value?

14. Distinguish vector quantities from scalar quantities. Give examples of each. How are vector quantities added; subtracted? Explain the parallelogram of forces and the polygon of forces.

15. Show that values of an alternating current, varying sinusoidally with time, may be determined at any instant by the projection of a rotating vector on a vertical axis. What determines the length of the rotating vector; its speed in rps?

16. Show the relation between the two rotating vectors which represent two currents differing in phase. How is the angle between the vectors determined?

17. In what fundamental manner may two currents in time phase with each other be added to find their sum? What is the relation of the resultant current to the component currents?

18. In what fundamental manner may two currents differing in phase be added to find their sum?

19. Show that if the two rotating vectors which determine the individual current curves are added vectorially the resultant rotating vector determines in phase and in magnitude the resultant curve found by adding the component current curves.

20. Why may the rms value of the resultant current be similarly determined, the rms values of the component currents being represented by vectors?

21. Show that when two alternators are operating in parallel the load current may be less than the scalar sum of the individual alternator currents.

22. Why is the vector method of adding currents also applicable to voltages?

PROBLEMS ON CHAPTER I
Alternating Current and Voltage

1. From sine tables (p. 436; see also p. 7), plot the values of sin x with values of x in degrees as abscissas. Take values of x at every 30°, that is, sin 30° = ? sin 60° = ? etc.

2. What is the maximum value of the sine function, $y = 20 \sin x$. Using the values of sines from Prob. 1, plot values of y as ordinates with values of x in degrees as abscissas.

3. Determine graphically the sine curve corresponding to $y = 20 \sin x$ (see Fig. 5, p. 7).

4. Plot the values of cos x from cosine tables (p. 436; see also p. 8) with values of x in degrees as abscissas. Take values of x at every 30°.

5. Plot values of y as ordinates for various values of x in the equation $y = 12 \cos x$.

6. Determine graphically the cosine curve corresponding to $y = 12 \cos x$ (see Fig. 6, p. 8).

7. Determine the values in radians of the following angles. (a) 30°; (b) 50°; (c) 90°; (d) 140°; (e) 490°.

8. Determine the values in degrees of the following angles expressed in radians r. (a) $0.4r$; (b) $1.5r$; (c) $3\pi/4r$; (d) $3.67r$.

9. A rotating vector completes 25 revolutions in 1 sec. (a) Through how many radians does it go in one revolution? (b) What is its angular velocity in radians per second? (c) What should be the angular velocity in radians per second of a vector to give 15 rps?

10. The maximum values of two sine curves A and B are 20 units and 30 units, and they differ in phase by 60°, curve B crossing the x-axis to the right of curve A. Plot the two curves, and determine graphically the resultant curve C and the maximum value of C.

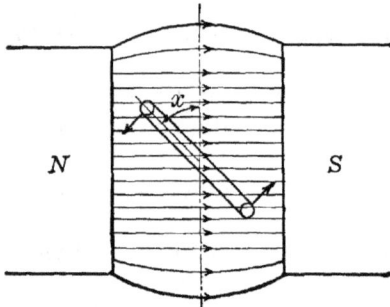

FIG. 11 A.

11. Figure 11A shows a single-turn coil, having an axial length of 30 cm and a breadth of 20 cm and rotating at a speed of 12 rps in a uniform magnetic field having a density of 2,000 gausses. Determine: (a) speed in centimeters per second of the 30-cm sides of the coil; (b) by means of equation $e = Blv10^{-8}$ volts, the induced emf per coil side when the plane of the coil is perpendicular to the magnetic field. Determine induced emf when: (c) plane of coil is parallel to magnetic field; (d) angle x is 45°; (e) angle x is 60°. (f) What is frequency of induced emf?

12. In Prob. 11, plot the emf as ordinates and values of the angle x as abscissas. Also mark a time scale in seconds along the axis of abscissas.

13. In Prob. 11, through how many radians does the emf go in each cycle? Each second? Repeat for a 25- and a 60-cycle emf.

14. The speed of a four-pole alternator is 750 rpm. What is its frequency? Repeat for a six-pole alternator whose speed is 1,200 rpm.

15. The speed of a 60-cycle, motor-driven alternator is 720 rpm. How many poles has it? Repeat for a 60-cycle, 90-rpm, slow-speed alternator.

16. What is the speed in rpm of a 20-pole, 25-cycle, water-wheel alternator? Of a 40-pole, 60-cycle alternator?

17. The flux density under the center of the pole of a 60-kva, 60-cycle, six-pole, 600-volt alternator is 8,000 gausses, the diameter of the armature is 50 cm, and the active length of conductor is 30 cm. The flux density is distributed sinusoidally along the air-gap. Determine: (a) the velocity in cm per sec at which the flux cuts each armature conductor; (*b*) the maximum induced emf per conductor; (*c*) the rms emf.

18. A 60-cycle, sine-wave emf has a maximum value of 120 volts. Determine its instantaneous value when it has gone through (*a*) $\pi/3$ radians after crossing the zero axis in a positive direction; (*b*) 2.5 radians; (*c*) 4.5 radians.

19. The equation of an alternating emf is $e = 250 \sin 377t$ volts. Determine: (*a*) radians per sec through which the emf goes; (*b*) frequency; (*c*) emf when time $t = \frac{1}{240}$ sec.

20. The equation of an alternating current is $i = 28.28 \sin 157t$. Determine: (*a*) frequency; (*b*) current when $t = \frac{1}{150}$ sec.; (*c*) current and time when current has gone through 0.5 radian after having passed through zero in a positive direction.

21. The equation of an alternating emf is $e = 325 \sin 314t$. Determine: (*a*) frequency; (*b*) value of emf when $t = 0.004$ sec; (*c*) value of emf when $t = 0.015$ sec; (*d*) time required for emf to reach value of 281 volts.

22. The maximum value of a sine-wave alternating current is 67.9 amp. Determine its effective, or rms, value. In Probs. 19, 20, and 21 the maximum values of voltage or current are 250 volts, 28.28 amp, and 325 volts. Determine the rms values.

23. Determine the maximum values of the sine-wave voltages whose rms values are 125, 230, and 550 volts.

24. Determine the average values over one-half cycle of the voltages and current in Probs. 19, 20, and 21 (250 volts, 28.28 amp, 325 volts).

25. Figure 25*A* shows a rectified, alternating sine wave of current (consisting of half sine waves), the maximum value of each half-wave being 34 amp (see p. 343). Such a rectified current is frequently used to charge batteries. In electrolytic work the chemical conversion is proportional to the coulombs and hence over any period of time is proportional to the *average* current. What value of steady direct current will have the same effect in charging the battery as the rectified current of Fig. 25*A*?

26. The instantaneous maximum value of a sine-wave current is 25.5 amp. Determine the average power that will be evolved in heating when this current flows in a 5-ohm resistance.

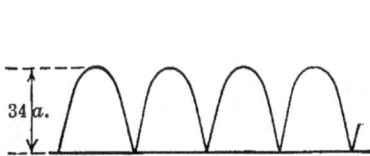

FIG. 25 *A*.

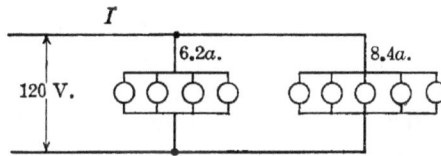

FIG. 28 *A*.

27. When a sine-wave alternating current flows in an 8-ohm resistor, the power is 4,608 watts. Determine the rms and the maximum value of the current. What value of direct current will give the same power?

28. Two lamp loads (Fig. 28*A*) are connected in parallel across 120-volt, 60-cycle mains. One lamp bank takes 6.2 amp (rms), and the other takes 8.4 amp (rms). Since incandescent lamps are practically pure resistance, the current in each lamp bank is in phase with the voltage. Determine:

(*a*) value of total current *I*, supplied by mains; (*b*) maximum value of this current.   (*c*) Sketch the three waves.

29. A resistance, when connected across 100-volt, 60-cycle mains, takes a current I_1 of 3.0 amp (rms) (Fig. 29*A*).   A capacitor connected in parallel

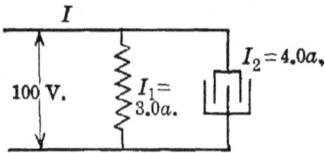

Fig. 29 *A*.

with this resistance takes a current I_2 of 4.0 amp (rms).   The current I_2 leads the current I_1 by 90°.   (*a*) Plot the two currents (see Fig. 29, p. 34).   (*b*) Find resultant current curve by adding ordinates, and determine maximum value of resultant current.   (*c*) Determine resultant rms current by adding vectors proportional in length to rms values of component currents.   (*d*) Compare (*c*) with rms value obtained from (*b*).

30. A resistance connected across 120-volt, 60-cycle mains takes a current I_1 of 4.2 amp (rms) (Fig. 30*A*), and an impedance in parallel with the resistance takes a current I_2 of 5.0 amp (rms) which lags the voltage by 60°.   The current in the resistance is in phase with the voltage.   (*a*) Plot currents as in Fig. 29 (p. 34).   (*b*) By adding ordinates, determine maximum value of resultant current *I*, and compute rms value from maximum value.   (*c*) Determine resultant rms current by finding resultant of rms component current vectors.   Compare (*b*) with (*c*).

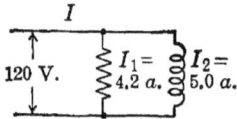

Fig. 30 *A*.

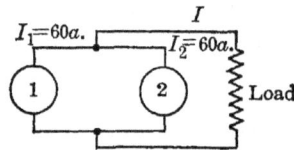

Fig. 31 *A*.

31. Each of two alternators 1 and 2 (Fig. 31*A*) delivers 60 amp (rms) to the load.   Determine rms load current *I* if the two currents differ in phase by 45°, I_1 leading I_2.

32. Determine the resultant rms current *I* when 1 (Fig. 31*A*) delivers 40 amp. (rms) and 2 delivers 60 amp (rms), the 40 amp leading the 60 amp by 45°.

33. Determine the resultant rms current (Prob. 31), assuming that the two currents differ in phase by 30°.

34. A lamp load and a single-phase induction motor are connected across 120-volt, 60-cycle mains (Fig. 34*A*).   The lamp load takes a current I_1 of 8.6 amp in phase with the voltage, and the single-phase motor takes a current I_2 of 7.0 amp which lags the voltage by 60°.   Determine the resultant current *I*.

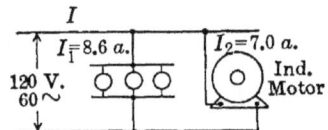

Fig. 34 *A*.

35. The two phases *A* and *B* of a two-phase alternator each generate an emf of 600 volts (rms) (Fig. 35*A*).   The voltage in phase *A* leads the voltage in phase *B* by 90°.   One end of each of the two phases is connected at *C*.   Find the voltage across the other two ends *DE*.

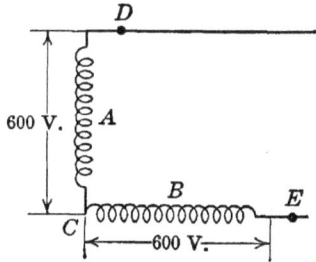

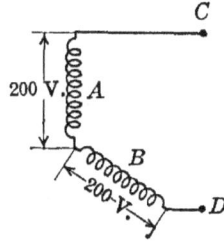

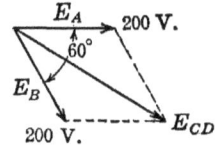

FIG. 35 *A*.   FIG. 36 *A*.

36. The emfs generated in two phases *A* and *B* (Fig. 36*A*) of a three-phase alternator are each equal to 200 volts (rms). These emfs differ in phase by 60°, the emf E_A in phase *A* leading the emf E_B in phase *B*, as shown by the vector diagram. Find the emf E_{CD} across the open ends *CD* of these phases.

37. Phase *B* is reversed (Prob. 36). This makes a phase difference of 120° between the two phases, the emf of phase *B* now leading that of *A*. Find resultant emf across open ends of phases.

HINT. Reverse the vector E_B (Fig. 36*A*).

QUESTIONS ON CHAPTER II

Single-phase Alternating-current Circuits

1. With resistance only in circuit, why must the impressed voltage be equal to the resistance drop? What phase relation exists between current and voltage with resistance only in circuit? What relation exists among current, voltage, and resistance? How are the foregoing relations shown vectorially?

2. Why does inductance have an effect on the flow of alternating current, whereas it has no effect on the flow of steady direct current? How may this be demonstrated experimentally?

3. Show that inductance in opposing change in current prevents an alternating current reaching the value which it would attain were inductance not present.

4. At what instants during the cycle is an alternating current not changing with time? At what instants is it changing at the maximum rate? At what instants is the emf of self-induction zero? A maximum?

5. What phase relation exists between the emf of self-induction and the current? Between the emf of self-induction and the impressed voltage with inductance only in circuit? Between impressed voltage and current?

6. How may the phase relation between current and voltage be shown vectorially?

7. How does current in an inductance vary with voltage? Frequency? Inductance? Give the equation which shows the relation among voltage, current, inductance, and frequency. Define "inductive reactance."

8. Describe a mechanical system which is analogous to an a-c circuit containing inductance only. What factor in the mechanical system corresponds to current? To inductance?

9. What hydraulic system is analogous to a capacitor connected across an alternating voltage? What factor corresponds to voltage? To current? Show that the flow of water *leads* the pressure.

10. A galvanometer is connected in series with a capacitance. Describe the effects that occur when the two are: (a) connected across a d-c power source; (b) short-circuited; (c) connected across the d-c source with reversed polarity; (d) again short-circuited.

11. Demonstrate that the current flowing into a pure capacitance leads the voltage by 90° when the impressed voltage varies sinusoidally with time. Compare this current flow with the flow of water (Question 9).

12. To what three factors is the flow of current in a capacitance proportional? Why?

13. Define "capacitive reactance."

14. Fundamentally, to what two factors is electrical power proportional? Show how this principle may be utilized in determining instantaneous values of power in a-c circuits.

15. Sketch a current, a voltage, and a power curve when current and voltage are in phase. How many times each current cycle is the power zero? What is the frequency of the power curve in terms of that of current and voltage?

16. To what is the average power over a cycle equal when current and voltage are in phase?

17. A battery is connected across a certain circuit and is acting as a source of energy. Under what conditions is the battery power positive? Negative?

18. If the battery is considered as being a translating device, that is, a device which receives energy, under what conditions is the power positive? Negative?

19. Sketch voltage, current, and power curves with inductance only in circuit. How many times each current cycle is the power zero? What is the frequency of the power curve in terms of the frequency of voltage and current?

20. Indicate the periods during which the circuit receives energy from the source (Question 19). Indicate the periods during which the circuit delivers energy to the source. What is the average power over a cycle?

21. Sketch voltage, current, and power curves with capacitance only in circuit. How many times each current cycle is the power zero? What is the frequency of the power curve in terms of the frequency of voltage and current?

22. Indicate the periods during which the circuit receives energy from the source (Question 21). Indicate the periods during which the circuit delivers energy to the source. What is the average power over a cycle?

23. Sketch voltage, current, and power curves when the current lags the voltage by an angle θ, where θ is greater than zero and less than 90°. What is the **frequency** of the power curve in terms of the frequency of voltage and current? How many times each cycle is the power zero?

24. Indicate the periods during which the power is positive and during which it is negative (Question 23). Which periods are of longer duration? Why cannot the average power over the cycle be equal to the product of volts and amperes? Why must it be greater than zero?

25. Define "power factor." Show its use in determining the power.

26. To what is the impedance equal for a circuit containing resistance and inductance in series? In what units is impedance expressed? If voltage and impedance are known, how is current found? Draw the vector diagram.

27. To what ratio is the tangent of the lag angle equal? The cosine?

28. With resistance and inductance in series, account for the power dissipated in the circuit. To what is this power equal in terms of current and resistance? Voltage, current, and power factor?

29. Write the expression which gives the impedance of a circuit containing resistance and capacitance in series. To the ratio of what factors is the tangent of the angle of current lead equal? To what is the cosine of this angle equal? Draw a vector diagram of the circuit.

30. With resistance and capacitance in series, account for the power dissipated in the circuit. To what is this power equal in terms of current and resistance? Voltage, current, and power factor?

31. Derive the general expression for the impedance of a circuit containing resistance, inductance, and capacitance in series, giving this expression in terms of resistance and simple *reactances*. Also, give this expression in terms of resistance, frequency, inductance, and capacitance.

32. To what ratio is the tangent of the phase angle equal? How is it determined whether the current lags or leads?

33. To what expression is the cosine of the phase angle equal?

34. Account for the power dissipated in a series circuit containing resistance, inductance, and capacitance in series. To what is this power equal in terms of current and resistance? Voltage, current, and power factor?

35. With a fixed resistance, under what conditions is the impedance a minimum for a circuit containing resistance, inductance, and capacitance in series? To what is the impedance equal when it is a minimum? How does the value of current when the circuit is in resonance compare with its value when the circuit is not in resonance?

36. Compare the voltage across the inductance with the voltage across the capacitance when the circuit is in resonance. Compare the voltage across the resistance with the voltage impressed across the circuit. What is the power factor of the circuit under these conditions?

37. What is the "natural frequency" of a circuit? To what is it equal? How is it utilized in tuning radio circuits? Show how current varies with frequency in a circuit containing fixed resistance, inductance, and capacitance in series, the voltage of the circuit remaining constant.

38. Explain why it is impossible practically to have pure inductance. Within how many degrees phase angle do practical inductance coils approach pure inductance? How may a practical inductance coil be considered so far as its effect in a-c circuits is concerned?

39. Why do practical capacitors usually come within a fraction of a degree to being pure capacitance?

40. Define "effective resistance." In a circuit consisting of resistance and impedance in series, the circuit voltage, the voltages across the resistance and impedance, and the current are all measured. Show how the vector diagram may be constructed from these data.

41. In a circuit consisting of resistance, inductive impedance, and capacitance in series, the circuit voltage, the voltages across the resistance, the impedance, the capacitance, and the current are all measured. Show how the vector diagram may be constructed from these data, stating any assumptions that may be necessary.

42. Why is the parallel circuit met more commonly in practice than the series circuit? What factor is common to all branches of the parallel circuit, and what position is it convenient to give this factor in vector diagrams of the circuit?

43. Describe a direct method of determining the currents when resistance, inductance, and capacitance are connected in parallel across constant-voltage supply.

44. With resistance and inductance in parallel, to what is the tangent of the power-factor angle equal? The cosine? Repeat for resistance and capacitance in parallel.

45. With resistance, inductance, and capacitance in parallel, what quantities in the vector diagram tend to cancel each other?

46. Under what conditions does antiresonance occur in the parallel circuit? Compare the value of current with the value occurring in the circuit when it is not in antiresonance.

47. Show that, with pure inductance and pure capacitance in parallel, current may flow in the inductance and capacitance and yet, in the line supplying this parallel circuit, the current may be zero.

48. A resistance and an impedance coil are connected in parallel across a constant-voltage source. Draw the vector diagram including line voltage and current, current to the resistance, and to the impedance coil.

49. A resistance, an impedance coil, and a capacitor are connected in parallel across a constant-voltage source. Draw the vector diagram including line voltage and current, current to the resistance, the impedance coil, and the capacitor.

50. Using a vector diagram, define "energy current"; "quadrature current." What is the relation of each to the power of the circuit?

51. Define "reactive power"; "vars"; "kilovars."

PROBLEMS ON CHAPTER II

Single-phase Alternating-current Circuits

38. A 100-watt, Mazda C lamp having a hot resistance of 140 ohms is connected across 120-volt (rms),[1] 60-cycle mains. Determine: (*a*) rms cur-

---

[1] Unless otherwise specified, all values of current and voltage are *effective*, or *rms*, values, and all voltage and current waves are sinusoidal.

rent; (b) maximum instantaneous value of current; (c) equation of current wave.

39. A resistance of 12 ohms and a resistance of 20 ohms are connected in parallel across 110-volt, 25-cycle a-c mains. Determine: (a) rms current to each resistance; (b) maximum instantaneous values of each current in (a); (c) equation of each current in (a); (d) maximum instantaneous value of total current and equation of current.

40. The two resistances of 12 and 20 ohms, Prob. 39, are connected in series across 110-volt, 25-cycle mains. Determine: (a) rms current and its maximum instantaneous value; (b) rms voltage across each resistance; (c) equation of voltage wave across each resistance. (d) Compare the sum of the equations in (c) with the equation for line voltage.

41. A coil having an inductance of 0.1 henry and negligible resistance is connected across a 100-volt, 60-cycle power source. Determine: (a) reactance; (b) current; (c) maximum instantaneous current. (d) Sketch the line emf wave and the current wave with their phase relation.

42. Repeat Prob. 41 with the coil connected across 100-volt, 25-cycle power source.

43. Determine the reactance of a 0.3-henry inductance at a frequency of 50 cycles per sec and at the telephonic frequency of 1,000 cycles per sec. Determine the voltage across the inductance at each frequency, when the current is 0.2 amp.

44. An inductance of 0.25 henry having negligible resistance takes 2.44 amp when connected across a 230-volt power circuit. Determine the frequency of the power circuit.

45. A capacitor, having a capacitance of 0.000050 farad, is connected across a 220-volt, 60-cycle power circuit. Determine: (a) rms current; (b) maximum instantaneous current; (c) equation of current. (d) Sketch the voltage and current waves with their phase relation.

46. A capacitor takes 3.77 amp when connected across a 200-volt, 50-cycle supply. It is then connected across a 200-volt a-c supply the frequency of which is unknown, and the current is now 9.05 amp. Determine the capacitance and the unknown frequency.

47. Determine the reactance of a 25-μf capacitor and a 40-μf capacitor at 60 cycles. What current will these capacitors take when connected in series across a 230-volt, 60-cycle circuit?

48. Determine the reactance of a 2-μf and a 3-μf capacitor at the telephonic frequency of 1,000 cycles. If a 1,000-cycle current of 0.2 amp flows through these capacitors in series, determine the voltage across each capacitor and across the two in series.

49. Determine the current to the two capacitors, Prob. 48, if the two are connected in parallel across 20 volts, 1,200 cycles.

50. In Prob. 38, determine the average power to the lamp. What is the frequency of the power curve and its maximum instantaneous value? Sketch the power wave.

51. An electric flatiron takes 800 watts from a 115-volt, 60-cycle outlet, the current being in phase with the voltage. Determine the current. What is the maximum instantaneous value of the power wave?

52. An electric heating unit takes 1,200 watts when connected to a 115-volt, 60-cycle supply, the current being in phase with the voltage.

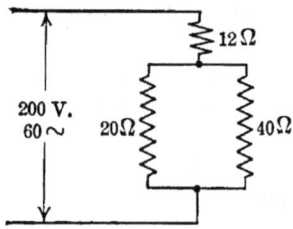

Fig. 53 *A*.

(*a*) Determine the power that the unit takes when connected to a 220-volt, 25-cycle source. (*b*) What is the maximum instantaneous value of the power wave in each case?

53. A resistance of 12 ohms is connected in series with resistances of 20 and 40 ohms in parallel (Fig. 53*A*). The entire circuit is connected across 200-volt, 60-cycle mains. Determine: (*a*) total current; (*b*) total power; (*c*) current in each resistance; (*d*) power dissipated in each resistance.

54. A single-phase, 60-cycle induction motor takes 8.1 amp and 800 watts at 120 volts. Determine; (*a*) power factor at which motor operates; (*b*) apparent power, or volt-amperes.

55. When the motor, Prob. 54, is taking 1,000 watts, the current is 10 amp and the voltage remains at 120 volts. Repeat (*a*) and (*b*), Prob. 54.

56. A 60-cycle, single-phase alternator is rated to deliver 10 kw at 250 volts and 0.8 power factor. Determine: (*a*) its current rating; (*b*) its volt-ampere rating; (*c*) its kva rating.

57. A transformer is taking 42 amp at 0.90 power factor from a 2,300-volt, 60-cycle power circuit. Determine: (*a*) power taken by transformer; (*b*) angle of lag of current.

58. An inductive single-phase load takes 5.6 amp at 230 volts, 25 cycles, and the current lags the voltage by an angle of 36.9°. Determine: (*a*) cosine of angle of lag; (*b*) load power factor; (*c*) power.

59. The current to a distribution feeder is 62 amp at 2,300 volts, 60 cycles, and the current lags the voltage by 31.2°. Determine: (*a*) power factor; (*b*) power; (*c*) kva.

60. A series circuit consisting of a 20-ohm noninductive resistance R and an inductive reactance X_L of 30 ohms at 60 cycles is connected across a 200-volt, 60-cycle supply (Fig. 60*A*). Determine: (*a*) impedance of circuit; (*b*) current; (*c*) voltage across resistance R; (*d*) voltage across reactance X_L; (*e*) vector sum of these two voltages; (*f*) power dissipated in resistance; (*g*) power dissipated in reactance; (*h*)

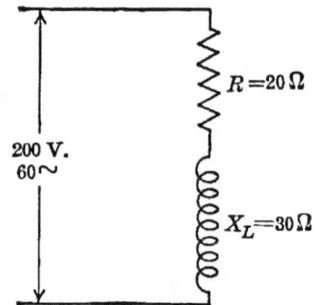

Fig. 60 *A*.

total power taken by circuit; (*i*) volt-amperes; (*j*) power factor; (*k*) angle of phase difference between voltage and current, using both cosine and tangent formulas. (*l*) Draw vector diagram.

61. Repeat Prob. 60 with $R = 30$ ohms and $X_L = 20$ ohms.

62. A resistance of 50 ohms and an inductance of 0.159 henry are connected in series across 120-volt, 60-cycle mains. Determine: (*a*) inductive reactance of circuit; (*b*) impedance of circuit; (*c*) current; (*d*) power; (*e*) power factor; (*f*) lag angle. (*g*) Draw vector diagram.

63. Repeat Prob. 62 with the circuit connected across 120-volt, 25-cycle mains.

64. A telephone relay having a resistance of 300 ohms and an inductance of 0.3 henry is connected in series with a 1,200-ohm resistance unit. The frequency is 796 cycles per sec. (ω = 5,000). Determine: (*a*) reactance of circuit; (*b*) impedance of circuit; (*c*) voltage which must be impressed across circuit to give the 5 ma necessary to operate the relay.

65. The voltage across a noninductive resistance R is 40 volts and that across an inductive reactance X_L is 30 volts, and the current is 2.4 amp when the resistance and reactance are connected in series across a 50-cycle voltage (Fig. 65*A*). Determine: (*a*) resistance of circuit; (*b*) reactance of circuit; (*c*) impedance; (*d*) power; (*e*) power factor; (*f*) power-factor angle; (*g*) circuit voltage. (*h*) Draw vector diagram.

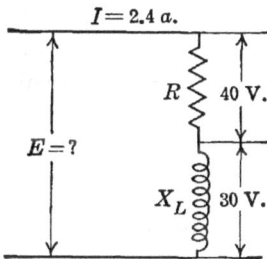

FIG. 65 *A*.

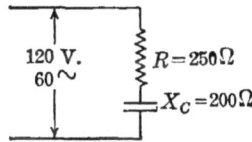

FIG. 67 *A*.

66. A resistance of 8 ohms and an inductance of 0.02 henry are connected in series. It is desired that the current be 12 amp at 60 cycles. Determine: (*a*) necessary voltage; (*b*) power-factor angle; (*c*) power factor; (*d*) power.

67. A 250-ohm resistance R and a capacitive reactance X_C of 200 ohms are connected in series across 120-volt, 60-cycle mains (Fig. 67*A*). Determine: (*a*) impedance of circuit; (*b*) current; (*c*) voltage across resistance; (*d*) voltage across reactance. (*e*) Draw vector diagram.

68. Repeat Prob. 67, (*a*) to (*e*), when this circuit is connected across 120-volt, 25-cycle mains, using the same capacitor.

69. A 40-μf capacitor and a 60-ohm resistance are connected in series; and a current of 2.5 amp, 60 cycles, flows in the circuit. Determine: (*a*) circuit impedance; (*b*) voltage across resistance; (*c*) voltage across capacitor; (*d*) voltage across circuit; (*e*) power-factor angle; (*f*) power; (*g*) power factor. (*h*) Draw vector diagram.

70. An 80-μf capacitor and a resistance of 40 ohms are connected in series across a 60-cycle power supply. The voltage across the resistance is 100 volts. Determine: (*a*) voltage across capacitor; (*b*) voltage at power supply.

71. A 1,000-ohm resistance is connected in series with a 2-μf capacitor across 20-volt, 60-cycle mains. Determine: (*a*) capacitive reactance; (*b*) impedance; (*c*) current.

72. Repeat Prob. 71, substituting 20-volt, 1,000-cycle mains for the 60-cycle mains.

73. An a-c circuit consisting of a capacitor and a resistance in series takes 2.0 amp and 120 watts when connected across 100-volt, 60-cycle mains. Determine: (a) resistance of circuit; (b) impedance; (c) reactance; (d) capacitance of capacitor; (e) power-factor angle.

74. A capacitive reactance of 8 ohms and a resistance of 6 ohms are connected in series across 120-volt, 60-cycle mains. Determine: (a) impedance; (b) current; (c) power; (d) power factor; (e) volt-amperes.

75. In a 2,300-volt, 60-cycle circuit, it is desired to obtain 60 amp, leading the voltage by 90°, by the use of a capacitor. Determine: (a) capacitance of capacitor; (b) volt-amperes; (c) kva.

76. An 80-ohm resistance is connected in series with a capacitive reactance across 25-cycle mains. The power is 1,280 watts, and the power factor is 0.60. Determine: (a) current; (b) reactance; (c) impedance; (d) circuit voltage; (e) volt-amperes. (f) Draw vector diagram.

77. A resistance of 100 ohms, an inductive reactance of 180 ohms, and a capacitive reactance of 80 ohms are all connected in series across 100-volt, 60-cycle mains (Fig. 77A). Determine: (a) impedance of circuit; (b) current; (c) voltage across resistance; (d) voltage across inductive reactance; (e) voltage across capacitive reactance; (f) power; (g) volt-amperes; (h) power factor; (i) power-factor angle. (j) Draw vector diagram.

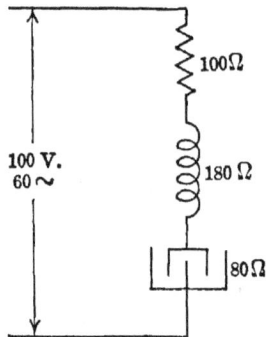

FIG. 77 A.

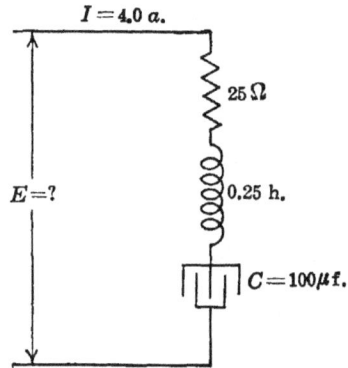

FIG. 79 A.

78. Repeat Prob. 77 with the frequency changed to 25 cycles, the resistance, inductance, and capacitance remaining unchanged.

79. When a resistance of 25 ohms, an inductance of 0.25 henry, and a capacitance of 100 μf are connected in series across a 60-cycle circuit, a current of 4.0 amp flows (Fig. 79A). Determine: (a) inductive reactance in circuit; (b) capacitive reactance; (c) impedance; (d) circuit voltage; (e) power; (f) power factor; (g) power-factor angle; (h) voltage across resistance, inductance, capacitance. (i) Draw vector diagram.

80. Repeat Prob. 79 for a 50-cycle circuit and the same current.

81. A series circuit consisting of 50 ohms resistance, 40 ohms inductive reactance, and a variable capacitive reactance is connected across 120-volt, 60-cycle mains. Determine: (a) value of capacitive reactance necessary to make circuit resonant; (b) current; (c) power; (d) voltage across inductive reactance; (*e*) voltage across capacitive reactance. (*f*) Draw vector diagram.

82. The circuit, Prob. 81, is connected across 120-volt, 25-cycle mains. Determine: (*a*) value of capacitance necessary to produce resonance in circuit. Also, determine: (*b*), (*c*), (*d*), (*e*), (*f*), Prob. 81.

83. A loudspeaker (Fig. 83*A*) has a resistance of 5,000 ohms and an inductance of 2.5 henrys at a frequency of 796 cycles per sec. ($\omega = 5,000$.) (*a*) What capacitance should be connected in series with it in order that the current taken by the loudspeaker at this frequency may be in phase with the voltage? (*b*) If the voltage *E* is 20 volts, what power is taken by the loudspeaker?

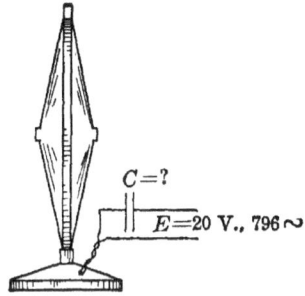

FIG. 83 *A.*

84. A series circuit consists of a resistance of 125 ohms, an inductance of 0.1 henry, and a capacitance of 70.4 µf. (*a*) At what frequency will this circuit be in resonance? If the circuit is connected across a 120-volt source whose frequency is the resonant frequency of the circuit, determine: (*b*) current; (*c*) voltage across inductance; (*d*) voltage across capacitance.

85. An impedance coil (Fig. 85*A*) when connected across 120-volt, 60-cycle mains takes 4.0 amp and 60 watts. Determine: (*a*) impedance;

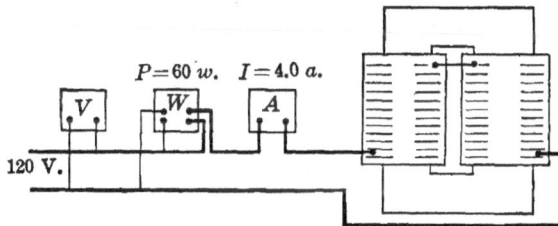

FIG. 85 *A.*

(*b*) power factor; (*c*) effective resistance; (*d*) reactance; (*e*) inductance; (*f*) phase angle by which current lags voltage. (The effective resistance includes iron losses. The iron losses may be considered, however, as increasing the resistance of the coil to give the effective resistance. See p. 82.)

86. When an impedance coil is connected across 230-volt, 25-cycle mains, the angle between voltage and current is 80° and the current is 4.2 amp. Determine: (*a*) volt-amperes; (*b*) power factor; (*c*) power; (*d*) effective resistance; (*e*) reactance; (*f*) inductance.

87. An impedance coil is connected across 208-volt, 60-cycle mains. The power is 150 watts and the power factor 0.12. Determine: (*a*) current; (*b*) volt-amperes; (*c*) effective resistance; (*d*) phase angle.

88. A resistance *R* and an impedance coil *Z′* are connected in series across a 60-cycle supply (Fig. 88*A*). The voltages of the system are measured and are as follows: line voltage, 120 volts; voltage E_R across *R*, 80 volts;

voltage $E_{Z'}$ across Z', 75 volts. The line current I is measured with an
ammeter and found to be 3.5 amp. (a) Draw
vector diagram to scale. Determine: (b) power
factor of circuit; (c) watts; (d) watts to resistance
R; (e) power to impedance Z'; (f) effective resistance of Z'; (g) impedance of Z'; (h) reactance of
Z'. (This problem may be solved analytically (p.
83) or graphically from (a).)

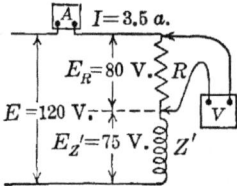

89. Repeat Prob. 88 with $E_R = 75$ volts and
$E_{Z'} = 80$ volts, the other quantities remaining unchanged.

90. In a circuit similar to that of Fig. 88A the data are as follows: $E = 200$
volts, 60 cycles; $E_R = 150$ volts; $E_{Z'} = 125$ volts; $I = 2.5$ amp. Determine:
(a) power factor of circuit; (b) watts; (c) watts to resistance R; (d) power to
impedance coil Z'; (e) effective resistance of Z'; (f) power factor of Z'; (g)
phase angle of Z'; (h) inductance of Z'.

91. A resistance R, a capacitance reactance X_C,
and an impedance coil Z' are connected in series
(Fig. 91A) across a 200-volt, 60-cycle supply. The
following four voltages are measured: $E = 200$ volts;
$E_R = 170$ volts; $E_C = 160$ volts; $E_{Z'} = 220$ volts.
The current is measured and found to be 4.0 amp.
(a) Draw vector diagram to scale. Determine: (b)
vector sum of voltages E_R and E_C; (c) power-factor
angle of circuit; (d) power factor; (e) watts to circuit; (f) watts to resistance
R; (g) watts to impedance Z'; (h) resistance of Z'; (i) impedance of Z'; (j)
reactance of Z'. (This problem may be solved analytically (p. 85) or
graphically from (a).)

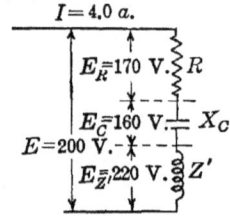

92. Repeat Prob. 91 with the following data: $E = 150$ volts, 60 cycles;
$E_R = 120$ volts; $E_{Z'} = 140$ volts; $E_C = 180$ volts; $I = 5.0$ amp.

93. A 20-ohm resistance and a 20-ohm inductive reactance are connected
in parallel across 100-volt, 60-cycle mains. Determine: (a) current in resistance; (b) current in reactance; (c) resultant current; (d) power to circuit;
(e) volt-amperes; (f) power factor of circuit; (g) power-factor angle. (h)
Draw vector diagram.

94. A resistance of 25 ohms and an inductance of 0.12 henry are connected
in parallel across 100-volt, 60-cycle mains. Determine: (a) current in resistance; (b) current in inductance; (c) resultant current; (d) power to circuit;
(e) volt-amperes; (f) power factor of circuit; (g) power-factor angle. (h)
Draw vector diagram.

95. Repeat Prob. 94 with the resistance and inductance connected in
parallel across 100-volt, 25-cycle mains.

96. A 100-ohm capacitive reactance and a 100-ohm resistance are connected in parallel across 220-volt, 50-cycle mains. Determine: (a) current in
capacitive reactance; (b) current in resistance; (c) resultant current; (d)
power to circuit; (e) volt-amperes to circuit; (f) power factor; (g) power-
factor angle. (h) Draw vector diagram.

97. A 50-ohm resistance and a 60-μf capacitor are connected in parallel
across 120-volt, 60-cycle mains. Determine: (a) current in resistance;

(b) current in capacitor; (c) resultant current; (d) power to circuit; (e) volt-amperes to circuit; (f) power factor; (g) power-factor angle. (h) Draw vector diagram.

98. Repeat Prob. 97 with the resistance and capacitor connected across 120-volt, 25-cycle mains.

99. An 80-ohm resistance R, a 50-ohm inductive reactance X_L, and a 40-ohm capacitive reactance X_C are connected in parallel across 120-volt, 60-cycle mains (Fig. 99A). Determine: (a) current in resistance, in inductive reactance, and in capacitive reactance; (b) resultant current; (c) power to circuit; (d) volt-amperes to circuit; (e) power factor of circuit; (f) power-factor angle. (g) Draw vector diagram.

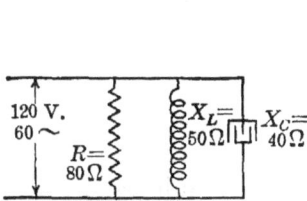

Fig. 99 *A*.

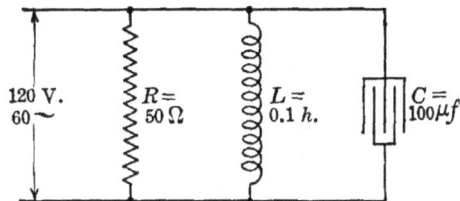

Fig. 100 *A*.

100. A 50-ohm resistance R, an inductance L of 0.1 henry, and a capacitance C of 100 μf are connected in parallel across 120-volt, 60-cycle mains (Fig. 100A). Determine: (a) reactance of inductive branch and current in this branch; (b) reactance of capacitive branch and current in this branch; (c) total circuit current; (d) power taken by circuit; (e) volt-amperes; (f) power factor; (g) power-factor angle. (h) Draw vector diagram.

101. Repeat Prob. 100 with the circuit connected across 120-volt, 25-cycle mains.

102. A resistance of 10 ohms, an inductance of 0.04 henry, and a capacitance of 150 μf are connected in parallel across 220-volt, 50-cycle mains. Determine: (a) current in inductance; (b) current in capacitance; (c) total current to circuit; (d) power; (e) volt-amperes; (f) power factor; (g) power-factor angle. (h) Draw vector diagram.

103. What value of the capacitance in Prob. 102 will cause the circuit to be in antiresonance? Determine the current and the power to the circuit under these conditions.

104. A 40-ohm resistance R and a 50-ohm inductive reactance X_L are connected in parallel across 240-volt, 60-cycle mains (Fig. 104A). It is

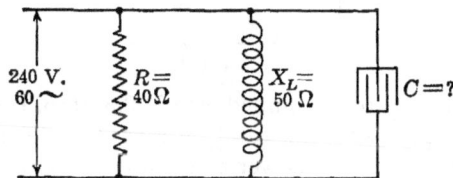

Fig. 104 *A*.

desired that the power factor of the entire circuit shall be made unity by connecting a capacitor C in parallel with the resistance and inductive reactance. Determine: (a) current to inductive reactance; (b) current that capacitor should take; (c) reactance of capacitor; (d) capacitance of capacitor; (e) power to circuit; (f) volt-amperes; (g) power factor. (h) Draw vector diagram.

105. A resistance of 20 ohms and an inductance of 0.4 henry are connected in parallel across 110-volt, 25-cycle mains. Determine: (a) capacitance connected in parallel which will make power factor of circuit unity; (b) current in resistance; (c) current in inductance; (d) current in capacitance; (e) total current; (f) power. (g) Draw vector diagram.

106. A resistance of 10 ohms and a capacitance of 50 µf are connected in parallel across a 120-volt, 50-cycle supply. What value of inductance in parallel with the resistance and capacitance will make the power factor of the circuit unity?

107. An inductance of 0.506 henry and a capacitance of 0.05 µf in parallel are bridged across the two wires of a telephone circuit. Determine: (a) impedance of each at 1,000 cycles; (b) current to each with 10 volts, 1,000 cycles, across the wires; (c) current to circuit; (d) impedance of circuit.

108. A resistance and an impedance coil are connected in parallel across a 110-volt, 60-cycle supply. The current to the resistance is 12 amp, that to the impedance is 12 amp, and the total current to the circuit is 20 amp. (a) Draw vector diagram. Determine: (b) power factor of circuit; (c) power to circuit; (d) power to resistance; (e) power to impedance coil; (f) power factor of impedance coil. (The problem may be solved analytically, p. 93, or graphically from (a).)

109. A resistance and an impedance coil are connected in parallel across a 240-volt, 60-cycle power source. The current to the resistance is 36 amp, the current to the impedance is 24 amp, and the total current is 48 amp. Determine: (a) power factor of circuit; (b) power to circuit; (c) power to resistance; (d) power to impedance coil; (e) power factor of impedance coil; (f) effective resistance of impedance coil; (g) reactance of impedance coil.

110. A resistance, an impedance coil, and a capacitance are connected in parallel across 120-volt, 60-cycle mains (Fig. 110A), and the line voltage and the four currents are measured. The current I_R to the resistance is

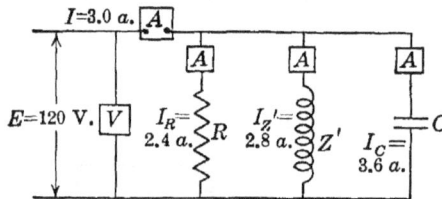

Fig. 110A.

2.4 amp, the current $I_{Z'}$ to the impedance is 2.8 amp, the current I_C to the capacitance is 3.6 amp, and the total current I is 3.0 amp. Determine: (a) power factor of circuit; (b) power to circuit; (c) power to resistance; (d) power to impedance; (e) effective resistance of impedance; (f) reactance of impedance. (g) Draw vector diagram. (The problem may be solved analytically or graphically by means of (g).)

111. Repeat Prob. 110 with the following data: $E = 200$ volts, 60 cycles: $I_R = 3.5$ amp; $I_{Z'} = 7.0$ amp; $I_C = 5.5$ amp; $I = 5.2$ amp.

112. A power load is taking 50 kw at 0.707 power factor, lagging current, from a 250-volt, 60-cycle supply. Determine: (*a*) total current; (*b*) energy current; (*c*) quadrature current; (*d*) vars (reactive volt-amperes).

113. An electric furnace takes 30 kw and 240 amp, lagging current, at 208 volts, 25 cycles. Determine: (*a*) power factor; (*b*) energy current; (*c*) quadrature current; (*d*) vars.

114. The current to a power load is 90 amp at 600 volts, 60 cycles, and the power factor is 0.65, lagging current. Determine: (*a*) power; (*b*) energy current; (*c*) quadrature current; (*d*) vars; (*e*) capacitance of a capacitor which will neutralize the quadrature current.

QUESTIONS ON CHAPTER III

Alternating-current Instruments and Measurements

1. Describe the Siemens dynamometer. Why does the moving coil tend to turn when current flows through the two coils in series? Explain the method of measuring the turning moment of the coil. Explain how the instrument may be adapted to measure voltage, and watts. Why is this type of instrument used only for very special measurements?

2. Describe the indicating electrodynamometer. What electromagnetic principle does its operation illustrate? What makes it more portable than the Siemens dynamometer?

3. How should the coils of an electrodynamometer be wound in order that the instrument may operate as a voltmeter? How should the coils and the instrument be connected in circuit? Explain the use of the instrument for direct current.

4. Why is the voltmeter-ammeter method of measuring power not practicable ordinarily with alternating current? Upon what principle does the wattmeter operate? Explain briefly why the torque of the instrument is nearly proportional to the power.

5. Show how a wattmeter is connected to measure the power taken by its current coil in addition to the measured power. Show how it is connected to measure the power taken by its potential circuit, in addition to the measured power. What is the order of magnitude of the power taken by each coil of the instrument, and how is correction made?

6. Explain how it is possible to burn out either the current coil or the potential circuit of a wattmeter, even if the instrument indication is well within the upper limit of the scale. What means may be used to determine whether or not the current coil or the potential circuit is overloaded?

7. Why is a polyphase wattmeter often desirable? Describe its construction and operation.

8. Describe the construction and operation of the Weston iron-vane type of instrument. How is it adapted to operate as a voltmeter? As an ammeter?

9. Describe the construction and operation of the General Electric inclined-coil, iron-vane type of instrument. In what way does the voltmeter differ from the ammeter?

10. On what principle do thermal-type instruments operate? Indicate the mode of operation of the hot-wire type of instrument. Of the vacuum thermocouple. Under what conditions is the use of thermal instruments advantageous?

11. Why is the d-c watthour meter operative with alternating current? What error is introduced in its use at low power factor?

12. Describe the construction and the connections of the induction watthour meter. Why is a power-factor-compensating, or lagging, coil necessary? What adjustment is made if the meter is slow with lagging current? If it is fast? Repeat for leading current.

13. In what manner is the error due to the friction torque compensated? Explain carefully the operation of the "shading coil."

14. Explain briefly the method by which the driving torque is obtained. The retarding torque.

15. Make a diagram of the connections which would be used in calibrating an a-c watthour meter. State the procedure to follow at unity power factor. At low power factor.

16. State the several advantages of the induction watthour meter over the d-c type.

17. On what mechanical principle does the vibrating-reed type of frequency meter operate? Why is the core of the electromagnet polarized?

18. For what purpose is a synchroscope used? How does the synchroscope indicate that the incoming alternator has the same frequency as the bus-bars? How does it indicate that the proper phase relation exists between the incoming alternator and the bus-bars for connecting in parallel?

19. On what simple principle does the magnetic-type oscillograph operate? What feature of its design makes it possible for it to follow rapid changes in the electric circuit? How is the deflection of the vibrator indicated? In what manner is the time element introduced?

20. Describe the general construction and operation of a laboratory type of oscillograph. How is a photographic record obtained? Show the method of connecting the two vibrators into an electric circuit.

21. Under what conditions does the magnetic type of oscillograph fail to respond accurately to electric-circuit conditions? Why is it possible for the cathode-ray oscillograph to overcome this difficulty?

22. What is the function of the "electron gun"? What is the source of the electrons? What are the functions of the anodes? The control electrode?

23. Explain the electrostatic method of deflecting the cathode-ray beam. What is the "sweep circuit"?

24. In what way may the beam be made to respond to current? What is the limitation of this method? How may very high frequency currents be made to deflect the beam?

QUESTIONS ON CHAPTER IV

Polyphase Systems

1. Give three reasons for the use of polyphase rather than single-phase power. Describe an elementary generator capable of generating three-phase emfs. Why do these three emfs differ in phase by 120 time-degrees? Show that at every instant the algebraic sum of the ordinates of the three emf curves is zero.

2. How may each of the three-phase emfs be connected to the external circuit?

3. Describe briefly symbolic notation, giving the relation of the order of subscripts to the direction of flow of energy. What is the order of the subscripts when two voltages are added?

4. Two coils of a three-phase generator are connected in series, the outer end of one coil being connected to the inner end of the other. What is the relation of the resultant voltage to the voltage of the individual coils?

5. Repeat Question 4 with the two inner ends of the coils connected.

6. Show by means of vector addition (or subtraction) that with a Y-system the line voltage is $\sqrt{3}$ times the coil voltage and also that there is a phase difference of 30° between line and coil voltage.

7. In the Y-system, what relation exists between the line currents and the individual coil currents? Why must the instantaneous algebraic sum and the vector sum of the three coil currents equal zero?

8. At unity power factor, what is the phase relation between coil voltage and coil current? If the power factor is other than unity, what phase relation exists between coil voltage and coil current?

9. At unity power factor, what phase relation exists between coil current and line voltage? Between line voltage and line current? If the power factor is other than unity, what phase relation exists between line voltage and line current?

10. At unity power factor, what power does the Y-system develop in terms of coil voltage and coil current? In terms of line voltage and line current?

11. Repeat Question 10 for power factors other than unity.

12. To what are the volt-amperes of a three-phase system equal? The kva? Give the expression for the power factor in terms of volts, amperes, and watts. What is the relation of the power factor of the system to the power factor of the individual coils?

13. Develop the delta-connection from three coils connected in Y. In the delta-system, what relation exists between the line voltage and the coil voltage?

14. In the delta-system, what is the numerical relation between the coil currents and the line currents? What phase relation exists between any one coil current and a corresponding line current?

15. In the delta-system, what phase relation exists between line currents and line voltages at unity power factor?

16. Derive the equation giving the power in watts of a delta-system in terms of line voltage, line current, and coil phase angle. To what is the kva equal? The power factor?

17. Why is it ordinarily better to connect alternators of large output in Y rather than in delta? Why is this not ordinarily true with small alternators? Give other uses of the Y- and delta-connections.

18. Make a wiring diagram, showing the use of the three-wattmeter method of measuring three-phase power when a neutral is and is not accessible. Under what conditions do all three wattmeters read the same?

19. Make a wiring diagram showing the use of the two-wattmeter method for measuring three-phase power. To what is the total power equal? Under what conditions are the readings of the two wattmeters equal? Under what conditions does one wattmeter read negative?

20. How may the power factor of a balanced three-phase system be determined from the wattmeter readings alone?

21. In what type of three-phase system can the two-wattmeter method *not* be used?

22. How are the emfs of a two-phase system generated? What is their phase relation?

23. What is meant by a "two-phase isolated system?" How are the loads applied? How many different values of voltage are available?

24. How may an interconnected, five-wire, quarter-phase system be obtained from a two-phase system? How many different values of voltage are available?

25. Show the connections of the two-phase, three-wire system. What relation exists among the three line voltages? The three line currents? Why is this system little used?

26. Show a mesh-connected, two-phase system.

27. Make a diagram showing the manner of measuring power in an isolated two-phase system. In an interconnected, two-phase system. In a two-phase, three-wire system.

28. Show diagrammatically a star-connected, six-phase system and a mesh-connected, six-phase system. Give the relations among voltages in the two systems.

PROBLEMS ON CHAPTER IV

Polyphase Systems

115. In (a) (Fig. 115A) are two transformer coils oa and ob connected at o. In (b) is shown the vector diagram giving E_{oa} and E_{ob}, each equal to

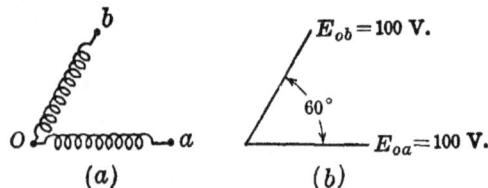

FIG. 115 A.

100 volts, 60 cycles, E_{ob} leading E_{oa} by 60°. (a) By means of subscript notation determine the emf E_{ab} between the open ends ab. (b) Reverse the connection of coil ob, and repeat.

116. In (a) (Fig. 116A) are shown the three coils of a three-phase, 60-cycle alternator, o_1a, o_2b, o_3c, in each of which 300 volts is being induced. The three emfs are shown in the vector diagram in (b). (a) Ends o_1 and o_2

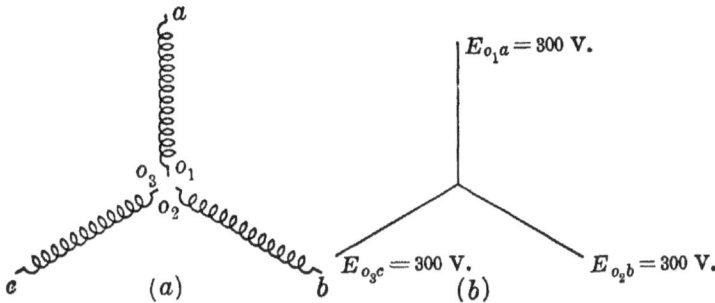

Fig. 116 *A*.

are connected. Determine the emf between ends a and b. (b) End o_3 of coil o_3c is connected to o_1 and o_2. Determine the emf between ends b and c. (c) Under the conditions of (b), determine the emf between ends c and a. (This is the Y-connection.)

117. In Fig. 116A the end o_1 of coil o_1a is connected to end b of coil o_2b. Determine the emf E_{ao_2} between the free ends a and o_2. (This gives the V-connection. See p. 254.)

118. In Fig. 116A the end o_2 of coil o_2b is connected to end c of coil o_3c. Determine the emf E_{bo_3} between the free ends b and o_3.

119. In Fig. 116A the three coils are connected in series, end o_1 being connected to b and end o_2 being connected to c. Determine the emf E_{o_3a} between the free ends o_3 and a. Show that the end o_3 may be connected to end a and form the delta-connection.

120. In Fig. 116A the three coils are connected in series, end o_1 being connected to b and end o_2 connected to o_3. Determine the emf E_{ca} between the open ends c and a.

121. It is desired that a 20,000-kva, three-phase, 60-cycle alternator shall have a rated voltage of 13,800 volts between its terminals. This alternator is to be Y-connected. Determine: (a) voltage rating of each of its coil circuits; (b) kva per coil; (c) ampere rating of each coil and the ampere rating per terminal of the alternator.

122. If the three coils, Prob. 121, are connected in delta, determine: (a) rated voltage of alternator; (b) rated current; (c) rated kva.

123. In Fig. 123A are shown three coils oa, ob, and oc of a 2,000-kva, 25-cycle alternator. In each coil an emf of 1,326 volts is induced, and the phase sequence is E_{oa}, E_{ob}, E_{oc}, E_{ob} lagging E_{oa} by 120° and leading E_{oc} by 120°. (a) With vector E_{oa} horizontal to the right or along the X-axis, draw the vector diagram, and determine E_{ab}, E_{bc}, and E_{ca} both in phase and magnitude. Determine the phase angle between: (*b*) E_{ob} and E_{ab}; (*c*) E_{ob} and E_{bc}. (*d*) Determine the current rating of the alternator.

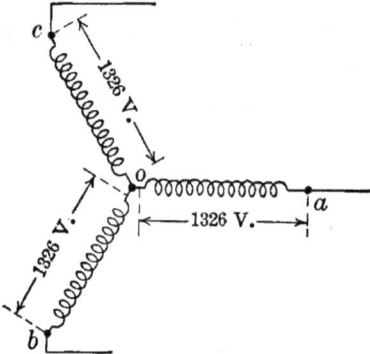

FIG. 123 *A*.

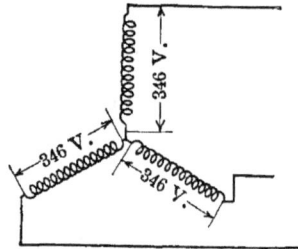

FIG. 124 *A*.

124. Each of three three-phase, Y-connected alternator coils (Fig. 124*A*) generates 346 volts, and the rated current per coil is 83.4 amp. Determine: (*a*) no-load terminal voltage of alternator; (*b*) kva rating per coil; (*c*) kva rating of alternator. (Check by means of Eq. (55), p. 132, using line values.) (*d*) If the alternator operates at unity power factor with the foregoing values of voltage and current, what is its kilowatt output? (Neglect the voltage drop in the armature coils in (*c*) and (*d*).)

125. If the current in each coil lags its voltage by 45° (Prob. 124), determine: (*a*) coil power factor; (*b*) system power factor; (*c*) system output in kilowatts; (*d*) system kva output.

126. In Fig. 126*A* is shown a three-phase power load of 80 kw connected across a 600-volt, three-phase, 60-cycle system. The power factor of the load is 0.75, lagging current. Determine: (*a*) current to the load; (*b*) voltage between each line and neutral *O*; (*c*) angle between each voltage to neutral and current.

127. Repeat Prob. 126 for the same power load with the load power factor equal to 0.80, leading current.

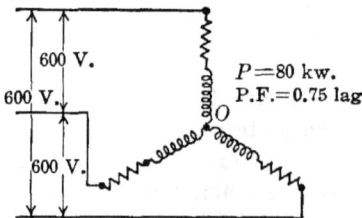

FIG. 126 *A*.

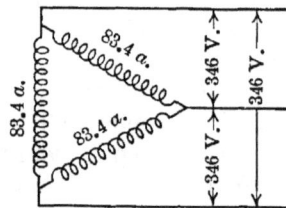

FIG. 129 *A*.

128. A three-phase, 6,000-kva, 11,000-volt, 25-cycle alternator is Y-connected. Determine: (*a*) rated voltage per coil circuit; (*b*) rated current per coil circuit. (*c*) If the coil power factor at rated kva load is 0.8, determine the kilowatt output under these conditions.

129. The three coils, Prob. 124, are connected in delta (Fig. 129*A*), and each coil still delivers 83.4 amp at 346 volts. Determine: (*a*) line voltage;

(*b*) line current; (*c*) kva output of system. (*d*) If the coil current lags its terminal voltage by 45°, what is the kilowatt output of the system?

130. The alternator coils, Prob. 128 are reconnected in delta. Determine: (*a*) rated coil voltage and current; (*b*) rated line voltage and current; (*c*) kva rating. (*d*) At 0.80 power factor, lagging current, what is the kilowatt rating of the alternator?

131. A three-phase delta-connected induction motor is delivering 12 hp, the efficiency is 0.89, and the power factor is 0.85. The input voltage is 208 volts, three-phase, 60 cycles. Determine: (*a*) input to motor in kilowatts; (*b*) input to motor in kva; (*c*) line current; (*d*) coil current.

132. A three-phase, 200-kva, 600-volt, 25-cycle alternator is delta-connected. Determine: (*a*) rated current per coil; (*b*) power factor of alternator if coil current lags coil voltage by 30° at rated kva load; (*c*) kilowatt output of alternator under these conditions.

133. Three 25-ohm resistors are connected in delta across 120-volt, three-phase, 60-cycle mains (Fig. 133*A*). Determine: (*a*) current in each resistor; (*b*) line current; (*c*) total power to three resistors.

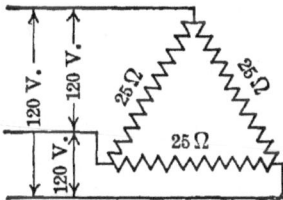

Fig. 133 *A*.

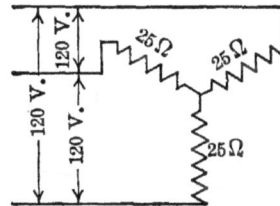

Fig. 134 *A*.

134. The three 25-ohm resistors, Prob. 133, are connected in Y across the same 120-volt, three-phase, 60-cycle mains (Fig. 134*A*). Determine: (*a*) voltage across each resistor; (*b*) current to each resistor; (*c*) total power to the three resistors; (*d*) ratio of power in (*c*) to that in (*c*), Prob. 133.

135. It is desired that the three 25-ohm resistors connected in Y, Prob 134, take the same power as when connected in delta (Prob. 133). What should be the line voltage under these conditions?

136. A 60-cycle, 230-volt, delta-connected alternator (Fig. 136*A*) supplies power to three 5-ohm resistors connected in Y. Determine: (*a*) voltage across each resistor; (*b*) current to each resistor; (*c*) current in each coil circuit of alternator; (*d*) power to load.

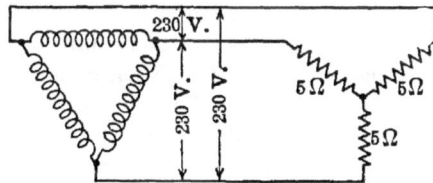

Fig. 136 *A*.

137. Repeat Prob. 136 (*a*) to (*d*) with the three alternator coils connected in Y, each coil generating 230 volts as before.

138. The voltage of each coil of a three-phase, Y-connected alternator (Fig. 138*A*) is 254 volts, 25 cycles, when a load consisting of three delta-connected impedances, each of 8 ohms, is connected across the generator terminals. The ratio of resistance to reactance in each of the impedances is such that each current lags its voltage by 36.9°.　Determine: (*a*) line voltage of system; (*b*) current in each impedance; (*c*) line current; (*d*) power factor of each impedance; (*e*) power to each impedance; (*f*) system power; (*g*) kva of system.

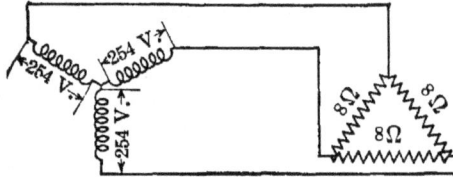

FIG. 138 *A*.

139. The rating of the alternator, Prob. 138, is 250 kva.　It is desired to apply a delta-connected resistance load so that the alternator is loaded to its rating (now in kilowatts).　Determine: (*a*) rated line current; (*b*) current in each delta-connected resistor; (*c*) resistance of each resistor.

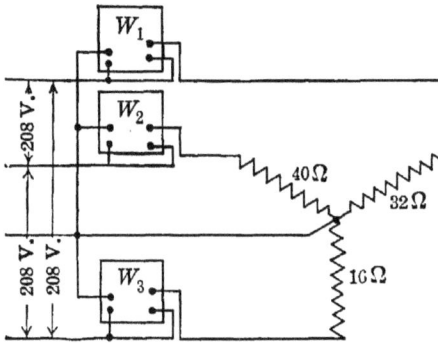

FIG. 140 *A*.

140. Figure 140*A* shows a three-phase, four-wire, 60-cycle system operating with 208 volts between the three line wires.　The voltages are balanced.　Three resistors of 32, 40, and 16 ohms are connected between line wires and neutral.　Determine: (*a*) voltage to neutral; (*b*) current in the 32-, 40- and 16-ohm resistors; (*c*) reading of W_1, W_2, W_3.

141. Repeat Prob. 140 with the three resistors connected to the neutral of a balanced, symmetrical, three-phase, four-wire system in which there is 440 volts between the three line wires.

142. The two-wattmeter method is used to measure the power taken by a balanced Y-connected load connected across 220-volt, 60-cycle mains (Fig. 142*A*).　Each unit of the load consists of a 20-ohm non-inductive resistor.　Determine: (*a*) voltage to neutral; (*b*) power taken by each of resistor units; (*c*) total power taken by load; (*d*) reading of each wattmeter.

143. In Prob. 142, with the three resistors connected in delta across the same three-phase mains, determine: (*a*) current to each resistor; (*b*) power to each resistor; (*c*) total power; (*d*) reading of each wattmeter.

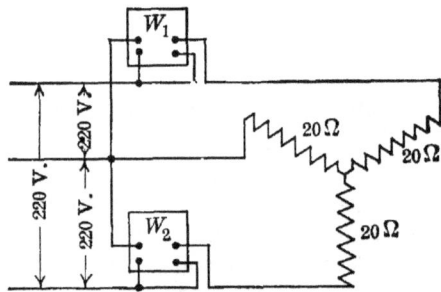

FIG. 142 *A*.

144. In Fig. 144*A* a delta-connected lamp load (resistance) is connected across 120-volt, 60-cycle mains, and the power is measured by the two wattmeters W_1 and W_2. The current to each load is 32 amp. Determine: (a) line current; (b) reading of each of the two wattmeters W_1 and W_2.

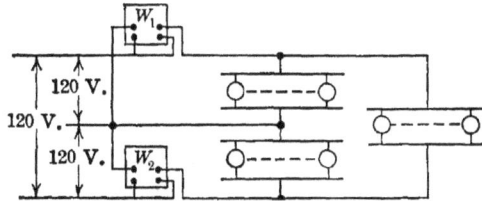

FIG. 144 A.

145. The input to a 230-volt, three-phase induction motor is measured by the two-wattmeter method (Fig. 145A). The load is balanced. Near rated load on the motor, wattmeter W_1 reads 6,900 and wattmeter W_2 reads 2,900 watts. Both instruments are known to be reading upscale. Determine: (a) total power taken by motor; (b) power factor of motor using the curve of Fig. 115 (p. 143); (c) volt-amperes; (d) reading of the three ammeters.

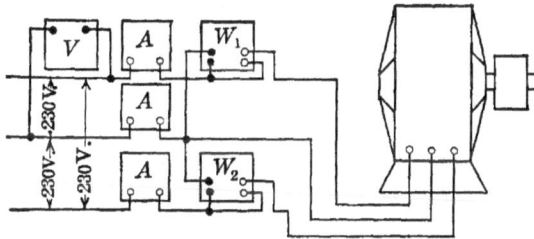

FIG. 145 A.

146. In Prob. 145 the load on the motor is reduced, and W_1 now reads 3,720 watts and W_2 reads 520 watts. In order that W_2 may read upscale, it is found necessary to reverse its current-coil connection. Determine: (a) power factor of motor, using Fig. 115 (p. 143); (b) volt-amperes; (c) line current. The motor is delta-connected. (d) Determine phase angle between coil voltage and coil current.

147. In Probs. 145 and 146 the load on the motor changes so that W_1 reads 4,500 watts and W_2 reads zero. Determine: (a) power factor of the motor, using Fig. 115 (p. 143); (b) volt-amperes; (c) line current; (d) phase angle between coil voltage and coil current.

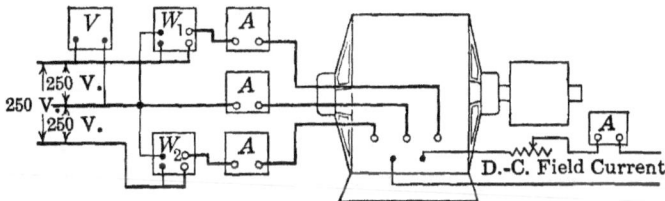

FIG. 148 A.

148. A synchronous motor is an alternator operated as a motor. The power factor of the motor may be controlled by varying the d-c field excitation, underexcitation causing the motor to take a lagging current and over-excitation causing the motor to take a leading current. With normal excitation, which is intermediate, the power factor is unity. (See p. 319.) In Fig. 148*A* is shown a delta-connected, 250-volt, three-phase, 60-cycle synchronous motor with two wattmeters W_1 and W_2, the three ammeters *A*, and a voltmeter *V* to measure the input.

The motor is underexcited; that is, the current lags. Wattmeter W_1 reads +8,600 watts, and W_2 reads +5,160 watts. From the curve, Fig. 115 (p. 143), determine: (*a*) power factor at which motor operates; (*b*) total watts; (*c*) current; (*d*) coil current.

149. The field of the motor, Prob. 148, is weakened, and the reading of W_1 becomes +11,460 watts and that of W_2 becomes +2,290 watts. Repeat (*a*) to (*d*), Prob. 148.

150. The field of the motor, Prob. 148, is weakened still further. The reading of W_1 now becomes +19,670 watts; that of W_2, −5,910 watts. Repeat (*a*) to (*d*), Prob. 148.

151. The motor, Prob. 148, is overexcited and the current leads. The reading of W_1 now becomes +3,930 watts; that of W_2, +9,840 watts. Repeat (*a*) to (*d*), Prob. 148.

152. With the motor, Prob. 148, still overexcited, W_1 reads −3,440 watts, and W_2 reads +17,200 watts. Repeat (*a*) to (*d*), Prob. 148.

153. A two-phase, three-wire system is shown in Fig. 153*A*. Phase *AB* and phase *BC* are connected at *B*. The voltage of each phase is 220 volts, the current in each phase is 40 amp, and the power factor of each phase is unity. The voltage and current in phase *AB* are in quadrature with the voltage and current in phase *BC*. Determine: (*a*) voltage across outer wires *A*-*C*; (*b*) current in common wire *BB'*; (*c*) volt-amperes of system.

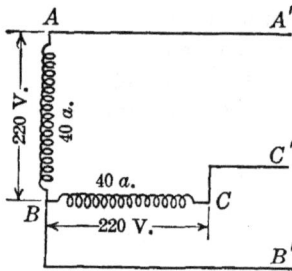

FIG. 153 *A*.

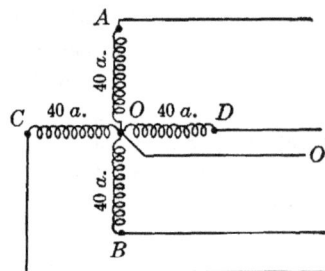

FIG. 155 *A*.

154. Repeat Prob. 153, with the voltage across each phase *AB* and *BC* equal to 120 volts and with the current in each phase equal to 60 amp.

155. Phase *AB* and phase *BC*, Prob. 153, are connected at their neutral points *O* (Fig. 155*A*), and a neutral wire *OO'* is also connected. This constitutes a quarter-phase, five-wire system. The voltages across *AB* and *CD* are each equal to 220 volts. Determine: (*a*) voltage from each outer wire *A*, *D*, *B*, *C* to neutral *O*; (*b*) voltage across adjacent outer wires *A*-*D*, *D*-*B*, *B*-*C*, *C*-*A*; (*c*) kva output of system when current in each phase is equal to 40 amp.

156. Repeat Prob. 155, with the voltages across AB and CD each equal to 120 volts and the current in each phase equal to 60 amp.

157. A two-phase, mesh-connected system is shown in Fig. 157A. Voltage E_{AB} is in phase with E_{DC}, and voltage E_{BC} is in phase with E_{AD}. The voltages E_{AB} and E_{DC} are in quadrature with voltages E_{BC} and E_{AD}. The voltage per coil is 110 volts, 60 cycles, and the current per coil is 40 amp. Determine: (a) voltage across the diametrical connections A-C and B-D; (b) current in each line wire; (c) volt-amperes of system.

158. Repeat Prob. 157, with voltages E_{AP}, E_{BC}, E_{CD}, E_{DA} each equal to 162.5 volts and current per coil equal to 61.5 amp.

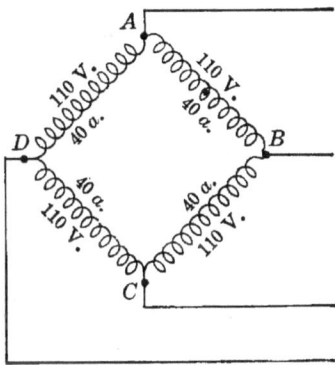

FIG. 157 A.

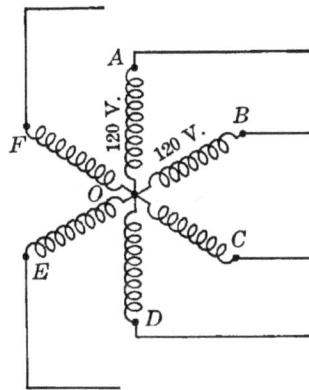

FIG. 159 A.

159. Figure 159A shows a star-connected, six-phase system. There is 120 volts, 25 cycles, from each coil terminal A, B, C, D, E, F to neutral O. Determine: (a) voltage between adjacent line terminals A-B, B-C, etc.; (b) voltage between diametrically opposite terminals A-D, B-E, etc.; (c) voltage between alternate terminals A-C, C-E, etc.

160. Repeat Prob. 159, with 300 volts, 25 cycles, from each coil terminal A, B, C, D, E, F to neutral O.

QUESTIONS ON CHAPTER V

The Alternator

1. In what way does the relation of armature to field in an alternator differ from the relation in a d-c generator? Why is this relation possible with the alternator whereas to have the field rotate is not practicable ordinarily with d-c dynamos?

2. Give three advantages of a stationary over a rotating armature.

3. Describe the method by which the field induces alternating emfs in the stator, or armature, coils. On what fundamental laws are these emfs based? Apply Fleming's right-hand rule to the stationary-armature type of alternator.

4. What two fundamental principles govern the design of alternator windings?

5. Why are simple single-phase windings little used in alternators?

6. Sketch a four-pole, single-phase lap winding in which there are four slots per pole. Repeat for a wave winding. Compare the emfs of the two windings, the number of conductors, etc., in each being the same.

7. Show the difference between half-coil and whole-coil windings. Why are armatures having but one slot per pole rarely used?

8. In what way does the spiral winding differ from the lap and the wave winding? What is meant by "single range"?

9. Show how a simple two-phase winding may be evolved from two single-phase windings. Sketch such a winding.

10. Sketch a four-pole, two-phase spiral winding having four slots per pole per phase. How many ranges must such a winding have?

11. Sketch a four-pole, two-phase, full-pitch lap winding having eight slots per pole or four slots per pole per phase.

12. In what way does the three-phase winding differ from the single-phase winding? What relation exists between the coil sides of adjacent phases?

13. In what way does a fractional-pitch winding differ from a full-pitch winding? How do the coil sides in the top layer of a fractional-pitch winding compare with those of a similar full-pitch winding. How do the coil sides in the bottom layers of the two windings compare?

14. What are the advantages of fractional-pitch windings?

15. Into what two general classes are rotating-field types of alternator divided? What are the advantages of the cylindrical type, and what is the range of speed at which it operates?

16. Why is it necessary that the stators of alternators be made of laminations? Describe the construction of a slow-speed alternator. How are the stator stampings held to the frame?

17. In what way does the construction of turbine-driven alternators differ from that of low-speed alternators? Why are ventilating ducts through the stator iron necessary? How are the coil ends braced, and why is this bracing necessary?

18. Sketch two types of alternator slot. How are the conductors held in the slot? What is the advantage of each type, and under what conditions is each used?

19. How are the field poles of low-speed alternators constructed? What two methods are used to hold them to the spider?

20. Describe the construction of moderate-speed rotors. How are the poles held to the rotor structure?

21. Why is it not possible to use salient poles with high-speed turbine-driven alternators? Of what material are the rotors of high-speed turbine-driven alternators made?

22. Compare the parallel-slot with the radial-slot rotor. Where is each used?

23. What is meant by a distributed-field winding? How does the flux distribution with this type of rotor compare with that of the ordinary salient-pole rotor?

24. Why does the construction of turbine-driven alternators make it particularly difficult to secure adequate ventilation? Describe the modern method of cooling turbine-driven alternators. What are the advantages of a totally enclosed system?

25. State the advantages of hydrogen as a cooling medium.

26. To what three factors is the induced emf in an alternator proportional? What is meant by "pitch-factor"? "Belt-factor"?

27. Describe in detail the manner of "phasing" alternator coils in Y. In delta.

28. What factor ordinarily determines the rating of electric machinery? Why is the output of an alternator determined by its *current* output rather than by its *kilowatt* output? What factor determines the rating of the prime mover?

PROBLEMS ON CHAPTER V

The Alternator

161. In Fig. 161*A* is shown a stator and a four-pole rotor. There are six slots per pole on the stator. Sketch a single-layer lap winding: (*a*) with the two conductor belts in series; (*b*) with the two conductor belts in parallel. (See Fig. 129 (*a*), p. 156.)

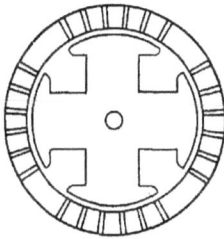

FIG. 161 *A*.

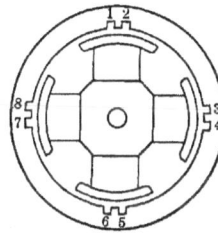

FIG. 163 *A*.

162. Sketch a single-layer wave winding adapted to the stator of Fig. 161*A*. (See Fig. 129(*b*), p. 156.)

163. Figure 163*A* shows the stator and field structure of a four-pole alternator. There are eight slots, 1 to 8. Draw a half-coil, single-phase winding, adapted to this stator, in which there is a total of four coils.

164. Draw a whole-coil, single-layer winding adapted to the stator, Fig. 163*A*, in which there is a total of four coils.

165. Draw a two-layer, whole-coil winding with a total of eight coils for the stator of Fig. 163*A*.

166. In Fig. 166*A* is shown a stator similar to that in Fig. 163*A*, except that there are four slots per pole. Draw a two-layer, whole coil or lap winding that will give two coil sides in every slot.

167. In the stator, Fig. 163*A*, draw a two-layer wave winding that will give two coil sides in every slot.

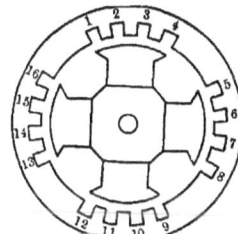

FIG. 166 *A*.

168. In the stator, Fig. 163*A*, draw a single-range spiral winding in which there are two coils per pole. (See Fig. 133, p. 159.)

169. In a stator, similar to that shown in Fig. 163*A*, except that the eight slots are spaced uniformly, draw a two-phase, whole-coil, two-layer winding.

170. Figure 170*A* shows 23 slots, each containing two coil sides of an alternator winding having 8 slots per pole. Connect these coil sides so as to give a portion of a two-phase, half-coil, full-pitch lap winding. The

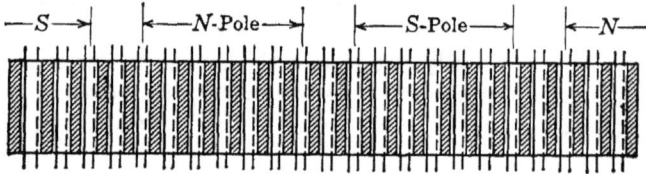

Fig. 170 *A*.

conductors shown solid lie in the tops of the slots, and the conductors shown dotted lie in the bottoms of the slots.

171. Repeat Prob. 170, with connections to give a wave winding.

172. Draw a portion of a two-phase, two-layer, ¾-pitch lap winding for an alternator having eight slots per pole as in Fig. 170*A* (see Figs. 136 and 140, pp. 162 and 165).

173. In a stator similar to that shown in Fig. 163*A*, there are 12 slots equally spaced around the stator. Draw a three-phase, two-layer lap-winding. (See Fig. 137, p. 163.)

174. Draw a portion of a three-phase, two-layer, full-pitch lap winding for an alternator having nine slots per pole. (See Fig. 138, p. 164.)

175. Repeat Prob. 174, making the winding ⅞ pitch.

176. Draw a portion of a three-phase, two-layer, ¾-pitch lap winding for an alternator having 12 slots per pole.

Note.—In the following problems the flux may be considered as sinusoidal along the air-gap.

177. Each of the coils of an alternator, Fig. 177*A*, is linked by 8,000,000 maxwells when the pole centers are directly under the coil centers as in Fig. 177*A*. Consider coil *a* which is being linked by the flux from the

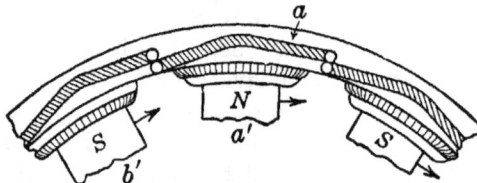

Fig. 177 *A*.

N-pole *a'*. In a half-cycle an *S*-pole *b'* will be directly under this same coil, and the flux linking coil *a* has been completely reversed in direction. Hence, in a quarter-cycle, when each pole center is directly under a coil side, *no* flux links the coil *a* (also see Part I, p. 208, Fig. 181). This is a 60-cycle generator so that in $\frac{1}{240}$ sec after the maximum flux links the coil, no flux links the coil. Determine: (*a*) total change of flux linking coil during each quarter-cycle; (b) average induced emf in coil during quarter-cycle; (c) total change of flux linking coil a during each half-cycle; (d) average emf induced in coil a during each half-cycle. Ratio of rms to average emf is 1.11 (see p. 25). (e) Determine rms emf in coil a during each half-cycle. (f) Will the induced average emf over a complete cycle differ from that over a quarter-cycle? A half-cycle? Explain your answer.

178. Repeat Prob. 177 for a 25-cycle alternator, the value of the flux remaining unchanged.

179. In a four-pole, single-phase, 50-cycle alternator, there is one slot per pole, there are 20 conductors per slot, and a half-coil winding is used. The resulting two armature coils are connected in series. A flux of 2,800,000 maxwells enters the armature from each N-pole. Determine the emf induced in the armature.

180. In a four-pole, two-phase, 60-cycle alternator, there is one slot per pole per phase. There are 24 conductors per slot, and a flux of 2,000,000 maxwells enters the armature from each N-pole. (a) Determine induced emf per phase. (b) If the two phases are connected to give a two-phase, three-wire system, what is the emf across the two free coil ends (see Fig. 119, p. 146)?

181. In a six-pole, two-phase, 50-cycle alternator, there are three slots per pole per phase and there are eight conductors per slot. The three slots per pole give a belt factor k_b of 0.91. The winding is full-pitch. A flux of 2,700,000 maxwells enters the armature from each N-pole. Determine: (a) induced emf per phase; (b) emfs available between alternator terminals, if phases are star-connected.

182. In an eight-pole, two-phase, 25-cycle alternator, there are four slots per pole per phase, and 24 conductors per slot. The winding is full-pitch. A flux of 6,000,000 maxwells enters the armature from each N-pole. The belt factor k_b is 0.91. Determine induced emf per phase.

183. Determine induced emf (Prob. 182) for a $\frac{5}{6}$-pitch winding, the pitch factor k_p being 0.92.

184. In a three-phase, six-pole, 60-cycle, delta-connected alternator, there are 12 slots per pole, or 4 slots per pole per phase, and there are 16 conductors per slot. A $\frac{5}{6}$-pitch winding is used. A flux of 4,800,000 maxwells enters the armature from each N-pole. The belt factor k_b is 0.958, and the pitch factor k_p is 0.966. Determine induced emf between each pair of terminals.

185. In a three-phase, Y-connected, 6,600-volt, 6,000-kva, water-wheel-driven alternator there are 30 poles, and the speed is 240 rpm. There are 360 armature slots with a full-pitch, two-layer winding and two conductors per slot. The belt factor k_b is 0.958. The flux is 12,600,000 maxwells per pole. Determine induced emf between: (a) line terminals and neutral; (b) line terminals.

186. Repeat Prob. 185, with a $\frac{5}{6}$-pitch winding and a flux of 13,200,000 maxwells per pole. $k_p = 0.966$.

187. Repeat Prob. 185, with a $\frac{3}{4}$-pitch winding and a flux of 14,000,000 maxwells per pole. $k_p = 0.924$.

188. The rating of a 25-cycle, three-phase, 6,600-volt, Y-connected, two-pole alternator is 4,000 kw at 0.8 power factor. Determine: (*a*) kva rating; (*b*) rated current per terminal; (*c*) rated coil voltage; (*d*) speed in rpm.

189. A Y-connected, 60-cycle, three-phase, four-pole, turbine-driven alternator is rated at 13,800 volts and 839 amp per terminal on an 0.8 power-factor basis. Determine: (*a*) kva rating; (*b*) power rating; (*c*) horsepower of driving turbine if alternator efficiency, exclusive of field loss, is 96 per cent; (*d*) speed in rpm.

190. A three-phase, 60-cycle, Y-connected, turbine-driven alternator delivers 12,000 kw at 13,200 volts and 0.8 power factor. Determine: (*a*) kva rating; (*b*) current rating; (*c*) turbine output in horsepower when alternator delivers rated load if alternator efficiency is 0.955, exclusive of field loss.

QUESTIONS ON CHAPTER VI

Alternator Characteristics and Operation

1. Why is the variation of voltage with load in alternators ordinarily much greater, proportionately, than the variation of voltage with load in d-c generators? Why does the selection of system apparatus make it desirable to know something of the characteristics of the alternators connected to the system? What three factors cause the terminal voltage of alternators to drop with increase of load if the current is not leading?

2. Make a sketch showing that considerable flux links the conductors embedded in the slots of an alternator armature. Show that this flux also links the armature coils. What flux, other than that which crosses the slot, links the armature coils?

3. How is armature inductance affected by slot dimensions? How is armature reactance affected by slot dimensions? How is armature reactance affected by frequency?

4. Give three reasons why the resistance of the alternator armature to alternating current is greater than it is to direct current. What is meant by "effective resistance"? Approximately what is the ratio of effective to ohmic resistance in commercial alternators?

5. Show with a diagram that when the armature current is in phase with the no-load induced emf, its mmf distorts the field of the alternator but alters but slightly the magnitude of the field. In what direction, relative to the rotation of the armature coil, is the field crowded?

6. Show with a diagram that when the armature current lags the no-load induced emf by 90°, its mmf weakens the alternator field.

7. Show with a diagram that when the armature current leads the no-load induced emf by 90°, its mmf strengthens the alternator field.

8. What voltage drop, other than the resistance drop, exists in the alternator armature? How is the armature impedance drop found?

9. If the armature terminal voltage, the current, the phase difference between current and terminal voltage, and the resistance and reactance of the armature are known, how may the *induced* emf be found?

10. By means of vector diagrams, compare the relative magnitudes of terminal voltage and induced emf when the current is in phase with the terminal voltage. When the current lags the terminal voltage. When the current leads the terminal voltage.

11. Why is the induced emf in an alternator not equal to the no-load emf, even with constant field excitation? Define "alternator regulation." What two effects combine to give poor regulation with lagging current?

12. Give two reasons why the rated-load terminal voltage may exceed in magnitude the no-load emf when the current leads.

13. Make a diagram of connections which would be used to determine the saturation curve of a three-phase, delta-connected alternator. A three-phase, Y-connected alternator. Sketch the saturation curve.

14. Make a diagram of connections which would be used for determining the voltage-load characteristics of a three-phase alternator. Discuss the methods which can be used in connecting the load. What difficulties are encountered in making such load tests?

15. Show with sketch the approximate voltage-load characteristics of an alternator when the current is in phase with the terminal voltage. When the current lags the terminal voltage. When the current leads the terminal voltage.

16. Why is a method, such as the synchronous-impedance method for determining the operating characteristics of an alternator, desirable? Show that armature reaction and armature reactance have the same general effect on the alternator terminal voltage.

17. Draw the synchronous-impedance-method vector diagram, and indicate the part of the diagram which replaces armature reaction.

18. Describe the method of determining synchronous impedance, that is, the short-circuit and the open-circuit tests. Why is it desirable that the armature current should be approximately twice its rated value?

19. In what manner does the application of the method to a delta-connected alternator differ from its application to a Y-connected alternator?

20. Why is the synchronous-impedance method called the "pessimistic" method? State briefly the factors that give values of regulation that are too high.

21. Sketch the field-discharge switch with its connections, and state the reason for its use.

22. Discuss the use of the Tirrill regulator in maintaining constant the terminal voltage of alternators.

23. Show that a change in field excitation cannot change appreciably the division of power load between alternators operating in parallel. What factors do determine the division of power load between alternators in parallel?

24. How may the division of kilowatt load between alternators operating in parallel be changed? Why are alternators operating in parallel in stable equilibrium? What occurs when the driving torque of one alternator is entirely removed?

25. Describe in general the effects which occur with two alternators operating in parallel if the field of one is strengthened and the field of the other is weakened.   In which machine does the current lead more?   Lag more?

26. What effects occur in each of the two alternators (a) which tend to equalize their *induced* emfs; (b) which bring their terminal voltages to equality?

27. What two conditions must be fulfilled before alternators can be connected in parallel?   How may the equality of terminal voltages be determined?   How may the correct polarity be determined?   Make a diagram of connections.

28. What does the flickering of the synchronizing lamps show?   What are the disadvantages of the "three-dark" method of synchronizing? Explain how these difficulties are eliminated in the "two-light-and-one-dark" method.   What is a synchroscope, and what does it show?

29. What is meant by the "hunting" of alternators?   What causes hunting?   What methods are employed to minimize and even to eliminate hunting?

PROBLEMS ON CHAPTER VI

Alternator Characteristics and Operation

191. Each of the coils *ab*, *bc*, *ca*, of a delta-connected alternator, Fig. 191*A*, has an ohmic resistance of 0.16 ohm.   (a) Determine the ohmic resistance measured between terminals *a* and *b*. (b) If the ratio of effective to ohmic resistance is 1.5, what is the effective resistance of each coil?

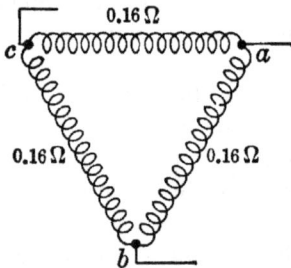

FIG. 191 *A*.

192. The ohmic resistance between each pair of terminals of a three-phase, delta-connected alternator is measured and found to average 0.08 ohm.   The ratio of effective to ohmic resistance is 1.4.   Determine: (a) ohmic resistance per coil; (b) effective resistance per coil.

193. The effective resistance per coil of a delta-connected alternator is 0.084 ohm, and the ratio of effective to ohmic resistance is 1.4.   If the ohmic resistance is measured between each pair of terminals (Fig. 191*A*), what will be its value?

194. In a 50-kva, single-phase, 600-volt, 60-cycle alternator the effective armature resistance is 0.20 ohm and the armature reactance is 1.0 ohm. Determine: (a) rated current of alternator; (b) resistance drop at rated current; (c) reactance drop at rated current; (d) induced emf at rated current and unity power factor.

195. In Prob. 194, determine induced emf at rated current when power factor is 0.8, lagging current.

196. In Prob. 194, determine induced emf at rated current when power factor is 0.8, leading current.

197. In a 20-kva, single-phase, 250-volt, 60-cycle alternator the effective armature resistance is 0.12 ohm and the armature reactance is 0.72 ohm.

Determine: (a) rated current of alternator; (b) armature resistance drop; (c) armature reactance drop. Determine induced emf at rated kva load: (d) when load power factor is unity; (e) when load power factor is 0.80, lagging current; (f) when load power factor is 0.80, leading current.

198. The terminal voltage of a 15-kva, three-phase, 230-volt, 60-cycle alternator is adjusted to its rated value at rated current under each of the following conditions, and the load is then removed: (a) unity power factor; (b) 0.8 power factor, lagging current; (c) 0.8 power factor, leading current. In (a) the no-load voltage is found to be 250 volts; in (b), it is found to be 298 volts; in (c), it is found to be 218 volts. Determine the regulation under each condition. (d) Determine kilowatt output in (a), (b), and (c).

199. In a 2,500-kva, 2,300-volt, 60-cycle, three-phase, Y-connected alternator the regulation at rated voltage and kva at unity power factor is 0.052; at 0.8 power factor, lagging current, 0.145; at 0.8 power factor, leading current, −0.06. When the field current is adjusted to give rated terminal voltage at rated current, determine no-load emf at: (a) unity power factor; (b) 0.8 power factor, lagging current; (c) 0.8 power factor, leading current.

200. In Prob. 199, determine: (a) no-load emf per coil in (a), (b), and (c); (b) power output in (a), (b), and (c).

201. The terminal voltage of a 400-kva, 600-volt, three-phase, delta-connected, 60-cycle alternator is adjusted to its rated value when the load power factor is unity, when the load power factor is 0.85, lagging current, and when the load power factor is 0.85, leading current. When the load is removed under each of these conditions, the regulation is 0.076, 0.18, and −0.04. When the field current is adjusted to give rated terminal voltage at rated current, determine no-load emf with power factor: (a) unity; (b) 0.85, lagging current; (c) 0.85, leading current. (d) Determine the power output under the conditions of (a), (b), and (c).

202. In a 25-kva, 250-volt, 60-cycle, single-phase alternator the effective armature resistance is 0.1 ohm and the synchronous reactance is 1.2 ohms. The alternator is delivering rated current at rated voltage at unity power factor. Determine: (a) no-load emf; (b) regulation.

203. In Prob. 202 with rated voltage and current at 0.8 power factor, lagging current, determine: (a) no-load emf; (b) regulation. The vector diagram is shown in Fig. 203.4.

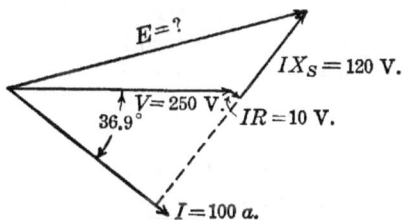 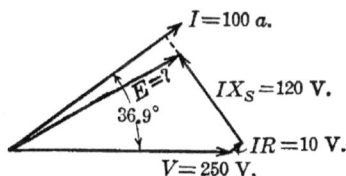

FIG. 203 *A*.   FIG. 204 *A*.

204. In Prob. 202, with rated voltage and current at 0.8 power factor, leading current, determine: (a) no-load emf; (b) regulation. The vector diagram is shown in Fig. 204*A*.

205. In a 15-kva, 230-volt, 60-cycle, six-pole, single-phase alternator the effective resistance of the armature is 0.24 ohm and the synchronous reactance is 2.4 ohms. When load is adjusted to rated voltage and current at unity power factor, determine: (*a*) no-load emf; (*b*) regulation.

206. In Prob. 205, with rated voltage and current at 0.8 power factor, lagging current, determine: (*a*) no-load emf; (*b*) regulation.

207. In Prob. 205, with rated voltage and current at 0.8 power factor, leading current, determine: (*a*) no-load emf; (*b*) regulation.

208. In a 100-kva, 600-volt, 60-cycle, three-phase, Y-connected alternator the effective resistance per coil is 0.20 ohm and the synchronous reactance is 1.8 ohms per coil. Determine: (*a*) rated current; (*b*) rated terminal voltage of coil; (*c*) no-load emf when load is adjusted to rated current and terminal voltage at unity power factor; (*d*) regulation in (*c*).

209. In the alternator, Prob. 208, at 0.8 power factor, lagging current, at rated kva, determine: (*a*) no-load emf; (*b*) regulation.

210. In the alternator, Prob. 208, at 0.8 power factor, leading current, at rated kva, determine: (*a*) no-load emf; (*b*) regulation.

211. In an 800-kva, three-phase, 25-cycle, 6,600-volt, Y-connected alternator, the effective resistance per coil is 0.9 ohm and the synchronous reactance is 11 ohms per coil. At unity power factor, determine: (*a*) rated coil current; (*b*) rated terminal voltage of coil; (*c*) no-load emf; (*d*) regulation.

212. In the alternator, Prob. 211, at 0.8 power factor, lagging current, at rated kva, determine: (*a*) no-load emf; (*b*) regulation.

213. In the alternator, Prob. 211, at 0.8 power factor, leading current, at rated kva, determine: (*a*) no-load emf; (*b*) regulation.

214. If, in Prob. 211, the determination of the effective resistance and the synchronous reactance is made on the assumption that the alternator is delta-connected, the effective resistance per coil is 2.7 ohms and the synchronous reactance per coil 33 ohms. Repeat (*a*), (*b*), (*c*), and (*d*).

215. Repeat Prob. 212, assuming a delta-connection.

216. Repeat Prob. 213, assuming a delta-connection.

217. A 1,000-kva, 6,600-volt, three-phase, Y-connected alternator is short-circuited and the field current increased from 0 to 150 amp. The current in each of the three short-circuited line wires is then 166 amp. (See Fig. 172, p. 203.) The short circuit is then removed, and with the field current still equal to 150 amp the emf between line terminals is 3,600 volts. The effective resistance per coil is 0.5 ohm. Determine: (*a*) synchronous impedance per coil; (*b*) synchronous reactance per coil; (*c*) rated current: (*d*) with rated kva and terminal voltage and unity power factor, the no-load emf; (*e*) regulation.

218. In the alternator, Prob. 217, at 0.8 power factor, lagging current, and at rated kva, determine: (*a*) no-load emf; (*b*) regulation.

219. In the alternator, Prob. 217, at 0.8 power factor, leading current, and at rated kva, determine: (*a*) no-load emf; (*b*) regulation.

220. Solve Prob. 217, assuming the alternator to be delta-connected.

221. Solve Prob. 218, assuming the alternator to be delta-connected.

222. Solve Prob. 219, assuming the alternator to be delta-connected.

223. A 1,500-kva, 2,300-volt, three-phase, delta-connected alternator is short-circuited, the connections being shown in Fig. 223A. The field current is gradually increased from 0 to 180 amp, and each of the line ammeters then reads 800 amp. The short circuit is then removed, the field

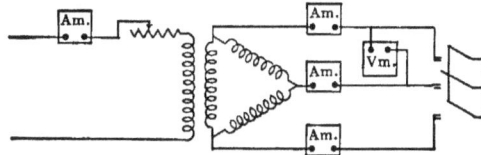

Fig. 223 *A.*

current remaining constant, and the three line emfs are found to be 1,700 volts. The effective resistance per coil is 0.28 ohm. Determine: (*a*) synchronous impedance per coil; (*b*) synchronous reactance per coil; (*c*) with rated kva and terminal voltage and unity power factor, the no-load emf; (*d*) regulation.

224. In the alternator, Prob. 223, at 0.85 power factor, lagging current, at rated kva, determine: (*a*) no-load emf; (*b*) regulation.

225. In the alternator, Prob. 224, at 0.85 power factor, leading current, at rated kva, determine: (*a*) no-load emf; (*b*) regulation.

226. Two three-phase, 25-cycle alternators, having ratings of 250 and 150 kva, operate in parallel. The governor characteristics of the engines driving these alternators are such that the speed of the 250-kva unit drops at a uniform rate from 25 to 24 cycles when the kilowatts change from 0 to 250 kw and the speed of the 150-kva unit drops at a uniform rate from 25 to 23

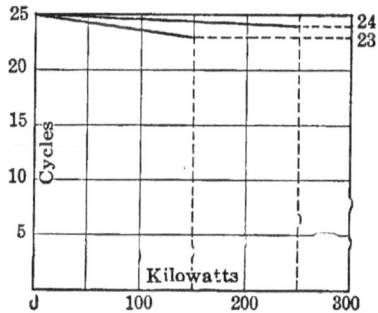

Fig. 226 *A.*

cycles when the kilowatts change from 0 to 150 kw (see Fig. 226A). Determine the kilowatts delivered by the 150-kva unit when the 250-kva unit delivers 250 kw.

227. In Prob. 226, determine: (*a*) power delivered by 250-kva unit when 150-kva unit delivers 50 kw; (*b*) total power delivered to load.

228. In Prob. 226, determine power delivered by 250-kva unit when 150-kva unit delivers 100 kw.

229. Two 1,500-kva, 2,300-volt, three-phase, 60-cycle turbine-driven alternator units 1 and 2, operating in parallel, have such speed-load characteristics that the speed of 1 increases from 60 to 62 cycles when the load on the alternator drops from 1,500 kw to zero and the speed of 2 increases from 60 to 63 cycles under the same change of load. The speed—load characteristic of each unit is a straight line. Determine the load which 2 delivers when 1 delivers 750 kw.

230. In Prob. 229, determine the load which 1 delivers when 2 delivers 750 kw.

231. In Prob. 229, determine the load which 2 delivers when 1 delivers 1,000 kw.

232. In Prob. 229, determine the load which 1 delivers when 2 delivers 1,000 kw.

233. In Prob. 229 the turbine governor of 2 is adjusted so that the two alternators divide the load equally when the total load is 2,000 kw. The frequency—load characteristic of 1 remains unchanged: that of 2 is parallel to its former characteristic, the frequency rising three cycles from 1,500 kw to no load. Determine: (a) load on 2 when 1 delivers 1,500 kw; (b) load on 2 when 1 delivers 500 kw.

234. In Prob. 233, determine: (a) frequency at which 2 operates when it delivers rated load of 1,500 kw; (b) power which 1 delivers under the conditions of (a), on the assumption that its frequency—load characteristic continues as a straight line to overload.

235. Two 50-kva, 230-volt, single-phase, 25-cycle alternators 1 and 2 operate in parallel to supply 75 kw at 230 volts to a lamp load. With the fields of the two alternators adjusted so that each operates at unity power factor, 1 delivers 30 kw and 2 delivers 45 kw. The field of 1 is weakened and the field of 2 is strengthened so that the power factor of 1 is 0.85. The governors of the prime movers are not changed. Before the readjustment of the two fields, determine: (a) current delivered by each alternator. After the readjustment of the fields determine: (b) current delivered by 1; (c) quadrature current of 1; (d) quadrature current of 2; (e) phase angle of 2; (f) total current delivered by 2; (g) power factor of 2; (h) current to the load.

236. Two 1,000-kva, 2,300-volt, three-phase, Y-connected, 60-cycle, Diesel-engine-driven alternators 1 and 2 operate in parallel across 2,300-volt bus-bars, and together are supplying 1,600 kw to a three-phase load of unity power factor. Alternator 1 delivers 850 kw and 2 delivers 750 kw, each at unity power factor. The field of 1 is increased and of 2 is weakened until the power factor of 1 is 0.80. The governors of the Diesel engines are not changed. Before the change in field excitation, determine: (a) current delivered by each alternator. After the change in field excitation, determine: (b) current delivered by 1; (c) quadrature current of 1 and of 2; (d) phase angle of 2; (e) total current of 2; (f) power factor of 2; (g) current to load. Draw vector diagram.

237. Repeat Prob. 236, with the field of 2 increased so that its power factor is 0.85, the field of 1 being weakened simultaneously to maintain constant bus-bar voltage.

QUESTIONS ON CHAPTER VII

The Transformer

1. Define "static transformer." Through what medium is the energy transferred? Why is the transformer a very useful piece of power apparatus?

2. Upon what fundamental principle does the induced emf in the secondary depend? In what manner does the production of emf in the transformer differ from that in the generator?

3. Define "primary." "Secondary."

4. To what three factors is the induced emf proportional? Show that the induced emf in the primary and the induced emf in the secondary are proportional to their respective number of turns. Why are the terminal voltages practically proportional to their respective number of turns?

5. What two functions does the no-load current perform? What is the order of magnitude of the no-load current? The no-load power factor? The no-load phase angle?

6. Show that the flux in the core of the ordinary transformer remains practically constant over the operating range of the transformer.

7. Why do the secondary ampere-turns *oppose* the flux? Analyze the sequence of reactions which follow the application of load to the secondary and which cause more energy to enter the primary from the line.

8. Why must the secondary ampere-turns be practically equal to the increase of primary ampere-turns over the no-load ampere-turns? To what is the ratio of primary to secondary current equal, practically?

9. What mmf produces the mutual flux? The primary leakage flux? Discuss the effect of the primary leakage flux on the operation of the transformer. Define "primary leakage reactance."

10. To what is the secondary leakage flux proportional? Discuss its effect on the operation of the transformer. Define "secondary leakage reactance."

11. What is the general effect of leakage flux on the operation of the transformer as a whole? When are small and when are large leakage reactances desirable?

12. Why is it difficult to determine transformer efficiency and regulation accurately by measurements of output and input?

13. To what are the core losses due, and how are they reduced? Sketch a diagram of connections which would be used in measuring such losses. How do these losses vary with the emf impressed on the transformer winding? In the constant-potential transformer, why are core losses practically constant irrespective of the load?

14. What d-c measurements are made in order to determine the copper losses, and what precautions are necessary in making these measurements?

15. Draw the diagram of connections used in making the short-circuit test. Account for the power input to the transformer. In a 2,300/230-volt transformer, why would such a measurement be made on the high side?

16. How is the equivalent effective resistance of the transformer computed, referred to both primary and secondary? Why do the losses computed from the short-circuit test differ from those computed from d-c measurements?

17. Write three equations giving the efficiency of a transformer in terms of output and losses. What is the order of magnitude of transformer efficiencies?

18. Show how the equivalent impedance of a transformer referred to the primary, for example, is determined from the short-circuit test. How is the equivalent reactance determined?

19. Given the equivalent resistance and reactance, draw vector diagrams showing how the regulation at any value of power factor is computed. What is the order of magnitude of transformer regulation?

20. Describe core- and shell-type transformers. How are the cores built up? To what type of service is the core-type well adapted, and why? In the core type, why are a portion of the primary and a portion of the secondary placed on each of the two legs?

21. Describe the type-H transformer of the General Electric Co. Discuss the disposition of the two windings and of the insulation. What are oil channels?

22. Why does the problem of cooling transformers become more difficult with increase in rating? Name six methods for cooling transformers.

23. What types of transformer are cooled by natural circulation of air? Forced circulation of air? Why is the voltage rating of air-cooled transformers limited to relatively low values?

24. Describe the mechanism of cooling by natural circulation of oil and air. What type of case may be used with transformers of the smaller ratings? As the rating is increased, how is the heat-dissipating surface increased?

25. What is Pyranol, and what are its advantages as a cooling medium for transformers?

26. Describe tubular-cooled transformers. What limits the size of such transformers?

27. What type of transformer employs natural circulation of oil and forced circulation of air? Describe the mechanism of oil cooling in both the tubular and the radiator construction.

28. Describe the method of cooling involving the natural circulation of oil and its artificial cooling.

29. Describe the phenomenon of "breathing" of transformers and the injurious effects which accompany it. How are these injurious effects prevented?

30. Sketch the core and windings of a three-phase, core-type transformer. Compare a three-phase transformer with three single-phase transformers of the same total rating with reference to weight, space, and reliability.

31. In what manner do autotransformers differ from the ordinary transformer? Where would autotransformers be used? Why is it not feasible to use them for large ratios of transformation? Sketch the connections of a two-to-one, three-phase, step-down autotransformer. Show the connections which would be used to obtain a 230- to 115-volt, three-wire system from a 230-volt, two-wire system. What is a "balance coil"?

32. Why is it highly undesirable to connect at random two primary coils of a transformer in series directly across a power source? Describe two methods by which the correct phase relation of such coils for both series and parallel connection may be determined without danger of short circuit. Describe methods which may be used to phase high-voltage coils.

33. Sketch typical series and parallel connections of both the primary and the secondary coils of a distribution transformer. How is a three-wire secondary system obtained?

34. State the advantages of the delta-delta transformer connection. Of a Y-Y connection. Under what conditions is it not practicable to use a Y-Y connection?

35. Under what conditions is the delta-Y connection used? The Y-delta connection? What is the particular advantage of using Y-connected transformers for high voltage? When must the phase relations among primaries be determined before connecting them to the three-phase power source? How are proper phase relations among secondaries determined?

36. Show, by means of the delta-connection, that a V-V connection gives three-phase to three-phase transformation of power. What is the ratio of the kva rating of a delta-bank to a V-bank, if units of the same kva rating are used for each?

37. In the T-transformer connection, why do the emfs in the two halves of the main transformer differ in phase by 180°? With equal ratios of transformation, why are the secondary emfs unequal? How may equality of secondary emfs be obtained? Where is the Scott- , or T- , connection used?

38. Show that when the load on a constant-current transformer is changed, reactions develop which tend to maintain the current at a constant value. What part does leakage flux play in maintaining constant current, and why? What precautions should be taken in the operation of constant-current transformers?

39. Give three reasons for the necessity of using instrument transformers. Describe a potential transformer. At what standard secondary voltage do such transformers operate?

40. In what manner does the current transformer differ from the constant-potential transformer? At what current are the secondaries ordinarily rated? What precautions should be taken in the operation of such transformers?

PROBLEMS ON CHAPTER VII

The Transformer

238. In a 100-kva, 13,800/2,300-volt, 60-cycle step-down transformer, there are 2,100 turns on the high side (primary). Determine: (*a*) ratio of transformation; (*b*) volts per turn; (*c*) turns on low side (secondary); (*d*) rated current of high side and of low side.

239. In a 2,000-kva, 66,000/6,600-volt, 60-cycle transformer, the induced emf per turn is 50 volts. Determine: (*a*) high-side turns; (*b*) low-side turns; (*c*) rated high-side and low-side currents.

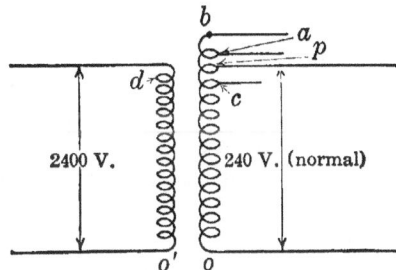

Fig. 240 A.

240. A 30-kva, 10-to-1, 60-cycle transformer (Fig. 240*A*) operates with 2,400 volts across its primary, or high side. There are 1,000 high-side turns. It is specified that the low-side *op* shall be provided with three taps; *a* and *b* give voltages 5 and 10 per cent above normal, and c gives 5 per cent below normal.   Determine turns in portion of secondary winding: (a) op; (b) oc; (c) oa; (d) ob.

241. In Prob. 240, if the high side, or primary, is provided with a tap d such that there are 960 turns in o'd, determine the four secondary emfs E_{oc}, E_{op}, E_{oa}, E_{ob} when the upper 2,400-volt wire is connected to d.

242. In a 100-kva, 6,900/460-volt, 60-cycle distribution transformer, there are 1,725 high-side turns, and the transformer operates with 1,565,000 maxwells, maximum instantaneous, in the core.   Determine: (a) actual high-side emf; (b) actual low-side emf; (c) volts per turn.   (Use Eq. (75), p. 220.)

243. In a 250-kva, 10-to-1, 60-cycle transformer, there are 2,200 high-side turns, and the maximum instantaneous flux is 2,500,000 maxwells.   Determine: (a) low-side turns; (b) induced emf per turn; (c) induced emf in high side; (d) induced emf in low side.   (Use Eq. 75, p. 220.)

244. In a 15-kva, 2,300/230-volt, 60-cycle transformer, there are 460 high-side turns.   Determine: (a) volts per turn; (b) low-side turns; (c) maximum instantaneous flux in core.

245. A 50-kva, 4,400/550 volt, 25-cycle transformer operates with a flux having a maximum instantaneous value of 3,960,000 maxwells.   Determine: (a) volts per turn; (b) number of turns required for high and for low side.

246. The high side of the transformer (Prob. 245) operates at 4,400 volts, 60 cycles.   (a) Determine maximum instantaneous flux in core.   (b) Repeat (a) with the high side operating at 10,000 volts, 60 cycles.

247. In a 500-kva, 25-cycle, 2,400/240-volt transformer, there are 200 primary, or high-side, turns.   Determine: (a) rated high-side current; (b) rated low-side current; (c) primary ampere-turns, neglecting exciting current; (d) secondary ampere-turns.

248. A transformer identical with that of Prob. 247 is provided with a secondary tap which gives 95 per cent rated voltage.   If the rated kva output can be obtained at this reduced voltage, determine: (a) secondary current; (b) secondary turns; (c) secondary ampere-turns; (d) primary ampere-turns, neglecting magnetizing current.

249. In the 30-kva transformer (Prob. 240), determine: (a) the primary ampere-turns when the secondary delivers rated current of 125 amp at taps c, p, a, b; (b) primary current in each case.   The entire high-side winding is utilized.

250. In a 5,000-kva, 25-cycle transformer which steps down the voltage from 13,800 to 2,300 volts, there are 620 turns in the high-side winding.   Determine: (a) rated high-side current; (b) high-side ampere-turns at rated current, neglecting exciting current; (c) low-side ampere-turns; (d) low-side current.   (Links ab and cd together are connected across bc (Fig. 251A), connecting coils a'c and d'd in series.)

251. The high-side winding (Prob. 250) is in two sections, each rated at 6,900 volts.   These sections are connected in parallel by means of the links ab and cd (Fig. 251A) so that the transformer high side then operates at 6,900 volts and rated current.   Under these conditions, determine: (a) total rated high-side current; (*b*) high-side ampere-turns; (*c*) low-side ampere-turns. Compare these results with those obtained in Prob. 250.

252. In a 25-kva, 2,200/550-volt, 60-cycle transformer, the no-load, or core, loss is 120 watts when the transformer is operating at rated voltage. The rated high-side current is 25,000/2,200, or 11.36, amp. The effective resistance of the high side is 1.90 ohms, and the effective resistance of the low side is 0.11 ohm. Determine: (*a*) low-side current; (*b*) high-side copper loss; (*c*) low-side copper loss; (*d*) total loss at rated kva load; (*e*) efficiency at rated kva load and unity power factor.

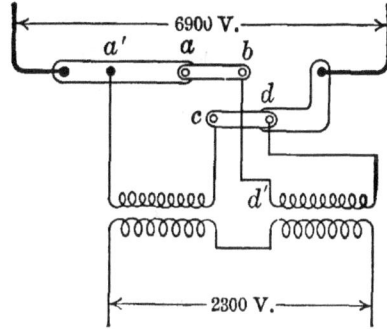

Fig. 251 *A*.

253. Determine the efficiency of the transformer (Prob. 252) at half its rated kva load and unity power factor.

254. Determine the efficiency of the transformer (Prob. 252) at rated kva load and 0.7 power factor, lagging current.

255. In a 75-kva, 2,400/240-volt, 25-cycle transformer, the core loss at rated voltage is 415 watts, the effective resistance of the high side is 0.95 ohm, and the effective resistance of the low side is 0.0093 ohm. Determine: (*a*) rated high-side current; (*b*) rated low-side current; (*c*) primary copper loss at rated current; (*d*) secondary copper loss at rated current; (*e*) efficiency at rated kva and unity power factor; (*f*) efficiency at rated voltage and one-half rated current at unity power factor.

256. In Prob. 255, determine the efficiency at 0.8 power factor, lagging current, at: (*a*) 1.25 per cent rated kva load; (*b*) at rated kva load; (*c*) at one-half rated kva load.

257. In a 1,000-kva, 22,000/2,200-volt, 60-cycle transformer, the core loss is 2,900 watts at rated voltage. The high-side effective resistance is 1.76 ohms, and the low-side effective resistance is 0.018 ohm. Determine: (*a*) rated-load high-side and low-side currents; (*b*) high-side copper loss at rated current; (*c*) low-side copper loss at rated current; (*d*) total loss at rated kva load; (*e*) efficiency at rated kva load and unity power factor.

258. Determine efficiency of transformer (Prob. 257) at rated kva load and: (*a*) 0.8 power factor, lagging current; (*b*) 0.8 power factor, leading current.

259. Determine efficiency of transformer (Prob. 257) at three-fourths rated kva load at unity and 0.7 power factor, lagging current.

260. In Prob. 257, compute and plot the efficiency from 1/16 to 1 1/4 rated load, unity power factor. See Fig. 190, p. 234.

261. In an open-circuit test of a 25-kva, 2,400/240-volt, 60-cycle transformer, the input at 240 volts corrected for instrument losses is 140 watts. The low side is short-circuited; and the voltage, current, power input to the high side corrected for instrument losses are 92 volts, 11.5 amp and 350 watts.

(See Figs. 185 and 188, pp. 227 and 229.)   Determine: (a) equivalent impedance Z_{01}, referred to high side; (b) equivalent effective resistance R_{01} referred to high side; (c) equivalent reactance X_{01} referred to high side; (d) rated high-side and low-side currents; (e) copper loss at rated current computed from (b) and from equivalent effective resistance referred to low side; (f) efficiency at rated kva load and unity power factor.

262. In Prob. 261, determine no-load emf and regulation at rated kva load and unity power factor: (a) using high-side constants; (b) using low-side constants.

263. Repeat Prob. 262 with rated kva load and 0.8 power factor, lagging current.

264. Repeat Prob. 262 with rated kva load and 0.8 power factor, leading current.

265. A short-circuit test is made on the 1,000-kva, 22,000/2,200-volt, 60-cycle transformer (Prob. 257).   The 22,000-volt side is short-circuited, and measurements on the 2,200-volt side are 108 volts, 454 amp, and 7,330 watts.   Determine: (a) equivalent impedance Z_{02} referred to low side; (b) equivalent resistance R_{02} referred to low side; (c) equivalent reactance X_{02} referred to low side; (d) no-load emf at rated kva and unity power factor, using low-side constants; (e) regulation corresponding to (d).

266. In Prob. 265, determine: (a) equivalent impedance Z_{01}, equivalent effective resistance R_{01}, and equivalent reactance X_{01}, all referred to high side; (b) no-load emf and regulation at rated kva and unity power factor, using constants in (a).   (c) Repeat (b) for 0.8 power factor, lagging current.

267. In Prob. 265, repeat (d) and (e) for 0.8 power factor, leading current.

268. A 250-kva, 13,200/575-volt, 60-cycle transformer is tested for regulation and efficiency.   In the open-circuit test the input to the low side at 575 volts is 1,120 watts.   The low side is short-circuited, and measurements on the high side are 600 volts, 20 amp and 2,830 watts.   Determine: (a) equivalent impedance Z_{01} referred to high side; (b) equivalent resistance R_{01} referred to high side; (c) equivalent reactance X_{01} referred to high side; (d) high-side no-load emf at rated kva and unity power factor; (e) regulation corresponding to (d); (f) efficiency corresponding to (d).

269. In Prob. 268, determine: (a) low-side constants Z_{02}, R_{02}, X_{02}; (b) low-side emf at rated kva and unity power factor; (c) regulation and efficiency corresponding to (b).

270. In Prob. 268, determine: (a) high-side emf at rated kva and 0.8 power factor, lagging current; (b) regulation and efficiency corresponding to (a).

271. Repeat (d) and (e), Prob. 268, for 0.8 power factor, leading current.

272. Repeat (d) and (e), Prob. 268, using low-side constants of Prob. 269.

273. The transformer, Prob. 252, is connected to a 2,200-volt, 60-cycle power line 24 hr per day.   It delivers at unity power factor; ¼ load, 4 hr; ½ load, 2 hr; rated load, 8 hr.   Determine all-day efficiency.

274. The transformer, Prob. 255, is connected to a 2,400-volt, 25-cycle power line 24 hr per day.   It delivers at unity power factor: ⅛ load, 8 hr; ¼ load, 4 hr; ½ load, 2 hr; rated load, 6 hr; 1¼ load, 2 hr.   Determine all-day efficiency.

275. The transformer, Probs. 257, 265, is disconnected from the 22,000-volt, 60-cycle supply line 4 hr each day and is energized the remaining 20 hr. It operates at ½ load, 6 hr; rated load, 6 hr; 1¼ load, 2 hr., the power factor being unity throughout. Determine all-day efficiency.

276. An autotransformer (Fig. 276*A*) is used to transform from 550 to 440 volts. The load at $b'c'$ is 25 kw at unity power factor. Neglecting all

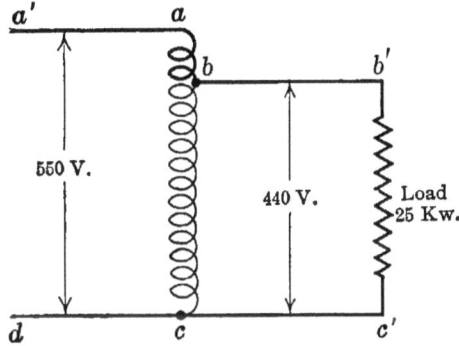

Fig. 276 *A*.

losses and magnetizing current, determine: (*a*) current in bb'; (*b*) current in ab; (*c*) current in cb.

277. An autotransformer (Fig. 277*A*) is used to boost the voltage on a 2,200-volt feeder to 2,400 volts. The load is 60 kw at unity power factor.

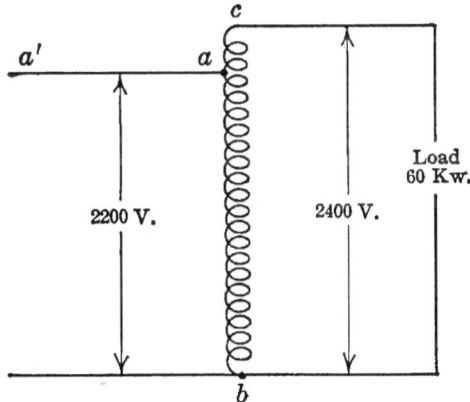

Fig. 277 *A*.

Neglecting all losses and magnetizing current, determine: (*a*) current in ac; (*b*) current in $a'a$; (*c*) current in ba.

278. The autotransformer (Fig. 276*A*) is inverted to step up the voltage from 440 to 550 volts. The 440-volt, a-c supply replaces the load $b'c'$, and a 20-kw, unity-power-factor load is connected between a' and d. Neglecting all losses and magnetizing current, determine: (*a*) current in aa'; (*b*) current in $b'b$; (*c*) current in cb.

279. A 230-volt starting compensator, or autotransformer, used for starting an a-c motor at reduced voltage, supplies the motor on starting from 40 per cent taps (Fig. 279A), giving 92 volts across the motor terminals. Determine line current when motor takes 100 amp. (See Fig. 236, p. 289.)

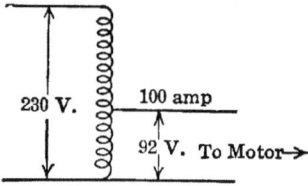

FIG. 279 A.

280. A Y-Y transformer bank (Fig. 280A) is used to transform three-phase, unity-power-factor load of 40 amp at 13,800 volts to 2,300 volts, three-phase. Determine: (a) voltage ratings of transformer primaries; (b) voltage ratings of transformer secondaries; (c) secondary currents; (d) kva input and kva output of transformer bank.

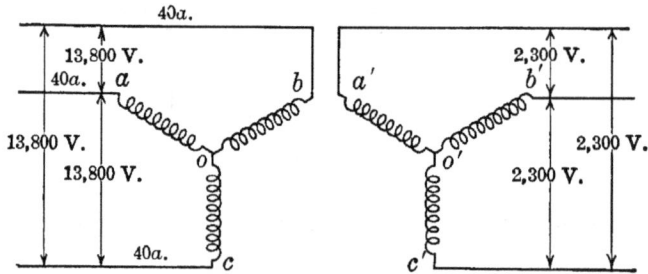

FIG. 280 A.

281. A 40-kw, 600-volt, three-phase, 60-cycle load at 0.8 power factor is supplied from a 1,100-volt, three-phase system by a Y-Y connected transformer bank. Determine: (a) voltage rating of each transformer secondary; (b) current rating of each transformer secondary; (c) voltage rating of each transformer primary; (d) current rating of each transformer primary; (e) kva rating of each transformer; (f) kva rating of bank.

282. A delta-Y transformer bank (Fig. 282A) is used to transform 3,000 kva at 0.9 power factor, lagging current, at 13,800 volts to 4,000 volts to supply a 4,000-volt, three-phase, four-wire distribution system. Determine:

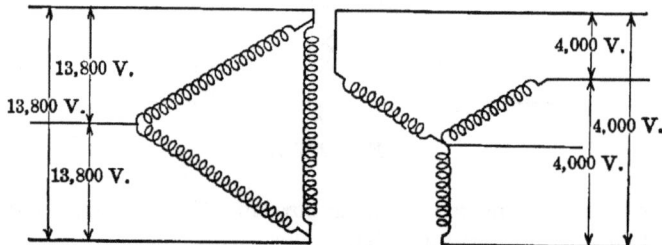

FIG. 282 A.

(a) voltage ratio of transformer bank; (b) voltage ratio of individual transformer; (c) primary coil currents; (d) secondary coil currents; (e) voltage to neutral on secondary.

283. Repeat Prob. 282, with the secondary load equal to 3,600 kva at 0.8 power factor.

284. A three-phase, 60-cycle transformer bank in a generating station is connected delta-Y and steps up the voltage from 13,200 to 138,000 volts, three-phase, for transmission. The power is 30,000 kw at 0.9 power factor, lagging current. Determine: (*a*) primary line current; (*b*) current in transformer primaries; (*c*) voltage ratings of secondaries; (*d*) ratio of transformation in each transformer; (*e*) current in secondaries; (*f*) kva delivered by each transformer.

285. Three Y-delta-connected transformers, each rated at 16,667 kva, 60 cycles (Fig. 285*A*), are installed to step down the voltage at a substation

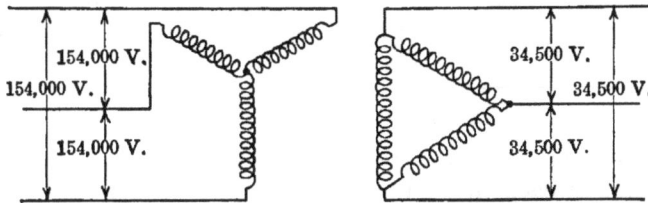

Fig. 285 *A.*

from a three-phase, 154,000-volt transmission line to 34,500 volts, three-phase, for local distribution. Determine: (*a*) voltage rating of each primary; (*b*) ratio of transformation in each transformer; (*c*) current rating of each secondary; (*d*) rated secondary line current.

286. In Prob. 285, when the secondary load is 40,000 kw at 0.92 power factor, lagging current, determine: (*a*) primary line current; (*b*) secondary coil current; (*c*) secondary line current.

287. Figure 287*A* shows a delta-Y transformer bank used for stepping up the voltage in a hydroelectric station from 13,200 to 66,000 volts, 60

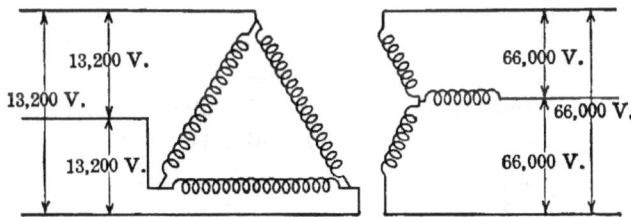

Fig. 287 *A.*

cycles, for transmission. The generators are supplying the primaries with 24,000 kw, 0.9 power factor, lagging current, 13,200 volts. Determine: (*a*) primary line current; (*b*) current in each transformer primary; (*c*) current in each transformer secondary; (*d*) voltage rating of each secondary; (*e*) individual transformer ratio; (*f*) secondary line current; (*g*) kva output of each transformer.

288. In a factory the demand is 250 kw at 0.707 power factor, lagging current, 440 volts, 60 cycles, three-phase. Power is supplied by a 2,300 volt, three-phase distribution line and transformed by a delta-delta transformer bank. Determine: (a) kva rating of transformer bank; (b) kva rating of individual transformer; (c) primary line current; (d) primary current of each transformer; (e) secondary current of each transformer; (f) secondary line current.

289. Repeat Prob. 288, when power is taken from a 6,600-volt, three-phase source.

290. It is desired to transform 120 kw at unity power factor and 25 cycles from 23,000 to 2,300 volts by means of V-connected transformers (Fig.

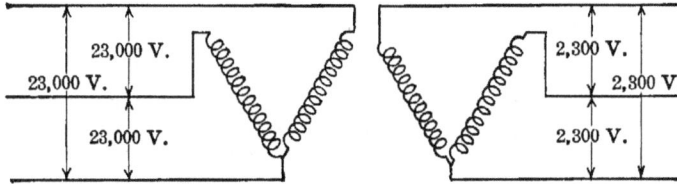

FIG. 290 *A*.

290*A*). Determine: (a) primary line current; (b) secondary line current; (c) kva rating of each transformer; (d) kilowatt load on each transformer.

291. If a third transformer similar to the other two is added to the V-bank (Prob. 290) to give a delta-connection, determine: (a) kva rating of the delta-delta transformer bank; (b) ratio of delta-delta to V-V rating.

292. Two transformers, each rated at 2,300/600 volts and 27.6 amp, primary current, are connected in V to supply power to a machine shop. Assuming unity-power-factor load, determine: (a) total kilowatts that these transformers can deliver without exceeding their kva rating; (b) secondary line currents.

293. The 2,300-volt system (Prob. 292) also supplies a load of 160 kw at 0.8 power factor, lagging current, by means of another V-connected transformer bank. Determine: (a) primary current; (b) secondary current; (c) kva rating of each transformer.

294. If the load, Prob. 293, were supplied by a delta-delta bank, what would be the kva rating of each transformer?

295. A T-connected transformer bank (Fig. 295*A*) transforms from 2,300 volts, 25 cycles, three-phase, to 230 volts, two-phase. Determine: (a)

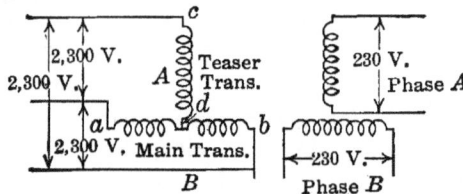

FIG. 295 *A*.

emfs E_{ad}, E_{bd}, E_{cd}; (b) transformation ratio of main transformer; (c) transformation ratio of teaser transformer.

296. It is desired to secure a 2,300-volt, two-phase system from a 13,800-volt, 50-cycle, three-phase system by means of the T-connection. (See

Fig. 295*A*.) Determine: (*a*) emfs E_{ad}, E_{bd}, E_{cd}; (*b*) transformation ratio of main transformer; (*c*) transformation ratio of teaser transformer.

297. A low-voltage, three-phase system requires 300 kva, 60 cycles, three-phase, 208 volts between line wires. The available power source consists of a two-phase, 60-cycle, 4,000-volt system; that is, there is 4,000 volts across phase *A* and across phase *B*. (See Fig. 295*A*.) Make a wiring diagram showing the necessary transformers and the transformer and system connections. Determine: (*a*) primary voltage of each transformer; (*b*) secondary voltage of main transformer; (*c*) secondary voltage of teaser transformer; (*d*) secondary current in each transformer; (*e*) primary current in each transformer.

QUESTIONS ON CHAPTER VIII

The Polyphase Induction Motor

1. Describe a simple experiment which shows that if a flux is made to cut a conducting body the currents induced in that body have such a direction that the body tends to follow the direction of motion of the flux.

2. Under these circumstances, why must a force exist between the flux and the conducting body? Why can the body never attain the speed of the flux?

3. By means of sketches, show that a rotating magnetic field may be produced by two-phase currents flowing in a two-phase, four-pole winding. How many space-degrees does the field advance for each time-degree advance of the current? What is the relation of the speed of the field in rps to the frequency of the currents?

4. Repeat Question 3 for a three-phase, four-pole winding.

5. What is meant by the "synchronous speed" of an induction motor? Derive the equation which gives the synchronous speed in terms of frequency and number of poles.

6. What is meant by the "revolutions slip" of an induction motor? To what is the slip equal?

7. Show that the rotor frequency f_2 is equal to the product of the stator frequency and the slip. To what is the rotor frequency equal at starting? Of what order of magnitude is the rotor frequency under ordinary operating conditions?

8. In what important respect does the change of torque with load for the induction motor differ from that of the d-c shunt motor?

9. Sketch a typical torque—slip curve for a motor having the usual type of rotor. At low values of slip, how does the torque vary with increase of load? What occurs when the breakdown torque is reached? With higher values of slip, why does the torque decrease even though the flux remains constant and the current increases to high values? What is the order of magnitude of the starting torque?

10. Show that for any given value of slip the torque varies with the square of the impressed voltage. Explain why it is important to keep this fact in mind.

11. Describe the construction of the squirrel-cage rotor. In what different ways are the squirrel-cage bars fastened to the end rings? What type of slot is used, and why?

12. Why must the slip of the motor increase with increase of load? Why is the slip of the squirrel-cage motor low? State the order of magnitude of the full-load slip for small and for large motors.

13. Sketch curves of current, efficiency, slip, speed, and power factor for an induction motor, using horsepower output as abscissas. Discuss each of these curves. Why is the no-load current comparatively large, and the no-load power factor low? Why does the power factor increase with load?

14. Name one disadvantage of the squirrel-cage induction motor. Why is the no-load torque low? State the advantages of this type of motor. For what classes of work is it best adapted? For what classes of work is it not adapted?

15. Analyze the effect on the operation of an induction motor of introducing resistance in the rotor circuit when the motor is carrying a load which requires a definite torque. Sketch the torque—slip curves with three different values of resistance in the rotor circuit.

16. How does the introduction of resistance in the rotor circuit affect the relation of breakdown torque to slip? Show that large starting torque may be obtained by the introduction of resistance in the rotor circuit. Name some industrial applications of the wound-rotor motor.

17. Describe a type of rotor that combines both high starting torque and good speed-load characteristic.

18. Why do squirrel-cage motors take relatively large currents when connected directly across the line? Sketch five different rotor designs, indicating the starting current, starting torque, and torque—speed characteristic that each design produces.

19. Why are modern motors classified by a letter in accordance with National Electrical Code standards?

20. Make a wiring diagram of an across-the-line starter with push-button control for both a wound-rotor and a squirrel-cage induction motor. What are thermal overload protective heaters?

21. Make a wiring diagram of a common type of starting compensator. Show the details of the connections of the overload relays and the undervoltage release.

22. Discuss speed control of the induction motor by the introduction of resistance in the rotor circuit. State two distinct disadvantages of this method of speed control.

23. Make a diagram showing the connections which are used when resistance is introduced in the rotor circuit.

24. How may a rotor be designed so that both high starting torque and good running torque are obtained?

25. Enumerate some of the industrial applications of the wound-rotor induction motor. How does its cost compare with that of the squirrel-cage motor?

26. What three factors completely determine the speed of an induction motor?

27. Discuss change of slip as a method of speed control. Change of frequency. Change of the number of poles. Where is each of these methods used?

28. How may an induction motor be made to act as a generator? What determines the frequency of the induction generator? Why is it called *asynchronous?*

29. Why must the induction generator always take lagging current from the line or deliver leading current? Give two reasons which make it necessary for the induction generator to operate always in parallel with a synchronous machine.

30. State the disadvantages of the induction generator. Give its industrial applications.

31. Describe briefly the principle underlying the stroboscopic method for measuring slip. Describe two other methods.

32. Describe the operation of the single-phase induction regulator, showing how "boost" and "buck" are obtained. Why is a short-circuited winding necessary, and where is it placed? Show with a diagram the method of connecting the regulator to a feeder circuit.

33. Describe the three-phase regulator.

PROBLEMS ON CHAPTER VIII

The Polyphase Induction Motor

298. A three-phase, four-pole induction motor is connected to a 230-volt, three-phase, 50-cycle power system; and its speed is 1,425 rpm. Determine: (a) speed of its rotating field in rps; (b) synchronous speed in rpm; (c) revolutions slip; (d) slip.

299. A 25-hp, 10-pole, two-phase, 60-cycle induction motor is connected to a two-phase, 440-volt, 60-cycle power system; and its rated-load speed is 684 rpm. Determine: (a) speed of rotating field in rps; (b) its synchronous speed in rpm; (c) revolutions slip; (d) slip.

300. Power for a three-phase, 440-volt, 25-cycle system is obtained from a 60-cycle system through a frequency changer, a 10-pole synchronous motor driving a four-pole alternator, so that actually the frequency is but 24 cycles per sec. A 10-hp, two-pole, three-phase induction motor is connected to the 24-cycle system; and its speed is 1,360 rpm. Determine: (a) speed of rotating field in rps; (b) synchronous speed in rpm; (c) revolutions slip; (d) slip.

301. Repeat (a), (b), (c), (d), Prob. 300 when a six-pole, 50-hp motor is connected to the 24-cycle system, the speed of the motor being 465 rpm.

302. A planer for woodworking machinery is to be driven at a speed of 8,100 rpm by a direct-connected, two-pole, three-phase, 230-volt induction motor. At rated load the speed is 8,100 rpm, and the slip is 0.10. Determine: (a) synchronous speed of motor; (b) frequency of power system. (Such higher frequencies are obtained from the 60-cycle system by means of a frequency changer.)

303. The speed of a 50-hp, three-phase induction motor when connected across three-phase, 440-volt, 60-cycle mains is 423 rpm. The slip at rated load is 0.06. Determine: (a) synchronous speed; (b) number of poles; (c) rotor frequency.

304. The speed of a 100-hp, six-pole, 25-cycle, 600-volt, three-phase induction motor is 480 rpm. The slip is 0.04. Determine: (a) synchronous speed; (b) number of poles; (c) rotor frequency.

305. The rotor speed of a 5-hp, 208-volt, three-phase, four-pole, induction motor is 1,710 rpm when it operates from a three-phase, 60-cycle power source. Determine: (a) slip; (b) rotor frequency.

306. A 10-hp, 208-volt, six-pole, three-phase, 60-cycle induction motor develops 46.2 lb-ft torque at rated load. If rated voltage were applied at standstill, the motor would develop 40 lb-ft torque with six times the rated current of 26.2 amp. The motor is started at 50 per cent rated voltage by means of an autotransformer starter or compensator. Determine: (a) starting current to motor; (b) line starting current; (c) starting torque; (d) ratio of starting to rated-load torque.

307. Repeat Prob. 306, with motor being started with 40 per cent autotransformer taps.

308. Repeat Prob. 306, with motor being started with 60 per cent autotransformer taps.

309. The breakdown torque of the motor, Prob. 306, is 120 lb-ft at the rated voltage of 208 volts. Determine breakdown torque: (a) when motor is operated on a 230-volt, three-phase, 60-cycle system; (b) if system voltage drops to 200 volts.

310. The starting torque of a 200-hp, 600-volt, three-phase, six-pole, 25-cycle, squirrel-cage motor is 320 lb-ft when one-third rated voltage is applied by means of an autotransformer and the current to the motor is 483 amp or 276 per cent rated current. Determine: (a) line current at starting. Determine starting torque were the following autotransformer taps used: (b) 40 per cent; (c) 60 per cent; (d) 80 per cent.

311. In Prob. 310, determine motor current and line current when the following taps are used: (a) 40 per cent; (b) 60 per cent.

312. Rated current and torque at rated voltage of the motor, Prob. 310, are 175 amp and 2,180 lb-ft. With the 50 per cent autotransformer tap being used, determine: (a) ratio of starting to rated current; (b) ratio of starting to rated torque.

313. The following data were taken at rated load during the brake test of a 5-hp, 220-volt, four-pole, three-phase, 60-cycle, squirrel-cage induction motor. The power input was measured by means of the two-wattmeter method. A prony brake having a 2-ft arm was used to measure the output (see Part I, p. 292, Fig. 254). The tare was +2.6 lb; line, 220 volts, 60 cycles; 13.3 amp; watts, $W_1 = +2,920$, $W_2 = +1,460$; slip, 90 rpm; balance reading, 10.3 lb. Determine: (a) slip; (b) speed; (c) torque; (d) efficiency; (e) power factor.

314. With the motor, Prob. 313, at approximately rated load, the following data were taken: line, 220 volts, 60 cycles; 10.63 amp; watts, $W_1 = +2,330$, $W_2 = +1,000$; slip, 68 rpm; balance reading, 8.30 lb. Determine: (a) to (e), Prob. 313.

315. With the motor, Prob. 313, at approximately one-half rated load, the following data were taken: line, 220 volts, 60 cycles; 8.12 amp; watts,

$W_1 = +1740$; $W_2 = +510$; slip, 45 rpm; balance reading, 6.35 lb. Determine: (a) to (e), Prob. 313.

316. In a test of a 7.5-hp, 220-volt, three-phase, four-pole, squirrel-cage induction motor the following data were obtained. The length of the brake arm is 2 ft and the tare of the balance +8.1 lb. For each set of prony-brake readings, compute: (a) torque; (b) slip; (c) horsepower output; (d) watts input; (e) efficiency; (f) power factor. As functions of hp output plot: (g) current; (h) power factor; (i) efficiency; (j) slip. (See Fig. 228, p. 278.)

Volts	Cycles per sec.	Average amp per wire	Watts		Balance, lb	Slip, rpm
			W_1	W_2		
220	25	31.1	6,830	3,100	48.75	56
220	25	28.4	6,240	2,870	45.0	51
220	25	22.1	4,870	2,230	36.3	39
220	25	18.8	4,120	1,740	31.1	31
220	25	15.6	3,380	1,150	25.2	24
220	25	13.4	2,800	600	20.0	17
220	25	11.5	2,250	120	15.0	10
220	25	10.1	1,650	−460	10.0	2
220	25	9.4	1,300	−750	8.1	1.5

NOTE.—The electrical efficiency of the rotor of an induction motor is the ratio of the mechanical power developed within the rotor to the power transferred across the air-gap. The mechanical power at the pulley is slightly less than that developed in the rotor by the amount of the mechanical losses. For example, when the slip of an induction motor is 0.07, the electrical efficiency of the rotor is 0.93.

317. A 50-hp, 550-volt, three-phase, 60-cycle, eight-pole, wound-rotor induction motor is operating at 855 rpm with the rotor resistance cut out. (a) Determine electrical efficiency of rotor. The speed is reduced by cutting resistance into the rotor circuit. Determine electrical efficiency of rotor when speed is: (b) 800 rpm; (c) 450 rpm.

318. A 75-hp, 550-volt, three-phase, 25-cycle, four-pole, wound-rotor induction motor is operating at 710 rpm with the rotor resistance cut out. (a) Determine electrical efficiency of rotor. The speed is reduced by cutting resistance into the rotor circuit. Determine electrical efficiency of rotor when speed is: (b) 600 rpm; (c) 400 rpm.

QUESTIONS ON CHAPTER IX

Single-phase Induction Motors

1. Why do both the d-c shunt motor and the d-c series motor develop torque in one direction when supplied with alternating current? What factor prevents the shunt motor from becoming a commercial a-c motor?

2. Why has the series motor possibilities as an a-c motor?

3. Before the series motor becomes operative with alternating current, what changes must be made in the following and why: field structure; series-field turns; frequency; air-gap?

4. Why is a compensating winding necessary? Explain "conductive compensation" and "inductive compensation."

5. What commutating difficulty exists in the a-c motor that is not present in the d-c motor? What measures are taken to overcome the difficulty?

6. Sketch the characteristic curves of the series type of motor, including speed, torque, horsepower, efficiency, power factor.

7. Give the industrial applications of this type of motor. What are its limitations?

8. What is meant by a "universal motor"? What are the advantages of this type of motor? Why do such motors ordinarily operate at high speed? What is the order of their horsepower ratings? Give some typical industrial applications of this type of motor.

9. What type of armature is used with the repulsion motor? Show that the armature of the repulsion motor behaves like a transformer secondary.

10. Between what two points on the armature does zero potential difference exist, and why? Between what two points does maximum potential difference exist?

11. Why does the motor develop no torque when the brushes are in the geometrical neutral? In the plane of the pole axis? For what positions of the brush axis does the motor develop torque? Give reasons.

12. How are the stator and rotor ordinarily constructed? Why? Describe briefly the operating characteristics of the repulsion motor. State its industrial applications.

13. Discuss the reactions which occur in the single-phase induction motor when its stator is supplied with single-phase current and the short-circuited rotor is stationary. Show that the motor develops no torque under these conditions.

14. What reaction occurs when the armature is rotating, which causes it to develop a rotating field? In what direction does this field rotate?

15. Compare the characteristics of the single-phase induction motor with those of the polyphase induction motor. Compare its output and size with those of the polyphase motor.

16. Discuss the reactions which follow the opening of one phase of the three-phase induction motor, when it is rotating. How is the rating of the motor affected by the opening of one phase?

17. Show that a polyphase motor may by accident operate single-phase. How is such operation detected?

18. Describe two split-phase methods of starting single-phase induction motors. How is the direction of rotation reversed in one of these methods?

19. Describe in detail the "shaded-pole" method of starting a single-phase induction motor. In what types of motor is this method used? Name one other industrial application of the shaded pole.

20. Describe the repulsion-motor method of starting single-phase induction motors. With what type of winding is the rotor wound? How is the rotor converted to induction-motor operation?

21. Make a diagram of connections of the capacitor motor. What is its principle of operation? Why are two capacitors desirable?

QUESTIONS ON CHAPTER X
Synchronous Motors and Converters

1. Compare the construction of the usual synchronous motor with that of the alternator.

2. Describe the principle on which the synchronous motor operates. Show that the motor must operate at synchronous speed or not operate at all.

3. Show that the rotor of the synchronous motor may be considered as being a salient-pole magnet dragged around by a synchronously rotating magnetic field. Give a mechanical analogue.

4. When load is applied to the rotor of the synchronous motor, why can it not take additional energy from the line by a decrease in the magnitude of its counter emf? What actually does occur when the mechanical load is increased? Illustrate with a vector diagram, and give a mechanical analogue. How may the effect of applying load be observed visually?

5. Why can not the synchronous motor act like the shunt motor when its field excitation is increased? What two reactions result when the field of the synchronous motor is overexcited? Show that a leading current in a synchronous motor weakens the field and also produces a counter emf which may exceed the impressed voltage in magnitude.

6. Why can not the synchronous motor act like the shunt motor when its field excitation is decreased? What two reactions result when the field of the synchronous motor is underexcited? Show that a lagging current in a synchronous motor strengthens the field and at the same time produces a counter emf which is less than the impressed voltage. Where does the synchronous motor obtain a portion of its excitation when it is operating underexcited?

7. Sketch the connections, with all instruments, which may be used in testing a synchronous motor.

8. Show by a graph the variation of armature current and power factor as the field current of the synchronous motor is increased from small to large values. What are V-curves?

9. What is meant by a "synchronous condenser"? How may the synchronous condenser be used to improve power factor? Define "energy current" and "quadrature current." Show that if a synchronous condenser running light takes current equal to the quadrature current of the inductive load the power factor of the system is brought practically to unity. Show how a synchronous motor may be used to improve power factor.

10. Under what conditions can a synchronous motor or condenser regulate the voltage at the end of a transmission line?

11. What is the purpose of amortisseur windings, or dampers? Analyze their operation.

12. Describe the procedure which is followed in starting a synchronous motor when it is connected to a d-c generator which may be operated as a motor. How is the synchronous motor, operating as a generator when starting, made to operate as a motor after being connected to the bus-bars?

13. Why will the synchronous motor start from rest and come up to speed when polyphase currents are supplied to its stator? Why can the motor pull into synchronism, even without its d-c excitation? Show that the application of excitation may cause a line disturbance. How may this disturbance be reduced?

14. To what danger is the field winding subjected during starting? How is this danger reduced?

15. Why is the starting torque of a synchronous motor inherently low? Describe two methods whereby high starting torque is secured.

16. What are the inherent disadvantages of the synchronous motor? Its advantages? Name several industrial applications.

17. Describe the construction and mode of operation of the Warren Telechron clock motor. Of the Holtz subsynchronous motor.

18. Show that the principle of the synchronous converter is related to the generation of emf in the armature of a d-c dynamo. How is the converter operated ordinarily? Give six other methods of operation.

19. How are the armature connections of the slip-rings made ordinarily? What relation exists among the number of phases, number of poles, and number of slip-ring taps?

20. In a general way, how does the rating of a converter vary as the number of its phases is increased? How is the rating affected by power factor?

21. Unless the shape of the flux-distribution curve changes, why is the ratio of a-c to d-c *induced* emf in a converter armature constant under all conditions? Show that the ratio of d-c emf to single-phase emf (rms) must be $\sqrt{2}$. Give a graphical method for determining the ratio of d-c to a-c emf for different numbers of phases.

22. How does change of excitation affect the ratio of a-c to d-c induced emf? Why does change of excitation change the ratio of slip ring to commutator voltage? Why does change of excitation ordinarily change the slip-ring voltage also? State the objections to voltage control by change of excitation.

23. Discuss the use of transformer taps for controlling the d-c voltage.

24. Describe the series booster. How is it used to control the d-c voltage?

25. What is meant by an "inverted synchronous converter"? What precaution is necessary when it is in operation, and why?

26. Show that the converter armature starts as an induction motor when polyphase alternating currents are supplied to its slip-rings. Describe the precautions which must be taken as regards the shunt and series fields and the brushes, when starting in this manner. How may the converter be caused to come up to speed with the correct d-c polarity?

27. What is the procedure followed when the converter is started from the d-c side? What difficulty appears when an attempt is made to synchronize?

28. What are dampers? Analyze their operation.

29. Make a wiring diagram showing the connections which would be used for making a laboratory test of a six-phase synchronous converter.

30. What is the order of magnitude of the ratings of large commercial synchronous converters? What is the order of magnitude of their efficiencies?

31. Compare the converter with the motor-generator set. What type of rectifying equipment has reduced the number of new installations of synchronous converters?

PROBLEMS ON CHAPTER X

Synchronous Motors and Converters

319. Determine the no-load and the rated-load speed of an 800-hp, 25-cycle, three-phase, 10-pole synchronous motor which is direct-connected to an ammonia compressor operating in an artificial ice-making plant.

320. It is desired to drive an ammonia compressor at 300 rpm by means of a 300-hp synchronous motor operating from 2,300-volt, three-phase, 60-cycle supply. How many poles must the motor have?

321. The rated speed of a 1,200-hp, 6,600-volt, 25-cycle synchronous motor is 250 rpm. Determine the number of poles.

322. The rated speed of a 2,000-hp, 5,000-volt, 50-cycle synchronous motor is $41\frac{2}{3}$ rpm. How many poles has it?

323. The rated speed of an 800-hp, 2,300-volt, 64-pole synchronous motor is $93\frac{3}{4}$ rpm. Determine the frequency.

324. A 500-hp, 2,300-volt, 12-pole, 60-cycle, three-phase synchronous motor is driving a d-c generator which requires 380 hp. The efficiency of the motor, disregarding field loss, is 0.92. The synchronous motor is operating at 2,300 volts and unity power factor. Determine: (a) power taken by motor armature; (b) current; (c) torque at coupling. (d) If the efficiency of the d-c generator is 0.93 at this load, what current does it deliver at 240 volts?

325. The excitation of the synchronous motor, Prob. 324, is increased so that the motor operates at a power factor of 0.80, leading current. The efficiency is now 0.915. Determine the current to the motor armature.

326. When the motor, Prob. 324, is operating near its rated load, the line voltage is 2,300 volts, the current is 100 amp, leading, and the power factor is 0.85. The generator delivers 1,216 amp at 240 volts at an efficiency of 0.935. Determine: (a) generator input in watts; (b) motor input in watts; (c) efficiency of synchronous motor, disregarding field loss; (d) torque at coupling.

327. It is desired to drive a 1,200-kw, 600-volt, 750-rpm railway generator by a synchronous motor which must take its power from a 6,900-volt, three-phase, 25-cycle line.   The generator efficiency at rated load is 0.94.   For the synchronous motor, determine: (*a*) horsepower rating; (*b*) number of poles.

328. The efficiency of the motor, Prob. 327, is 0.930, disregarding field loss.   Determine: (*a*) ampere input at unity power factor when generator delivers rated load; (*b*) ampere input as in (*a*) except that power factor is 0.80, leading current; (*c*) torque at coupling.

329. At half rated load on the generator, Prob. 328, the generator efficiency is 0.92 and the motor efficiency is 0.91.   Determine the ampere input to the synchronous motor at: (*a*) unity power factor; (*b*) 0.75 power factor, leading current.

330. A 12-hp, 230-volt, 60-cycle, single-phase induction motor operates at 0.80 power factor and delivers 10.0 hp when it takes 47.7 amp at 230 volts. Determine: (*a*) energy current of motor; (*b*) quadrature current; (*c*) leading quadrature current necessary to bring system power factor to unity; (*d*) vars (reactive volt-amperes) necessary to bring system power factor to unity.

331. In Prob. 330, determine: (*a*) induction-motor input in watts; (*b*) motor efficiency.   (*c*) Draw vector diagram.

332. The load of a machine shop, consisting almost entirely of induction motors, takes 120 kw at 600 volts, 60 cycles, and 0.60 power factor.   Determine: (*a*) total current; (*b*) energy current; (*c*) quadrature current; (*d*) reactive kva (kilovars) which synchronous condenser, located in shop, must take in order to bring power factor of system to unity.   Draw vector diagram.   (For simplicity, assume that load is single-phase.)

333. Repeat Prob. 332 for three-phase system.

334. A factory takes 300 kw at 2,300 volts, 25 cycles, three-phase, and 0.707 power factor, lagging current.   Determine: (*a*) total current; (*b*) energy current; (*c*) quadrature current; (*d*) synchronous-condenser kva necessary to raise system power factor to unity.

335. Repeat Prob. 334, with factory power factor 0.80, lagging current.

336. A six-phase, 25-cycle, 250-volt direct current, 750-rpm synchronous converter is rated at 750 kw.   Determine its rated output if it becomes necessary to operate it three-phase (see table, Par. 190, p. 333).

337. If the converter, Prob. 336, were direct-connected to a prime mover and operated as a 250-volt, d-c generator, what would its rating be?

338. A four-phase, 60-cycle, 600-volt synchronous converter is rated at 600 kw.   Determine maximum load which it can safely carry if one of the two phases, supplying energy to two of the four slip-rings, becomes disabled and the converter operates single phase (see table, p. 333).

339. Determine the rating of the converter, Prob. 338, if it is operated as a d-c generator.

340. Determine the rating of the converter, Prob. 336, when it operates: (*a*) six-phase and 0.9 power factor; (*b*) three-phase and 0.9 power factor.

341. Determine the rating of the 600-kw converter, Prob. 338, when it operates four-phase and 0.9 power factor.

342. In Prob. 336 the d-c voltage of the 750-kw synchronous converter is 250 volts.   When it operates six-phase, determine: (*a*) rated d-c amperes output; (*b*) diametrical, or single-phase, slip-ring voltage; (*c*) six-phase voltage to neutral or across adjacent slip-ring taps; (*d*) rated alternating current per slip-ring. Assume unity power factor, and neglect losses.

343. The efficiency of a 1,500-kw, six-phase, 600-rpm, 600-volt, railway synchronous converter is 94.6 per cent. Determine: (*a*) rated direct current of converter; (*b*) six-phase voltage to neutral at slip-rings; (*c*) current per slip-ring. Power factor is unity. Neglect armature voltage drops.

344. The efficiency of a 1,000-kw, six-phase, 25-cycle, 750-rpm, 250-volt synchronous converter is 0.958. Determine: (*a*) rated direct current; (*b*) six-phase voltage to neutral at slip-rings; (*c*) current per slip-ring tap at unity power factor. Neglect armature voltage drops.

345. The converter, Prob. 344, is supplied at 6,900 volts, three-phase, 25 cycles. The transformers are connected with the primaries in delta and the secondaries in six-phase, star. The d-c side is a three-wire system. Make a complete wiring diagram, and show: (*a*) rated direct-current output; (*b*) current per slip-ring; (*c*) voltage between adjacent slip-rings; (*d*) transformer secondary voltages to neutral; (*e*) transformer secondary currents; (*f*) transformer primary currents; (*g*) primary line current. Neglect voltage drops and transformer losses.

QUESTIONS ON CHAPTER XI

Rectifiers

1. Into what four general types may rectifiers be divided? What is "inversion"?

2. Describe the rectifying commutator and the vibrating-contact type of rectifier. What limits their output?

3. Differentiate between half-wave and full-wave rectification. Draw the connections of the bridge circuit with which full-wave rectification may be obtained.

4. On what principle do electrolytic rectifiers operate? Make a diagram of connections.

5. Describe the principle of hot-cathode rectification, explaining the effects of the electrons and the positive ions.

6. On what principle does the Tungar operate? Why does the Tungar or Rectigon conduct current in one direction only? Make a diagram of connections. What are the approximate efficiencies and outputs of this type of rectifier?

7. On what principle does the mercury-arc rectifier operate? Why are two or more anodes used? Why is smoothing inductance necessary?

8. In the glass-tube mercury-arc rectifier, describe the operation of the starting anode. Make a diagram of connections. What is the approximate power limit of this type of rectifier?

9. What type of rectifier is used for large power output, and what difficulties was it necessary to overcome in its design? Why is inductance not necessary to maintain the arc? Why is inductance used nevertheless?

10. In a multianode rectifier, explain how current is transferred from one anode to the next.

11. Explain the operation of the grid in a vapor-type rectifier, and compare with the grid in a vacuum tube. Why cannot the grid in a vapor-type rectifier interrupt the current flow from anode to cathode?

12. Describe the construction of the thyratron tube. Why should the cathode heater be operated for at least 5 or 6 min before putting the tube in service?

13. Describe two methods of thyratron-tube control. Make a diagram of connections of a thyratron tube operating as a full-wave rectifier.

14. To what maximum ratings is it now possible to construct multianode metal-tank rectifiers. Cite the advantages of such rectifiers over motor-generator sets and synchronous converters.

15. Explain the relation of voltage rating to efficiency in mercury-arc rectifiers. What are the causes of arc-back, or backfiring?

16. State the advantages of the ignitron over the multianode steel-tank rectifier. How is ignition effected? How is output controlled?

QUESTIONS ON CHAPTER XII

Transmission and Distribution of Electrical Energy

1. Name some of the factors which determine the location of power stations. What makes it possible to transmit and utilize energy at places which are located many miles from the generating plant?

2. In a typical large power system, at what voltages is the energy ordinarily generated? If it is to be transmitted a long distance, what voltage might be used for transmission? Give reasons.

3. Why is the transmission voltage reduced before entering the more thickly populated districts? What is the function of the distributing substation?

4. At what voltages and frequencies are the following services usually delivered: (a) trolley cars; (b) a-c lighting; (c) d-c lighting; (d) street lighting; (e) a-c power; (f) downtown light and power.

5. What types of apparatus and what connections would be used to give each class of service in Question 4?

6. How does the weight of conductor vary with the transmission voltage, other factors being the same? What factors limit the transmission voltage?

7. Show that a transmission line has self-inductance. Capacitance.

8. What three types of structures are used most frequently for transmission-line supports? State the field of use of each.

9. Why is porcelain the most satisfactory material for line insulators? Under what conditions is the pin-type insulator used, and what are its limitations? State the advantages of the suspension-type insulator.

10. What are the functions of lightning arresters? Why must an arrester cut off at a voltage considerably above normal line voltage? Describe the Thyrite type of arrester.

11. Show how the 208–120-volt, three-phase, four-wire network is obtained. Under what conditions is this network used, and what are its advantages?

12. Where is low-voltage direct current used, and what are its advantages? Why is not this system generally used in new installations?

13. Distinguish between feeders and mains. How is the voltage at the end of a feeder measured at the station? How is it maintained constant at the desired value?

14. In street-railway systems, how are trolley and rail connected with the generator? What is meant by the "ladder system" of feeding? Make a diagram showing the sectionalized-trolley method of feeding. What are the advantages and disadvantages of this last method of feeding?

15. State the underlying cause of electrolysis of water mains, gas mains, etc. What measures are taken to minimize electrolysis?

16. What factors were considered in determining the value of 110 to 120 volts for lighting and domestic service? State the advantages and disadvantages of the series-parallel system.

17. Describe the Edison three-wire system, giving its advantages. Why should every precaution be taken to prevent the neutral being opened?

18. Describe the two-generator and the balancer-set methods of maintaining the neutral in a three-wire system. What connections of the machines of the balancer set give better voltage balancing?

19. Show the connections of the three-wire generator, and explain how it makes a neutral connection available.

20. Why are storage batteries seldom used in large power systems to give a more uniform load curve? For what purposes are storage batteries used? How are their charge and discharge controlled?

21. State the advantages of series distribution. For what service are such systems used? How are the loads connected in and removed from service? What is a "film cutout"? Make a diagram showing the parallel-loop system of series distribution.

PROBLEMS ON CHAPTER XII

Transmission and Distribution of Electrical Energy

346. Figure 346A shows a load taking 100 amp direct current at 110 volts (11 kw) over a 300-ft (91.5 m), No. 0 A.W.G. copper feeder, having resistance of 0.03 ohm per conductor. Determine: (a) bus-bar voltage; (b) power lost in feeder; (c) efficiency of transmission; (d) weight of copper in feeder. No. 0 copper weighs 319 lb per 1,000 ft.

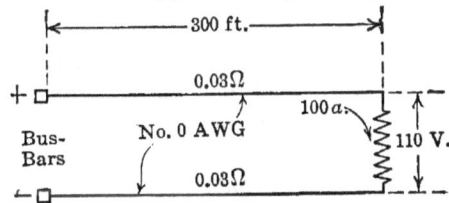

Fig. 346 A.

347. Assume that the load voltage, Prob. 346, is doubled and the power at the load remains 11 kw. Determine: (a) current; (b) bus-bar voltage; (c) power lost in feeder; (d) efficiency of transmission; (e) weight of copper.

348. In Prob. 347, No. 6 A.W.G. wire having four times the resistance per unit length of the No. 0 wire is used for the feeder. Determine: (a) bus-bar voltage; (b) power lost in feeder; (c) efficiency of transmission; (d) weight of copper.

349. In Fig. 349A is shown a 10-hp motor located 1,000 ft from 240-volt bus-bars and taking 40 amp over a No. 4 A.W.G. feeder, the resistance of

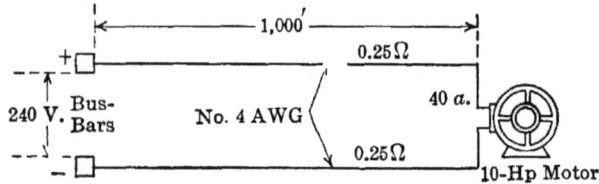

FIG. 349 A.

which is 0.25 ohm per 1,000 ft. Determine: (a) voltage drop in each wire; (b) voltage at motor; (c) efficiency of transmission; (d) weight of feeder. Number 4 wire weighs 126 lb per 1,000 ft.

350. Repeat Prob. 349, with bus-bar voltage being 120 volts, feeder No. 000 A.W.G. having four times the cross-section of No. 4, and current 80 amp because the voltage rating of the motor is now halved.

351. Repeat Prob. 349 with the motor running overload at 12 hp taking 48 amp.

352. A 50-hp motor is located 500 ft from 250-volt bus-bars; and when it takes rated current of 175 amp, it is desired that the voltage at its terminals shall be not less than 240 volts. Determine: (a) resistance per wire of necessary feeder; (b) next larger size wire A.W.G. (p. 442); (c) weight of wire; (d) efficiency of transmission.

353. Repeat Prob. 352, with bus-bar voltage twice 250 volts, or 500 volts, and voltage at the motor terminals not less than 480 volts. In order that the power to the motor shall not change, the current is now 87.5 amp.

354. A 4-mile, single-phase distribution line consists of two No. 0000 A.W.G. solid conductors spaced 30 in. on centers. The diameter of the wire is 0.460 in. Using Eqs. (95) and (96) (p. 364), determine: (a) loop inductance of line; (b) 60-cycle reactance.

355. Compute capacitance in microfarads of line, Prob. 354, using Eq. (97), p. 365.

356. The resistance of a single-phase, 60-cycle distribution line is 2.14 ohms per conductor. A 100-kw, 2,300-volt, unity-power-factor load is connected at the far end. Determine: (a) current; (b) copper loss; (c) power input to line; (d) efficiency of transmission.

357. Repeat Prob. 356, with power factor of the 100-kw load reduced to 0.707, lagging current.

358. In Prob. 356 the wire is No. 000 A.W.G. solid copper the cross-section of which is 168,000 cir mils. If the voltage at the load were raised to 4,000 volts, the power at the load remaining at 100 kw, determine: (a) cross-section of wire to give same line loss; (b) nearest A.W.G. wire number to give this cross-section.

359. It is desired to transmit 200 kw single-phase at unity power factor, 60 cycles, 8 miles (12.9 km), with 4,600 volts at the load. The resistance of each conductor is 4.2 ohms. Determine: (*a*) current; (*b*) copper loss; (*c*) input to line; (*d*) efficiency of transmission.

360. Repeat Prob. 359, with power factor 0.8, lagging current. Compare (*d*) in Probs. 359 and 360.

361. It is desired to transmit 300 kw, three-phase, 60 cycles, a distance of 10 miles (16.1 km) with 6,900 volts at the load. The load power factor is unity. The resistance per conductor is 15 ohms. Determine: (*a*) current per conductor; (*b*) power loss per conductor; (*c*) total transmission loss; (*d*) efficiency of transmission.

362. Repeat Prob. 361, with power factor of 0.75, other factors remaining unchanged.

363. In Fig. 363*A* is shown a single-phase distribution line delivering 150 kw at unity power factor and 2,300 volts, 60 cycles. The resistance

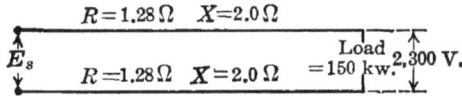

FIG. 363 *A*.

of each conductor is 1.28 ohms and the reactance 2.0 ohms. Determine: (*a*) current; (*b*) resistance drop; (*c*) reactance drop; (*d*) sending-end voltage E_s; (*e*) line power loss; (*f*) efficiency of transmission.

364. Repeat Prob. 363, with load of 150 kw at 0.8 power factor, lagging current.

365. Repeat Prob. 363, with load of 150 kw at 0.8 power factor, leading current.

366. A single-phase transmission line 30 miles long consists of two No. 0000 A.W.G. stranded copper conductors spaced 5 ft (1.525 m) between centers. A load of 2,500 kw at 26,000 volts, 60 cycles, and unity power factor is connected to the receiver end. Determine: (*a*) total resistance of line; (*b*) total reactance of line (see Appendixes H and I, pp. 442 and 443); (*c*) voltage at sending end; (*d*) line power loss; (*e*) efficiency of transmission.

367. Repeat Prob. 366 for same kilowatt load but with power factor 0.707, lagging current.

368. Repeat Prob. 366 for same kilowatt load but with power factor 0.707, leading current.

369. A single-phase transmission line, 25 miles long, consists of two No. 00 A.W.G. stranded copper conductors spaced 4 ft between centers. The voltage at the receiver is 33,000 volts, 60 cycles; and the load of 3,200 kw at unity power factor is connected to this end. Determine: (*a*) resistance per conductor; (*b*) reactance per conductor; (*c*) current; (*d*) *IR* and *IX* drops; (*e*) sending-end voltage; (*f*) line loss; (*g*) efficiency of transmission. (See Appendixes H and I, pp. 442 and 443.)

370. Repeat Prob. 369 with same kilowatt load but with power factor 0.80, lagging current.

371. The following lamp load (Fig. 371*A*) is supplied at 115 volts d-c by a 200-ft No. 8 A.W.G. feeder: eight 100-watt lamps; eight 150-watt lamps;

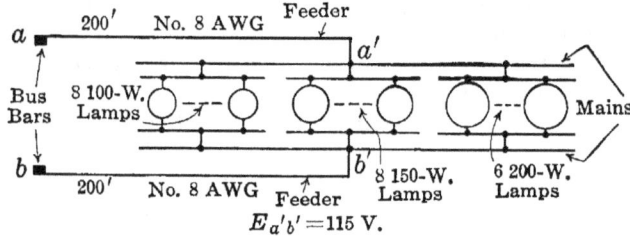

Fig. 371 *A*.

six 200-watt lamps. The resistance of the feeder is 0.64 ohm per 1,000 ft. The resistance of the mains may be neglected. Determine: (*a*) the current to load; (*b*) resistance of each wire *aa'*, *bb'*, of feeder; (*c*) voltage at bus-bars *ab*; (*d*) efficiency of transmission.

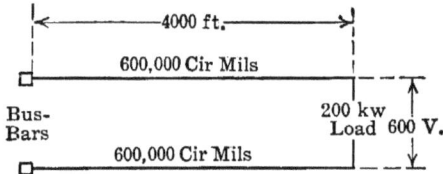

Fig. 372 *A*.

372. The voltage at a 200-kw d-c load supplied by a 4,000-ft (1.22 km) feeder is 600 volts (Fig. 372*A*). The feeder consists of two 600,000-cir-mil cables, each having a resistance of 0.018 ohm per 1,000 ft. Determine: (*a*) current to load; (*b*) feeder resistance; (*c*) bus-bar voltage; (*d*) power loss in feeder; (*e*) efficiency of transmission.

373. In Prob. 372, with same power of 200 kw, voltage at load is 300 volts. Determine: (*a*) current; (*b*) feeder resistance which would produce same loss (16.0 kw) found in (*d*); (*c*) resistance per 1,000 ft of such a feeder; (*d*) size in circular mils of feeder. (*e*) Compare weight of copper in two cases.

374. A 1,000-ft (305-m) length of 10,000-cir-mil copper conductor weighs 31.4 lb (see Part I, p. 19). It is desired to transmit 200 kw a distance of 2,000 ft (610 m). With 120 volts at the load and with a line drop equal to 10 per cent of the load voltage, determine: (*a*) size of necessary copper conductor; (*b*) weight of copper. (*c*) With copper costing $0.12 per lb ($0.264 per kg), determine cost of copper in feeder.

375. A trolley system extends 4.0 miles from the station (Fig. 375*A*), a No. 0000 A.W.G. trolley having a resistance of 0.27 ohm per mile being used. The station bus-bar voltage is 600 volts.

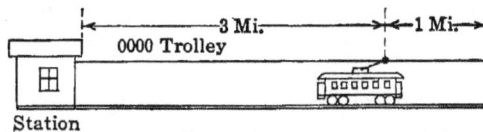

Fig. 375 *A*.

Resistance of rail and ground return is 0.10 ohm per mile. A car 3.0 miles from station takes 65 amps. Determine voltage: (*a*) at car; (*b*) at end of line.

376 In Fig. 375*A*, assume the car to be at the far end of the line, taking a starting current of 80 amp. Determine: (*a*) voltage at car; (*b*) efficiency of transmission.

377. In Fig 375*A*, assume the car to be 2 miles from the station, taking 70 amp. Determine: (*a*) voltage at car; (*b*) efficiency of transmission.

378. In the Edison three-wire system (Fig. 378*A*), there are loads of 70 and 40 amp on the positive side and a load of 80 amp on the negative side. Determine current and its direction at points *a*, *b*, *c* in neutral.

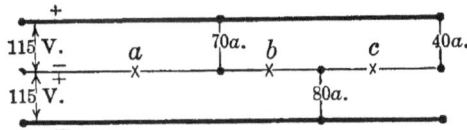

Fɪɢ. 378 *A*.

379. Repeat Prob. 378, with the 70-amp load reduced to 20 amp.

380. Figure 380*A* shows an Edison three-wire system with loads of 50, 60, and 40 amp on the positive side and loads of 70 and 30 amp on the

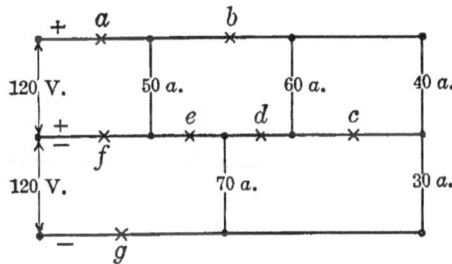

Fɪɢ. 380 *A*.

negative side. Determine magnitudes and directions of currents at points *a*, *b*, *c*, *d*, *e*, *f*, *g*.

381. Repeat Prob. 380, with the 70- and 30-amp loads changed to 100 and 60 amp.

382. The loop length of a 6.6-amp series arc system is 9.6 miles (15.4 km), and the circuit consists of No. 6 A.W.G. underground cable, the resistance of which is 0.442 ohm per 1,000 ft (305 m) at 50°C, the operating temperature in the ducts. There are forty-five 510-watt, 8,960-lumen, magnetite lamps connected to the system. Determine: (*a*) voltage drop in loop system; (*b*) circuit voltage; (*c*) transmission efficiency of system.

383. Repeat Prob. 382 for similar system having loop length of 11.4 miles (18.3 km) and 50 magnetite lamps.

QUESTIONS ON CHAPTER XIII

Vacuum Tubes

1. Discuss the nature of electrons, the order of magnitude of their mass, their relation to the a\ ,m, and their collisions.

2. What conditions are necessary for a free emission of electrons? What is the effect of the electrons in the space outside a body on the number of electrons that remain in this space? What is meant by "critical velocity"?

3. State and analyze Richardson's law. Define "thermionic efficiency." Compare the thermionic efficiencies of tungsten, oxide-coated platinum, and thoriated tungsten. What is meant by "space charge"?

4. Analyze electron emission in the two-electrode tube. What is meant by "space-charge saturation," and what is its effect on the emission characteristic of the tube?

5. State and analyze Child's three-halves power law. At any one temperature, why does the current become essentially constant with increase in voltage, after the voltage reaches a certain value?

6. Describe the "Edison effect."

7. Describe the operation of the Fleming valve and its use as a rectifier. Draw the connections for a full-wave rectifier, and explain the function of the filter. Describe the construction and operation of an X-ray tube.

8. Describe the construction and analyze the operation of a three-electrode vacuum tube. How does the grid act to control the output of the tube?

9. State the equivalent plate-circuit theorem. Give the formula for amplification factor, plate resistance, and transconductance; and state the relation connecting them.

10. Draw the connections used for determining the static characteristics of the three-electrode tube. Define the three tube coefficients, "amplification factor," "plate resistance," and "transconductance." State the relation connecting them. What is the effect of the geometry of the tube on these quantities?

11. Sketch and analyze the three following static characteristics: I_p–E_g for different values of E_p; I_p–E_p for different values of E_g; E_p–E_g for different values of I_p.

12. Sketch the circuit connections which are used for measuring by dynamic methods the three tube coefficients, amplification factor μ, plate resistance r_p, and transconductance g_m. What are the relations in the measuring circuits which determine these coefficients?

13. Describe the construction of three-electrode receiving tubes. What special feature is used with a-c tubes, and why is a directly heated filament sometimes used?

14. Draw the connections and analyze the operation of the three-electrode tube as an amplifier. What are the effects of the steady voltages and currents in the plate and grid circuits on the operation of the tube as an amplifier?

15. Define class A, class B, and class C amplification.

16. With load on the tube, why is the actual voltage amplification always less than the amplification factor? Draw the circuit connections of a resistance-coupled amplifier. What is the order of magnitude of voltage amplification in a two-stage resistance-coupled amplifier? What limits amplification obtainable in a multistage amplifier? What factor normally limits the voltage gain in a single stage, and how may the gain be increased?

17. Compare resistance-coupled and transformer-coupled amplifiers as to over-all gain and frequency characteristics.

18. Explain why the screen-grid tube eliminates regenerative effects.

19. In what manner does the four-electrode tube, or tetrode, differ from the three-electrode tube? Describe the operation of the space-charge and screen-grid types of tubes.

20. What is the object of the pentode tube? Show the tube connections, and sketch its I_p–E_p characteristic. For what purposes are multi-electrode tubes sometimes used?

21. Explain why a pentode introduces considerable distortion in the plate current.

22. Make a diagram of circuit connections for regeneration, and analyze the principle of operation.

23. On what principle do oscillators operate? Draw wiring diagrams of their four circuits.

24. Discuss the relative merits of the various oscillator circuits: tuned grid, tuned plate, Hartley, Colpitts.

25. In what manner do power tubes differ from receiving tubes? How is the rating of power tubes increased? Describe the construction of tubes whose rating exceeds 1 kw.

26. What is meant by "modulation"? What three frequencies exist in a modulated carrier current? Distinguish between "side frequencies" and "side bands."

27. Expand Eq. (115) (p. 419), and indicate the audiofrequency and radiofrequency components.

28. Make a diagram of circuit connections, and describe plate-circuit modulation.

29. What is meant by "detection"? Show how detection may be accomplished with a two-electrode tube. How is the sensitivity increased?

30. Describe detection with the three-electrode tube with polarized grid. Why is the grid polarized negatively? Where is this type of detection used?

31. Analyze detection with three-electrode tubes in which grid resistance is employed. What are the functions of the grid resistance and the grid capacitor? Show how detection and regeneration may be combined in a single tube.

32. Analyze heterodyne, or beat, reception. What is meant by "beat frequency"? Make a diagram of connections of separate heterodyne reception.

33. Describe the general principles on which most receivers operate. Why are filters used with a-c sets? Discuss the advantages of the super-heterodyne receiver.

INDEX

www.ingramcontent.com/pod-product-compliance
Lightning Source LLC
Chambersburg PA
CBHW080406270326
41929CB00018B/2922